Material											
Magnesium, extrusion, AZ80X	0.066	35	26	49	b	21	19	6.5	2.4	12	14.4
Magnesium, sand cast, AZ63-HT	0.066	14	14	40	b	19	14	6.5	2.4	12	14.4
Monel, wrought, hot-rolled	0.319	50	b	90	b		40	26	9.5	35	7.8
Red brass, cold-rolled	0.316	60		75				15	5.6	4	9.8
Red brass, annealed	0.316	15	b	40	b			15	5.6	50	9.8
Bronze, cold-rolled	0.320	75		100				15	6.5	3	9.4
Bronze, annealed	0.320	20	b	50	b			15	6.5	50	9.4
Titanium alloy, annealed	0.167	135	b	155	b			14	5.3	13	9.4
Invar, annealed	0.292	42	b	70	b			21	8.1	41	0.6
Nonmetallic materials											
Douglas fir, green[e]	0.022	4.8	3.4		3.9	0.9		1.6			
Douglas fir, air dry[e]	0.020	8.1	6.4		7.4	1.1		1.9			
Red oak, green[e]	0.037	4.4	2.6		3.5	1.2		1.4			1.9
Red oak, air dry[e]	0.025	8.4	4.6		6.9	1.8		1.8			
Concrete, medium strength	0.087		1.2		3.0			3.0			6.0
Concrete, fairly high strength	0.087		2.0		5.0			4.5			6.0

[a] Elastic strength may be represented by proportional limit, yield point, or yield strength at a specified offset (usually 0.2 percent for ductile metals).

[b] For ductile metals (those with an appreciable ultimate elongation), it is customary to assume the properties in compression have the same values as those in tension.

[c] Rotating beam.

[d] Elongation in 8 in.

[e] All timber properties are parallel to the grain.

INTRODUCTION TO
MECHANICS OF MATERIALS

INTRODUCTION TO
MECHANICS OF MATERIALS

WILLIAM F. RILEY

Iowa State University

LOREN W. ZACHARY

Iowa State University

WILEY

JOHN WILEY & SONS   New York   Chichester   Brisbane   Toronto   Singapore

Cover: *Frank Stella,* **Damascus Gate Variation I,** 1969;
Private collection, Courtesy of M. Knoedler & Co., Inc., New York

Text and cover design by Madelyn Lesure
Production supervised by Lucille Buonocore
Illustrations supervised by John Balbalis
Copy editing supervised by Richard Koreto

Library of Congress Cataloging in Publication Data:

Riley, William F. (William Franklin), 1925-
   Introduction to mechanics of materials/by William F. Riley and
Loren W. Zachary.
      p.   cm.
   Includes index.
   ISBN 0-471-84933-2
   1. Strength of materials.   2. Structures, Theory of.   I. Zachary,
Loren W.   II. Title.

TA405.R54 1989
620.1'12—dc 19   88-21587
                                                                        CIP

Printed in the United States of America
10 9 8 7 6 5 4 3 2 1

PREFACE

This textbook is an outgrowth of the book *Mechanics of Materials* by A. Higdon, E.H. Ohlsen, and W.B. Stiles that was first published in 1960. This abbreviated edition retains many of the features of the previous four editions of this earlier work.

The primary objectives of a course in mechanics of materials are:

1. to develop a working knowledge of the relations between the loads applied to a nonrigid body made of a given material and the resulting deformations of the body
2. to develop a thorough understanding of the relations between the loads applied to a nonrigid body and the stresses produced in the body
3. to develop a clear insight into the relations between stresses and strains for a wide variety of conditions and materials, and
4. to develop adequate procedures for finding the required dimensions of a member of a specified material to carry a given load subject to stated specifications of stress and deflection.

These objectives involve the concepts and skills that form the foundation of all structural analysis and machine design.

The principles and methods used to meet the objectives are drawn largely from prerequisite courses in mechanics, physics, and mathematics. Some basic concepts of the theory of elasticity and some knowledge of the mechanical properties of engineering materials are also used. This book is designed to emphasize fundamental principles. Instead of deriving numerous formulas for specialized states of stress, we emphasize use of free-body diagrams, application of the equations of equilibrium, visualization and use of the geometry of the deformed body, and use of the relations between stresses and strains for the material being used. This approach makes it possible to develop all of the commonly used *Mechanics of Materials* formulas in a rational and logical manner, and to clearly indicate the conditions under which they may be safely applied to the analysis and design of actual engineering structures and machine components. Since extensive use is made of course material from statics and calculus, a working knowledge of these subjects is considered an essential prerequisite to a successful study of mechanics of materials as presented in this book.

This abbreviated edition is designed for a first course in mechanics of deformable bodies. The extensive subdivision of the book into different topics will provide flexibility in the choice of assignments to cover courses of different length and content. The topics progress logically from a study of stresses and deformations in members subjected to simple tension, compression, or shear, to a study of stresses and deformations in shafts under torsion, beams under bending, and structural members and machine components under combined loadings. Statically indeterminate members are introduced throughout the text in connection with those topics for which the subject is appropriate. The variation of stress (and strain) at a point is presented at a place in the book where the topic can be treated fully and correctly and where students have sufficient understanding of the subject to appreciate the concepts of principal stress and maximum shearing stress at a point.

Throughout the book, strong emphasis is placed on the engineering significance of the subject, and not merely on mathematical methods of analysis. The course aims to develop in students the ability to analyze a given problem in a simple and logical manner, to apply a few fundamental principles to its solution, and to present results in a clear, logical, and neat manner. We believe that students gain mastery of a subject once they find that they can apply the basic theory to the solution of a problem that appears somewhat difficult. They learn less by solving a large number of simple but similar problems. A conscientious effort has been made to present the material in a simple and direct manner, with a student's point of view constantly in mind.

Students are usually more enthusiastic about a subject if they can see and appreciate its value as they proceed into the subject. The practical side of design is shown wherever possible, and there are frequent reminders that safety is a prime consideration in all engineering design. To help students acquire an intuitive feeling for the sizes, shapes, proportions, and load-carrying capacities of real machine components and structural members, we have chosen only realistic sizes and loads for the members used in illustrative examples and homework problems. Also, the materials used in the fabrication of the members are chosen from those materials that are commonly used in engineering practice, and the stresses to which the materials are subjected are realistic.

Many illustrative example problems have been integrated into the main body of the text at places where the presentation of a method can best be reinforced by the immediate use of the method in an illustrative example problem. The illustrative examples and homework problems have been carefully selected, and in general, require an understanding of the principles of mechanics of materials without demanding excessive time for computational work.

The answers to about half the problems are included at the end of the book. We feel that the first assignment on a given topic should include some problems for which the answers are given. Since the simpler problems are usually reserved for this first assignment, the answers are provided for the first four or five problems of each article and thereafter are given in general for alternate problems. We feel that answers for certain problems should not be given; hence, we have not followed the flat rule of answers for all even-numbered problems. Since the convenient designation of problems for which answers are provided is of great value to those who make up assignment sheets, the problems for which answers are provided are indicated by means of an asterisk after the number.

In general, all given data are assumed to be composed of three significant figures regardless of the number of figures shown. Answers are therefore given to three significant figures, unless the number lies between 1 and 2 or any decimal multiple thereof, in which case four significant figures are reported. Some of the problems involve material properties, which are quite variable, and possibly stress concentration factors, which are difficult to obtain to more than two significant figures from graphs. We hesitate to give results to more than two significant figures in these instances; however, in order to be consistent and to avoid confusion, the answers will be given to three significant figures.

Use of the International System of Units (SI) is slowly gaining acceptance in the United States, and as a result, engineers must be proficient during the transition period in both the SI system and the U.S. Customary System (USCS) in common use today. In response to this need, both U.S. customary units and SI units are used in approximately equal proportions in the text for both illustrative examples and homework problems. Since a large number of homework problems are provided, the instructor can assign homework problems in whatever proportion he finds desirable for his class. A discussion of the SI system together with a table of conversion factors is presented in an appendix of the book.

We are grateful for comments and suggestions received from colleagues and from users of the earlier editions of this book. We appreciate also the careful appraisal made by the publisher's consultant and have incorporated those recommended changes that we considered appropriate. We will be pleased to receive comments from readers of this edition of the book and will attempt to acknowledge all such communications.

William F. Riley
Loren W. Zachary

ABOUT THE AUTHORS

William F. Riley holds the rank of Distinguished Professor of Engineering Science and Mechanics at Iowa State University, Ames, Iowa. He received his B.S. in mechanical engineering from the Carnegie Institute of Technology in 1951 and his M.S. in mechanics from the Illinois Institute of Technology in 1958. He worked from 1951 to 1954 as a mechanical engineer for the Mesta Machine Company in the design and development of steel-mill equipment and heavy machine tools. From 1954 to 1966 he engaged in experimental stress analysis activities at IIT Research Institute as a research engineer, as manager of the experimental stress analysis group, and finally as science advisor to the mechanics research division. During the summers of 1966 and 1970 he served as a consultant for USAID/NSF in India in their summer institute program for college teachers. Since 1966 he has been an engineering educator at Iowa State University. Professor Riley was elected a Fellow of the Society for Experimental Mechanics in 1977 and an Honorary Member in 1984. In 1977, he received the M.M. Frocht Award of the Society for Experimental Mechanics for Excellence in Engineering Education. Professor Riley has published approximately fifty technical papers dealing with stress analysis topics in national and international journals and has coauthored six books.

Loren W. Zachary, Professor of Engineering Science and Mechanics at Iowa State University since 1976, received his B.S. degree in Aerospace Engineering in 1966 from Iowa State University. He worked for Martin Marietta Corporation in Denver, Colorado, from 1966 to 1972. He conducted vibration studies on Titan launch vehicles and Skylab.

In 1972, Dr. Zachary returned to Iowa State and received his M.S. and Ph.D. degrees in engineering mechanics. He has been active in experimental stress analysis and nondestructive evaluation. He has authored and coauthored several journal papers in these areas. He is a member of the Society for Experimental Mechanics and is a licensed professional engineer.

CONTENTS

CHAPTER 6   361

STATICALLY INDETERMINATE BEAMS

CHAPTER 7   391

STRESS AND STRAIN TRANSFORMATION EQUATIONS

CHAPTER 8   453

COMBINED STATIC LOADINGS

SYMBOLS AND ABBREVIATIONS*

A	area
avg.	average
b	breadth, width
C	compression, Celsius
c	distance from neutral axis or from center of twist to extreme fiber
E	modulus of elasticity in tension or compression
e	eccentricity
Eq.	equation
F	force, or body force, Fahrenheit
f	normal force per unit of length
ft	foot or feet
ft-lb	foot-pound
G	modulus of rigidity (modulus of elasticity in shear)
g	acceleration of gravity
h	height, depth of beam
hp	horsepower
I	moment of inertia or second moment of area
in.	inch
in.-lb	inch-pound
J	polar moment of inertia of area
J	joule ($N \cdot m$)
K	stress concentration factor
k	modulus of spring, factor of safety
kg	kilogram
kip	kilopound (1000 lb)
ksi	kilopounds (or kips) per square inch
L	length
lb	pound

*The symbols and abbreviations used in this book conform essentially with those approved by the International Standards Organization.

M	mega
M	bending moment
m	meter
max	maximum
min	minimum
N	normal force
N	newton
N · m	newton · meter
n	ratio of elastic moduli
P	concentrated load, force
p	fluid pressure
Pa	pascal (N/m^2)
psi	pounds per square inch
Q	first moment of area
q	shearing force per unit of length
R	radius, modulus of rupture, resultant force
r	radius of gyration
rad	radian
rpm	revolutions per minute
rps	revolutions per second
S	surface force, resultant stress, stress vector, section modulus ($S = I/c$)
sec	second
T	torque, temperature, tension
t	thickness, tangential deviation
U	strain energy
u	strain energy per unit volume
u, v, w	components of displacement
V	shearing force
W	weight
W	watt (N·m/s)
W_k	work
w	load per unit of length
x, y, z	cartesian coordinates
y	deflection
α (alpha)	coefficient of thermal expansion
γ (gamma)	shearing strain, specific weight
δ (delta)	total deformation
ϵ (epsilon)	normal strain

θ (theta)	total angle of twist in radians, slope of deflected beam
μ (mu)	micro (10^{-6})
ν (nu)	Poisson's ratio
ρ (rho)	radius, radius of curvature
σ (sigma)	normal stress
τ (tau)	shearing stress
ω (omega)	angular velocity
$\circ$	degree
C12 $\times$ 30	a 12-in. rolled channel section with a weight of 30 lb/ft
C305 $\times$ 45	a 305-mm rolled channel section with a mass of 45 kg/m
L6 $\times$ 4 $\times$ 1/2	a rolled angle section with 6-in. and 4-in. legs 1/2 in. thick
L102 $\times$ 89 $\times$ 13	a rolled angle section with 102-mm and 89-mm legs 13 mm thick
S10 $\times$ 35	a 10-in. American standard rolled section (I-beam) with a weight of 35 lb/ft
S127 $\times$ 22	a 127-mm American standard rolled section (I-beam) with a mass of 22 kg/m
W16 $\times$ 100	a 16-in. wide-flange rolled section with a weight of 100 lb/ft
W914 $\times$ 342	a 914-mm wide-flange rolled section with a mass of 342 kg/m
WT8 $\times$ 50	an 8-in. structural T-section (cut from a W-section) with a weight of 50 lb/ft
WT457 $\times$ 171	a 457-mm structural T-section (cut from a W-section) with a mass of 171 kg/m

INTRODUCTION TO
MECHANICS OF MATERIALS

CHAPTER 1

INTRODUCTION TO STRESS, STRAIN, AND THEIR RELATIONSHIPS

1-1

INTRODUCTION

The primary objective of a course in mechanics of materials is the development of relationships between the loads applied to a nonrigid body and the internal forces and deformations induced in the body. Ever since the time of Galileo Galilei (1564–1642), scientists and engineers have studied the problem of the load-carrying capacity of structural members and machine components, and have developed mathematical and experimental methods of analysis for determining the internal forces and the deformations induced by the applied loads. The experiences and observations of these scientists and engineers of the last three centuries are the heritage of the engineer of today. The fundamental knowledge gained over the last three centuries, together with the theories and analysis techniques developed, permit the modern engineer to design, with complete competence and assurance, structures and machines of unprecedented size and complexity.

The subject matter of this book forms the basis for the solution of three general types of problems:

1. Given a certain function to perform (the transporting of traffic over a river by means of a bridge, conveying scientific instruments to

Mars in a space vehicle, the conversion of water power into electric power), of what materials should the machine or structure be constructed, and what should the sizes and proportions of the various elements be? This is the designer's task, and obviously there is no single solution to any given problem.

2. Given the completed design, is it adequate? That is, does it perform the function economically and without excessive deformation? This is the checker's problem.

3. Given a completed structure or machine, what is its actual load-carrying capacity? The structure may have been designed for some purpose other than the one for which it is now to be used. Is it adequate for the proposed use? For example, a building may have been designed as an office building, but is later found to be desirable for use as a warehouse. In such a case, what maximum loading may the floor safely support? This is the rating problem.

Since the complete scope of these problems is obviously too comprehensive for mastery in a single course, this book is restricted to a study of individual members and very simple structures or machines. The design courses that follow will consider the entire structure or machine, and will provide essential background for the complete analysis of the three problems.

The principles and methods used to meet the objective stated at the beginning of this chapter depend to a great extent on prerequisite courses in mathematics and mechanics, supplemented by additional concepts from the theory of elasticity and the properties of engineering materials. The equations of equilibrium from *statics* are used extensively, with one major change in the free-body diagrams; namely, most free bodies are isolated by *cutting through* a member instead of removing a pin or some other connection. The forces transmitted by the cut sections are *internal* forces. The intensities of these internal forces (force per unit area) are called *stresses*.

It will frequently be found that the equations of equilibrium (or motion) are not sufficient to determine all the unknown loads or reactions acting on a body. In such cases it is necessary to consider the geometry (the change in size or shape) of the body after the loads are applied. The *deformation* per unit length in any direction or dimension is called *strain*. In some instances, the specified maximum deformation and not the specified maximum stress will govern the maximum load that a member may carry.

Some knowledge of the physical and mechanical properties of materials is required in order to create a design, to properly evaluate a given design, or even to write the correct relation between an applied load and the resulting deformation of a loaded member. Essential information will be introduced as required, and more complete information can be obtained from textbooks and handbooks on properties of materials.

1-2

LOAD CLASSIFICATION

Certain terms are commonly used to describe applied loads; their definitions are given here so that the terminology will be clearly understood.

Loads may be classified with respect to time:

1. A *static load* is a gradually applied load for which equilibrium is reached in a relatively short time.
2. A *sustained load* is a load that is constant over a long period of time, such as the weight of a structure (called *dead load*). This type of load is treated in the same manner as a static load; however, for some materials and conditions of temperature and stress, the resistance to failure may be different under short-time loading and under sustained loading.
3. An *impact load* is a rapidly applied load (an energy load). Vibration normally results from an impact load, and equilibrium is not established until the vibration is eliminated, usually by natural damping forces.
4. A *repeated load* is a load that is applied and removed many thousands of times. The helical springs that close the valves on automobile engines are subjected to repeated loading.

Loads may also be classified with respect to the area over which the load is applied:

1. A *concentrated load* is a load or force applied at a point. Any load applied to a relatively small area compared with the size of the loaded member is assumed to be a concentrated load; for example, a truck wheel load on the longitudinal members of a bridge.
2. A *distributed load* is a load distributed along a length or over an area. The distribution may be uniform or nonuniform. The weight of a concrete bridge floor of uniform thickness is an example of a uniformly distributed load.

Loads may be classified with respect to the location and method of application:

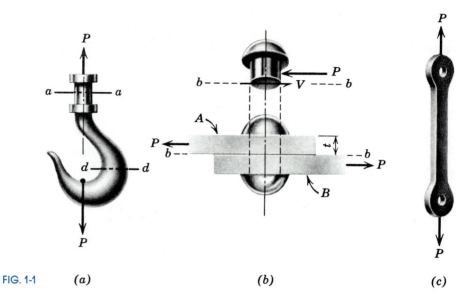

FIG. 1-1   *(a)*   *(b)*   *(c)*

1. A *centric load* is one in which the resultant force passes through the centroid of the resisting section. If the resultant passes through the centroids of all resisting sections, the loading is termed *axial*. In Fig. 1-1*a*, the line of action of the load P passes through the centroid of section *a–a*; therefore, the load is centric for section *a–a*. The force P in Fig. 1-1*b* is applied through the plate to the shank of the rivet and is resisted by the force V on the shaded section *b–b*. If the thickness t is not too large (the usual case for riveted joints), the force P can be assumed to be collinear with V, thus satisfying the definition of a centric load. The loading illustrated in this figure is called *single shear*, and the force V is a shearing force. If the plate A was sandwiched between two plates B with the rivet extending through all three plates, the rivet would be subjected to two shearing forces. Such loading is called *double shear*. Figure 1-1*c* represents an eyebar used for trusses, tie rods, and linkages. This is an example of axial loading.

Resultant of bolt forces is a couple

FIG. 1-2

FIG. 1-3

2. A *torsional load* is one that subjects a shaft or some other member to couples that twist the member. If the couples lie in planes transverse to the axis of the member, as in Fig. 1-2, the member is subjected to pure torsion.

3. A *bending* or *flexural load* is one in which the loads are applied transversely to the longitudinal axis of the member. The applied load may include couples that lie in planes parallel to the axis of the member. A member subjected to bending loads bends or bows along its length. Figure 1-3 illustrates a beam subjected to flexural loading consisting of a concentrated load, a uniformly distributed load, and a couple.

4. A *combined loading* is a combination of two or more of the previously defined types of loading.

1-3

NORMAL STRESS UNDER AXIAL LOADING

In every subject area there are certain fundamental concepts that assume paramount importance for a satisfactory comprehension of the subject matter. The basic concept of paramount importance in the subject area of "mechanics of materials" is that of *stress*. In the simplest qualitative terms, *stress is the intensity of internal force*. Force is a vector quantity and as such has both magnitude and direction. Intensity implies an area over which the force is distributed. Thus,

$$\text{Stress} = \frac{\text{Force}}{\text{Area}} \qquad (1\text{-}1)$$

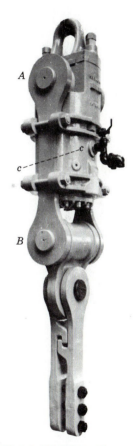

The term stress is sometimes used to denote the resultant force on a section; in other instances, it is used to indicate force per unit area. In this book, *stress will always refer to force per unit area.*

The eyebar *AB* from a pile extractor (see Fig. 1-4), which is used to reclaim steel sheet piles from cofferdams and similar installations, will be used to introduce the concept of a *normal stress*. Although the bar is subjected to dynamic loading in use, an equivalent static load (a static load that will produce the same stresses as the dynamic load) can be used for the design of the eyebar. Since the eyebar is a two-force member, the pin reactions at *A* and *B* are collinear with the axis

FIG. 1-4 (*Courtesy of Vulcan Iron Works Inc., Chicago, Illinois*)

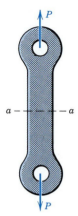

FIG. 1-5

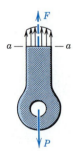

FIG. 1-6

FIG. 1-7

of the bar and produce an axial tensile loading of the bar as shown in Fig. 1-5. A review of two-force members and their application in trusses, as presented in a course in statics, is strongly recommended. When the eyebar is cut by a transverse plane, such as plane *a–a* of Fig. 1-5, a free-body diagram of the bottom half of the bar can be drawn as shown in Fig. 1-6. Equilibrium of this portion of the bar is obtained with a distribution of internal force that develops on the exposed cross section. This distribution of internal force has a resultant F that is normal to the exposed surface, is equal in magnitude to P, and has a line of action that is collinear with the line of action of P. An average intensity of internal force, which is also known as the *average normal stress* σ_{avg} on the cross section, can be computed as

$$\sigma_{avg} = \frac{F}{A} \qquad (1\text{-}2)$$

where A is the cross-sectional area of the eyebar.

The Greek letter sigma (σ) will be used to denote a normal stress in this book. A positive sign will be used to indicate a *tensile normal stress* (member in tension), and a negative sign will be used to indicate a *compressive normal stress* (member in compression).

Consider now a small area ΔA on the exposed cross section of the bar, as shown in Fig. 1-7, and let ΔF represent the resultant of the internal forces transmitted by this small area. The average intensity of internal force being transmitted by area ΔA is obtained by dividing ΔF by ΔA. If the internal forces transmitted across the section are assumed to be continously distributed, the area ΔA can be made smaller and smaller and will approach a point on the exposed surface in the limit. The corresponding force ΔF also becomes smaller and smaller. The stress at the point on the cross section to which ΔA converges is defined as

$$\sigma = \lim_{\Delta A \to 0} \frac{\Delta F}{\Delta A} \qquad (1\text{-}3)$$

In general, the stress σ at a given point on a transverse cross section of an axially loaded bar will not be the same as the average stress computed by dividing the force F by the cross-sectional area A. A necessary (but not sufficient) condition for the distribution of stress to be uniform is that the loading be centric. For long, slender, axially loaded members, such as those found in trusses and similar structures, it is generally assumed that the normal stresses are uniformly distributed except in the vicinity of the points of application of the loads. The subject of nonuniform stress distributions under centric loading will be discussed in a later section on stress concentrations. *Unless otherwise indicated, a uniform stress distribution is to be assumed for all centric (or assumed centric) loading in this book.*

1-4

SHEARING STRESSES UNDER CENTRIC LOADING

Loads applied to a structure or machine are generally transmitted to the individual members through connections that use rivets, bolts, pins, nails, or welds. In all of these connections, one of the most significant stresses induced is a *shearing stress*; hence, these members are often said to be subjected to centric shear loading. The bolted and pinned connection shown in Fig. 1-8 will be used to introduce the concept of a *cross shearing stress*.

The method by which loads are transferred from one member of the connection to another is by means of a distribution of shearing force on a transverse cross section of the bolt or pin used to effect the connection. A free-body diagram of the left member of the connection of Fig. 1-8 is shown in Fig. 1-9. In this diagram, the distribution of shearing force on the transverse cross section of the bolt has been replaced by a resultant shear force V. Since only one cross section of the bolt is used to effect load transfer between the members, the bolt is said to be in *single shear*; therefore, equilibrium requires that the resultant shear force V equal the applied load P. A similar free-body diagram for the threaded eyebar at the right end of the connection of Fig. 1-8 is shown in Fig. 1-10. In this case, two cross sections of the pin are used to effect load transfer between members of the connection, and the pin is said to be in *double shear*. As a result, equilibrium requires that the resultant shear force V on each cross section of the pin equals one-half of the applied load P.

From the definition of stress given by Eq. 1-1, an average shearing stress on the transverse cross section of the bolt or pin can be computed as

$$\tau_{\text{avg}} = \frac{V}{A_s} \qquad (1\text{-}4)$$

where A_s is the cross-sectional area of the bolt or pin.

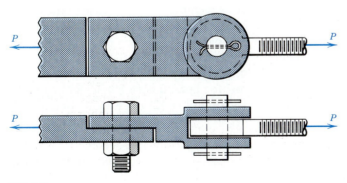

FIG. 1-8

FIG. 1-9

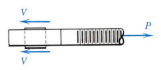

FIG. 1-10

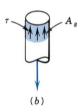

(a)

(b)

FIG. 1-11

The Greek letter tau (τ) will be used to denote shearing stress in this book. A sign convention for shearing stress will be presented in a later section of the book.

The stress at a point on the transverse cross section of the bolt or pin can be obtained by using the same type of limit process that was used to obtain Eq. 1-3 for the normal stress at a point. Thus,

$$\tau = \lim_{\Delta A_s \to 0} \frac{\Delta V}{\Delta A_s} \tag{1-5}$$

It will be shown later in this text that the shearing stresses cannot be uniformly distributed over the transverse cross section of a pin or bolt and that the *maximum shearing stress* on the transverse cross section may be much larger than the average shearing stress obtained by using Eq. 1-4. The design of simple connections, however, is usually based on average stress considerations, and this procedure will be followed in this book.

Another type of centric shear loading is termed *punching shear*. Examples of this type of loading include the action of a punch in forming rivet holes in a metal plate, the tendency of building columns to punch through footings, and the tendency of a tensile axial load on a bolt to pull the shank of the bolt through the head. Under a punching shear load, the significant stress is the average shearing stress on the surface described by the periphery of the punching member and the thickness of the punched member; for example, the shaded cylindrical area A_s shown extending through the head of the bolt in Fig. 1-11b.

1-5

BEARING STRESSES

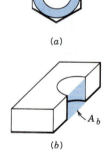

(a)

(b)

FIG. 1-12

Bearing stresses (compressive normal stresses) occur on the surface of contact between two interacting members. For the case of the bolted connection shown in Fig. 1-8, bearing stresses occur on the surfaces of contact between the head of the bolt and the top plate and between the nut and the bottom plate. The force producing the stress is the axial tensile internal force F developed in the shank of the bolt as the nut is tightened. The area of interest for bearing stress calculations is the area A_b of the bolt head or nut (see Fig. 1-12a) that is in contact with the plate. Thus, the *average bearing stress* σ_b is expressed as

$$\sigma_b = \frac{F}{A_b} \tag{1-6}$$

Bearing stresses also develop on the contact surfaces between the plates and the shank of the bolt or the pin. Since the distribution of these forces is quite complicated, an average bearing stress σ_b computed by dividing the force F transmitted across the surface of contact by the projected area A_b shown in Fig. 1-12b, instead of the actual contact area, is often used for design purposes.

The following example will serve to illustrate the concepts discussed in Sections 1-3, 1-4, and 1-5.

> *Note:* In all problems and examples in this textbook, the weights of the members will be neglected unless otherwise specified.

EXAMPLE 1-1

A hoist is made from a 4 × 4-in. wood post BC and a 1-in.-diameter steel eyebar AB as shown in Fig 1-13a. The hoist supports a load W of 7000 lb. Determine the axial stresses in members AB and BC and the cross shearing stress in the 1-in.-diameter bolt at A, which is in double shear.[1]

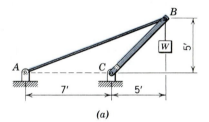

(a)

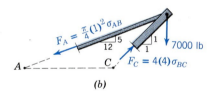

(b)

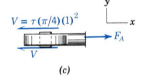

(c)

FIG. 1-13

Solution

The members AB and BC are two-force members subjected to axial loading. When the equations of equilibrium are applied to the free-body diagram of Fig. 1-13b, the following results are obtained:

$$\circlearrowleft+ \qquad \Sigma M_C = 0$$

$$7000(5) - (5/13)(F_A)(7) = 0, \qquad F_A = 13{,}000 \text{ lb } T$$

$$\sigma_{AB} = \frac{F_A}{A} = \frac{13{,}000}{(\pi/4)(1)^2} = \underline{16{,}550 \text{ psi } T} \qquad \text{Ans.}$$

[1] Structural timbers specified as rough sawn will be approximately full size. Cross-sectional dimensions of dressed timbers will be between $\frac{3}{8}$ and $\frac{1}{2}$ in. smaller than nominal. For convenience, all timbers in this book will be considered full size

Observe that it is not necessary to solve the equation explicitly for the force F_A in the member if only the stress is required. Hence, for the next equation this step will be omitted; thus,

$$\circlearrowleft+ \qquad \Sigma M_A = 0$$

$$7000(12) - \frac{1}{\sqrt{2}}(16)(\sigma_{BC})(7) = 0$$

$$\sigma_{BC} = \underline{1061 \text{ psi } C} \qquad \text{Ans.}$$

From the free-body diagram of Fig 1-13c, which is a free-body diagram of the connection at A, and using the value of force F_A as determined above, the cross shearing stress is obtained as follows:

$$\xrightarrow{+} \qquad \Sigma F_x = 0$$

$$F_A - 2V = 0$$

$$V = \tau\pi(1)^2/4 = 6500$$

$$\tau = \underline{8280 \text{ psi}} \qquad \text{Ans.}$$

Stress, being the intensity of internal force, has the dimensions of force per unit area (FL^{-2}). Until recently, the commonly used unit for stress in the United States was the pound per square inch, abbreviated as psi. Since metals can sustain stresses of several thousand pounds per square inch, the unit ksi (kips per square inch) is also frequently used (1 ksi = 1000 psi). With the advent of the International System of Units (usually called SI units), units of stress based on the metric system are beginning to be used in the United States and will undoubtedly come into wider use in the next few years.

During the transition, both systems will be encountered, so some problems in this book are given using the English system (pounds and inches) and others with SI units (newtons and meters). For problems with SI units, forces will be given in newtons (N) or kilonewtons (kN), dimensions in meters (m) or millimeters (mm), and masses in kilograms (kg). The SI unit for stress or pressure is a newton per square meter (N/m^2), also known as a pascal (Pa). Stress magnitudes normally encountered in engineering applications are expressed in meganewtons per square meter (MN/m^2), or megapascals (MPa).

Data will usually be given in either the English or the SI system; a combination of the two systems will not be used for a given problem. When it is necessary to convert units from one system to the other, the following conversion factors will be useful:

$$
\begin{array}{ll}
1\ \text{lb} = 4.448\ \text{N} & 1\ \text{N} = 0.2248\ \text{lb} \\
1\ \text{in.} = 25.40\ \text{mm} & 1\ \text{mm} = 0.03937\ \text{in.} \\
1\ \text{ft} = 0.3048\ \text{m} & 1\ \text{m} = 3.281\ \text{ft} \\
1\ \text{ksi} = 6.895\ \text{MPa} & 1\ \text{MPa} = 0.1450\ \text{ksi}
\end{array}
$$

Note that one MPa is approximately one-seventh of one ksi. This one-seventh "rule of thumb" is convenient for interpreting the physical significance of answers in SI units for those more accustomed to English units.

Acceleration due to gravity is 9.81 meters per second per second. From Newton's second law, the force due to gravitational attraction on a mass of 1 kg is 9.81 N. Consequently, when a 1-kg mass is supported by a cord in the earth's gravitational field, the tension in the cord is 9.81 N. Similarly the tension in a cable sustaining a mass of m kg is 9.81 m N. A more detailed discussion of SI units is given in Appendix E.

EXAMPLE 1-2

The bar with circular cross section shown in Fig. 1-14a has steel, brass, and aluminum sections. Axial loads are applied at cross sections A, B, C, and D. If the allowable axial stresses in tension or compression are 125 MPa in the steel, 70 MPa in the brass, and 85 MPa in the aluminum, determine the diameters required for each of the sections.

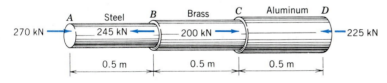

FIG. 1-14a

Solution

The forces transmitted by cross sections in intervals AB, BC, and CD of the bar shown in Fig. 1-14a are obtained by using the three free-body diagrams shown in Fig. 1-14b. Summing forces along the axis of the bar yields:

$$
\begin{array}{ll}
F_s + 270 = 0 & F_s = -270\ \text{kN} = 270\ \text{kN}\ C \\
F_b + 270 - 245 = 0 & F_b = -25\ \text{kN} = 25\ \text{kN}\ C \\
F_a + 270 - 245 + 200 = 0 & F_a = -225\ \text{kN} = 225\ \text{kN}\ C
\end{array}
$$

A pictorial representation of the distribution of force in bar $ABCD$ is shown in Fig. 1-14c. This type of representation is known as a *load*

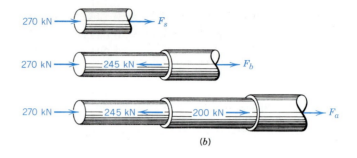

(b)

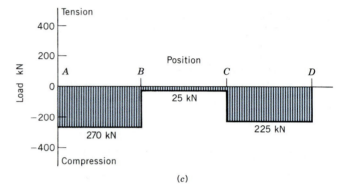

FIG. 1-14b,c

(c)

diagram and has been shown to be useful in solving many problems involving axial-force distributions.

The cross-sectional areas of the bar required to limit the stresses to the specified values are obtained from Eq. 1-2. Thus,

$$A = F/\sigma_{avg}$$

$$\frac{\pi}{4}d_s{}^2 = \frac{270(10^3)}{125(10^6)} \qquad d_s = 52.4(10^{-3})\text{ m} = \underline{52.4\text{ mm}} \qquad \text{Ans.}$$

$$\frac{\pi}{4}d_b{}^2 = \frac{25(10^3)}{70(10^6)} \qquad d_b = 21.3(10^{-3})\text{ m} = \underline{21.3\text{ mm}} \qquad \text{Ans.}$$

$$\frac{\pi}{4}d_a{}^2 = \frac{225(10^3)}{85(10^6)} \qquad d_a = 58.1(10^{-3})\text{ m} = \underline{58.1\text{ mm}} \qquad \text{Ans.}$$

PROBLEMS

Note: Unless otherwise specified, all members are assumed to have negligible weight, and all pins used for connections are assumed to be smooth. An asterisk after the problem number indicates that the answer is given in the back of the book.

1-1* The round bar shown in Fig. P1-1 has steel, brass, and aluminum sections. Axial loads are applied at cross sections *A, B, C,* and *D.* If the allowable axial stresses are 40 ksi in the steel, 30 ksi in the brass, and 35 ksi in the aluminum, determine the diameters required for each of the sections.

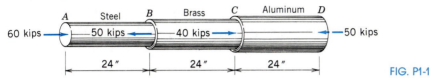

FIG. P1-1

1-2* A flat steel bar has axial loads applied at points *A, B, C,* and *D* as shown in Fig. P1-2. If the bar has a cross-sectional area of 2000 mm², determine the axial stress in the bar

  (a) on a cross section 200 mm to the right of point *A.*

  (b) on a cross section 1500 mm to the right of point *A.*

  (c) on a cross section 2500 mm to the right of point *A.*

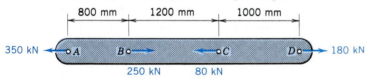

FIG. P1-2

1-3* Axial loads are applied with rigid bearing plates to the steel pipes shown in Fig. P1-3. The cross-sectional areas are 16 in.² for *AB,* 4 in.² for *BC,* and 12 in.² for *CD.* Determine the axial stress in the pipe

  (a) on a cross section in the interval *AB.*

  (b) on a cross section in the interval *BC.*

  (c) on a cross section in the interval *CD.*

1-4* An aluminum bar is loaded and supported as shown in Fig. P1-4. If the axial stress in the bar must not exceed 150 MPa *T*, determine the minimum cross-sectional area required for each of the sections.

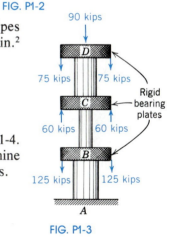

FIG. P1-3

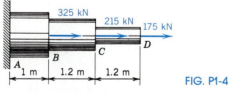

FIG. P1-4

1-5 A 2024-T4 aluminum tube with an outside diameter of 1.000 in. will be used to support a 10-kip load. If the axial stress in the member must be limited to 30 ksi *T* or *C,* determine the wall thickness required for the tube.

1-6 A stainless-steel tube with an outside diameter of 50 mm and a wall thickness of 5 mm is used as a compression member. If the axial stress in the member must be limited to 500 MPa, determine the maximum load that the member can support.

1-7* A brass tube with an outside diameter of 2.00 in. and a wall thickness of 0.375 in. is connected to a steel tube with an inside diameter of 2.00 in. and a wall thickness of 0.250 in. by using a 0.750-in. diameter pin as shown in Fig. P1-7. Determine

  (a) the cross shearing stress in the pin when the joint is carrying an axial load of 10 kips.

  (b) the length of joint required if the pin is replaced by a glued joint and the shearing stress in the glue must be limited to 250 psi.

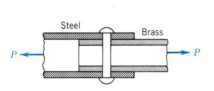

FIG. P1-7

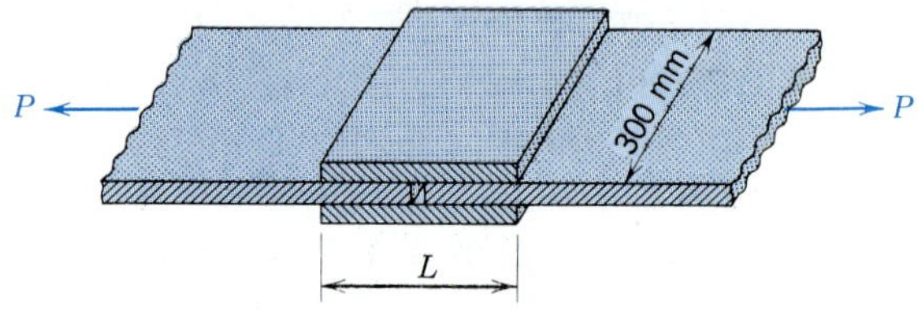

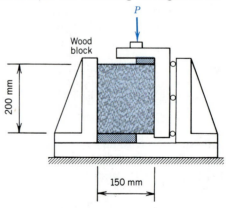

FIG. P1-8

1-8* Two strips of a plastic material are bonded together as shown in Fig. P1-8. The average shearing stress in the glue must be limited to 950 kPa. What length L of splice plate is needed if the axial load carried by the joint is 50 kN?

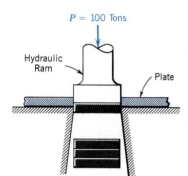

FIG. P1-9

1-9 A 100-ton hydraulic punch press is used to punch holes in a 0.50-in.-thick steel plate, as illustrated schematically in Fig. P1-9. If the average punching shear resistance of the steel plate is 40 ksi, determine the maximum diameter hole that can be punched.

1-10 A device for determining the shearing strength of wood is shown in Fig. P1-10. The dimensions of the wood specimen are 150 mm wide by 200 mm high by 50 mm thick. If a load of 75 kN is required to fail the specimen, determine the shearing strength of the wood.

FIG. P1-10

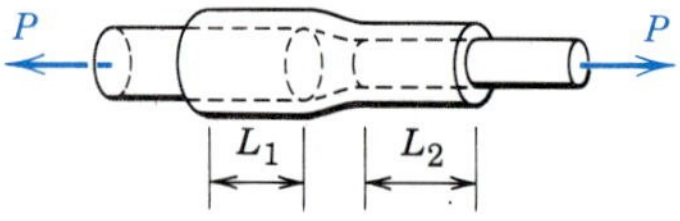

1-11* A coupling is used to connect 2-in.-diameter plastic rods to 1.5-in.-diameter rods as shown in Fig. P1-11. If the average shearing stress in the adhesive must be limited to 500 psi, determine the minimum

FIG. P1-11

lengths L_1 and L_2 required for the joint if the applied axial load P is 8000 lb.

1-12* The steel pipe column shown in Fig. P1-12 has an outside diameter of 150 mm and a wall thickness of 15 mm. The load imposed on the column by the timber beam is 150 kN. Determine

(a) the average bearing stress at the surface between the column and the steel bearing plate.

(b) the diameter of a circular bearing plate if the average bearing stress between the steel plate and the wood beam is nct to exceed 3.25 MPa.

1-13 The tie rod shown in Fig. P1-13 has a diameter of 1.50 in. and is used to resist the lateral pressure against the walls of a grain bin. If the tensile stress in the rod is 10 ksi, determine

(a) the diameter of the washer that must be used if the bearing stress on the wall is not to exceed 350 psi.

(b) the thickness of the head of the rod if the punching shear stress in the head is not to exceed 5 ksi.

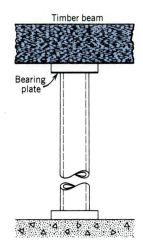

FIG. P1-12

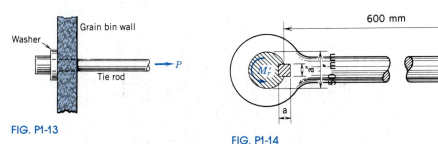

FIG. P1-13

FIG. P1-14

1-14 A lever is attached to the shaft of a steel gate valve with a square key as shown in Fig. P1-14. If the shearing stress in the key must not exceed 125 MPa, determine the minimum dimension "a" that must be used if the key is 20 mm long.

1-15* A lever with two protruding pins, as shown in Fig. P1-15, is used to operate a locking mechanism. The maximum load P required to operate the lock is 25 lb. Determine the minimum diameter needed for each of the plastic pins if the average shearing stress in the plastic must not exceed 6 ksi.

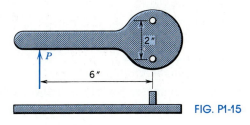

FIG. P1-15

1-16* A vertical shaft is supported by a thrust collar and bearing plate, as shown in Fig. P1-16. Determine the maximum axial load that can be applied to the shaft if the average punching shear stress in the collar and the average bearing stress between the collar and the plate are limited to 75 and 100 MPa, respectively.

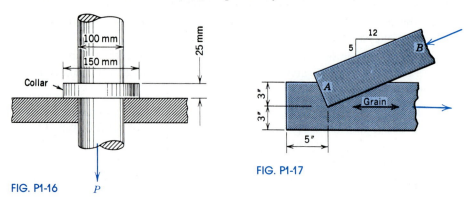

FIG. P1-16

FIG. P1-17

1-17 The inclined member AB of a timber truss is framed into a 4×6-in. bottom chord, as shown in Fig. P1-17. Determine the axial compressive force in member AB when the average shearing stress parallel to the grain at the end of the bottom chord is 225 psi.

1-18 Member AD of the timber truss shown in Fig. P1-18 is framed into the 100×150-mm bottom chord ABC, as shown in the insert. Determine the dimension "a" that must be used if the average shearing stress parallel to the grain at the ends of chord ABC is not to exceed 2.25 MPa.

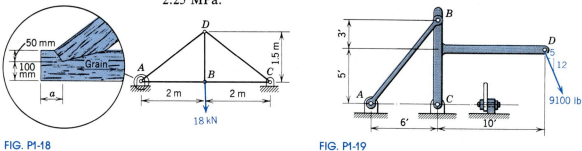

FIG. P1-18

FIG. P1-19

1-19* A simple structure is used to support a 9100-lb load as shown in Fig. P1-19. Determine

    (a) the minimum diameter for tie rod AB if the axial stress in the rod is limited to 20 ksi.

    (b) the minimum diameters for pins A and B if the cross shearing stress in the pins is limited to 10 ksi.

    (c) the minimum diameter for pin C if the cross shearing stress in the pin is limited to 12 ksi.

1-20* The axial stresses are 10 MPa *C* in the wood post *B* and 200 MPa *T* in the steel bar *A* of Fig. P1-20. Determine

(a) the load *P*.

(b) the minimum diameter for pin *C* if it is in single shear and the cross shearing stress is limited to 75 MPa.

(c) the minimum diameter for pin *D* if it is in double shear and the cross shearing stress is limited to 85 MPa.

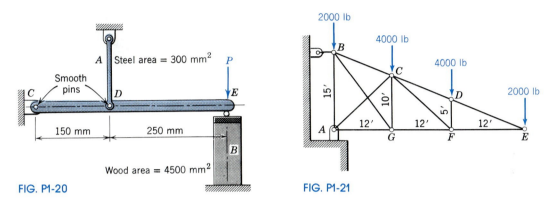

FIG. P1-20

FIG. P1-21

1-21 A pin-connected truss is loaded and supported as shown in Fig. P1-21. Determine

(a) the axial stress in member *GF* if it has a cross-sectional area of 2.50 in.²

(b) the minimum cross-sectional area for member *CD* if the axial stress is limited to 7500 psi.

1-22 A pin-connected truss is loaded and supported as shown in Fig. P1-22. Determine

(a) the axial stress in member *GF* if it has a cross-sectional area of 4750 mm².

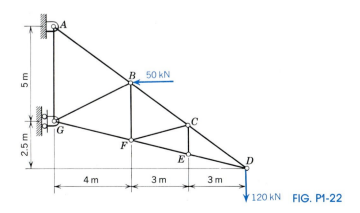

FIG. P1-22

(b) the minimum cross-sectional area for member *BC* if the axial stress is limited to 70 MPa.

(c) the minimum cross-sectional area for member *BG* if the axial stress is limited to 35 MPa.

1-23* The turnbuckles in Fig. P1-23 are tightened until the compression block *D* exerts a force of 28,000 lb on the beam *AC* at *B*. Member *D* is a hollow tube with an inside diameter of 1.00 in. and an outside diameter of 2.00 in. Determine

(a) the axial stress in member *D*.

(b) the minimum diameter for rod *AE* if the axial stress is limited to 15 ksi.

(c) the minimum diameter for the pin at joint *A* if the cross shearing stress is limited to 12 ksi.

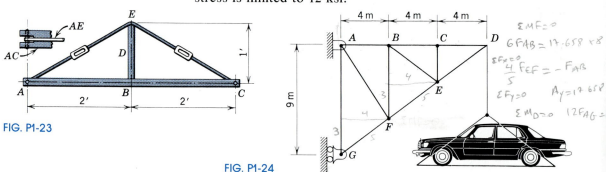

FIG. P1-23

FIG. P1-24

1-24* The truss shown in Fig. P1-24 supports an automobile whose mass is 1800 kg. Determine

(a) the axial stress in member *AB* if it has a cross-sectional area of 200 mm².

(b) the minimum cross-sectional area for member *EF* if the axial stress is limited to 50 MPa.

(c) the minimum cross-sectional area for member *AG* if the axial stress is limited to 150 MPa.

1-25 A flat steel bar 4 in. wide by 1 in. thick has axial loads applied with 1.50-in.-diameter pins in double shear at points *A*, *B*, *C*, and *D*, as shown in Fig. P1-25. Determine

(a) the axial stress in the bar on a cross section at pin *B*.

(b) the average bearing stress on the bar at pin *B*.

(c) the cross shearing on the pin at *A*.

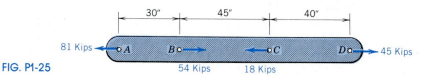

FIG. P1-25

1-26 Three plates are joined with a 12-mm-diameter pin, as shown in Fig. P1-26. Determine the maximum load P that can be transmitted by the joint if

(a) the maximum axial stress on a cross section at the pin must be limited to 350 MPa.

(b) the maximum bearing stress between a plate and the pin must be limited to 650 MPa.

(c) the maximum shearing stress on a cross section of the pin must be limited to 240 MPa.

(d) the punching shear resistance of the material in the top and bottom plates is 300 MPa.

$P_m = \sigma \times A_m \left(25(25-12) \right)$
$P_u = 25 \times 10 \times (30-12)$
$P_m = \sigma \times 25 \times 12$
$P_u = 25 \times 30 \times 12$
$P = 2\tau \left(\frac{\pi}{4} \times 12^2 \right)$
$\frac{P}{2} = 2\tau (10 \times 12)$

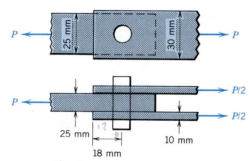

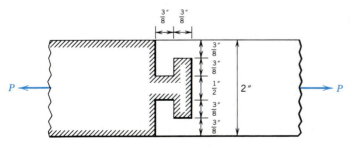

FIG. P1-26

1-27* The joint shown in Fig. P1-27 is used in a steel tension member which has a 2×1-in. rectangular cross section. If the allowable normal, bearing, and punching-shear stresses in the joint are 13.5 ksi, 18.0 ksi, and 6.50 ksi, respectively, determine the maximum load P that can be carried by the joint.

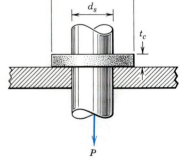

FIG. P1-27

FIG. P1-28

1-28 A thrust collar is press-fitted onto a shaft as shown in Fig. P1-28. Develop an expression for the maximum load P that the thrust collar can support in terms of the dimensions of the shaft and collar d_c, d_s, t_c, and the normal stress σ and the coefficient of friction μ at the interface between the shaft and the collar.

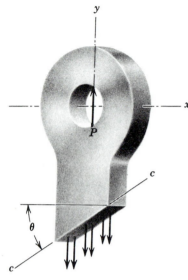

FIG. 1-15

FIG. 1-16

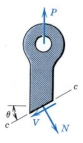

FIG. 1-17

1-6

STRESSES ON AN INCLINED PLANE IN AN AXIALLY LOADED MEMBER

In previous sections, normal, shear, and bearing stresses on planes parallel and perpendicular to the axes of centrically loaded members were introduced. Stresses on planes inclined to the axes of axially loaded bars will now be considered. When the eyebar AB shown in Figs. 1-4 and 1-5 is cut by an inclined plane c–c, a free-body diagram of the upper portion of the bar would appear as shown in Fig. 1-15. Equilibrium of the upper portion of the bar is established by placing a distribution of internal force on the cut section as shown in Fig. 1-15. The resultant F of this distribution of internal force is equal in magnitude to the applied load P and has a line of action that is coincident with the axis of the bar as shown in Fig. 1-16. An average total stress S_{avg} on the inclined surface can be computed by using Eq. 1-2. This total stress conveys very little information that is useful for design purposes. The resultant F, however, can be replaced by normal and tangential components N and V, as shown in Fig. 1-17. These components can then be used to compute normal and shear stresses on the inclined surface by using Eqs. 1-2 and 1-4. The area of the inclined surface A_n equals $A/\cos\theta$, where A is the cross-sectional area of the axially loaded member, and $N = P\cos\theta$, and $V = P\sin\theta$. Therefore,

$$\sigma_n = \frac{N}{A_n} = \frac{P\cos\theta}{A/\cos\theta} = \frac{P}{A}\cos^2\theta = \frac{P}{2A}(1 + \cos 2\theta) \qquad (1\text{-}7)$$

$$\tau_n = \frac{V}{A_n} = \frac{P\sin\theta}{A/\cos\theta} = \frac{P}{A}\sin\theta\cos\theta = \frac{P}{2A}\sin 2\theta \qquad (1\text{-}8)$$

These stresses, when evaluated, can be compared with experimentally determined values for the strength of the material in tension, compression, and shear, and thus yield significant design information.

In the preceding discussion, the assumption is made that the stress is uniformly distributed over the inclined surface. A necessary (but not sufficient) condition for the distribution to be uniform is that the loading be centric. Nonuniform stress distributions under centric loading will be discussed later in the book. *In this book, unless otherwise indicated, a uniform stress distribution is to be assumed for all centric (or assumed centric) loading.*

Since both the area of the inclined surface A_n and the values for the normal and shear forces N and V on the surface depend on the angle of inclination θ of the inclined plane with respect to the applied load, the normal and shear stresses σ_n and τ_n on the inclined plane also depend on the angle of inclination θ of the plane. This dependence of

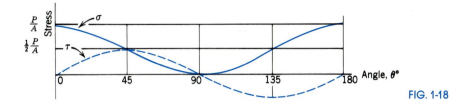

FIG. 1-18

stress on both force and area means that stress is not a vector quantity; therefore, the laws of vector addition do not apply to stresses that act on different planes. This need not be cause for concern if, in the application of the equations of equilibrium (or motion), one always replaces a stress with a total force (stress multiplied by the appropriate area), thus reducing the problem to one involving ordinary force vectors. However, stresses that act on a particular plane can be treated as vectors, since they all are associated with the same area.

A graph showing the magnitudes of σ_n and τ_n as a function of θ is shown in Fig. 1-18. These results indicate that σ_n is maximum when θ is 0° or 180°, that τ_n is maximum when θ is 45° or 135°, and also that $\tau_{max} = \sigma_{max}/2$. Therefore, the maximum normal and shearing stresses for *uniaxial* centric tensile or compressive loading are

$$\sigma_{max} = P/A \qquad (1\text{-}9)$$

$$\tau_{max} = P/2A \qquad (1\text{-}10)$$

Note that the normal stress is either maximum or minimum on planes for which the shearing stress is zero. It can be shown that the shearing stress is always zero on the planes of maximum or minimum normal stress. The concepts of maximum and minimum normal stress and maximum shearing stress for more general cases will be treated in later sections of the book.

Laboratory experiments indicate that both normal and shearing stresses under axial loading are important, since a brittle material loaded in uniaxial tension will fail in tension on a transverse plane, whereas a ductile material loaded in uniaxial tension will fail in shear on a 45° plane.

The plot of normal and shear stresses for axial loading, shown in Fig. 1-18, indicates that the sign of the shearing stress changes when θ is greater than 90°. The magnitude of the shearing stress for any angle θ, however, is the same as that for 90° + θ. The sign change merely indicates that the shear force V changes sense, being directed down to the right on plane 90° + θ instead of down to the left on plane θ, as shown in Fig. 1-17.

The equality of shearing stresses on orthogonal planes can be demonstrated by applying the equations of equilibrium to the free-body

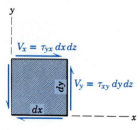

FIG. 1-19

diagram of a small rectangular block of thickness dz shown in Fig. 1-19. If a shearing force[2] $V_x = \tau_{yx}\, dx\, dz$ is applied to the top surface of the block, the equation $\Sigma F_x = 0$ will dictate the application of an oppositely directed force V_x to the bottom of the block, thus leaving the block subjected to a clockwise couple. This clockwise couple must be balanced by a counterclockwise couple composed of the oppositely directed forces V_y applied to the vertical faces of the block. Finally, application of the equation $\Sigma M_z = 0$ yields

$$\tau_{yx}\,(dx\;dz)\,dy \;=\; \tau_{xy}\,(dy\;dz)\,dx$$

from which

$$\tau_{yx} \;=\; \tau_{xy} \tag{1-11}$$

Therefore, *if a shearing stress exists at a point on any plane, there must also exist at this point a shearing stress of the same magnitude on an orthogonal plane*. This statement is also valid when normal stresses are acting on the planes, since the normal stresses occur in collinear but oppositely directed pairs, and thus have zero moment with respect to any axis.

EXAMPLE 1-3

The two parts of the eyebar shown in Fig. 1-20a are connected with $\frac{1}{2}$-in.-diameter bolts (one on each side). Specifications for the bolts require that the axial tensile stress not exceed 12.0 ksi and the cross shearing stress not exceed 8.0 ksi. Determine the maximum load P that can be applied to the eyebar without exceeding either specification. The cross-sectional area of the bolts at the root of the threads is 0.126 in.[2].

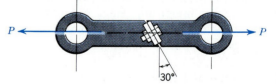

FIG. 1-20a

Solution

The forces transmitted between the two parts of the eyebar by the bolts can be obtained by using a free-body diagram of a part of the eyebar

---

[2]The double subscript on the shearing stress is used to designate both the plane on which the stress acts and the direction of the stress. The first subscript indicates the plane (or rather the normal to the plane), and the second subscript indicates the direction of the stress.

as shown in Fig. 1-20b. Summing forces in directions normal n and tangent t to the mating surface of the joint yields

FIG. 1-20b

$$N - P \cos 30° = 0 \quad \text{or} \quad P = N/\cos 30° \qquad (a)$$
$$V - P \sin 30° = 0 \quad \text{or} \quad P = V/\sin 30°$$

The cross-sectional areas of the bolts transmitting the forces N and V are

$$A_N = 2(0.1260) = 0.2520 \text{ in.}^2$$

since the area at the root of the threads limits the tensile force that can be transmitted by a bolt, and

$$A_V = 2 \left(\frac{\pi}{4}\right)\left(\frac{1}{2}\right)^2 = 0.3927 \text{ in.}^2$$

since the full diameter of the shank of the bolt is effective in transmitting shearing forces.

The maximum permissible values for forces N and V are obtained from Eqs. 1-2 and 1-4 as

$$N_{max} = \sigma_{max} A_N = 12.00(0.2520) = 3.024 \text{ kips}$$
$$V_{max} = \tau_{max} A_V = 8.00(0.3927) = 3.142 \text{ kips}$$

The maximum load P that can be transmitted by the eyebar is obtained from Eqs. (a) as

$$P = \frac{N}{\cos 30°} = \frac{3.024}{\cos 30°} = 3.49 \text{ kips}$$

$$P = \frac{V}{\sin 30°} = \frac{3.124}{\sin 30°} = 6.28 \text{ kips}$$

Since both conditions must be satisfied,

$$P_{max} = \underline{3.49 \text{ kips}} \qquad\qquad \text{Ans.}$$

PROBLEMS

1-29* A steel rod of circular cross section will be used to carry an axial tensile load of 50 kips. The maximum stresses in the rod must be limited to 25 ksi in tension and 15 ksi in shear. Determine the required diameter d for the rod.

1-30* A concrete cylinder 75 mm in diameter and 150 mm high failed along a plane making an angle of 57° with the horizontal when subjected to an axial vertical compressive load of 80 kN. Determine the normal and shearing stresses on the failure plane.

1-31 A structural steel eyebar with a 2 × 6-in. rectangular cross section is subjected to an axial tensile load of 270 kips. Determine

    (a) the normal and shearing stresses on a plane through the bar that makes an angle of 40° with the direction of the load.

    (b) the maximum normal and shearing stresses in the bar.

1-32 A structural steel bar 25 mm in diameter is subjected to an axial tensile load of 55 kN. Determine

    (a) the normal and shearing stresses on a plane through the bar that makes an angle of 30° with the direction of the load.

    (b) the maximum normal and shearing stresses in the bar.

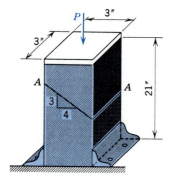

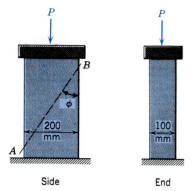

FIG. P1-33

1-33* Specifications for the 3 × 3 × 21-in. block shown in Fig. P1-33 require that the normal and shearing stresses on plane $A–A$ not exceed 800 psi and 500 psi, respectively. Determine the maximum load P that can be applied without exceeding the specifications.

1-34* The normal stress on plane AB of the rectangular block shown in Fig. P1-34 is 12 MPa C when the load P is applied. If angle ϕ is 36°, determine

    (a) the load P.

    (b) the shearing stress on plane AB.

    (c) the maximum normal and shearing stresses in the block.

FIG. P1-34

1-35 A plastic bar with a circular cross section will be used to support an axial load of 1000 lb, as shown in Fig. P1-35. The allowable normal and shearing stresses in the adhesive joint used to connect the two parts of the bar are 675 psi and 350 psi, respectively. Determine the required diameter d for the bar.

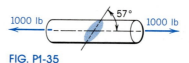

FIG. P1-35

1-36 Determine the maximum axial load P that can be applied to the wood compression block shown in Fig. P1-36 if specifications require that the shearing stress parallel to the grain not exceed 5.25 MPa, the compressive stress perpendicular to the grain not exceed 13.60 MPa, and the maximum shearing stress in the block not exceed 8.75 MPa.

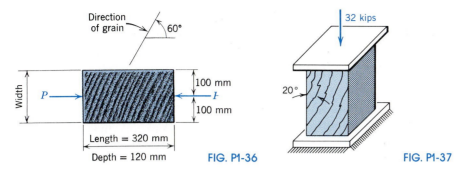

FIG. P1-36   FIG. P1-37

1-37* A timber block with a square cross section will be used to support a compressive load of 32 kips as shown in Fig. P1-37. Determine the size of the block required if the shearing stress parallel to the grain is not to exceed 200 psi and if the normal stress parallel to the grain is not to exceed 2000 psi.

1-38* A steel bar with a butt-welded joint, as shown in Fig. P1-38, will be used to carry an axial tensile load of 400 kN. If the normal and shearing stresses on the plane of the butt weld must be limited to 70 MPa and 45 MPa, respectively, determine the minimum thickness required for the bar.

$$A = \frac{0.1 \, t}{\sin 57°}$$

$$\frac{P}{A} \left[\frac{57°}{\tau_n A} \right.$$
$$\sigma_n A$$

FIG. P1-38

1-39 A steel eyebar with a 4 × 1-in. rectangular cross section has been designed to transmit an axial tensile load. The length of the eyebar must be increased by welding a new center section in the bar ($45° \leq \phi \leq 90°$) as shown in Fig. P1-39. The stresses in the weld material must be limited to 12,000 psi in tension and 9000 psi in shear. Determine

(a) the optimum angle ϕ for the joint.

(b) the maximum safe load P for the redesigned member.

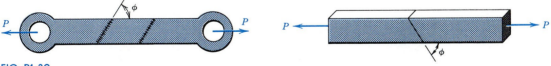

FIG. P1-39

FIG. P1-40

1-40 A wood tension member with a 50 × 100-mm rectangular cross section will be fabricated with an inclined glued joint ($45° \leq \phi \leq 90°$) at its midsection, as shown in Fig. P1-40. If the allowable working stresses for the glue are 5 MPa in tension and 3 MPa in shear, determine

(a) the optimum angle ϕ for the joint.

(b) the maximum safe load P for the member.

1-7

STRESS AT A GENERAL POINT IN AN ARBITRARILY LOADED MEMBER

In Sections 1-3, 1-4, and 1-6 the concept of stress was introduced by considering the internal force distribution required to satisfy equilibrium in a portion of a bar under centric load. The nature of the force distribution led to uniformly distributed normal and shearing stresses on transverse planes through the bar. In more complicated structural members or machine components the stress distributions will not be uniform on arbitrary internal planes; therefore, a more general concept of the state of stress at a point is needed.

Consider a body of arbitrary shape that is in equilibrium under the action of a system of applied forces. The nature of the internal force distribution at an arbitrary interior point O can be studied by exposing an interior plane through O as shown in Fig. 1-21a. The force distribution required on such an interior plane to maintain equilibrium of the isolated part of the body, in general, will not be uniform; however, any distributed force acting on the small area ΔA surrounding the point of interest O can be replaced by a statically equivalent resultant force ΔF_n through O and a couple ΔM_n. The subscript n indicates that the

FIG. 1-21

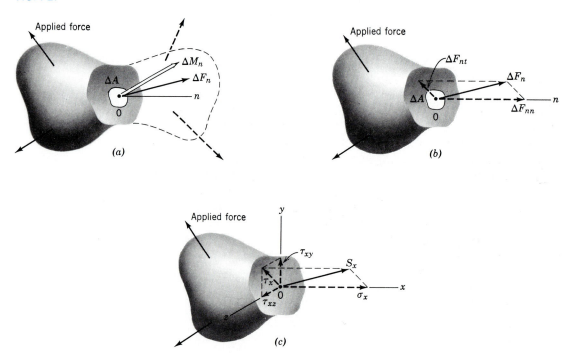

(a)

(b)

(c)

resultant force and couple are associated with a particular plane through O—namely, the one having an outward normal in the n direction at O. For any other plane through O the values of ΔF and ΔM could be different. Note that the line of action of ΔF_n or ΔM_n may not coincide with the direction of n. If the resultant force ΔF_n is divided by the area ΔA, an average force per unit area (average resultant stress) is obtained. As the area ΔA is made smaller and smaller, the couple ΔM_n vanishes as the force distribution becomes more and more uniform. In the limit a quantity known as the *stress vector*[3] or resultant stress is obtained. Thus,

$$S_n = \lim_{\Delta A \to 0} \frac{\Delta F_n}{\Delta A}$$

In Section 1-6 it was pointed out that materials respond to components of the stress vector rather than the stress vector itself. In particular, the components normal and tangent to the internal plane were important. As shown in Fig. 1-21b, the resultant force ΔF_n can be resolved into the components ΔF_{nn} and ΔF_{nt}. A normal stress σ_n and a shearing stress τ_n are then defined as

$$\sigma_n = \lim_{\Delta A \to 0} \frac{\Delta F_{nn}}{\Delta A}$$

and

$$\tau_n = \lim_{\Delta A \to 0} \frac{\Delta F_{nt}}{\Delta A}$$

For purposes of analysis it is convenient to reference stresses to some coordinate system. For example, in a Cartesian coordinate system the stresses on planes having outward normals in the x, y, and z directions are usually chosen. Consider the plane having an outward normal in the x direction. In this case the normal and shear stresses on the plane will be σ_x and τ_x, respectively. Since τ_x, in general, will not coincide with the y or z axes, it must be resolved into the components τ_{xy} and τ_{xz}, as shown in Fig. 1-21c.

Unfortunately the state of stress at a point in a material is not completely defined by these three components of the stress vector, since the stress vector itself depends on the orientation of the plane with which it is associated. An infinite number of planes can be passed through the point, resulting in an infinite number of stress vectors being

[3] The component of a tensor on a plane is a vector; therefore, on a particular plane, the stresses can be treated as vectors.

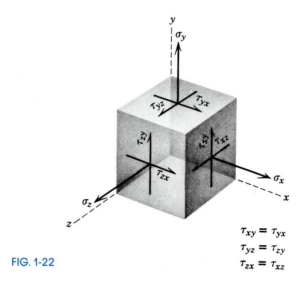

$$\tau_{xy} = \tau_{yx}$$
$$\tau_{yz} = \tau_{zy}$$
$$\tau_{zx} = \tau_{xz}$$

FIG. 1-22

associated with the point. Fortunately it can be shown[4] that the spec-ification of stresses on three mutually perpendicular planes is sufficient to describe completely the state of stress at the point. The rectangular components of stress vectors on planes having outward normals in the coordinate directions are shown in Fig. 1-22. The six faces of the small element are denoted by the directions of their outward normals so that the positive x face is the one whose outward normal is in the direction of the positive x axis. The coordinate axes x, y, and z are arranged as a right-hand system.

The sign convention for stresses is as follows. Normal stresses (in-dicated by the symbol σ and a single subscript to indicate the plane on which the stress acts) are positive if they point in the direction of the outward normal. Thus, normal stresses are positive if tensile. Shearing stresses are denoted by the symbol τ followed by two subscripts; the first subscript designates the plane on which the shearing stress acts and the second the coordinate axis to which it is parallel. Thus, τ_{xz} is the shearing stress on an x plane parallel to the z axis. A positive shearing stress points in the positive direction of the coor-dinate axis of the second subscript if it acts on a surface with an out-ward normal in the positive direction. Conversely, if the outward normal of the surface is in the negative direction, then the positive shearing stress points in the negative direction of the coordinate axis of the second subscript. The stresses shown on the element in Fig. 1-22 are all positive.

[4]See Section 7-10.

1-8

TWO-DIMENSIONAL OR PLANE STRESS

Considerable insight into the nature of stress distributions can be gained by considering a state of stress known as two-dimensional or *plane stress*. For this case, two parallel faces of the small element shown in Fig. 1-22 are assumed to be free of stress. For purposes of analysis, let these faces be perpendicular to the z axis. Thus,

$$\sigma_z = \tau_{zx} = \tau_{zy} = 0$$

From Eq. 1-11, however, this also implies that

$$\tau_{xz} = \tau_{yz} = 0$$

The components of stress present for plane stress analysis will be σ_x, σ_y, and $\tau_{xy} = \tau_{yx}$. For convenience this state of stress is usually represented by the two-dimensional sketch shown in Fig. 1-23. The three-dimensional element of which the sketch is a plane projection should be kept in mind at all times. Normal and shearing stresses on an arbitrary plane such as plane $A–A$ in Fig 1-23 can be obtained by using the free-body diagram method discussed for centric loading in Section 1-6. The solution to the following example illustrates this method of approach under plane stress conditions.

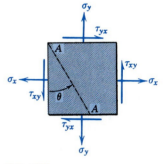

FIG. 1-23

EXAMPLE 1-4

At a given point in a machine element, the following stresses were evaluated: 100 MPa T and zero shear on a horizontal plane and 50 MPa C on a vertical plane. Determine the stresses at this point on a plane having a slope of 3 vertical to 4 horizontal.

Solution

As an aid to visualization of the data, it is suggested that the differential block of Fig. 1-24*a* be drawn (a stress picture, not a free-body diagram). The next step is to draw a free-body diagram subjected to ordinary force vectors. Various satisfactory free-body diagrams can be used, such as the wedge-shaped element defined by the three given planes in Fig. 1-24*b*, in which the shaded area indicates the plane on which the stresses to be evaluated are acting. To this area is assigned arbitrarily the magnitude dA, and the corresponding areas of the horizontal and vertical faces of the element are 0.8 dA and 0.6 dA, respectively. The forces acting on these areas will be as indicated in the diagram.

FIG. 1-24

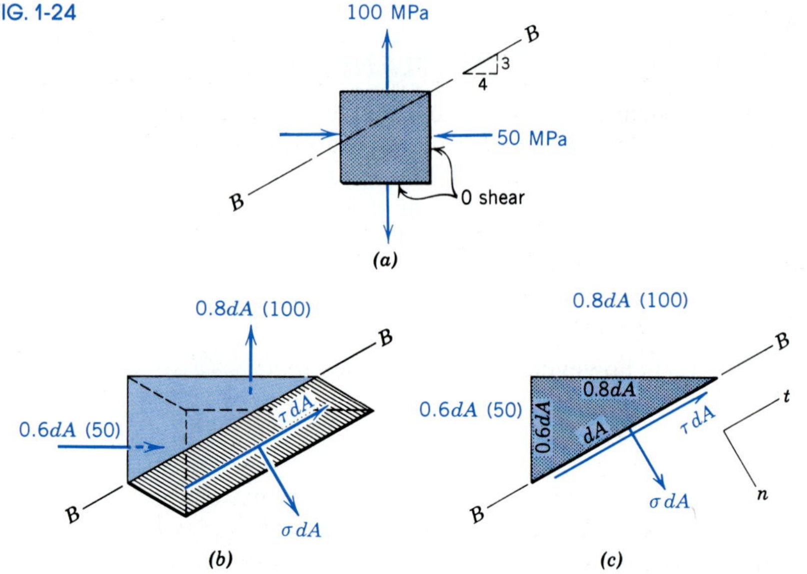

(a)

(b)

(c)

It must be emphasized that *force vectors* act on these areas. For practical purposes, the two-dimensional free-body diagram of Fig. 1-24c will be adequate. Summing forces in the *n* direction yields the following:

$$\overset{+\searrow}{\bigcirc} \qquad \Sigma F_n = 0$$

$$\sigma \, dA + 0.6 \, dA \, (50)(0.6) - 0.8 \, dA \, (100)(0.8) = 0$$

from which

$$\sigma = \underline{46.0 \text{ MPa } T} \qquad \text{Ans.}$$

When forces are summed in the *t* direction, the shearing stress is found to be

$$\tau = \underline{-72.0 \text{ MPa}} \qquad \text{Ans.}$$

Note that the normal stress should be designated as *tension* or *compression*. The presence of shearing stresses on the horizontal and vertical planes, had there been any, would merely have required two more forces on the free-body diagram: one parallel to the vertical face and one parallel to the horizontal face. Note, however, that the magnitude of the shearing stresses (not the forces) must be the same on any two orthogonal planes (see Section 1-6) and that the vectors should be directed as in Fig. 1-19 or both vectors reversed, depending on the given data.

PROBLEMS

1-41* At a point in a stressed body, there are normal stresses of 16 ksi T on a vertical plane and 12 ksi T on a horizontal plane, as shown in Fig. P1-41. Determine the normal and shearing stresses at this point on the inclined plane shown in the figure.

1-42* At a point in a stressed body, there are normal stresses of 95 MPa T on a vertical plane and 125 MPa T on a horizontal plane, as shown in Fig. P1-42. Determine the normal and shearing stresses at this point on the inclined plane shown in the figure.

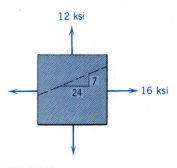

FIG. P1-41

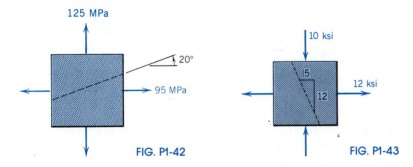

FIG. P1-42   FIG. P1-43

1-43 The stresses shown in Fig. P1-43 act at a point in a stressed body. Determine the normal and shearing stresses at this point on the inclined plane shown in the figure.

1-44 The stresses shown in Fig. P1-44 act at a point in a stressed body. Determine the normal and shearing stresses at this point on the inclined plane shown in the figure.

1-45* The stresses shown in Fig. P1-45 act at a point in a stressed body. Determine the normal and shearing stresses at this point on the inclined plane AB shown in the figure.

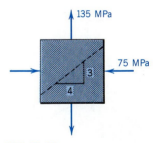

FIG. P1-44

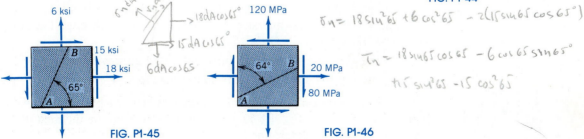

FIG. P1-45   FIG. P1-46

1-46* The stresses shown in Fig. P1-46 act at a point in a stressed body. Determine the normal and shearing stresses at this point on the inclined plane AB shown in the figure.

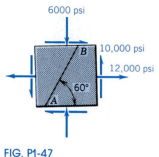

FIG. P1-47

1-47 The stresses shown in Fig. P1-47 act at a point in a stressed body. Determine the normal and shearing stresses at this point on the inclined plane *AB* shown in the figure.

1-48 The stresses shown in Fig. P1-48 act at a point in a stressed body. Determine the normal and shearing stresses at this point on the inclined plane *AB* shown in the figure.

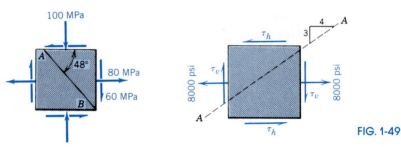

FIG. P1-48   FIG. 1-49

1-49* The stresses on horizontal and vertical planes at a point in a stressed body are shown in Fig. P1-49. The normal stress on plane *A–A* at this point is 6000 psi *T*. Determine

    (a) the magnitude of the shearing stresses τ_h and τ_v.

    (b) the magnitude and direction of the shearing stress on the inclined plane *A–A*.

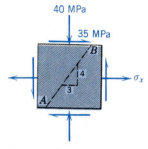

FIG. P1-50

1-50* At a point in a structural member, there are stresses on horizontal and vertical planes as shown in Fig. P1-50. The normal stress on the inclined plane *AB* is 15 MPa *T*. Determine

    (a) the normal stress σ_x on the vertical plane.

    (b) the magnitude and direction of the shearing stress on the inclined plane *AB*.

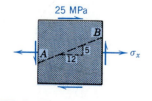

FIG. P1-51

1-51 At a point in a machine component, the normal and shearing stresses on an inclined plane are 4800 psi *T* and 1500 psi, respectively, as shown in Fig. P1-51. The normal stress on a vertical plane through the point is zero. Determine

    (a) the shearing stresses on horizontal and vertical planes.

    (b) the normal stress on a horizontal plane.

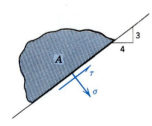

FIG. P1-52

1-52 The stresses on horizontal and vertical planes at a point in a stressed body are shown in Fig. P1-52. The normal stress on the inclined plane *AB* is 15 MPa *T*. Determine

    (a) the normal stress σ_x on the vertical plane.

    (b) the magnitude and direction of the shearing stress on the inclined plane *AB*.

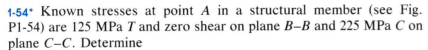

1-53 At a point in a structural member, there are stresses on horizontal and vertical planes, as shown in Fig. P1-53. The magnitude of the compressive stress σ_c is three times the magnitude of the tensile stress σ_t. Specifications require that the shearing stress on plane AB not exceed 3500 psi, and the normal stress on AB not exceed 7500 psi. Determine the maximum value of stress σ_c that will satisfy the specifications.

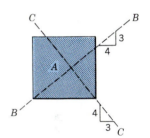

FIG. P1-53

1-54* Known stresses at point A in a structural member (see Fig. P1-54) are 125 MPa T and zero shear on plane $B–B$ and 225 MPa C on plane $C–C$. Determine

    (a) the stresses on a vertical plane through the point.

    (b) the stresses on a horizontal plane through the point.

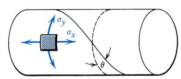

FIG. P1-54

1-55 The thin-walled cylindrical pressure vessel shown in Fig. P1-55 was constructed by wrapping a thin steel plate into a helix that forms an angle $\theta = 35°$ with respect to a transverse plane through the cylinder and butt-welding the resulting seam. In a thin-walled cylindrical pressure vessel, the normal stress σ_y on a horizontal plane through a point on the surface of the vessel is twice as large as the normal stress σ_x on a vertical plane through the point, and the shearing stresses on both horizontal and vertical planes are zero. If the stresses in the weld material on the plane of the weld must be limited to 10 ksi in tension and 7 ksi in shear, determine the maximum normal stress σ_x permitted in the vessel.

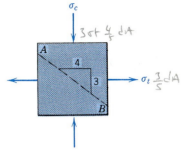

FIG. P1-55

1-56 The thin-walled cylindrical pressure vessel shown in Fig. P1-55 was constructed by wrapping a thin steel plate into a helix that forms an angle θ with respect to a transverse plane through the cylinder and butt-welding the resulting seam. The normal stresses σ_x and σ_y on horizontal and vertical planes through a point on the surface of the cylinder are 50 MPa and 100 MPa, respectively, and the shearing stresses are zero. Prepare a plot, similar to Fig. 1-18, showing the variation of normal and shearing stresses on the plane of the weld as angle θ is varied from 0° to 180°.

1-57 In a thin-walled spherical pressure vessel, the normal stresses σ_x and σ_y on vertical and horizontal planes through a point on the surface of the vessel are 10 ksi and the shearing stresses are zero, as shown in Fig. P1-57. Prepare a plot, similar to Fig. 1-18, showing the variation of normal and shearing stresses on plane $A–A$ through the point as angle θ is varied from 0° to 180°.

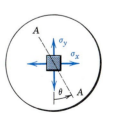

FIG. P1-57

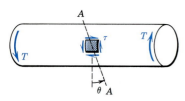

1-58 In a solid circular shaft subjected to a torsional form of loading, the normal stresses σ_x and σ_y on vertical and horizontal planes through a point on the surface of the shaft are zero and the shearing stresses are 75 MPa, as shown in Fig. P1-58. Prepare a plot, similar to Fig. 1-18, showing the variation of normal and shearing stresses on plane A–A through the point as angle θ is varied from 0° to 180°.

1-9

DISPLACEMENT, DEFORMATION, AND THE CONCEPT OF STRAIN

The relationships between stresses on planes having different aspects at a point were developed using equilibrium considerations. No assumptions involving deformations or materials used in fabricating the body were made; therefore, the results are valid for an idealized rigid body or for a real deformable body. In the design of structural elements or machine components, the deformations experienced by the body as a result of the applied loads often represent as important a design consideration as the stresses previously discussed. For this reason, the nature of the deformations experienced by a real deformable body as a result of internal force or stress distributions will be studied, and methods to measure or compute deformations will be established.

When a system of loads is applied to a machine component or structural element, individual points of the body generally move. This movement of a point with respect to some convenient reference system of axes is a vector quantity known as a *displacement*. In some instances displacements are associated with a translation and/or rotation of the body as a whole. Since the size or shape of the body is not changed by this type of displacement (referred to as a rigid-body displacement), they have no significance in the analysis of deformations. When displacements are induced by an applied load or a temperature change, individual points of the body move relative to each other; therefore, the size and/or shape of the body is altered. The change in any dimension associated with these load- or temperature-induced displacements is known as a *deformation* and will be designated by the Greek letter δ.

Under general conditions of loading, deformations will not be uniform throughout the body. Some line segments will experience extensions while others will experience contractions. Different segments (of the same length) along the same line may experience different amounts of extension or contraction. Similarly, angle changes between line segments may vary with position and orientation in the body. This nonuniform nature of load-induced deformations requires an analysis of deformations similar to the analysis of internal forces developed in previous sections of this chapter.

Strain is a quantity used to provide a measure of the intensity of a deformation (deformation per unit length) just as stress is used to provide a measure of the intensity of an internal force (force per unit area). In Sections 1-3 and 1-4, two types of stresses were defined: normal stresses and shearing stresses. This same classification is used for strains. *Normal strain,* designated by the Greek letter ϵ, is used to provide a measure of the elongation or contraction of an arbitrary line segment in a body during deformation. *Shearing strain,* designated by the Greek letter γ, is used to provide a measure of angular distortion (change in angle between two lines that are orthogonal in the undeformed state). The deformation or strain may be the result of a change in temperature, of a stress, or of other physical phenomena such as grain growth or shrinkage. In this book, only strains resulting from changes in temperature or stress are considered. The term *strain* is sometimes used to denote total deformation δ; however, *in this book, strain will always refer to deformation per unit length for normal strains or changes in angle for shearing strains.*

1-10

NORMAL AND SHEARING STRAINS

The deformation (change in length and width) of a simple bar under an axial load (see Fig. 1-25a) can be used to illustrate the idea of a normal strain. The average axial strain (a normal strain, hereafter called *axial strain*) ϵ_{avg} over the length of the bar is obtained by dividing the axial deformation δ_n by the length L of the bar. Thus,

$$\epsilon_{\text{avg}} = \frac{\delta_n}{L} \qquad (1\text{-}12)$$

In those cases in which the deformation is nonuniform along the length of the bar (a long bar hanging under its own weight, for example), the average axial strain given by Eq. 1-12 may be significantly different from the true axial strain at an arbitrary point P along the bar. The true axial strain at a point can be determined by making the length over which the axial deformation is measured smaller and smaller. In the limit a quantity defined as the axial strain at the point, $\epsilon(P)$, is obtained. This limit process is indicated by the expression

$$\epsilon(P) = \lim_{\Delta L \to 0} \frac{\Delta \delta_n}{\Delta L} = \frac{d\delta_n}{dL} \qquad (1\text{-}13)$$

In a similar manner a deformation involving a change in shape (distortion) can be used to illustrate a shearing strain. An average shearing strain γ_{avg} associated with two reference lines that are orthogonal in the undeformed state (two edges of the element shown in Fig. 1-25b)

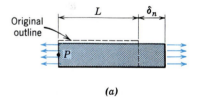

(a)

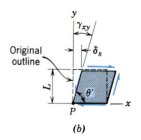

(b)

FIG. 1-25

can be obtained by dividing the shearing deformation δ_s (displacement of the top edge of the element with respect to the bottom edge) by the distance L between these two edges. The shearing strain is defined here for small deformations only (i.e., $\sin \gamma = \tan \gamma = \gamma$ and $\cos \gamma \approx 1$). Thus,

$$\gamma_{\text{avg}} = \frac{\delta_s}{L} \tag{1-14}$$

Again, for those cases in which the deformation is nonuniform, the shearing strain at a point, $\gamma_{xy}(P)$, associated with two orthogonal reference lines x and y, is obtained by measuring the shearing deformation as the size of the element is made smaller and smaller. In the limit

$$\gamma_{xy}(P) = \lim_{\Delta L \to 0} \frac{\Delta \delta_s}{\Delta L} = \frac{d\delta_s}{dL} \tag{1-15}$$

Since shearing strain is the tangent of the angle of distortion, which for small angles is equal to the angle in radians, an equivalent expression for shearing strain that is sometimes useful for calculations is

$$\gamma_{xy}(P) = \frac{\pi}{2} - \theta' \tag{1-16}$$

In this expression θ' is the angle in the deformed state between the two initially orthogonal reference lines.

Equations 1-12 through 1-16 indicate that both normal and shearing strains are dimensionless quantities; however, normal strains are frequently expressed in units of in./in. or microin./in. while shearing strains are expressed in radians (rad) or microradians. The symbol μ is frequently used to indicate mirco (10^{-6}).

From the definition of normal strain given by Eq. 1-12 or 1-13 it is evident that normal strain is positive when the line elongates and negative when the line contracts. In general, if the axial stress is tensile, the axial deformation will be an elongation. Therefore, positive normal strains are referred to as *tensile strains*. The reverse will be true for compressive axial stresses; therefore, negative normal strains are referred to as *compressive strains*. From Eq. 1-16 it is evident that shearing strains will be positive if the angle between the reference lines decreases. If the angle increases, the shearing strain is negative. Positive and negative shearing strains are not given special names. Normal and shearing strains for most engineering materials in the elastic range (see Section 1-13) seldom exceed values of 0.2 percent, which is equivalent to 0.002 in./in. or 0.002 rad.

EXAMPLE 1-5

A rigid bar C is supported by two steel rods as shown in Fig. 1-26a. There is no strain in the vertical bars before load P is applied. After load P is applied, the axial strain in bar B is 0.0006 in./in. Determine

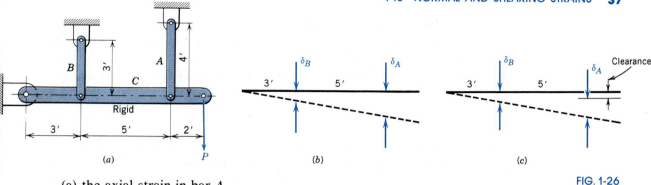

FIG. 1-26

(a) the axial strain in bar A.

(b) the axial strain in bar A if there is a 0.001-in. clearance in the connection between A and C before the load is applied.

Solution

The change in length of bar B is given by Eq. 1-12 as

$$\delta_B = \epsilon_B L_B = 0.0006(3)(12) = 0.0216 \text{ in.}$$

The relationship between deformations in bars A and B is controlled by the rigid bar C, since it does not deform but simply rotates about the support as shown in Figs. 1-26b and c. Sketches of the type shown in Figs. 1-26b and c are referred to as *deformation diagrams*.

(a) From the geometry of Fig. 1-26b, the change in length of bar A is

$$\delta_A = \frac{8}{3}\delta_B = \frac{8}{3}(0.0216) = 0.0576 \text{ in.}$$

The strain in bar A is given by Eq. 1-12 as

$$\epsilon_A = \frac{\delta_A}{L_A} = \frac{0.0576}{4(12)}$$

$$= 0.001200 \text{ in./in.} = \underline{1200 \ \mu\text{in./in.}} \qquad \text{Ans.}$$

(b) From the geometry of Fig. 1-26c, the relationship between deformations in bars A and B is

$$\delta_A + \text{clearance} = \frac{8}{3}\delta_B$$

Thus

$$\delta_A = \frac{8}{3}\delta_B - \text{clearance}$$

$$= \frac{8}{3}(0.0216) - 0.001 = 0.0566 \text{ in.}$$

The strain in bar A is given by Eq. 1-12 as

$$\epsilon_A = \frac{\delta_A}{L_A} = \frac{0.0566}{4(12)}$$

$$= 0.001179 \text{ in./in.} = \underline{1179 \ \mu\text{in./in.}} \qquad \text{Ans.}$$

PROBLEMS

1-59* A 50-ft length of steel wire is subjected to a tensile load that produces a change in length of 1.25 in. Determine the axial strain in the wire.

1-60* Compression tests of concrete indicate that concrete fails when the axial compressive strain is 1200 μm/m. Determine the maximum change in length that a 200-mm-diameter by 400-mm-long concrete test specimen can tolerate before failure occurs.

1-61 A 1.00-in.-diameter steel bar is 8 ft long. The diameter is reduced to $\frac{1}{2}$ in. in a 2-ft central portion of the bar. When an axial load is applied to the ends of the bar, the axial strain in the central portion of the bar is 960 μin./in., and the total elongation of the bar is 0.04032 in. Determine

(a) the elongation of the central portion of the bar.

(b) the axial strain in the end portions of the bar.

1-62 A structural steel bar was loaded in tension to fracture. The 200-mm gage length of the bar was marked off in 25-mm lengths before loading. After the bar broke, the 25-mm segments were found to have lengthened to 30.0, 30.5, 31.5, 34.0, 44.5, 32.0, 31.0, and 30.0 mm, consecutively. Determine

(a) the average strain over the 200-mm gage length.

(b) the maximum average strain over any 50-mm gage length.

1-63* Mutually perpendicular axes in an unstressed member were found to be oriented at 89.92° when the member was stressed. Determine the shearing strain associated with these axes in the stressed member.

1-64* The shear force V shown in Fig. P1-64 produces an average shearing strain γ_{xy} of 1000 μm/m in the block of material. Determine the horizontal movement of point A resulting from application of the shear force V.

1-65 The 0.5 × 2.0 × 4.0-in. rubber mounts shown in Fig. P-165 are used to isolate the vibrational motion of a machine from its supports. Determine the average shearing strain in the rubber mounts if the rigid frame displaces 0.01 in. vertically relative to the support.

1-66 A steel sleeve is connected to a steel shaft with a flexible rubber insert, as shown in Fig. P1-66. The insert has an inside diameter of 85 mm and an outside diameter of 110 mm. When the unit is subjected to a torque T, the shaft rotates 1.5° with respect to the sleeve. Assume that radial lines in the unstressed state remain straight as the rubber deforms. Determine the shearing strain $\gamma_{r\theta}$ in the rubber insert

(a) at the inside surface.

(b) at the outside surface.

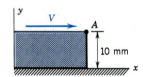

FIG. P1-64

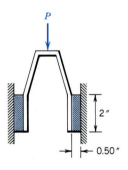

FIG. P1-65

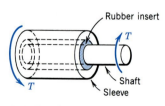

FIG. P1-66

1-67 A thin-walled cylindrical pressure vessel increases in diameter from 80.00 in. to 80.10 in. when the pressure inside the vessel is increased. Determine the change in circumferential strain produced in the vessel by the change in pressure.

1-68 The sanding-drum mandrel shown in Fig. P1-68 is made for use with a hand drill. The mandrel is made from a rubberlike material that expands when the nut is tightened to secure the sanding drum placed over the outside surface. If the diameter of the mandrel increases from 50 mm to 55 mm as the nut is tightened, determine

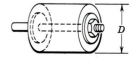

FIG. P1-68

  (a) the average normal strain along a diameter of the mandrel.

  (b) the circumferential strain at the outside surface of the mandrel.

1-69* A rigid steel plate A is supported by three rods, as shown in Fig. P1-69. There is no strain in the rods before the load P is applied. After load P is applied, the axial strain in rod C is 900 μin./in. Determine

  (a) the axial strain in rods B.

  (b) the axial strain in rods B if there is a 0.006-in. clearance in the connections between A and B before the load is applied.

  (c) the axial strain in rods B if there is a 0.004-in. clearance in the connection between A and C before the load is applied.

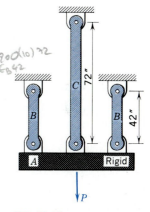

FIG. P1-69

1-70* A rigid bar AD is supported by two rods as shown in Fig. P1-70. There is no strain in the vertical bars before load P is applied. After load P is applied, the axial strain in rod BF is 400 μm/m. Determine

  (a) the axial strain in rod CE.

  (b) the axial strain in rod CE if there is a 0.25-mm clearance in the connection at pin C before the load is applied.

  (c) the axial strain in rod CE if there is a 0.15-mm clearance in the connection at pin B before the load is applied.

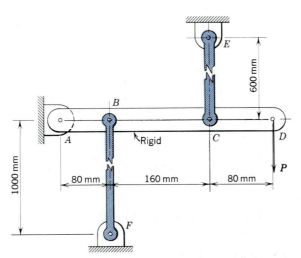

FIG. P1-70

1-71 The load P produces an axial strain in the brass post B of Fig. P1-71 of 950 μin./in. Determine

(a) the axial strain in the aluminum alloy rod A.

(b) the axial strain in the aluminum alloy rod A if there is a 0.006-in, clearance in the connection between A and C in addition to the 0.009-in. clearance between B and C.

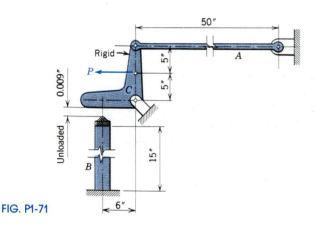

FIG. P1-71

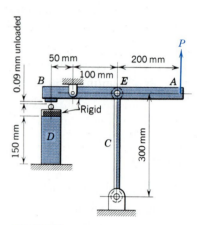

FIG. P1-72

1-72 The load P produces an axial strain in the steel post D of Fig. P1-72 of 450 μm/m. Determine

(a) the axial strain in the aluminum alloy rod C.

(b) the axial strain in the aluminum alloy rod C if there is a 0.08-mm clearance in the connection at E in addition to the 0.09-mm clearance between B and D before the load P is applied.

1-73* The axial strain in a suspended bar of material of varying cross section due to its own weight is given by the expression $(\gamma y)/(3E)$ where γ is the specific weight of the material, y is the distance from the free (bottom) end of the bar, and E is a material constant. Determine, in terms of γ, L, and E,

(a) the change in length of the bar due to its own weight.

(b) the average axial strain over the length L of the bar.

(c) the maximum axial strain in the bar.

a) $\delta = \int_0^L \epsilon \, dy = \dfrac{\gamma L^2}{6E}$

b) $\epsilon_{avg} = \dfrac{\delta}{L} = \dfrac{\gamma L}{6E}$

c) $\epsilon_{max} = \dfrac{\gamma L}{3E}$ at $y = L$

1-74* A steel rod is subjected to a nonuniform heating that produces an extensional (axial) strain that is proportional to the square of the distance from the unheated end ($\epsilon = kx^2$). If the strain is 1250 μm/m at the midpoint of a 3.00-m rod, determine

(a) the change in length of the rod.

(b) the average axial strain over the length L of the rod.

(c) the maximum axial strain in the rod.

1-75 A steel cable is used to support an elevator cage at the bottom of a 2000-ft-deep mineshaft. A uniform axial strain of 250 μin./in. is produced in the cable by the weight of the cage. At each point the weight of the cable produces an additional axial strain that is proportional to the length of the cable below the point. If the total axial strain in the cable at the cable drum (upper end of the cable) is 700 μin./in., determine

(a) the strain in the cable at a depth of 500 ft.

(b) the total elongation of the cable.

1-76 A steel cable is used to tether an observation balloon. The force exerted on the cable by the balloon is sufficient to produce a uniform strain of 500 μm/m in the cable. In addition, at each point in the cable, the weight of the cable reduces the axial strain by an amount that is proportional to the length of the cable between the balloon and the point. When the balloon is directly overhead at an elevation of 300 m, the axial strain at the midlength of the cable is 350 μm/m. Determine

(a) the total elongation of the cable.

(b) the maximum height that the balloon could achieve.

1-11

STRAIN AT A GENERAL POINT IN AN ARBITRARILY LOADED MEMBER

The material of Section 1-10 serves to convey the concept of strain as a unit deformation, but it is inadequate for other than one-directional loading. The extension of the concept to biaxial loading is essential because of the important role played by the strain in experimental methods of stress evaluation. In many practical problems involving the design of structural or machine elements, the configuration and loading are too complicated to permit stress determination solely by mathematical analysis; hence, this technique is supplemented by laboratory measurements.

Strains can be measured by several methods but, except for the simplest cases, stresses cannot be obtained directly. Consequently, the usual procedure in experimental stress analysis is to measure strains and calculate the state of stress from the stress-strain equations presented in Section 1-14.

The complete state of strain at an arbitrary point P in a body under load can be determined by considering the deformation associated with a small volume of material surrounding the point. For convenience the volume is normally assumed to have the shape of a rectangular parallelepiped with its faces oriented perpendicular to the reference axes

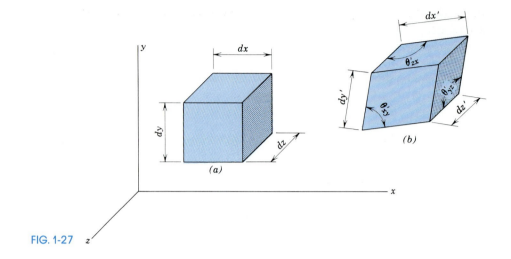

FIG. 1-27

x, y, and z in the undeformed state, as shown in Fig. 1-27a. Since the element of volume is very small, deformations are assumed to be uniform; therefore, parallel planes remain plane and parallel and straight lines remain straight in the deformed element, as shown in Fig. 1-27b. The final size of the deformed element is determined by the lengths of the three edges dx', dy', and dz'. The distorted shape of the element is determined by the angles θ'_{xy}, θ'_{yz}, and θ'_{zx} between faces.

The Cartesian components of strain at the point can be expressed in terms of the deformations by using the definitions of normal and shearing strain presented in Section 1-10. These are the strain components associated with the Cartesian components of stress discussed in Section 1-7 and shown in Fig. 1-22. Thus,

$$\epsilon_x = \frac{dx' - dx}{dx} = \frac{d\delta_x}{dx} \qquad \gamma_{xy} = \frac{\pi}{2} - \theta'_{xy}$$

$$\epsilon_y = \frac{dy' - dy}{dy} = \frac{d\delta_y}{dy} \qquad \gamma_{yz} = \frac{\pi}{2} - \theta'_{yz} \qquad (1.17a)$$

$$\epsilon_z = \frac{dz' - dz}{dz} = \frac{d\delta_z}{dz} \qquad \gamma_{zx} = \frac{\pi}{2} - \theta'_{zx}$$

In a similar manner the normal strain component associated with a line oriented in an arbitrary n direction and the shearing strain component associated with two arbitrary orthogonal lines oriented in the n and t directions in the undeformed element are given by

$$\epsilon_n = \frac{dn' - dn}{dn} = \frac{d\delta_n}{dn} \qquad \gamma_{nt} = \frac{\pi}{2} - \theta'_{nt} \qquad (1.17b)$$

Alternative forms of Eq. 1-17, which will be useful in later developments, are

$$dx' = (1 + \epsilon_x)\, dx \qquad \theta'_{xy} = \frac{\pi}{2} - \gamma_{xy}$$

$$dy' = (1 + \epsilon_y)\, dy \qquad \theta'_{yz} = \frac{\pi}{2} - \gamma_{yz}$$

$$dz' = (1 + \epsilon_z)\, dz \qquad \theta'_{zx} = \frac{\pi}{2} - \gamma_{zx} \qquad \text{(1-18)}$$

$$dn' = (1 + \epsilon_n)\, dn \qquad \theta'_{nt} = \frac{\pi}{2} - \gamma_{nt}$$

1-12

TWO-DIMENSIONAL OR PLANE STRAIN

Considerable insight into the nature of strain distributions can be gained by considering a state of strain known as two-dimensional or *plane strain*. For this case, if the xy plane is taken as the reference plane, the length dz shown in Fig. 1-27 does not change and the angles θ_{yz} and θ_{zx} remain 90°. Thus, from Eqs. 1-17, for the conditions of plane strain $\epsilon_z = \gamma_{yz} = \gamma_{zx} = 0$. Therefore, the change in size of the deformed element is determined by the lengths dx' and dy', and the distorted shape of the element is determined by the angle θ'_{xy}. The state of plane strain can be represented by the two-dimensional element shown in Fig. 1-28. The law of sines or the law of cosines can be used to establish relationships between the different strains, as illustrated in the following example.

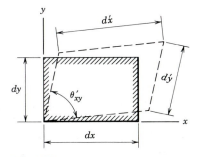

FIG. 1-28

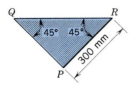

FIG. 1-29

EXAMPLE 1-6

The thin triangular plate shown in Fig. 1-29 is uniformly deformed such that $\epsilon_{PQ} = +0.0030$ m/m, $\epsilon_{QR} = 0$, and $\epsilon_{PR} = +0.0050$ m/m. Compute the shearing strain at P associated with the two edges (PQ and PR) that were orthogonal in the undeformed plate.

Solution

The lengths of the sides of the triangle in the deformed state can be determined by using Eq. 1-12. Thus,

$$L'_{PQ} = L_{PQ} + \delta_{PQ} = L_{PQ} + \epsilon_{PQ}L_{PQ}$$
$$= 300 + 0.0030(300) = 300.90 \text{ mm}$$
$$L'_{PR} = L_{PR} + \delta_{PR} = L_{PR} + \epsilon_{PR}L_{PR}$$
$$= 300 + 0.0050(300) = 301.50 \text{ mm}$$
$$L'_{QR} = L_{QR} = L_{PR}/\cos 45°$$
$$= 300/\cos 45° = 424.26 \text{ mm}$$

Once the lengths of the sides of the deformed triangle are known, the angle θ' between sides PQ and PR in the deformed state can be determined by using the law of cosines. Thus

$$\cos \theta' = \frac{(L'_{PQ})^2 + (L'_{PR})^2 - (L'_{QR})^2}{2(L'_{PQ})(L'_{PR})}$$
$$= \frac{(300.90)^2 + (301.50)^2 - (424.26)^2}{2(300.90)(301.50)}$$
$$= 0.007972$$

Therefore $\theta' = 89.5432°$. The shearing strain $\gamma(P)$ is obtained by using Eq. 1-16. Thus,

$$\gamma(P) = 90° - \theta' = 90° - 89.5432°$$
$$= 0.4568° = 0.007972 \text{ rad} = \underline{7972 \ \mu\text{rad}} \quad \text{Ans.}$$

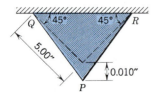

FIG. P1-77

PROBLEMS

1.77* A thin triangular plate is uniformly deformed as shown in Fig. P1-77. Determine the shearing strain at P associated with the two edges (PQ and PR) that were orthogonal in the undeformed plate.

1-78* A thin rectangular plate is uniformly deformed as shown in Fig. P1-78. Determine the shearing strain γ_{xy} at P.

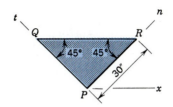

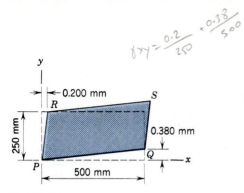

FIG. P1-78

FIG. P1-79

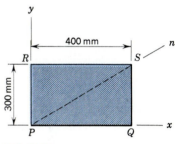

FIG. P1-80

1-79 The thin triangular plate shown in Fig. P1-79 is uniformly deformed such that $\epsilon_n = +1575\ \mu\text{in./in.}$, $\epsilon_t = +1350\ \mu\text{in./in.}$, and $\epsilon_x = +1250\ \mu\text{in./in.}$ Determine the shearing strain γ_{nt} at P.

1-80 The thin rectangular plate shown in Fig. P1-80 is uniformly deformed such that $\epsilon_x = +1950\ \mu\text{m/m}$, $\epsilon_y = -1625\ \mu\text{m/m}$, and $\epsilon_n = -1275\ \mu\text{m/m}$. Determine the shearing strain γ_{xy} at P.

1-81* The thin rectangular plate shown in Fig. P1-81 is uniformly deformed such that $\epsilon_x = +2500\ \mu\text{in./in.}$, $\epsilon_y = +1250\ \mu\text{in./in.}$, and $\gamma_{xy} = -1000\ \mu\text{rad}$. Determine the normal strain ϵ_n in the plate.

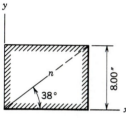

FIG. P1-81

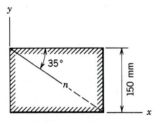

FIG. P1-82

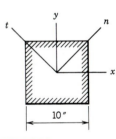

FIG. P1-83

1-82* The thin rectangular plate shown in Fig. P1-82 is uniformly deformed such that $\epsilon_x = -2000\ \mu\text{m/m}$, $\epsilon_y = -1500\ \mu\text{m/m}$, and $\gamma_{xy} = +1250\ \mu\text{rad}$. Determine the normal strain ϵ_n in the plate.

1-83 The thin square plate shown in Fig. P1-83 is uniformly deformed such that $\epsilon_x = +3200\ \mu\text{in./in.}$, $\epsilon_y = +1500\ \mu\text{in./in.}$, and $\gamma_{xy} = 0$. Determine

(a) the normal strain ϵ_n in the plate.

(b) the shearing strain γ_{nt} in the plate.

1-84 The thin square plate shown in Fig. P1-84 is uniformly deformed such that $\epsilon_x = +1750\ \mu\text{m/m}$, $\epsilon_y = -2200\ \mu\text{m/m}$, and $\gamma_{xy} = 0$. Determine

(a) the normal strain ϵ_n in the plate.

(b) the shearing strain γ_{nt} in the plate.

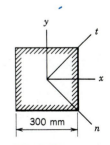

FIG. P1-84

1-13

PROPERTIES OF MATERIALS

The satisfactory performance of a structure frequently is determined by the amount of deformation or distortion that can be permitted. A deflection of a few thousandths of an inch might make a boring machine useless, whereas the boom on a dragline might deflect several inches without impairing its usefulness. It is often necessary to relate the loads on a structure, or on a member in a structure, to the deflection the loads will produce. Such information can be obtained by plotting diagrams showing loads and deflections for each member and type of loading in a structure, but such diagrams will vary with the dimensions of the members, and it would be necessary to draw new diagrams each time the dimensions were varied. A more useful diagram is one showing the relation between the stress and strain. Such diagrams are called stress-strain diagrams.

Data for stress-strain diagrams are usually obtained by applying an axial load to a test specimen and measuring the load and deformation simultaneously. A testing machine (Fig. 1-30) is used to strain the

FIG. 1-30
Hydraulic testing machine set up for a tension test (*courtesy of MTS Systems Corporation*).

specimen and to measure the load required to produce the strain. The stress is obtained by dividing the load by the initial cross-sectional area of the specimen. The area will change somewhat during the loading, and the stress obtained using the initial area is obviously not the exact stress occurring at higher loads. It is the stress most commonly used, however, in designing structures. The stress obtained by dividing the load by the actual area is frequently called the *true stress* and is useful in explaining the fundamental behavior of materials. Strains are usually relatively small in materials used in engineering structures, often less than 0.001, and their accurate determination requires special measuring equipment. Normal strain can be obtained by measuring the deformation δ in a length L and dividing δ by L. Instruments for measuring the deformation δ are called strain gages, or sometimes extensometers or compressometers, and obtain the desired accuracy by multiplying levers, dial indicators, beams of light, or other means. The electrical resistance strain gage described in Section 7-9 is widely used for this type of measurement.

True strain, like true stress, is computed on the basis of the actual length of the test specimen during the test and is used primarily to study the fundamental properties of materials. The difference between nominal stress and strain, computed from initial dimensions of the specimen, and true stress and strain is negligible for stresses usually encountered in engineering structures, but sometimes the difference becomes important with larger stresses and strains.

A more complete discussion of the experimental determination of stress and strain will be found in various books on experimental stress analysis.[5,6]

Figures 1-31*a* and *b* show tensile stress-strain diagrams for structural steel (a low-carbon steel) and for a magnesium alloy. These diagrams will be used to explain a number of properties useful in the study of mechanics of materials.

The initial portion of the stress-strain diagram for most materials used in engineering structures is a straight line. The stress-strain diagrams for some materials, such as gray cast iron and concrete, show a slight curve even at very small stresses, but it is common practice to draw a straight line to average the data for the first part of the diagram and neglect the curvature. The proportionality of load to deflection was first recorded by Robert Hooke who observed in 1678, "*Ut tensio sic vis*" (as the stretch so the force); this is frequently referred to as Hooke's law. Thomas Young, in 1807, suggested what amounts to using the ratio of stress to strain to measure the stiffness of a material. This

---

[5] *Handbook of Experimental Stress Analysis,* M. Hetenyi, Wiley, New York, 1950.

[6] *Experimental Stress Analysis,* J. W. Dally and W. F. Riley, McGraw-Hill, New York, 1978.

ratio is called *Young's modulus* or the *modulus of elasticity* and is the slope of the straight-line portion of the stress-strain diagram. Young's modulus is written as

$$E = \sigma/\epsilon \qquad \text{or} \qquad G = \tau/\gamma \qquad (1\text{-}19)$$

where E is used for normal stress and strain and G (sometimes called the modulus of rigidity) is used for shearing stress and strain. The maximum stress for which stress and strain are proportional is called the *proportional limit* and is indicated by the ordinates at points A on Fig. 1-31a or b.

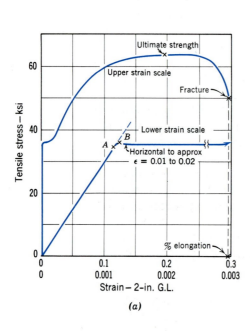

(a)

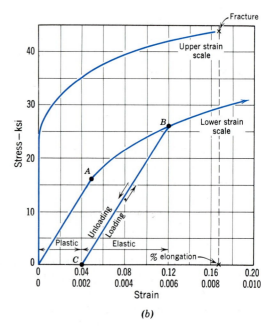

(b)

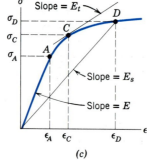

(c)

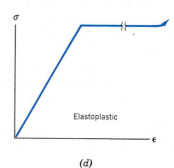

(d)

FIG. 1-31 Stress-strain diagrams for structural steel and a magnesium alloy.

For points on the stress-strain curve beyond the proportional limit (such as point C on Fig. 1-31c), other quantitites such as the *tangent modulus* and the *secant modulus* are used as measures of the stiffness of a material. The tangent modulus E_t is defined as the slope of the stress-strain diagram at a particular stress level. Thus, the tangent modulus is a function of the stress (or strain) for stresses greater than the proportional limit. For stresses less than the proportional limit, the tangent modulus is the same as Young's modulus. The secant modulus E_s is the ratio of the stress to the strain at any point on the diagram. Young's modulus E, the tangent modulus E_t, and the secant modulus E_s are all illustrated in Fig. 1-31c.

The action is said to be *elastic* if the strain resulting from loading disappears when the load is removed. The *elastic limit* is the maximum stress for which the material acts elastically. For most materials it is found that the stress-strain diagram for unloading is aproximately parallel to the loading portion (see line BC in Fig. 1-31b). If the specimen is again loaded, the stress-strain diagram will usually follow the unloading curve until it reaches a stress a little less than the maximum stress attained during the initial loading, at which time it will start to curve in the direction of the initial loading curve. As indicated in Fig. 1-31b, the proportional limit for the second loading is greater than that for the initial loading. This phenomenon is called *strain hardening* or *work hardening*.

When the stress exceeds the elastic limit (or proportional limit for practical purposes), it is found that a portion of the deformation remains after the load is removed. The deformation remaining after an applied load is removed is called *plastic deformation*. Plastic deformation independent of the time duration of the applied load is known as *slip*. *Creep* is plastic deformation that continues to increase under a constant stress. In many instances creep continues until fracture occurs; however, in other instances the rate of creep decreases and approaches zero as a limit. Some materials are much more susceptible to creep than are others, but most materials used in engineering exhibit creep at elevated temperatures. The total strain is thus made up of elastic strain, possibly combined with plastic strain that results from slip, creep, or both. When the load is removed, the elastic portion of the strain is recovered, but the plastic part (slip and creep) remains as permanent set.[7]

A precise value for the proportional limit is difficult to obtain, particularly when the transition of the stress-strain diagram from a straight

---

[7]In some instances a portion of the strain that remains immediately after the stress is removed may disappear after a period of time. This reduction of strain is sometimes called *recovery*.

line to a curve is gradual. For this reason, other measures of stress that can be used as a practical elastic limit are required. The yield point and the yield strength for a specified offset are frequently used for this purpose.

The *yield point* is the stress at which there is an appreciable increase in strain with no increase in stress, with the limitation that, if straining is continued, the stress will again increase. This latter specification indicates that there is a kink or "knee" in the stress-strain diagram, as indicated in Fig. 1-31*a*. The yield point is easily determined without the aid of strain-measuring equipment because the load indicated by the testing machine drops (or if a dial indicator is used, the needle halts) at the yield point. Unfortunately, few materials possess this property, the most common examples being the low-carbon steels.

The *yield strength* is defined as the stress that will induce a specified permanent set, usually 0.05 to 0.3 percent, which is equivalent to a strain of 0.0005 to 0.003. The yield strength is particularly useful for materials with no yield point. The yield strength can be conveniently determined from a stress-strain diagram by laying off the specified offset (permanent set) on the strain axis as *OC* in Fig. 1-31*b* and drawing a line *CB* parallel to *OA*. The stress indicated by the intersection of *CB* and the stress-strain diagram is the yield strength.

The maximum stress, based on the original area, developed in a material before rupture is called the *ultimate strength* of the material, and the term may be modified as the ultimate tensile, compressive, or shearing strength of the material. Ductile materials undergo considerable plastic tensile or shearing deformation before rupture. When the ultimate strength of a ductile material is reached, the cross-sectional area of the test specimen starts to decrease or neck down, and the resultant load that can be carried by the specimen decreases. Thus, the stress based on the original area decreases beyond the ultimate strength of the material, although the true stress continues to increase until rupture.

Many engineering structures are designed so that the stresses are less than the proportional limit, and Young's modulus provides a simple and convenient relationship between the stress and strain. When the stress exceeds the proportional limit, the deformation becomes partially plastic, and no simple relation exists between the stress and strain. Various empirical equations have been proposed relating the stress and strain beyond the proportional limit. The following equation, known as the Ramberg-Osgood[8] formula, is one that has been used in plastic analysis:

$$\epsilon = \epsilon_0 \left[\frac{\sigma}{\sigma_0} + \frac{3}{7} \left(\frac{\sigma}{\sigma_0} \right)^n \right] \tag{1-20}$$

[8]"Description of Stress-Strain Curves by Three Parameters," W. Ramberg and W. R. Osgood, National Advisory Committee for Aeronautics, Tech. Note 902, 1943.

in which ϵ and σ are the strain and stress to be related, and ϵ_0, σ_0, and n are experimentally determined constants. Corresponding values of stress and strain can also be read from stress-strain diagrams. A stress-strain diagram similar to the one in Fig. 1-31d is frequently assumed for mild steel or other material with similar properties in order to simplify the calculations. In this book, material with such a stress-strain diagram will be designated *elastoplastic,* and the proportional limit and yield points will be assumed to be the same. For mild steel the plastic strain that occurs at the yield point with no increase in stress is 15–20 times the elastic strain at the proportional limit.

Strength and stiffness are not the only properties of interest to a design engineer. Another important property is *ductility,* defined as the capacity for plastic deformation in tension or shear. This property controls the amount of cold forming to which a material may be subjected. The forming of automobile bodies, the bending of concrete-reinforcing bars, and the manufacture of fencing and other wire products all require ductile materials. Ductility is also an important property of materials used for fabricated structures. Under static loading the presence of stress concentration in the region of rivet holes or welds may be ignored, since the ductility permits considerable plastic action to take place in the region of high stresses, with a resulting redistribution of stress and the establishment of equilibrium. Two commonly used quantitative indices of ductility are the *ultimate elongation* (expressed as a percentage elongation of the gage length at rupture) and the *reduction of cross-sectional area* at the section where rupture occurs (expressed as a percentage of the original area).

The property indicating the resistance of a material to failure by creep is known as the *creep limit* and is defined as the maximum stress for which the plastic strain will not exceed a specified amount during a specified time interval at a specified temperature. This property is important when designing parts to be fabricated with polymeric materials (commonly known as plastics) and when designing metal parts that will be subjected to high temperatures and sustained loads (for example, the turbine blades in a turbojet engine).

A material loaded in one direction will undergo strains perpendicular to the direction of the load as well as parallel to it. The ratio of the lateral or perpendicular strain (ϵ_{lat} or ϵ_t) to the longitudinal or axial strain (ϵ_{long} or ϵ_a) for a uniaxial state of stress is called Poisson's ratio, after Simeon D. Poisson, who identified the constant in 1811. Poisson's ratio is a constant for stresses below the proportional limit and has a value between $\frac{1}{4}$ and $\frac{1}{3}$ for most metals. The symbol ν is used for Poisson's ratio, which is given by the equation

$$\nu = -\frac{\epsilon_{\text{lat}}}{\epsilon_{\text{long}}} = -\frac{\epsilon_t}{\epsilon_a} \qquad (1\text{-}21)$$

The ratio $\nu = -\epsilon_t/\epsilon_a$ is valid only for a uniaxial state of stress. It will be shown in Section 1-14 that Poisson's ratio is related to E and G by the formula

$$E = 2(1 + \nu)G$$

The properties discussed in this section are primarily concerned with static or continuous loading. Other properties are important when impact loads or repeated loads are encountered. A complete discussion of properties of materials can be found in textbooks[9,10] devoted to the subject.

PROBLEMS

1-85* At the proportional limit, an 8-in. gage length of a $\frac{1}{2}$-in.-diameter alloy bar has elongated 0.012 in., and the diameter has been reduced 0.00025 in. The total axial load carried was 4650 lb. Determine the following properties of the material:

   (a) the modulus of elasticity.
   (b) Poisson's ratio.
   (c) the proportional limit.

1-86* At the proportional limit, a 200-mm gage length of a 15-mm-diameter alloy bar has elongated 0.90 mm and the diameter has been reduced 0.022 mm. The total axial load carried was 62.6 kN. Determine the following properties of the material:

   (a) the modulus of elasticity. $E = \frac{\sigma}{\epsilon_a}$
   (b) Poisson's ratio. $\nu = \frac{\epsilon_t}{\epsilon_a}$
   (c) the proportional limit. $\sigma_{PL} = \sigma$

1-87 A $1 \times 4 \times 90$-in. bar elongates 0.09 in. under an axial load of 120 kips. After the load is applied, the 4-in. side measures 3.9988 in. Determine the following properties of the material:

   (a) the modulus of elasticity.
   (b) Poisson's ratio.
   (c) the modulus of rigidity.

[9]*The Structure and Properties of Materials*, Vol. III, *Mechanical Behavior*, H. W. Hayden, W. G. Moffatt, and John Wulff, Wiley, New York, 1965.

[10]*Elements of Material Science*, L. H. VanVlack, 5th ed., Addison-Wesley, Reading, Mass., 1985.

1-88 A 50-mm-diameter rod 6 m long elongates 12 mm under an axial load of 375 kN. The diameter of the rod decreases 0.035 mm during the loading. Determine the following properties of the material:

(a) the modulus of elasticity.

(b) Poisson's ratio.

(c) the modulus of rigidity.

1-89 A tensile test specimen having a diameter of 0.505 in. and a gage length of 2.000 in. was tested to fracture. Load and deformation data obtained during the test were as follows:

Load (lb)	Change in Length (in.)	Load (lb)	Change in Length (in.)
0	0	12,600	0.0600
2,200	0.0008	13,200	0.0800
4,300	0.0016	13,900	0.1200
6,400	0.0024	14,300	0.1600
8,200	0.0032	14,500	0.2000
8,600	0.0040	14,600	0.2400
8,800	0.0048	14,500	0.2800
9,200	0.0064	14,400	0.3200
9,500	0.0080	14,300	0.3600
9,600	0.0096	13,800	0.4000
10,600	0.0200	13,000	Fracture
11,800	0.0400		

Determine

(a) the modulus of elasticity.

(b) the proportional limit.

(c) the ultimate strength.

(d) the yield strength (0.05 percent offset).

(e) the yield strength (0.20 percent offset).

(f) the fracture stress.

(g) the true fracture stress if the final diameter of the specimen at the location of the fracture was 0.425 in.

(h) the tangent modulus at a stress level of 46,000 psi.

(i) the secant modulus at a stress level of 46,000 psi.

1-90 A tensile test specimen having a diameter of 11.28 mm and a gage length of 50 mm was tested to fracture. Load and deformation data obtained during the test were as follows:

Load (kN)	Change in Length (mm)	Load (kN)	Change in Length (mm)
0	0		
7.6	0.02	43.8	1.50
14.9	0.04	45.8	2.00
22.2	0.06	48.3	3.00
28.5	0.08	49.7	4.00
29.9	0.10	50.4	5.00
30.6	0.12	50.7	6.00
32.0	0.16	50.4	7.00
33.0	0.20	50.0	8.00
33.3	0.24	49.7	9.00
36.8	0.50	47.9	10.00
41.0	1.00	45.1	Fracture

Determine

    (a) the modulus of elasticity.

    (b) the proportional limit.

    (c) the ultimate strength.

    (d) the yield strength (0.05 percent offset).

    (e) the yield strength (0.20 percent offset).

    (f) the fracture stress.

    (g) the true fracture stress if the final diameter of the specimen at the location of the fracture was 9.50 mm.

    (h) the tangent modulus at a stress level of 315 MPa.

    (i) the secant modulus at a stress level of 315 MPa.

1-14

GENERALIZED HOOKE'S LAW FOR ISOTROPIC MATERIALS

Hooke's law (see Eq. 1-19) can be extended to include the biaxial (see Fig. 1-23) and triaxial (see Fig. 1-22) states of stress often encountered in engineering practice. Consider, for example, Fig. 1-32, which shows a differential element of material subjected to a biaxial state of normal stress. Shearing stresses have not been shown on the faces of the element, since they produce distortion of the element (angle changes), but do not produce changes in the lengths of the sides of the element that would contribute to the normal strains. The deformations of the element in the direction of the normal stresses, for a combined loading,

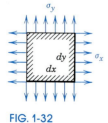

FIG. 1-32

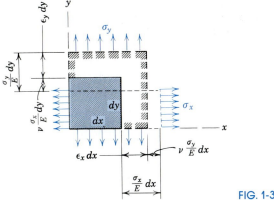

FIG. 1-33

can be determined by computing the deformations resulting from the individual stresses separately and adding the values obtained algebraically. This procedure is based on the *principle of superposition*, which states that the effects of separate loadings can be added algebraically if two conditions are satisfied:

1. Each effect is linearly related to the load that produced it.
2. The effect of the first load does not significantly change the effect of the second load.

The first condition is satisifed if the stresses do not exceed the proportional limit for the material. The second condition is satisfied if the deformations are small, so that the small changes in the areas of the faces of the element do not produce significant changes in the stresses.

The deformations of the element of Fig. 1-32, associated with the stresses σ_x and σ_y, are shown in Fig. 1-33. The shaded square in Fig. 1-33 indicates the original or unstrained configurations of the element. Under the action of the stress σ_x, the element extends in the x direction and contracts in the y direction to the configuration indicated by the dashed lines (deformations are greatly exaggerated). Then, under the action of the stress σ_y superimposed on the stress σ_x, the element assumes the outline shown by the hatched dashed lines. If the material is isotropic, Young's modulus has the same value for all directions and the final deformation in the x direction is

$$d\delta_x = \epsilon_x dx = \frac{\sigma_x}{E} dx - \nu \frac{\sigma_y}{E} dx$$

and the three normal strains are

$$\epsilon_x = \frac{1}{E}(\sigma_x - \nu\sigma_y)$$

$$\epsilon_y = \frac{1}{E}(\sigma_y - \nu\sigma_x) \tag{1-22}$$

$$\epsilon_z = -\frac{\nu}{E}(\sigma_x + \sigma_y)$$

The analysis above is readily extended to triaxial normal stresses, and the expressions for strain become

$$\epsilon_x = \frac{1}{E}[\sigma_x - \nu(\sigma_y + \sigma_z)]$$

$$\epsilon_y = \frac{1}{E}[\sigma_y - \nu(\sigma_z + \sigma_x)] \tag{1-23}$$

$$\epsilon_z = \frac{1}{E}[\sigma_z - \nu(\sigma_x + \sigma_y)]$$

In these expressions, tensile stresses and strains are considered positive, and compressive stresses and strains are considered negative.

When Eqs. 1-22 are solved for the stresses in terms of the strains, they give

$$\sigma_x = \frac{E}{1 - \nu^2}(\epsilon_x + \nu\epsilon_y)$$

$$\sigma_y = \frac{E}{1 - \nu^2}(\epsilon_y + \nu\epsilon_x) \tag{1-24}$$

Equations 1-24 can be used to calculate normal stresses from measured or computed normal strains. When Eqs. 1-23 are solved for stresses in terms of strains, they give

$$\sigma_x = \frac{E}{(1 + \nu)(1 - 2\nu)}[(1 - \nu)\epsilon_x + \nu(\epsilon_y + \epsilon_z)]$$

$$\sigma_y = \frac{E}{(1 + \nu)(1 - 2\nu)}[(1 - \nu)\epsilon_y + \nu(\epsilon_z + \epsilon_x)] \tag{1-25}$$

$$\sigma_z = \frac{E}{(1 + \nu)(1 - 2\nu)}[(1 - \nu)\epsilon_z + \nu(\epsilon_x + \epsilon_y)]$$

Torsion test specimens are used to study material behavior under pure shear, and it is observed that a shearing stress produces only a single corresponding shearing strain. Thus, Hooke's law extended to shearing stresses is simply

$$\tau = G\gamma \tag{1-26}$$

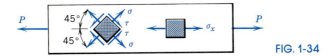

FIG. 1-34

Equations 1-25 and 1-26 seem to indicate that three elastic constants, E, ν, and G, are required to determine the deformations (and strains) in a material resulting from an arbitrary state of stress. In fact, only two of these constants need be determined experimentally for a given material.

The relationship between the elastic constants E, ν, and G can be determined by considering the stresses and strains produced by an axial tensile load in a bar of the material. Equations 1-7 and 1-8 indicate that both normal and shearing stresses are produced on different inclined planes through the bar by the axial tensile load. For the discussion which follows, consider the stresses that develop on the faces of the two small square elements shown in Fig. 1-34. From Eqs. 1-7 and 1-8:

$$\sigma_x = \frac{P}{2A}(1 + \cos 0°) = \frac{P}{A}$$

$$\sigma = \frac{P}{2A}(1 + \cos 90°) = \frac{P}{2A} = \frac{\sigma_x}{2} \qquad (a)$$

$$\tau = \frac{P}{2A}\sin 90° = \frac{P}{2A} = \frac{\sigma_x}{2}$$

The strains produced by these stresses are given by Eqs. 1-22 and 1-26 as:

$$\epsilon_x = \frac{\sigma_x}{E}$$

$$\epsilon_y = -\frac{\nu\sigma_x}{E} \qquad (b)$$

$$\gamma = \frac{\tau}{G} = \frac{\sigma_x}{2G}$$

(handwritten annotations: $P = -\dfrac{E}{1-2\nu}\,\epsilon_x$ pressure)

Under the actions of the axial tensile load, the two square elements shown in Fig. 1-34 deform into the shapes shown in Fig. 1-35. In this figure, the deformations of the two elements have been superimposed to show their relationships. The relationship between the shearing strain γ and the two normal strains ϵ_x and ϵ_y is determined from the geometry of Fig. 1-35. Thus,

$$\text{Tan}\left(\frac{\pi}{4} - \frac{\gamma}{2}\right) = \frac{(1 + \epsilon_y)(dL/2)}{(1 + \epsilon_x)(dL/2)} \qquad (c)$$

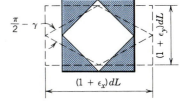

FIG. 1-35

The tangent of the difference between the angles can be expressed as

$$\frac{\text{Tan } \dfrac{\pi}{4} - \text{Tan } \dfrac{\gamma}{2}}{1 + \text{Tan } \dfrac{\pi}{4} \text{Tan } \dfrac{\gamma}{2}} = \frac{1 + \epsilon_y}{1 + \epsilon_x} = \frac{1 - \nu\epsilon_x}{1 + \epsilon_x} \tag{d}$$

Since all of the strains are small, Eq. (d) reduces to

$$\frac{1 - \dfrac{\gamma}{2}}{1 + \dfrac{\gamma}{2}} = \frac{1 - \nu\epsilon_x}{1 + \epsilon_x} \tag{e}$$

Solving for the shearing strain γ yields

$$\gamma = \frac{(1 + \nu)\epsilon_x}{1 + (\frac{1}{2})(1 - \nu)(\epsilon_x)} \approx (1 + \nu)\epsilon_x \tag{f}$$

Finally, if Eqs. (b) are substituted into Eq. (f)

$$\frac{\tau}{G} = \frac{\sigma_x}{2G} = (1 + \nu)\frac{\sigma_x}{E} \tag{g}$$

Solving Eq. (g) for G yields the desired relationship between G, E, and ν. Thus,

$$G = \frac{E}{2(1 + \nu)} \tag{1-27}$$

If Eq. 1-27 is substituted into Eq. 1-26, an alternate form of generalized Hooke's law for shearing stress and strain in isotropic materials is obtained. Thus,

$$\tau_{xy} = G\gamma_{xy} = \frac{E}{2(1 + \nu)} \gamma_{xy}$$

$$\tau_{yz} = G\gamma_{yz} = \frac{E}{2(1 + \nu)} \gamma_{yz} \tag{1-28}$$

$$\tau_{zx} = G\gamma_{zx} = \frac{E}{2(1 + \nu)} \gamma_{zx}$$

Equations 1-22 through 1-28 are widely used for experimental stress determinations. The following example will illustrate the method of application.

EXAMPLE 1-7

At a point on the surface of an alloy steel ($E = 30,000$ ksi and $\nu = 0.30$) machine part subjected to a biaxial state of stress, the measured strains were: $\epsilon_x = +1394 \ \mu$in./in., $\epsilon_y = -660 \ \mu$in./in., and $\gamma_{xy} = 2054$ μrad. Determine the stresses σ_x, σ_y, and τ_{xy} at the point.

Solution

The normal stresses σ_x and σ_y are obtained by using Eqs. 1-24. Thus,

$$\sigma_x = \frac{30(10^6)}{1 - 0.09} [1394 + 0.3(-660)](10^{-6})$$

$$= +39,400 \text{ psi} = \underline{39,400 \text{ psi } T} \qquad \text{Ans.}$$

$$\sigma_y = \frac{30(10^6)}{1 - 0.09} [-660 + 0.3(1394)](10^{-6})$$

$$= -7980 \text{ psi} = \underline{7980 \text{ psi } C} \qquad \text{Ans.}$$

The shearing stress τ_{xy} is obtained by using Eqs. 1-28. Thus,

$$\tau_{xy} = \frac{30(10^6)}{2(1 + 0.3)} (2054)(10^{-6}) = \underline{23,700 \text{ psi}} \qquad \text{Ans.}$$

Once the stresses σ_x, σ_y, and τ_{xy} are known, the equilibrium method presented in Section 1-8 can be used to determine the stresses on any other plane of interest through the point.

PROBLEMS

1-91* An aluminum alloy ($E = 10,000$ ksi) bar with a circular cross section will be subjected to an axial tensile load of 10 kips. Determine the minimum diameter required for the bar if the axial stress must not exceed 20 ksi and the axial strain must not exceed 1500 μin./in.

1-92* Two rigid plates are loaded and supported with steel ($E = 200$ GPa and $\nu = 0.30$) pipe columns, as shown in Fig. P1-92. The columns have outside diameters of 100 mm and inside diameters of 75 mm. Determine

  (a) the axial compressive strain in columns A.
  (b) the axial compressive strain in columns B.
  (c) the change in outside diameter of columns B.

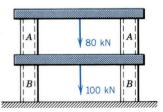

FIG. P1-92

1-93 A 2 × 2 × 6-in.-long steel (E = 29,600 ksi and ν = 0.30) bar is subjected to an axial load. When the load is applied, the lateral dimensions of the bar increase by 0.0011 in. Determine

(a) the magnitude and direction of the applied axial load.

(b) the change in length of the bar.

(c) the change in volume of the bar.

1-94 A 25 × 25 × 200-mm-long cold-rolled bronze (E = 100 GPa and G = 45 GPa) bar is subjected to an axial tensile load of 120 kN. Determine

(a) the axial strain in the bar.

(b) the lateral strain in the bar.

(c) the change in volume of the bar.

1-95* A 2 × 4 × ¼-in-thick steel (E = 30,000 ksi and ν = 0.30) plate is subjected to a horizontal shear force V, as shown in Fig. P1-95. If the average shearing stress τ_{xy} produced in the plate by application of the force is 2 ksi, determine

(a) the magnitude of the applied shear force V.

(b) the horizontal displacement of the top edge of the plate with respect to its no-load position.

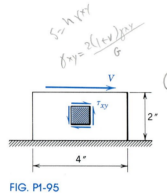

$$\gamma = h \gamma + \gamma$$
$$\gamma_{xy} = \frac{2(1+\nu)\tau_{xy}}{G}$$

FIG. P1-95

1-96* A rigid plate is shock-isolated from a foundation with a rubber (G = 0.50 GPa and ν = 0.48) block as shown in Fig. P1-96. The length and width of the block are 300 mm and 50 mm, respectively. If the horizontal movement of the rigid plate must not exceed 5 mm and if the average shearing stress in the rubber block must not exceed 24 MPa, determine

(a) the maximum thickness h for the rubber block.

(b) the maximum load P that can be applied to the plate.

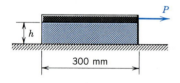

FIG. P1-96

1-97* At a point on the surface of an aluminum alloy (E = 10,000 ksi and G = 3800 ksi) machine part subjected to a biaxial state of stress, the measured strains were: ϵ_x = +900 μin./in., ϵ_y = −300 μin./in., and γ_{xy} = −400 μrad. Determine the stresses σ_x, σ_y, and τ_{xy} at the point.

1-98* At a point on the surface of a structural steel (E = 200 GPa and G = 76 GPa) machine part subjected to a biaxial state of stress, the measured strains were: ϵ_x = +750 μm/m, ϵ_y = +350 μm/m, and γ_{xy} = −560 μrad. Determine the stresses σ_x, σ_y, and τ_{xy} at the point.

1-99 At a point on the surface of a titanium alloy (E = 14,000 ksi and G = 5300 ksi) aircraft part subjected to a biaxial state of stress, the measured strains were: ϵ_x = +1250 μin./in., ϵ_y = +600 μin./in., and γ_{xy} = 650 μrad. Determine the stresses σ_x, σ_y, and τ_{xy}.

1-100 At a point on the surface of a stainless steel (E = 190 GPa and G = 76 GPa) machine part subjected to a biaxial state of stress, the measured strains were: ϵ_x = +1175 μm/m, ϵ_y = −1250 μm/m, and γ_{xy} = +850 μrad. Determine the stresses σ_x, σ_y, and τ_{xy} at the point.

1-101* Determine the state of strain that corresponds to the following state of stress at a point in a steel (E = 30,000 ksi and ν = 0.30) machine part; σ_x = 15,000 psi, σ_y = 5000 psi, σ_z = 7500 psi, τ_{xy} = 5500 psi, τ_{yz} = 4750 psi, and τ_{zx} = 3200 psi.

$$G = \frac{E}{2(1+\nu)}$$
$$\epsilon_x = \frac{\sigma_x - \nu(\sigma_y + \sigma_z)}{E}$$

1-102* Determine the state of strain that corresponds to the following state of stress at a point in an aluminum alloy (E = 73 GPa and ν = 0.33) machine part: σ_x = 120 MPa, σ_y = −85 MPa, σ_z = 45 MPa, τ_{xy} = 35 MPa, τ_{yz} = 48 MPa, and τ_{zx} = 76 MPa.

1-103 A 10 × 10 × 10-in. cube of steel (E = 30,000 ksi and ν = 0.30) is loaded with a uniformly distributed pressure of 30,000 psi on the four faces having outward normals in the x and y directions. Rigid constraints limit the deformation of the cube in the z direction to 0.002 in. Determine the normal stress σ_z that develops as the pressure is applied.

1-104 A 10 × 10 × 25.4-mm block of rubberlike material (E = 1.40 GPa and ν = 0.40) is to be pushed into a lubricated 10 × 10 × 25-mm flat-bottomed hole in a rigid material as shown in Fig. P1-104. Determine the load P required to push the block into the hole until its top surface is flush with the top of the hole.

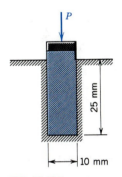

FIG. P1-104

1-105 An elastomer (G = 100 ksi) is bonded to a rigid circular shaft and to a rigid circular sleeve as shown in Fig. P1-105 to form a spring. An axial load P of 10 kips is applied to the shaft. Determine

    (a) the shearing strain at the outer surface of the elastomer.

    (b) the shearing strain at the inner surface of the elastomer.

    (c) the deflection of the end of the shaft with respect to its no-load position.

1-106 A thick-walled cylindrical pressure vessel will be used to store gas under a pressure of 100 MPa. During initial pressurization of the vessel, axial and hoop components of strain were measured on the inside and outside surfaces. On the inside surface the axial strain was 500 μm/m and the hoop strain was 750 μm/m. On the outside surface the axial strain was 500 μm/m and the hoop strain was 100 μm/m. Determine the axial and hoop components of stress associated with these strains if E = 200 GPa and ν = 0.30.

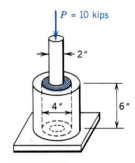

FIG. P1-105

1-15

TEMPERATURE EFFECTS

Most engineering materials when unrestrained expand when heated and contract when cooled. The strain due to a 1° temperature change is designated by α and is known as the coefficient of thermal expansion. The strain due to a temperature change of $\Delta T°$ is

$$\epsilon_T = \alpha(\Delta T) \qquad (a)$$

The coefficient of thermal expansion is approximately constant for a considerable range of temperatures (in general, the coefficient increases with an increase of temperature). For a homogeneous, isotropic material, the coefficient applies to all dimensions (all directions). Values of the coefficient of expansion for several materials are included in Appendix A.

When a member is restrained (free movement prevented) while a temperature change takes place, stresses (referred to as *thermal stresses*) are induced in the member. If the action is elastic, thermal stresses are easily computed by (1) assuming that the restraining influence has been removed and the member permitted to expand or contract freely, (2) applying forces that cause the member to assume the configuration dictated by the restraining influence. For example, the bar AB of Fig. 1-36a is securely fastened to rigid supports at the ends. If the end B is assumed to be cut loose from the wall and the temperature drops, the end will move to B' a distance δ_T, as indicated in Fig. 1-36b. Next, a force P of sufficient magnitude, shown in Fig. 1-36c, is applied to B to move end B through a distance δ_P so that the length of the bar is again L, the distance between the walls. Since the walls do not move, $\delta_T = \delta_P$, where $\delta_T = \alpha L \Delta T$, and $\delta_P = L\epsilon$; if Hooke's law applies, $\epsilon = \sigma/E$, where σ is the induced axial stress.

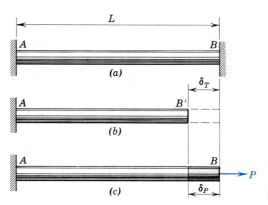

FIG. 1-36

In case a stress exists before the temperature change takes place (the rod of Fig. 1-36a might have been tightened by means of a turn-buckle or nuts at the ends), the stress due to the temperature change may be added algebraically to the original stress by using the *principle of superposition*. The principle states that stresses (at a point on a given plane) due to different loads may be computed separately and added algebraically, provided that the sum of the stresses does not exceed the proportional limit of the material and that the structure remains stable. In explanation of this last limitation, consider a member with a length considerably greater than the lateral dimension subjected to axial compression loads. This member may successfully carry either of two applied loads, and the sum of the axial stresses for both loads may be within the proportional limit; yet, upon application of both loads, the member may collapse (this action is discussed further in Chapter 9).

The effects of temperature change under more complicated states of stress (plane stress or the three-dimensional state of stress) can be evaluated by observing that an arbitrary normal strain ϵ_n will be composed of two parts: namely, a strain ϵ_σ associated with the stresses and a strain ϵ_T associated with the temperature change. Thus,

$$\epsilon_n = \epsilon_\sigma + \epsilon_T \qquad\qquad (b)$$

In cases in which the material is completely restrained, the strains due to the stresses must be the negative of the strains due to the temperature change, as the total strain must be zero. The strains resulting from stresses are given by generalized Hooke's law (Eq. 1-23). The strains resulting from a temperature change are given by Eq. *a*. When Eqs. 1-23 and *a* are substituted in Eq. *b*, the stress-strain-temperature equations for a linear elastic isotropic material are obtained. The results are as follows:

$$\epsilon_x = \frac{1}{E}[\sigma_x - \nu(\sigma_y + \sigma_z)] + \alpha(\Delta T)$$

$$\epsilon_y = \frac{1}{E}[\sigma_y - \nu(\sigma_z + \sigma_x)] + \alpha(\Delta T) \qquad\qquad (1\text{-}29)$$

$$\epsilon_z = \frac{1}{E}[\sigma_z - \nu(\sigma_x + \sigma_y)] + \alpha(\Delta T)$$

Often it is more convenient to have equations for stress in terms of strain. When Eqs. 1-29 are solved for stresses in terms of strains, they yield

$$\sigma_x = \frac{E}{(1 + \nu)(1 - 2\nu)} [(1 - \nu)\epsilon_x + \nu(\epsilon_y + \epsilon_z)] - \frac{E\alpha(\Delta T)}{(1 - 2\nu)}$$

$$\sigma_y = \frac{E}{(1 + \nu)(1 - 2\nu)} [(1 - \nu)\epsilon_y + \nu(\epsilon_z + \epsilon_x)] - \frac{E\alpha(\Delta T)}{(1 - 2\nu)} \quad (1\text{-}30)$$

$$\sigma_z = \frac{E}{(1 + \nu)(1 - 2\nu)} [(1 - \nu)\epsilon_z + \nu(\epsilon_x + \epsilon_y)] - \frac{E\alpha(\Delta T)}{(1 - 2\nu)}$$

Many of the problems considered later in this text can be classified as plane stress problems. For the case of plane stress in the xy plane ($\sigma_z = \tau_{zx} = \tau_{zy} = 0$), Eqs. 1-29 reduce to

$$\epsilon_x = \frac{1}{E} (\sigma_x - \nu\sigma_y) + \alpha(\Delta T)$$

$$\epsilon_y = \frac{1}{E} (\sigma_y - \nu\sigma_x) + \alpha(\Delta T) \quad (1\text{-}31)$$

$$\epsilon_z = -\frac{\nu}{E} (\sigma_x + \sigma_y) + \alpha(\Delta T)$$

In a similar manner, Eqs. 1-30 reduce to

$$\sigma_x = \frac{E}{1 - \nu^2} (\epsilon_x + \nu\epsilon_y) - \frac{E\alpha(\Delta T)}{1 - \nu}$$

$$\sigma_y = \frac{E}{1 - \nu^2} (\epsilon_y + \nu\epsilon_x) - \frac{E\alpha(\Delta T)}{1 - \nu} \quad (1\text{-}32)$$

Additional effects that an increase (decrease) in temperature can produce include a decrease (increase) in modulus of elasticity, a decrease (increase) in yield strength, a decrease (increase) in tensile strength, an increase (decrease) in ductility, and an increase (decrease) in creep rate. The changes in material properties for most engineering materials (polymer materials excepted) are small for a temperature change of a few hundred degrees Fahrenheit near room temperature.

EXAMPLE 1-8

A 6-m-long by 50-mm-diameter rod of aluminum alloy ($E = 70$ GPa, $\nu = 0.346$, and $\alpha = 22.5(10^{-6})/C°$) is attached at the ends to supports that yield to permit a change in length of 1.00 mm in the rod when stressed. When the temperature is 35°C, there is no stress in the rod. After the temperature of the rod drops to $-20°C$, determine

(a) the axial stress in the rod.

(b) the change in diameter of the rod.

Solution

(a) As a result of yielding of the supports, the final strain in the rod, as given by Eq. 1-12, after the temperature drops and the axial stress develops is

$$\epsilon_x = \frac{\delta}{L} = \frac{-0.001}{6} = -166.67(10^{-6}) \ \text{m/m} = -166.67 \ \mu\text{m/m}$$

The axial stress σ_x in the rod, resulting from the combined effects of the temperature change and yielding of the supports, is obtained by using Eq. 1-31. Since the stress σ_y is zero, Eq. 1-31 yields

$$\sigma_x = E\epsilon_x - E\alpha(\Delta T)$$
$$= 70(10^9)(-166.67)(10^{-6}) - 70(10^9)(22.5)(10^{-6})(-55)$$
$$= 74.96(10^6) \ \text{N/m}^2 = \underline{75.0 \ \text{MPa} \ T} \qquad \text{Ans.}$$

(b) The change in diameter of the rod results from the combined effects of the axial stress and the temperature change. Thus, from Eqs. 1-12 and 1-31

$$\delta_D = \epsilon_D \ D = [-\nu\sigma_x/E + \alpha(\Delta T)] \ D$$
$$= [-0.346(74.96)(10^6)/(70)(10^9)$$
$$+ 22.5(10^{-6})(-55)](50)(10^{-3})$$
$$= -80.4(10^{-6}) \ \text{m} = \underline{-0.0804 \ \text{mm}} \qquad \text{Ans.}$$

PROBLEMS

1-107* A bridge has a span of 4800 ft. Determine the change in length of a steel [$E = 30{,}000$ ksi and $\alpha = 6.5(10^{-6})/°F$] longitudinal floor support beam (that must be provided for with expansion joints) due to a seasonal temperature variation from $-30°F$ to $100°F$.

1-108* An airplane has a wing span of 40 m. Determine the change in length of the aluminum alloy [$E = 73 \ \text{GPa}$ and $\alpha = 22.5(10^{-6})/°C$] wing spar if the plane leaves the ground at a temperature of $40°C$ and climbs to an altitude where the temperature is $-40°C$.

1-109* A large cement kiln has a length of 225 ft and a diameter of 12 ft. Determine the change in length and diameter of the structural steel [$E = 30{,}000$ ksi and $\alpha = 6.5(10^{-6})/°F$] shell caused by an increase in temperature of $250°F$.

1-110* A steel [$E = 200$ GPa and $\alpha = 11.9(10^{-6})/°C$] micrometer designed for measuring diameters of approximately 2 m was calibrated at 20°C. What percent error will be introduced if the micrometer is used to make a 2-m measurement at a temperature of 40°C without correcting for temperature?

1-111 A steel [$E = 30,000$ ksi and $\alpha = 6.5(10^{-6})/°F$] surveyor's tape 1/2-in. wide by 1/32-in. thick is exactly 100 ft long at 72°F and under a pull of 10 lb. What percent error would be introduced if the tape is used to make a 100-ft measurement at a temperature of 100°F and under a pull of 25 lb?

1-112 A 25-mm-diameter aluminum [$\alpha = 22.5(10^{-6})/°C$, $E = 73$ GPa, and $\nu = 0.33$] rod hangs vertically while suspended from one end. A 2500-kg mass is attached at the other end. After the load is applied, the temperature decreases 50°C. Determine

(a) the axial stress in the rod. $\sigma = \dfrac{P}{A}$

(b) the axial strain in the rod. $\epsilon_a = \dfrac{\sigma}{E}$

(c) the change in diameter of the rod. $\epsilon_d = \dfrac{-\nu\sigma}{E} + \alpha\Delta T$ $\Delta D = \epsilon_d D$

1-113* Determine the movement of the pointer of Fig. P1-113 with respect to the scale zero, when the temperature increases 80°F. Assume that the coefficients of thermal expansion are $6.6(10^{-6})/°F$ for the steel and $12.5(10^{-6})/°F$ for the aluminum.

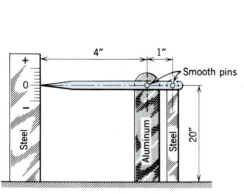

FIG. P1-113

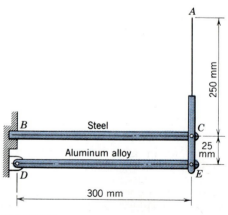

FIG. P1-114

1-114* Determine the horizontal movement of point A of Fig. P1-114 due to a temperature increase of 75°C. Assume that member AE has an insignificant coefficient of thermal expansion. The coefficients of thermal expansion are $11.9(10^{-6})/°C$ for the steel and $22.5(10^{-6})/°C$ for the aluminum alloy.

1-115 A 20-ft section of steel [$E = 30,000$ ksi and $\alpha = 6.5(10^{-6})/°F$] rail has a cross-sectional area of 12 in.^{2}. Both ends of the rail are tight against adjacent rails that, for this problem, can be assumed to be rigid. The rail is supported against lateral movement. For an increase in temperature of 80°F, determine

(a) the normal stress in the rail.

(b) the internal force on a cross section of the rail.

1-116 A steel [$E = 200$ GPa and $\alpha = 11.9(10^{-6})/°C$] rod containing a turnbuckle has its ends attached to rigid walls. During the summer when the temperature is 28°C, the turnbuckle is tightened to produce a stress in the rod of 25 MPa. Determine the stress in the rod in the winter when the temperature is $-30°C$.

1-117* At a temperature of 100°F, a cold-rolled bronze [$E = 15,000$ ksi and $\alpha = 9.4(10^{-6})/°F$] sleeve has an inside diameter of 6.000 in. and an outside diameter of 6.500 in. At a temperature of 0°F, determine

(a) the inside diameter of the sleeve.

(b) the uniaxial stress in the sleeve if the inside diameter of the sleeve is prevented from contracting with the change in temperature.

1-118* At a temperature of 25°C, a cold-rolled red brass [$E = 100$ GPa and $\alpha = 17.6(10^{-6})/°C$] sleeve has an inside diameter of 299.75 mm and an outside diameter of 310 mm. The sleeve is to be placed on a steel shaft with an outside diameter of 300 mm. Determine

(a) the temperature at which the sleeve will slip over the shaft with a clearance of 0.05 mm.

(b) the stress in the sleeve after it has cooled to 25°C if the shaft is assumed to be rigid.

1-119 A 1-in.-diameter steel [$\alpha = 6.5(10^{-6})/°F$, $E = 30,000$ ksi, and $\nu = 0.30$] bar is subjected to a temperature decrease of 150°F. The ends of the bar are supported by two walls that displace a small amount during the temperature change. If the measured strain in the bar is -600 μin./in. after the temperature change, determine the load being transmitted to the walls.

1-120 A prismatic bar, free of stress at room temperature, is fastened to rigid walls at its ends. One end of the bar is heated to 100°C above room temperature, while the other end is maintained at room temperature. The change in temperature ΔT along the bar is proportional to the square of the distance from the unheated end. Determine the axial stress in the bar if it is made of an aluminum [$E = 70$ GPa and $\alpha = 22.5(10^{-6})/°C$] alloy.

1-121* At a point on the free surface of an aluminum alloy [$E = 10,000$ ksi, $\nu = 0.33$, and $\alpha = 12.5(10^{-6})/°F$] plate, the measured strains resulting from both a temperature increase of 100°F and externally applied loads are: $\epsilon_x = +500\ \mu\text{in./in.}$, $\epsilon_y = +1000\ \mu\text{in./in.}$, and $\gamma_{xy} = 0$. Determine the stresses σ_x and σ_y at the point.

1-122* At a point on the free surface of a steel [$E = 200\ \text{GPa}$, $\nu = 0.30$, and $\alpha = 12(10^{-6})/°\text{C}$] machine part, the measured strains resulting from both a temperature decrease of 100°C and externally applied loads are: $\epsilon_x = -1250\ \mu\text{m/m}$, $\epsilon_y = +200\ \mu\text{m/m}$, and $\gamma_{xy} = 0$. Determine the stresses σ_x and σ_y at the point.

$$\sigma_x = \frac{E}{1-\nu^2}(\epsilon_x + \delta_y)$$

1-16

HOOKE'S LAW FOR UNIDIRECTIONAL COMPOSITE MATERIALS

The word *composite* is used to describe a material formed by combining two or more unlike material components. Composite materials have a long history of usage; the use of straw to strengthen mud bricks precedes recorded history. More recently, glass fibers in combination with thermosetting resins (epoxies and polyesters) have been used for car bodies, boat hulls, large tanks, and pipes. Composites utilizing high-modulus fibers, such as boron, carbon, and Kevlar, are currently used in aircraft and some aerospace structures. The sporting goods industry uses carbon-fiber-reinforced plastics for fishing poles and golf clubs.

The three common types of composite material are *fibrous composites* (continuous high-strength fibers in a resin matrix), *laminated composites* (layers of a fibrous composite or other material), and *particulate composites* (chopped fibers or particles in a resin matrix).

Composite materials have many characteristics that are different from the conventional engineering materials (steel, aluminum, brass, etc.). Most of the common engineering materials can be considered *homogeneous* (properties are not a function of position in the body) and *isotropic* (properties are the same in all directions at a point in the body). In contrast, composite materials can be *nonhomogeneous* (properties are a function of position in the body) and *anisotropic* (properties are different in all directions at a point in the body). A material with properties that are different in three mutually perpendicular directions at a point in the body but which has three mutually perpendicular planes of material symmetry is known as an orthotropic material. The discussion of composite materials in this text will be limited to the case of the simple unidirectional fiber-reinforced orthotropic material shown in Fig. 1-37. The 1 and 2 directions are principal material directions.

The material properties required to describe the behavior of this material include Young's modulus in the fiber direction, E_1; Young's modulus in the transverse direction, E_2; major Poisson's ratio, ν_{12};

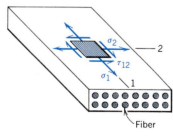

FIG. 1-37

minor Poisson's ratio, ν_{21}; and the shear modulus, G_{12}. These five
properties are defined as follows:

$$E_1 = \frac{\sigma_1}{\epsilon_1} \qquad \text{when} \quad \sigma_2 = \tau_{12} = 0 \qquad (1\text{-}33)$$

$$E_2 = \frac{\sigma_2}{\epsilon_2} \qquad \text{when} \quad \sigma_1 = \tau_{12} = 0 \qquad (1\text{-}34)$$

$$\nu_{12} = -\frac{\epsilon_2}{\epsilon_1} \qquad \text{when} \quad \sigma_2 = \tau_{12} = 0 \qquad (1\text{-}35)$$

$$\nu_{21} = -\frac{\epsilon_1}{\epsilon_2} \qquad \text{when} \quad \sigma_1 = \tau_{12} = 0 \qquad (1\text{-}36)$$

$$G_{12} = \frac{\tau_{12}}{\gamma_{12}} \qquad \text{when} \quad \sigma_1 = \sigma_2 = 0 \qquad (1\text{-}37)$$

For orthotropic materials, a tensile normal stress in a principal material direction produces an extension in the direction of the stress and
a contraction perpendicular to the stress. The contraction can be larger
or smaller than the contraction experienced by a similarly loaded
isotropic material with the same elastic modulus in the direction of
the load. A shear stress causes a shearing deformation, but the magnitude of the deformation is independent of the Young's moduli and the
Poisson's ratios. In other words, the shear modulus of an orthotropic
material is independent of the other material properties.

For anisotropic materials, application of a tensile normal stress
produces an extension in the direction of the stress, a contraction
perpendicular to the direction of the stress, and a shearing deformation.
Conversely, shearing stresses produce extensions and contractions in
addition to the shearing deformation. This behavior is also observed
in orthotropic materials when the normal stress is applied in a non-
principal material direction as shown in Fig. 1-38. This previous discussion illustrates a few of the problems encountered when composite
materials are used.

Fiber-reinforced composite materials are usually treated as linear
elastic materials, since the fibers provide most of the strength and
stiffness. As a result, the strains can be combined using the principle
of superposition. Thus, the stress-strain relationships for plane stress
in the principal material directions are:

$$\epsilon_1 = \frac{\sigma_1}{E_1} - \nu_{21}\frac{\sigma_2}{E_2}$$

$$\epsilon_2 = \frac{\sigma_2}{E_2} - \nu_{12}\frac{\sigma_1}{E_1} \qquad (1\text{-}38)$$

$$\gamma_{12} = \frac{\tau_{12}}{G_{12}}$$

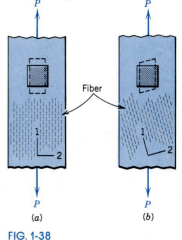

FIG. 1-38

TABLE 1-1 Material Properties for Two Unidirectional Composites

Constant	Glass/Epoxy	Graphite/Epoxy
E_1	8000 ksi (55 GPa)	21,000 ksi (145 GPa)
E_2	3000 ksi (21 GPa)	1400 ksi (9.7 GPa)
ν_{12}	0.25	0.29
ν_{21}	0.09	0.02
G_{12}	1400 ksi (9.7 GPa)	1000 ksi (6.9 GPa)

Solving Eqs. 1-38 for stresses in terms of strains yields

$$\sigma_1 = \frac{E_1}{1 - \nu_{12}\nu_{21}} (\epsilon_1 + \nu_{21}\epsilon_2)$$

$$\sigma_2 = \frac{E_2}{1 - \nu_{12}\nu_{21}} (\epsilon_2 + \nu_{12}\epsilon_1) \qquad (1\text{-}39)$$

$$\tau_{12} = G_{12}\gamma_{12}$$

Equations 1-38 and 1-39 will reduce to Hooke's law for isotropic materials as given by Eqs. 1-22, 1-24, and 1-26, when $E_1 = E_2$ and $\nu_{12} = \nu_{21}$.

Some typical material properties for a unidirectional glass-fiber/epoxy composite and a unidirectional graphite-fiber/epoxy composite are listed in Table 1-1. The properties of a particular composite material are a function of the volume fraction of fibers, that is, the volume of fibers relative to the total volume of material. Typically, modulus E_1, which depends primarily on the fibers, is much larger than modulus E_2, which depends mainly on the matrix material. Similarly, the major Poisson's ratio is much larger than the minor Poisson's ratio. If the tensile and compressive elastic properties for the composite are the same, the minor Poisson's ratio can be written in terms of the other material properties as

$$\nu_{21} = \frac{E_2}{E_1} \nu_{12} \qquad (1\text{-}40)$$

A temperature change in a composite material produces effects that are similar to those encountered in the conventional homogeneous and isotropic engineering materials and these are described in Section 1-15. When a temperature change occurs, the stress-strain relationships must be adjusted to account for the free thermal strains. If α_1 and α_2 are the thermal coefficients of expansion in the 1 and 2 directions, respectively, then

$$\epsilon_1 = \frac{\sigma_1}{E_1} - \nu_{21}\frac{\sigma_2}{E_2} + \alpha_1 \Delta T$$

$$\epsilon_2 = \frac{\sigma_2}{E_2} - \nu_{12}\frac{\sigma_1}{E_1} + \alpha_2 \Delta T \qquad (1\text{-}41)$$

$$\gamma_{12} = \frac{\tau_{12}}{G_{12}}$$

Solving for the stresses in terms of the strains yields

$$\sigma_1 = \frac{E_1}{1 - \nu_{12}\nu_{21}}[(\epsilon_1 - \alpha_1 \Delta T) + \nu_{21}(\epsilon_2 - \alpha_2 \Delta T)]$$

$$\sigma_2 = \frac{E_2}{1 - \nu_{12}\nu_{21}}[(\epsilon_2 - \alpha_2 \Delta T) + \nu_{12}(\epsilon_1 - \alpha_1 \Delta T)] \qquad (1\text{-}42)$$

$$\tau_{12} = G_{12}\gamma_{12}$$

The stress-strain relationships developed in this section are valid only for the principal material directions. A general stress-strain relationship for any direction relative to the principal material directions can be developed, but is beyond the scope of this introductory treatment of composite materials and their behavior.

1-17
DESIGN LOADS, WORKING STRESSES, AND FACTORS OF SAFETY

As mentioned in Section 1-1, the designer is required to select a material and properly proportion a member therefrom to perform a specified function without failure. Failure is defined as the state or condition in which a member or structure no longer functions as intended. Several types of failure warrant discussion at this point. *Elastic failure* is characterized by excessive elastic deformation—a bridge may deflect elastically under traffic to an extent that may result in discomfort to the vehicle passengers; a close-fitting machine part may deform sufficiently under load to prevent proper operation of the machine. When a structure is designed to avoid elastic failure, the stiffness of the material, indicated by Young's modulus, is the significant property. *Slip failure* is characterized by excessive plastic deformation due to slip. Yield strength, yield point, and proportional limit are used as indices of strength with respect to failure by slip for members subjected to static loads. *Creep failure* is characterized by excessive plastic deformation over a long period of time under constant stress. For machines or structures that are to be subjected to relatively high stress, high temperature, or both over a long period of time, creep is a design consideration, and the creep limit is the strength index to be used. The creep limit normally decreases as the temperature increases. *Failure by fracture* is a complete separation of the material. The ultimate strength of a material is the index of resistance to failure by fracture under static loads in which creep is not involved.

Most design problems involve many unknown variables. The load that the structure or machine must carry is usually estimated; the actual load may vary considerably from the estimated, especially when loads at some future time must be considered. Since testing usually damages the material, the properties of the material used in the structure cannot be evaluated directly but are normally determined by testing specimens of a similar material. Furthermore, the actual stresses that will exist in a structure are unknown because the calculations are based on assumptions about the distribution of stresses in the material. Because of these and other unknown variables, it is customary to design for the load required to produce failure, which is larger than the estimated actual load, or to use a working (or design) stress below the stress required to produce failure.

A *working stress* or *allowable stress* is defined as the maximum stress permitted in the design computation. The *factor of safety* may be defined as the ratio of a failure-producing load to the estimated actual load. Factor of safety may also be defined as the ratio of the strength of a material to the maximum computed stress in the material. The latter factor of safety may be used to determine an allowable (or working) stress by dividing the strength of a material by the factor of safety; however, because actual sections of structural and machine elements, in general, are not the same as the minimum computed sections, the actual (computed) factor of safety is the strength divided by the maximum computed stress based on actual sections of members in the design.

For a given design, the factor of safety based on the ultimate strength is, in general, quite different from that based on the elastic strength (proportional limit, yield point, or yield strength); hence, the term *strength* is ambiguous without a qualifying statement regarding the pertinent mode of failure. The following example illustrates the use of a factor of safety.

EXAMPLE 1-9

The improvised system of cables, shown schematically in Fig. 1-39a, is used to remove the pile Q. A tractor with a maximum drawbar pull of 16 kips is available to apply the force P. Cable A is a 1-in.-diameter iron hoisting cable that has a breaking load of 29.0 kips. Cables B, C,

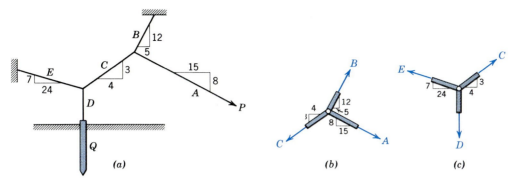

FIG. 1-39

and D are parts of a continuous $\frac{3}{4}$-in.-diameter carbon steel aircraft cable with a breaking strength of 49.6 kips. Cable E is a $\frac{7}{8}$-in.-diameter steel hoisting cable with a nominal area of 0.60 in.^{2} and an ultimate tensile strength, based on the nominal area, of 71.6 ksi. The minimum factor of safety for each cable, based on fracture, is to be 2.0. Determine

(a) the maximum allowable force that can be exerted on the pile.

(b) the factor of safety for each of the five cables when the load in part (a) is applied.

Solution

(a) The two free-body diagrams needed to solve the problem are shown in Figs. 1-39b and c. The maximum allowable tension in each cable can be obtained from the strength of the cable and the factor of safety, because for axial loading the factor of safety based on load is the same as that based on stress. The allowable tensions are:

$$A = 29.0/2 = 14.5 \text{ kips}$$
$$B = C = D = 49.6/2 = 24.8 \text{ kips}$$
$$E = 0.60(71.6)/2 = 43.0/2 = 21.5 \text{ kips}$$

The maximum tractor pull (16.0 kips) is greater than the allowable force in cable A; therefore, the maximum tractor pull cannot be utilized without exceeding the specifications. Assuming that the force in A is 14.5 kips, the equations of equilibrium applied to Fig. 1-39b give

$$(5/13)B + (15/17)(14.50) - (4/5)C = 0$$
$$(12/13)B - (8/17)(14.50) - (3/5)C = 0$$

from which $B = 25.9$ kips and $C = 28.4$ kips. Since the maximum allowable load on cables B and C is only 24.8 kips, it is apparent that the force in A must be less than the allowable load. The force in B is less than that in C; therefore, cable C is the controlling member for Fig. 1-39b. The equations of equilibrium for this free-body diagram with $C = 24.8$ kips give the results

$$A = 12.65 \text{ kips} \quad \text{and} \quad B = 22.6 \text{ kips}$$

The equations of equilibrium for Fig. 1-39c are

$$(4/5)C - (24/25)E = 0$$
$$(3/5)C + (7/25)E - D = 0$$

When $C = 24.8$ kips is substituted in these equations, the results are

$$D = E = 20.7 \text{ kips}$$

Both of these forces are less than the maximum allowable loads in the corresponding cables and are therefore safe loads. The strength of cable C is thus the limiting factor. The maximum pull on the pile is the force in cable D and is

$$D = \underline{20.7 \text{ kips}}$$

Ans.

(b) The factor of safety in each of the cables is

$$k_A = 29.0/12.65 = \underline{2.29} \qquad \text{Ans.}$$
$$k_B = 49.6/22.6 \ = \underline{2.19} \qquad \text{Ans.}$$
$$k_C = 49.6/24.8 \ = \underline{2.00} \qquad \text{Ans.}$$
$$k_D = 49.6/20.7 \ = \underline{2.40} \qquad \text{Ans.}$$
$$k_E = 43.0/20.7 \ = \underline{2.08} \qquad \text{Ans.}$$

PROBLEMS

1-123* A 1-in.-diameter structural steel (see Appendix A for properties) bar is supporting an axial tensile load of 8 kips. Determine

(a) the factor of safety with respect to failure by slip.
(b) the factor of safety with respect to failure by fracture.

1-124* A structural steel (see Appendix A for properties) bar with a 25 × 50-mm rectangular cross section is supporting an axial tensile load of 125 kN. Determine

(a) the factor of safety with respect to failure by slip.
(b) the factor of safety with respect to failure by fracture.

1-125 A 2-in.-wide by 6-in.-long strip of SAE 4340 heat-treated steel (see Appendix A for properties) will be used to transmit an axial tensile load of 15 kips. Determine the minimum thickness required if

(a) a factor of safety of 4 with respect to failure by slip is specified.

(b) a factor of safety of 3 with respect to failure by fracture is specified.

1-126 Two aluminum alloy plates are connected by using a lap joint with three aluminum alloy rivets. The yield and ultimate strengths of the rivets in shear are 190 MPa and 280 MPa, respectively. The plate will be used to transmit an axial tensile load of 100 kN. Determine the minimum diameter for the rivets if

(a) a factor of safety of 2.5 with respect to failure by slip is specified.

(b) a factor of safety of 2 with respect to failure by fracture is specified.

1-127* A 10-kip load is supported by two tie rods as shown in Fig. P1-127. Rod *AC* is made of annealed red brass (see Appendix A for properties), has a cross-sectional area of 1.00 in.², and has a working stress specification of 10 ksi. Rod *BC* is made of annealed bronze, has a cross-sectional area of 1.25 in.², and has a working stress specification of 14 ksi. Determine

(a) the factor of safety with respect to failure by slip for each of the rods based on the working stress specifications.

(b) the actual factor of safety with respect to failure by fracture for each of the rods.

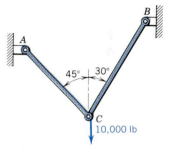

FIG. P1-127

1-128* A 50-kN load is supported by a tie rod and a pipe column, as shown in Fig. P1-128. Tie rod *AB* is made of aluminum alloy 2024-T4 (see Appendix A for properties), has a cross-sectional area of 650 mm², and has a working stress specification of 160 MPa. Pipe column *BC* is made of structural steel, has a cross-sectional area of 925 mm², and has a working stress specification of 125 MPa. Determine

(a) the factor of safety with respect to failure by slip for each of the members based on the working stress specifications.

(b) the actual factor of safety with respect to failure by slip for each of the members.

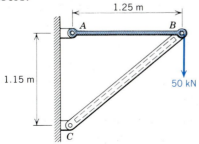

FIG. P1-128

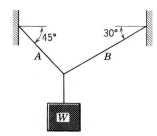

FIG. P1-129

1-129 The load W of Fig. P1-29 is supported by two cables. Cable A has a breaking strength of 3.5 kips. Cable B has a breaking strength of 3 kips. Determine the maximum safe load that can be supported if a factor of safety of 2.5 with respect to failure by fracture is specified.

1-130 The load P of Fig. P1-30 is supported by a continuous, carbon steel aircraft cable having a length L and a breaking strength of 40 kN. Determine

   (a) the equilibrium position of load P (distance a in terms of cable length L).

   (b) the maximum load P that can be supported if a factor of safety of 2 with respect to failure by fracture is specified.

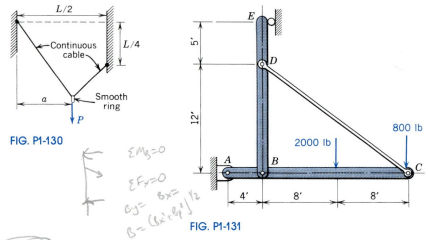

FIG. P1-130

FIG. P1-131

1-131* Two bars and a tie rod are used to support an 800-lb load and a 2000-lb load, as shown in Fig. P-131. The tie rod CD and the rivets at B, C, and D (which are in single shear) are made of aluminum alloy 2014-T4 (see Appendix A for properties). Determine

   (a) the minimum diameter D for tie rod CD if a factor of safety of 3 with respect to failure by slip is specified.

   (b) the minimum diameter D for the rivet at B if a factor of safety of 4 with respect to failure by fracture is specified.

   (c) the minimum diameter D for the rivets at C and D if a factor of safety of 4 with respect to failure by fracture is specified.

1-132* The tie rod AB and the bolts at A, B, and C of the structure shown in Fig. P1-132 are made of steel, for which the elastic strengths are 250 MPa in tension and 150 MPa in shear, and the ultimate strengths are 450 MPa in tension and 270 MPa in shear. Determine

   (a) the minimum diameter D for tie rod AB if a factor of safety of 2.5 with respect to failure by slip is specified.

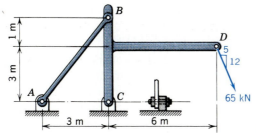

FIG. P1-132

(b) the minimum diameter D for the bolt at C if a factor of safety of 3.5 with respect to failure by fracture is specified.

(c) the minimum diameter D for the bolts at A and B if a factor of safety of 3.5 with respect to failure by fracture is specified.

1-133 A pin-connected truss is loaded and supported as shown in Fig. P1-133. All members of the truss are two-force members. The weights of the members are negligible with respect to the applied loads. Member CF is to be fabricated by using two structural steel (see Appendix A for properties) channels fastened back to back. A factor of safety of 4 with respect to failure by slip is specified. Determine

(a) the cross-sectional area needed for the member.

(b) the lightest pair of channels from Appendix C that will be acceptable.

(c) the actual factor of safety for the channels chosen.

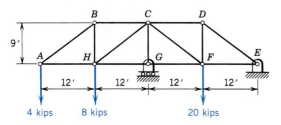

FIG. P1-133

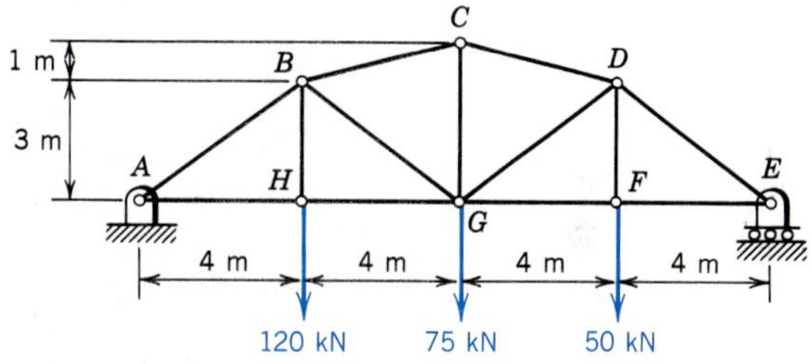

FIG. P1-134

1-134 A pin-connected truss is loaded and supported as shown in Fig. P1-134. All members of the truss are two-force members. The weights of the members are negligible with respect to the applied loads. Member GH is to be fabricated by using two structural steel (see Appendix A for properties) angles fastened back to back. A factor of safety of 3 with respect to failure by slip is specified. Determine

(a) the cross-sectional area needed for the member.

(b) the lightest pair of angles from Appendix C that will be acceptable.

(c) the actual factor of safety for the angles chosen.

1-18

REPEATED LOADING—FATIGUE

The discussion in previous sections of this chapter has been limited to static loading, for which the behavior of a structure under load can be predicted on the basis of the static properties of the material. However, many machine components and structural elements are subjected to a type of loading known as *repeated loading*, under which the load is applied and removed, or the magnitude is changed, many thousands of times during the life of the member. For example, springs, push rods, connecting rods, and crankshafts, are subjected to repeated loading. In other instances, the repeated loading is accidental, as a result of vibrations induced by unpredictable or uncontrollable circumstances.

If the applied load changes from any magnitude in one direction to the same magnitude in the opposite direction, the loading is termed *completely reversed*, whereas if the load changes from one magnitude to another the load is said to be a *fluctuating load*. The physical effect of a repeated load on a material is different from that of a static load. With a repeated load, the failure is always a fracture (or brittle type of failure) for both brittle and ductile materials. Under repeated loading, a small crack forms in a region of high localized stress. As the load fluctuates, the crack opens and closes and progresses across the section. This crack propagation continues until there is insufficient cross section left to carry the load and the member ruptures. This type of failure is known as a *fatigue failure*. It is unfortunate that fatigue failures occur at stress levels well below the static elastic strength of most materials.

The index of resistance to failure under a completely reversed repeated load is known as the *endurance limit*. The endurance limit of a material is defined as the maximum completely reversed stress to which the material can be subjected millions of times without fracture. The magnitude of the endurance limit is determined from repeated load (or

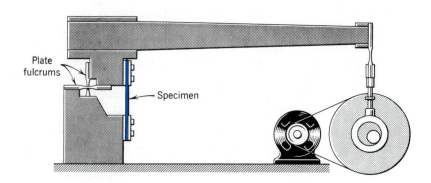

FIG. 1-40

Plate fulcrums

Specimen

fatigue) tests. Many types of fatigue testing machines are used; a common type is shown schematically in Fig. 1-40. With this machine, the load is applied by means of a crank and connecting rod and can be adjusted to produce different ranges of load in the specimen. Endurance limit determinations require the testing of a series of identical specimens at completely reversed loads of different magnitudes. The number of cycles of stress (or load) required to fracture each specimen is recorded. The data are then plotted as a stress-cycle ($\sigma - n$) diagram, and the endurance limit is determined from the diagram. It is customary to plot the data on semilogarithmic paper, the stress being plotted to a linear scale and the cycles to failure to a logarithmic scale, as shown in Fig. 1-41. For some materials (particularly the ferrous materials) the curve seems to break sharply, as in curves A and B of Fig. 1-41. Thus, if the stress level is less than the endurance limit σ_{EA}, material A could probably be subjected to an unlimited number of cycles without failure. On the other hand, the alloys of aluminum and magnesium exhibit curves such as C and D of Fig. 1-41, which are apparently asymptotic to the n axis. For such materials, the stress corresponding to some arbitrary number of cycles, such as $5(10^8)$ for the aluminum alloys, is taken as the endurance limit.

Fatigue failure data—dependent as they are on local imperfections, grain structure, and precision of machining identical specimens—show considerable scatter. Thus, results can be better represented by a series

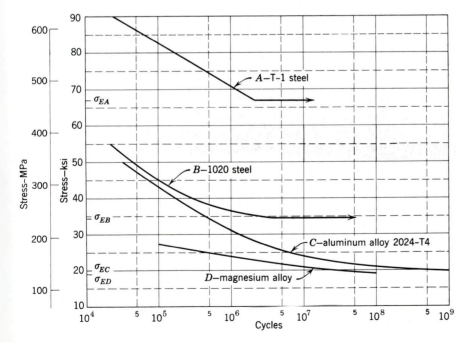

FIG. 1-41

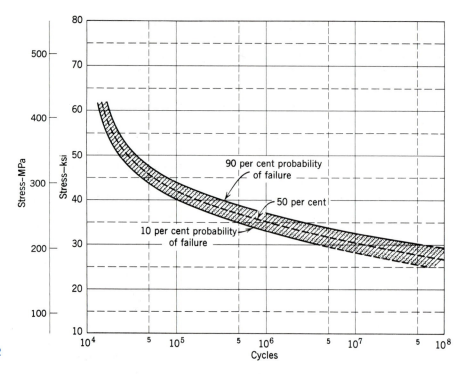

FIG. 1-42

of curves that represent different probabilities of failure or as a band such as that of Fig.1-42. The $\sigma - n$ diagram shown in Fig. 1-41 is quite likely an approximate 50 percent probability of failure curve; that is, a specimen tested at a particular maximum stress will have a 50–50 chance of failure when subjected to the number of cycles indicated by the intersection of the line representing the test stress with the 50 percent curve.

If a machine part is to be designed for a limited life, a strength index higher than the endurance limit can be used. Theoretically, one can estimate the number of cycles of stress to which the member will probably be subjected during its lifetime, enter the $\sigma - n$ diagram with this value, and read the corresponding stress from the stress scale. If the member was stressed to this value, one could expect a fracture failure at the end of the expected lifetime, and the design of the machine part could be based on this value reduced by an appropriate factor of safety. In this connection, however, the statistical nature of the $\sigma - n$ data should be emphasized since the life of any particular manufactured part may be much less than that indicated by the ordinary $\sigma - n$ diagram.

The maximum stress necessary to cause fracture under a fluctuating load is usually larger than the endurance limit. Considerable data have been accumulated from tests performed in machines similar to that shown schematically in Fig. 1-40, in which the stress in the specimen may vary from some minimum to a different maximum value. However, the task of experimentally determining the ranges of stress that will cause fatigue failure for all possible (or reasonable) combinations of maximum and minimum stresses and for the numerous materials in use today would be unending. Hence, some method of predicting the strength based on available data is essential. The approach to the problem is made through the interpretation of data from a limited number of tests in various ranges and is based on the concept that a fluctuating stress can be considered the sum of a steady stress and a completely reversed (or alternating) stress, as indicated for normal stresses in Fig. 1-43, where σ_a, the steady stress, is one-half the sum of σ_{max} and σ_{min}, and where σ_r, the alternating stress, is one-half the difference between these two stresses (compressive stresses are negative), The steady and alternating stresses may not be the same kind of stress; for example, one could have a steady torsional shearing stress combined with an alternating flexural stress. However, for the restricted scope of this book, the discussion will be limited to the case in which the steady and alternating stresses at a point are both normal (or both shearing) stresses on the same plane.

The data from tests can be plotted on a diagram, and a curve can be drawn to represent the strength of the material for various combinations of steady and alternating stresses. Two curves that have received widespread attention are shown in Fig. 1-44, in which σ_E is the endurance limit, σ_Y is the static elastic[11] tensile strength, and σ_U is the static ultimate tensile strength. Any point with coordinates (σ_a, σ_r), such as A on the diagram, would represent a particular combination of steady and repeated stresses, and if the point is above the straight line (Soderberg[12] or Goodman,[13] depending on which line is accepted

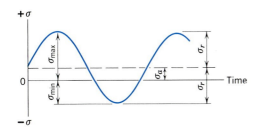

FIG. 1-43

FIG. 1-44

[11] Yield point, yield strength, or proportional limit.

[12] See "Working Stresses," C. R. Soderberg, *Trans. ASME*, Vol. 55, APM-55-16, 1933, p. 131.

[13] Goodman's diagram does not resemble the one of Fig. 1-44. However, since the two diagrams yield the same results, it has become customary to identify the upper line of Fig. 1-44 as the Goodman criterion.

as the failure criterion), failure is presumed to result. Most of the available test data plot above the two lines shown, indicating that both curves are conservative.

When a design problem involves machine elements that will not function properly if the static elastic strength of the material is exceeded (a slip failure), the Soderberg diagram should be regarded as the criterion of failure. The reason for using the more conservative diagram is that σ_{max} might exceed the static elastic strength on the first application of the maximum load, whereupon the usefulness of the mechanism would be impaired or destroyed. If fracture is the only mode of failure to be considered, slip is of no concern, and the less conservative Goodman diagram should be used. Needless to say, the approximate criteria of strength are to be used only if fluctuating strength data from mechanical tests are unavailable. For designs that warrant the expense, fatigue tests should be conducted on structural components loaded to simulate actual service conditions. When the approximate failure criteria are applied, if the endurance limit is unknown *and the material is ferrous*, one-half of the ultimate tensile strength will give a reasonable value for the endurance limit. The use of the Soderberg diagram is illustrated in Example 1-10. The choice of diagrams is usually indicated by a statement concerning the mode of failure. For example, a factor of safety with respect to failure by slip or fracture means that both slip (due to the application of the maximum load) and fracture (due to the alternating load) must be avoided; hence, the Soderberg criterion should be used.

EXAMPLE 1-10

A machine element of heat-treated SAE 4340 steel is subjected to stresses that fluctuate from 350 MPa T to 70 MPa C. Determine the factor of safety with respect to failure by slip or fracture.

Solution

Failure by slip or fracture indicates that the Soderberg diagram should be used. The diagram is shown in Fig. 1-45, for which the properties were obtained from Appendix A. The first step in the solution is to compute the ratio of the working stresses σ_{rw} to σ_{aw} as follows:

$$\sigma_{aw} = \frac{350 + (-70)}{2} = 140 \text{ MPa } T$$

and

$$\sigma_{rw} = \frac{350 - (-70)}{2} = 210 \text{ MPa } T \text{ and } C$$

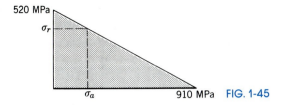

FIG. 1-45

from which

$$\sigma_r = \frac{3\sigma_a}{2}$$

The equation of the line in Fig. 1-45 is

$$\sigma_r = 520 - \left(\frac{520}{910}\right)\sigma_a$$

and since

$$\sigma_r = \frac{3\sigma_a}{2}$$

$$\frac{3\sigma_a}{2} = 520 - \left(\frac{520}{910}\right)\sigma_a$$

from which

$$\sigma_a = 251 \text{ MPa}$$

This value represents the limit of the steady stress. Because of the fixed ratio between σ_a and σ_r, 251 MPa can be used as a strength index; hence, the factor of safety is

$$k = \frac{251 \text{ MPa}}{140 \text{ MPa}} = \underline{1.79} \qquad \text{Ans.}$$

Observe that this value is obtained regardless of whether one uses the steady, the alternating, the maximum, or the minimum stress as the strength index; for example,

$$\sigma_{\max} = \sigma_a + \sigma_r = \sigma_a + \frac{3\sigma_a}{2} = \frac{5\sigma_a}{2}$$

and the factor of safety is

$$k = \frac{(251)}{(140)} = \frac{(3/2)(251)}{(3/2)(140)} = \frac{(5/2)(251)}{(5/2)(140)} = 1.79$$

PROBLEMS

1-135* Estimate the probable service life of a machine part made of 1020 steel (see Fig. 1-41) if the part is subjected to a completely reversed repeated stress of

(a) 45 ksi at 250 cycles per min.

(b) 40 ksi at 300 cycles per min.

(c) 38 ksi at 350 cycles per min.

(d) 32 ksi at 500 cycles per min

1-136 Estimate the probable service life of an aluminum alloy 2024-T4 (see Fig. 1-41) panel in an airplane if the panel is subjected to a completely reversed repeated stress (due to vibration) of

(a) 295 MPa at 100 cycles per min.

(b) 210 MPa at 200 cycles per min.

(c) 180 MPa at 150 cycles per min.

(d) 100 MPa at 250 cycles per min.

1-137* Determine a suitable working stress with a factor of safety of 1.25 for a machine part made of T-1 steel (see Fig. 1-41) if the part is subjected to a completely reversed repeated load and the estimated service life of the part is

(a) 100,000 cycles.

(b) 500,000 cycles.

(c) 5,000,000 cycles.

1-138 A machine part made of 1020 steel (see Fig. 1-41) is subjected to a completely reversed repeated stress of 150 MPa. Determine the factor of safety if the required service life of the part is

(a) 100,000 cycles.

(b) 500,000 cycles.

(c) 5,000,000 cycles.

1-139* Determine the factor of safety with respect to failure by slip or fracture for a machine part for which the stress varies from 10 ksi *T* to 6 ksi *C*. The material is aluminum alloy 2024-T4 (see Appendix A).

1-140* A machine component is subjected to an axial load that varies from 3000-lb compression to 5000-lb tension. Determine the combination of maximum and minimum stresses to which the member could probably be subjected millions of times without failure by slip or fracture if the proposed material is hardened and tempered 0.93 percent carbon spring-steel with an endurance limit of 56 ksi, a proportional limit of 85 ksi, and an ultimate strength of 97 ksi.

COMPUTER PROBLEMS

Note: The following problems have been designed to be solved with a programmable calculator, microcomputer, or mainframe computer. The problems require application of the basic principles developed in this chapter; however, most problems must be solved a number of times to study the effects of varying a particular parameter. Such solutions permit characteristics of the problem to be displayed that are not evident by solving the problem with a single value of the parameter. Appendix F contains a description of a few numerical methods, together with a few simple programs in BASIC and FORTRAN, that can be modified for use in the solution of these problems.

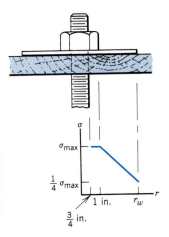

FIG. PC1-1

C1-1 Because the washer of Problem 1-13 is not rigid, the stress between the washer and the grain bin wall will not be uniform. Assume that the stress varies as shown in Fig. PC1-1; a uniform value of σ_{max} under the restraining nut (which has a 2-in. diameter), decreasing linearly to $\sigma_{max}/4$ at the outside edge r_w of the washer. Using the data from Problem 1-13,

(a) compute and plot σ_{max} versus the radius r_w of the washer for 1 in. $\leq r_w \leq 5$ in.

(b) compute the percent decrease in σ_{max} for a 4-in.-diameter washer compared to no washer.

(c) compute the percent decrease in σ_{max} for a 6-in.-diameter washer compared to no washer.

C1-2 If the bearing plate of Problem 1-12 is not rigid, the stress between the bearing plate and the wood beam will not be uniform. Assume that the stress varies as shown in Fig. PC1-2; a uniform value of σ_{max} above the column, decreasing linearly to $\sigma_{max}/5$ at the outside edge r_p of the bearing plate. Using the data from Problem 1-12,

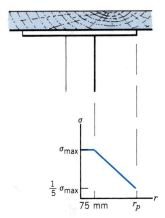

FIG. PC1-2

(a) compute and plot σ_{max} versus the radius r_p of the bearing plate for 75 mm $\leq r_p \leq 500$ mm.

(b) compute the percent decrease in σ_{max} for a 400-mm-diameter bearing plate compared to a 150-mm-diameter bearing plate.

(c) compute the percent decrease in σ_{max} for a 600-mm-diameter bearing plate compared to a 150-mm-diameter bearing plate.

C1-3 Repeat Problem C1-1 for the stress distribution shown in Fig. PC1-3.

$$\sigma = \sigma_{max}; \qquad 0.75 \text{ in.} \leq r \leq 1.00 \text{ in.}$$

$$\sigma = \frac{1}{r}\sigma_{max}; \qquad 1.00 \text{ in.} \leq r \leq r_w$$

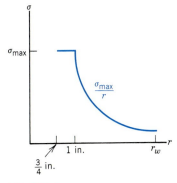

FIG. PC1-3

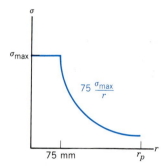

FIG. PC1-4

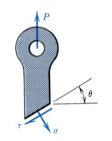

FIG. PC1-5

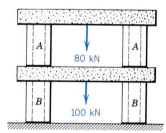

FIG. PC1-6

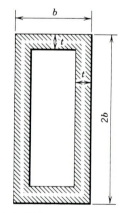

FIG. PC1-7

C1-4 Repeat Problem C1-2 for the stress distribution shown in Fig. PC1-4.

$$\sigma = \sigma_{max}; \qquad 0 \text{ mm} \leqslant r \leqslant 75 \text{ mm}$$

$$\sigma = \frac{75}{r}\sigma_{max}; \qquad 75 \text{ mm} \leqslant r \leqslant r_p$$

C1-5 The bar shown in Fig. PC1-5 has a cross-sectional area of 3/in.^{2} and carries an axial load of 12,000 lb.

(a) Compute and plot the normal stress σ versus the angle θ for $-90° \leqslant \theta \leqslant 90°$.

(b) Compute and plot the shear stress τ versus the angle θ for $-90° \leqslant \theta \leqslant 90°$.

(c) Plot the shear stress τ versus the normal stress σ for $-90° \leqslant \theta \leqslant 90°$. Identify the points on the graph corresponding to $\theta = -60°, -45°, -30°, -15°, 0°, 15°, 30°, 45°, 60°$.

C1-6 Two rigid plates are loaded and supported with steel ($E = 200$ GPa and $\nu = 0.30$) pipe columns as shown in Fig. PC1-6. Let the outside diameter of the pipes vary (50 mm $\leqslant 2r_o \leqslant 400$ mm), while keeping the pipe wall thickness constant ($r_o - r_i = 10$ mm). Compute and plot:

(a) the axial compressive strain in columns A versus r_o.

(b) the axial compressive strain in columns B versus r_o.

(c) the change in outside diameter of columns B versus r_o.

C1-7 A hollow rectangular, steel ($E = 29,600$ ksi and $\nu = 0.30$) bar with the cross section shown in Fig. PC1-7 is 36 in. long. The bar is subjected to an axial tensile load of 10 kips. Let the width b of the bar vary (1.5 in. $\leqslant b \leqslant 4$ in.), while keeping the wall thickness t constant ($t = 0.5$ in.). Compute and plot:

(a) the axial tensile strain ϵ in the bar versus the width b of the bar.

(b) the change in width Δb of the bar versus the width b of the bar.

C1-8 A hollow rectangular bronze ($E = 100$ Gpa and $\nu = 0.30$) bar with the cross section shown in Fig. PC1-7 is 1 m long. The bar is subjected to an axial compressive load of 120 kN. Assume that the width b of the bar is fixed ($b = 100$ mm), while the wall thickness t of the bar varies (5 mm $\leqslant t \leqslant 50$ mm). Compute and plot:

(a) the axial compressive strain ϵ in the bar versus the thickness t of the bar.

(b) the change in width Δb of the bar versus the thickness t of the bar.

C1-9 For a state of plane stress, write a program to convert the strains ϵ_x, ϵ_y, and γ_{xy} into the stress σ_x, σ_y, and τ_{xy}. Use the program to solve problems 1-97, 1-98, 1-99, and 1-100.

CHAPTER 2

AXIAL LOADING: APPLICATIONS AND PRESSURE VESSELS

2-1

INTRODUCTION

The problem of determining forces and deformations at all points within a body subjected to external forces is extremely difficult when the loading or geometry of the body is complicated. A refined analytical method of analysis that attempts to obtain general solutions to such problems is known as the *theory of elasticity*. The number of problems solved by such methods has been limited; therefore, practical solutions to most design problems are obtained by what has become known as the *mechanics of materials approach*. With this approach, real structural elements are analyzed as idealized models subjected to simplified loadings and restraints. The resulting solutions are approximate, since they consider only the effects that significantly affect the magnitudes of stresses, strains, and deformations.

In Chapter 1, the concepts of stress and strain were developed and a discussion of material behavior led to the development of equations relating stress to strain. In the remaining chapters of the book, the stresses and deformations produced in a wide variety of structural members by axial, torsional, and flexural loadings will be considered. The "mechanics of material analyses," as presented here, are somewhat less rigorous than the "theory of elasticity approach," but experience indicates that the results obtained are quite satisfactory for most engineering problems.

2-2

DEFORMATIONS IN AXIALLY LOADED BARS

When a bar of uniform cross section is axially loaded by forces applied at the ends (two-force member), the axial strain along the length of the bar is assumed to have a constant value, and the elongation (or contraction) of the bar resulting from the axial load P may be expressed as $\delta = \epsilon L$ by definition. If Hooke's law applies, Eq. 1-22 of Section 1-14 reduces to $\epsilon = \sigma/E$, where σ is the axial stress P/A induced by the load. Thus, the axial deformation may be expressed in terms of stress, or load, as follows:

$$\delta = \epsilon L = \frac{\sigma L}{E} \tag{2-1a}$$

or

$$\delta = \frac{PL}{AE} \tag{2-1b}$$

The first form will frequently be found convenient in elastic problems in which limiting axial stress and axial deformation are both specified. The stress corresponding to the specified deformation can be obtained from Eq. 2-1a and compared to the specified working stress, the smaller of the two values being used to compute the unknown load or cross-sectional area. In general, Eq. 2-1a is the preferred form when the problem involves the determination or comparison of stresses. If a bar is subjected to a number of axial loads applied at different points along the bar, or if the bar consists of parts having different cross-sectional areas or parts composed of different materials, then the change in length of each part can be computed by using Eq. 2-1b. These individual changes in length of the various parts of the bar can then be added algebraically to give the total change in length of the complete bar.

For those cases in which the axial force or the cross-sectional area varies continuously along the length of the bar, Eq. 2-1 is not valid. In Section 1-10 the axial strain at a point for the case of nonuniform deformation was defined as $\epsilon = d\delta/dL$. Thus, the increment of deformation associated with a differential element of length $dL = dx$ may be expressed as $d\delta = \epsilon\, dx$. If Hooke's law applies, the strain may again be expressed as $\epsilon = \sigma/E$, where $\sigma = P_x/A_x$. The subscripts indicate that both the applied load P and the cross-sectional area A may be functions of position x along the bar. Thus,

$$d\delta = \frac{P_x\, dx}{A_x\, E} \tag{a}$$

Integrating Eq. a yields the following expression for the total elongation (or contraction) of the bar:

$$\delta = \int_0^L d\delta = \int_0^L \frac{P_x \, dx}{A_x E} \qquad (2\text{-}2)$$

Equation 2-2 gives acceptable results for tapered bars, provided the angle between the sides of the bar does not exceed 20°.

EXAMPLE 2-1

The rigid yokes B and C of Fig. 2-1a are securely fastened to the 2-in. square steel ($E = 30,000$ ksi) bar AD. Determine

(a) the maximum normal stress in the bar.

(b) the change in length of the complete bar.

Solution

Since bar AD is subjected to a number of axial loads applied at different points along the bar, the different sections AB, BC, and CD of the bar will transmit different levels of load. An internal force (load) diagram, such as the one shown in Fig. 2-1b, provides a pictorial representation of the levels of internal force in each of the sections and serves as an aid for stress and deformation calculations.

The load diagram is easily constructed by isolating different portions of the bar with imaginary cuts and computing the internal force that must be transmitted by the cross section exposed by the cut to maintain equilibrium of the isolated portion of the bar. For example, an imaginary cut anywhere in section AB will expose a cross section that must transmit 82 kips to maintain equilibrium of the portions of the bar on either side of the cut.

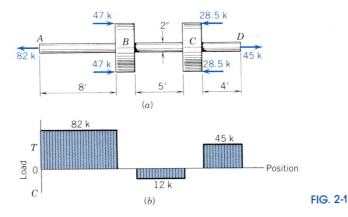

(a)

(b)

FIG. 2-1

(a) Examination of the load diagram indicates that the maximum load transmitted by any section of the bar is 82 kips; therefore, since the bar has a uniform cross section,

$$\sigma_{max} = \frac{P_{max}}{A} = \frac{82}{4} = \underline{20.5 \text{ ksi } T} \qquad \text{Ans.}$$

(b) Since different parts of the bar are transmitting different levels of load, Eq. 2-1 must be used to determine the changes in length associated with each of the different parts of the bar. These individual values are then added algebraically to give the change in length of the complete bar. Thus,

$$\delta_{AD} = \Sigma \frac{PL}{AE} = \delta_{AB} + \delta_{BC} + \delta_{CD}$$

$$= \frac{82(8)(12)}{4(30,000)} - \frac{12(5)(12)}{4(30,000)} + \frac{45(4)(12)}{4(30,000)}$$

$$= 0.0656 - 0.0060 + 0.0180$$

$$= \underline{0.0776\text{-in. increase}} \qquad \text{Ans.}$$

EXAMPLE 2-2

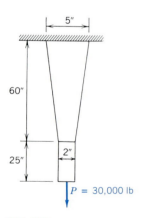

FIG. 2-2

A steel ($E = 30,000$ ksi) bar of rectangular cross section consists of uniform and tapered sections as shown in Fig. 2-2. The width of the tapered section varies linearly from 2 in. at the bottom of the section to 5 in. at the top of the section. The bar has a constant thickness of $\frac{1}{2}$ in. Determine the elongation of the bar resulting from application of the 30-kip load. Neglect the weight of the bar.

Solution

The elongation of the bar can be determined by using Eq. 2-1 for the uniform portion and Eq. 2-2 for the tapered portion. Since the width w of the bar varies linearly with position in the tapered section, the cross-sectional area in the tapered section can be expressed as

$$A = wt = (2 + 0.05y)(1/2) = 1 + 0.025y$$

where y is measured from the bottom of the section. The elongations in the two parts of the bar are

For the uniform portion:

$$\delta_U = \frac{PL}{AE} = \frac{30(25)}{2(1/2)(30,000)} = 0.0250 \text{ in.}$$

For the tapered portion:

$$\delta_T = \int_0^L \frac{Pdy}{A_yE} = = \int_0^{60} \frac{30 \ dy}{(1 + 0.025y)(30,000)}$$

$$= 0.04 \ \text{Log}_e(1 + 0.025y)]_0^{60}$$

$$= 0.0367 \text{ in.}$$

Thus

$$\delta = \delta_U + \delta_T = 0.0250 + 0.0367 = \underline{0.0617 \text{ in.}} \qquad \text{Ans.}$$

PROBLEMS

2-1* A $\frac{3}{4}$-in.-diameter by 3-ft-long structural steel rod (see Appendix A for properties) is supporting an axial tensile load of 5 kips. Determine

(a) the elongation of the bar.

(b) the factor of safety with respect to failure by slip.

2-2* A 2024-T4 aluminum alloy bar (see Appendix A for properties) 3 m long has a 25-mm-square cross section over 1 m of its length and a 25-mm-diameter circular cross section over the other 2 m of its length. The bar is supporting an axial tensile load of 50 kN. Determine

(a) the elongation of the bar.

(b) the factor of safety with respect to failure by fracture.

2-3* Specifications for a steel ($E = 30,000$ ksi) bar with a circular cross section require that the axial stress not exceed 20 ksi and that the total elongation in a 3-ft length not exceed 0.025 in. Determine the minimum diameter required if the bar must transmit an axial tensile load of 40 kips.

2-4* A steel ($E = 200$ GPa) rod with a circular cross section is 3 m long. Determine the minimum diameter required if the rod must transmit a tensile force of 20 kN without exceeding an allowable stress of 175 MPa or stretching more than 3 mm.

2-5 A hollow brass ($E = 15,000$ ksi) tube A with an outside diameter of 4 in. and an inside diameter of 2 in. is fastened to a 2-in.-diameter steel ($E = 30,000$ ksi) rod B and loaded and supported as shown in Fig. P2-5. Determine

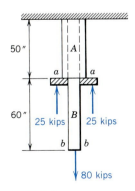

FIG. P2-5

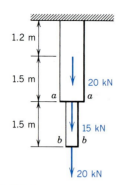

FIG. P2-6

(a) the deflection of cross section a–a.

(b) the deflection of cross section b–b.

2-6 An aluminum ($E = 73$ GPa) bar is loaded and supported as shown in Fig. P2-6. The diameters of the top and bottom sections of the bar are 25 and 15 mm, respectively. Determine

(a) the deflection of cross section a–a.

(b) the deflection of cross section b–b.

2-7* A hollow steel ($E = 30,000$ ksi) tube A with an outside diameter of 2.5 in. and an inside diameter of 2 in. is fastened to an aluminum ($E = 10,000$ ksi) bar B that has a 2-in. diameter over one-half its length and a 1-in. diameter over the other half. The bar is loaded and supported as shown in Fig. P2-7. Determine

(a) the change in length of the steel tube.

(b) the overall elongation of the member.

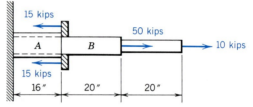

FIG. P2-7

2-8* A steel ($E = 200$ GPa) bar is 3 m long and has a 50-mm diameter over one-half its length and a 30-mm diameter over the other half.

(a) How much will the bar elongate under a tensile load of 125 kN?

(b) How much will a bar of uniform cross section, which has the same weight as the bar of part (a), elongate under the action of the 125-kN load?

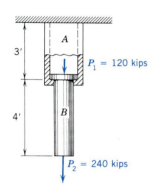

FIG. P2-9

2-9 The tension member shown in Fig. P2-9 consists of a steel pipe A, which has an outside diameter of 6 in. and an inside diameter of 4.5 in., and a solid aluminum alloy bar B, which has an outside diameter of 4 in. The moduli of elasticity for the steel and aluminum are 29,000 ksi and 10,600 ksi, respectively. Determine the overall elongation of the member under the action of the indicated loads.

2-10 The compression member shown in Fig. P2-10 consists of a solid aluminum bar A, which has an outside diameter of 100 mm, a brass tube B, which has an outside diameter of 150 mm and an inside diameter of 100 mm, and a steel pipe C, which has an outside diameter of 200 mm and an inside diameter of 125 mm. The moduli of elasticity of the aluminum, brass, and steel are 73, 100, and 210 GPa, respectively. Determine the overall shortening of the member under the action of the indicated loads.

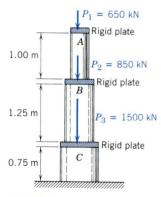

FIG. P2-10

2-11* An aluminum alloy ($E = 10,000$ ksi) tube A with an outside diameter of 3 in. is used to support a 1-in.-diameter steel ($E = 30,000$ ksi) rod B, as shown in Fig. P2-11. Determine the minimum thickness required for the tube if the maximum deflection of the loaded end of the rod must be limited to 0.015 in.

2-12* A control rod in a mechanism must stretch 5 mm when a load of 5 kN is applied. The rod is made from an aluminum alloy with an elastic strength of 330 MPa and a modulus of elasticity of 73 GPa. If a factor of safety of 2 with respect to failure by slip is specified, determine the diameter and length of the lightest rod that can be used.

2-13 A 1-in.-diameter by 16-ft-long cold-rolled bronze bar ($E = 15,000$ ksi and $\gamma = 0.320$ lb/in.^{3}) hangs vertically while suspended from one end. Determine the change in length of the bar due to its own weight.

2-14 An aluminum alloy ($E = 70$ GPa) bar of circular cross section consists of uniform and tapered sections as shown in Fig. P2-14. The diameter of the tapered section varies linearly from 50 mm at the top to 20 mm at the bottom. The uniform section at the top is hollow and has a wall thickness of 10 mm. Determine the elongation of the bar resulting from application of the 75-kN load. Neglect the weight of the bar.

2-15* Determine the extension, due to its own weight, of the conical bar shown in Fig. P2-15. The bar is made of aluminum alloy ($E = 10,600$ ksi and $\gamma = 0.100$ lb/in.^{3}). The bar has a radius of 2 in. and a length L of 20 ft. Assume the taper of the bar is slight enough for the assumption of a uniform axial stress distribution over a cross section to be valid.

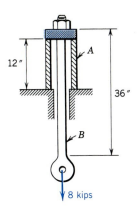

FIG. P2-11

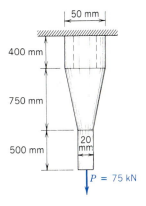

FIG. P2-14

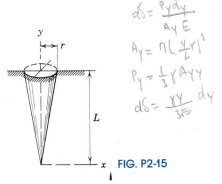

FIG. P2-15

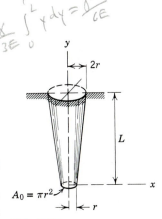

FIG. P2-16

2-16* Determine the extension, due to its own weight, of the 20-m-long stainless-steel ($E = 190$ GPa and $\rho = 7.92$ Mg/m^3) bar shown in Fig. P2-16. The radii of the bar are 20 mm at the bottom and 40 mm at the top. Assume that the taper of the bar is slight enough for the assumption of a uniform axial stress distribution over a cross section to be valid.

2-3

DEFORMATIONS IN A SYSTEM OF AXIALLY LOADED BARS

It is sometimes necessary to determine axial deformations and strains in a loaded system of pin-connected deformable bars (two-force members). The problem is approached through a study of the geometry of the deformed system, from which the axial deformations δ of the various bars in the system are obtained. Suppose, for example, one is interested in the axial deformations of bars *AB, BC,* and *BD* of Fig. 2-3, in which the solid lines represent the unstrained (unloaded) configuration of the system and the dashed lines represent the configuration due to a force applied at *B*. From the Pythagorean theorem, the axial deformation in bar *AB* is

$$\delta_{AB} = \sqrt{(L + y)^2 + x^2} - L$$

Transposing the last term and squaring both sides gives

$$\delta_{AB}^2 + 2L\delta_{AB} + L^2 = L^2 + 2Ly + y^2 + x^2$$

If the displacements are small (the usual case for stiff materials and elastic action), the terms involving the squares of the displacements may be neglected; hence,

$$\delta_{AB} \approx y$$

In a similar manner,

$$\delta_{BD} \approx x$$

The axial deformation in bar *BC* is

$$\delta_{BC} = \sqrt{(R \cos \theta - x)^2 + (R \sin \theta + y)^2} - R$$

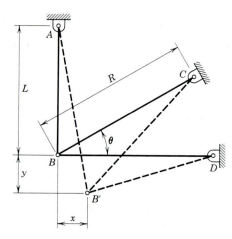

FIG. 2-3

Transposing the last term and squaring both sides gives

$$\delta_{BC}^2 + 2R\delta_{BC} + R^2 = R^2 \cos^2 \theta - 2Rx \cos \theta \\ + x^2 + R^2 \sin^2 \theta + 2Ry \sin \theta + y^2$$

Neglecting small second-degree terms and noting that $\sin^2 \theta + \cos^2 \theta = 1$, one obtains

$$\delta_{BC} \approx y \sin \theta - x \cos \theta$$

or in terms of the deformations of the other two bars,

$$\delta_{BC} \approx \delta_{AB} \sin \theta - \delta_{BD} \cos \theta$$

The geometric interpretation of this equation is indicated by the shaded right triangles of Fig. 2-4.

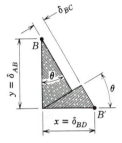

FIG. 2-4

The general conclusion that may be drawn from the above discussion is that, *for small displacements,* the axial deformation in any bar may be assumed equal to the component of the displacement of one end of the bar (relative to the other end) taken in the direction of the unstrained orientation of the bar. Any rigid members of the system will change orientation or position, but will not be deformed in any manner. For example, if bar BD of Fig. 2-3 was rigid and subjected to a small downward rotation, point B could be assumed to be displaced vertically through a distance y, and δ_{BC} would be equal to $y \sin \theta$.

EXAMPLE 2-3

A tie rod and a pipe strut are used to support a 50-kN load, as shown in Fig. 2-5a. The cross-sectional areas are 650 mm² for tie rod AB and 925 mm² for pipe strut BC. Both members are made of structural steel that has a modulus of elasticity of 200 GPa. Determine

(a) the axial stresses in tie rod AB and pipe strut BC.

(b) the lengthening or shortening of tie rod AB and pipe strut BC.

(c) the horizontal and vertical displacements of point B.

(d) the angles through which members AB and BC rotate.

Solution

(a) The forces in members AB and BC can be determined by using the free-body diagram of joint B shown in Fig. 2-5b. Thus,

$$F_{AB} = 50(\cos 42.61°)/\sin 42.61° = 54.36 \text{ kN } T$$

$$F_{BC} = 50/\sin 42.61° = 73.85 \text{ kN } C$$

The stresses in members AB and BC are determined by using Eq. 1-2. Thus,

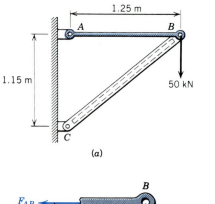

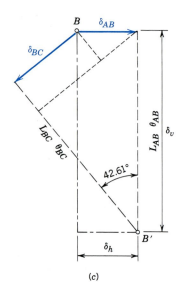

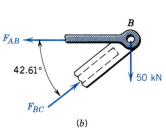

FIG. 2-5

(a)

(b)

(c)

$$\sigma_{AB} = \frac{F_{AB}}{A_{AB}} = \frac{+54.36(10^3)}{650(10^{-6})}$$

$$= +83.63(10^6) \text{ N/m}^2 = \underline{83.6 \text{ MPa } T} \quad \text{Ans.}$$

$$\sigma_{BC} = \frac{F_{BC}}{A_{BC}} = \frac{-73.85(10^3)}{925(10^{-6})}$$

$$= -79.84(10^6) \text{ N/m}^2 = \underline{79.8 \text{ MPa } C} \quad \text{Ans.}$$

(b) The changes in length of the members are determined by using Eq. 2-1a. Thus,

$$\delta_{AB} = \frac{\sigma_{AB}L_{AB}}{E} = \frac{+83.63(10^6)(1.25)}{200(10^9)}$$

$$= +0.5227(10^{-3})\text{m} = \underline{+0.523 \text{ mm}} \quad \text{Ans.}$$

$$\delta_{BC} = \frac{\sigma_{BC}L_{BC}}{E} = = \frac{-79.84(10^6)(1.70)}{200(10^9)}$$

$$= -0.6786(10^{-3}) \text{ m} = \underline{-0.679 \text{ mm}} \quad \text{Ans.}$$

(c) The horizontal and vertical displacements δ_h and δ_v of point B are indicated on the deformation diagram shown in Fig. 2-5c. The deformations have been greatly exaggerated in this diagram and the arcs through which the members rotate have been replaced by straight lines drawn perpendicular to the unloaded positions of the members. From the diagram it is observed that

$$\delta_h = \delta_{AB} = +0.5227 \text{ mm} = \underline{0.523 \text{ mm}} \rightarrow \quad \text{Ans.}$$

$$\delta_v = \frac{-\delta_{AB}\cos 42.61° + \delta_{BC}}{\sin 42.61°}$$

$$= \frac{-0.5227\cos 42.61° - 0.6786}{\sin 42.61°}$$

$$= -1.5706 \text{ mm} = \underline{1.571 \text{ mm}} \downarrow \qquad \text{Ans.}$$

(d) The angles through which members AB and BC rotate (see deformation diagram) are

$$\theta_{AB} = \tan^{-1}\frac{\delta_v}{L_{AB}}$$

$$= \tan^{-1}\frac{-0.0015706}{1.25}$$

$$= -0.0012565 \text{ rad} = \underline{0.0720°}\ \curvearrowright \qquad \text{Ans.}$$

$$\theta_{BC} = \tan^{-1}\frac{\delta_v\cos 42.61° - \delta_{h_o}\sin 42.61°}{L_{BC}}$$

$$= \tan^{-1}\frac{-0.0015706\cos 42.61° - 0.0005227\sin 42.61°}{1.700}$$

$$= -0.0008881 \text{ rad} = -0.0509° = \underline{0.0509°}\ \curvearrowleft \qquad \text{Ans.}$$

PROBLEMS

> *Note:* For the problems in this section, assume small deformations and elastic action.

2-17* A rigid bar CD is loaded and supported as shown in Fig. P2-17. Bars A and B are unstressed before the load P is applied. Bar A is

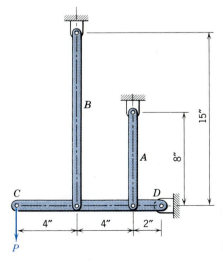

FIG. P2-17

made of steel (E = 30,000 ksi) and has a cross-sectional area of 2 in.^{2}. Bar B is made of brass (E = 15,000 ksi) and has a cross-sectional area of 1.5 in.^{2}. After the load P is applied, the strain in bar B is found to be 1500 μin./in. Determine

(a) the stresses in bars A and B.
(b) the vertical displacement (deflection) of pin C.
(c) the load P.

2-18* A rigid bar CD is loaded and supported as shown in Fig. P2-18. Bars A and B are unstressed before the load P is applied. Bar A is made of stainless steel (E = 190 GPa) and has a cross-sectional area of 750 mm^2. Bar B is made of an aluminum alloy (E = 73 GPa) and has a cross-sectional area of 1250 mm^2. After the load P is applied, the strain in bar B is found to be 1200 μm/m. Determine

(a) the stresses in bars A and B.
(b) the vertical displacement (deflection) of pin D.
(c) the load P.

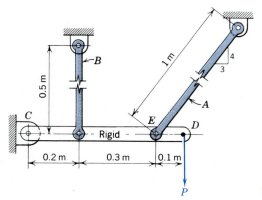

FIG. P2-18

2-19 Solve Problem 2-17 when there is a clearance of 0.015 in. in the pin connection at the bottom of bar B.

2-20 Solve Problem 2-18 when there is a clearance of 0.50 mm in the pin connection at the bottom of bar B.

2-21* The three bars shown in Fig. P2-21 will be used to support a load P of 150 kips. Bar A will be made of aluminum alloy (E = 10,500 ksi) and will have a circular cross section. Bars B will be made of structural steel (E = 29,000 ksi) and will have 2.5 × 1.0-in. rectangular cross sections. If the stress in bar A must be limited to 15 ksi, determine the minimum satisfactory diameter for bar A.

2-22* The three bars shown in Fig. P2-22 will be used to support a load P of 425 kN. All of the bars have the same circular cross section and are 4 m long. Bar A is made of Monel (E = 180 GPa), bar B is made

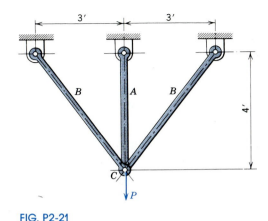

FIG. P2-21

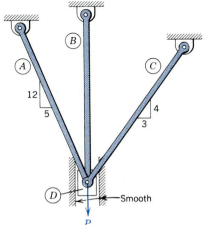

FIG. P2-22

of a magnesium alloy ($E = 45$ GPa), and bar C is made of structural steel ($E = 200$ GPa). If the stress in bar C must be limited to 150 MPa, determine the minimum satisfactory diameter for the bars.

2-23 Solve Problem 2-21 when there is an initial clearance of 0.025 in. between the pin at C and bar A. The load is applied to pin C, which fits snugly in bars B and moves downward 0.025 in. before contacting bar A.

2-24 Solve Problem 2-22 when there is an initial clearance of 1.50 mm between the pin at D and bar B. The load is applied to pin D, which fits snugly in bars A and C and moves downward 1.50 mm before contacting bar B.

2-25* Two tie rods are used to support a 10-kip load as shown in Fig. P2-25. Rod AC is 10 ft long and is made of an aluminum alloy with a modulus of elasticity of 10,600 ksi and an elastic strength of 41 ksi. Rod BC is 15 ft long and is made of structural steel with a modulus of elasticity of 29,000 ksi and a yield point of 36 ksi. Determine

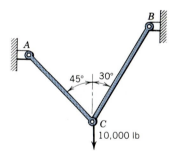

FIG. P2-25

   (a) the minimum cross-sectional area required for each of the rods if a factor of safety of 2.5 with respect to failure by slip is specified.

   (b) the lengthening of each of the rods when the areas of part (a) are used.

   (c) the horizontal and vertical displacements of pin C when the areas of part (a) are used.

2-26* Two tie rods are used to support a 75-kN load as shown in Fig. P2-26. Rod AB is 2 m long and is made of an aluminum alloy with a modulus of elasticity of 73 GPa and an elastic strength of 280 MPa. Rod BC is 4 m long and is made of structural steel with a modulus of elasticity of 200 GPa and a yield point of 250 MPa. Determine

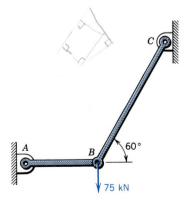

FIG. P2-26

(a) the minimum cross-sectional area required for each of the rods if a factor of safety of 2.25 with respect to failure by slip is specified.

(b) the lengthening of each of the rods when the areas of part (a) are used.

(c) the horizontal and vertical displacements of pin B when the areas of part (a) are used.

2-27 A tie rod and a pipe strut are used to support a 25-kip load as shown in Fig. P2-27. Tie rod AB is made of stainless steel with a modulus of elasticity of 28,000 ksi and an elastic strength of 165 ksi and pipe strut BC is made of structural steel with a modulus of elasticity of 29,000 ksi and a yield point of 36 ksi. Determine

(a) the minimum cross-sectional areas required for the rod and strut if a factor of safety of 2 with respect to failure by slip is specified.

(b) the lengthening or shortening of the rod and strut when the areas of part (a) are used.

(c) the horizontal and vertical displacements of pin B when the areas of part (a) are used.

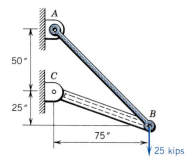

FIG. P2-27

2-28 A tie rod and a pipe strut are used to support a 100-kN load as shown in Fig. P2-28. Tie rod AB is made of a titanium alloy with a modulus of elasticity of 96 GPa and an elastic strength of 930 MPa, and pipe strut AC is made of structural steel with a modulus of elasticity of 200 GPa and a yield point of 250 MPa. Determine

(a) the minimum cross-sectional areas required for the rod and strut if a factor of safety of 1.75 with respect to failure by slip is specified.

(b) the lengthening or shortening of the rod and strut when the areas of part (a) are used.

(c) the horizontal and vertical displacements of pin A when the areas of part (a) are used.

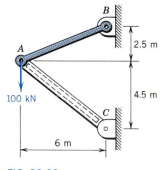

FIG. P2-28

2-4

STATICALLY INDETERMINATE AXIALLY LOADED MEMBERS

In many simple structures (and mechanical systems) constructed with axially loaded members, it is possible to find the reactions at supports and the forces in the individual members by drawing free-body diagrams and solving equilibrium equations. Such structures (and systems) are referred to as being *statically determinate*.

For many other structures (and mechanical systems), the equations of equilibrium are not sufficient for the determination of axial forces in the members and reactions at the supports; these structures (and systems) are referred to as being *statically indeterminate*. Problems of this type can be analyzed by supplementing the equilibrium equations with additional equations involving the geometry of the deformations in the members of the structure or system. The following outline of procedure will be helpful in the analysis of problems involving statically indeterminate situations.

1. Draw a free-body diagram.
2. Note the number of unknowns involved (magnitudes and positions).
3. Recognize the type of force system on the free-body diagram and note the number of independent equations of equilibrium available for this system.
4. If the number of unknowns exceeds the number of equilibrium equations, a deformation equation must be written for each extra unknown.
5. When the number of independent equilibrium equations and deformation equations equals the number of unknowns, the equations can be solved simultaneously. Deformations and forces must be related in order to solve the equations simultaneously.

Hooke's law (Eq. 1-19) and the definitions of stress (Eq. 1-2) and strain (Eq. 1-12) can be used to relate deformations and forces when all stresses are less than the corresponding proportional limits of the materials used in the fabrication of the members. If some of the stresses exceed the proportional limits of the materials, stress-strain diagrams can be used to relate the loads and deformations. *In this chapter, the problems will be limited to the region of elastic action of the materials.* Problems involving inelastic behavior of materials are discussed in Chapter 10. It is recommended that a displacement diagram be drawn showing deformations to assist in obtaining the correct deformation equation. The displacement diagram should be as simple as possible (a line diagram), with the deformations indicated with exaggerated magnitudes and clearly dimensioned. Loading members (members assumed to be rigid), especially, should be indicated by single lines. Note that an equilibrium equation and the corresponding deformation equation *must be compatible;* that is, when a tensile force is assumed for a member in the free-body diagram, a tensile deformation must be indicated for the same member in the deformation diagram. If the diagrams are compatible, a negative result will indicate that the assumption was wrong; however, the magnitude of the result will be correct.

In all the following problems and examples, the loading members, pins, and supports are assumed to be rigid, and the mechanism is so constructed that the force system is coplanar. The procedure outlined above is illustrated in the following examples.

EXAMPLE 2-4

A rigid plate C is used to transfer a 20-kip load P to a steel ($E = 30{,}000$ ksi) rod A and to an aluminum alloy ($E = 10{,}000$ ksi) pipe B, as shown in Fig. 2-6a. The supports at the top of the rod and bottom of the pipe are rigid and there are no stresses in the rod or pipe before the load P is applied. The cross-sectional areas of rod A and pipe B are 0.800 in.^{2} and 3.00 in.^{2}, respectively. Determine

(a) the axial stresses in rod A and pipe B.

(b) the displacement of plate C.

Solution

(a) A free-body diagram of plate C and portions of rod A and pipe B is shown in Fig. 2-6b. The free-body diagram contains two unknown forces, P_A and P_B. Since only one equation of equilibrium, $\Sigma F_y = 0$, is available, the problem is statically indeterminate. The additional equation needed to obtain a solution to the problem is obtained from the deformation diagram shown in Fig. 2-6c. As the load P is applied to plate C, it moves downward an amount δ, which represents the deflection experienced by both rod A and pipe B. The relationship between load and deflection for axial loading is given by Eq. 2-1a. Thus, the two equations needed to solve the problem are:

Equilibrium equation

$$\Sigma F_y = 0$$
$$P_A + P_B - 20 = 0$$

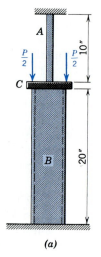

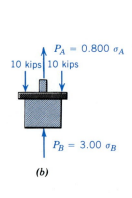

(b)

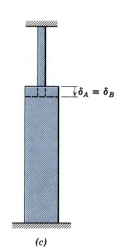

FIG. 2-6   (a)

or in terms of stresses

$$0.800\,\sigma_A + 3.00\,\sigma_B = 20 \qquad (a)$$

Deformation equation

$$\delta_A = \delta_B$$

$$\frac{\sigma_A L_A}{E_A} = \frac{\sigma_B L_B}{E_B}$$

$$\frac{\sigma_A(10)}{30,000} = \frac{\sigma_B(20)}{10,000}$$

from which

$$\sigma_A = 6\sigma_B \qquad (b)$$

Solving Eqs. a and b simultaneously yields

$$\sigma_A = \underline{15.38\ \text{ksi}\ T} \qquad \text{and} \qquad \sigma_B = \underline{2.56\ \text{ksi}\ C} \qquad \text{Ans.}$$

(b) The displacement of plate C is the same as the deflection of rod A or the deflection of pipe B. Thus, from Eq. b

$$\delta_C = \delta_A = \delta_B = \frac{\sigma_A(10)}{30,000}$$

$$= \frac{15.38(10)}{30,000} = \underline{0.00513\ \text{in.}} \qquad \text{Ans.}$$

EXAMPLE 2-5

A pin-connected structure is loaded and supported as shown in Fig. 2-7a. Member CD is rigid and is horizontal before the load P is applied.

FIG. 2-7

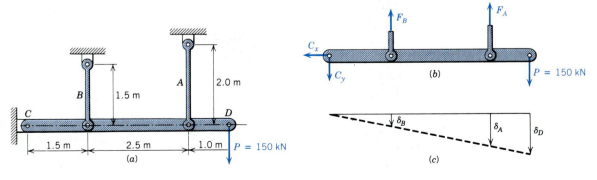

Member A is an aluminum alloy bar with a modulus of elasticity of 75 GPa and a cross-sectional area of 1000 mm². Member B is a structural steel bar with a modulus of elasticity of 200 GPa and a cross-sectional area of 500 mm². Determine

(a) the axial stresses in bars A and B.

(b) the deflection of pin D.

Solution

(a) A free-body diagram of member CD and portions of members A and B is shown in Fig. 2-7b. The free body-diagram contains four unknown forces: C_x, C_y, F_A, and F_B; therefore, since only three equilibrium equations are available, the problem is statically indeterminate. As the load P is applied to member CD, it will tend to rotate clockwise about pin C and produce deformations in members A and B as shown in Fig. 2-7c. The extensions shown in Fig. 2-7c are compatible with the tensile forces shown in members A and B in Fig. 2-7b. The unknown reaction at C is not needed to complete the solution of the problem and can be eliminated from further consideration by summing moments about pin C. The equilibrium and deformation equations needed to solve for F_A and F_B are:

Equilibrium equation

$$\curvearrowleft+ \qquad \Sigma M_C = 0$$

$$P(5) - F_A(4) - F_B(1.5) = 0$$

from which

$$4F_A + 1.5F_B = 150(10^3)(5) = 750(10^3) \qquad (a)$$

Deformation equation

$$\frac{\delta_A}{4} = \frac{\delta_B}{1.5}$$

$$\frac{F_A L_A}{4A_A E_A} = \frac{F_B L_B}{1.5 A_B E_B}$$

$$= \frac{F_A(2)}{4(1000)(10^{-6})(75)(10^9)} = \frac{F_B(1.5)}{1.5(500)(10^{-6})(200)(10^9)}$$

or

$$F_A = 1.5 F_B \qquad (b)$$

Solving Eqs. a and b simultaneously yields

$$F_A = 150 \text{ kN} \qquad F_B = 100 \text{ kN}$$

The stresses in the two bars are

$$\sigma_A = \frac{F_A}{A_A} = \frac{150(10^3)}{1000(10^{-6})} = 150(10^6) \ \text{N/m}^2 = \underline{150 \ \text{MPa} \ T} \quad \text{Ans.}$$

$$\sigma_B = \frac{F_B}{A_B} = \frac{100(10^3)}{500(10^{-6})} = 200(10^6) \ \text{N/m}^2 = \underline{200 \ \text{MPa} \ T} \quad \text{Ans.}$$

(b) Since bar CD rotates as a rigid body, the deflection of pin D is

$$\delta_D = \frac{5}{4} \delta_A = \frac{5(150)(10^3)(2)}{4(1000)(10^{-6})(75)(10^9)}$$

$$= 5.00(10^{-3}) \ \text{m} = \underline{5.00 \ \text{mm downward}} \quad \text{Ans.}$$

EXAMPLE 2-6

After the load $P = 150$ kN was applied to the pin-connected structure shown in Fig. 2-7, the temperature increased 100°C. The thermal coefficients of expansion are $22(10^{-6})$/°C for the aluminum alloy rod A and $12(10^{-6})$/°C for the steel rod B. Determine

(a) the axial stresses in bars A and B.

(b) the deflection of pin D.

Solution

(a) The free-body diagram shown in Fig. 2-7b remains valid; therefore, the equilibrium equation $\Sigma M_C = 0$ applies. The deformations in bars A and B are functions of both load and temperature change, as shown in Fig. 2-8. Thus

$$\delta_A = \delta_{AP} + \delta_{AT} \tag{a}$$
$$\delta_B = \delta_{BP} + \delta_{BT}$$

where the subscripts P and T refer to load and temperature, respectively. The equilibrium and deformation equations needed to solve for F_A and F_B are:

Equilibrium equation

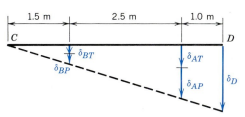

FIG. 2-8

$$\circlearrowleft+ \quad \Sigma M_C = 0$$

$$P(5) - F_A(4) - F_B(1.5) = 0$$

from which

$$4F_A + 1.5 F_B = 750(10^3) \qquad (b)$$

Deformation equation

$$\frac{\delta_A}{4} = \frac{\delta_B}{1.5}$$

$$\frac{F_A L_A}{4 A_A E_A} + \frac{\alpha_A L_A (\Delta T)}{4} = \frac{F_B L_B}{1.5 A_B E_B} + \frac{\alpha_B L_B (\Delta T)}{1.5}$$

$$\frac{F_A(2)}{4(1000)(10^{-6})(75)(10^9)} + \frac{22(10^{-6})(2)(100)}{4}$$

$$= \frac{F_B(1.5)}{1.5(500)(10^{-6})(200)(10^9)} + \frac{12(10^{-6})(1.5)(100)}{1.5}$$

from which

$$F_A = 1.5 F_B + 15(10^3) \qquad (c)$$

Solving Eqs. *b* and *c* simultaneously yields

$$F_A = 153 \text{ kN} \qquad F_B = 92 \text{ kN}$$

The stresses in the two bars are

$$\sigma_A = \frac{F_A}{A_A}$$

$$= \frac{153(10^3)}{1000(10^{-6})} = 153(10^6) \text{ N/m}^2 = \underline{153 \text{ MPa } T} \quad \text{Ans.}$$

$$\sigma_B = \frac{F_B}{A_B}$$

$$= \frac{92(10^3)}{500(10^{-6})} = 184(10^6) \text{ N/m}^2 = \underline{184 \text{ MPa } T} \quad \text{Ans.}$$

(b) Since bar *CD* rotates as a rigid body, the deflection of pin *D* is

$$\delta_D = \frac{5}{4} \delta_A = \frac{5 F_A L_A}{4 A_A E_A} + \frac{5}{4} \alpha_A L_A (\Delta T)$$

$$= \frac{5(153)(10^3)(2)}{4(1000)(10^{-6})(75)(10^9)} + \frac{5}{4}(22)(10^{-6}(2)(100)$$

$$= 10.60(10^{-3}) \text{ m} = \underline{10.60 \text{ mm downward}} \quad \text{Ans.}$$

EXAMPLE 2-7

A $\frac{1}{2}$-in.-diameter alloy-steel bolt (E = 30,000 ksi) passes through a cold-rolled brass sleeve (E = 15,000 ksi), as shown in Fig. 2-9a. The cross-sectional area of the sleeve is 0.375 in.². Determine the axial stresses produced in the bolt and sleeve by tightening the nut a quarter turn (0.020 in.).

Solution

A free-body diagram of the nut and parts of the bolt and sleeve is shown in Fig. 2-9b. The free-body diagram contains two unknown forces F_B and F_S. Since the only equilibrium equation available is $\Sigma F_x = 0$, the problem is statically indeterminate. The additional equation needed to obtain a solution to the problem is obtained from deformation considerations. As the nut is turned it would move a distance Δ = 0.020 in., as shown in Fig. 2-9c, if the sleeve were not present; however, the sleeve is present and the movement is resisted. As a result, tensile stresses develop in the bolt and compressive stresses develop in the sleeve. These stresses produce the extension δ_B of the bolt and contraction δ_S of the sleeve shown in Figs. 2-9d and e. The deformation equation obtained from the final positions of the nut and sleeve is $\delta_S + \delta_B = \Delta$. The two equations needed to solve the problem are

Equilibrium equation

$$\Sigma F_x = 0$$
$$F_S - F_B = 0$$

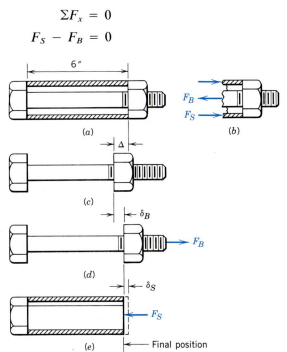

(a)

(b)

(c)

(d)

(e)   Final position   **FIG. 2-9**

or in terms of stresses

$$0.375\,\sigma_S = \frac{\pi}{4}\left(\frac{1}{2}\right)^2 \sigma_B$$

from which

$$\sigma_S = 0.5236\sigma_B \qquad (a)$$

Deformation equation

$$\delta_S + \delta_B = \Delta$$

$$\frac{\sigma_B L_B}{E_B} + \frac{\sigma_S L_S}{E_S} = \Delta$$

$$\frac{\sigma_B(6)}{30,000} + \frac{\sigma_S(6)}{15,000} = 0.020$$

from which

$$\sigma_B + 2\sigma_s = 100 \qquad (b)$$

Solving Eqs. a and b simultaneously yields

$$\sigma_B = \underline{48.8\ \text{ksi}\ T} \quad \sigma_S = \underline{25.6\ \text{ksi}\ C} \qquad \text{Ans.}$$

PROBLEMS

> *Note:* For the problems in this section, assume small deformations and elastic action. All members are assumed to have negligible weight, and all pins used for connection are assumed to be smooth. The temperature remains constant unless otherwise specified.

2-29* A steel (E = 30,000 ksi) pipe column with an outside diameter of 3 in. and an inside diameter of 2.5 in. is attached to unyielding supports at the top and bottom, as shown in Fig. P2-29. A rigid collar C is used to apply a 50-kip load P. Determine

(a) the stresses in the top and bottom portions of the pipe.

(b) the deflection of collar C.

2-30* A hollow brass (E = 100 GPa) tube A with an outside diameter of 100 mm and an inside diameter of 50 mm is fastened to a 50-mm-diameter steel (E = 200 GPa) rod B, as shown in Fig. P2-30. The supports at the top and bottom of the assembly and the collar C used to apply the 500-kN load are rigid. Determine

(a) the stresses in each of the members.

(b) the deflection of collar C.

2-31 Solve Problem 2-29 if the lower support in Fig. P2-29 yields and displaces 0.030 in. as the load P is applied.

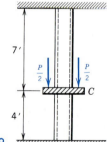

FIG. P2-29

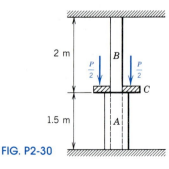

FIG. P2-30

2-32 Solve Problem 2-30 if the top support in Fig. P2-30 yields and displaces 0.75 mm as the load P is applied.

2-33* A hollow steel (E = 30,000 ksi) tube A with an outside diameter of 2.5 in. and an inside diameter of 2 in. is fastened to an aluminum (E = 10,000 ksi) bar B that has a 2-in. diameter over one-half its length and a 1-in. diameter over the other half. The assembly is attached to unyielding supports at the left and right ends and is loaded as shown in Fig. P2-33. Determine

(a) the stresses in all parts of the bar.

(b) the deflection of cross section $a–a$.

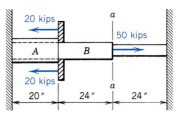

FIG. P2-33

2-34* A 3-mm-diameter cord (E = 7 GPa) that is covered with a 0.5-mm-thick plastic sheath (E = 14 GPa) is subjected to an axial tensile load. The elastic strengths for the cord and sheath are 15 MPa and 56 MPa, respectively. Determine the maximum allowable load if a factor of safety of 3 with respect to failure by slip is specified.

2-35* Nine $\frac{3}{4}$-in.-diameter steel (E = 30,000 ksi) reinforcing bars are used in the short concrete (E = 4500 ksi) pier shown in Fig. P2-35. A load P of 150 kips is applied to the pier through a rigid capping plate. Determine

(a) the stresses in the concrete and in the steel bars.

(b) the shortening of the pier.

2-36* The 200 × 200 × 500-mm oak (E = 12 GPa) block shown in Fig. P2-36 was reinforced by bolting two 50 × 200 × 500-mm steel (E = 200 GPa) plates to opposite sides of the block. If the stresses in the wood and the steel are to be limited to 32 MPa and 150 MPa, respectively, determine

(a) the maximum compressive load that can be applied to the reinforced block.

(b) the shortening of the block when the load of part (a) is applied.

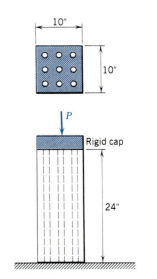

FIG. P2-35

2-37 A column is being designed to carry a compressive load of 1000 kips. The column, which will have a 20 × 20-in. square cross section, will be made of concrete (E = 3000 ksi) and will be reinforced with 2-in.-diameter steel (E = 30,000 ksi) bars. If the maximum allowable stresses are 18 ksi in the steel and 1.2 ksi in the concrete, determine

(a) the number of steel bars required.

(b) the stresses in the steel and concrete when the bars of part (a) are used.

(c) the change in length of a 10-ft column when the bars of part (a) are used.

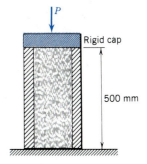

FIG. P2-36

2-38 Five 25-mm-diameter steel (E = 200 GPa) reinforcing bars will be used in a 1-m-long concrete (E = 31 GPa) pier with a square cross section similar to the one shown in Fig. P2-35. The elastic strengths in compression for concrete and steel are 14 MPa and 200 MPa, respectively. Determine the minimum size of pier required to support a 900 kN load with a factor of safety of 4 with respect to failure by slip.

2-39* After the steel-reinforced concrete pier described in Problem 2-35 was constructed and loaded, the temperature increased 100°F. The coefficients of thermal expansion for steel and concrete are 6.6(10^{-6})/°F and 6.0(10^{-6})/°F, respectively. Determine

(a) the stresses in the concrete and in the steel bars after the temperature increase.

(b) the change in length of the pier resulting from the combined effects of the temperature change and the load.

2-40* A 150-mm-diameter by 200-mm-long polymer (E = 2.10 GPa) cylinder will be attached to a 45-mm-diameter by 400-mm-long brass (E = 100 GPa) rod by using the flange type of connection shown in Fig. P2-40. A 0.15-mm clearance exists between the parts as a result of a machining error. If the bolts are inserted and tightened, determine

(a) the axial stresses produced in each of the members.

(b) the final position of the flange–polymer interface after assembly, with respect to the left support.

2-41 The assembly shown in Fig. P2-41 consists of a steel bar A (E = 30,000 ksi and A = 1.25 in.^{2}), a rigid bearing plate C that is securely fastened to bar A, and a bronze bar B (E = 15,000 ksi and A = 3.75 in.^{2}). The elastic strengths of the steel and bronze are 62 ksi and 75 ksi, respectively. A clearance of 0.015 in. exists between the bearing plate C and bar B before the assembly is loaded. After a load of 95 kips is applied to the bearing plate, determine

(a) the axial stresses in bars A and B.

(b) the factors of safety with respect to failure by slip for each of the members.

(c) the vertical displacement of bearing plate C.

2-42 The assembly shown in Fig. P2-42 consists of a steel (E = 210 GPa) cylinder A, a rigid bearing plate C, and an aluminum alloy (E = 71 GPa) bar B. The elastic strengths of the steel and aluminum are 430 MPa and 170 MPa, respectively. An axial load of 600 kN is applied. If a factor of safety of 2 with respect to failure by slip is specified, determine

(a) the minimum acceptable cross-sectional area for cylinder A if bar B has a cross-sectional area of 2500 mm^2.

(b) the displacement of plate C when the area of part (a) is used.

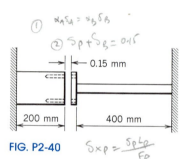

FIG. P2-40

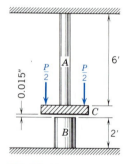

FIG. P2-41

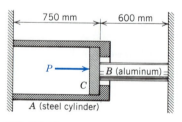

FIG. P2-42

2-43* Solve Problem 2-41 if the temperature of bar A increases 100°F and the temperature of bar B increases 50°F after the load P is applied. The coefficients of thermal expansion are $6.6(10^{-6})$/°F and $9.4(10^{-6})$/°F for the steel and bronze, respectively.

2-44* Solve Problem 2-42 if the temperature of cylinder A decreases 50°C and the temperature of bar B increases 25°C after the load P is applied. The coefficients of thermal expansion are $11.9(10^{-6})$/°C and $22.5(10^{-6})$/°C for the steel and aluminum, respectively.

2-45* A pin-connected structure is loaded and supported as shown in Fig. P2-45. Member CD is rigid and is horizontal before the load P is applied. Member A is an aluminum alloy bar with a modulus of elasticity of 10,600 ksi and a cross-sectional area of 2.25 in.². Bar B is a stainless-steel bar with a modulus of elasticity of 28,000 ksi and a cross-sectional area of 1.75 in.². After the 30-kip load is applied to the structure, determine

(a) the axial stresses in bars A and B.

(b) the vertical displacement of pin D.

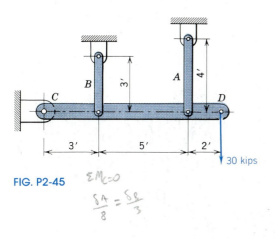

FIG. P2-45

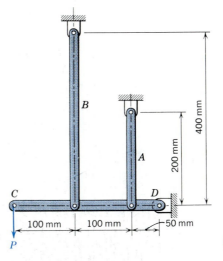

FIG. P2-46

2-46* A pin-connected structure is loaded and supported as shown in Fig. P2-46. Member CD is rigid and is horizontal before the load P is applied. Member A is a hot-rolled steel bar with a modulus of elasticity of 210 GPa and a cross-sectional area of 200 mm². Bar B is a cold-rolled brass bar with a modulus of elasticity of 100 GPa and a cross-sectional area of 225 mm². After the 15-kN load is applied to the structure, determine

(a) the axial stresses in bars A and B.

(b) the vertical displacement of pin C.

2-47 Solve Problem 2-45 if a clearance of 0.025 in. exists in the connection at the bottom of bar B before the 30-kip load is applied.

2-48 Solve Problem 2-46 if a clearance of 0.50 mm exists in the connection at the bottom of bar B before the 15-kN load is applied.

2-49* Solve Problem 2-45 if the temperature of bar A increases 75°F and the temperature of bar B increases 100°F after the 30-kip load is applied. The coefficients of thermal expansion are $9.6(10^{-6})/°F$ for the stainless steel and $12.5(10^{-6})/°F$ for the aluminum alloy.

2-50* Solve Problem 2-46 if the temperature of bar A increases 50°C and the temperature of bar B increases 70°C after the 15-kN load is applied. The coefficients of thermal expansion are $11.9(10^{-6})/°C$ for the steel and $17.6(10^{-6})/°C$ for the brass.

2-51 A pin-connected structure is loaded and supported as shown in Fig. P2-51. Member CD is rigid and is horizontal before the 20-kip load P is applied. Bar A is made of an aluminum alloy ($E = 10,600$ ksi) and bar B is made of steel ($E = 30,000$ ksi). If the allowable stresses are 10 ksi for the aluminum alloy and 20 ksi for the steel, determine

(a) the minimum acceptable cross-sectional area for bar A if bar B has a cross-sectional area of 1.75 in.²

(b) the vertical displacement of the pin used to apply the load.

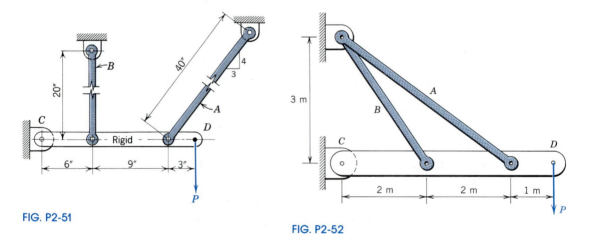

FIG. P2-51

FIG. P2-52

2-52 A pin-connected structure is loaded and supported as shown in Fig. P2-52. Member CD is rigid and is horizontal before the 75-kN load P is applied. Bar A is made of structural steel ($E = 200$ GPa) and bar B is made of an aluminum alloy ($E = 73$ GPa). If the allowable stresses are 125 MPa for the steel and 70 MPa for the aluminum alloy, determine

(a) the minimum acceptable cross-sectional area for bar B if bar A has a cross-sectional area of 625 mm².

(b) the vertical displacement of the pin used to apply the load.

2-53* The pin-connected structure shown in Fig. P2-53 consists of a cold-rolled bronze ($E = 15,000$ ksi) bar A that has a cross-sectional area of 3.00 in.² and two 0.2%C hardened steel ($E = 30,000$) bars B that have cross-sectional areas of 2.50 in.². The bars are supporting a load P of 200 kips. Determine

(a) the axial stresses in the bars.

(b) the displacement of pin C.

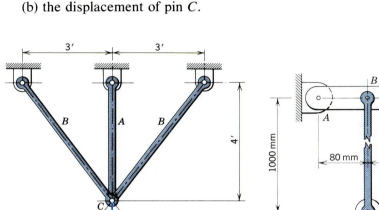

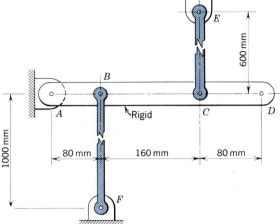

FIG. P2-53   FIG. P2-54

2-54* Bar BF of Fig. P2-54 is made of steel ($E = 210$ GPa) and bar CE is made of aluminum alloy ($E = 73$ GPa). The cross-sectional areas are 1200 mm² for bar BF and 900 mm² for bar CE. As a result of a misalignment of the pinholes at A, B, and C, a force of 50 kN must be applied upward at D, after pins A and B are in place, to permit insertion of pin C. Determine

(a) the axial stress in bar CE when the force P is removed with all pins in place.

(b) the displacement of pin D from its no-load position.

2-55 Solve Problem 2-53 when there is a clearance of $\frac{1}{32}$-in. between the pin at C and bar A that must be closed before bar A carries any load.

2-56 Solve Problem 2-53 if the tempereature of bar A decreases 50°F and the temperature of bars B increases 30°F after the 200-kip load is applied. The coefficients of thermal expansion are 9.4(10^{-6})/°F for the bronze and 6.6(10^{-6})/°F for the steel.

2-57* The structural steel ($E = 29,000$ ksi) post B of Fig. P2-57 is 2 by 2 in. square, and the aluminum alloy ($E = 10,600$ ksi) bar C is 1.5 in. wide by 1 in. thick. Bar A and the bearing block on B are to be considered rigid. The clearance between A and B is 0.002 in. before the load P is applied. If the axial stresses are not to exceed 30 ksi for the steel and 18 ksi for the aluminum alloy, determine

(a) the maximum load P that can be applied.

(b) the vertical displacement of the pin used to apply the load.

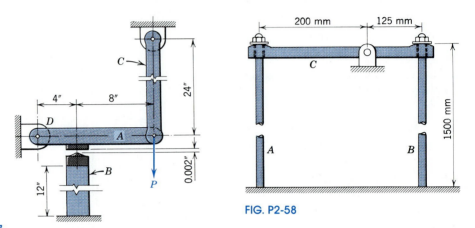

FIG. P2-57

FIG. P2-58

2-58* The mechanism of Fig. P2-58 consists of a structural steel ($E = 200$ GPa) rod A with a cross-sectional area of 350 mm², a cold-rolled brass ($E = 100$ GPa) rod B with a cross-sectional area of 750 mm², and a rigid bar C. The nuts at the top ends of rods A and B are initially tightened to the point where all slack is removed from the mechanism but the bars remain free of stress. Determine

(a) the axial stress induced in rod A by advancing the nut at the top of rod B one turn (2.5 mm).

(b) the vertical displacement of the nut at the top end of rod A.

2-59 The pin-connected structure shown in Fig. P2-59 occupies the position shown when unloaded. When the loads $D = 16$ kips and $E = $

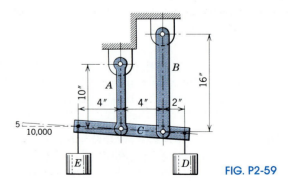

FIG. P2-59

8 kips are applied to the structure, the rigid bar C must become horizontal. Bar A is made of an aluminum alloy (E = 10,600 ksi) and bar B is made of bronze (E = 15,000 ksi). If the stresses in the bars must be limited to 20 ksi in the aluminum alloy and 15 ksi in the bronze, determine

> (a) the minimum cross-sectional areas that will be satisfactory for the bars.
>
> (b) the changes in length of rods A and B.

2-60 Solve Problem 2-59 if the temperature of bars A and B increases 50°F before the loads D and E are applied. The coefficients of thermal expansion are 12.5(10^{-6})/°F for the aluminum alloy bar and 9.4(10^{-6})/°F for the bronze bar.

2-61* The bronze (E = 100 GPa) post D of Fig. P2-61 has a cross-sectional area of 2500 mm², and the high-strength steel (E = 200 GPa) bar C has a cross-sectional area of 600 mm². Bar AB and the bearing block on post D are to be considered rigid. The clearance between post D and bar AB is 0.09 mm before the load P is applied. If the axial stresses are not to exceed 215 MPa for the steel and 95 MPa for the bronze, determine

> (a) the maximum load P that can be applied.
>
> (b) the vertical displacement of the point of application of the load P.

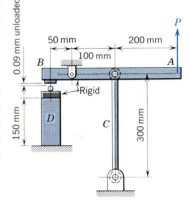

FIG. P2-61

2-62* Solve Problem 2-61 if the temperature of the bronze post D increases 30°C and the temperature of the steel bar C increases 20°C before the load P is applied. The coefficients of thermal expansion are 16.9(10^{-6})/°C for the bronze and 11.9(10^{-6})/°C for the steel.

2-63 A high-strength steel (E = 30,000 ksi) bolt passes through a brass (E = 15,000 ksi) sleeve as shown in Fig. P2-63. After the unit is assembled, the nut is tightened a quarter turn (0.03125 in.) and the temperature is raised 60°F. The coefficients of thermal expansion are 6.6(10^{-6})/°F for the steel bolt and 9.8(10^{-6})/°F for the brass sleeve. The cross-sectional areas are 0.785 in.² for the bolt and 1.767 in.² for the sleeve. Determine

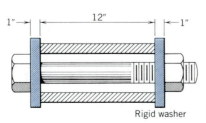

Rigid washer   FIG. P2-63

(a) the axial stresses that develop in the bolt and in the sleeve.

(b) the change in length of the brass sleeve.

2-64 The two faces of the clamp, shown in Fig. P2-64, are 250 mm apart when the two stainless-steel ($E = 190$ GPa) bolts connecting them are unstretched. A force P is applied to separate the faces of the clamp so that an aluminum alloy ($E = 73$ GPa) bar with a length of 251 mm can be inserted as shown. After the load P is removed, the temperature is raised 100°C. The coefficients of thermal expansion are $17.3(10^{-6})$/°C for the stainless steel and $22.5(10^{-6})$/°C for the aluminum alloy. Each of the bolts has a cross-sectional area of 120 mm² and the bar has a cross-sectional area of 625 mm². Determine

(a) the axial stresses in the bolts and in the bar.

(b) the change in length of the aluminum alloy bar.

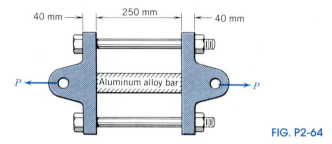

FIG. P2-64

2-5

WELDED CONNECTIONS UNDER CENTRIC LOADING

Connections between metal plates, angles, pipes, and other structural elements are frequently made by welding. Fusion welds are made by melting portions of the materials to be joined with an electric arc, a gas flame, or with Thermit. In fusion welds, additional welding material is usually added to the melted metal to fill the space between the two parts to be joined or to form a fillet. A gas shield is provided when welding certain metals to prevent rapid oxidation of the molten metal. A good welded joint will usually develop the full strength of the material being joined unless the high tempereature necessary for the process changes the properties of the materials.

Metals may also be joined by resistance welding in which a small area or spot is heated under high localized pressure. The material is not melted with this type of welding. Other joining methods for metals include brazing and soldering, in which the joining metal is melted but the parts to be connected are not melted. Such connections are usually much weaker than the materials being connected. Fusion welding is the most effective method when high strength is an important factor, and it will be discussed more in detail.

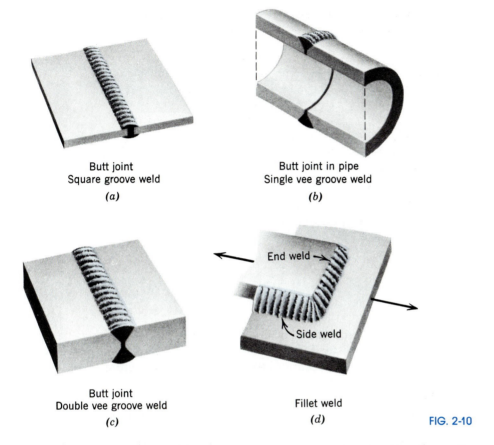

Butt joint
Square groove weld

(a)

Butt joint in pipe
Single vee groove weld

(b)

Butt joint
Double vee groove weld

(c)

Fillet weld

(d)

FIG. 2-10

Figure 2-10 illustrates the two general types of fusion-welded joints. The joints illustrated in Figs. 2-10a, b, and c are called *butt* joints and are made with groove welds. The weld may be square for fairly thin materials, or there may be one or two vees for thicker metals. When approved welding techniques and materials are used and complete penetration of the material is obtained, the strength of the joint is assumed to be the same as that of the material being joined.

A fillet weld is illustrated by Fig. 2-10d and is used to join a plate, angle, or other structural shape to another structural member. The end weld may be omitted in some applications. Ideally, the fillet should be slightly concave, rather than convex or straight as shown, to reduce possible stress concentrations at reentrant corners.

The shearing stress in the side weld cannot be uniformly distributed along the weld as long as the action is elastic, and tests[1] indicate the distribution is somewhat as indicated in Fig. 2-11. However, if the

[1]"Distribution of Shear in Welded Connections," H. W. Troelsch, *Trans. Am. Soc. of Civil Engrs.*, Vol. 99, 1934, p. 409.

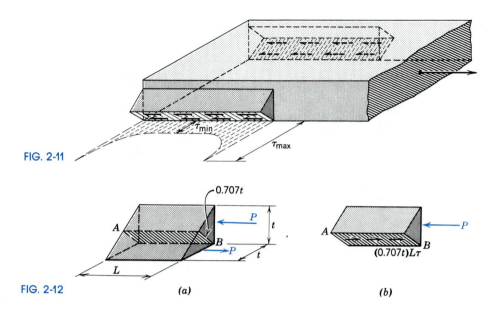

FIG. 2-11

FIG. 2-12   *(a)*   *(b)*

material is ductile (the usual condition for welded connections), a slight amount of inelastic action at the ends of the welds will result in essentially a uniform shearing stress distribution; therefore, in the design of welded connections, a uniform distribution is assumed. Fillet welds are designed on the assumption that failure will occur by shearing the minimum section of the weld, which is called the throat of the weld (section *AB* in Fig. 2-12*a*), whether the weld is a side (longitudinal) weld or an end (transverse) weld. The strength of side fillet welds can be determined from Figs. 2-12*a* and *b*. Although it is desirable to make fillet welds slightly concave, calculations are usually based on the assumption that the weld surface is straight and makes an angle of 45° with the two legs. The maximum allowable load *P* that can be carried by a side fillet weld with a length *L* and legs *t* (it is common practice to designate the size of fillet welds by the length of leg *t*) can be obtained from the free-body diagram in Fig. 2-12*b* and is

$$P = 0.707tL\tau \tag{2-3}$$

where τ is the allowable shearing stress permitted in the weld material.

Tests indicate that the strength of a transverse fillet weld (end weld in Fig. 2-10*d*) is about 30 percent greater than that of a side fillet weld with the same dimensions. It is common practice, however, to assume that the strength of a transverse weld is the same as that of a side weld with the same dimensions. This practice will be followed in this book.

The following example illustrates the analysis of a centrically loaded fillet weld.

EXAMPLE 2-8

An $8 \times 6 \times \frac{1}{2}$-in. steel angle (see Appendix C for additional data) is to be welded to a flat plate with the long side of the angle against the plate, as shown in Figs. 2-13a and b. The allowable tensile stress for the material in the angle is 18,000 psi, and the allowable shearing stress in the weld material is 13,600 psi. Each leg of the weld is 0.50 in. Determine

  (a) the minimum lengths L_1 and L_2 that will permit the angle to carry the maximum allowable axial load.
  (b) the minimum lengths L_1' and L_2' that will permit the angle to carry the maximum allowable axial load if a transverse weld is added across the end of the angle.

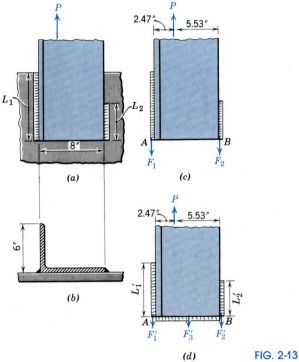

FIG. 2-13

Solution

  (a) The maximum longitudinal force in the angle will be developed when the tensile stress is uniformly distributed and will act through the centroid of the cross section of the angle. The area of the cross section and the location of its centroid can be obtained from the table of properties of sections of structural shapes in Appendix C. The area is 6.75 sq. in., and the location of the centroid is shown on Fig. 2-13c, which is a free-body diagram of a part of the angle. The forces F_1 and F_2 are assumed to act at the edges of the angle. The force P is

$$P = \sigma A = 18,000(6.75) = 121,500 \text{ lb}$$

The equation $\Sigma M_A = 0$, applied to the free-body diagram, gives

$$(\circlearrowleft+) \qquad \Sigma M_A = 8F_2 - 2.47P = 0$$

from which

$$F_2 = 37,500 \text{ lb as shown}$$

Similarly,

$$F_1 = 84,000 \text{ lb as shown}$$

The shear area of the throat that must carry the load F_2 is $L_2(0.707)(0.50)$, and the shearing stress multiplied by this area must equal the applied load. Thus,

$$F_2 = 37,500 = \tau A = 13,600 L_2(0.707)(0.50)$$

from which

$$L_2 = \underline{7.80 \text{ in.}} \qquad\qquad \text{Ans.}$$

In the same manner

$$L_1 = \underline{17.45 \text{ in.}} \qquad\qquad \text{Ans.}$$

(b) When a weld is placed across the end of the angle, an additional force must be shown on the free-body diagram, as in Fig. 2-13d. The Force F_3' is

$$F_3' = \tau A = 13,600(8)(0.707)(0.50) = 38,500 \text{ lb}$$

Note that the transverse weld is assumed to carry the same load per inch of weld as the side weld. The equation $\Sigma M_A = 0$ applied to Fig. 2-13d gives

$$(\circlearrowleft+) \qquad \Sigma M_A = 8F_2' + 4(38,500) - 2.47P = 0$$

from which

$$F_2' = 18,250 \text{ lb} = 13,600 L_2'(0.707)(0.50)$$

as in part (a) of the solution. This equation gives

$$L_2' = \underline{3.80 \text{ in.}} \qquad\qquad \text{Ans.}$$

Similarly,

$$L_1' = \underline{13.45 \text{ in.}} \qquad\qquad \text{Ans.}$$

PROBLEMS

2-65* A steel tube with an outside diameter of 4 in. and a wall thickness of $\frac{1}{4}$ in. is to be attached to a $\frac{1}{2}$-in. steel plate by machining a slot in the tube and using a $\frac{3}{16}$-in. fillet weld, as shown in Fig. P2-65. If the maximum shearing stress in the weld must be limited to 12 ksi, determine the length of weld required to develop an axial tensile working stress in the tube of 15 ksi.

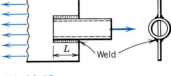

FIG. P2-65

2-66* A steel bar 100 mm wide by 10 mm thick will be connected to a large steel plate by means of two side welds as shown in Fig. P2-66. Determine the required length L of the 8-mm welds if the working stress for the weld material in shear is 85 MPa.

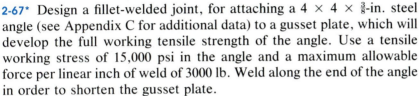

125 kN

FIG. P2-66

2-67* Design a fillet-welded joint, for attaching a $4 \times 4 \times \frac{3}{8}$-in. steel angle (see Appendix C for additional data) to a gusset plate, which will develop the full working tensile strength of the angle. Use a tensile working stress of 15,000 psi in the angle and a maximum allowable force per linear inch of weld of 3000 lb. Weld along the end of the angle in order to shorten the gusset plate.

2-68* Design a fillet-welded joint, for attaching a $127 \times 127 \times 19.1$-mm steel angle (see Appendix C for additional data) to a gusset plate, which will develop the full working tensile strength of the angle. Use a tensile working stress of 100 MPa in the angle and a maximum allowable force per linear millimeter of weld of 1000 N. Weld along the end of the angle in order to shorten the gusset plate.

2-69 A $6 \times 4 \times \frac{3}{8}$-in. steel angle (see Appendix C for additional data) is to be welded to a gusset plate with the 6-in. side against the plate. The tensile working strength of the angle is 24 ksi, and the allowable shearing stress in the $\frac{5}{16}$-in. weld is 13.6 ksi. Determine

(a) the minimum lengths L_1 and L_2 (see Fig. 2-13a) of the side welds that will permit the angle to carry the maximum allowable axial load.

(b) The minimum lengths L_1' and L_2' (see Fig. 2-13d) of the side welds that will permit the angle to carry the maximum allowable axial load if a transverse weld is added across the end of the angle.

2-70 A $127 \times 76 \times 12.7$-mm steel angle (see Appendix C for additional data) is to be welded to a gusset plate with the 127-mm side against the plate. The tensile working strength of the angle is 165 MPa, and the allowable shearing stress in the 11-mm weld is 85 MPa. Determine

(a) the minimum lengths L_1 and L_2 (see Fig. 2-13a) of the side welds that will permit the angle to carry the maximum allowable axial load.

(b) the minimum lengths L_1' and L_2' (see Fig. 2-13d) of the side welds that will permit the angle to carry the maximum allowable axial load if a transverse weld is added across the end of the angle.

2-71* Design a fillet-welded joint, for attaching a $4 \times 3 \times \frac{1}{2}$-in. steel angle (see Appendix C for additional data) to a gusset plate, which will develop the full working tensile strength of the angle. The specifications permit a tensile working stress of 20 ksi in the angle and a maximum shearing stress of 12.4 ksi in the $\frac{3}{8}$-in. weld. Weld along the end of the angle in order to shorten the gusset plate. Assume that

(a) the 4-in. side of the angle is against the plate.

(b) the 3-in. side of the angle is against the plate.

2-72* Design a fillet-welded joint, for attaching a $127 \times 89 \times 12.7$-mm steel angle (see Appendix C for additional data) to a gusset plate, that will develop the full working tensile strength of the angle. The specifications permit a tensile working stress of 150 MPa in the angle and a maximum shearing stress of 85 MPa in the 11-mm weld. Weld along the end of the angle in order to shorten the gusset plate. Assume that

(a) the 127-mm side of the angle is against the plate.

(b) the 89-mm side of the angle is against the plate.

2-73 The flanges of the steel cantilever beam AB shown in Fig. P2-73 are to be connected to the column with 1×14-in. butt welds. The web connection is to be $\frac{3}{8} \times 2$-in. intermittent fillet welds on both sides of the web. Recall that the 80-kip load can be replaced with a vertical force at A and a couple. Assume that the couple is resisted only by the flange welds and that the vertical force is resisted only by the web welds. Determine

(a) the average tensile stress in the flange weld (on the vertical plane).

(b) the number of $\frac{3}{8} \times 2$-in. welds to be used on each side of the

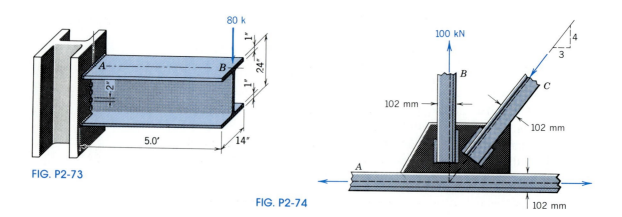

FIG. P2-73

FIG. P2-74

web if the maximum shearing stress in the fillet welds must not exceed 12.6 ksi.

2-74 A joint from a truss is shown in Fig. P2-74. Each member of the truss consists of two 102 × 76 × 9.5-mm angles that will be fillet welded to a gusset plate with side welds as shown in Fig. P2-74. If the maximum shearing stress in the 8-mm welds must be limited to 80 MPa, determine

    (a) the lengths of the side welds for member *B* of the joint.

    (b) the lengths of the side welds for member *C* of the joint.

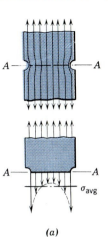

(a)

2-6

STRESS CONCENTRATIONS

In the foregoing sections, it was assumed that the average stress, as determined by the expression σ or τ equals P/A, is the significant or critical stress. For many problems, this is true; for other problems, however, the maximum normal stress on a given section may be considerably greater than the average; for certain combinations of loading and material, the maximum rather than the average is the important stress. If there exists in the structural or machine element a discontinuity that interrupts the stress path (called a stress trajectory),[2] the stress at the discontinuity may be considerably greater than the nominal (average, in the case of centric loading) stress on the section; thus, there is a *stress concentration* at the discontinuity. This is illustrated in Fig. 2-14, in which a type of discontinuity is shown in the upper figure and the approximate distribution of normal stress on a transverse plane is shown in the accompanying lower figure. The ratio of the maximum stress to the nominal stress on the section is known as the *stress concentration factor.* Thus, the expression for the maximum normal stress in a centrically loaded member becomes

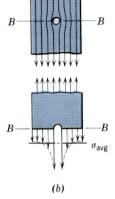

(b)

$$\sigma = K(P/A) \qquad (2\text{-}4)$$

where *A* is either the gross area or the net area (area at the reduced section) depending on the value used for *K*, the stress concentration factor. Curves, similar to the ones shown in Fig. 2-15, can be found in numerous design handbooks. It is important that the user of such curves (or tables of factors) ascertain whether the factors are based on the gross or net section. The factors K_t shown in Figs. 2-15*a, b,* and *c* are based on the net section. Sometimes the solution of a problem is expedited by use of the factor K_g based on the gross section, and for this purpose conversion expressions are given with the various curves.

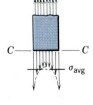

(c)

FIG. 2-14

  [2]A stress trajectory is a line everywhere parallel to the maximum normal stress.

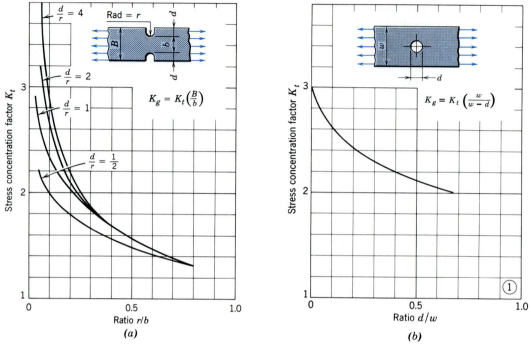

FIG. 2-15 Stress concentration factors for grooves, holes, and fillets.

A classic example of the solution of a problem involving a localized redistribution of stress occurs in the case of a small circular hole in a wide plate under uniform unidirectional tension.[3] The theory of elasticity solution is expressed in terms of a radial stress σ_r, a tangential stress σ_θ, and a shearing stress $\tau_{r\theta}$, as shown in Fig. 2-16. The equations are

$$\sigma_r = \frac{\sigma}{2}\left(1 - \frac{a^2}{r^2}\right) - \frac{\sigma}{2}\left(1 - \frac{4a^2}{r^2} + \frac{3a^4}{r^4}\right)\cos 2\theta$$

$$\sigma_\theta = \frac{\sigma}{2}\left(1 + \frac{a^2}{r^2}\right) + \frac{\sigma}{2}\left(1 + \frac{3a^4}{r^4}\right)\cos 2\theta$$

$$\tau_{r\theta} = \frac{\sigma}{2}\left(1 + \frac{2a^2}{r^2} - \frac{3a^4}{r^4}\right)\sin 2\theta$$

On the boundary of the hole (at $r = a$) these equations reduce to

$$\sigma_r = 0$$

$$\sigma_\theta = \sigma(1 + 2\cos 2\theta)$$

$$\tau_{r\theta} = 0$$

[3]This solution was obtained by G. Kirsch; see *Z. Ver. deut. Ing.*, Vol. 42, 1898. See also *Elasticity in Engineering Mechanics*, A. P. Boresi and P. P. Lynn, Prentice-Hall, Englewood Cliffs, N.J., 1974, pp. 304–309.

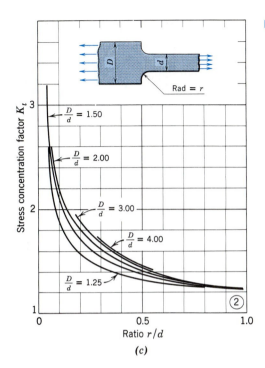

FIG. 2-15 *(continued)*

(c)

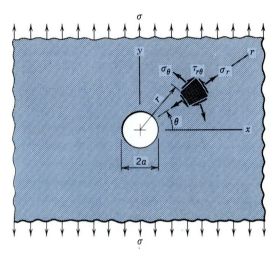

FIG. 2-16

At $\theta = 0°$, the tangential stress σ_θ equals 3σ, where σ is the uniform tensile stress in the plate in regions far removed from the hole. Thus, the stress concentration factor associated with this type of discontinuity is 3.

The localized nature of a stress concentration can be evaluated by considering the distribution of the tangential stress σ_θ along the x axis ($\theta = 0°$). Here

$$\sigma_\theta = \frac{\sigma}{2}\left(2 + \frac{a^2}{r^2} + \frac{3a^4}{r^4}\right)$$

At a distance $r = 3a$ (one hole diameter from the hole boundary) this equation yields $\sigma_\theta = 1.074\sigma$. Thus, the stress that began as three times the nominal at the boundary of the hole has decayed to a value only 7 percent greater than the nominal at a distance of one diameter from the hole. This rapid decay is typical of the redistribution of stress in the neighborhood of a discontinuity.

Stress concentration is not significant in the case of static loading of a ductile material (defined in Section 1-13) because the material will yield inelastically in the region of high stress and, with the accompanying redistribution of stress, equilibrium may be established and no harm done. However, if the load is an impact or repeated load, instead of the action described above, the material may fracture. Also, if the material is brittle, even a static load may cause fracture. Therefore, in the case of impact or repeated loading on any material or static loading on a brittle material, the presence of stress concentration should not be ignored. Before we leave the subject of stress concentration, it should be noted that in regions of support and load application, the stress distribution varies from the nominal (defined as the stress obtained from elementary theories of stress distribution—uniform for centric loading). This fact was discussed in 1864 by Barre de Saint-Venant (1797–1886), a French mathematician. Saint-Venant observed that although localized distortions in such regions produced stress distributions different from the theoretical distributions, these localized effects disappeared at some distance (the implication being that the distance is not of great magnitude[4]) from such locations. This statement is known as *Saint-Venant's principle* and is constantly used in engineering design.

EXAMPLE 2-9

The machine part shown in Fig. 2-17 is 20 mm thick and is made of 0.4 percent carbon hot-rolled steel (see Appendix A for properties). Determine the maximum safe load P if a factor of safety of 2.5 with respect to failure by slip is specified.

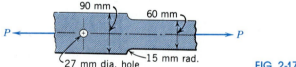

FIG. 2-17

Solution

The yield strength of 0.4 percent carbon hot-rolled steel, obtained from Appendix A, is 360 MPa. The working stress, based on a factor of safety of 2.5, is 360/2.5 = 144 MPa. The maximum stress in the machine part will occur either in the fillet between the two sections or on the boundary of the circular hole. Thus:

At the fillet

$$D/d = 90/60 = 1.5$$
$$r/d = 15/60 = 0.25$$

---

[4]For example, it can be shown mathematically that the localized effect of a concentrated load on a beam may disappear at a section slightly greater than the depth of the beam away from the load. See *Theory of Elasticity,* S. Timoshenko and J. N. Goodier, 3rd ed., McGraw-Hill, New York, 1970.

From Fig. 2-15c

$$K_t = 1.62$$

Thus, from Eq. 2-4

$$P = \sigma_w A_t/K_t = 144(10^6)(60)(20)(10^{-6})/1.62$$
$$= 106.7(10^3) \text{ N} = 106.7 \text{ kN}$$

At the hole

$$d/w = 27/90 = 0.3$$

From Fig. 2-15b

$$K_t = 2.30$$

Thus, from Eq. 2-4

$$P = \sigma_w A_t/K_t = 144(10^6)(90-27)(20)(10^{-6})/2.30$$
$$= 78.9(10^3) \text{ N} = 78.9 \text{ kN}$$

Therefore,

$$P_{max} = \underline{78.9 \text{ kN}} \qquad \text{Ans.}$$

PROBLEMS

2-75* The machine part shown in Fig. P2-75 is $\frac{1}{4}$ in. thick and is made of SAE 4340 heat-treated steel (see Appendix A for properties). Determine the maximum safe load P if a factor of safety of 2 with respect to failure by slip is specified.

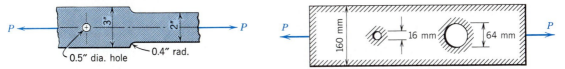

FIG. P2-75   FIG. P2-76

2-76* The machine part shown in Fig. P2-76 is 10 mm thick and is made of cold-rolled 18-8 stainless steel (see Appendix A for properties). Determine the maximum safe load P if a factor of safety of 1.5 with respect to failure by slip is specified.

2-77 A $\frac{1}{8}$-in.-thick by 4-in.-wide steel bar is transmitting an axial tensile load of 500 lb. After the load is applied, a $\frac{1}{64}$-in.-diameter hole is drilled through the bar as shown in Fig. P2-77.

   (a) Determine the stress at point A (on the edge of the hole) in the bar before and after the hole is drilled.

(b) Does the axial stress at point B on the edge of the bar increase or decrease as the hole is drilled? Explain.

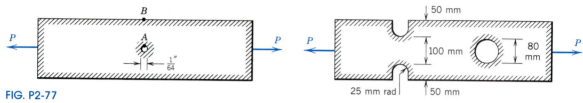

FIG. P2-77

FIG. P2-78

2-78 The machine part shown in Fig. P2-78 is 200 mm wide by 25 mm thick and is made of 2024-T4 Aluminum (see Appendix A for properties). Determine the maximum safe load P if a factor of safety of 2.25 with respect to failure by slip is specified.

2-79 The machine part shown in Fig. P2-79 is $\frac{1}{4}$- in. thick and is made of structural steel (see Appendix A for properties). Determine the maximum safe load P if a factor of safety of 3 with respect to failure by slip is specified.

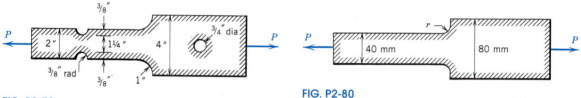

FIG. P2-79

FIG. P2-80

2-80* The machine part shown in Fig. P2-80 is 20 mm thick, is made of cold-rolled red brass (see Appendix A for properties), and is subjected to a tensile load P of 100 kN. Determine the minimum radius r that can be used between the two sections if a factor of safety of 2 with respect to failure by slip is specified.

2-81 The $\frac{1}{2}$-in.-thick bar with semicircular ($d/r = 1$) edge grooves, shown in Fig. P2-81, is made of structural steel (see Appendix A for properties) and will be subjected to an axial tensile load P of 10 kips. Determine the minimum safe width B for the bar if a factor of safety of 1.8 with respect to failure by slip must be maintained.

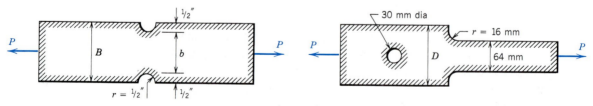

FIG. P2-81

FIG. P2-82

2-82 The stepped bar with a circular hole, shown in Fig. P2-82, is made of annealed 18-8 stainless steel (see Appendix A for properties). The bar is 25 mm thick and will be subjected to an axial tensile load P of 125 kN. Determine the minimum safe width D for the bar if a factor of safety of 2 with respect to failure by slip must be maintained.

2-7

THIN-WALLED PRESSURE VESSELS

A pressure vessel is described as *thin-walled* when the ratio of the wall thickness to the radius of the vessel (an approximately cylindrical or spherical shape is implied) is so small that the distribution of normal stress on a plane perpendicular to the surface of the shell is essentially uniform throughout the thickness of the shell. Actually, this stress varies from a maximum value at the inside surface to a minimum value at the outside surface of the shell, but it can be shown that if the ratio of the wall thickness to the inner radius of the vessel is less than 0.1, the maximum normal stress is less than 5 percent greater than the average. Boilers, gas storage tanks, pipelines, metal tires, and hoops are normally analyzed as thin-walled elements. Gun barrels, certain high-pressure vessels in the chemical processing industry, and cylinders and piping for heavy hydraulic presses need to be treated as thick-walled vessels.

Problems involving thin-walled vessels subjected to liquid (or gas) pressure p are readily solved with the aid of free-body diagrams of sections of the vessels together with the fluid contained therein. In the following subsections, spherical, cylindrical, and other thin shells of revolution are considered.

1. Spherical Pressure Vessels. A typical thin-walled spherical pressure vessel used for gas storage is shown in Fig. 2-18. If the weights of the gas and vessel are negligible (a common situation), symmetry of loading and geometry requires that stresses on sections that pass through the center of the sphere be equal. Thus, on the small element shown in Fig. 2-19a,

$$\sigma_x = \sigma_y = \sigma_n$$

Furthermore, there are no shearing stresses on any of these planes, since there are no loads to induce them. The normal stress component in a sphere is known as a meridional or axial stress and is commonly denoted as σ_m or σ_a.

The free-body diagram shown in Fig. 2-19b can be used to evaluate the stress $\sigma_x = \sigma_y = \sigma_n = \sigma_a$ in terms of the pressure p, and the

FIG. 2-18 Hortonsphere for gas storage in Superior, Wisconsin. (*Courtesy of Chicago Bridge and Iron Co., Chicago, Ill.*)

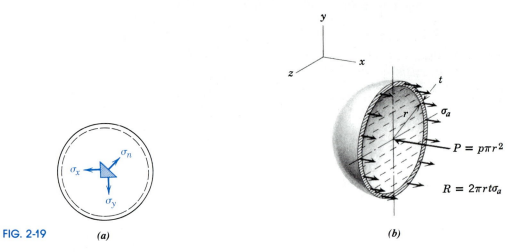

FIG. 2-19   *(a)*   *(b)*

radius r and thickness t of the spherical vessel. The force R is the resultant of the internal forces that act on the cross-sectional area of the sphere that is exposed by passing a plane through the center of the sphere. The force P is the resultant of the fluid forces acting on the fluid remaining within the hemisphere. From a summation of forces in the x direction,

$$R - P = 0$$

or

$$2\pi r t \sigma_a = p\pi r^2$$

from which

$$\sigma_a = \frac{pr}{2t} \qquad (2\text{-}5)$$

2. Cylindrical Pressure Vessels. A typical thin-walled cylindrical pressure vessel used for gas storage is shown in Fig. 2-20. Normal stresses, such as those shown on the small element of Fig. 2-21a, are easy to evaluate by using appropriate free-body diagrams. The normal stress

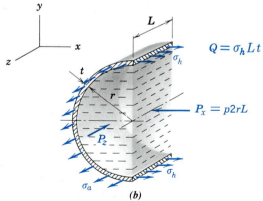

FIG. 2-20
Cylindrical tanks for gas storage in Northlake, Illinois. (*Courtesy of Chicago Bridge and Iron Co., Chicago, Ill.*)

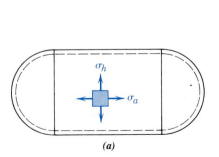

(a)

$Q = \sigma_h L t$

$P_x = p2rL$

(b)

FIG. 2-21

component on a transverse plane is known as an axial or meridional stress and is commonly denoted as σ_a or σ_m. The normal stress component on a longitudinal plane is known as a hoop, tangential, or circumferential stress and is denoted as σ_h, σ_t, or σ_c. There are no shearing stresses on transverse or longitudinal planes.

The free-body diagram used for axial stress determinations is similar to Fig. 2-19b, which was used for the sphere, and the results are the same. The free-body diagram used for the hoop stress determination is shown in Fig. 2-21b. The force P is the resultant of the fluid forces acting on the fluid remaining within the portion of the cylinder isolated by the longitudinal plane and two transverse planes. The forces Q are the resultant of the internal forces on the cross-sectional area exposed by a longitudinal plane containing the axis of the cylinder. From a summation of forces in the x direction,

$$2Q - P = 0$$

or

$$2\sigma_h Lt = p2rL$$

from which

$$\sigma_h = \frac{pr}{t} \tag{2-6a}$$

Also (see Eq. 2-5),

$$\sigma_a = \frac{pr}{2t} \tag{2-6b}$$

The previous analysis of the stresses in a cylindrical vessel subjected to uniform internal pressure indicates that the stress on a longitudinal plane is twice the stress on a transverse plane. Consequently, a longitudinal joint needs to be twice as strong as a transverse (or girth) joint.

3. Thin Shells of Revolution. The discussion in the two previous subsections was limited to thin-walled cylindrical and spherical vessels under uniform internal pressure. The theory can be extended, however, to include other shapes and other loading conditions. Consider, for example, the thin shell of revolution shown in Fig. 2-22. Such shells are generated by rotating a plane curve, called the meridian, about an axis lying in the plane of the curve. Shapes that can be formed in this manner include the sphere, hemisphere, torus (doughnut), cylinder, cone, and ellipsoid. In shells of revolution, the two unknown principal stresses are a meridional stress σ_m that acts on a plane perpendicular to the meridian and a tangential stress σ_t that acts on a plane perpen-

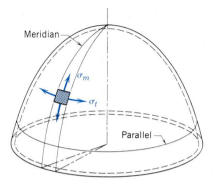

Meridian

σ_m

σ_t

Parallel

FIG. 2-22

dicular to a parallel. Both of these stresses are shown on the small element of Fig. 2-22. The two stresses can be evaluated by using two equilibrium equations.

The small element of Fig. 2-22 is shown enlarged in Fig. 2-23. The element has a uniform thickness, is subjected to an internal pressure p, and has different curvatures in two orthogonal directions. The resultant forces on the various surfaces are shown on the diagram. Summing forces in the n direction gives

$$P - 2F_m \sin d\theta_m - 2F_t \sin d\theta_t = 0$$

from which

$$p(2r_t d\theta_t)(2r_m d\theta_m) = 2\sigma_m(2tr_t d\theta_t)\sin d\theta_m + 2\sigma_t(2tr_m d\theta_m)\sin d\theta_t$$

and since, for small angles, $\sin d\theta \approx d\theta$, the above equation becomes

$$\frac{\sigma_m}{r_m} + \frac{\sigma_t}{r_t} = \frac{p}{t} \tag{2-7}$$

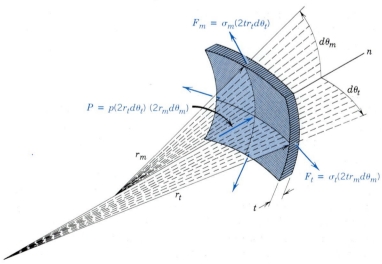

$F_m = \sigma_m(2tr_t d\theta_t)$

$d\theta_m$

n

$d\theta_t$

$P = p(2r_t d\theta_t)\,(2r_m d\theta_m)$

r_m

r_t

t

$F_t = \sigma_t(2tr_m d\theta_m)$

FIG. 2-23

Since Eq. 2-7 contains two unknown stresses, an additional independent equation is needed. Such an equation can be obtained by considering equilibrium of a portion of the vessel above or below the parallel that passes through the point of interest. Application of Eq. 2-7 is illustrated in Example 2-11.

EXAMPLE 2-10

A cylindrical pressure vessel 1.50 m in diameter is constructed by wrapping a 15-mm-thick steel plate into a spiral and butt-welding the mating edges of the plate as shown in Fig. 2-24a. The butt-welded seams form an angle of 30° with a transverse plane through the cylinder. Determine the normal stress σ perpendicular to the weld and the shearing stress τ parallel to the weld when the internal pressure in the vessel is 1500 kPa.

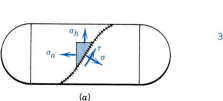
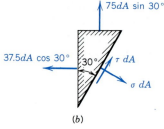

FIG. 2-24   (a)   (b)

Solution

The hoop stress σ_h and the axial stress σ_a in the cylinder can be determined by using Eqs. 2-6a and 2-6b. Thus

$$\sigma_h = \frac{pr}{t} = \frac{1500(10^3)(0.75)}{0.015} = 75.0(10^6) \text{ N/m}^2 = 75.0 \text{ MPa}$$

$$\sigma_a = \frac{pr}{2t} = \frac{1500(10^3)(0.75)}{2(0.015)} = 37.5(10^6) \text{ N/m}^2 = 37.5 \text{ MPa}$$

The normal stress σ perpendicular to the weld and the shearing stress τ parallel to the weld can be determined by using the free-body diagram shown in Fig. 2-24b.

From a summation of forces in the n direction:

$$\sigma dA - 37.5 dA \cos 30° \cos 30° - 75 dA \sin 30° \sin 30° = 0$$

$$\sigma = \underline{46.9 \text{ MPa}} \qquad \text{Ans.}$$

From a summation of forces in the t direction:

$$\tau dA - 37.5 dA \cos 30° \sin 30° + 75 dA \sin 30° \cos 30° = 0$$

$$\tau = \underline{-16.24 \text{ MPa}} \qquad \text{Ans.}$$

EXAMPLE 2-11

A pressure vessel of $\frac{1}{4}$-in. steel plate has the shape of a paraboloid closed by a thick flat plate, as shown in Fig. 2-25a. The equation of the generating parabola is $y = x^2/4$, where x and y are in inches. Determine the meridional and tangential stresses σ_m and σ_t in the shell at a point 16 in. above the bottom of the vessel due to an internal gas pressure of 250 psi gage.

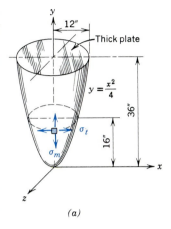

(a)

Solution

Determining σ_m:

The meridional stress σ_m, which must be tangent to the shell, can be determined with the aid of the free-body diagram shown in Fig. 2-25b. This free-body diagram represents a thin slice of the vessel and gas, with the forces perpendicular to the slice omitted. From the equation of the parabola, the radius x and the slope of the shell dy/dx at $y = 16$ in. are determined to be 8 in. and 4/1, respectively.

Summing forces in the y direction gives

$$-\int_{A_p} dP + \int_{A_\sigma} dF \cos \alpha = 0$$

$$-p\pi x^2 + \sigma_m 2\pi x t \cos \alpha = 0$$

Substituting the given data yields

$$-250(\pi)(8^2) + \sigma_m(2\pi)(8)(1/4)(4/\sqrt{17}) = 0$$

from which

$$\sigma_m = 4123 = \underline{4120 \text{ psi } T} \qquad \text{Ans.}$$

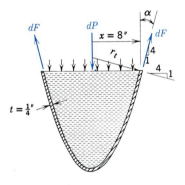

(b)

FIG. 2-25

Determining σ_t:

In order to find σ_t from Eq. 2-7, the radii r_m and r_t at the point must be determined. The radius of curvature r_m of the shell in the xy plane is determined from the expression

$$r_m = \frac{[1 + (dy/dx^2)]^{1.5}}{d^2y/dx^2} = \frac{(1 + 4^2)^{1.5}}{1/2} = 140.19 \text{ in.}$$

and the perpendicular radius r_t is found from the geometry of Fig. 2-25b as

$$r_t = 8(\sqrt{17}/4) = 8.246 \text{ in.}$$

Then, from Eq. 2-7

$$p/t = (\sigma_m/r_m) + (\sigma_t/r_t)$$

$$250(4) = (4123/140.19) + (\sigma_t/8.246)$$

from which

$$\sigma_t = 8003 = \underline{8000 \text{ psi } T} \qquad \text{Ans.}$$

PROBLEMS

2-83* Determine the maximum normal stress in a 12-in.-diameter basketball that has a $\frac{1}{16}$-in. wall thickness after it has been inflated to a gage pressure of 15 psi.

2-84* A spherical gas-storage tank 50 ft in diameter is being constructed to store gas under an internal pressure of 200 psi. Determine the thickness of structural steel (see Appendix A for properties) plate required if a factor of safety of 3 with respect to failure by slip is specified.

2-85* A steel pipe with an inside diameter of 12 in. will be used to transmit steam under a pressure of 1000 psi. If the hoop stress in the pipe must be limited to 10 ksi because of a longitudinal weld in the pipe, determine the minimum satisfactory thickness for the pipe.

2-86* A cylindrical propane tank, similar to the ones shown in Fig. 2-20, has an outside diameter of 3.25 m and a wall thickness of 22 mm. If the allowable hoop stress is 100 MPa and the allowable axial stress is 45 MPa, determine the maximum internal pressure that can be applied to the tank.

2-87 A spherical shell, similar to the one shown in Fig. 2-18, has an outside diameter of 10 ft, a wall thickness of $\frac{7}{8}$ in., and is made of structure steel (see Appendix A for properties). Determine

(a) the normal stress produced by an internal pressure of 275 psi.

(b) the maximum allowable pressure if a factor of safety of 3 with respect to failure by slip is specified.

2-88 A spherical shell, similar to the one shown in Fig. 2-18, is being designed to store gas at a pressure of 500 kPa. The shell will be made of structural steel (see Appendix A for properties) plates and will have an outside diameter of 8 m. If a factor of safety of 3 with respect to failure by slip is specified, determine the minimum thickness of plate that will be satisfactory for the shell.

2-89* A cylindrical boiler with hemispherical ends has an outside diameter of 6 ft. If the hoop and axial stresses in the boiler must be limited to 15 ksi and 8 ksi, respectively, determine the thickness of steel plate required if the internal pressure in the boiler will be 200 psi.

2-90* A cylindrical boiler with an outside diameter of 3 m and a wall thickness of 50 mm is made of structural steel (see Appendix A for properties). Determine

(a) the maximum normal stress produced by an internal pressure of 2 MPa.

(b) the maximum allowable pressure if a factor of safety of 2.75 with respect to failure by slip is specified.

2-91 The force transmitted across a 10-in. length of the joint uniting the two halves of a spherical pressure vessel 20 ft in diameter must be limited to 3000-lb tension. Determine the maximum internal pressure that can be safely applied to the vessel.

2-92 A spherical pressure vessel 12 m in diameter contains gas at a gage pressure of 300 kPa. Determine the force transmitted across a 500-mm length of the joint uniting the two halves of the vessel.

2-93* A standpipe 12 ft in diameter and 50 ft tall is being constructed for use as a storage tank for water ($\gamma = 62.4$ lb/ft^3). Determine the minimum thickness of steel plate that can be used if the hoop stress in the standpipe must be limited to 5000 psi.

2-94* A cylindrical tank 30 m in diameter and 10 m high is being constructed as a storage tank for oil ($\rho = 850$ kg/m^3). If the hoop stresses in the tank must be limited to 75 MPa, determine

(a) the required wall thickness for the steel plate in the bottom half of the tank.

(b) the required wall thickness for the steel plate in the top half of the tank.

2-95 The axial strain measured on the outside surface of an aluminum alloy ($E = 10,000$ ksi and $\nu = \frac{1}{3}$) spherical pressure vessel that has an outside diameter of 20 in. and a wall thickness of $\frac{1}{2}$ in. is 500 μin./in. Determine

(a) the axial stress in the vessel.

(b) the internal pressure being applied to the vessel.

2-96 A spherical pressure vessel 3 m in diameter is being designed to withstand a maximum internal pressure of 500 kPa. The material being used in its construction has an elastic strength of 430 MPa, a modulus of elasticity of 210 GPa, and a Poisson's ratio of 0.30. If a factor of safety of 4 with respect to failure by slip is specified, determine

(a) the minimum satisfactory wall thickness.

(b) the circumferential normal strain at maximum pressure when the wall thickness of part (a) is used.

(c) the change in diameter of the pressure vessel at maximum pressure when the wall thickness of part (a) is used.

2-97 A 2-ft-diameter by 10-ft-long cylindrical pressure vessel will be subjected to a maximum internal pressure of 250 psi. The material used in the vessels fabrication has an elastic strength of 60 ksi, a modulus of elasticity of 30,000 ksi, and a Poisson's ratio of 0.30. If a factor of safety of 4 with respect to failure by slip is specified, determine

(a) the minimum wall thickness required for the vessel.

(b) the hoop and axial stresses at maximum pressure when the thickness of part (a) is used.

(c) the change in diameter and the change in length of the vessel at a pressure of 200 psi when the thickness of part (a) is used.

2-98* A steel boiler 1 m in diameter is welded using a spiral seam that makes an angle of 30° with the longitudinal direction (axis of the boiler). Determine the magnitudes of the normal and shearing forces transmitted across a 150-mm length of the seam due to an internal pressure of 950 kPa.

2-99 A cylindrical tank, similar to the one shown in Fig. 2-20, has an outside diameter of 8 ft and a wall thickness of $\frac{3}{4}$ in. The tank is subjected to an internal pressure of 150 psi.

(a) Determine the axial and hoop stresses in the tank.

(b) Prepare a plot, similar to Fig. 1-18, showing the variation of normal and shearing stresses on planes through a point in the wall of the tank as angle θ, measured counterclockwise from a transverse plane, varies from 0° to 90°.

2-100* A thin-walled cylindrical pressure vessel has an outside diameter of 2 m and a wall thickness of 10 mm. The vessel is made of steel with a modulus of elasticity of 200 GPa and a Poisson's ratio of 0.30. During proof testing of the vessel, an axial strain of 300 μm/m is recorded. Determine

(a) the internal pressure applied to the vessel.

(b) the axial and hoop stresses in the vessel.

(c) the hoop strain present when the axial strain was measured.

2-101* The cylindrical pressure tank, shown in Fig. P2-101, is made of 10-mm steel plate. The magnitude of the shearing stress at point A on plane $B–B$ (which is perpendicular to the surface of the plate at A) is 45 MPa. Determine the air pressure in the tank.

2-102 A thin-walled tube with a 1-in. outside diameter and a 0.032-in. wall thickness is subjected to an internal pressure of 1 ksi. Prepare a sketch showing the stresses acting on a small cube of material at a point

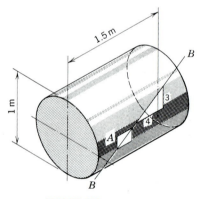

FIG. P2-101

(a) on the outside surface of the tube.

(b) on the inside surface of the tube.

2-103 The strains measured on the outside surface of the cylindrical pressure vessel shown in Fig. P2-103 are $\epsilon_1 = +619$ μin./in. and $\epsilon_2 = +330$ μin./in. The angle $\theta = 30°$. The outside diameter of the vessel is 20 in. and the wall thickness is $\frac{1}{8}$ in. The vessel is made of 0.4 percent carbon hot-rolled steel (see Appendix A for properties). Determine

(a) the stresses σ_1 and σ_2 in the vessel.

(b) the internal pressure applied to the vessel.

(c) the factor of safety with respect to failure by fracture based on the hoop stress.

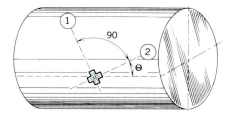

FIG. P2-103

2-104 A cylindrical pressure tube with an outside diameter of 20 mm and a wall thickness of 0.50 mm is 1 m long. The tube is made of cold-rolled bronze (see Appendix A for properties) and is subjected to an internal pressure of 5 MPa. Prepare a sketch showing the strains experienced by a small cube of material at a point

(a) on the outside surface of the shell.

(b) on the inside surface of the shell.

(c) Explain any inconsistencies in the results of parts (a) and (b).

2-105* A hemispherical tank of radius r and thickness t is supported by a flange, as shown in the cross section of Fig. P2-105. The tank is filled with a fluid having a specific weight γ. Determine, in terms of γ, r, and t, the meridional and tangential stresses σ_m and σ_t at a depth $y = r/2$.

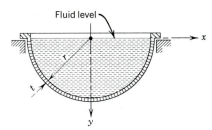

FIG. P2-105

2-106* The thin-walled spherical water tank shown in Fig. P2-106 is completely full of water and open to the atmosphere at the top. Determine the meridional and tangential stresses σ_m and σ_t at a point on the equator of the sphere in terms of γ, r, and t, where γ is the specific weight of the water, r is the radius of the sphere, and t is the wall thickness of the sphere.

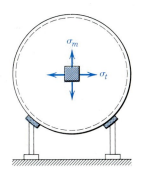

FIG. P2-106

2-107 The conical water tank shown in Fig. P2-107 was fabricated from $\frac{1}{8}$-in. steel plate. When the tank is completely full of water (specific weight $\gamma = 62.4$ lb/ft³), determine the axial and hoop stresses σ_a and σ_h at a point in the wall 8 ft below the apex of the cone.

2-108 A reducer in a pipeline (see Fig. P2-108) has a wall thickness of 4 mm. Assume that the flange bolts at the 150-mm end take all the end thrust (no compression in the flange at the 75-mm end). The pipeline

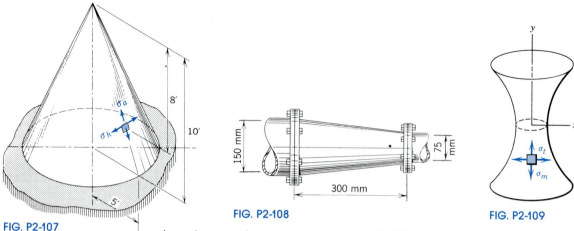

FIG. P2-107

FIG. P2-108

FIG. P2-109

is under a static water pressure of 420 kPa (gage). Determine the axial and hoop stresses σ_a and σ_h at a point midway between the flanges.

2-109* The closed glass vial shown in Fig. P2-109 has the shape of a hyperboloid of revolution. The equation of the generating curve is $x^2 = \frac{1}{4} + y^2$, where x and y, are in inches. The wall thickness is 0.02 in. and the internal pressure is 2 psi. Determine the meridional and tangential stresses σ_m and σ_t at a point where $y = 1$ in.

COMPUTER PROBLEMS

> *Note:* The following problems have been designed to be solved with a programmable calculator, microcomputer, or mainframe computer. Appendix F contains a description of a few numerical methods, together with a few simple programs in BASIC and FORTRAN, that can be modified for use in the solution of these problems.

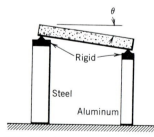

FIG. PC2-1

C2-1 The rigid platform shown in Fig. PC2-1 is 4 ft long and weighs 25,000 lb. The platform rests on two solid circular posts that are initially 30 in. long. The mass of the platform is uniformly distributed. The steel ($E = 30,000$ ksi) post has a diameter d_s of 4 in., while the diameter d_a of the aluminum ($E = 10,600$ ksi) post varies, 3 in. $\leq d_a \leq 8$ in. Compute and plot:

(a) the deflection of the aluminum end of the platform relative to the deflection of the steel end of the platform as a function of the diameter d_a of the aluminum post.

(b) the angle θ as a function of the diameter d_a of the aluminum post.

C2-2 The rigid platform shown in Fig. PC2-1 is 2 m long and weighs

150 kN. The platform rests on two posts that are initially 1 m long. The center of mass of the platform is $\frac{1}{2}$ m from the right end. The steel (E = 200 GPa) post has a diameter d_s of 100 mm, while the diameter d_a of the aluminum (E = 73 GPa) post varies, 50 mm $\leqslant d_a \leqslant$ 250 mm. Compute and plot:

(a) the deflection of the aluminum end of the platform relative to the deflection of the steel end of the platform as a function of the diameter d_a of the aluminum post.

(b) the angle θ as a function of the diameter d_a of the aluminum post.

C2-3 The 10-in.-diameter structural steel (E = 30,000 ksi) pile shown in Fig. PC2-3 is being extracted from the ground. Assume that the entire 30-ft length of the pile is in the ground and that the horizontal normal stresses σ_n and the vertical shearing stresses τ (which are a function of the type of soil surrounding the pile) are uniformly distributed over the surface of the cylinder. The magnitudes of the normal and shearing stresses are 40 psi and 20 psi, respectively. Compute and plot:

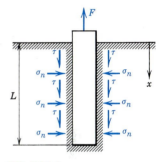

FIG. PC2-3

(a) The axial stress $\sigma_x(x)$ in the pile as a function of the distance x from the ground surface.

(b) The axial deformation of the pile δ_x as a function of the distance x from the ground surface.

C2-4 The 300-mm-diameter timber (E = 13 GPa and ν = 0.30) pile shown in Fig. PC2-3 is being extracted from the ground. Assume that the entire 8-m length of the pile is in the ground and that the distribution of horizontal normal stress σ_n and vertical shearing stresses τ (which are a function of the type of soil surrounding the pile) can be approximated by the expressions

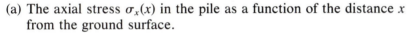

$$\sigma_n(x) = \gamma x (1 - \sin \phi) \qquad \tau(x) = \sigma_n \tan \phi$$

where γ = 400 N/m³ is the specific weight of the surrounding material, ϕ = 28° is the friction angle of the soil, and x is the distance from the ground surface. Compute and plot:

(a) the axial stress $\sigma_x(x)$ in the pile as a function of the distance x from the ground surface.

(b) the axial deformation of the pile δ_x as a function of the distance x from the ground surface.

C2-5 For the pile of Problem C2-3, compute and plot the force F required to extract the pile from the ground as a function of the length L' of the pile that remains in the ground (0 ft $\leqslant L' \leqslant$ 30 ft).

C2-6 For the pile of Problem C2-4, compute and plot the force F required to extract the pile from the ground as a function of the length L' of the pile that remains in the ground (0 m $\leqslant L' \leqslant$ 8 m).

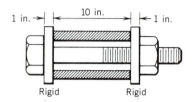

FIG. PC2-7

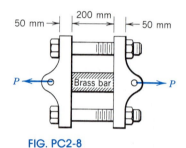

FIG. PC2-8

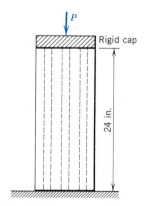

FIG. PC2-9

C2-7 A high-strength steel bolt (E_s = 30,000 ksi and A_s = 0.785 in.^{2}) passes through a brass sleeve (E_b = 15,000 ksi and A_b = 1.767 in.^{2}) as shown in Fig. PC2-7. As the nut is tightened, it advances a distance of 0.125-in. along the bolt for each complete turn of the nut. Compute and plot:

(a) the axial stress σ_s in the steel bolt and the axial stress σ_b in the brass sleeve as functions of the angle of twist θ of the nut ($0° \leqslant \theta \leqslant 180°$).

(b) the elongation δ_s of the steel bolt and the elongation δ_b of the brass sleeve as functions of the angle of twist θ of the nut ($0° \leqslant \theta \leqslant 180°$).

(c) the distance L between the two washers as a function of the angle of twist θ of the nut ($0° \leqslant \theta$ angle $180°$).

C2-8 The two faces of the clamp shown in Fig. PC2-8 are 200 mm apart when the two steel bolts (E_s = 190 GPa, A_s = 115 mm^2 each, and α_s = 17.3 $\times$ 10^{-6}/°C) connecting them are unstretched. A force P is applied to separate the faces of the clamp so that a brass bar (E_b = 100 GPa, A_b = 625 mm^2, and α_b = 17.6 $\times$ 10^{-6}/°C) with a length of 200.50 mm can be inserted as shown. After the load P is removed, the temperature of the system is slowly raised. Compute and plot:

(a) the axial stress σ_s in the steel bolt and the axial stress σ_b in the brass bar as functions of the temperature rise T ($0°C \leqslant T \leqslant 100°C$).

(b) the elongation δ_s of the steel bolts and the elongation δ_b of the brass bar as functions of the temperature rise T ($0°C \leqslant T \leqslant 100°C$).

(c) The distance L between the faces of the clamps as a function of the temperature rise ($0°C \leqslant T \leqslant 100°C$).

C2-9 The short pier shown in Fig. PC2-9 is reinforced with nine steel (E_s = 3000 ksi) reinforcing bars. A compressive load P is applied to the pier through the rigid capping plate. The axial load carried by the pier is a function of R, the percentage of the cross section taken up by the steel reinforcement bars. The load is also a function of the modulus ratio E_R/E_M, where E_R and E_M are Young's modulus for the reinforcement material and the matrix material, respectively. For the three matrix–reinforcement combinations listed below, compute and plot the percentage of the load carried by the matrix as a function of R ($0 \leqslant R \leqslant 100$ percent).

Matrix	Reinforcement	E_R/E_M
(a) Rubber	Steel	50,000
(b) Wood	Steel	20
(c) Concrete	Steel	7.5

CHAPTER 3
TORSIONAL LOADING

3-1
INTRODUCTION

In the design of machinery (and some structures), the problem of transmitting a torque (a couple or twisting moment) from one plane to a parallel plane is frequently encountered. The simplest device for accomplishing this function is the circular shaft, and as a result, a circular shaft is commonly used as the connecting member between an electric motor and a pump, compressor, or similar mechanism. A modified free-body diagram of a typical installation is shown in Fig. 3-1. The weight and bearing reactions are not shown on this modified diagram, since they do not contribute useful information to the torsion problem. The resultant of the electromagnetic forces applied to armature A of the motor is a couple that is resisted by the resultant of the bolt forces (another couple) acting on the flange coupling B. The circular shaft transmits the torque from the armature to the coupling. The torsion problem is concerned with the determination of stresses in the shaft and deformation of the shaft.

For purposes of this elementary analysis, a segment of the shaft between transverse planes $a–a$ and $b–b$ of Fig. 3-1 will be used. By limiting the analysis to this portion of the shaft, the complicated states of stress at the locations of the torque-applying devices (armature and flange coupling) can be avoided. Recall that Saint-Venant's principle

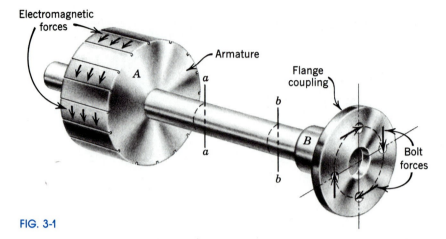

FIG. 3-1

states that the effects introduced by attaching the armature and coupling to the shaft will cease to be felt in the shaft at a distance of approximately one shaft diameter from the devices. A free-body diagram of the segment of the shaft between planes a–a and b–b is shown in Fig. 3-2. The resultant of the electromagnetic forces applied by the armature to the shaft is shown on the left end as torque T. The resisting torque T_r, at the right end of the shaft, is the resultant of the differential forces dF acting on the transverse plane b–b. The force dF is equal to $\tau_\rho \, dA$, where τ_ρ is the shearing stress on the transverse plane at a distance ρ from the center of the shaft, and dA is a differential area. In keeping with the notation outlined in Section 1-7, the shearing stress τ_ρ should be designated $\tau_{z\theta}$ to indicate that it acts on the z face in the direction of increasing θ. For the elementary theory of torsion of circular sections discussed in this book, the shearing stress on any transverse plane is always perpendicular to the radius to the point; therefore, the formal

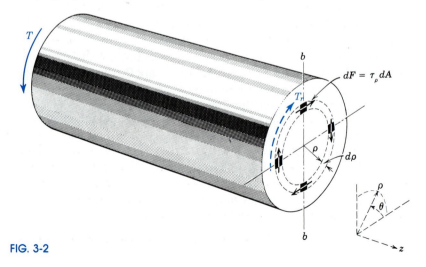

FIG. 3-2

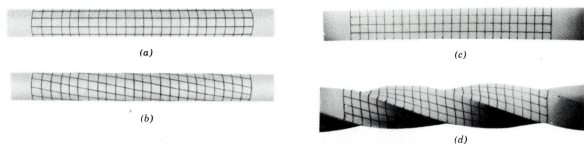

(a)

(b)

(c)

(d)

FIG. 3-3

subscript notation for shearing stress is not needed for accuracy and can be omitted. In this chapter, a single subscript will be used with the symbol τ to indicate the radial position on the cross section where the shearing stress is acting. Thus, the symbols τ_ρ and τ_c will be used to indicate the shearing stresses at an arbitrary radial position ρ and at the outside surface ($\rho = c$) of the shaft, respectively.

If the shaft is in equilibrium, the summation of moments about the axis of the shaft indicates that

$$T = T_r = \int_{\text{area}} \rho \tau_\rho \, dA \tag{3-1}$$

The law of variation of the shearing stress on the transverse plane must be known before the integral of Eq. 3-1 can be evaluated. The approach to the problem of stress distribution will be prefaced by a brief reference to the past.

In 1784 C. A. Coulomb, a French engineer, developed experimentally the relationship between the applied torque and the angle of twist for circular bars.[1] A. Duleau, another French engineer, in a paper published in 1820[1] derived analytically the same relationship by making the assumption that *a plane section before twisting remains plane after twisting and a diameter remains a straight line.* Visual examination of twisted models indicates that these assumptions are apparently correct for circular sections either solid or hollow (provided the hollow section is circular and symmetrical with respect to the axis of the shaft), but incorrect for any other shape. Compare, for example, the distortions of the rubber models with circular and square cross sections shown in Fig. 3-3. Figures 3-3a and b show a circular shaft before and after a torque is applied. Each cross section remains plane and undistorted as it rotates with respect to a neighboring cross section. Figures 3-3c and d show a square shaft before and after a torque is applied. Figure 3-3d clearly shows that planes do not remain plane, but warp as the torque is applied. The behavior exhibited by the square shaft is characteristic of all but circular sections; therefore, the analysis that follows is valid only for solid or hollow circular shafts.

[1] From *History of Strength of Materials,* S. P. Timoshenko, McGraw-Hill, New York, 1953.

3-2

TORSIONAL SHEARING STRAIN

If the assumptions noted above are made, the distortion of the shaft will be as indicated in Fig. 3-4a, where points B and D on a common radius in a plane section move to B' and D' in the same plane and still on the same radius. The angle θ is called the *angle of twist*. The surface ABB' of Fig. 3-4a is shown developed in Fig. 3-4b, and a differential cube of the material at B assumes at B' the familiar distortion due to shearing stress. At this point the assumption will be made that *all longitudinal elements have the same length* (which limits the results to straight shafts of constant diameter). From Figs. 3-4b and c, the following strain relations may be written:

$$\tan \gamma_c = \frac{BB'}{L} = \frac{c\theta}{L}$$

and

$$\tan \gamma_\rho = \frac{DD'}{L} = \frac{\rho\theta}{L}$$

or, *if the strain is small,*

$$\gamma_c = \frac{c\theta}{L} \quad \text{and} \quad \gamma_\rho = \frac{\rho\theta}{L} \tag{3-2}$$

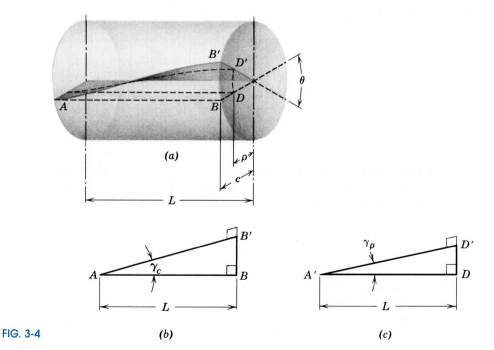

FIG. 3-4

(a)

(b)

(c)

From Eq. 3-2,

$$\theta = \frac{\gamma_c L}{c} = \frac{\gamma_\rho L}{\rho}$$

and as the longitudinal elements have the same length, the expression reduces to

$$\gamma_\rho = \frac{\gamma_c}{c}\rho \qquad\qquad (3\text{-}3)$$

which indicates that the shearing strain ($\gamma_\rho = \gamma_{z\theta}$) is proportional to the distance from the axis of the shaft. Note that Eq. 3-3 is valid for elastic or inelastic action and for homogeneous or heterogeneous materials, provided the strains are not too large ($\tan \gamma = \gamma$). Problems and examples in this book will be assumed to satisfy this requirement.

3-3

TORSIONAL SHEARING STRESSES: THE ELASTIC TORSION FORMULA

If the assumption is now made that Hooke's law applies (the accompanying limitation is that the stresses must be below the proportional limit of the material), shearing stress τ is related to shearing strain γ by the expression $\tau = G\gamma$ and Eq. 3-3 can be expressed in terms of stresses as

$$\tau_\rho = \frac{\tau_c}{c}\rho \qquad\qquad (3\text{-}4)$$

When Eq. 3-4 is substituted into Eq. 3-1, the result is

$$T_r = \frac{\tau_c}{c}\int_{\text{area}} \rho^2\, dA = \frac{\tau_\rho}{\rho}\int_{\text{area}} \rho^2\, dA \qquad\qquad (a)$$

The integral in Eq. a is the polar second moment of area J (called the polar moment of inertia), with respect to the longitudinal axis of the shaft (see Appendix G). Equation a can then be written as

$$T_r = \frac{\tau_\rho J}{\rho}$$

or in terms of the unknown shearing stress as

$$\tau_\rho = \frac{T\rho}{J} \qquad\qquad (3\text{-}5)$$

The polar second moment of area J (polar moment of inertia) for a hollow circular cross section is

$$J = (\pi/2)(R_o^4 - R_i^4) = (\pi/32)(D_o^4 - D_i^4)$$

where the subscripts o and i are used with the radius R or diameter D of the cross section to indicate outside and inside, respectively. For a solid shaft, $R_i = 0$ and $J = \pi R^4/2 = \pi c^4/2$. The units of J are in.^{4} in the U.S. Customary System and mm^4 in the SI System.

Equation 3-5 is known as the *elastic torsion formula,* in which τ_ρ is the shearing stress on a transverse plane at a distance ρ from the axis of the shaft, and T is the resisting torque, which, in general, is obtained from a free-body diagram and an equilibrium equation. Note that Eq. 3-5 applies only for linearly elastic action in homogeneous and isotropic materials. Note also that $\tau_{\max}$ occurs at the outside surface of the shaft (at $\rho = R_o$).

3-4

STRESSES ON OBLIQUE PLANES

As stated in Section 3-3, the torsion formula can be used to evaluate the shearing stress on a transverse plane. It is necessary to ascertain if the transverse plane is a plane of maximum shearing stress and if there are other significant stresses induced by torsion. For this study, the stresses at point A in the shaft of Fig. 3-5a will be analyzed. Figure

FIG. 3-5

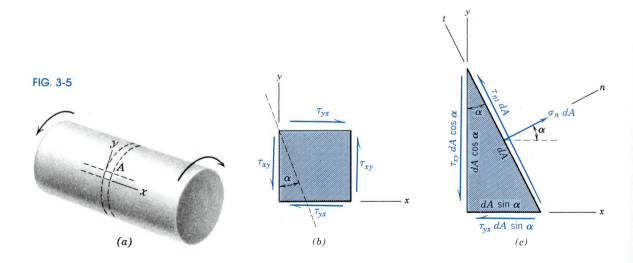

(a) $\qquad$ (b) $\qquad$ (c)

3-5*b* shows a differential element taken from the shaft at *A* and the stresses acting on transverse and longitudinal planes. The stress τ_{xy} may be determined by means of the torsion formula, and $\tau_{yx} = \tau_{xy}$ (see Eq. 1-11). If the equations of equilibrium are applied to the free-body diagram of Fig. 3-5*c*, the following results are obtained:

$$\overset{\curvearrowleft}{+} \qquad \Sigma F_t = 0$$

$$\tau_{nt}dA - \tau_{xy}(dA \cos \alpha)\cos \alpha + \tau_{yx}(dA \sin \alpha)\sin \alpha = 0$$

from which

$$\tau_{nt} = \tau_{xy}(\cos^2 \alpha - \sin^2 \alpha) = \tau_{xy} \cos 2\alpha \qquad (3\text{-}6)$$

and

$$\overset{\nearrow}{+} \qquad \Sigma F_n = 0$$

$$\sigma_n dA - \tau_{xy}(dA \cos \alpha)\sin \alpha - \tau_{yx}(dA \sin \alpha)\cos \alpha = 0$$

from which

$$\sigma_n = 2\tau_{xy} \sin \alpha \cos \alpha = \tau_{xy} \sin 2\alpha \qquad (3\text{-}7)$$

These results are shown in the graph of Fig. 3-6, from which it is apparent that the maximum shearing stress occurs on transverse and longitudinal (diametral) planes. The graph also shows that the maximum normal stresses occur on planes oriented at 45° with the axis of the bar and perpendicular to the surface of the bar. On one of these planes ($\alpha = 45°$ for Fig. 3-5) the stress is tension, and on the other ($\alpha = 135°$) the stress is compression. Furthermore, *all these stresses have the same magnitude;* hence, the torsion formula will give the magnitude of the maximum normal and shearing stresses at a point in a circular shaft subjected to pure torsion.

FIG. 3-6

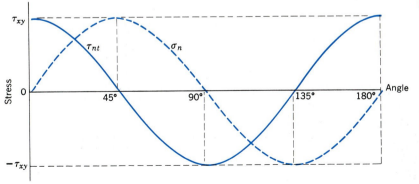

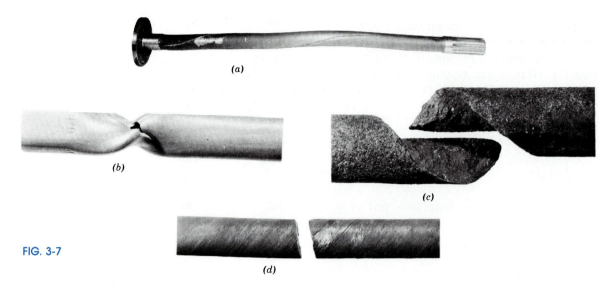

(a)

(b)

(c)

FIG. 3-7

(d)

Any of the stresses discussed in the preceding paragraph may be significant in a particular problem. Compare, for example, the failures shown in Fig. 3-7. In Fig. 3-7a, the steel truck rear axle split longitudinally. One would also expect this type of failure to occur in a shaft of wood with the grain running longitudinally. In Fig. 3-7b, the compressive stress caused the thin-walled aluminum alloy tube to buckle along one 45° plane, while the tensile stress caused tearing on the other 45° plane. Buckling of thin-walled tubes (and other shapes) subjected to torsional loading is a matter of paramount concern to the designer. In Fig. 3-7c, the tensile stresses caused the gray cast iron to fail in tension—typical of any brittle material subjected to torsion. In Fig. 3-7d, the low-carbon steel failed in shear on a plane that is almost transverse—a typical failure for ductile material. The reason the fracture in Fig. 3-7d did not occur on a transverse plane is that under the large plastic twisting deformation before rupture (note the spiral lines indicating elements originally parallel to the axis of the bar), longitudinal elements were subjected to axial tensile loading because the grips of the testing machine would not permit the bar to shorten as the elements were twisted into spirals. This axial tensile stress (not shown in Fig. 3-5) changes the plane of maximum shearing stress from a transverse to an oblique plane (resulting in a warped surface of rupture).[2]

[2]The tensile stress is not entirely due to the grips because the plastic deformation of the outer elements of the bar is considerably greater than that of the inner elements. This results in a spiral tensile stress in the outer elements and a similar compressive stress in the inner elements.

3-5

TORSIONAL DISPLACEMENTS

Frequently the amount of twist in a shaft (or structural element) is of paramount importance, making the computation of the angle of twist a necessary accomplishment of the machine or structural designer. The fundamental approach to the problem, involving members satisfying the limitations of Section 3-3, is provided by the following equations:

$$\gamma_\rho = \frac{\rho\theta}{L} \quad \text{or} \quad \gamma_\rho = \frac{\rho d\theta}{dL} \tag{3-2}$$

$$\tau_\rho = \frac{T\rho}{J} \tag{3-5}$$

$$\tau_\rho = G\gamma_\rho \tag{1-26}$$

The second form of Eq. 3-2 is useful when the torque or the cross section varies as a function of position along the length of the shaft. Equation 3-2 is valid for both elastic and inelastic action. Equation 3-5 is the elastic torsion formula that provides the shearing stress τ_ρ on a transverse plane at a distance ρ from the axis of the shaft. Equation 1-26 is Hooke's law for shearing stresses. The last two expressions are limited to stresses below the proportional limit of the material (elastic action). The three equations can be combined to give several different relationships; for example,

$$\theta = \frac{\gamma_\rho L}{\rho} = \frac{\tau_\rho L}{\rho G} = \frac{\tau_c L}{cG} \tag{3-8a}$$

or

$$\theta = \frac{TL}{JG} \tag{3-8b}$$

The units of θ are radians in both the U.S. Customary and SI Systems. The first form of Eq. 3-8 will often be most useful in dual-specification problems (limiting θ and τ are specified).

The angle of twist determined from the above expressions is for a length of shaft of constant diameter sufficiently removed from sections where pulleys, couplings, or other mechanical devices are attached to make Saint-Venant's principle applicable. However, for practical purposes, it is customary to neglect local distortion at all connections and compute angles as though there were no discontinuities, and as though the torques applied were distributed over the cross sections in exactly the same manner in which the stresses given by the torsion formula are distributed. The following examples illustrate the application of the principles developed.

EXAMPLE 3-1

A hollow steel shaft with an outside diameter of 100 mm and an inside diameter of 75 mm is subjected to a pure torque of 7.5 kN·m. The modulus of rigidity G (shear modulus) for the steel is 80 GPa. Determine

(a) the maximum shearing stress in the shaft.

(b) the shearing stress on a transverse cross section at the inside surface of the shaft.

(c) the magnitude of the angle of twist in a 2-m length.

Solution

Equations 3-5 and 3-8 for the shearing stresses and angle of twist in a circular shaft subjected to a pure torque form of loading both contain the polar moment of inertia J of the cross section. Thus

$$J = \left(\frac{\pi}{32}\right)(D_o{}^4 - D_i{}^4) = \left(\frac{\pi}{32}\right)(100^4 - 75^4)$$

$$= 6.711(10^6) \text{ mm}^4 = 6.711(10^{-6}) \text{ m}^4$$

(a) The maximum shearing stress occurs on a transverse cross section at the outer surface of the shaft and is given by Eq. 3-5 as

$$\tau_{max} = \tau_c = \frac{Tc}{J} = \frac{7.5(10^3)(50)(10^{-3})}{6.711(10^{-6})}$$

$$= 55.9(10^6) \text{ N/m}^2 = \underline{55.9 \text{ MPa}} \qquad \text{Ans.}$$

(b) The shearing stress on a transverse cross section at the inner surface of the shaft is given by Eq. 3-5 as

$$\tau_\rho = \frac{T\rho}{J} = \frac{7.5(10^3)(37.5)(10^{-3})}{6.711(10^{-6})}$$

$$= 41.9(10^6) \text{ N/m}^2 = \underline{41.9 \text{ MPa}} \qquad \text{Ans.}$$

(c) The angle of twist in a 2-m length if given by Eq. 3-8b as

$$\theta = \frac{TL}{JG} = \frac{7.5(10^3)(2)}{6.711(10^{-6})(80)(10^9)}$$

$$= \underline{0.0279 \text{ rad}} \qquad \text{Ans.}$$

EXAMPLE 3-2

A solid steel shaft 14 ft long has a diameter of 6 in. for 9 ft of its length and a diameter of 4 in. for the remaining 5 ft. The shaft is in equilibrium when subjected to the three torques shown in Figure 3-8a. The modulus of rigidity (shear modulus) of the steel is 12,000 ksi. Determine

   (a) the maximum shearing stress in the shaft.

   (b) the rotation of end B of the 6-in. segment with respect to end A.

   (c) the rotation of end C with respect to end A.

Solution

   (a) In general, free-body diagrams should be drawn in order to eval-uate correctly the resisting torque. Such diagrams are shown in Figs. 3-8b and c, where in part b the shaft is cut by any transverse plane through the 6-in. segment, and T_6 is the resisting torque on this section. Similarly, in Fig. 3-8c the plane is passed through the 4-in. section, and T_4 is the resisting torque on this section. A torque diagram, such as the one shown in Fig. 3-8d, provides a pictorial representation of the levels of torque being transmitted by each of the sections and serves as an aid for stress and de-

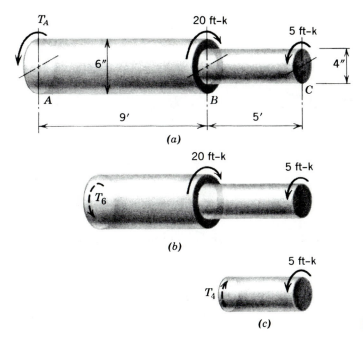

(a)

(b)

(c)

FIG. 3-8

FIG. 3-8*d–h*

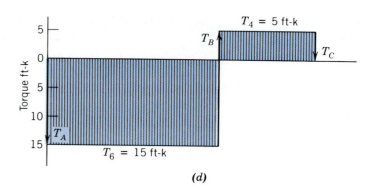

(d)

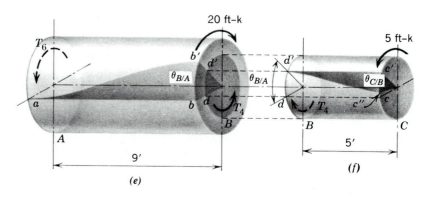

(e)   *(f)*

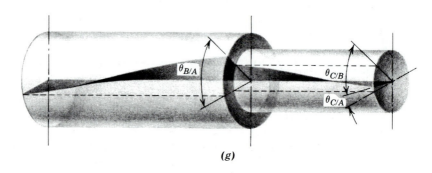

(g)

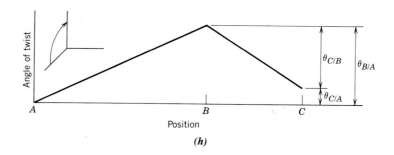

(h)

formation calculations. The location of the maximum shearing stress is not apparent; hence, the stress must be checked at both sections. Thus, from Fig. 3-8b or from Fig. 3-8d,

$$\Sigma M = 0, \quad T_6 = 20 - 5 = 15 \text{ ft-kips} = 15,000(12) \text{ in.-lb}$$

and

$$\tau_{max} = \frac{Tc}{J} = \frac{15,000(12)(3)}{(\pi/2)(3^4)} = \frac{40,000}{3\pi}$$

From Fig. 3-8c or from Fig. 3-8d,

$$\Sigma M = 0, \quad T_4 = 5 \text{ ft-kips} = 5000(12) \text{ in.-lb}$$

and

$$\tau_{max} = \frac{5000(12)(2)}{(\pi/2)(2^4)} = \frac{15,000}{\pi} > \frac{40,000}{3\pi}$$

Therefore, the maximum shearing stress is

$$\tau_{max} = 15,000/\pi = \underline{4770 \text{ psi}} \qquad \text{Ans.}$$

This stress is less than the shearing proportional limit of any steel; hence, the torsion formula applies. Note that had the larger torque been carried by the smaller section, the maximum stress would obviously occur in the small section, and only one stress determination would have been required.

(b) As an aid to visualizing the distortion of the shaft, the segments AB and BC and the torques acting on them are drawn separately in Figs. 3-8e and f, with the distortions greatly exaggerated. As the resisting torque of 15 ft-kips is transmitted from section to section along the shaft in segment AB, the section at B twists relative to the section at A by an amount $\theta_{B/A}$ and points b and d move to b' and d', respectively. Segment BC would rotate as a rigid body through the same angle, and point c would move to c', if there were no resisting torque being transmitted by the segment. However, a resisting torque of 5 ft-kips is transmitted by segment BC and this torque causes the part of the shaft between B and C to twist so that the section at C rotates counterclockwise relative to the section at B by an amount $\theta_{C/B}$. As segment BC of the shaft deforms, point c' moves back to c''. The resultant distortion for the entire shaft is pictorially shown in Fig. 3-8g. This distortion of the shaft can also be shown on an angle-of-twist diagram, as shown in Fig. 3-8h. The slope of the angle-of-twist diagram is constant in both segments of the shaft, since the term T/JG in Eq. 3-8b is constant in both segments. The direction of the twist is determined by noting that the torque of 15 ft-kips in segment AB causes the section at B to rotate clockwise relative to the section at A. Thus, a torque plotted below the line in Fig. 3-8d causes a clockwise rotation and a torque plotted above the line causes a counterclockwise rotation of a section with respect to a section to its left (for example, $\theta_{B/A}$).

When computing distortions, it is recommended that the two fundamental relationships of Eqs. 3-2 and 3-5 plus Hooke's law be used; thus,

$$\gamma_c = \frac{c\theta}{L} = \frac{\tau_c}{G} = \frac{Tc}{JG}$$

When this expression is applied to Fig. 3-8e, the twist in segment AB is obtained as

$$\theta_{B/A} = \frac{(20 - 5)(12)(9)(12)}{(\pi/2)(3^4)(12)(10^3)}$$

$$= \frac{40}{\pi(10^3)} = 12.73(10^{-3}) \text{ rad} \quad \text{Ans.}$$

(c) From Figs. 3-8f and g,

$$\theta_{C/A} = \theta_{B/A} - \theta_{C/B}$$

$$= \frac{40}{\pi(10^3)} - \frac{5(12)(5)(12)}{(\pi/2)(2^4)(12)(10^3)}$$

$$= \frac{40 - 37.5}{\pi(10^3)} = 0.796(10^{-3}) \text{ rad} \quad \text{Ans.}$$

EXAMPLE 3-3

Two 35-mm-diameter steel ($G = 80$ GPa) shafts from a factory drive system are connected with gears, as shown in Fig. 3-9. The diameters of gears B and C are 250 mm and 150 mm, respectively. If an input torque of 1.5 kN·m is applied at section A of shaft AB, determine

(a) the maximum shearing stress produced at section D of shaft CD.

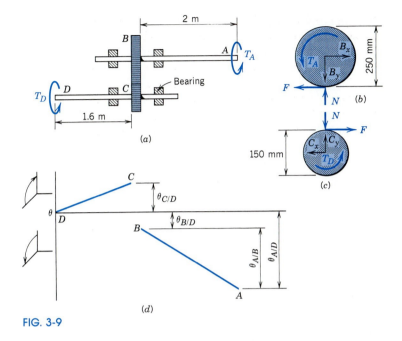

FIG. 3-9

(b) the rotation of section A of shaft AB relative to section D of shaft CD resulting from torsional deformations of the shafts.

Solution

(a) The output torque at section D of shaft CD can be determined from equilibrium considerations for the two shafts. As an aid for these considerations, free-body diagrams for gears B and C with their shafts attached are shown in Figs. 3-9b and c, respectively. The input torque T_A in shaft AB is transferred to shaft BC by means of the gear tooth force F shown in the two diagrams. Thus, from $\Sigma M = 0$ about the axis of each of the shafts:

For shaft AB

$$T_A - R_B F = 0 \qquad\qquad (a)$$

For shaft CD

$$T_D - R_C F = 0 \qquad\qquad (b)$$

Since the force F in Eqs. a and b must be equal,

$$T_D = (R_C/R_B)T_A = (75/125)(1.5)(10^3) = 900\text{N·m}$$

The maximum shearing stress at section D, which occurs on a transverse cross section at the outside surface of the shaft, can be determined by using Eq. 3-5. Thus

$$\tau_{max} = \tau_c = \frac{Tc}{J}$$

$$= \frac{900(35/2)(10^{-3})}{(\pi/32)(35^4)(10^{-12})}$$

$$= 106.9(10^6) \text{ N/m}^2 = \underline{106.9 \text{ MPa}} \qquad \text{Ans.}$$

(b) Since the required rotation $\theta_{A/D}$ is measured relative to the section at D, it is convenient to consider section D fixed. The quantities required for the determination of $\theta_{A/D}$ are illustrated on the angle of twist diagram shown in Fig. 3-9d. The rotation of the section at C relative to the section at D in shaft CD can be determined by using Eq. 3-8b. Thus

$$\theta_{C/D} = \frac{TL}{JG} = \frac{900(1.6)}{(\pi/32)(35^4)(10^{-12})(80)(10^9)}$$

$$= 0.12218 \text{ rad}$$

The teeth on gears B and C must move through the same arc length. Therefore,

$$S = R_B\theta_{B/D} = R_C\theta_{C/D}$$

From which

$$\theta_{B/D} = (R_C/R_B)(\theta_{C/D}) = (75/125)(0.12218) = 0.07331 \text{ rad}$$

The rotation of the section at A relative to the section at B in shaft AB can be determined by using Eq. 3-8b. Thus

$$\theta_{A/B} = \frac{TL}{JG} = \frac{1.5(10^3)(2)}{(\pi/32)(35^4)(10^{-12})(80)(10^9)}$$

$$= 0.2545 \text{ rad}$$

Finally,

$$\theta_{A/D} = \; = \theta_{B/D} + \theta_{A/B}$$

$$= 0.07331 + 0.2545$$

$$= 0.328 \text{ rad} = \underline{18.78°} \qquad \text{Ans.}$$

EXAMPLE 3-4

The solid circular tapered shaft of Fig. 3-10 is subjected to end torques applied in transverse planes. Determine the magnitude of the angle of twist in terms of T, L, G, and r. Assume elastic action and a slight taper.

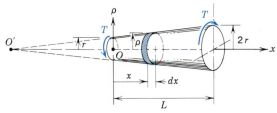

FIG. 3-10

Solution

Note that Eq. 3-2 was developed on the assumption that plane cross sections remain plane and all longitudinal elements have the same length. Neither of these assumptions is strictly valid for the tapered shaft, but if the taper is slight, the error involved is negligible; hence, Eq. 3-2 will be used as follows:

$$d\theta = \frac{\gamma}{\rho} \, dx$$

also

$$\gamma = \frac{\tau}{G}$$

and

$$\tau = \frac{T\rho}{J} = \frac{2T\rho}{\pi\rho^4} = \frac{2T}{\pi\rho^3}$$

Therefore,

$$d\theta = \frac{\tau}{G\rho} \, dx = \frac{2T}{G\pi\rho^4} \, dx$$

The radius can be expressed as a function of x; thus,

$$\rho = r + \frac{2r - r}{L} \, x = \frac{r}{L} \, (L + x)$$

which, substituted into the expression for $d\theta$, gives

$$d\theta = \frac{2TL^4}{G\pi r^4 (L + x)^4} \, dx$$

Integration gives

$$\theta = \frac{2TL^4}{G\pi r^4} \int_0^L \frac{dx}{(L + x)^4} = -\frac{2TL^4}{3G\pi r^4} \left(\frac{1}{8L^3} - \frac{1}{L^3} \right)$$

$$= \frac{7TL}{12G\pi r^4} \qquad\qquad \text{Ans.}$$

An alternate setup is to place the origin of coordinates at a distance L to the left of 0 in Fig. 3-10 (point $0'$). The function for ρ then becomes

$$\rho = \frac{r}{L}x$$

and

$$\theta = \frac{2TL^4}{G\pi r^4}\int_L^{2L}\frac{dx}{x^4} = \frac{7TL}{12G\pi r^4}$$

PROBLEMS

Note: Assume linearly elastic action for all of the following problems.

3-1* A solid circular steel shaft 2 in. in diameter is subjected to a pure torque of 18,000 in.-lb. The modulus of rigidity (shear modulus) for the steel is 12,000,000 psi. Determine

(a) the maximum shearing stress in the shaft.

(b) the magnitude of the angle of twist in a 6-ft length.

3-2* A hollow steel shaft has an outside diameter of 150 mm and an inside diameter of 100 mm. The shaft is subjected to a pure torque of 35 kN·m. The modulus of rigidity (shear modulus) for the steel is 80 GPa. Determine

(a) the shearing stress on a transverse cross section at the outside surface of the shaft.

(b) the shearing stress on a transverse cross section at the inside surface of the shaft.

(c) the magnitude of the angle of twist in a 2.5-m length.

3-3 A solid circular steel ($G = 12,000$ ksi) shaft that is 9 ft long will be subjected to a pure torque of 75 in.-kips. Determine the minimum diameter required if the shearing stress must not exceed 12,000 psi and the angle of twist must not exceed 0.075 rad.

3-4 A hollow circular steel ($G = 80$ GPa) shaft with an outside diameter of 100 mm will be subjected to a pure torque of 8 kN·m. Determine the maximum inside diameter that can be used if the shearing stress must not exceed 50 MPa and the angle of twist in a 3-m length must not exceed 0.04 rad.

3-5* A torque of 75 in.-kips is supplied to the steel ($G = 12,000$ ksi) factory drive shafts of Fig. P3-5 by a belt that drives pulley A. A torque of 45 in.-kips is taken off by pulley C. Shafts AB and BC are 4 ft and 3 ft long, respectively. For an allowable shearing stress of 10 ksi, determine

(a) the minimum permissible diameters for the two shafts.
(b) the rotation of pulley B with respect to pulley A if shaft AB has a 3.5-in. diameter.
(c) the rotation of pulley C with respect to pulley A if both shafts have 3-in. diameters.

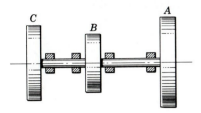

FIG. P3-5

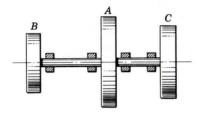

FIG. P3-6

3-6* A torque of 10 kN·m is supplied to the steel ($G = 80$ GPa) factory drive shaft of Fig. P3-6 by a belt that drives pulley A. A torque of 6.5 kN·m is taken off by pulley B and the remainder by pulley C. Shafts AB and AC are 1.75 m and 1.25 m long, respectively. For an allowable shearing stress of 75 MPa, determine

(a) the minimum permissible diameters for the two shafts.
(b) the rotation of pulley B with respect to pulley A if shaft AB has a diameter of 75 mm.
(c) the rotation of pulley C with respect to pulley B if both shafts have diameters of 80 mm.

3-7 A solid circular steel ($G = 12,000$ ksi) shaft with diameters, as shown in Fig. P3-7 is subjected to a torque T. The allowable shearing stress is 12,000 psi and the maximum allowable angle of twist in the 7-ft length is 0.06 rad. Determine the maximum allowable value of T.

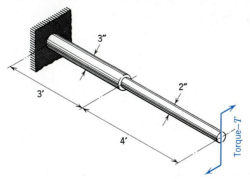

FIG. P3-7

3-8 A solid circular steel ($G = 80$ GPa) shaft is fastened securely to a solid circular bronze ($G = 40$ GPa) shaft, as indicated in Fig. P3-8. The allowable shearing stress is 70 MPa and the maximum allowable angle of twist in the 3.5-m length is 0.04 rad. Determine the maximum allowable value of T.

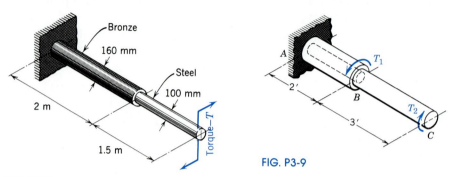

FIG. P3-8

FIG. P3-9

3-9* The shaft shown in Fig. P3-9 consists of a brass ($G = 5600$ ksi) tube AB that is securely connected to a solid stainless-steel ($G = 12,500$ ksi) bar BC. The tube AB has an outside diameter of 5 in. and an inside diameter of 2.5 in. Bar BC has an outside diameter of 3.5 in. Torques T_1 and T_2 are 100 in.-kips and 40 in.-kips, respectively, in the directions shown. Determine

(a) the maximum shearing stress in the shaft.

(b) the rotation of a section at C with respect to its no-load position.

3-10 The shaft shown in Fig. P3-9 consists of a hollow steel ($G = 12,000$ ksi) tube AB that is securely connected to a solid bronze ($G = 6500$ ksi) bar BC. Tube AB has an outside diameter of 4 in. and an inside diameter of 2 in. Bar BC has an outside diameter of 3 in. With the torques T_1 and T_2 applied as shown, the shearing stress in bar BC at a radius of 0.5 in. is 2 ksi. Determine

(a) the torque T_2.

(b) the torque T_1 if the rotation of a cross section at B with respect to its no-load position is 0.012 rad in the direction of torque T_1.

3-11* The solid circular shaft and the hollow tube shown in Fig. P3-11 are both attached to a rigid circular plate at the left end. A torque T_A applied to the right end of the shaft is resisted by a torque T_B at the right end of the tube. Both the shaft and the tube are made of an aluminum alloy ($G = 28$ GPa). If the shaft has a diameter of 50 mm and the tube has an outside diameter of 70 mm and an inside diameter of 60 mm, determine

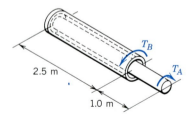

FIG. P3-11

(a) the maximum shearing stress in the shaft when $T_A = 2$ kN·m.

(b) the maximum shearing stress in the tube when $T_A = 2.5$ kN·m.

(c) the rotation of the right end of the shaft with respect to the right end of the tube when $T_A = 2.25$ kN·m.

3-12 The solid circular shaft and the hollow tube shown in Fig. P3-11 are both attached to a rigid circular plate at the left end. A torque $T_A = 2$ kN·m applied to the right end of the shaft is resisted by a torque T_B at the right end of the tube. The shaft is made of steel ($G = 80$ GPa) and the tube is made of an aluminum alloy ($G = 28$ GPa). If the shaft has a diameter of 50 mm and the tube has an outside diameter of 80 mm, determine

(a) the maximum inside diameter that can be used for the tube if the maximum shearing stress in the tube must be limited to 50 MPa.

(b) the maximum inside diameter that can be used for the tube if the rotation of the right end of the shaft with respect to the right end of the tube must be limited to 0.25 rad.

3-13* A torque of 40 ft-lb is applied through gear A to the left end of the gear train shown in Fig. P3-13. The diameters of gears B and C are 5 in. and 2 in., respectively. If the maximum shearing stresses in the aluminum alloy ($G = 3800$ ksi) shafts AB and CD are limited to 15 ksi, determine

(a) the minimum permissible diameter for shaft AB.

(b) the minimum permissible diameter for shaft CD.

(c) the maximum length for shaft CD if the rotation of a section at D with respect to a section at A must not exceed 1.5 rad.

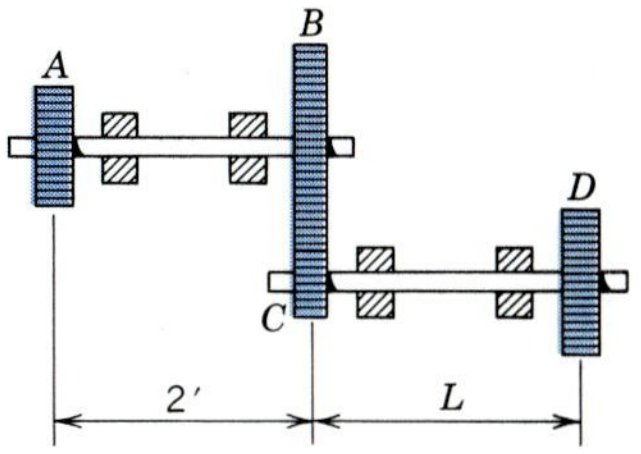

FIG. P3-13

3-14* A motor supplies a torque of 5.5 kN·m to the constant-diameter steel ($G = 80$ GPa) line shaft shown in Fig. P3-14. Three machines are driven by gears B, C, and D on the shaft and they require torques of 3 kN·m, 1.5 kN·m, and 1 kN·m, respectively. Determine

(a) the minimum diameter required if the maximum shearing stress in the shaft is limited to 100 MPa.

(b) the rotation of gear D with respect to the coupling at A if the coupling and gears are spaced at 2-m intervals and the shaft diameter is 75 mm.

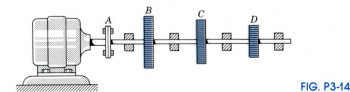

FIG. P3-14

3-15 A motor supplies a torque of 450 in.-kips to the steel (G = 12,000 ksi) line shaft shown in Fig. P3-14. Three machines are driven by gears B, C, and D on the shaft and they require torques of 200 in.-kips, 150 in.-kips, and 100 in.-kips, respectively. Determine

   (a) the diameter required in interval BC of the shaft if the maximum shearing stress must be limited to 20 ksi.

   (b) the rotation of gear D with respect to the coupling at A if the coupling and gears are spaced at 3-ft intervals and the shaft has a constant diameter of 5 in.

3-16 The elastic strength in shear for the aluminum alloy (G = 28 GPa) shaft ABC shown in Fig. P3-16 is 160 MPa. Torque is applied to the shaft through gear C and is removed through gears A and B. If the torque applied to gear C by the motor is 9 kN·m and the torque removed through gear A is 2 kN·m, determine,

   (a) the minimum permissible diameter for each section of the shaft if a factor of safety of 2 with respect to failure by slip is specified.

   (b) the rotation of gear A with respect to gear C if a shaft with a constant diameter of 75 mm is used and lengths L_1 and L_2 are

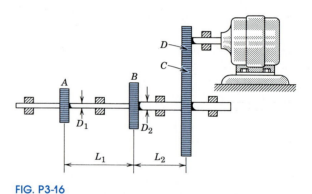

FIG. P3-16

3-17* A motor applies a torque to the shaft shown in Fig. P3-16 through gear C, which has a diameter of 8 in. Gears A and B remove torque from the shaft, with the torque at A being 12 in.-kips. The shafts are made of steel (G = 12,000 ksi) and D_1 = 2 in., D_2 = 3 in., L_1 = 4 ft, and L_2 = 2.5 ft. When the force between the teeth of gears C and D is 20 kips, determine

   (a) the maximum shearing stress in shaft AB.

   (b) the maximum shearing stress in shaft BC.

   (c) the rotation of gear B relative to gear C.

   (d) the rotation of gear A relative to gear C.

3-18* Torque is applied to the steel (G = 80 GPa) shaft shown in Fig. P3-16 through gear C and is removed through gears A and B. If the torque applied to gear C by the motor is 20 kN·m and the torque removed through gear B is 12 kN·m, determine

(a) the minimum permissible diameter for each section of the shaft if the maximum shearing stresses must not exceed 125 MPa.
(b) the minimum permissible uniform diameter for a shaft with L_1 = 1.5 m and L_2 = 1.25 m if the rotation of gear A relative to gear C must be less than 0.15 rad.

3-19 An aluminum alloy (G = 12,000 ksi) tube is to be used to transmit a torque in a control mechanism. The tube has an outside diameter of 1.50 in. and a wall thickness of 0.075 in. If the maximum compressive stress must be limited to 8000 psi, determine

(a) the maximum torque that can be applied.
(b) the angle of twist in a 3-ft length when a torque of 1000 in.-lb is applied.

3-20 The hollow circular aluminum alloy (G = 28 GPa) shaft shown in Fig. P3-20 has an outside diameter of 150 mm and an inside diameter of 75 mm. The 5-kN forces are applied with a rigid bar that is securely fastened to the right end of the shaft. Determine

(a) the maximum compressive stress in the shaft.
(b) the distance point B moves as the torque is applied.
(c) the maximum compressive stress in the shaft after the inside diameter is increased to 100 mm.

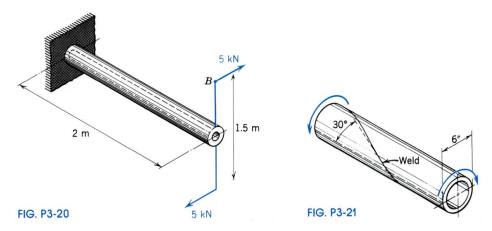

FIG. P3-20   5 kN   FIG. P3-21

3-21* A cylindrical tube is fabricated by butt-welding a ¼-in. steel plate along a spiral seam, as shown in Fig. P3-21. If the maximum compressive stress in the tube must be limited to 12,000 psi, determine

(a) the maximum torque T that can be applied to the tube.

(b) the factor of safety with respect to failure by fracture for the weld, when a torque of 200 in.-kips is applied, if the ultimate strengths of the weld metal are 30 ksi in shear and 50 ksi in tension.

3-22* The hollow circular steel ($G = 80$ GPa) shaft of Fig. P3-22 is in equilibrium under the torques indicated. Determine

(a) the minimum permissible outside diameter d if the maximum shearing stress in the shaft is not to exceed 100 MPa.

(b) the rotation of a section at D with respect to a section at A for a shaft with an outside diameter of 120 mm.

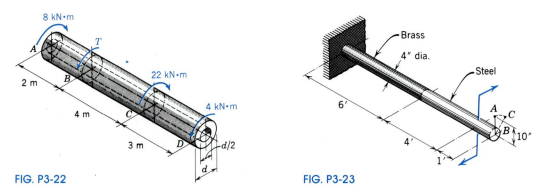

FIG. P3-22

3-23 The 4-in. diameter shaft shown in Fig. P3-23 is composed of brass ($G = 5000$ ksi) and steel ($G = 12,000$ ksi) sections that are rigidly connected. Determine the maximum allowable torque, applied as shown, if the maximum shearing stresses in the brass and the steel are not to exceed 8000 psi and 12,000 psi, respectively, and the distance AC, through which the end of the 10-in. pointer AB moves, is not to exceed 0.75 in.

3-24 A stepped steel ($G = 80$ GPa) shaft has the dimensions and is subjected to the torques shown in Fig. P3-24. Determine

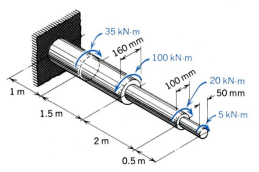

FIG. P3-24

(a) the maximum shearing stress on a section 2 m from the left end.

(b) the rotation of a section 3 m from the left end with respect to its no-load position.

(c) the rotation of the section at the right end of the shaft with respect to its no-load position.

3-25* The maximum shearing stress in the stepped steel ($G = 12,000$ ksi) shaft shown in Fig. P3-25 is 18,000 psi. Determine

(a) the magnitude of the torque T_3.

(b) the rotation of a section at B with respect to its no-load position.

(c) the rotation of a section at D with respect to its no-load position.

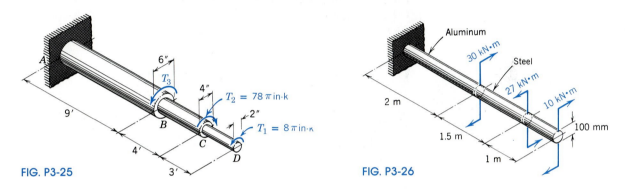

FIG. P3-25   FIG. P3-26

3-26* The 100-mm diameter shaft shown in Fig. P3-26 is composed of aluminum alloy ($G = 28$ GPa) and steel ($G = 80$ GPa) sections that are rigidly connected. Determine

(a) the maximum shearing stress in the shaft.

(b) the rotation of a section 3 m from the left end with respect to its no-load position.

(c) the rotation of a section at the right end with respect to its no-load position.

3-27 A steel ($G = 12,000$ ksi) shaft is loaded and supported as shown in Fig. P3-27. Determine

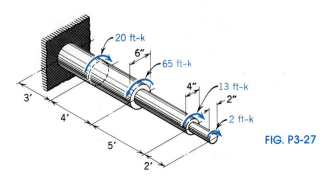

FIG. P3-27

(a) the maximum shearing stress in the shaft.

(b) the rotation of the right end with respect to its no-load position.

(c) the rotation of a section 5 ft from the left end with respect to its no-load position.

3-28 When the two torques are applied to the steel ($E = 200$ GPa and $G = 80$ GPa) shaft of Fig. P3-28, point A moves 3 mm in the direction indicated by torque T_1. A strain gage bonded to the surface of the 50-mm shaft at an angle of 45° with the axis of the shaft indicates a strain of 750 μ. Determine

(a) the maximum tensile stress in the 50-mm shaft.

(b) the magnitudes of the two torques.

(c) the rotation of the right end of the 100-mm shaft with respect to its no-load position.

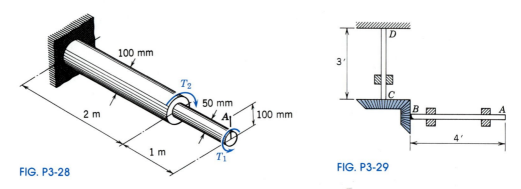

FIG. P3-28   FIG. P3-29

3-29* A torque of 10 in.-kips is applied to the right end of shaft AB of Fig. P3-29. The mean diameter of bevel gear C is twice that of bevel gear B. If the shafts are made of steel ($G = 12,000$ ksi) and if the maximum shearing stress must not exceed 15 ksi in either shaft, determine

(a) the minimum diameters required for the shafts.

(b) the rotation of a section at A relative to its no-load position when the diameters of part (a) are used.

3-30* The motor shown in Fig. P3-30 supplies a torque of 45 kN·m to shaft AB. Two machines are powered by gears D and E. The torque delivered by gear E to the machine is 8 kN·m. Shafts AB and DCE are made of steel ($G = 80$ GPa) and have 150-mm and 80-mm diameters, respectively. If the diameters of gears B and C are 450 mm and 150 mm, respectively, determine

(a) the maximum shearing stress in shaft AB.

(b) the maximum shearing stress in shaft DCE.

(c) the rotation of gear E relative to gear D.

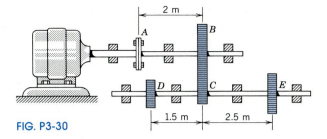

FIG. P3-30

(d) the rotation of gear D relative to the coupling at A.

3-31 A motor supplies a torque of 31.4 in.-kips to the steel (G = 12,000 ksi) shaft AB shown in Fig. P3-31. Two machines are powered by gears D and E with the torque removed at gear E being 5.7 in.-kips. The diameters of gears B and C are 10 in. and 5 in., respectively. If the maximum shearing stresses in shafts AB and DCE must be limited to 20 ksi, determine

(a) the minimum satisfactory diameter for shaft AB.

(b) the minimum satisfactory diameter for shaft DCE.

(c) the rotation of gear B with respect to the coupling at A if a shaft with a 2.5 in. diameter is used.

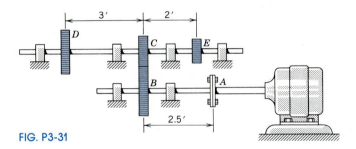

FIG. P3-31

3-32 The motor shown in Fig. P3-31 supplies a torque of 50 in.-kips to shaft AB. Two machines are powered by gears D and E. The torque delivered by gear E to the machine is 10 in.-kips. Shafts AB and DCE are made of steel (G = 12,000 ksi) and have 3-in. diameters. If the diameters of gears B and C are 9 in. and 6 in., respectively, determine

(a) the rotation of gear E relative to gear D.

(b) the rotation of gear D relative to the coupling at A.

3-33* The hollow tapered shaft of Fig. P3-33 has a constant wall thickness t that is small with respect to the radius r. Determine the angle of twist for a constant torque T in terms of T, L, G, t, and r. The approximate expression for the polar moment of inertia of a hollow thin-walled circular section ($J = 2\pi r^3 t$) may be used.

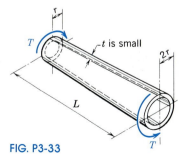

FIG. P3-33

3-34 The hollow circular tapered shaft of Fig. P3-34 is subjected to a constant torque T. Determine the angle of twist in terms of T, L, G, and r.

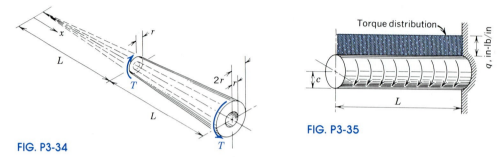

FIG. P3-34

FIG. P3-35

3-35* The solid cylindrical shaft of Fig. P3-35 is subjected to a uniformly distributed torque of q in.-lb per inch of length. Determine, in terms of q, L, G, and c, the rotation of the left end under the applied torque.

3-36 The solid cylindrical shaft of Fig. P3-36 is subjected to a distributed torque that varies linearly from zero at the left end to q N·m per meter of length at the right end. Determine, in terms of q, L, G, and c, the rotation of the left end under the applied torque.

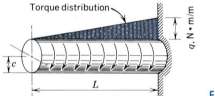

FIG. P3-36

3-6

POWER TRANSMISSION

Almost anyone will recognize that one of the most common uses for the circular shaft is the transmission of power; therefore, no discussion of torsion would be adequate without including this topic. Power is the time rate of doing work, and the basic relationship for work done by a constant torque is $W_k = T\phi$, where W_k is work and ϕ is the angular displacement of the shaft in radians. The time derivative of this expression gives

$$\frac{dW_k}{dt} = T\frac{d\phi}{dt} = T\omega \qquad (3-9)$$

where dW_k/dt is power in foot-pounds per minute (or similar units), T is a constant torque in foot-pounds, and ω is the angular velocity (of the shaft) in radians per minute. All units, of course, may be changed to any other consistent set of units. Since the angular velocity is usually given in revolutions per minute (rpm), the conversion of revolutions to radians will often be necessary. Also, in the English system, power is usually given in units of horsepower, and the relation 1 hp = 33,000 ft-lb per min will be found useful. In the metric (SI) system, power is given in watts (N·m/sec). An example of shaft design follows.

EXAMPLE 3-5

A diesel engine for a small commercial boat operates at 200 rpm and delivers 800 hp through a gearbox with a ratio of 4 to 1 to the propeller shaft as shown in Fig. 3-11. Both the shaft from engine to gearbox and the propeller shaft are to be solid and made of heat-treated alloy steel. Determine the minimum permissible diameters for the two shafts if the allowable stress is 20 ksi and the angle of twist in a 10-ft length of the propeller shaft is not to exceed 4°. Neglect power loss in the gearbox and assume (incorrectly because of thrust stresses) that the propeller shaft is subjected to pure torsion.

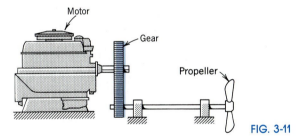

FIG. 3-11

Solution

The first step is the determination of the torques to which the shafts are to be subjected. By means of the expression, power = $T\omega$, the torques are obtained as follows:

$$800(33,000) = T_1(200)(2\pi)$$

from which

$$T_1 = 21008 \text{ ft-lb}$$

which is the torque at the crankshaft of the engine. Because the propeller shaft speed is four times that of the crankshaft and power loss in the gearbox is to be neglected, the torque on the propeller shaft is one-fourth that on the crankshaft and is equal to 5252 ft·lb. The torsion formula can be used to determine the shaft sizes necessary to satisfy the stress specification. For the main shaft,

$$\frac{J}{c} = \frac{T}{\tau} = \frac{21008(12)}{20(10^3)} = \frac{(\pi/2)c_1^4}{c_1}$$

$$c_1^3 = 8.02 \quad \text{and} \quad c_1 = 2.002 \text{ in.}$$

or the shaft from engine to gearbox should be

$$d_m = 2c_1 = \underline{4.00 \text{ in.}} \qquad \text{Ans.}$$

The torque on the propeller shaft is one-fourth that on the main shaft, and this is the only change in the expression for c_1^3; therefore,

$$c_2^3 = 8.02/4 \quad \text{and} \quad c_2 = 1.261 \text{ in.}$$

The size of propeller shaft necessary to satisfy the distortion specification will be determined as follows:

$$\gamma = \frac{c\theta}{L} = \frac{\tau}{G} = \frac{Tc}{JG}$$

from which

$$\frac{\theta}{L} = \frac{T}{JG}$$

$$\frac{4(\pi/180)}{10(12)} = \frac{5252(12)}{(\pi c_3^4/2)(12)(10^6)}$$

$$c_3^4 = 5.75 \quad \text{and} \quad c_3 = 1.548 > 1.261$$

Therefore, the propeller shaft must be

$$d_p = 2c_3 = \underline{3.10 \text{ in.}} \qquad \text{Ans.}$$

PROBLEMS

> *Note:* In the following problems, the specified allowable stresses are all below the shearing proportional limits of the materials.

3-37* Determine the horsepower that a 10-in.-diameter shaft can transmit at 200 rpm if the maximum shearing stress in the shaft must be limited to 15,000 psi.

3-38* The shaft of a diesel engine is being designed to transmit 250 kW at 240 rpm. Determine the minimum diameter required if the maximum shearing stress in the shaft is not to exceed 80 MPa.

3-39* A steel ($G = 12,000$ ksi) shaft with a 4-in. diameter must not twist more than 0.06 rad in a 20-ft length. Determine the maximum power that the shaft can transmit at 270 rpm.

3-40* A solid circular steel (G = 80 GPa) shaft transmits 225 kW at 180 rpm. Determine the minimum diameter required if the angle of twist in a 3-m length must not exceed 0.035 rad.

3-41 A solid circular steel (G = 12,000 ksi) shaft is 3 in. in diameter and 4 ft long. If the maximum shearing stress must be limited to 12.5 ksi and the angle of twist over the 4-ft length must be limited to 0.030 rad, determine the maximum horsepower that this shaft can deliver

(a) when rotating at 225 rpm.

(b) when rotating at 500 rpm.

3-42 A hollow steel (G = 80 GPa) shaft 3 m long has an outside diameter of 100 mm and an inside diameter of 60 mm. The maximum shearing stress in the shaft must be limited to 80 MPa. Determine

(a) the maximum power that can be transmitted by the shaft if the speed must be limited to 200 rpm.

(b) the magnitude of the angle of twist in a 2-m length of the shaft when 150 kW is being transmitted at 150 rpm.

3-43* The hydraulic turbines in a water-power plant rotate at 60 rpm and are rated at 20,000 hp with an overload capacity of 25,000 hp. The vertical shaft between the turbine and the generator has a 30-in. diameter and is 20 ft long. Determine

(a) the maximum shearing stress in the shaft at rated load.

(b) the maximum shearing stress in the shaft at maximum overload.

(c) the angle of twist in the steel (G = 12,000 ksi) shaft at maximum overload.

3-44* A solid circular steel (G = 80 GPa) shaft 3 m long transmits 250 kW at a speed of 500 rpm. If the allowable shearing stress is 70 MPa and the allowable angle of twist is 0.045 rad, determine

(a) the minimum permissible diameter for the shaft.

(b) the speed at which this power can be delivered if the stress is not to exceed 50 MPa in a shaft with a diameter of 75 mm.

3-45 The engine of an automobile supplies 162 hp at 3800 rpm to the drive shaft. If the maximum shearing stress in the drive shaft must be limited to 5 ksi, determine

(a) the minimum diameter required for a solid drive shaft.

(b) the maximum inside diameter permitted for a hollow drive shaft if the outside diameter is 3 in.

(c) the percent savings in weight realized if the hollow shaft is used instead of the solid shaft.

3-46 A rear axle of an automobile transmits 28 kW at 900 rpm to a rear wheel. Determine

(a) the minimum diameter required if the maximum shearing stress in the axle is not to exceed 60 MPa.

(b) the angle of twist in a 1-m length if the rear axle is made of steel (G = 80 GPa) and has a diameter of 30 mm.

3-47* A motor delivers 200 hp at 250 rpm to gear B of the factory drive shaft shown in Fig. P3-47. Gears A and C transfer 120 hp and 80 hp, respectively, to operating machinery in the factory. Determine

(a) the maximum shearing stress in shaft AB.

(b) the maximum shearing stress in shaft BC.

(c) the rotation of gear C with respect to gear A if the shafts are both made of steel (G = 12,000 ksi).

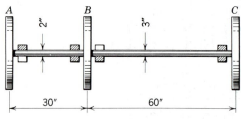

FIG. P3-47

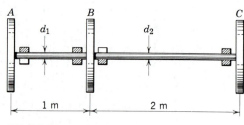

FIG. P3-48

3-48* A motor supplies 200 kW at 250 rpm to gear A of the factory drive shaft shown in Fig. P3-48. Gears B and C transfer 125 kW and 75 kW, respectively, to operating machinery in the factory. For an allowable shearing stress of 75 MPa, determine

(a) the minimum permissible diameter d_1 for shaft AB.

(b) the minimum permissible diameter d_2 for shaft BC.

(c) the rotation of gear C with respect to gear A if both shafts are made of steel (G = 80 GPa) and have diameters of 75 mm.

3-49 A hollow shaft of aluminum alloy (G = 4000 ksi) is to transmit 1200 hp at 1500 rpm. The shearing stress is not to exceed 15 ksi and the angle of twist is not to exceed 0.20 rad in a 10-ft length. Determine the minimum permissible outside diameter if the inside diameter is to be three-fourths of the outside diameter.

3-50 A hollow shaft of aluminum alloy (G = 28 GPa) is to transmit 1200 kW at 1800 rpm. The maximum shearing stress must not exceed 100 MPa and the angle of twist in a 3-m length must not exceed 0.20 rad. Prepare a design curve showing maximum inside diameter as a function of outside diameter ($d_o < 150$ mm) for the cross sections that are satisfactory for this application.

3-51* A motor delivers 350 hp at 1800 rpm to a gear box that reduces the speed to 200 rpm to drive a ball mill. If the maximum shearing stress in the shafts is not to exceed 15 ksi, determine

   (a) the required diameter for the shaft between the motor and the gear box.

   (b) the required diameter for the shaft between the gear box and the ball mill.

   (c) the rotation, resulting from the transmitted torques, of a cross section of the shaft at the ball mill with respect to a cross section of the shaft at the motor if the shafts are made of steel ($G = 12,000$ ksi) and each shaft is 10 ft long and has the minimum permissible diameter.

3-52* A motor delivers 225 kW at 180 rpm to a gear box that increases the speed to 900 rpm to drive a blower. If the maximum shearing stress in the shafts is not to exceed 70 MPa, determine

   (a) the minimum permissible diameter for the shaft between the motor and the gear box.

   (b) the minimum permissible diameter for the shaft between the gear box and the blower.

   (c) the rotation, resulting from the transmitted torques, of a cross section of the shaft at the blower with respect to a cross section of the shaft at the motor if the shafts are made of steel ($G = 80$ GPa) and each shaft is 2 m long and has the minimum permissible diameter.

3-53 The motor shown in Fig. P3-53 develops 100 hp at a speed of 360 rpm. Gears A and B deliver 40 hp and 60 hp, respectively, to operating units in a factory. If the maximum shearing stress in the shafts must be limited to 12 ksi, determine

   (a) the minimum satisfactory diameter for the motor shaft.

   (b) the minimum satisfactory diameter for the power shaft.

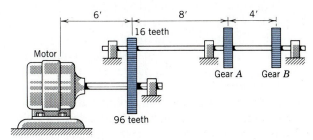

FIG. P3-53

(c) the rotation, resulting from the transmitted torques, of gear *B* with respect to the motor when the shafts are made of steel (*G* = 12,000 ksi) and both have the minimum satisfactory diameters.

3-54 A motor provides 180 kW of power at 400 rpm to the drive shafts shown in Fig. P3-54. The maximum shearing stress in the three solid steel (*G* = 80 GPa) shafts must not exceed 70 MPa. Gears *A, B,* and *C* supply 40 kW, 60 kW, and 80 kW, respectively, to operating units in the plant. Determine

(a) the minimum satisfactory diameter for shaft *A*.

(b) the minimum satisfactory diameter for shaft *B*.

(c) the minimum satisfactory diameter for shaft *C*.

(d) the rotation of gear *A* with respect to gear *C* when 75-mm diameter shafts are used.

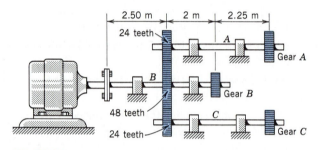

FIG. P3-54

3-7

STATICALLY INDETERMINATE MEMBERS

In all problems discussed in the preceding sections of this chapter, the resisting torque at any cross section of the member could be determined by using an equilibrium equation. Occasionally, torsionally loaded members are so constructed and loaded that the number of independent equilibrium equations is less than the number of unknowns. When this occurs, the member is statically indeterminate and it becomes necessary to develop a sufficient number of distortion equations so that the total number of independent equations agrees with the number of unknowns to be determined.

Since the distortion equations will involve the angles of twist, a simplified twist diagram will often be of assistance in obtaining the correct equations. The following examples will serve to illustrate the application of the principles.

EXAMPLE 3-6

The circular shaft AC of Fig. 3-12a is fixed to rigid walls at A and C. The solid section AB is made of annealed bronze and the hollow section BC is made of aluminum alloy 2024-T4. There is no stress in the shaft before the 30-kN·m torque T is applied. Determine the maximum shearing stresses in both the bronze and aluminum portions of the shaft after the torque T is applied.

Solution

A free-body diagram of the shaft is shown in Fig. 3-12b. The torques T_A and T_C at the supports are unknown. A summation of moments about the axis of the shaft, as shown in the torque diagram of Fig. 3-12c, gives

$$T_A + T_C = 30(10^3) \qquad (a)$$

This is the only independent equation of equilibrium; therefore, the problem is statically indeterminate. A second equation can be obtained from the deformation of the shaft, since the left and right portions of the shaft undergo the same angle of twist as shown in Fig. 3-12d. Thus,

$$\theta_{B/A} = \theta_{B/C}$$

Since Eq. a is expressed in terms of T_A and T_C, the convenient form of the angle of twist equation for use in this example is Eq. 3-8b. Thus,

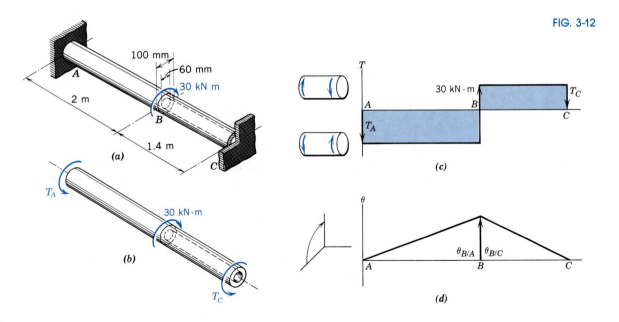

FIG. 3-12

$$\frac{T_A L_{AB}}{G_{AB} J_{AB}} = \frac{T_C L_{BC}}{G_{BC} J_{BC}}$$

From Appendix A

$$G_{AB} = 45 \text{ GPa} \quad \text{and} \quad G_{BC} = 28 \text{ GPa}$$

Also

$$J_{AB} = (\pi/32)(100^4) = 9.817(10^6) \text{ mm}^4 = 9.817(10^{-6}) \text{ m}^4$$
$$J_{BC} = (\pi/32)(100^4 - 60^4) = 8.545(10^6) \text{ mm}^4 = 8.545(10^{-6}) \text{ m}^4$$

Therefore,

$$\frac{T_A(2)}{45(10^9)(9.817)(10^{-6})} = \frac{T_C(1.4)}{28(10^9)(8.545)(10^{-6})}$$

From which

$$T_C = 0.7737 T_A \qquad\qquad (b)$$

Solving Eqs. a and b simultaneously yields

$$T_A = 16914 \text{ N·m} = 16.914 \text{ kN·m}$$
$$T_C = 13086 \text{ N·m} = 13.086 \text{ kN·m}$$

The stresses in the two portions of the shaft can then be obtained by using Eq. 3-5. Thus

$$\tau_{AB} = \frac{T_A c_{AB}}{J_{AB}} = \frac{16.914(10^3)(50)(10^{-3})}{9.817(10^{-6})}$$
$$= 86.1(10^6) \text{ N/m}^2 = \underline{86.1 \text{ MPa}} \qquad \text{Ans.}$$

$$\tau_{BC} = \frac{T_C c_{BC}}{J_{BC}} = \frac{13.086(10^3)(50)(10^{-3})}{8.545(10^{-6})}$$
$$= 76.6(10^6) \text{ N/m}^2 = \underline{76.6 \text{ MPa}} \qquad \text{Ans.}$$

These maximum stresses, which occur at the surface of the shaft, do not exceed the proportional limits of the materials; therefore, only elastic behavior is involved and the results are valid.

EXAMPLE 3-7

A hollow circular aluminum alloy 2024-T4 cylinder has a 0.4%C hot-rolled steel core as shown in Fig. 3-13a. The steel and aluminum parts are securely connected at the ends. If the working stresses in the steel and aluminum must be limited to 14 ksi and 10 ksi, respectively, determine

   (a) the maximum torque T that can be applied to the right end of the composite shaft.
   (b) the rotation of the right end of the composite shaft when the torque of part (a) is applied.

Solution

A free-body diagram of the shaft is shown in Fig. 3-13b. Since torques and stresses will be related by the torsion formula, which is limited to cross sections of homogeneous material, two unknown torques, the torque in the aluminum T_a and the torque in the steel T_s have been placed on the left end of the shaft. Summing moments with respect to the axis of the shaft yields

$$T_a + T_s = T \qquad (a)$$

Since Eq. a is the only independent equation of equilibrium, the problem is statically indeterminate. A second equation can be obtained from the deformation of the shaft. The fact that the shaft is nonhomogeneous does not invalidate the assumptions of plane cross sections remaining plane and diameters remaining straight. As a result, strains remain proportional to the distance from the axis of the shaft; however, stresses are not proportional to the radii throughout the entire cross section, since G is not single-valued. The steel and aluminum parts of the shaft experience the same angle of twist because of the secure connections at the ends. Thus,

$$\theta_s = \theta_a$$

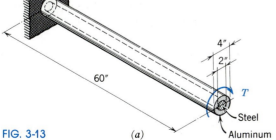

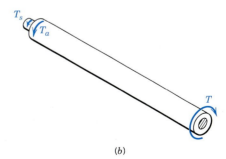

FIG. 3-13   (a)   (b)

Since maximum shearing stresses are specified, the convenient form of the angle of twist equation for use in this example is Eq. 3-8a. Thus,

$$\frac{\tau_s L_s}{G_s c_s} = \frac{\tau_a L_a}{G_a c_a}$$

From Appendix A

$$G_s = 11.6(10^6) \text{ psi} \quad \text{and} \quad G_a = 4.00(10^6) \text{ psi}$$

Therefore,

$$\frac{\tau_s(60)}{11.6(10^6)(1)} = \frac{\tau_a(60)}{4.0(10^6)(2)}$$

From which

$$\tau_s = 1.45 \, \tau_a \qquad\qquad (b)$$

It is obvious from Eq. b that the shearing stress in the steel controls; therefore,

$$\tau_s = 14 \text{ ksi}$$

$$\tau_a = 14/1.45 = 9.655 \text{ ksi}$$

(a) Once the maximum shearing stresses in the steel and aluminum portions of the shaft are known, Eq. 3-5 can be used to determine the torques transmitted by the two parts of the shaft. Thus

$$T_s = \frac{\tau_s J_s}{c_s} = \frac{14000(\pi/32)(2^4)}{1}$$

$$= 21,990 \text{ in.-lb}$$

$$T_a = \frac{\tau_a J_a}{c_a} = \frac{9655(\pi/32)(4^4 - 2^4)}{2}$$

$$= 113,750 \text{ in.-lb}$$

From Eq. a

$$T = T_a + T_s = 21,990 + 113,750$$

$$= 135,740 \text{ in.-lb} = \underline{135.7 \text{ in.-kip}} \qquad \text{Ans.}$$

(b) The rotation of the right end of the shaft with respect to its no-load position can be determined by using either Eq. 3-8a or Eq. 3-8b. If Eq. 3-8a is used,

$$\theta = \theta_a = \theta_s = \frac{\tau_s L_s}{G_s c_s}$$

$$= \frac{14000(60)}{11.6(10^6)(1)}$$

$$= \underline{0.0724 \text{ rad}} \qquad \text{Ans.}$$

EXAMPLE 3-8

The torsional assembly shown in Fig. 3-14a consists of a solid cold-rolled bronze shaft CD and a hollow 2024-T4 aluminum alloy shaft EF, which has a 0.4%C hot-rolled steel core. The ends C and F are fixed to rigid walls and the steel core of shaft EF is connected to the flange at E, so that the aluminum and steel parts act as a unit. The two flanges D and E are bolted together, and the bolt clearance permits flange D to rotate through 0.03 rad before EF carries any of the load. Determine the maximum shearing stress in each of the shaft materials when the torque $T = 54$ kN·m is applied to flange D.

Solution

A free-body diagram for the assembly is shown in Fig. 3-14b. An unknown torque T_B is shown at the left support and two unknown torques T_A and T_S are shown at the right support. Summing moments with respect to the axis of the shaft, as shown in the torque diagram of Fig. 3-14c, gives

$$T_B + T_A + T_S = 54(10^3) \qquad (a)$$

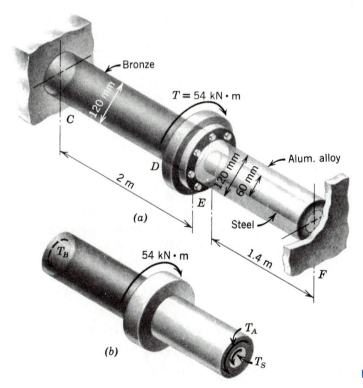

(a)

(b)

FIG. 3-14a,b

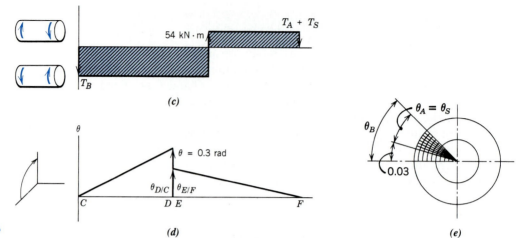

FIG. 3-14c,d,e

Equation a is the only independent equilibrium equation that can be written for this problem. Since there are three unknown torques, two distortion equations are needed to solve the problem. Two different types of angle-of-twist diagrams are shown in Figs. 3-14d and 3-14e. Angles of rotation for all cross sections of the shafts are shown in Fig. 3-14d. Relationships between angles $\theta_{D/C}$, $\theta_{E/F}$, and the rotation $\theta = 0.03$ rad permitted by the bolt clearance are required for solution of this example. The same quantities are shown in a polar form of representation in Fig. 3-14e. The angle of twist in the bronze shaft θ_B is the same as the rotation of coupling D with respect to the support at C. The angles of twist in the aluminum and steel shafts θ_A and θ_S, respectively, are equal and the same as the rotation of coupling E with respect to the support at F. Thus, the two distortion equations required for solution of the problem are

$$\theta_B = \theta_A + 0.03 \qquad (b)$$

and

$$\theta_A = \theta_S \qquad (c)$$

Equations a, b, and c can be written in terms of the same three unknowns (torque, angle, or stress) and solved simultaneously. Since maximum stresses are required, Eqs. a, b, and c will be written in terms of the maximum stress in each material by using Eqs. 3-5 and 3-8b.

From Appendix A

$$G_A = 28 \text{ GPa}$$
$$G_S = 80 \text{ GPa}$$
$$G_B = 45 \text{ GPa}$$

Also

$$J_A = (\pi/32)(120^4 - 60^4) = 19.085(10^6) \text{ mm}^4 = 19.085(10^{-6}) \text{ m}^4$$
$$J_B = (\pi/32)(120^4) = 20.36(10^6) \text{ mm}^4 = 20.36(10^{-6}) \text{ m}^4$$
$$J_S = (\pi/32)(60^4) = 1.2723(10^6) \text{ mm}^4 = 1.2723(10^{-6}) \text{ m}^4$$

Thus,

$$\frac{\tau_B J_B}{c_B} + \frac{\tau_A J_A}{c_A} + \frac{\tau_S J_S}{c_S} = 54(10^3)$$

$$\frac{\tau_B(20.36)(10^{-6})}{60(10^{-3})} + \frac{\tau_A(19.085)(10^{-6})}{60(10^{-3})} + \frac{\tau_S(1.2723)(10^{-6})}{30(10^{-3})} = 54(10^3)$$

from which

$$16\tau_B + 15\tau_A + 2\tau_S = 2.546(10^9) \qquad (d)$$

Similarly,

$$\frac{\tau_B L_B}{G_B c_B} = \frac{\tau_A L_A}{G_A c_A} + 0.03$$

$$\frac{\tau_B(2)}{45(10^9)(60)(10^{-3})} = \frac{\tau_A(1.4)}{28(10^9)(60)(10^{-3})} + 0.03$$

from which

$$8\tau_B = 9\tau_A + 324(10^6) \qquad (e)$$

And,

$$\frac{\tau_A L_A}{G_A c_A} = \frac{\tau_S L_S}{G_S c_S}$$

$$\frac{\tau_A(1.4)}{28(10^9)(60)(10^{-3})} = \frac{\tau_S(1.4)}{80(10^9)(30)(10^{-3})}$$

from which

$$7\tau_S = 10\tau_A \qquad (f)$$

Solving Eqs. d, e, and f simultaneously yields

$$\tau_A = 52.93(10^6) \text{ N/m}^2 = \underline{52.9 \text{ MPa}} \qquad \text{Ans.}$$
$$\tau_B = 100.1(10^6) \text{ N/m}^2 = \underline{100.1 \text{ MPa}} \qquad \text{Ans.}$$
$$\tau_S = 75.62(10^6) \text{ N/m}^2 = \underline{75.6 \text{ MPa}} \qquad \text{Ans.}$$

The elastic strengths in shear for most engineering materials are approximately six-tenths the elastic strengths in tension. The above stresses are all below these limits; therefore, the elastic solution is valid and the maximum shearing stress in the assembly is at the surface of the bronze shaft.

PROBLEMS

Note: In the following problems, the specified allowable stresses are all below the shearing proportional limits of the materials.

3-55* A hollow circular brass (G = 5600 ksi) tube with an outside diameter of 4 in. and an inside diameter of 2 in. is attached at the ends to a solid 2-in.-diameter steel (G = 12,000 ksi) core as shown in Fig. P3-55. Determine the maximum shearing stress in the tube when the composite shaft is transmitting a torque of 10 ft-kips.

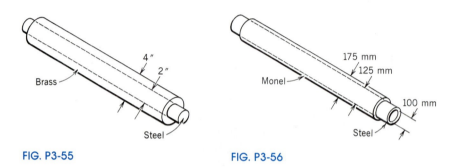

FIG. P3-55   FIG. P3-56

3-56* A steel (G = 80 GPa) tube with an inside diameter of 100 mm and an outside diameter of 125 mm is encased in a Monel (G = 65 GPa) tube with an inside diameter of 125 mm and an outside diameter of 175 mm as shown in Fig. P3-56. The tubes are connected at the ends to form a composite shaft. The shaft is subjected to a torque of 15 kN·m. Determine

(a) the maximum shearing stress in each material.

(b) the angle of twist in a 5-m length.

3-57 A 3-in.-diameter cold-rolled steel (G = 11,600 ksi) shaft, for which the maximum allowable shearing stress is 15 ksi, exhibited severe corrosion in a certain installation. It is proposed to replace the shaft with one in which an aluminum alloy (G = 4000 ksi) tube $\frac{1}{4}$-in. thick is bonded to the outer surface of the cold-rolled steel shaft to produce a composite shaft. If the maximum allowable shearing stress in the aluminum alloy shell is 12 ksi, determine

(a) the maximum torque that the original shaft can transmit.

(b) the maximum torque that the replacement shaft can transmit.

3-58 A composite shaft consists of a bronze (G = 45 GPa) sleeve with an outside diameter of 80 mm and an inside diameter of 60 mm over a solid aluminum (G = 28 GPa) rod with an outside diameter of 60 mm.

$J_n = \frac{\pi}{32}(d_{oi}^4 - d_{in}^4)$

$J_s =$ - - - - - -

① $T = 15 = T_m + T_s$

$\theta_m = \theta_s$

$\frac{T_m L}{GJ} = \frac{T_s L}{GJ}$ ②

a) $T_s = \frac{Tc}{J}$

b) $\theta = \theta_m = \theta_s = \frac{T_s L}{GJ}$

If the allowable shearing stress in the bronze is 150 MPa, determine

(a) the maximum torque T that can be transmitted by the composite shaft.

(b) the maximum shearing stress in the aluminum rod when the maximum torque is being transmitted,

3-59* The 2-in.-diameter steel (G = 12,000 ksi) shaft shown in Fig. P3-59 is fixed to rigid walls at both ends. When a torque of 3500 ft-lb is applied as shown, determine

(a) the maximum shearing stress in the shaft.

(b) the angle of rotation of the section where the torque is applied with respect to its no-load position.

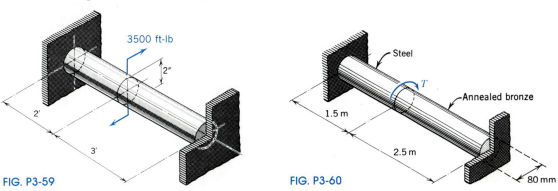

FIG. P3-59   FIG. P3-60

3-60* Two 80-mm-diameter solid circular steel (G = 80 GPa) and bronze (G = 45 GPa) shafts are rigidly connected and supported, as shown in Fig. P3-60. A torque T is applied at the junction of the two shafts as indicated. The allowable shearing stresses are 125 MPa for the steel and 40 MPa for the bronze. Determine

(a) the maximum torque T that can be applied.

(b) the angle of rotation of the section where the torque is applied with respect to its no-load position.

3-61 The composite shaft shown in Fig. P3-61 consists of a solid brass (G = 5600 ksi) core with an outside diameter of 2 in. covered by a steel (G = 11,600 ksi) tube with an inside diameter of 2 in. and a wall thickness of $\frac{1}{2}$ in., which is in turn covered by an aluminum alloy (G = 4000 ksi) sleeve with an inside diameter of 3 in. and a wall thickness of $\frac{1}{4}$ in. The three materials are bonded so that they act as a unit. Determine

(a) the maximum shearing stress in each material when the assembly is transmitting a torque of 10 ft-kips.

(b) the angle of twist in a 10-ft length when the assembly is transmitting a torque of 8 ft-kips.

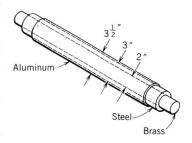

FIG. P3-61

3-62 A composite shaft, similar to the one shown in Fig. P3-61, consists of a solid steel ($G = 80$ GPa) core with an outside diameter of 40 mm covered by a brass ($G = 39$ GPa) tube with an inside diameter of 40 mm and a wall thickness of 20 mm, which is in turn covered by an aluminum alloy ($G = 28$ GPa) sleeve with an inside diameter of 80 mm and a wall thickness of 10 mm. The three materials are bonded so that they act as a unit. Determine

(a) the maximum shearing stress in each material when the assembly is transmitting a torque of 15 kN·m.

(b) the angle of twist in a 3-m length when the assembly is transmitting a torque of 10 kN·m.

3-63* The composite shaft shown in Fig. P3-63 is used as a torsional spring. The solid circular polymer ($G = 150$ ksi) portion of the shaft is encased in and firmly attached to a steel ($G = 12,000$ ksi) sleeve for part of its length. If a torque of 1.00 in.-kip is being transmitted by the composite shaft, determine

(a) the rotation of a cross section at C.

(b) the rotation of a cross section at C if the steel shell is assumed to be rigid.

(c) the percent error introduced by assuming the steel shell to be rigid.

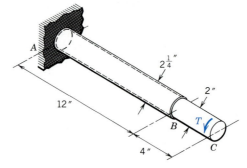

FIG. P3-63

3-64* A hollow steel ($G = 80$ GPa) tube with an outside diameter of 100 mm and an inside diameter of 50 mm is covered with a Monel ($G = 65$ GPa) tube that has an outside diameter of 125 mm and an inside diameter of 102 mm. The tubes are connected at the ends to form a composite shaft. If the maximum allowable shearing stress in the steel is 70 MPa and the maximum allowable shearing stress in the Monel is 85 MPa, determine

(a) the maximum torque T that the composite shaft can transmit.

(b) the angle of twist in a 2.5-m length when the composite shaft is transmitting the maximum torque.

3-65 A solid aluminum alloy ($G = 4000$ ksi) rod with an outside diameter of 2 in. is used as a shaft. A hollow steel (G = 12,000 ksi) tube with an inside diameter of 2 in. is placed over the rod to increase the torque transmitting capacity of the shaft. The tube and the rod are attached at the ends to form a composite shaft. Determine the minimum tube thickness required to permit the torque transmitting capacity of the shaft to be increased by 50 percent.

3-66 A solid Monel (G = 65 GPa) rod with an outside diameter of 60 mm is used as a shaft. A hollow stainless-steel (G = 86 GPa) tube with an inside diameter of 60 mm is placed over the rod to increase the torsional stiffness of the shaft. The tube and rod are attached at the ends to form a composite shaft. Determine the minimum tube thickness required if the angle of twist for a given torque must be decreased by 50 percent.

3-67* A hollow steel (G = 11,600 ksi) tube with an inside diameter of 2 in., an outside diameter of 2.5 in., and a length of 12 in. is encased with a brass (G = 5600 ksi) tube that has an inside diameter of 2.5 in. and an outside diameter of 3.25 in. The brass and steel tubes, while unstressed, are brazed together at one end. A couple is then applied to the other end of the brass tube, which twists one degree with respect to the steel tube. The tubes are then brazed together in that position. Determine the maximum shearing stresses in the steel tube and in the brass tube after the twisting couple is removed from the brass tube.

3-68* A composite shaft consists of a bronze (G = 45 GPa) shell that has an outside diameter of 100 mm bonded to a solid steel (G = 80 GPa) core. Determine the diameter of the steel core when the torque resisted by the steel core is equal to the torque resisted by the bronze shell.

3-69 The steel (G = 12,000 ksi) shaft shown in Fig. P3-69 is attached to rigid walls at both ends. The right 10 ft of the shaft is hollow, having an inside diameter of 2 in. Determine

(a) the maximum shearing stress in the shaft.

(b) the angle of rotation of the section where the torque is applied with respect to its no-load position.

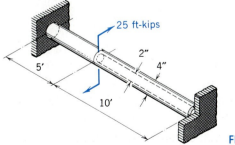

25 ft-kips

2"

4"

5'

10'

FIG. P3-69

3-70 A disk and two circular shafts are connected and supported between rigid walls, as shown in Fig. P3-70. Shaft *A* is made of brass (G = 39 GPa) and has a diameter of 100 mm and a length of 400 mm. Shaft *B* is made of Monel (G = 65 GPa) and has a diameter of 80 mm and a length of 600 mm. If a torque of 20 kN·m is applied to the disk, determine

    (a) the maximum shearing stress in each of the shafts.

    (b) the angle of rotation of the disk with respect to its no-load position.

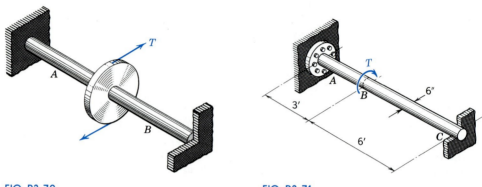

FIG. P3-70   FIG. P3-71

3-71* The solid steel (G = 12,000 ksi) shaft shown in Fig. P3-71 is fixed to the wall at *C*. The bolt holes in the flange at *A* have an angular misalignment of 0.0018 rad with respect to the holes in the wall. Determine

    (a) the torque, applied at *B*, required to align the bolt holes.

    (b) the maximum shearing stress in the shaft after the bolts are inserted and tightened and the torque at *B* is removed.

    (c) the maximum torque that can be applied at section *B* after the bolts are tightened if the maximum shearing stress in the shaft is not to exceed 10 ksi.

3-72* The torsional assembly of Fig. P3-72 consists of an aluminum alloy (G = 28 GPa) segment *AB* securely connected to a steel (G = 80 GPa) segment *BCD* by means of a flange coupling with four bolts. The diameters of both segments are 75 mm, the cross-sectional area of each bolt is 150 mm², and the bolts are located 75 mm from the center of the shaft. If the cross shearing stress in the bolts must be limited to 60 MPa, determine

    (a) the maximum torque *T* that can be applied at section *C*.

    (b) the maximum shearing stress in the steel.

    (c) the maximum shearing stress in the aluminum.

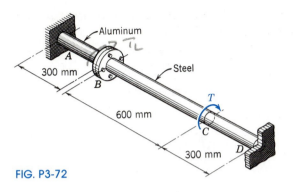

FIG. P3-72

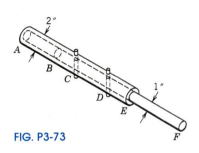

FIG. P3-73

3-73 The steel (G = 12,000 ksi) shaft shown in Fig. P3-73 will be used to transmit a torque of 1000 in.-lb. The hollow portion AE of the shaft is connected to the solid portion BF with two pins at C and D as shown. If the average cross shearing stress in the pins must be limited to 25 ksi, determine the minimum satisfactory diameter for each of the pins.

3-74 A stainless-steel (G = 86 GPa) shaft 2.5 m long extends through and is attached to a hollow brass (G = 39 GPa) shaft 1.5 m long, as shown in Fig. P3-74. Both shafts are fixed at the wall. When the two couples shown are applied to the shaft, determine

(a) the maximum shearing stress in the steel.

(b) the maximum shearing stress in the brass.

(c) the rotation of the right end of the shaft.

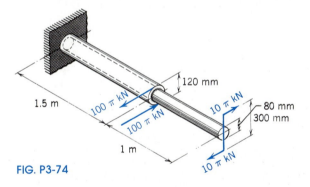

FIG. P3-74

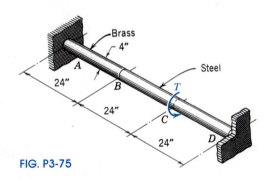

FIG. P3-75

3-75* The 4-in.-diameter shaft shown in Fig. P3-75 is composed of brass (G = 6000 ksi) and steel (G = 12,000 ksi) segments. Determine the maximum permissible magnitude for the torque T, applied at C, if the allowable shearing stresses are 5 ksi for the brass and 12 ksi for the steel.

3-76* The circular shaft shown in Fig. P3-76 consists of a steel (G = 80 GPa) segment ABC securely connected to a bronze (G = 40 GPa) segment *CD*. Ends *A* and *D* of the shaft are fastened securely to rigid supports. Determine

(a) the maximum shearing stress in the bronze segment.

(b) the maximum shearing stress in the steel segment.

(c) the rotation of a section at *B* with respect to its no-load position.

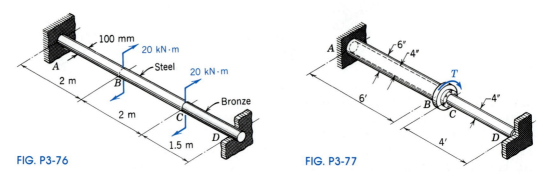

FIG. P3-76   FIG. P3-77

3-77 The shaft shown in Fig. P3-77 consists of a 6-ft hollow steel ($G = 12{,}000$ ksi) section *AB* and a 4-ft solid aluminum alloy ($G = 4000$ ksi) section *CD*. The torque *T* of 40 ft-kips is applied initially only to the steel section *AB*. Section *CD* is then connected and the torque *T* is released. When the torque is released, the connection slips 0.010 rad before the aluminum section takes any load. Determine

(a) the maximum shearing stress in the aluminum alloy.

(b) the maximum shearing stress in the steel after the torque *T* is released.

(c) the final rotation of the collar at *B* with respect to its no-load position.

3-78 A torque *T* of 10 kN·m is applied to the steel ($G = 80$ GPa) shaft shown in Fig. P3-78 without the brass ($G = 40$ GPa) shell. The brass shell is then slipped into place and attached to the steel. After the original torque is released, determine

(a) the maximum shearing stress in the brass shell.

(b) the maximum shearing stress in the steel shaft.

(c) the final rotation of the right end of the steel shaft with respect to the left end.

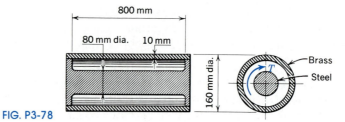

FIG. P3-78

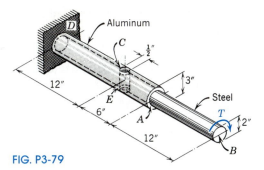

FIG. P3-79

3-79* The inner surface of the aluminum alloy (G = 4000 ksi) sleeve A and the outer surface of the steel (G = 12,000 ksi) shaft B of Fig. P3-79 are smooth. Both the sleeve and the shaft are rigidly fixed to the wall at D. The 0.500-in.-diameter pin C fills a hole drilled completely through a diameter of the sleeve and shaft. If the average shearing stress on the cross-sectional area of the pin at the interface between the shaft and the sleeve must not exceed 5000 psi, determine

(a) the maximum torque T that can be applied to the right end of the steel shaft B.

(b) the maximum shearing stress in the aluminum sleeve A when the torque of part (a) is applied.

(c) the rotation of the right end of the shaft when the maximum torque T is applied.

3-8

STRESS CONCENTRATIONS IN CIRCULAR SHAFTS UNDER TORSIONAL LOADINGS

In Section 2-6 it was shown that the introduction of a circular hole or other geometric discontinuity into an axially loaded member can cause a significant increase in the magnitude of the stress (stress concentration) in the immediate vicinity of the discontinuity. This is also the case for circular shafts under torsional forms of loading.

In the previous sections of this chapter it was assumed that the maximum shearing stress in a circular shaft of uniform cross section and made of a linearly elastic material is given by Eq. 3-5 as

$$\tau_{max} = \tau_c = \frac{Tc}{J} \tag{3-5}$$

Equation 3-5 can also be used to determine the maximum shearing stress in a tapered shaft if the change in diameter occurs in a gradual manner. For stepped shafts, however, large increases in stress (stress

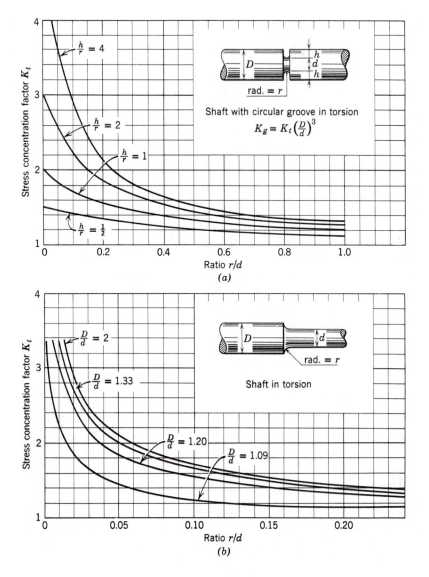

FIG. 3-15 Stress
concentration factors for
torsional loading.

concentrations) occur in the vicinity of the abrupt changes in diameter. These large stresses can be reduced by using a fillet between the parts of the shaft with the different diameters. The maximum shearing stress in the fillet can be expressed in terms of a stress-concentration factor K as

$$\tau_{\max} = K \frac{Tc}{J} \qquad (3\text{-}10)$$

The stress-concentration factor K depends upon the ratio of the diameters of the two portions of the shaft (D/d) and the ratio of the radius of the fillet to the diameter of the smaller shaft (r/d). Stress-concentration factors K_t (based on the net section) for stepped circular shafts and for circular shafts with U-shaped grooves are shown in Fig. 3-15. A careful examination of Fig. 3-15a shows that a generous fillet radius r should be used wherever a change in shaft diameter occurs. Equation 3-10 can be used to determine localized maximum shearing stresses in stepped shafts as long as the value of τ_{max} does not exceed the proportional limit of the material.

Stress concentrations also occur in circular shafts at oil holes and grooves (see Fig. 3-15b), and at keyways used for attaching pulleys and gears to the shaft. Each of these types of discontinuity require special consideration during the design process.

EXAMPLE 3-9

A stepped shaft has a 4-in. diameter for one-half of its length and a 2-in. diameter for the other half. If the maximum shearing stress in the shaft must be limited to 8 ksi when the shaft is transmitting a torque of 6280 in.-lb, determine the minimum radius needed at the junction between the two portions of the shaft.

Solution

The maximum shearing stress produced by the torque of 6280 in.-lb in the portion of the shaft with the 2-in. diameter is given by Eq. 3-5 as

$$\tau_{max} = \frac{Tc}{J} = \frac{6280(1)}{(\pi/32)(2^4)}$$

$$= 3998 \text{ psi} \approx 4.00 \text{ ksi}$$

Since the maximum shearing stress in the fillet between the two portions of the shaft must be limited to 8 ksi, the maximum permissible value for the stress-concentration factor K_t based on the maximum shearing stress in the small (net) section is

$$K_t = 8.00/4.00 = 2.00$$

The stress-concentration factor K_t depends on two ratios (D/d) and (r/d). For the 4-in.-diameter shaft with the 2-in.-diameter turned section, the ratio $D/d = 4.00/2.00 = 2.00$. From the curves presented in Fig. 3-15a, a ratio $r/d = 0.06$ together with a ratio $D/d = 2.00$ will produce a stress-concentration factor $K_t = 2.00$. Thus, the minimum permissible radius for the fillet between the two portions of the shaft is:

$$r = 0.06d = 0.06(2.00) = \underline{0.12 \text{ in.}} \qquad \text{Ans.}$$

PROBLEMS

Note: In the following problems, the specified allowable stresses are all below the shearing proportional limits of the materials.

3-80* A fillet with a radius of 0.15 in. is used at the junction in a stepped shaft where the diameter is reduced from 4.00 in. to 3.00 in. Determine the maximum shearing stress in the fillet when the shaft is transmitting a torque of 4000 ft-lb.

3-81* A fillet with a radius of 12 mm is used at the junction in a stepped shaft where the diameter is reduced from 135 mm to 100 mm. Determine the maximum shearing stress in the fillet when the shaft is transmitting a torque of 10 kN·m.

3-82 A fillet with a radius of $\frac{1}{8}$ in. is used at the junction in a stepped shaft where the diameter is reduced from 8.00 in. to 6.00 in. Determine the maximum torque that the shaft can transmit if the maximum shearing stress in the fillet must be limited to 12 ksi.

3-83 A stepped shaft has a 5-in. diameter for one-half of its length and a 4-in. diameter for the other half. If the maximum shearing stress in the fillet between the two portions of the shaft must be limited to 12 ksi when the maximum shearing stress in the 4-in. portion is 8 ksi, determine the minimum radius needed at the junction between the two portions of the shaft.

3-84* The small portion of a stepped shaft has a diameter of 50 mm. The radius of the fillet at the junction between the large and small portions is 4.5 mm. If the maximum shearing stress in the fillet must be limited to 40 MPa when the shaft is transmitting a torque of 614 N·m, determine the maximum diameter that can be used for the large portion of the shaft.

3-85* A 2-in.-diameter shaft contains a $\frac{1}{2}$-in.-deep U-shaped groove that has a $\frac{1}{4}$-in. radius at the bottom of the groove. The shaft must transmit a torque of 500 in.-lb. If a factor of safety of 3 with respect to failure by slip is specified, determine the minimum elastic strength in shear required for the shaft material.

3-86 A semicircular groove with a 5-mm radius is required in a 110-mm-diameter shaft. If the maximum allowable shearing stress in the shaft must be limited to 60 MPa, determine the maximum torque that can be transmitted by the shaft.

3-87 A shallow crack has been located in a 100-mm-diameter shaft. The crack will be removed by turning down a 200-mm length of the shaft surrounding the crack with a tool bit that has a 5-mm radius. If the maximum shearing stress in the 5-mm fillet must be limited to 60 MPa when the shaft is transmitting a torque of 3.27 kN·m, determine the minimum allowable diameter for the reduced section.

3-9

TORSION OF NONCIRCULAR SECTIONS

Prior to 1820, when A. Duleau published experimental results to the contrary, it was thought that the shearing stresses in any torsionally loaded member were proportional to the distance from its axis. Duleau proved experimentally that this is not true for rectangular cross sections. An examination of Fig. 3-16 will verify Duleau's conclusion. If the stresses in the rectangular bar were proportional to the distance from its axis, the maximum stress would occur at the corners. However, if there was a stress of any magnitude at the corner, as indicated in Fig. 3-16a, it could be resolved into the components indicated in Fig. 3-16b. If these components existed, the two components shown dashed would also exist. These last components cannot exist, since the surfaces on which they are shown are free boundaries. Therefore, the shearing stresses at the corners of the rectangular bar must be zero.

The first correct analysis of the torsion of a prismatic bar of noncircular cross section was published by Saint Venant in 1855; however, the scope of this analysis is beyond the elementary discussions of this

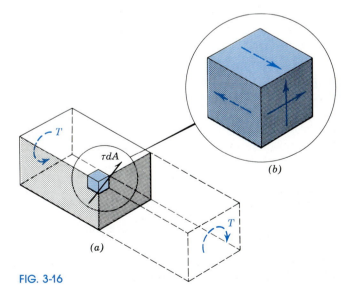

(a)

(b)

FIG. 3-16

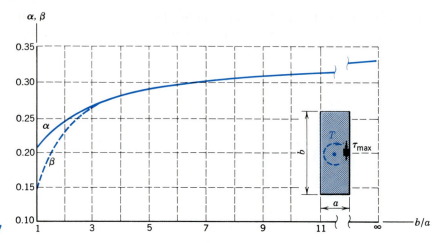

FIG. 3-17

book.[3] The results of Saint Venant's analysis indicate that, in general, *except for members with circular cross sections*, every section will warp (not remain plane) when the bar is twisted.

For the case of the rectangular bar shown in Fig. 3-3*d*, the distortion of the small squares is maximum at the midpoint of a side of the cross section and disappears at the corners. Since this distortion is a measure of shearing strain, Hooke's law requires that the shearing stress be maximum at the midpoint of a side of the cross section and zero at the corners. Equations for the maximum shearing stress and angle of twist for a rectangular section obtained from Saint Venant's theory are

$$\tau_{max} = \frac{T}{\alpha a^2 b}$$

$$\theta = \frac{TL}{\beta a^3 b G}$$

where *a* and *b* are the lengths of the short and long sides of the rectangle, respectively. The numerical factors α and β can be obtained from Fig. 3-17.

3-10

TORSION OF THIN-WALLED TUBES: SHEAR FLOW

Although the elementary torsion theory presented in Sections 3-1, 3-2, and 3-3 is limited to circular sections, one class of noncircular

[3] A complete discussion of this theory is presented in various books, such as *Mathematical Theory of Elasticity*, I. S. Sokolnikoff, 2nd ed., McGraw-Hill, New York, 1956, pp. 109–134.

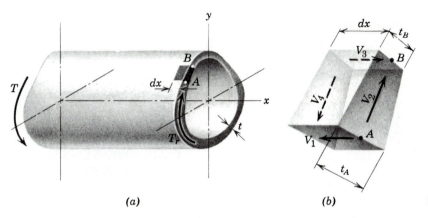

(a) (b) FIG. 3-18

sections that can be readily analyzed by elementary methods is the thin-walled section, such as the one illustrated in Fig. 3-18a, that represents a noncircular section with a wall of variable thickness (t varies).

A useful concept associated with the analysis of thin-walled sections is *shear flow q,* defined as internal shearing force per unit of length of the thin section. Typical units for q are pounds per inch or newtons per meter. In terms of stress, q equals $\tau(1)(t)$, where τ is the average shearing stress across the thickness t. It will be demonstrated that the shear flow on a cross section is constant even though the thickness of the section wall varies. Figure 3-18b shows a block cut from the member of Fig. 3-18a between A and B, and as the member is subjected to pure torsion, the shear forces $V_1 \cdots V_4$ alone (no normal forces) are necessary for equilibrium. Summing forces in the x direction gives:

$$V_1 = V_3$$

or

$$q_1 dx = q_3 dx$$

from which

$$q_1 = q_3$$

and, as $q = \tau(t)$,

$$\tau_1 t_A = \tau_3 t_B \qquad (a)$$

The shearing stresses at point A on the longitudinal and transverse planes have the same magnitude; likewise, the shearing stresses at point B have the same magnitude on the two orthogonal planes; hence, Eq. *a* may be written

$$\tau_A t_A = \tau_B t_B$$

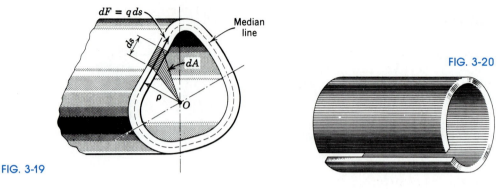

FIG. 3-20

FIG. 3-19

or

$$q_A = q_B$$

which was to be proved.

To develop the expression for the resulting torque on the section, consider the force dF acting through the center of a differential length of perimeter ds, as shown in Fig. 3-19. The resisting torque T_r is the resultant of the moments of the forces dF; that is

$$T_r = \int(dF)\rho = \int(q\ ds)\rho = q\int\rho\ ds$$

This integral may be difficult to evaluate by formal calculus; however, the quantity $\rho\ ds$ is twice the area of the triangle shown shaded in Fig. 3-19, which makes the integral equal to twice the area A *enclosed by the median line*. The resulting expression is

$$T_r = q(2A) \qquad (3\text{-}11)$$

or, in terms of stress,

$$\tau = \frac{T}{2At} \qquad (3\text{-}12)$$

where τ is the *average* shearing stress across the thickness t (and tangent to the perimeter) and is reasonably accurate when t is relatively small. For example, in a round tube with a diameter to wall thickness ratio of 20, the stress as given by Eq. 3-12 is 5 percent less than that given by the torsion formula. It must be emphasized that Eq. 3-12 applies only to "closed" sections—that is, sections with a continuous periphery. If the member were slotted longitudinally (see, for example, Fig. 3-20), the resistance to torsion would be diminished considerably from that for the closed section.

EXAMPLE 3-10

A rectangular box section of aluminum alloy has outside dimensions 100×50 mm. The plate thickness is 2 mm for the 50-mm sides and 3 mm for the 100-mm sides. If the maximum shearing stress must be limited to 95 MPa, determine the maximum torque that can be applied to the section.

Solution

The maximum stress will occur in the thinnest plate; therefore,

$$q = \tau t = 95(10^6)(2)(10^{-3}) = 190(10^3) \text{ N/m}$$

The torque that can be transmitted by the section is given by Eq. 3-11 as

$$
\begin{aligned}
T &= 2qA \\
&= 2(190)(10^3)(100 - 2)(50 - 3)(10^{-6}) \\
&= 1750 \text{ N·m} \qquad\qquad \text{Ans.}
\end{aligned}
$$

PROBLEMS

> *Note:* In the following problems, assume that buckling of the shell is prevented and that the shape of the section is maintained with ribs, diaphragms, or stiffeners. The specified allowable stresses are all below the shearing proportional limits of the materials.

3-88* A 24-in.-wide by 0.100-in.-thick by 100-in.-long steel sheet is to be formed into a hollow section by bending through 360° and welding (butt-weld) the long edges together. Assume a median length of 24 in. (no stretching of the sheet due to bending). If the maximum shearing stress must be limited to 12 ksi, determine the maximum torque that can be carried by the hollow section if

(a) the shape of the section is a circle.

(b) the shape of the section is an equilateral triangle.

(c) the shape of the section is a square.

(d) the shape of the section is an 8 × 4-in. rectangle.

3-89* A 500-mm-wide by 3-mm-thick by 2-m-long aluminum sheet is to be formed into a hollow section by bending through 360° and welding

FIG. P3-90

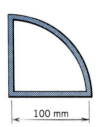

FIG. P3-91

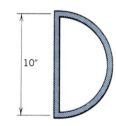

FIG. P3-92

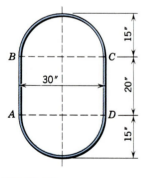

FIG. P3-95

(butt-weld) the long edges together. Assume a median length of 500 mm (no stretching of the sheet due to bending). If the maximum shearing stress must be limited to 75 MPa, determine the maximum torque that can be carried by the hollow section if

(a) the shape of the section is a circle.

(b) the shape of the section is an equilateral triangle.

(c) the shape of the section is a square.

(d) the shape of the section is a 150×100-mm rectangle.

3-90 A torque of 150 in.-kips will be applied to the hollow, thin-walled aluminum alloy section shown in Fig. P3-90. If the maximum shearing stress must be limited to 10 ksi, determine the minimum thickness required for the section.

3-91 A torque of 2.5 kN·m will be applied to the hollow, thin-walled aluminum section shown in Fig. P3-91. If the maximum shearing stress must be limited to 50 MPa, determine the minimum thickness required for the section.

3-92* A torque of 100 in.-kips will be applied to the hollow, thin-walled aluminum section shown in Fig. P3-92. If the section has a uniform thickness of 0.100 in., determine the magnitude of the maximum shearing stress developed in the section.

3-93* A torque of 2.75 kN·m will be applied to the hollow, thin-walled aluminum section shown in Fig. P3-93. If the section has a uniform thickness of 4 mm, determine the magnitude of the maximum shearing stress developed in the section.

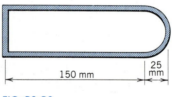

FIG. P3-93

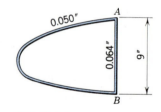

FIG. P3-94

3-94 A cross section of the leading edge of an airplane wing is shown in Fig. P3-94. The enclosed area is 82 in.^{2}. Sheet thicknesses are shown on the diagram. For an applied torque of 100 in.-kips, determine the magnitude of the maximum shearing stress developed in the section.

3-95* A cross section of an airplane fuselage made of aluminum alloy is shown in Fig. P3-95. For an applied torque of 1250 in.-kips and an allowable shearing stress of 7.5 ksi, determine the minimum thickness of sheet (constant for the entire periphery) required to resist the torque.

3-96* A cross section of an airplane fuselage made of aluminum alloy is shown in Fig. P3-96. For an applied torque of 200 kN·m and an allowable shearing stress of 50 MPa, determine the minimum thickness of sheet (constant for the entire periphery) required to resist the torque.

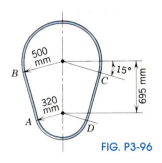

FIG. P3-96

COMPUTER PROBLEMS

Note: The following problems have been designed to be solved with a programmable calculator, microcomputer, or mainframe computer. Appendix F contains a description of a few numerical methods, together with a few simple programs in BASIC and FORTRAN, that can be modified for use in the solution of these problems.

C3-1 A hollow circular steel ($G = 11{,}000$ ksi) shaft 3-ft long is being designed to transmit a torque T of 3000 ft-lb. The outer radius r_o of the shaft can vary (1 in. $\leqslant r_o \leqslant$ 4 in.), but the cross-sectional area A of the shaft must remain constant ($A = 3$ in.²). Compute and plot:

(a) the angle of twist θ for the 3-ft length as a function of the outer radius r_o.

(b) the maximum shearing stress τ_{max} in the shaft as a function of the outer radius r_o.

C3-2 A hollow circular brass ($G = 40$ GPa) shaft 2-m long is being designed to transmit a torque T of 7500 N-m. The outer radius r_o of the shaft must be fixed ($r_o = 50$ mm); however, the inner radius r_i of the shaft can vary (0 mm $\leqslant r_i \leqslant$ 40 mm). Compute and plot:

(a) the angle of twist θ for the 2-m length as a function of the radius ratio r_i/r_o.

(b) the maximum shearing stress τ_{max} in the shaft as a function of the radius ratio r_i/r_o.

C3-3 The solid circular steel ($G = 12{,}000$ ksi) shaft shown in Fig. PC3-3 is subjected to a torque T of 4000 ft-lb. The diameter d_c of the

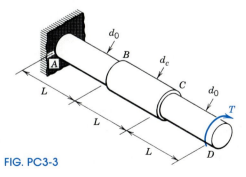

FIG. PC3-3

2-ft center section of the shaft can vary (2 in. $\leqslant d_c \leqslant$ 5 in.), while the diameters d_o of the 2-ft end sections of the shaft remain fixed at 3 in. Compute and plot the angle of twist θ as a function of position x along the shaft (0 ft $\leqslant x \leqslant$ 6 ft) when the center section diameter d_c equals 2 in., 3 in., 4 in., and 5 in.

C3-4 The hollow circular brass (G = 40 GPa) shaft shown in Fig. PC3-4 is 3 m long and is attached to rigid supports at both ends. An 8-kN-m torque T is applied at section C, which is located 1 m from the left end. The outer radius r_o of the shaft can vary (25 mm $\leqslant r_o \leqslant$ 100 mm); but, the cross-sectional area A of the shaft must remain constant at 1500 mm^2. Compute and plot:

    (a) the rotation θ of a section at C with respect to its no-load position as a function of the outer radius r_o.

    (b) the maximum shearing stress τ_{max} in the shaft as a function of the outer radius r_o.

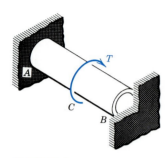

FIG. PC3-4

C3-5 The hollow circular aluminum alloy (G = 4000 ksi) shaft shown in Fig. PC3-4 is 5 ft long and is attached to rigid supports at both ends. A 7500-ft-lb torque T is applied at C, which is 2 ft from the right end. The outer radius r_o of the shaft must be 2 in.; however, the inner radius r_i can vary (0 $\leqslant r_i \leqslant$ 1.80 in.). Compute and plot:

    (a) the rotation θ of a section at C with respect to its no-load position as a function of the radius ratio r_i/r_o.

    (b) the maximum shearing stress τ_{max} in the shaft as a function of the radius ratio r_i/r_o.

C3-6 A composite shaft consists of a 2-m-long solid circular steel (G = 80 GPa) section securely fastened to a 2-m-long solid circular bronze (G = 40 GPa) section as shown in Fig. PC3-6. Both ends of the composite shaft are attached to rigid supports. The maximum shearing stress τ_{max} in the shaft must not exceed 60 MPa and the rotation θ of any cross section in the shaft must not exceed 0.04 rad. The ratio of the diameters d_b/d_s of the two sections can vary (1/2 $\leqslant d_b/d_s \leqslant$ 2), but the average diameter $(d_b + d_s)/2$ must be 100 mm. Compute and plot:

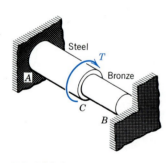

FIG. PC3-6

    (a) the maximum allowable torque T as a function of the diameter ratio d_b/d_s.

    (b) the rotation θ of a section at C as a function of the diameter ratio d_b/d_s.

    (c) the maximum shearing stress τ_b in the bronze shaft as a function of the diameter ratio d_b/d_s.

    (b) the maximum shearing stress τ_s in the steel shaft as a function of the diameter ratio d_b/d_s.

CHAPTER 4

FLEXURAL LOADING: STRESSES

4-1

INTRODUCTION

The beam, or flexural member, is frequently encountered in structures and machines, and its elementary stress analysis constitutes one of the more interesting facets of mechanics of materials. A beam is a member subjected to loads applied transverse to the long dimension, causing the member to bend. For example, Fig. 4-1 is a photograph of an aluminum alloy I-beam, *AB*, simply supported in a testing machine and loaded at the one-third points, and Fig. 4-2 depicts the shape (exaggerated) of the beam when loaded.

Before proceeding with a discussion of stress analysis for flexural members, it may be well to classify some of the various types of beams and loadings encountered in practice. Beams are frequently classified on the basis of the supports or reactions. A beam supported by pins, rollers, or smooth surfaces at the ends is called a *simple beam*. A simple support will develop a reaction normal to the beam but will not produce a couple. If either or both ends of a beam project beyond the supports, it is called a *simple beam with overhang*. A beam with more than two simple supports is a *continuous beam*. Figures 4-3a, b, and c show, respectively, a simple beam, a beam with overhang, and a

FIG. 4-1 Setup for measuring longitudinal strains in a beam.

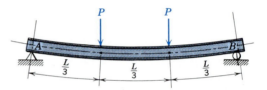

FIG. 4-2

continuous beam. A *cantilever beam* is one in which one end is built into a wall or other support so that the built-in end cannot move transversely or rotate. The built-in end is said to be *fixed* if no rotation occurs and *restrained* if a limited amount of rotation occurs. The supports shown in Figs. 4-3*d* and *e* represent fixed ends unless otherwise stated. The beams in Figs. 4-3*d, e,* and *f* are, in order, a cantilever beam, a beam fixed (or restrained) at the left end and simply supported near the other end (which has an overhang), and a beam fixed (or restrained) at both ends.

Cantilever beams and simple beams have only two reactions (two forces or one force and a couple), and these reactions can be obtained from a free-body diagram of the beam by applying the equations of equilibrium. Such beams are said to be *statically determinate,* since the reactions can be obtained from the equations of equilibrium. Continuous and other beams, with only transverse loads, with more than two reaction components are called *statically indeterminate,* since there are not enough equations of equilibrium to determine the reactions. Statically indeterminate beams are discussed in Chapter 6.

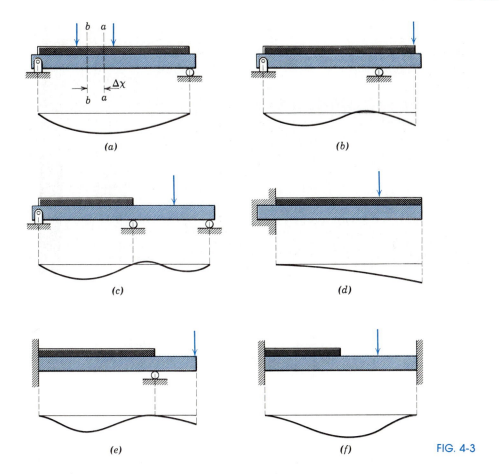

FIG. 4-3

The beams shown in Fig. 4-3 are all subjected to uniformly distributed loads and to concentrated loads, and, although shown as horizontal, they may have any orientation. Distributed loads will be shown on the side of the beam on which they are acting; that is, if drawn on the bottom of the beam, the load is pushing upward, and if drawn on the right side of a vertical beam, the load is pushing to the left. The deflection curves shown beneath the beams are greatly exaggerated to assist in visualizing the shape of the loaded beam.

A free-body diagram of the portion of the beam of Fig. 4-3a between the left end and plane a–a is shown in Fig. 4-4a. A study of this diagram reveals that a transverse force V_r and a couple M_r at the cut section and a force R (a reaction) at the left support are needed to maintain equilibrium. The force V_r is the resultant of the shearing stresses acting on the cut section (on plane a–a) and is called the resisting shear. The

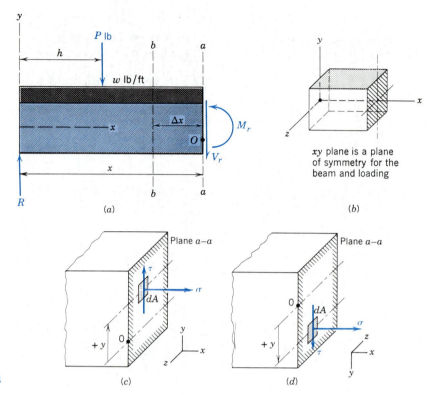

FIG. 4-4

couple M_r is the resultant of the normal stresses acting on the cut section (on plane a–a) and is called the resisting moment. The magnitudes and senses of V_r and M_r can be obtained from the equations of equilibrium $\Sigma F_y = 0$ and $\Sigma M_0 = 0$, where 0 is any axis perpendicular to the xy plane. The reaction R must be evaluated from a free body of the entire beam.

The normal and shearing stresses σ and τ on plane a–a are related to the resisting moment M_r and the shear V_r by equations of the form

$$V_r = -\int_{\text{Area}} \tau \, dA \qquad (a)$$

$$M_r = -\int_{\text{Area}} y\sigma \, dA \qquad (b)$$

The resisting shear and moment (V_r and M_r), as shown on Fig. 4-4a, will be defined as positive quantities later in Section 4-6. Previously, in Section 1-7, the normal and shearing stresses (σ and τ), as shown on Fig. 4-4c, were defined as positive stresses. The minus signs in Eqs. a and b are required to bring these two definitions into agreement.

Ideally, a positive shearing stress should result from a positive shear and a positive normal stress (a tensile stress) should result from a combination of a positive moment and a positive moment arm y. Both of these conditions can be satisfied by choosing the y direction to be positive downward as shown in Fig. 4-4d. Therefore, in this book, *the y direction will be positive downward for the computation of stresses in beam-type members*. In all other discussions and applications (beam deflections, column problems, and the like), the positive direction for y will be upward. When the downward positive direction for y is used, Eqs. *a* and *b* become

$$V_r = \int_{\text{Area}} \tau \, dA$$

$$M_r = \int_{\text{Area}} y\sigma \, dA \tag{4-1}$$

It is obvious from Eqs. 4-1 that the laws of variation of the normal and shearing stresses must be known before the integrals can be evaluated. For the present, the shearing stresses will be ignored while the normal stresses are studied.

The first recorded hypothesis on the normal stress distribution is that of Galileo Galilei who, possibly observing that beams of stone when subjected to bending loads broke somewhat as indicated in Fig. 4-5, concluded that the material at point A acted as a fulcrum and that

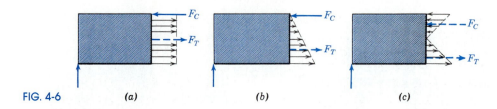

FIG. 4-6   (a)   (b)   (c)

the resultant of the tensile stresses was at the center of the cross section (as for axial loading).[1] This theory would lead to the stress distribution diagram of Fig. 4-6a, in which F_T is the resultant of the tensile stresses. About fifty years after Galileo's observations, E. Mariotte (1620–1684), a French physicist, still retaining the concept of the fulcrum at the compression surface of the beam, but reasoning that the extensions of the longitudinal elements of the beam (fibers) would be proportional to the distance from the fulcrum, suggested that the tensile stress distribution was as indicated in Fig. 4-6b. Mariotte later rejected the concept of the fulcrum and observed that part of the beam on the compression side was subjected to compressive stress having a triangular distribution; however, his expression for the ultimate load was still based on his original concept and was accepted by famous scientists, such as Jacob Bernoulli and Leonard Euler, until C. A. Coulomb (1736–1806), a French military engineer, in a paper published in 1773 discarded the fulcrum concept and proposed the distribution of Fig. 4-6c, in which both the tensile and compressive stresses have the same linear distribution.[2]

4-2

FLEXURAL STRAINS

The correctness of the Coulomb theory can be demonstrated as follows: A segment of the beam of Fig. 4-4, between planes a–a and b–b, is shown in Fig. 4-7 with the distortion greatly exaggerated. When Fig. 4-7 was drawn, the assumption was made that *a plane section before*

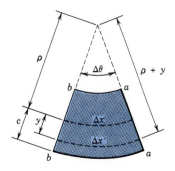

FIG. 4-7

[1] *Two New Sciences,* Galileo Galilei, 1638. Actually, Galileo's published work contains no record of test results; therefore, the manner in which he arrived at this conclusion is mere speculation.

[2] Sixty years before Coulomb's paper, Parent (1666–1716), a French mathematician, proposed the correct triangular stress distribution. However, his work remained in obscurity, and the majority of engineers continued to use formulas based on Mariotte's theory. For a more complete discussion of the history of the flexure problem, the reader is referred to the *History of Strength of Materials,* S. P. Timoshenko, McGraw-Hill, New York, 1953, from which most of the foregoing discussion is taken.

bending remains plane after bending. For this to be strictly true, it is necessary that the beam be bent only with couples (no shear on transverse planes). Also, the beam must be so proportioned that it will not buckle and the loads applied so that no twisting occurs (this last limitation will be satisfied if the loads are applied in a plane of symmetry—a sufficient though not a necessary condition). When a beam is bent only with couples, the deformed shape of all longitudinal elements (also referred to as fibers) is an arc of a circle.

Precise experimental measurements indicate that at some distance c above the bottom of the beam, the longitudinal elements undergo no change in length. The curved surface formed by these elements (at radius ρ in Fig. 4-7) is referred to as the *neutral surface* of the beam, and the intersection of this surface with any cross section is called the *neutral axis* of the section. All elements (fibers) on one side of the neutral surface are compressed, and those on the opposite side are elongated. As shown in Fig. 4-7, the fibers above the neutral surface of the beam of Fig. 4-4 are compressed and the fibers below the neutral surface are elongated.

Finally, the assumption is made that all longitudinal elements have the same initial length. This assumption imposes the restriction that the beam be initially straight and of constant cross section; however, in practice, considerable deviation from these last restrictions is often tolerated.

The longitudinal strain ϵ_x experienced by a fiber, which is located a distance y from the neutral surface of the beam, can be determined by using the definition of normal strain as expressed by Eq. 1-17a. Thus,

$$\epsilon_x = \frac{\Delta L}{L} = \frac{L_f - L_i}{L_i}$$

where

L_f = the final length of the fiber after the beam is loaded

L_i = the initial length of the fiber before the beam is loaded

From the geometry of the beam segment shown in Fig. 4-7,

$$\epsilon_x = \frac{\Delta x' - \Delta x}{\Delta x}$$

$$= \frac{(\rho + y)(\Delta \theta) - \rho(\Delta \theta)}{\rho(\Delta \theta)}$$

$$= \frac{1}{\rho} y \qquad\qquad (4\text{-}2)$$

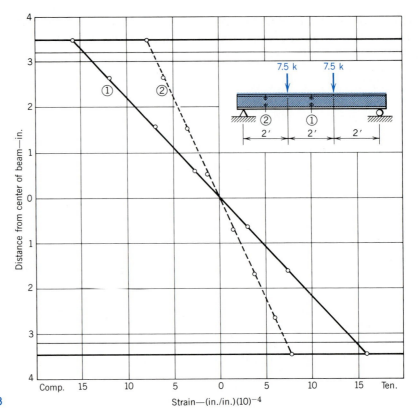

FIG. 4-8

Equation 4-2 indicates that the strain developed in any fiber is directly proportional to the distance of the fiber from the neutral surface of the beam. This variation can be demonstrated experimentally by means of strain gages attached to a beam, as shown in Fig. 4-1. The strains, as measured by gages on two different sections, are plotted against the vertical position of the gages on the beam in Fig. 4-8. Curve 1 represents strains on a section at the center of the beam where pure bending occurs (no transverse shear), and curve 2 shows strains at a section near one end of the beam where both fiber stresses and transverse shearing stresses exist. These curves are both straight lines within the limits of the accuracy of the measuring equipment.[3]

[3] A more exact analysis using principles developed in the theory of elasticity indicates that curve 2 should be curved slightly. *Note:* Other experiments indicate that a plane section of an initially curved beam will also remain plane after bending and that the deformations will still be proportional to the distance of the fiber from the neutral surface. The strain, however, will not be proportional to this distance, since each deformation must be divided by a different original length.

Note that Eq. 4-2 is valid for elastic or inelastic action so long as the beam does not twist or buckle and the transverse shearing stresses are relatively small. Problems in this book will be assumed to satisfy these restrictions.

4-3

FLEXURAL STRESSES: LINEARLY ELASTIC ACTION

With the acceptance of the premise that the longitudinal strain ϵ_x is proportional to the distance of the fiber from the neutral surface of the beam, the law of variation of the normal stress σ_x on the transverse plane can be determined by using a tensile–compressive stress-strain diagram for the material used in fabricating the beam. For most real materials, the tension and compression stress-strain diagrams are identical in the elastic range. Although the diagrams may differ somewhat in the inelastic range, the differences can be neglected for most real problems. *For beam problems in this book, the compressive stress-strain diagram will be assumed to be identical to the tensile diagram.*

For the special case of elastic action (which in this book always means linearly elastic action), the relationship between strains ϵ_x and the three normal stresses σ_x, σ_y, and σ_z is given by generalized Hooke's law (Eq. 1-23) as

$$\epsilon_x = \frac{1}{E} [\sigma_x - \nu(\sigma_y + \sigma_z)] \qquad (1\text{-}23)$$

Since no load is applied in the z direction, σ_z is zero throughout the beam. Because of loads applied in the y direction, σ_y is not zero throughout the beam; however, for the elementary solution considered here, σ_y is small enough to neglect. Thus, Eq. 1-23 reduces to

$$\epsilon_x = \frac{\sigma_x}{E} \qquad (a)$$

Substituting Eq. 4-2 into Eq. a and solving for σ_x yields

$$\sigma_x = E\epsilon_x = \left(\frac{E}{\rho}\right)y \qquad (4\text{-}3)$$

Equation 4-3 shows that the normal stress σ_x on the transverse cross section of the beam varies linearly with distance y from the neutral surface. Also, since plane cross sections remain plane, the normal stress σ_x is uniformly distributed in the z direction.

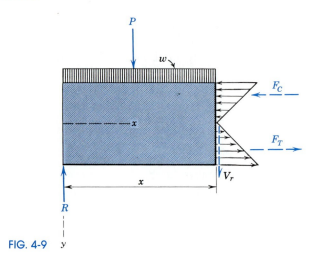

FIG. 4-9

With the law of variation of flexural stress known, Fig. 4-4 can now be redrawn as shown in Fig. 4-9. The vectors F_C and F_T are the resultants of the compressive and tensile flexural stresses. Since the sum of the forces in the x direction must be zero, F_C is equal to F_T; hence, they form a couple of magnitude M_r.

The resisting moment M_r developed by the normal stresses in a typical beam of arbitrary cross section, such as the one shown in Fig. 4-10, is given by Eq. 4-1 as

$$M_r = \int_A y\sigma_x \, dA = \int_A y \, dF \qquad (b)$$

Since y is measured from the neutral surface, it is first necessary to locate this surface by means of the equilibrium equation $\Sigma F_x = 0$, which gives

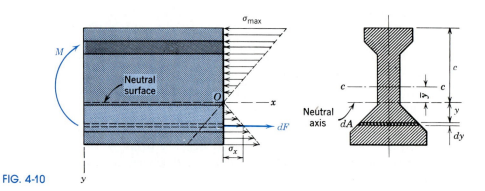

FIG. 4-10

$$\int_A \sigma_x \, dA = 0 \qquad\qquad (c)$$

Substituting Eq. 4-3 into Eq. c yields

$$
\begin{aligned}
\int_A \sigma_x \, dA &= \int_A \left(\frac{E}{\rho}\right) y \, dA \\
&= \left(\frac{E}{\rho}\right) \int_A y \, dA \qquad\qquad (4\text{-}4) \\
&= \left(\frac{E}{\rho}\right) \bar{y} A = 0
\end{aligned}
$$

where $\bar{y}$ is the distance from the neutral axis to the centroidal axis (c–c) of the cross section that is perpendicular to the plane of bending. Since neither (E/ρ) nor A is zero, $\bar{y}$ must equal zero. Thus, *for flexural loading and linearly elastic action, the neutral axis passes through the centroid of the cross section*. Instances in which the neutral axis (the line of zero longitudinal strain) does not pass through the centroid of the cross section include

1. a combined axial and flexural load (see Chapter 8),
2. a beam with cross section unsymmetrical with respect to the neutral axis and subjected to inelastic action, and
3. a curved beam.

Since the normal stress σ_x varies linearly with distance y from the neutral surface, the maximum normal stress σ_{max} on the cross section can be written as

$$\sigma_{max} = \left(\frac{E}{\rho}\right) c \qquad\qquad (4\text{-}5)$$

where c is the distance to the surface of the beam (top or bottom) farthest from the neutral surface. If the quantity (E/ρ) is eliminated from Eqs. 4-3 and 4-5, a useful relationship between the maximum stress σ_{max} on a transverse cross section and the stress σ_x at an arbitrary distance y from the neutral surface is obtained. Thus

$$\sigma_x = \left(\frac{y}{c}\right) \sigma_{max} \qquad\qquad (4\text{-}6)$$

Use of the concepts represented by Eqs. 4-1 through 4-6 will be demonstrated in the following two examples.

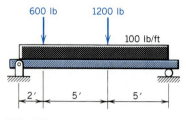

FIG. 4-11

EXAMPLE 4-1

The beam shown in Fig. 4-11 has a rectangular cross section 4 in. wide by 12 in. deep and a span of 12 ft. It is subjected to a uniformly distributed load of 100 lb/ft and concentrated loads of 600 lb and 1200 lb at 2 ft and 7 ft, respectively, from the left support. Assuming elastic action, determine the maximum fiber stress at the center of the span by computing the magnitudes and positions of the vectors F_T and F_C in terms of σ_{max} and substituting into the moment equilibrium equation.

Solution

The reactions are evaluated from a free-body diagram of the entire beam and moment equations with respect to each of the two reactions. By this method, the reactions are evaluated independently, and summation of forces may be used to check the results. The left reaction R_L is 1600 lb, and the right reaction R_R is 1400 lb, both upward. Figure 4-12a is a free-body diagram of the left half of the beam. The equation $\Sigma M_O = 0$ gives

$$(\curvearrowleft +) \qquad M_r + 600(4) + 100(6)(3) - 1600(6) = 0$$

from which $M_r = 5400$ ft-lb or 5400(12) in.-lb as shown. The moment M_r is the resultant of the flexural stresses indicated in Figs. 4-12b and c, in which F_C and F_T are the resultants of the compressive and tensile stresses, respectively. The resultants F_C and F_T (which act through the centroids of the wedge-shaped stress distribution diagram of Fig. 4-12c) are located through the centroids of the triangular stress distribution diagrams of Fig. 4-12b. The magnitude of the couple M_r equals $F_C(8)$ or $F_T(8)$, and $F_C = (\frac{1}{2})(\sigma_c)(6)(4) = 12(\sigma_c)$; therefore,

$$M_r = 8(12)\sigma_c = 5400(12) \text{ in.-lb}$$

FIG. 4-12

from which

$$\sigma_{max} = \sigma_c = \underline{675 \text{ psi } C \text{ at top and } T \text{ at bottom}} \qquad \text{Ans.}$$

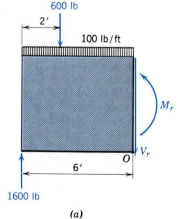

(a)

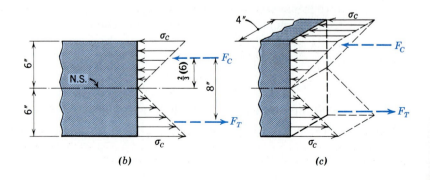

(b)   (c)

EXAMPLE 4-2

The beam shown in Fig. 4-11 is composed of two 2 × 10-in. timbers[4] securely fastened together to form a symmetrical T-section with the flange on top.

(a) Determine the maximum fiber stress at the center of the span by computing the magnitudes and positions of the vectors F_T and F_C in terms of σ_{max} and substituting into the moment equilibrium equation.

(b) Check the result of part (a) by integration of Eq. 4-1.

Solution

(a) The horizontal centroidal axis of the cross section is first determined to be 4 in. down from the top surface (see any text on statics). The fiber stress distribution diagrams are shown in Fig. 4-13. For convenience, the compression diagram is subdivided into three parts composed of two triangular distribution (wedge-shaped) diagrams and one uniform distribution diagram. This is done because the locations of the centroids of these diagrams are known. The values of the force components and their locations are shown in the figure. It will be noted that each component F_1, F_2, F_3, being the average stress multipled by the area on which the stress acts is equal to the volume of the corresponding part of the three-dimensional diagram of Fig. 4-13b. The moment of the component forces with respect to O is

$$M_r = F_1 \left(\frac{10}{3}\right) + F_2 \left(\frac{9}{3}\right) + F_3 \left(\frac{4}{3}\right) + F_4 \left(\frac{16}{3}\right)$$

which from the preceding example is equal to 5400(12); therefore,

$$\left(\frac{5}{2}\right) \sigma_c \left(\frac{10}{3}\right) + 5\sigma_c \left(\frac{9}{3}\right) + \left(\frac{1}{2}\right) \sigma_c \left(\frac{4}{3}\right) + 8\sigma_c \left(\frac{16}{3}\right) = 5400(12)$$

from which

$$\sigma_{max} = \sigma_c = 972 \text{ psi } T \text{ at bottom} \qquad \text{Ans.}$$

(b) A more direct solution can be obtained by integration of Eq. 4-1; thus,

$$M_r = \int_A y\,\sigma_x\, dA = \int_{-4}^{-2} (y)\left(\frac{\sigma_c}{8}y\right)(10\,dy) + \int_{-2}^{8} (y)\left(\frac{\sigma_c}{8}y\right)(2\,dy)$$

$$= \left[\frac{5\sigma_c y^3}{12}\right]_{-4}^{-2} + \left[\frac{\sigma_c y^3}{12}\right]_{-2}^{8} = 5400(12)$$

[4] Structural timbers specified as rough sawn will be approximately full size. Cross-sectional dimensions of dressed timbers will be from $\frac{1}{4}$ to $\frac{3}{4}$ in. smaller than nominal. For convenience, all timbers in this book will be considered full size.

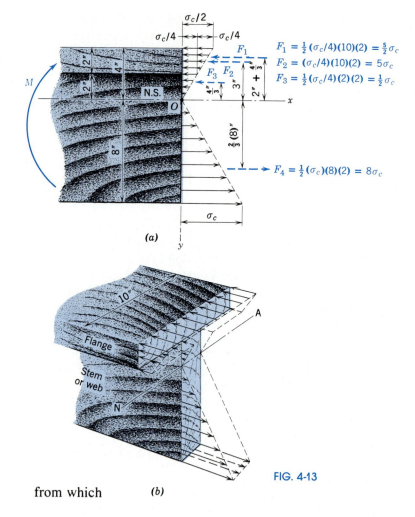

$F_1 = \frac{1}{2}(\sigma_c/4)(10)(2) = \frac{5}{2}\sigma_c$

$F_2 = (\sigma_c/4)(10)(2) = 5\sigma_c$

$F_3 = \frac{1}{2}(\sigma_c/4)(2)(2) = \frac{1}{2}\sigma_c$

$F_4 = \frac{1}{2}(\sigma_c)(8)(2) = 8\sigma_c$

(a)

FIG. 4-13

(b)

from which

$$\sigma_c = \sigma_{\max} = \underline{972 \text{ psi } T \text{ at bottom}} \qquad \text{Ans.}$$

Note: In order to simplify the calculations in beam problems in this book, the following additional assumptions are made.

1. The weight of the beam itself can be neglected unless otherwise stated. Obviously this would not, as a rule, be done in professional practice.

2. All beams are loaded in a plane of symmetry parallel to the largest cross-sectional dimension unless noted otherwise.

3. The specified allowable stresses are all below the proportional limits of the materials.

PROBLEMS

4-1* A structural steel (E = 29,000 ksi) bar with a rectangular cross section is bent over a rigid mandrel (R = 10 in.), as shown in Fig. P4-1. If the maximum fiber stress in the bar is not to exceed the yield strength (σ_y = 36 ksi) of the steel, determine the maximum allowable thickness h for the bar.

4-2* An aluminum alloy (E = 73 GPa) bar with a rectangular cross section is bent over a rigid mandrel as shown in Fig. P4-1. The thickness h of the bar is 25 mm. If the maximum fiber stress in the bar must be limited to 200 MPa, determine the minimum allowable radius R for the mandrel.

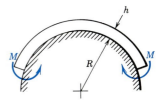

FIG. P4-1

4-3 A timber beam consists of four 2-by-8-in. planks fastened together to form a box section 8 in. wide by 12 in. deep as shown in Fig. P4-3. If the fiber stress at point A of the cross section is 1250 psi T, determine

(a) the fiber stress at point B of the cross section.

(b) the fiber stress at point C of the cross section.

(c) the fiber stress at point D of the cross section.

4-4 A timber beam consists of three 50 × 200-mm planks fastened together to form an I-beam 200 mm wide by 300 mm deep as shown in Fig. P4-4. If the fiber stress at point A of the cross section is 7.5 MPa C, determine

(a) the fiber stress at point B of the cross section.

(b) the fiber stress at point C of the cross section.

(c) the fiber stress at point D of the cross section.

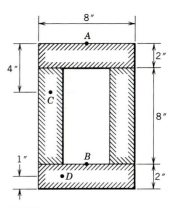

FIG. P4-3

4-5* The maximum fiber stress at a certain section in a rectangular beam 4 in. wide by 8 in. deep is 1500 psi. Determine the resisting moment M_r in the beam at that section by computing the magnitudes and positions of the vectors F_T and F_C in terms of σ_{max} and substituting into the moment equilibrium equation.

4-6* The maximum fiber stress at a certain section in a rectangular beam 150 mm wide by 300 mm deep is 15 MPa. Determine the resisting moment M_r in the beam at that section by computing the magnitudes and positions of the vectors F_T and F_C in terms of σ_{max} and substituting into the moment equilibrium equation.

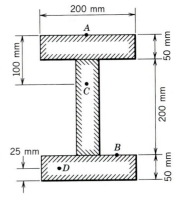

FIG. P4-4

4-7 A timber beam is made of three 2-by-6-in. planks fastened together to form an I-beam 6 in. wide by 10 in. deep. The maximum fiber stress must not exceed 1200 psi. Use the free-body diagram method (method of Example 4-1) to determine the maximum bending moment that the beam can support.

4-8 A timber beam is made of three 50 × 100-mm members fastened together to form an I-beam 100 mm wide by 200 mm deep. Use the free-body diagram method (method of Example 4-1) to determine the maximum fiber stress on a section where the bending moment is 8 kN·m.

4-9* The maximum fiber stress on a transverse cross section of a beam with a 2 × 4-in. rectangular cross section must not exceed 10 ksi. Determine the resisting moment M_r developed in the beam at the section if the neutral axis is

(a) parallel to the 4-in. side.

(b) parallel to the 2-in. side.

4-10* A 0.2%C hardened steel (E = 210 GPa) bar with a 50-mm-square cross section is subjected to a flexural form of loading that produces a fiber strain of 1200 μm/m at a point on the top surface of the beam. Determine

(a) the maximum fiber stress at the point.

(b) the resisting moment M_r developed in the beam on a transverse cross section through the point.

4-11 A structural steel (E = 29,000 ksi) beam, loaded in the vertical plane of symmetry, has the cross section shown in Fig. P4-11. If the fiber strain at point A of the cross section is −200 μin./in., determine

(a) the maximum fiber stress at the section.

(b) the resisting moment M_r developed at the section.

4-12 A steel bar with a rectangular cross section will be used as a beam. If the maximum stress on a cross section that develops a resisting moment of 75 kN·m must be limited to 100 MPa, determine the minimum depth required if the width of the bar (parallel to the neutral surface) is 50 mm.

4-13* An I-beam is fabricated by welding two 16 × 2-in. flange plates to a 24 × 1-in. web plate. The beam is loaded in the plane of symmetry parallel to the web. Use the free-body diagram method (method of Example 4-1) to determine the percentage of the bending moment carried by the flanges.

4-14* The maximum fiber stress at a certain section in a rectangular beam 100 mm wide by 200 mm deep is 15 MPa. Use the free-body diagram method (method of Example 4-1) to determine

(a) the resisting moment M_r developed at the section.

(b) the percentage decrease in M_r if the dotted central portion of the cross section shown in Fig. P4-14 is removed.

4-15 The load-carrying capacity of an S24 × 80 American standard beam (see Appendix C for dimensions) is to be increased by fastening

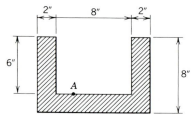

FIG. P4-11

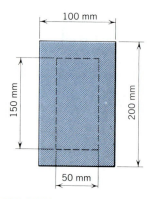

FIG. P4-14

two 8 × $\frac{3}{4}$-in. plates to the flanges of the beam. If the maximum fiber stress in both the original and modified beams must be limited to 15 ksi, determine

(a) the maximum moment that the original beam can support.

(b) the maximum moment that the modified beam can support.

4-16 Two L102 × 102 × 12.7-mm structural steel angles (see Appendix C for dimensions) are attached back-to-back to form a T-section. Determine the maximum moment that can be supported by the section if the maximum fiber stress must be limited to 125 MPa.

4-17* A beam is loaded so that the maximum fiber stress on the cross section shown in Fig. P4-17 is 8.5 ksi C. Use the free-body diagram method (method of Example 4-1) to determine the magnitude of the flexural force carried by the flange.

4-18* The beam of Fig. P4-18 is made of a material that has a yield point of 250 MPa. Use the free-body diagram method (method of Example 4-1) to determine the maximum moment that the beam can support if yielding must be avoided.

4-19 Use the free-body diagram method (method of Example 4-1) to determine the percentage of the bending moment carried by the flanges of a W36 × 160 wide-flange beam (see Appendix C for dimensions).

4-20 Use the free-body diagram method (method of Example 4-1) to determine the percentage of the bending moment carried by the flanges of W838 × 226 wide-flange beam (see Appendix for C dimensions).

4-21 Use the free-body diagram method (method of Example 4-1) to determine the percentage of the bending moment carried by the flanges of an S24 × 100 American standard beam (see Appendix C for dimensions).

4-22 Use the free-body diagram method (method of Example 4-1) to determine the percentage of the bending moment carried by the flanges of an S178 × 23 American standard beam (see Appendix C for dimensions).

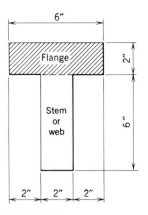

FIG. P4-17

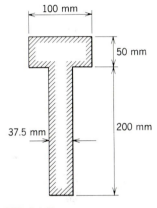

FIG. P4-18

4-4

FLEXURAL STRESSES: THE ELASTIC FLEXURE FORMULA

As the sophistication of structural and machine design increases, problems involving inelastic action are receiving increased attention. However, the design of beams on the basis of linearly elastic action is still a very important area of engineering activity; for such design, the fundamental approach outlined in Section 4-3 becomes too time-consuming. Therefore, a formula will be developed that has general applicability within the limitations noted in Sections 4-2 and 4-3.

Substitution of Eq. 4-6 into Eq. 4-1 after noting that Eq. 4-6 can be written as $\sigma_{max}/c = \sigma_x/y$ or alternatively as $\sigma_c/c = \sigma_x/y$ yields

$$M_r = \int_A y\sigma_x\, dA = \frac{\sigma_c}{c} \int_A y^2\, dA = \frac{\sigma_x}{y} \int_A y^2\, dA$$

The integral is the familiar second moment of the cross-sectional area, commonly called "moment of inertia" (see Appendix G), *with respect to the centroidal axis*. When this integral is replaced by the symbol I, the elastic flexure formula is obtained as

$$\sigma_x = \frac{M_r y}{I} \tag{4-7}$$

where σ_x is the normal stress at a distance y from the neutral surface and on a transverse plane, and M_r is the resisting moment of the section.

At any section of the beam, the fiber stress will be maximum at the surface farthest from the neutral axis, and Eq. 4-7 becomes

$$\sigma_{max} = \frac{M_r c}{I} = \frac{M_r}{S} \tag{4-8}$$

where $S = I/c$ is called the section modulus of the beam. Although the section modulus can be readily calculated for a given section, values of the modulus are often included in tables to simplify calculations. Observe that for a given area, S becomes larger as the shape is altered to concentrate more of the area as far as possible from the neutral axis. Commercial rolled shapes such as I- and H-beams and the various built-up sections are intended to optimize the area–section modulus relation.

Thus far, the discussion of flexural behavior has been limited to structural members with symmetric cross sections that are loaded in a plane of symmetry. Many other shapes are subjected to flexural loadings, and methods are needed to determine stress distributions in these nonsymmetric shapes. The flexure formula (Eq. 4-7) provides a means for relating the resisting moment M_r at a section of a beam to the normal stress at a point on the transverse cross section. Further insight into the applicability of the flexure formula to nonsymmetric sections can be gained by considering the requirements for equilibrium when the applied moment M does not have a component about the y axis of the cross section. Thus, from Eq. 4-6 and the equilibrium equation $\Sigma M_y = 0$,

$$\int_A z\sigma_x \, dA = \int_A z\frac{\sigma_c}{c}y \, dA = \frac{\sigma_c}{c}\int_A zy \, dA = \frac{\sigma_c}{c}I_{yz} = 0 \quad (a)$$

The quantity I_{yz} is commonly known as the "product of inertia" of the cross-sectional area with respect to the centroidal y and z axes. Obviously, Eq. a can be satisfied only if $I_{yz} = 0$. For symmetric cross sections, $I_{yz} = 0$ when the y and z axes coincide with the axes of symmetry. For nonsymmetric cross sections, $I_{yz} = 0$ when the y and z axes are centroidal principal axes (see Appendix G) for the cross section. Thus, the flexure formula is valid for any cross section, provided y is measured along the principal direction, I is a principal moment of inertia, and M_r is a moment about a principal axis.

EXAMPLE 4-3

Solve Example 4-2 by means of the flexure formula.

Solution

The neutral axis is horizontal and passes through the centroid of the cross section. The centroid is located by using the principle of moments as applied to areas. The total area A of the cross section is

$$A = 2(2)(10) = 40 \text{ in.}^2$$

The moment of the area M_A about the bottom edge of the cross section is

$$M_A = 2(10)(11) + 2(10)(5) = 320 \text{ in.}^3$$

The distance y_c from the bottom edge of the cross section to the centroid is

$$y_c = \frac{M_A}{A} = \frac{320}{40} = 8 \text{ in.}$$

The second moment of the area (moment of inertia) I of the cross section about the neutral axis is

$$I = (1/12)(10)(2)^3 + 10(2)(11 - 8)^2 + (1/12)(2)(10^3) + 2(10)(8 - 5)^2$$
$$= 533.3 \text{ in.}^4$$

The distance c from the neutral axis to the fiber farthest from the neutral axis is 8 in. The moment M_r at the center of the span is 5400 ft-lb. The fiber stress σ_{max} is given by Eq. 4-7 as

$$\sigma_{max} = \frac{M_r c}{I} = \frac{5400(12)(8)}{533.3}$$

$$= \underline{972 \text{ psi } T} \text{ at the bottom} \qquad \text{Ans.}$$

EXAMPLE 4-4

A timber beam, loaded and supported as shown in Fig. 4-14a, has the cross section shown in Fig. 4-14b. On a section 1.2 m from the left end, determine

(a) the fiber stress at point A of the cross section.
(b) the fiber stress at point B of the cross section.

Solution

The neutral axis is horizontal and passes through the centroid of the cross section. The centroid is located by using the principle of moments as applied to areas. The total area A of the cross section is

$$A = 2(50)(200) + 150(50) = 27\ 500 \text{ mm}^2$$

The moment of the area M_A about the bottom edge of the cross section is

$$M_A = 2(50)(200)(100) + 150(50)(25) = 2\ 187\ 500 \text{ mm}^3$$

The distance y_c from the bottom edge of the cross section to the centroid is

$$y_c = \frac{M_A}{A} = \frac{2187500}{27500} = 79.55 \text{ mm}$$

The second moment of the area (moment of inertia) I of the cross section about the neutral axis is

$$I = 2(1/12)(50)(200)^3 + 2(50)(200)(100 - 79.55)^2$$
$$+ (1/12)(150)(50)^3 + 150(50)(79.55 - 25)^2$$
$$= 98.91(10^6) \text{ mm}^4 = 98.91(10^{-6}) \text{ m}^4$$

The reaction at the left support can be determined by summing moments with respect to the right support. Thus

$$R_L(3) - 9000(4) - 15000(2)(1.5) = 0$$

from which

$$R_L = 27\ 000 \text{ N}$$

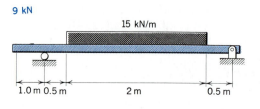

9 kN
15 kN/m
1.0 m   0.5 m   2 m   0.5 m

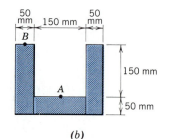

50 mm   150 mm   50 mm
B
150 mm
A
50 mm

FIG. 4-14   (a)   (b)

The moment M_r at a section 1.2 m from the left end of the beam is

$$M_r = 27000(0.2) - 9000(1.2) = -5400 \text{ N·m}$$

(a) The distance y_A from the neutral axis to point A is

$$y_A = 79.55 - 50 = 29.55 \text{ mm} = 29.55(10^{-3}) \text{ m}$$

The fiber stress at point A is given by Eq. 4-7 as

$$\sigma_{xA} = \frac{M_r y_A}{I} = \frac{-5400(29.55)(10^{-3})}{98.91(10^{-6})}$$
$$= -1.613(10^6) \text{ N/m}^2 = \underline{1.613 \text{ MPa } C} \quad \text{Ans.}$$

(b) The distance y_B from the neutral axis to point B is

$$y_B = 79.55 - 200 = -120.45 \text{ mm} = -120.45(10^{-3}) \text{ m}$$

The fiber stress at point B is given by Eq. 4-7 as

$$\sigma_{xB} = \frac{M_r y_B}{I} = \frac{-5400(-120.45)(10^{-3})}{98.91(10^{-6})}$$
$$= 6.58(10^6) \text{ N/m}^2 = \underline{6.58 \text{ MPa } T} \quad \text{Ans.}$$

EXAMPLE 4-5

An S152 × 19 steel beam (see Appendix C) is loaded and supported as shown in Fig. 4-15. On a section 2 m to the right of B, determine

(a) the fiber stress at a point 25 mm below the top of the beam.
(b) the maximum fiber stress on the section.

Solution

On a section 2 m to the right of B, the resisting moment M_r is

$$M_r = -8 + 3(2)(3) + 3(2) - 4(2)(1) = 8 \text{ kN·m}$$

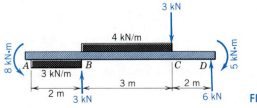

FIG. 4-15

For an S152 × 19 section (see Appendix C), $I = 9.20(10^6)$ mm⁴ = $9.20(10^{-6})$ m⁴, $S = 121(10^3)$ mm³ = $121(10^{-6})$ m³, and the depth of the beam is 152.4 mm = 0.1524 m.

(a) At a point 25 mm below the top of the beam, $y = -51.2$ mm = -0.0512 m; therefore, with all quantities expressed in newtons and meters, the stress is

$$\sigma_x = \frac{M_r y}{I} = \frac{8(10^3)(-0.0512)}{9.20(10^{-6})}$$

$$= -44.5(10^6) \text{ N/m}^2 = \underline{44.5 \text{ MPa } C} \qquad \text{Ans.}$$

(b) The maximum fiber stress is

$$\sigma_{max} = \frac{M_r}{S}$$

$$= \frac{8(10^3)}{121(10^{-6})} = 66.1(10^6) \text{ N/m}^2$$

$$= \underline{66.1 \text{ MPa } T \text{ on the bottom}}$$
$$\underline{C \text{ on the top}} \qquad \text{Ans.}$$

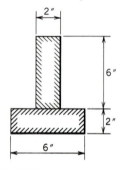

FIG. P4-23

PROBLEMS

> *Note:* In problems involving rolled shapes (see Appendix C), unless otherwise stated, the beam is oriented so that the maximum section modulus applies.

4-23* Determine the maximum fiber stress produced by a moment of + 5000 ft-lb if the beam has the the cross section shown in Fig. P4-23.

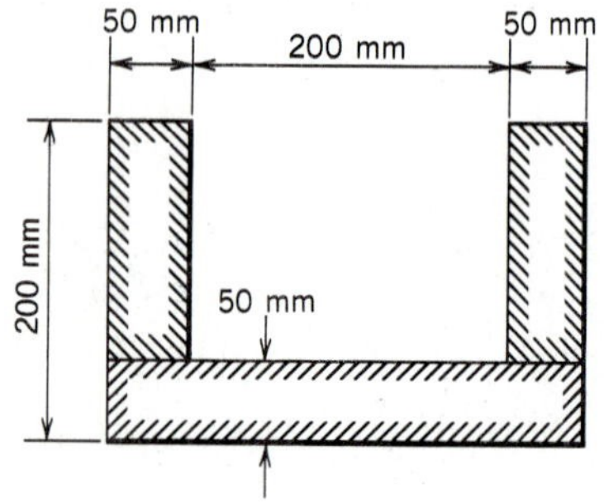

4-24* Determine the maximum fiber stress produced by a moment of − 15 kN·m if the beam has the cross section shown in Fig. P4-24.

4-25 Use the flexure formula to solve Problem 4-7.

4-26 Use the flexure formula to solve Problem 4-8.

4-27* Use the flexure formula to solve Problem 4-9.

4-28* Use the flexure formula to solve Problem 4-14.

4-29 A wide-flange beam will be used to resist a bending moment of 55,000 ft-lb. If the maximum fiber stress must not exceed 18,000 psi, select the most economical wide-flange section listed in Appendix C.

FIG. P4-24

4-30 A pair of channels fastened back-to-back will be used as a beam to resist a bending moment of 60 kN·m. If the maximum fiber stress must not exceed 120 MPa, select the most economical channel section listed in Appendix C.

4-31* Determine the maximum fiber stress produced by a moment of −30 ft-kips if the beam has the cross section shown in Fig. P4-31.

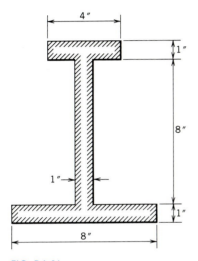

FIG. P4-31

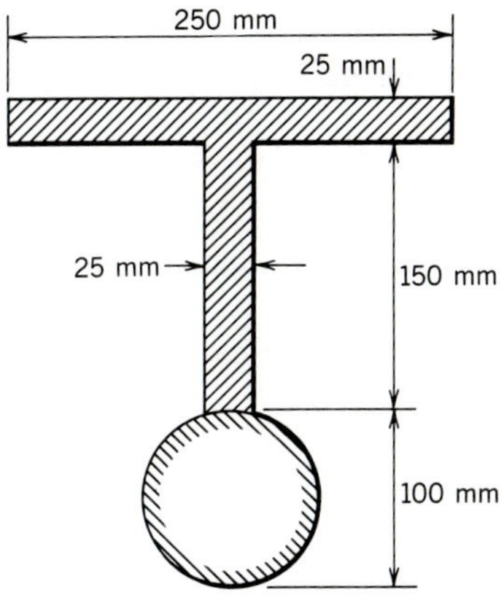

FIG. P4-32

4-32* Determine the maximum fiber stress produced by a moment of +100 kN·m if the beam has the cross section shown in Fig. P4-32.

4-33 A 4 × 8-in. structural timber 10 ft long is simply supported at the ends and carries a concentrated load of 2000 lb at the center of the span. Determine the maximum fiber stress

(a) on a cross section 2 ft from a support.

(b) on a cross section at the center of the span.

4-34 A 150 × 250-mm structural timber 4 m long is simply supported at the ends and carries a uniformly distributed load of 25 kN/m. Determine the maximum fiber stress

(a) on a cross section 0.75 m from a support.

(b) on a cross section at the center of the span.

4-35* A beam has the cross section shown in Fig. P4-35. On a section where the moment is +12.5 ft-kips, determine

(a) the maximum tensile fiber stress.

(b) the maximum compressive fiber stress.

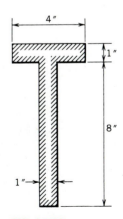

FIG. P4-35

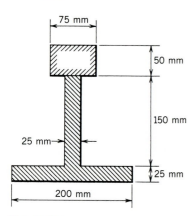

FIG. P4-36

4-36* A beam has the cross section shown in Fig. P4-36. On a section where the moment is -75 kN·m, determine

(a) the maximum tensile fiber stress.

(b) the maximum compressive fiber stress.

4-37 A beam has the cross section shown in Fig. P4-37. On a section where the moment is -30 ft-kips, determine

(a) the maximum tensile fiber stress.

(b) the maximum compressive fiber stress.

4-38 A beam has the cross section shown in Fig. P4-38. On a section where the moment is $+50$ kN·m, determine

(a) the maximum tensile fiber stress.

(b) the maximum compressive fiber stress.

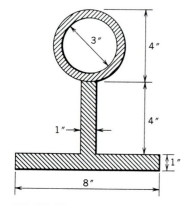

FIG. P4-37

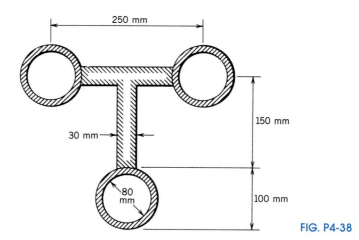

FIG. P4-38

4-39* A steel pipe with an outside diameter of 4 in. and an inside diameter of 3 in. is simply supported at the ends and carries two concentrated loads, as shown in Fig. P4-39. On a section 5 ft from the right support, determine

(a) the fiber stress at point A on the cross section.

(b) the fiber stress at point B on the cross section.

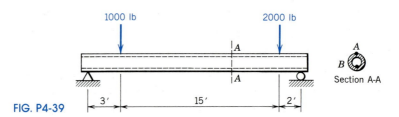

FIG. P4-39

4-40* Two 50 × 200-mm structural timbers are used to fabricate a beam with an inverted-T cross section (flange on the bottom). The beam is simply supported at the ends and is 4 m long. If the maximum fiber stress must be limited to 30 MPa, determine

   (a) the maximum moment that can be resisted by the beam.
   (b) the largest concentrated load P that can be supported at the center of the span.
   (c) the largest uniformly distributed load (over the entire span) that can be supported by the beam.

4-41 A beam is loaded and supported as shown in Fig. P4-41. If the allowable fiber stresses on the section at P are 4800 psi T and 7200 psi C, determine the maximum permissible value for the concentrated load P.

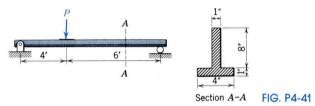

Section A–A   FIG. P4-41

4-42 A beam is loaded and supported as shown in Fig. P4-42. If the allowable fiber stresses on a section at the middle of the span are 22.5 MPa T and 15 MPa C, determine the maximum permissible value for the distributed load w.

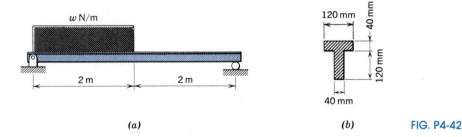

(a)   (b)   FIG. P4-42

4-43* An S18 × 70 steel beam (see Appendix C) is simply supported at the ends and carries a concentrated load of 25 kips at the center of a 20-ft span. On a section 7.5 ft from the left end of the beam, determine

   (a) the fiber stress at a point $\frac{1}{2}$ in. below the top surface of the beam.
   (b) the maximum fiber stress on the section.

4-44* A W610 × 155 steel beam (see Appendix C) is simply supported at the ends and carries a uniformly distributed load of 30 kN/m over an 8-m span. On a section 3 m from the left end of the beam, determine

(a) the fiber stress at a point 15 mm above the bottom surface of the beam.

(b) the maximum fiber stress on the section.

4-45 A simply supported timber beam with a rectangular cross section and a span of 20 ft must support a concentrated load of 2500 lb at the center of the span. If the depth of the cross section is 12 in., determine the minimum acceptable width for the cross section if the maximum fiber stress must be limited to 1250 psi.

4-46 A simply supported timber beam with a circular cross section and a span of 6 m must support a uniformly distributed load of 7.5 kN/m. Determine the minimum acceptable diameter for the timber if the maximum fiber stress must be limited to 15 MPa.

4-47* A steel pipe with an outside diameter of 6 in. and an inside diameter of 5 in. rests horizontally on simple supports near the ends of the pipe. Steel weighs approximately 490 lb/ft^3. If the maximum fiber stress in the pipe due to its own weight must be limited to 7500 psi, determine the maximum permissible distance between supports.

4-48* A simply supported timber beam with a rectangular cross section and a span of 6 m must support a concentrated load of 10 kN at the center of the span. If the width of the cross section is 200 mm, determine the minimum acceptable depth for the cross section if the maximum fiber stress must be limited to 10 MPa.

4-49 A beam with a hollow circular cross section is being designed to support a maximum moment of 100 in.-kips. The maximum fiber stress in the beam must not exceed 40 ksi. Prepare a design curve showing the acceptable combinations of inside diameter d_i and outside diameter d_o. The maximum diameter of the beam cannot exceed 5 in.

4-50 A beam with a solid rectangular cross section (width b and depth h) is being designed to support a maximum moment of 4 kN·m. The maximum fiber stress in the beam must not exceed 50 MPa. Prepare a design curve showing the acceptable combinations of width b and depth h. The width b of the beam cannot exceed 75 mm.

4-51 An S4 × 9.5 structural steel beam (see Appendix C) is simply supported at the ends and carries a concentrated load P at the center of the span. The load P must be increased by 75 percent, but the maximum fiber stress of 20 ksi cannot be increased. In order to strengthen the beam, two steel plates will be attached to the top and bottom flanges. Prepare a design curve showing the acceptable combinations of plate width b and thickness t. The thickness t of the plate cannot exceed $\frac{1}{2}$ in.

4-52 A beam with a hollow square cross section is being designed to support a maximum moment of 10 kN·m. The maximum fiber stress in the beam must not exceed 100 MPa. Prepare a design curve showing the acceptable combinations of outside dimension a_o and inside dimension a_i. The outside dimension a_o cannot exceed 120 mm.

4-5

SHEAR AND MOMENT IN BEAMS

The procedures for determining flexural stresses outlined in Sections 4-3 and 4-4 are quite adequate if one wishes to compute the stresses at any specified section in a beam. However, if it is necessary to determine the maximum fiber (flexural) stress in a given beam subjected to loading producing a bending moment that varies with position along the beam, it is desirable to have a convenient method for determining the maximum moment.

In Section 4-8 a relation will be developed between the resisting shear (V_r of Fig. 4-4) and the longitudinal and transverse shearing stresses at the section. Since the maximum transverse shearing stress will occur at the section at which V_r is maximum, a convenient method of determining such sections is likewise desirable.

When the equilibrium equation $\Sigma F_y = 0$ is applied to the free-body diagram of Fig. 4-4, the result can be written as

$$R - wx - P = V_r$$

or

$$V = V_r$$

where *V is defined as the resultant of the external transverse forces acting on the part of a beam to either side of a section and is called the transverse shear or just the shear at the section.*

As seen from the definitions of V and V_r, the shear is equal in magnitude and opposite in sense to V_r. Since these shear forces are always equal in magnitude, they are frequently treated as though they were identical. For simplicity the symbol V will be used henceforth to represent both the transverse shear and the resisting shear. The sign convention selected will have sufficient generality to apply to both quantities.

The resultant of the fiber stresses on any transverse section has been shown to be a couple (if only transverse loads are considered) and has been designated as M_r. When the equilibrium equation $\Sigma M_O = 0$ (where O is any axis parallel to the neutral axis of the section) is applied to the free-body diagram of Fig. 4-4, the result can be written as

$$Rx - wx^2/2 - P(x - h) = M_r$$

or

$$M = M_r$$

where *M is defined as the algebraic sum of the moments of the external forces, acting on the part of the beam to either side of the section, with respect to an axis in the section and is called the bending moment or just the moment at the section.*

As seen from the definitions of M and M_r, the bending moment is equal in magnitude and opposite in sense to M_r. Since these moments are always equal in magnitude, they are frequently treated as though they were identical. For simplicity the symbol M will be used henceforth to represent both the bending moment and the resisting moment. The sign convention selected will have sufficient generality to apply to both quantities.

The bending moment and the transverse shear are not normally shown on a free-body diagram. The usual procedure is to show each external force individually as indicated in Fig. 4-4. The variation of V and M along the beam can be shown conveniently by means of equations or shear and moment diagrams.

A sign convention is necessary for the correct interpretation of results obtained from equations or diagrams for shear and moment. The following convention will give consistent results regardless of whether one proceeds from left to right or from right to left. By definition, the shear at a section is positive when the portion of the beam to the left of the section (for a horizontal beam) tends to move upward with respect to the portion to the right of the section as shown in Fig. 4-16*a*. Also by definition, the bending moment in a horizontal beam is positive at sections for which the top of the beam is in compression and the bottom is in tension, as shown in Fig. 4-16*b*. Observe that the signs of the terms in the preceding equations for V and M agree with this convention.

Since M and V vary with x, they are functions of x, and equations for M and V can be obtained from free-body diagrams of portions of the beam. The procedure is illustrated in Example 4-6.

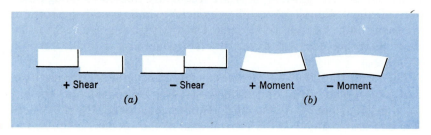

+ Shear   − Shear   + Moment   − Moment

(a)   (b)

FIG. 4-16

EXAMPLE 4-6

A beam is loaded and supported as shown in Fig. 4-17a. Write equations for the shear V and bending moment M for any section of the beam in the interval BC.

Solution

A free-body diagram, or load diagram, for the beam is shown in Fig. 4-17b. The reactions shown on the load diagram are determined from the equations of equilibrium. A free-body diagram of a portion of the beam from the left end to any section between B and C is shown in Fig. 4-17c. Note that the resisting shear V and the resisting moment M are shown as positive values. From the definition of V or from the equilibrium equation $\Sigma F_y = 0$,

$$V = 1400 - 400(x - 2) = \underline{2200 - 400x} \quad 2 < x < 6 \quad \text{Ans.}$$

From the definition of M, or from the equilibrium equation $\Sigma M_O = 0$,

$$M = 1400x - 400(x - 2)\left(\frac{x - 2}{2}\right)$$
$$= \underline{200x^2 + 2200x - 800} \quad 2 < x < 6 \quad \text{Ans.}$$

The equations for V and M in the other intervals can be determined in a similar manner.

FIG. 4-17

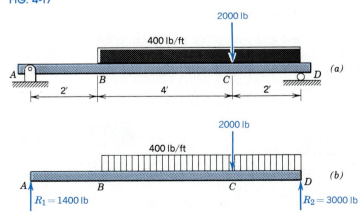

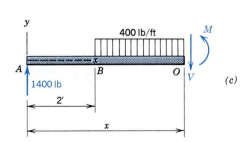

PROBLEMS

4-53* A beam is loaded and supported as shown in Fig. P4-53. Using the coordinate axes shown, write equations for the shear V and bending moment M for any section of the beam in the interval $4 < x < 8$.

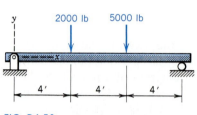

FIG. P4-53

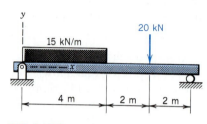

FIG. P4-54

4-54* A beam is loaded and supported as shown in Fig. P4-54. Using the coordinate axes shown, write equations for the shear V and bending moment M for any section of the beam in the interval $0 < x < 4$.

4-55 A beam is loaded and supported as shown in Fig. P4-55. Using the coordinate axes shown, write equations for the shear V and bending moment M for any section of the beam in the interval $2 < x < 8$.

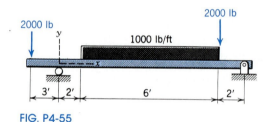

FIG. P4-55

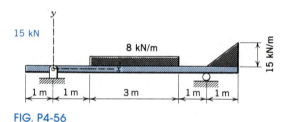

FIG. P4-56

4-56 A beam is loaded and supported as shown in Fig. P4-56. Using the coordinate axes shown, write equations for the shear V and bending moment M for any section of the beam in the interval $1 < x < 4$.

4-57* A beam is loaded and supported as shown in Fig. P4-57. Using the coordinate axes shown, write equations for the shear V and bending moment M for any section of the beam in the interval $0 < x < 10$.

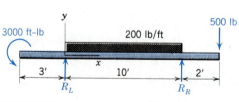

FIG. P4-57

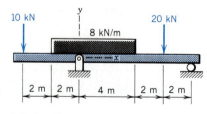

FIG. P4-58

4-58* A beam is loaded and supported as shown in Fig. P4-58. Using the coordinate axes shown, write equations for the shear V and bending moment M for any section of the beam in the interval $0 < x < 4$.

4-59 A beam is loaded and supported as shown in Fig. P4-59. Using the coordinate axes shown, write equations for the shear V and bending moment M for any section of the beam

(a) in the interval $5 < x < 10$.

(b) in the interval $10 < x < 20$.

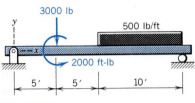

FIG. P4-59

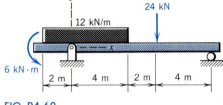

FIG. P4-60

4-60 A beam is loaded and supported as shown in Fig. P4-60. Using the coordinate axes shown, write equations for the shear V and bending moment M for any section of the beam

(a) in the interval $0 < x < 4$.

(b) in the interval $4 < x < 6$.

4-61* A beam is loaded and supported as shown in Fig. P4-61. Using the coordinate axes shown, write equations for the shear V and bending moment M for any section of the beam

(a) in the interval $4 < x < 8$.

(b) in the interval $8 < x < 20$.

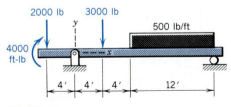

FIG. P4-61

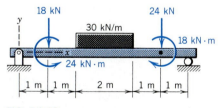

FIG. P4-62

4-62* A beam is loaded and supported as shown in Fig. P4-62. Using the coordinate axes shown, write equations for the shear V and bending moment M for any section of the beam

(a) in the interval $1 < x < 2$.

(b) in the interval $2 < x < 4$.

4-63 A beam is loaded and supported as shown in Fig. P4-63. Using the coordinate axes shown, write equations for the shear V and bending moment M for any section of the beam

(a) in the interval $0 < x < 6$.

(b) in the interval $6 < x < 16$.

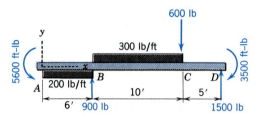

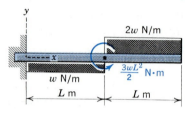

FIG. P4-63

FIG. P4-64

4-64 A beam is loaded and supported as shown in Fig. P4-64. Using the coordinate axes shown, write equations for the shear V and bending moment M for any section of the beam

(a) in the interval $0 < x < L$.

(b) in the interval $L < x < 2L$.

(c) Use the results of parts (a) and (b) to determine the magnitudes and locations of the maximum shear and the maximum bending moment in the beam.

4-65* A beam is loaded and supported as shown in Fig. P4-65. Using the coordinate axes shown,

(a) write equations for the shear V and bending moment M for any section of the beam.

(b) Use the results of part (a) to determine the magnitudes and locations of the maximum shear and the maximum bending moment in the beam.

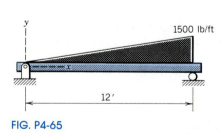

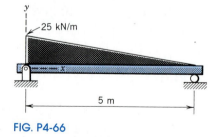

FIG. P4-65

FIG. P4-66

4-66* A beam is loaded and supported as shown in Fig. P4-66. Using the coordinate axes shown,

(a) write equations for the shear V and bending moment M for any section of the beam.

(b) Use the results of part (a) to determine the magnitudes and locations of the maximum shear and the maximum bending moment in the beam.

4-67 A beam is loaded and supported as shown in Fig. P4-67. Using the coordinate axes shown, write equations for the shear V and bending moment M for any section of the beam

   (a) in the interval $0 < x < L$.

   (b) in the interval $L < x < 2L$.

   (c) Use the results of parts (a) and (b) to determine the magnitudes and locations of the maximum shear and the maximum bending moment in the beam.

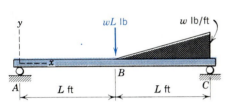

FIG. P4-67   FIG. P4-68

4-68 A beam is loaded and supported as shown in Fig. P4-68. Using the coordinate axes shown, write equations for the shear V and bending moment M for any section of the beam

   (a) in the interval $0 < x < L$.

   (b) in the interval $L < x < 2L$.

4-69* A beam is loaded and supported as shown in Fig. P4-69. Using the coordinate axes shown,

   (a) write equations for the shear V and bending moment M for any section of the beam.

   (b) Use the results of part (a) to determine the magnitudes and locations of the maximum shear and the maximum bending moment in the beam.

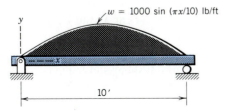

FIG. P4-69

4-70* A beam is loaded and supported as shown in Fig. P4-70. Using the coordinate axes shown,

(a) write equations for the shear V and bending moment M for any section of the beam.

(b) Use the results of part (a) to determine the magnitudes and locations of the maximum shear and the maximum bending moment in the beam.

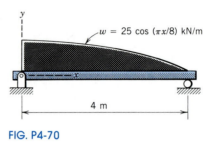

FIG. P4-70

4-71 A beam is loaded and supported as shown in Fig. P4-71. Using the coordinate axes shown,

(a) write equations for the shear V and bending moment M for any section of the beam.

(b) Use the results of part (a) to determine the magnitudes and locations of the maximum shear and the maximum bending moment in the beam.

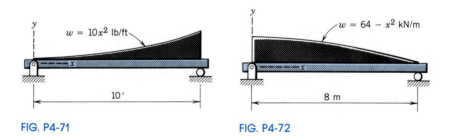

FIG. P4-71   FIG. P4-72

4-72 A beam is loaded and supported as shown in Fig. P4-72. Using the coordinate axes shown,

(a) write equations for the shear V and bending moment M for any section of the beam.

(b) Use the results of part (a) to determine the magnitudes and locations of the maximum shear and the maximum bending moment in the beam.

4-6

LOAD, SHEAR, AND MOMENT RELATIONSHIPS

The mathematical relationships between loads and shears, and between shears and moments, can be used to facilitate construction of shear and moment diagrams. The relationships can be developed from a free-body diagram of an elemental length of a beam as shown in Fig. 4-18, in which the upward direction is considered positive for the applied load w, and the shears and moments are shown as positive according to the sign convention established in Section 4-5. The element must be in equilibrium, and the equation $\Sigma F_y = 0$ gives

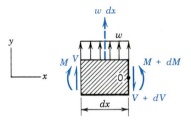

FIG. 4-18

$$(\uparrow +) \qquad \Sigma F_y = V + w\, dx - (V + dV) = 0$$

from which

$$dv = w\, dx \quad \text{or} \quad \frac{dV}{dx} = w \qquad (4\text{-}9)$$

The preceding equation indicates that at any section in the beam *the slope of the shear diagram is equal to the intensity of loading.* When w is known as a function of x, the equation can be integrated between definite limits as follows:

$$\int_{V_1}^{V_2} dV = \int_{x_1}^{x_2} w\, dx = V_2 - V_1 \qquad (4\text{-}10)$$

That is, *the change in shear between sections at x_1 and x_2 is equal to the area under the load diagram between the two sections.*

The equation $\Sigma M_O = 0$, applied to Fig. 4-18, gives

$$(\curvearrowleft +) \qquad \Sigma M_O = M + V\, dx + w\, dx(dx/2) - (M + dM) = 0$$

from which

$$dM = V\, dx + w(dx)^2/2$$

This expression reduces to

$$dM = V\, dx \quad \text{or} \quad V = \frac{dM}{dx} \qquad (4\text{-}11)$$

when second-order differentials are considered as being negligible compared with first-order differentials. The preceding equation indicates that at any section in the beam *the slope of the moment diagram is equal to the shear*. The equation can be integrated between definite limits to give

$$\int_{M_1}^{M_2} dM = \int_{x_1}^{x_2} = V\,dx = M_2 - M_1 \qquad (4\text{-}12)$$

Thus, the change in moment between sections at x_1 and x_2 is equal to the area under the shear diagram between the two sections.

Note that the equations in this section were derived with the x axis positive to the right, the applied loads positive upward, and the shear and moment signs as indicated in Fig. 4-16. If one or more of these assumptions is changed, the algebraic signs in the equation may need to be altered.

4-7

SHEAR AND MOMENT DIAGRAMS

As stated in Section 4-5, shear and moment diagrams provide a convenient method of obtaining maximum values of shear and moment and other information, to be discussed later. *A shear diagram is a graph in which abscissas represent distances along the beam and ordinates represent the transverse shear at the corresponding sections. A moment diagram is a graph in which abscissas represent distances along the beam and ordinates represent the bending moment at the corresponding sections.*

Shear and moment diagrams can be drawn by calculating values of shear and moment at various sections along the beam and plotting enough points to obtain a smooth curve. Such a procedure is rather time-consuming, and, although it may be desirable for graphic solutions of certain structural problems, other more rapid methods will be developed.

A convenient arrangement for constructing shear and moment diagrams is to draw a free-body diagram of the entire beam and construct shear and moment diagrams directly below. Two methods of procedure are presented in this section.

The first method consists of writing algebraic equations for the shear V and the moment M and constructing curves from the equations. This method has the disadvantage that unless the load is uniformly distributed or varies according to a known equation along the entire beam, no single elementary expression can be written for the shear V or the moment M, which applies to the entire length of the beam. Instead, it is necessary to divide the beam into intervals bounded by the abrupt changes in the loading. An origin should be selected (different origins may be used for different intervals); positive directions should be shown for the coordinate axes, and the limits of the abscissa (usually x) should be indicated for each interval.

Complete shear and moment diagrams should indicate values of shear and moment at each section where the load changes abruptly and at sections where they are maximum or minimum (negative maximum values). Sections where the shear and moment are zero should also be located.

The second method consists of drawing the shear diagram from the load diagram and the moment diagram from the shear diagram by means of the mathematical relationships developed in Section 4-6. The latter method, though it may not produce a precise curve, is less time-consuming than the first and it does provide the information usually required.

When all loads and reactions are known, the shear and moment at the ends of the beam can be determined by inspection. Both shear and moment are zero at the free end of a beam unless a force or a couple or both is applied there; in this case, the shear is the same as the force and the moment the same as the couple. At a simply supported or pinned end, the shear must equal the end reaction and the moment must be zero. At a built-in or fixed end, the reactions are the shear and moment values.

Once a starting point for the shear diagram is established, the diagram can be sketched by using the definition of shear and the fact that the slope of the shear diagram can be obtained from the load diagram. When positive directions are chosen as upward and to the right, a positive distributed load, one acting upward, will result in a positive slope on the shear diagram, and a negative load will give a negative slope. A concentrated force will produce an abrupt change in shear. The *change* in shear between any two sections is given by the area under the load diagram between the same two sections. The change of shear at a concentrated force is equal to the concentrated force.

The moment diagram is drawn from the shear diagram in the same manner. The slope at any point on the moment diagram is given by the shear at the corresponding point on the shear diagram, a positive shear representing a positive slope and a negative shear representing a negative slope, *when upward and to the right are positive*. The *change* in moment between any two sections is given by the area under the shear diagram between the corresponding sections. A couple applied to a beam will cause the moment to change abruptly by an amount equal to the moment of the couple. Examples 4-7 and 4-8 illustrate these two methods for construction of shear and moment diagrams.

EXAMPLE 4-7

A beam is loaded and supported, as shown in Fig. 4-19a.

   (a) Write equations for the shear V and the bending moment M for any section of the beam in the interval BC.

   (b) Draw complete shear and moment diagrams for the beam.

Solution

A free-body diagram, or load diagram, for the beam is shown in Fig. 4-19b. The reactions at C are computed from the equations of equilibrium. It is not necessary to compute the reactions on a cantilever beam in order to write equations for the shear and bending moment or to draw shear and moment diagrams from the load diagram, but the reactions provide a convenient check.

   (a) A free-body diagram of a portion of the beam from the left end to any section between B and C is shown in Fig. 4-19c. Note that the resisting shear V and the resisting moment M are shown as positive values. From the definition of V or from the equilibrium equation $\Sigma F_y = 0$,

$$V = -500 - 200(x - 4) = \underline{300 - 200x} \qquad 4 < x < 10 \quad \text{Ans.}$$

From the definition of M, or from the equilibrium equation $\Sigma M_O = 0$,

$$M = -500x - 200(x - 4)(x - 4)/2$$
$$= \underline{-100x^2 + 300x - 1600} \qquad 4 < x < 10 \quad \text{Ans.}$$

   (b) The equations for V and M in the interval AB can be written in the same manner as in part (a), after which the two equations for V can be plotted in the appropriate intervals to give the shear diagram in Fig. 4-19d. Likewise, the two equations for M can be plotted in the appropriate intervals to give the moment diagram in Fig. 4-19e.

   The shear and moment diagrams can also be drawn, working directly from the load diagram without writing the shear and moment equations.

   The shear diagram is drawn below the load diagram (Fig. 4-19d). The shear just to the right of the 500-lb load is -500 lb. The slope of the shear diagram is equal to the load, and since no load is applied between A and B, the slope of the diagram is zero. From B to C the load is uniform; therefore, the slope of the shear diagram is constant. The *change* of shear from B to C is equal to the applied load (1200 lb) from B to C, and the shear

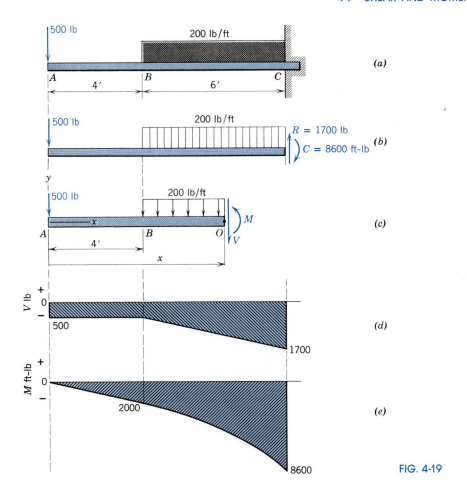

FIG. 4-19

is -1700 lb at C. This shear is the same as the reaction at C, which provides a check.

The moment diagram is drawn below the shear diagram as Fig. 4-19e. The moment is zero at the free end A. From A to B the shear is constant (-500); therefore, the slope of the moment diagram is constant, and the moment diagram is a straight line. The shear increases (negatively) from -500 lb at B to -1700 lb at C; therefore, the slope of the moment diagram is negative from B to C and increases uniformly in magnitude. The change in moment from A to B is equal to the area under the shear diagram and is $500(4) = 2000$ ft-lb. The area under the shear diagram from B to C is $(\frac{1}{2})(500 + 1700)(6) = 6600$ ft-lb, which represents the change in moment from B to C. Thus, the moment at C is $-2000 + (-6600) = -8600$ ft-lb.

EXAMPLE 4-8

A beam is loaded and supported, as shown in Fig. 4-20a.

(a) Write equations for the shear V and the bending moment M for any section of the beam in the interval CD.

(b) Draw complete shear and moment diagrams for the beam.

Solution

The reactions shown on the load diagram (Fig. 4-20b) are determined from the equations of equilibrium.

(a) A free-body diagram of a portion of the beam from the left end to any section between C and D is shown in Fig. 4-20c. From the definition of V, or from the equilibrium equation $\Sigma F_y = 0$,

$$V = 10 - 4x - 8 = \underline{-4x + 2} \qquad 3.5 < x < 5 \qquad \text{Ans.}$$

From the definition of M, or from the equilibrium equation $\Sigma M_O = 0$,

$$M = 10x - 4x(x/2) + 4 - 8(x - 3.5)$$
$$= \underline{32 + 2x - 2x^2} \qquad 3.5 < x < 5 \qquad \text{Ans.}$$

(b) The equations for V and M in the other intervals can be written and the shear and moment diagrams obtained by plotting these equations. In this example, the second method is used and the shear diagram is drawn directly from the load diagram.

   The shear just to the right of A is 10 kN and decreases at the rate of 4 kN each meter to -4 kN at C. The concentrated downward load at C causes the shear to change suddenly from -4 kN to -12 kN. The shear changes uniformly to -18 kN just to the left of D. The reaction at D changes the shear by 26 kN (from -18 kN to $+8$ kN). The shear decreases uniformly to zero at E. Note that the distributed load is uniform over the entire beam; consequently, the slope of the shear diagram is constant. Points of zero shear, such as F in Fig. 4-20d, are located from the geometry of the shear diagram. For example, the slope of the left portion of the shear diagram is 4 kN/m. Therefore,

$$\text{Slope} = 4 = 10/x_1$$

and

$$x_1 = 2.50 \text{ m}$$

   The moment is zero at A, and the slope of the moment diagram (equal to the shear) is 10 kN·m/m. From A to C the shear and,

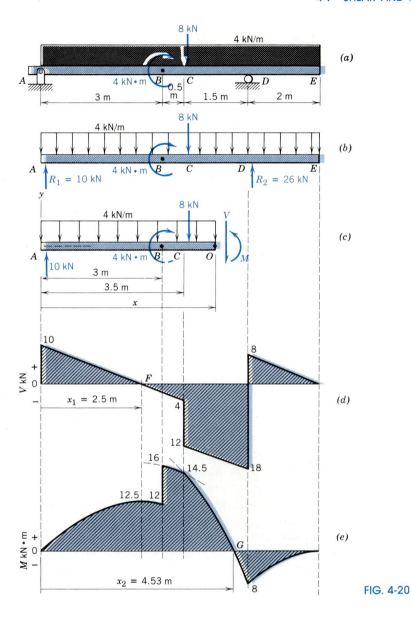

FIG. 4-20

hence, the slope of the moment diagram decreases uniformly to zero at F and to -4 at C. The abrupt change of shear at C indicates a sudden change of slope of the moment diagram; thus, the two parts of the moment diagram at C are not tangent. The slope of the moment diagram changes from -12 at C to -18 just to the left of D. From D to E the slope changes from $+8$ to zero.

The change of moment from A to F is equal to the area under the shear diagram from A to F and is

$$\Delta M = 10(2.5)/2 = 12.5 \text{ kN·m}$$

which is the moment at F, since the moment at A is zero. From F to B, $\Delta M = -0.5$ kN·m and $M_B = +12$ kN·m. The 4-kN·m couple (point moment) is applied at B. Since the moment just to the left of B is positive and the couple contributes an additional positive moment to sections to the right of B, the moment changes abruptly from $+12$ kN·m to 16 kN·m. Moments at C, D, and E can be determined from the shear diagram areas that give the changes in moment from section to section. If the moment at E is not equal to zero, it indicates that an error has occurred.

The point of inflection G, where the moment is zero, can be determined by setting the expression for the moment from part (a) equal to zero and solving for x. The result is

$$x_2 = 0.5 \pm \sqrt{16.25} = 4.531 = 4.53 \text{ m}$$

Note in the example that maximum and minimum moments may occur at sections where the shear curve passes through zero. In general, the shear curve may pass through zero at a number of points along the beam, and each such crossing indicates a point of *possible* maximum moment (in engineering, the moment with the largest absolute value is the maximum moment). It should be emphasized that the shear curve does not indicate the presence of abrupt discontinuities in the moment curve; hence, the maximum moment may occur where a couple is applied to the beam, not where the shear passes through zero. All possibilities should be examined to determine the maximum moment.

Sections where the bending moment is zero, called points of inflection or contraflexure, can be located by equating the expression for M to zero. The fiber stress is zero at such sections, and if a beam must be spliced, the splice should be located at or near a point of inflection, if there is one.

PROBLEMS

Note: In problems involving rolled shapes (see Appendix C), unless otherwise stated, the beam is oriented so that the maximum section modulus applies.

4-73* Draw complete shear and bending moment diagrams for the beam shown in Fig. P4-73.

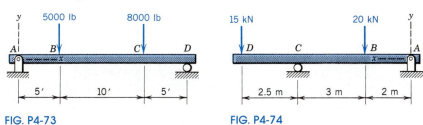

FIG. P4-73   FIG. P4-74

4-74* Draw complete shear and bending moment diagrams for the beam shown in Fig. P4-74.

4-75 Draw complete shear and bending moment diagrams for the beam shown in Fig. P4-75.

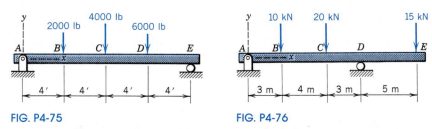

FIG. P4-75   FIG. P4-76

4-76 Draw complete shear and bending moment diagrams for the beam shown in Fig. P4-76.

4-77* A beam is loaded and supported, as shown in Fig. P4-77.

(a) Draw complete shear and bending moment diagrams for the beam.

(b) Write shear and bending moment equations for interval *AB* of the beam.

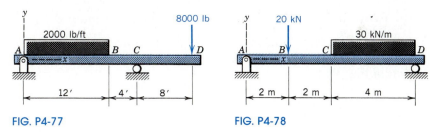

FIG. P4-77   FIG. P4-78

4-78* A beam is loaded and supported, as shown in Fig. P4-78.

(a) Draw complete shear and bending moment diagrams for the beam.

(b) Write shear and bending moment equations for interval *CD* of the beam.

4-79 Draw complete shear and bending moment diagrams for the beam shown in Fig. P4-79.

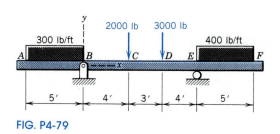

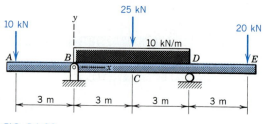

FIG. P4-79

FIG. P4-80

4-80 Draw complete shear and bending moment diagrams for the beam shown in Fig. P4-80.

4-81* A beam is loaded and supported, as shown in Fig. P4-81.

(a) Draw complete shear and bending moment diagrams for the beam.

(b) Write shear and bending moment equations for interval *BC* of the beam.

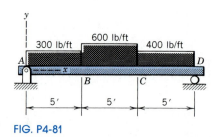

FIG. P4-81

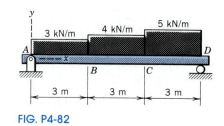

FIG. P4-82

4-82* A beam is loaded and supported, as shown in Fig. P4-82.

(a) Draw complete shear and bending moment diagrams for the beam.

(b) Write shear and bending moment equations for interval *CD* of the beam.

4-83 Draw complete shear and bending moment diagrams for the beam shown in Fig. P4-83.

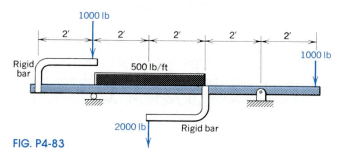

FIG. P4-83

4-84 Draw complete shear and bending moment diagrams for the beam shown in Fig. P4-84.

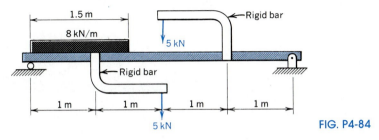

FIG. P4-84

4-85* Draw complete shear and bending moment diagrams for the beam shown in Fig. P4-59.

4-86* Draw complete shear and bending moment diagrams for the beam shown in Fig. P4-60.

4-87 Draw complete shear and bending moment diagrams for the beam shown in Fig. P4-61.

4-88 Draw complete shear and bending moment diagrams for the beam shown in Fig. P4-52.

4-89* Select the lightest steel wide-flange or American standard beam (see Appendix C) that may be used for the beam of Fig. P4-89 if the working fiber stress must be limited to 10 ksi.

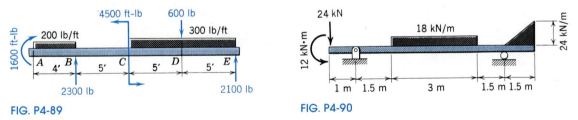

FIG. P4-89

FIG. P4-90

4-90* Select the lightest steel wide-flange or American standard beam (see Appendix C) that may be used for the beam of Fig. P4-90 if the working fiber stress must be limited to 120 MPa.

4-91 Select the lightest steel wide-flange or American standard beam (see Appendix C) that may be used for the beam of Fig. P4-91 if the working fiber stress must be limited to 12.5 ksi.

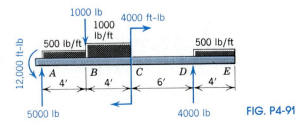

FIG. P4-91

4-92 Select the lightest steel wide-flange or American standard beam (see Appendix C) that may be used for the beam of Fig. P4-92 if the working fiber stress must be limited to 75 MPa.

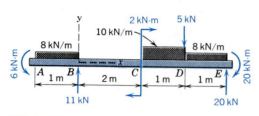

FIG. P4-92

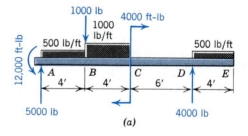

FIG. P4-93

4-93 Select the lightest pair of structural steel angles (see Appendix C) that may be used for the beam of Fig. P4-93 if the working fiber stress must be limited to 10 ksi. The angles will be fastened back-to-back to form a T-section.

4-94 Select the lightest pair of structural steel angles (see Appendix C) that may be used for the beam of Fig. P4-94 if the working fiber stress must be limited to 60 MPa. The angles will be fastened back-to-back to form a T-section.

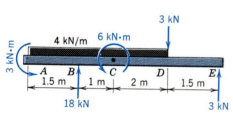

FIG. P4-94

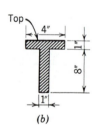

FIG. P4-95

4-95* The beam shown in Fig. P4-95a has the cross section shown in Fig. P4-95b. Determine

(a) the maximum tensile fiber stress in the beam

(b) the maximum compressive fiber stress in the beam.

4-96* The beam shown in Fig. P4-96a has the cross section shown in Fig. P4-96b. Determine

(a) the maximum tensile fiber stress is the beam.

(b) the maximum compressive fiber stress in the beam.

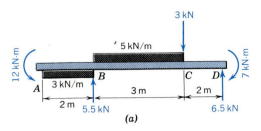
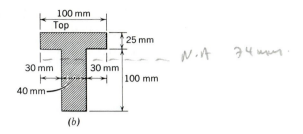

FIG. P4-96

4-97 A W8 × 40 steel beam (see Appendix C) is loaded and supported as shown in Fig. P4-97. The total length of the beam is 10 ft. If the maximum allowable fiber stress is 12,000 psi, determine the maximum permissible value for the distributed load w.

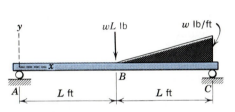

FIG. P4-97

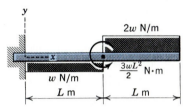

FIG. P4-98

4-98 An S203 × 34 steel beam (see Appendix C) is loaded and supported as shown in Fig. P4-98. The total length of the beam is 4 m. If the maximum allowable fiber stress is 125 MPa, determine the maximum permissible value for the distributed load w.

4-99* An S15 × 50 steel beam (see Appendix C) is loaded and supported as shown in Fig. P4-99. The total length of the beam is 15 ft. If the maximum allowable fiber stress is 15,000 psi, determine the maximum permissible value for the distributed load w.

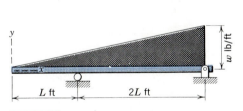

FIG. P4-99

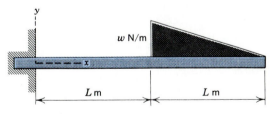

FIG. P4-100

4-100* A W152 × 24 steel beam (see Appendix C) is loaded and supported as shown in Fig. P4-100. The total length of the beam is 3 m. If the maximum allowable fiber stress is 100 MPa, determine the maximum permissible value for the distributed load w.

4-101 A timber beam with a rectangular cross section (width equal to one-half the depth) is to be loaded and supported as shown in Fig. P4-101. If the maximum allowable fiber stress in the wood is 1500 psi, determine the size timber required for the beam.

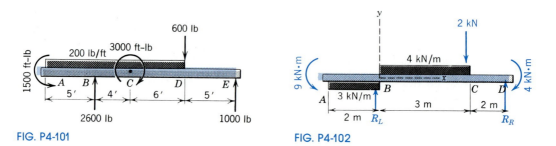

FIG. P4-101

FIG. P4-102

4-102 A timber beam with a rectangular cross section (width equal to one-half the depth) is to be loaded and supported as shown in Fig. P4-102. If the maximum allowable fiber stress in the wood is 12 MPa, determine the size timber required for the beam.

4-103* The beam shown in Fig. P4-103 has a pin-connected joint at C. The moment and shear at A are $+6000$ ft-lb and $+1200$ lb, respectively.

  (a) Draw complete shear and moment diagrams for the length $ABCD$ of the beam.

  (b) Write the moment equation for interval BCD of the beam if the origin of the coordinate axes is located at support B.

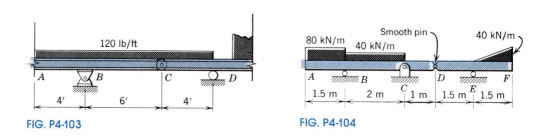

FIG. P4-103

FIG. P4-104

4-104* The beam shown in Fig. P4-104 has a pin-connected joint at D.

  (a) Draw complete shear and moment diagrams for the beam.

  (b) Write the moment equation for interval BC of the beam if the origin of the coordinate axes is located at the left support.

  (c) Write the moment equation for interval EF of the beam if the origin of the coordinate axes is located at the middle support.

4-105 The moment diagram for a beam is shown in Fig. P4-105. Construct the shear and load diagrams for the beam.

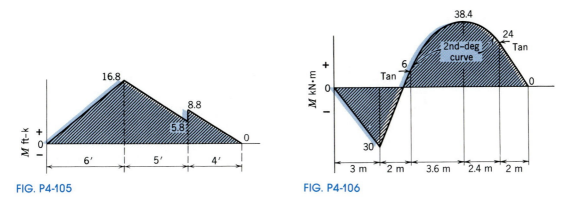

FIG. P4-105

FIG. P4-106

4-106 The moment diagram for a beam is shown in Fig. P4-106. Construct the shear and load diagrams for the beam.

4-107 The moment diagram for a beam is shown in Fig. P4-107. Construct the shear and load diagrams for the beam.

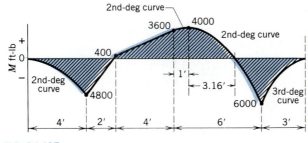

FIG. P4-107

4-108 The moment diagram for a beam is shown in Fig. P4-108. Construct the shear and load diagrams for the beam.

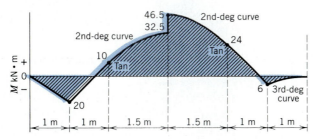

FIG. P4-108

4-8

SHEARING STRESSES BY EQUILIBRIUM

In Section 4-1, the discussion of shearing stresses in beams was bypassed while flexural stresses were studied. This procedure seems to be in keeping with the historical record on the study of beam stresses. From the time of Coulomb's paper, which contained the correct theory of distribution of flexural stresses, approximately seventy years elapsed before the Russian engineer D. J. Jourawski (1821–1891), while designing timber railroad bridges in 1844–1850, developed the elementary shear stress theory used today. In 1856 Saint-Venant developed a rigorous solution for beam shearing stresses, but the elementary solution of Jourawski, being much easier to apply and quite adequate, is the one in general use today by engineers and architects and the one that will be presented here. Note that the method, because it involves the elastic flexure formula, is limited to elastic action. This limitation need be no cause for concern because, for practical engineering problems, the shearing stress evaluation discussed in this and following sections is important only in (1) the evaluation of maximum normal and maxi-

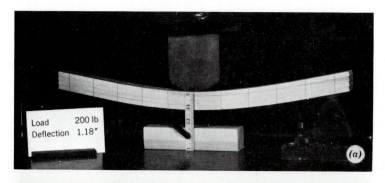

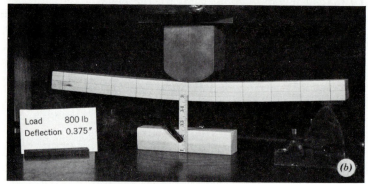

FIG. 4-21

mum shearing stresses at interior points in certain metallic beams (see Section 8-5), and (2) the design of timber beams, because of the longitudinal plane of low shear resistance. The allowable stresses for timber are always in the elastic range. In beams of elastoplastic steel designed on the basis of plastic theory, methods for evaluation of shear stresses are available;[5] however, this area is beyond the scope of this book.

If one constructs a beam by stacking flat slabs one on top of another without fastening them together, and then loads this beam in a direction normal to the surface of the slabs, the resulting deformation will appear somewhat like that in Fig. 4-21a. The same type of deformation can be observed by taking a pack of cards and bending them, noting the relative motion of the ends of the cards with respect to each other. The fact that a solid beam does not exhibit this relative movement of longitudinal elements (see Fig. 4-21b, in which the beam is identical to that of Fig. 4-21a, except that the layers are glued together) indicates the presence of shearing stresses on longitudinal planes. The evaluation of these shearing stresses will now be studied by means of the free-body diagram and equilibrium approach, as outlined in the following example.

EXAMPLE 4-9

A 12-in.-long segment of a beam having the T cross section shown in Fig. 4-22b is subjected to a moment of +5400 ft-lb at the left end and a constant shear of +900 lb. Determine the average shearing stress on a horizontal plane 4 in. above the bottom of the beam.

Solution

A free-body diagram of the segment is shown in Fig. 4-22a, where AB designates the plane on which the shearing stress is to be evaluated. The moment M_2, obtained from a moment equilibrium equation, is +6300 ft-lb. A free-body diagram of the portion of the beam segment below plane AB is shown in Fig. 4-22c. The force V_H is the resultant of the shearing stresses on plane AB. The forces F_1 and F_2 are the resultants of the normal stresses on the ends of the segment and are equal to the average fiber stress (which occurs 6 in. below the neutral surface) on the end area multiplied by the area; that is,

[5]See, for example, *Plastic Design in Steel*, A Guide and Commentary, Am. Soc. Civil Engineers Manual of Engineering Practice No. 41, 1971.

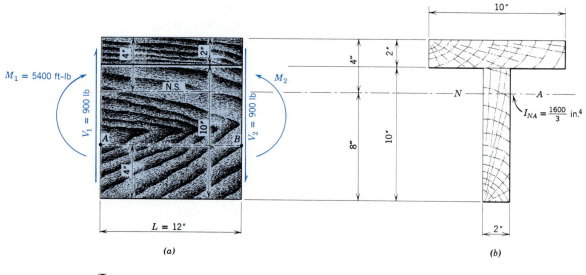

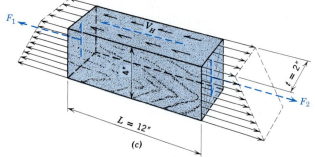

FIG. 4-22

$$F_1 = \frac{M_1 y}{I} A = \frac{5400(12)(6)}{1600/3}(2)(4) = 5832 \text{ lb}$$

and

$$F_2 = \frac{M_2 y}{I} A = \left(\frac{M_2}{M_1}\right) F_1 = \frac{63}{54}(5832) = 6804 \text{ lb}$$

Summing forces in the longitudinal direction in Fig. 4-22c gives

$$V_H = F_2 - F_1 = 6804 - 5832 = 972 \text{ lb}$$

The average shearing stress on plane AB is

$$\tau = \frac{V_H}{A_H} = \frac{972}{12(2)} = \underline{40.5 \text{ psi}} \qquad \text{Ans.}$$

In this example, the shearing stress on plane AB does not vary with the length; therefore, since the shearing stresses on orthogonal planes have the same magnitude, the shearing stresses on the transverse (vertical) planes at A and B are also 40.5 psi in the directions shown.

PROBLEMS

4-109* A 12-in. section of a 6-in.-deep by 3-in.-wide timber beam is shown in Fig. P4-109a. A free-body diagram of the bottom 2 in. of the beam is shown in Fig. P4-109b. Determine

(a) values for the moment M_2 and the forces F_1 and F_2.

(b) the resultant of the shearing stresses on area $ABCD$.

(c) the average shearing stress on area $ABCD$.

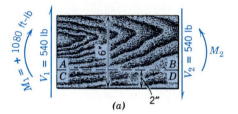

(a)

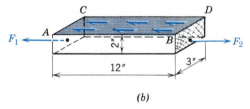

(b) FIG. P4-109

4-110* A beam is fabricated by using four 50×100-mm timbers to form the cross section shown in Fig. P4-110. In a region of constant shear, the two ends of a 300-mm segment of the beam are subjected to bending moments of $+3.5$ kN·m and $+4.1$ kN·m. Determine

(a) the average shearing stress in the glued joint 50 mm below the top surface of the beam.

(b) the average shearing stress in the glued joint 100 mm below the top surface of the beam.

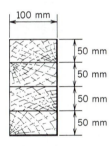

FIG. P4-110

4-111 A timber beam fabricated from 2-in. planks may have either of the cross sections shown in Fig. P4-111. An 8-in. length of the beam

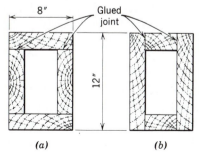

(a)   (b)   FIG. P4-111

for which the shear is constant will be subjected to bending moments of $+12$ ft-kips and $+18$ ft-kips at the ends. Determine

(a) the average shearing stress in each of the joints shown in Fig. P4-111a.

(b) the average shearing stress in each of the joints shown in Fig. P4-111b.

4-112 A timber T-beam consists of two 40×120-mm planks glued together as shown in Fig. P4-112a. A free-body diagram of the flange is shown in Fig. P4-112b. The vertical shear at sections 1–1 and 2–2 is $+4800$ N. The bending moment is $+2860$ N·m at section 1–1. Determine

(a) the bending moment at section 2–2.

(b) values for forces F_1 and F_2.

(c) the shear force transferred from the flange to the stem by the glue on surface $ABCD$.

(d) the average shearing stress at the joint.

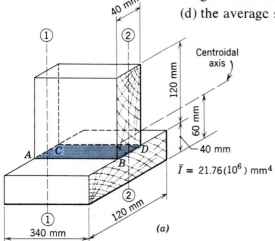

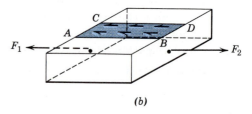

FIG. P4-112

4-113* A timber beam is fabricated from two 2×6-in. and one 2×12-in. pieces of lumber to form the cross section shown in Fig. P4-113. In a region of constant shear, the two ends of a 10-in. segment of the beam are subjected to bending moments of $+9.0$ ft-kips and $+16.5$ ft-kips, respectively. Determine

(a) the shear force V on a cross section of the beam.

(b) the average shearing stress in the glued joints between the horizontal and vertical parts of the cross section.

(c) the average shearing stress on a horizontal plane 2 in. from the bottom of the beam.

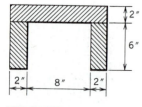

FIG. P4-113

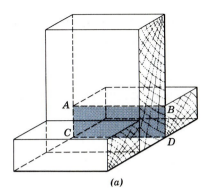

FIG. P4-114

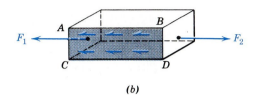

(a)   (b)

4-114* If the beam of Problem 4-112 consists of one 40 × 160-mm plank and two 40 × 40-mm sticks as shown in Fig. P4-114, determine

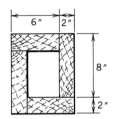

FIG. P4-115

(a) values for forces F_1 and F_2.

(b) the shear force transferred from the flange to the stem by the glue on surface *ABCD*.

(c) the average shearing stress at the joint.

4-115 A timber beam is fabricated from two 2 × 6-in. and two 2 × 8-in. pieces of lumber to form the cross section shown in Fig. P4-115. In a region of constant shear, the two ends of a 12-in. segment of the beam are subjected to bending moments of +18.0 ft-kips and +26.0 ft-kips, respectively. Determine

(a) the shear force V on a cross section of the beam.

(b) the average shearing stress in the horizontal glued joints.

(c) the average shearing stress in the vertical glued joints.

4-116 A timber beam is fabricated from one 40 × 80-mm and two 40 × 120-mm pieces of lumber to form the cross section shown in Fig. P4-116. In a region of constant shear, the two ends of a 300-mm segment of the beam are subjected to bending moments of −8.4 kN·m and −6.6 kN·m, respectively. Determine

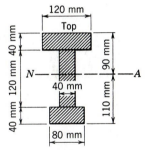

$I_{NA} = 56.75(10^6) \text{ mm}^4$

FIG. P4-116

(a) the shear force V on a cross section of the beam.

(b) the average shearing stress in the glued joint between the top flange and the web.

(c) the average shearing stress in the glued joint between the bottom flange and the web.

4-117* A timber beam is fabricated from one 2 × 8-in. and two 2 × 6-in. pieces of lumber to form the cross section shown in Fig. P4-117. The flanges of the beam are fastened to the web with nails that can safely transmit a shear force of 100 lb. If the beam is simply supported and carries a 1000-lb load at the center of a 12-ft span, determine

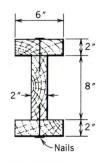

FIG. P4-117

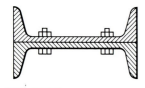

FIG. P4-118

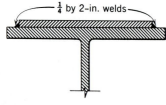

FIG. P4-119

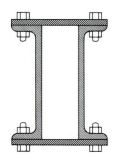

FIG. P4-120

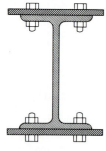

FIG. P4-121

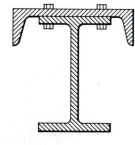

FIG. P4-122

(a) the shear force transferred by the nails from the flange to the web in a 12-in. length of the beam.

(b) the spacing required for the nails.

4-118   A cantilever beam is used to support a concentrated load of 20 kN. The beam is fabricated by bolting two C457 × 86 steel channels (see Appendix C) back-to-back to form the H-section shown in Fig. P4-118. If the pairs of bolts are spaced at 300-mm intervals along the beam, determine

(a) the shear force carried by each of the bolts.

(b) the bolt diameter required if the shear and bearing stresses for the bolts must be limited to 60 MPa and 125 MPa, respectively.

4-119*   A W18 × 97 steel beam (see Appendix C) will have $10 \times \frac{1}{2}$-in. cover plates welded to the top and bottom flanges as shown in Fig. P4-119. A 20-in. length of the beam for which the shear is constant will be subjected to bending moments of +4600 in.-kips and +2300 in.-kips at the ends. The fillet weld has an allowable load of 2400 lb per in. Determine the number of fillet welds, each 2-in. long, required on each side of the cover plate.

4-120*   A box beam will be fabricated by bolting two 15 × 260-mm steel plates to two C305 × 45 steel channels (see Appendix C) as shown in Fig. P4-120. The beam will be simply supported at the ends and will carry a concentrated load of 125 kN at the center of a 5-m span. Determine the bolt spacing required if the bolts have a diameter of 20 mm and an allowable shearing stress of 150 MPa.

4-121   A W21 × 101 steel beam (see Appendix C) is simply supported at the ends and carries a concentrated load at the center of a 20-ft span. The concentrated load must be increased to 125 kips, which requires that the beam be strengthened. It has been decided that two $\frac{3}{4} \times$ 16-in. steel plates will be bolted to the flanges, as shown in Fig. P4-121. Determine the bolt spacing required if the bolts have a diameter of $\frac{3}{4}$ in. and an allowable shearing stress of 17.5 ksi.

4-122   A W356 × 122 steel beam (see Appendix C) has a C381 × 74 channel bolted to the top flange, as shown in Fig. P4-122. The beam is simply supported at the ends and carries a concentrated load of 96 kN at the center of an 8-m span. If the pairs of bolts are spaced at 500-mm intervals along the beam, determine

(a) the shear force carried by each of the bolts.

(b) the bolt diameter required if the shear and bearing stresses for the bolts must be limited to 60 MPa and 125 MPa, respectively.

4-9

THE SHEARING STRESS FORMULA

The shearing stress as determined in Example 4-9 is reasonably accurate only if the width of the section on plane AB is not large compared to the depth of the beam and the transverse shear in the beam is constant throughout the length of the segment AB. If the transverse shear varies, the length L must be reduced to a short length Δx. The method outlined in the example, though excellent for emphasizing the concept of horizontal shear, is too laborious for general use; therefore, a formula will be developed that will be generally applicable within limits imposed by the assumptions. For this development refer to Fig. 4-23, which is a free-body diagram similar to that of Fig. 4-22c. The force dF_1 is the normal force acting on a differential area dA and is equal to $\sigma\, dA$. The resultant of these differential forces is F_1 (not shown). Thus, $F_1 = \int \sigma\, dA$ integrated over the shaded area of the cross section, where σ is the fiber stress at a distance y from the neutral surface and is given by the expression $\sigma = My/I$. When the two expressions are combined, the force F_1 becomes

$$F_1 = \frac{M}{I} \int y\, dA = \frac{M}{I} \int_h^c ty\, dy$$

Similarly, the resultant force on the right side of the element is

$$F_2 = \frac{(M + \Delta M)}{I} \int_h^c ty\, dy$$

The summation of forces in the horizontal direction in Fig. 4-23 yields

$$V_H = F_2 - F_1 = \frac{\Delta M}{I} \int_h^c ty\, dy$$

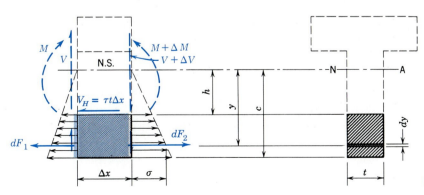

FIG. 4-23

The average shearing stress is V_H divided by the area, from which

$$\tau = \lim_{\Delta x \to 0} \frac{\Delta M}{\Delta x} \left(\frac{1}{It} \right) \int_h^c ty \, dy = \frac{dM}{dx} \left(\frac{1}{It} \right) \int_h^c ty \, dy \qquad (a)$$

and from Eq. 4-11, dM/dx equals V, the shear at the beam section where the stress is to be evaluated. Also, the integral is the first moment of that portion of the cross-sectional area between the transverse line where the stress is to be evaluated and the extreme fiber of the beam. The integral is designated Q, and when V and Q are substituted into Eq. a, the formula for horizontal (or longitudinal) shearing stress becomes

$$\tau = \frac{VQ}{It} \qquad (4\text{-}13)$$

Because the flexure formula was used in the derivation, Eq. 4-13 is subject to the same assumptions and limitations. Although the stress given by Eq. 4-13 is associated with a particular point in a beam, it is averaged across the thickness t and hence is accurate only if t is not too great. For a rectangular section having a depth twice the width, the maximum stress as computed by Saint-Venant's more rigorous method is about 3 percent greater than that given by Eq. 4-13. If the beam is square, the error is about 12 percent. If the width is four times the depth, the error is almost 100 percent, from which one must conclude that, if Eq. 4-13 were applied to a point in the flange of the beam of Example 4-9, the result would be worthless. Furthermore, if Eq. 4-13 is applied to sections where the sides of the beam are not parallel, such as a triangular section, the average stress is subject to additional error because the transverse variation of stress is greater when the sides are not parallel.

As noted previously, at each point in a beam, the horizontal (longitudinal) and vertical (transverse) shearing stresses have the same magnitude; hence, Eq. 4-13 gives the vertical shearing stress at a point in a beam (averaged across the width). The variation of the transverse shearing stress in a beam will be demonstrated for the T-shaped beam of Example 4-9. Note that in Eq. 4-13, V and I are constant for any section, and only Q and t vary for different points in the section. The transverse shearing stress at any point in the stem of the T-section a distance y_O from the neutral axis is, from Fig. 4-24a and Eq. 4-13,

$$\tau = \frac{V}{It_1} \int_{y_O}^8 yt_1 \, dy = \frac{V}{2I} (8^2 - y_O^2) \qquad (-2 \le y_O \le 8)$$

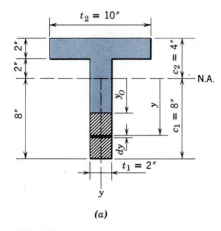

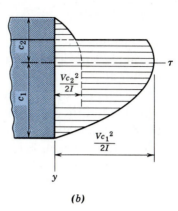

(a)   (b)

FIG. 4-24

Note that y_O is shown as positive downward. An expression for the average shearing stress in the flange can be written in a similar manner and is

$$\tau = \frac{V}{2I}(4^2 - y_O^2) \qquad (-4 \le y_O \le -2)$$

These are parabolic equations for the theoretical stress distribution, and the results are shown in Fig. 4-24b. The diagram has a discontinuity at the junction of the flange and stem because the thickness of the section changes abruptly. As pointed out previously, the distribution in the flange is fictitious, since the stress at the bottom of the flange must be zero (a free surface). Also, because of the abrupt change in section, a stress concentration will exist, and the maximum shearing stress at the junction of flange and stem will be more than the average stress. From Fig. 4-24b and Eq. 4-13, one may conclude that in general the maximum[6] longitudinal and transverse shearing stress occurs at the neutral surface at a section where the transverse shear is maximum. There may be some exceptions such as a beam with a cross section in the form of a triangle or a Greek cross with Q/t at the neutral surface less than the value some distance from the neutral surface. The following example illustrates the use of the shearing stress formula.

---

[6]In this book the term *maximum*, as applied to a longitudinal and transverse shearing stress, will mean the average stress across the thickness t at a point where such average has the maximum value.

EXAMPLE 4-10

A beam having the T cross section shown in Fig. 4-22b of Example 4-9 is simply supported and carries a concentrated load of 1800 lb at the center of a 15-ft span, as shown in Fig. 4-25a. Determine

(a) the average shearing stress on a horizontal plane 4 in. above the bottom of the beam and 6 ft from the left support.

(b) the maximum vertical shearing stress in the beam.

(c) the average shearing stress in the joint between the flange and the stem at a point 5 ft from the left support.

(d) the force transmitted from the flange to the stem by the glue in a 12-in. length of the joint centered 6 ft from the left support.

(e) the maximum tensile fiber stress in the beam.

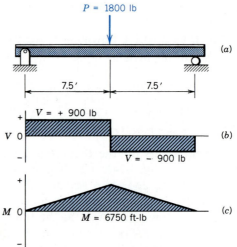

FIG. 4-25

Solution

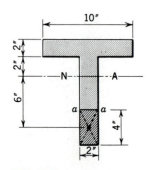

FIG. 4-26

(a) The moment of inertia for the cross section, as indicated in Fig. 4-22b, is 1600/3 in.⁴. The shear force V on a cross section 6 ft from the left support is $+900$ lb, as shown on Fig. 4-25b. The first moment Q for the shaded area below line a–a on Fig. 4-26 is

$$Q_{4''} = 4(2)(6) = 48 \text{ in.}^3$$

Finally,

$$\tau = \frac{VQ}{It} = \frac{900(48)}{(1600/3)(2)} = \underline{40.5 \text{ psi}} \qquad \text{Ans.}$$

(b) The maximum vertical shearing stress in the beam will occur at the neutral axis on the cross section supporting the largest shear force V. Thus

$$Q_{NA} = \overset{A\ \bar{y}}{8(2)(4)} = 64 \text{ in.}^3$$

Since the shear force V equals 900 lb on all cross sections of the beam,

$$\tau_{max} = \frac{V_{max}Q_{NA}}{It} = \frac{900(64)}{(1600/3)(2)} = \underline{54.0 \text{ psi}} \qquad \text{Ans.}$$

(c) The average shearing stress in the joint between the flange and the stem depends on the shear force V at the cross section 5 ft from the left support and the moment of the area of the flange about the neutral axis. Thus

$$Q_j = 2(10)(3) = 60 \text{ in.}^3$$

Alternatively, the moment of the area of the stem about the neutral axis can be used to calculate Q for the joint. Thus

$$Q_j = 2(10)(3) = 60 \text{ in.}^3$$

Since the shear force V at the cross section is 900 lb,

$$\tau_{joint} = \frac{VQ_j}{It} = \frac{900(60)}{(1600/3)(2)} = \underline{50.6 \text{ psi}} \qquad \text{Ans.}$$

(d) The force transmitted from the flange to the stem by the glue is

$$V_g = \tau_j A_j = 50.625(12)(2) = \underline{1215 \text{ lb}} \qquad \text{Ans.}$$

(e) The maximum tensile fiber stress in the beam will occur in a fiber at the bottom of the beam, since the resisting moment at all cross sections of the beam is positive. The largest moment occurs at midspan, as shown in Fig. 4-25c. Thus from Eq. 4-7,

$$\sigma_{max} = \frac{Mc}{I} = \frac{6750(12)(8)}{(1600/3)} = \underline{1215 \text{ psi}} \qquad \text{Ans.}$$

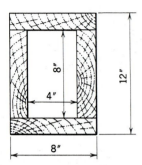

8"

4"

8"

12"

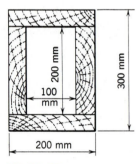

200 mm

100 mm

200 mm

300 mm

FIG. P4-124

PROBLEMS

4-123* The transverse shear V at a certain section of a timber beam is 7500 lb. If the beam has the cross section shown in Fig. P4-123, determine

(a) the horizontal shearing stress in the glued joint 2 in. below the top of the beam.

(b) the vertical shearing stress at a point 3 in. below the top of the beam.

(c) the magnitude and location of the maximum vertical shearing stress on the cross section.

4-124* The transverse shear V at a certain section of a timber beam is 25 kN. If the beam has the cross section shown in Fig. P4-124, determine

(a) the horizontal shearing stress in the glued joint 50 mm above the bottom of the beam.

(b) the vertical shearing stress at a point 80 mm above the bottom of the beam.

(c) the magnitude and location of the maximum vertical shearing stress on the cross section.

4-125* The cross section of a timber beam is a rectangle 6-in. wide by 12-in. deep. The beam is simply supported and carries a concentrated load of 18 kips at the center of a 12-ft span. Determine

(a) the horizontal shearing stress at a point 3 in. from the top of the beam and 3 ft from the left support.

(b) the maximum vertical shearing stress in the beam.

(c) the maximum tensile fiber stress in the beam.

4-126* The cross section of a timber beam is a rectangle 200-mm wide by 500-mm deep. The beam is simply supported and carries a uniformly distributed load of 40 kN/m over its entire 6-m span. Determine

(a) the maximum vertical shearing stress in the beam.

(b) the vertical shearing stress at a point 80 mm above the bottom of the beam and 1 m from the left support.

(c) the maximum compressive fiber stress in the beam.

4-127 A timber beam (simply supported) is needed to support a concentrated load of 10 kips at the center of a 10-ft span. The maximum tensile fiber stress in the beam must not exceed 1800 psi and the maximum horizontal shearing stress in the beam must not exceed 125 psi. Determine the width of beam required if the depth of the beam must be 12 in.

4-128 A timber beam (simply supported) is needed to support a uniformly distributed load of 25 kN/m over its entire 5-m span. The maximum tensile fiber stress in the beam must not exceed 10 MPa and the maximum horizontal shearing stress in the beam must not exceed 0.750 MPa. Determine the depth of beam required if the width of the beam must be 200 mm.

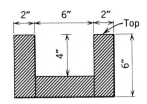

FIG. P4-129

4-129* A timber beam is simply supported and carries a uniformly distributed load of 320 lb/ft over its entire 18-ft span. If the beam has the cross section shown in Fig. P4-129, determine

- (a) the vertical shearing stress at a point 2 ft from the right end and 2 in. below the top surface of the beam.
- (b) the magnitude and location of the maximum horizontal shearing stress in the beam.
- (c) the maximum tensile fiber stress in the beam.

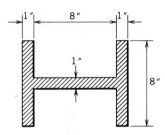

FIG. P4-130

4-130* A timber beam 4-m long is simply supported at its ends and carries a uniformly distributed load of 10 kN/m over its entire length. If the beam has the cross section shown in Fig. P4-130, determine

- (a) the vertical shearing stress at a point 500 mm from the left end and 40 mm below the neutral surface.
- (b) the magnitude and location of the maximum horizontal shearing stress in the beam.
- (c) the magnitude and location of the maximum horizontal shearing stress in the glued joints between the web and flanges of the beam.

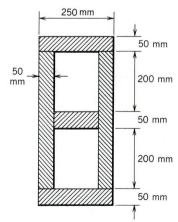

FIG. P4-131

4-131 A structural steel cantilever beam 6-ft long supports a concentrated load of 5 kips at the free end. If the beam has the cross section shown in Fig. P4-131, determine

- (a) the horizontal shearing stress at a point 1 in. from the top of the beam and 1 ft from the left support.
- (b) the maximum horizontal shearing stress in the beam.
- (c) the maximum tensile fiber stress in the beam.

4-132 A timber beam is simply supported and carries a concentrated load of 50 kN at the center of a 4-m span. If the beam has the cross section shown in Fig. P4-132, determine

- (a) the horizontal shearing stress in the glued joint 50 mm below the top of the beam and 1 m from the left support.
- (b) the maximum horizontal shearing stress in the beam.
- (c) the maximum tensile fiber stress in the beam.

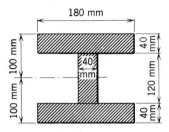

FIG. P4-132

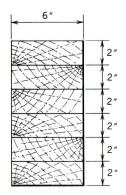

FIG. P4-133

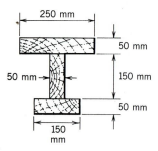

FIG. P4-134

4-133* A laminated wood beam consists of six 2 × 6-in. planks glued together to form a section 6 in. wide by 12 in. deep as shown in Fig. P4-133. If the strength of the glue in shear is 120 psi, determine

(a) the maximum uniformly distributed load w that can be applied over the full length of the beam if the beam is simply supported and has a span of 15 ft.

(b) the horizontal shearing stress in the glued joint 2 in. below the top of the beam and 2 ft from the left support when the load of part (a) is applied.

(c) the maximum tensile fiber stress in the beam when the load of part (a) is applied.

4-134* A timber beam is simply supported and carries a uniformly distributed load of 10 kN/m over the full length of the beam. If the beam has the cross section shown in Fig. P4-134 and a span of 6 m, determine

(a) the horizontal shearing stress in the glued joint 50 mm below the top of the beam and 1 m from the left support.

(b) the horizontal shearing stress in the glued joint 50 mm above the bottom of the beam and $\frac{1}{2}$ m from the left support.

(c) the maximum horizontal shearing stress in the beam.

(d) the maximum tensile fiber stress in the beam.

4-135 The segment of a timber beam shown in Fig. P4-135a has the cross section shown in Fig. P4-135b. At section b–b, the shear is −3000 lb, and the bending moment is +7500 ft-lb. Determine

(a) the fiber stress and the vertical shearing stress at point A of the cross section at section a–a.

(b) the magnitude and location of the maximum horizontal shearing stress in the beam segment.

(c) the magnitude and location of the maximum compressive fiber stress in the beam segment.

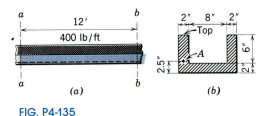

FIG. P4-135

4-136 The segment of a timber beam shown in Fig. P4-136a has the cross section shown in Fig. P4-136b. At section a–a, the shear is +10 kN, and the bending moment is −4 kN·m. Determine

(a) the fiber stress and the vertical shearing stress 40 mm below the top of the beam at section *b–b*.

(b) the magnitude and location of the maximum horizontal shearing stress in the glued joint between the flange and the stem in the beam segment.

(c) the magnitude and location of the maximum tensile fiber stress in the beam segment.

4-137 The beam shown in Fig. P4-137*a* is composed of two 1 × 6-in. and two 1 × 3-in. hard maple boards that are glued together, as shown in Fig. P4-137*b*. Determine the magnitude and location of

(a) the maximum tensile fiber stress in the beam.

(b) the maximum horizontal shearing stress in the beam.

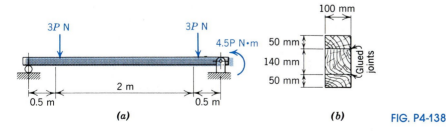

(*a*)

(*b*)

FIG. P4-137

4-138* The beam shown in Fig. P4-138*a* is composed of three pieces of timber that are glued together, as shown in Fig. P4-138*b*. If the maximum horizontal shearing stresses are not to exceed 500 kPa in the glued joints or 750 kPa in the wood and if the maximum tensile and compressive fiber stresses in the beam are not to exceed 15 MPa, determine the maximum permissible value for *P*.

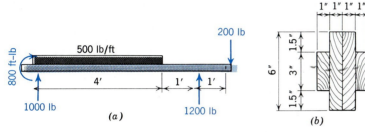

(*a*)

(*b*)

FIG. P4-138

4-139 The timber beam shown in Fig. P4-139*a* is fabricated by glueing two 1 × 5-in. and two 1 × 4-in. boards together, as shown in Fig. P4-139*b*. If the maximum horizontal shearing stresses are not to exceed 175 psi in the glued joints or 250 psi in the wood, and if the maximum

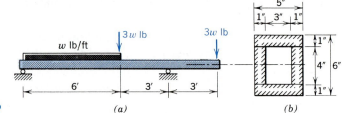

FIG. P4-139   (a)   (b)

tensile and compressive fiber stresses in the beam are not to exceed 2000 psi, determine the maximum permissible value for the distributed load w.

4-140 The cross section of the simple beam shown in Fig. P4-140 is a T-section formed by glueing two 40 × 120-mm boards together. If the maximum horizontal shearing stresses are not to exceed 750 kPa in the glued joint or 800 kPa in the wood, and if the maximum tensile and compressive fiber stresses in the beam are not to exceed 12.5 MPa, determine the maximum permissible value for the distributed load w.

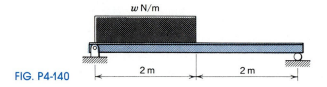

FIG. P4-140

4-141* A WT8 × 20 structural tee (see Appendix C) is loaded and supported as a beam (with the flange on top), as shown in Fig. P4-141. Determine

   (a) the maximum tensile fiber stress in the beam.
   (b) the maximum compressive fiber stress in the beam.
   (c) the maximum vertical shearing stress in the beam.
   (d) the vertical shearing stress at a point in the stem just below the flange on the cross section where the maximum vertical shearing stress occurs.

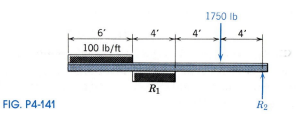

FIG. P4-141

4-142* A W610 × 92 steel beam (see Appendix C) is loaded and supported, as shown in Fig. P4-142. Determine

(a) the maximum tensile fiber stress in the beam.

(b) the maximum vertical shearing stress in the beam.

(c) the vertical shearing stress at a point in the web just below the flange on the cross section where the maximum vertical shearing stress occurs.

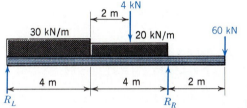

FIG. P4-142

4-143 A WT12 × 47 structural tee (see Appendix C) is loaded and supported as a beam (with the flange on bottom), as shown in Fig. P4-143. Determine

(a) the maximum tensile fiber stress in the beam.

(b) the maximum compressive fiber stress in the beam.

(c) the maximum vertical shearing stress in the beam.

(d) the vertical shearing stress at a point in the stem just above the flange on the cross section where the maximum vertical shearing stress occurs.

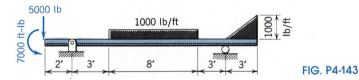

FIG. P4-143

4-144 A WT305 × 77 structural tee (see Appendix C) is loaded and supported as a beam (with the flange on top) as shown in Fig. P4-144. Determine

(a) the maximum tensile fiber stress in the beam.

(b) the maximum compressive fiber stress in the beam.

(c) the maximum vertical shearing stress in the beam.

(d) the vertical shearing stress at a point in the stem just below the flange on the cross section where the maximum vertical shearing stress occurs.

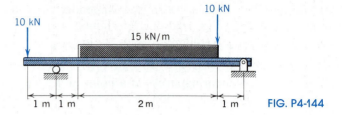

FIG. P4-144

4-10

STRESS CONCENTRATIONS UNDER FLEXURAL LOADINGS

In Section 2-6, it was shown that the introduction of a circular hole or other geometric discontinuity into an axially loaded member can cause a significant increase in the magnitude of the stress (stress concentration) in the immediate vicinity of the discontinuity. This is also the case for circular shafts under torsional forms of loading (see Section 3-8) and for flexural members.

In Section 4-4 of this chapter it was shown that the fiber stress in a beam of uniform cross section in a region of pure bending is given by Eq. 4-7 as:

$$\sigma = \frac{My}{I} \tag{4-7}$$

If a notch, hole, or other type of discontinuity is introduced into the flexural member, the magnitude of the fiber stress in the vicinity of the discontinuity will increase significantly. Similarly, the magnitude of the fiber stress in regions of abrupt change in cross section will increase significantly unless adequate fillets are introduced to smooth the transition from one size to the other. In these cases, the magnitude of the fiber stress at the notch, hole, or fillet can be expressed in terms of a stress-concentration factor K as

$$\sigma = K\left(\frac{My}{I}\right) \tag{4-14}$$

Since the factor K depends only upon the geometry of the member, curves can be developed that show the stress-concentration factor K as a function of the ratios of the parameters involved. Such curves (based on a net section) for fillets, holes, and grooves in flexural members are shown in Fig. 4-27.

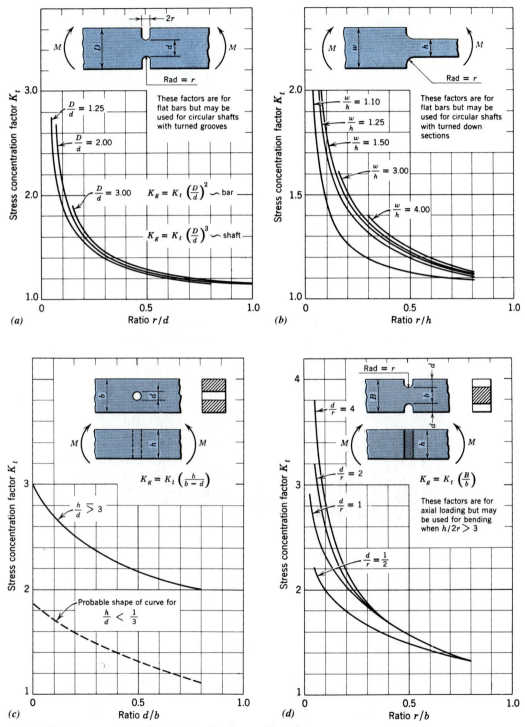

FIG. 4-27 Stress-concentration factors for flexural loading.

EXAMPLE 4-11

The cantilever spring shown in Fig. 4-28 is made of SAE 4340 heat-treated steel and is 50 mm wide. Determine the maximum safe moment M if a factor of safety k of 2.5 with respect to failure by fracture is specified and

(a) the radius r is 5 mm.

(b) the radius r is 10 mm.

(c) the radius r is 15 mm.

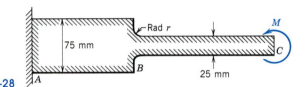

FIG. 4-28

Solution

The ultimate strength σ_u for heat-treated SAE 4340 steel (see Appendix A) is 1030 MPa. Thus, the allowable working stress σ_w is

$$\sigma_w = \frac{\sigma_u}{k} = \frac{1030}{2.5} = 412 \text{ MPa}$$

The moment of inertia I about the neutral axis for bending of the spring is

$$I = \frac{bh^3}{12} = \frac{50(25^3)}{12} = 65.10(10^3) \text{ mm}^4$$

The stress-concentration factor K_t for the fillet is obtained from Fig. 4-27b.

For $r = 5$ mm: $w/h = 75/25 = 3$, $r/h = 5/25 = 0.20$, which gives $K_t = 1.51$

For $r = 10$ mm: $w/h = 75/25 = 3$, $r/h = 10/25 = 0.40$, which gives $K_t = 1.28$

For $r = 15$ mm: $w/h = 75/25 = 3$, $r/h = 15/25 = 0.60$, which gives $K_t = 1.18$

Thus, from Eq. 4-14:

$$M = \frac{\sigma I}{K_t y} = \frac{412(10^6)(65.10)(10^{-9})}{(25/2)(10^{-3})(K_t)} = \frac{2146}{K_t}$$

(a) $$M = \frac{2146}{K_t} = \frac{2146}{1.51} = \underline{1421 \text{ N·m}}$$ Ans.

(b) $$M = \frac{2146}{K_t} = \frac{2146}{1.28} = \underline{1677 \text{ N·m}}$$ Ans.

(c) $$M = \frac{2146}{K_t} = \frac{2146}{1.18} = \underline{1819 \text{ N·m}}$$ Ans.

PROBLEMS

4-145* A stainless-steel spring, similar to the one shown in Fig. 4-28, has a width of $\frac{3}{4}$ in. and a change in depth at section B from $\frac{3}{8}$ in. to $\frac{1}{4}$ in. Determine the minimum acceptable radius for the fillet if the stress-concentration factor must not exceed 1.40.

4-146* An alloy-steel spring, similar to the one shown in Fig. 4-28, has a width of 25 mm and a change in depth at section B from 75 mm to 60 mm. If the radius of the fillet between the two sections is 6 mm, determine the maximum moment that the spring can resist if the maximum fiber stress in the spring must not exceed 100 MPa.

4-147 A stainless-steel bar $\frac{3}{4}$-in. wide by $\frac{3}{8}$-in. deep has a pair of semi-circular grooves cut in the edges of the bar (from top to bottom). If the grooves have a $\frac{1}{16}$-in. radius, determine the percent reduction in strength for flexural-type loadings.

4-148 A timber beam 150 mm wide by 200 mm deep has a 25-mm-diameter hole drilled from top to bottom of the beam on the centerline of a cross section. Determine the percent reduction in strength produced by the presence of the hole.

4-149* A 3-in.-diameter 0.4%C hot-rolled steel (see Appendix A) shaft has a reduced diameter of 2.73 in. for 12 in. of its length as shown in Fig. P4-149. If the tool used to turn down the section had a radius of 0.25 in., determine the maximum allowable bending load P that can be applied to the end of the shaft if a factor of safety of 3 with respect to failure by slip is specified.

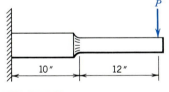

FIG. P4-149

4-150* A 0.4%C hot-rolled steel bar (see Appendix A) with a rectangular cross section will be loaded as a cantilever beam. The bar has a depth h of 200 mm and has a 25-mm-diameter hole drilled from top to bottom of the beam on the centerline of a cross section where a bending moment of 60 kN·m must be supported. If a factor of safety of 4 with respect to failure by slip is specified, determine the minimum acceptable width b for the bar.

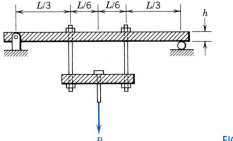

FIG. P4-151

4-151 A load of 5000 lb is supported by the beams shown in Fig. P4-151. The beams are 2 in. wide by 4 in. deep. The holes for the threaded rods have a $\frac{5}{8}$-in. diameter. If the maximum tensile fiber stresses in the beams must not exceed 20 ksi, determine the maximum permissible span for the top beam.

4-152 A steel bar with a rectangular cross section will be loaded as a cantilever beam. The bar has a 10-mm-diameter hole drilled from top to bottom of the beam on the centerline of a cross section where a moment of 10 kN·m must be supported without exceeding a stress of 200 MPa. Prepare a curve showing the acceptable combinations of beam width b and beam depth h. Limit the minimum depth of the beam to 30 mm, so that $h/d > 3$.

4-153 Two beams with rectangular cross sections, similar to the ones shown in Fig. P4-151, support a concentrated load P of 4.5 kips. The top beam has a span of 42 in. If the maximum tensile fiber stress in the top beam must not exceed 15 ksi, prepare a curve showing the acceptable combinations of beam width b and beam depth h if the hole diameters are $\frac{3}{4}$ in. Limit the minimum depth of the beam to 2.5 in., so that $h/d > 3$.

4-11

FLEXURAL STRESSES IN BEAMS OF TWO MATERIALS

The method of fiber stress computation covered in Section 4-3 is sufficiently general to cover symmetrical beams composed of longitudinal elements (layers) of different materials. However, for many real beams of two materials (often referred to as reinforced beams), a method can be developed to allow the use of the elastic flexure formula, thus reducing the computational labor involved. The method is applicable to elastic design only.

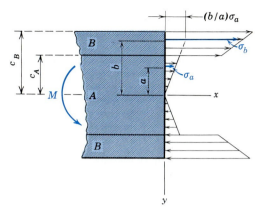

FIG. 4-29

The assumption of the plane section remaining plane is still valid, provided the different materials are securely bonded together to provide the necessary resistance to longitudinal shearing stresses. Therefore, the usual linear transverse distribution of longitudinal strains is valid.

The beam of Fig. 4-29, composed of a central portion of material A and two outer layers of material B, will serve as a model for the development of the stress distribution. The section is assumed to be symmetrical with respect to the xy and xz planes, and the moment is applied in the xy plane. As long as neither material is subjected to stresses above the proportional limit, Hooke's law applies and the strain relation

$$\epsilon_b = \epsilon_a \frac{b}{a}$$

becomes

$$\frac{\sigma_b}{E_B} = \frac{\sigma_a}{E_A}\left(\frac{b}{a}\right)$$

or

$$\sigma_b = \sigma_a \left(\frac{E_B}{E_A}\right)\left(\frac{b}{a}\right) \qquad (a)$$

From this relation it is evident that, at the junction between the two materials where distances a and b are equal, there is an abrupt change in stress determined by the ratio $n = E_B/E_A$ of the two moduli. If Eq. a is used, the normal force on a differential end area of element B is given by the expression

$$dF_B = \sigma_b\, dA = n\frac{b}{a}\sigma_a\, dA = \left(\frac{b}{a}\sigma_a\right)(nt)\, dy \qquad c_A \le y \le c_B$$

where t is the width of the beam at a distance b from the neutral surface. The first factor in parentheses represents the linear stress distribution in a homogeneous material A. The second factor in parentheses may be interpreted as the extended width of the beam from $y = c_A$ to $y = c_B$ if material B were replaced by material A, thus resulting in an equivalent or transformed *cross section* for a beam of homogeneous material. The transformed section is obtained by replacing either material by an equivalent amount of the other material as determined by the ratio n of their elastic moduli. The method is not limited to two materials; however, the use of more than two materials in one beam would be unusual. The method is illustrated by the following example.

EXAMPLE 4-12

A timber beam 4 in. wide by 8 in. deep has an aluminum alloy plate with a net section $\frac{10}{3}$ in. wide by $\frac{1}{2}$ in. deep securely fastened to its bottom face. The working stresses in the timber and aluminum alloy are 1200 and 16,000 psi, respectively. The moduli of elasticity for the timber and aluminum alloy are 1250 and 10,000 ksi, respectively. Determine the maximum allowable resisting moment for the beam.

Solution

The ratio of the moduli of the aluminum and timber is

$$n = 10{,}000/1250 = 8$$

The actual cross section (timber A and aluminum B) and the transformed timber cross section are shown in Figs. 4-30a and b. The neutral axis of the transformed section is located by the principle of moments as

$$\bar{y} = \frac{\Sigma M}{\Sigma A} = \frac{(40/3)(1/4) + 32(9/2)}{(40/3) + 32} = 3.25 \text{ in.}$$

above the bottom of the aluminum plate. The moment of inertia of the cross-sectional area with respect to the neutral axis is

$$I_c = \frac{1}{12}\left(\frac{80}{3}\right)\left(\frac{1}{2}\right)^3 + \frac{40}{3}(3)^2 + \frac{1}{12}(4)(8)^3 + 32\left(\frac{3}{2}\right)^2 = 341 \text{ in.}^4$$

The stress relation

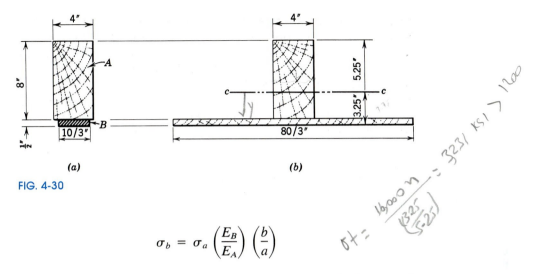

FIG. 4-30

$$\sigma_b = \sigma_a \left(\frac{E_B}{E_A}\right)\left(\frac{b}{a}\right)$$

will be used to determine which working stress controls the capacity of the section. From Fig. 4-30 it is seen that the distance to the outermost timber fiber is 5.25 in. and that the distance to the outermost aluminum fiber is 3.25 in. With 1200 psi for σ_a the maximum stress in the aluminum is

$$\sigma_b = 1200(8)(3.25/5.25) = 5940 \text{ psi}$$

which is well below the allowable stress. Therefore, the stress in the timber is the controlling stress. From the flexure formula

$$M = \frac{\sigma I}{c} = \frac{1200(341)}{5.25} = 77,900 \text{ in.-lb} \qquad \text{Ans.}$$

PROBLEMS

4-154* A timber beam 150 mm wide by 350 mm deep has a 150-mm-wide by 15-mm-deep steel plate fastened securely to its top face. The moduli of elasticity for the timber and steel are 10 GPa and 200 GPa, respectively. Determine the maximum fiber stress in the timber when the maximum fiber stress in the steel is 75 MPa *T*.

4-155* A timber beam 6 in. wide by 12 in. deep has a 6-in.-wide by ½-in.-deep steel plate fastened securely to its bottom face. The moduli of elasticity for the timber and steel are 1500 ksi and 30,000 ksi, respectively. Determine the maximum fiber stress in the steel when the maximum fiber stress in the timber is 1250 psi *C*.

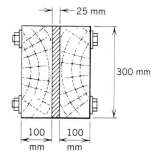

FIG. P4-156

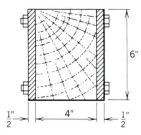

FIG. P4-157

4-156 A composite beam 225 mm wide by 300 mm deep by 4 m long is made by bolting two 100-mm-wide by 300-mm-deep timber planks to the sides of a 25-mm-wide by 300-mm-deep structural aluminum plate, as shown in Fig. P4-156. The moduli of elasticity of the timber and aluminum are 8 GPa and 73 GPa, respectively. Determine the maximum tensile fiber stress in each of the materials if the composite beam is simply supported and carries a uniformly distributed load of 7.5 kN/m over its entire length.

4-157 A 4-in.-wide by 6-in.-deep timber cantilever beam 8 ft long is reinforced by bolting two $\frac{1}{2}$-in.-wide by 6-in.-deep structural steel plates to the sides of the timber beam, as shown in Fig. P4-157. The moduli of elasticity of the timber and steel are 1600 ksi and 29,000 ksi, respectively. Determine the maximum tensile fiber stress in each of the materials when a static load of 1250 lb is applied to the free end of the beam.

4-158* A 150-mm-wide by 300-mm-deep timber beam 6 m long is reinforced with 150-mm-wide by 15-mm-thick steel plates on the top and bottom faces. The beam is simply supported and carries a uniformly distributed load of 25 kN/m over its entire length. The moduli of elasticity for the timber and steel are 13 GPa and 200 GPa, respectively. Determine the maximum tensile fiber stresses in the timber and in the steel.

4-159* A cantilever beam 8 ft long carries a concentrated load of 5000 lb at the free end. The beam consists of a 4-in.-wide by 10-in.-deep timber section reinforced with 4-in.-wide by $\frac{3}{4}$-in.-thick steel plates on the top and bottom surfaces. The moduli of elasticity for the wood and steel are 1600 ksi and 30,000 ksi, respectively. Determine the maximum fiber stresses in the wood and in the steel.

4-160 A timber beam 200 mm wide by 350 mm deep has a 200-mm-wide by 16-mm-thick steel plate securely fastened to its bottom face. The moduli of elasticity for the wood and steel are 12 GPa and 200 GPa, respectively. If the allowable fiber stresses are 10 MPa in the wood and 75 MPa in the steel, determine the maximum load P that can be applied at the center of a simply supported beam having a span of 6 m.

4-161 A timber beam 8 in. wide by 15 in. deep has an 8-in.-wide by $\frac{1}{2}$-in.-thick steel plate securely fastened to its bottom face. The beam will be simply supported, will have a span of 20 ft, and will carry a uniformly distributed load over its entire length. The moduli of elasticity for the wood and steel are 1600 ksi and 30,000 ksi, respectively. If the allowable fiber stresses are 1600 psi in the wood and 18,000 psi in the steel, determine the maximum allowable magnitude for the distributed load.

4-162* A 50 mm wide by 80 mm deep wood (E_w = 10 GPa) beam will be reinforced with 3-mm-thick structural aluminum (E_a = 70 GPa) plates on its top and bottom faces. A maximum bending moment of 3 kN·m must be resisted by the composite beam. If the allowable fiber stresses are 15 MPa in the wood and 135 MPa in the aluminum, determine the minimum width required for the aluminum plates.

4-163* A 2 in. wide by 3 in. deep polymer (E_p = 300 ksi) beam will be reinforced with $\frac{1}{8}$-in.-thick structural aluminum (E_a = 10,000 ksi) plates on its top and bottom faces. A maximum bending moment of 10 in.-kips must be resisted by the composite beam. If the allowable fiber stresses are 1 ksi in the polymer and 20 ksi in the aluminum, determine the minimum width required for the aluminum plates.

4-164 A 50 mm wide by 125 mm deep polymer (E_p = 1.400 GPa) beam will be reinforced with a 6-mm-thick brass (E_b = 100 GPa) plate on its bottom face. If the allowable fiber stresses are 6 MPa in the polymer and 60 MPa in the brass, determine the width of brass plate required to have the allowable stresses in the two materials occur simultaneously.

4-165 A 6 in. wide by 12 in. deep timber (E_w = 1500 ksi) beam will be reinforced with steel (E_s = 30,000 ksi) plates on its top and bottom faces. The beam will be simply supported, will have a span of 20 ft, and will carry a concentrated load of 5000 lb at the center of the span. If the allowable fiber stresses are 1 ksi in the wood and 10 ksi in the steel, prepare a curve showing the acceptable combinations of plate width and plate thickness. Limit the plate thicknesses to values less than $\frac{3}{4}$ in.

4-12

FLEXURAL STRESSES IN REINFORCED CONCRETE BEAMS

Concrete is widely used in beam construction because it is economical, readily available, fireproof, and exhibits a reasonable compressive strength. However, concrete has relatively little tensile strength; therefore, concrete beams must be reinforced with another material, usually steel, that can resist the tensile forces.

The transformed section method of Section 4-11 provides a satisfactory procedure for analyzing reinforced concrete beam problems. The transformed section used for these problems consists of the actual concrete on the compression side of the neutral axis plus the equivalent amount of hypothetical concrete (which is able to develop tensile stresses) on the tension side of the neutral axis required to replace the steel reinforcing rods. The actual concrete on the tension side of the beam is assumed to crack to the neutral surface; therefore, it has no tensile load-carrying ability and is neglected.

The solution for maximum stresses in a given beam or for the maximum bending moment with given allowable stresses consists of three steps.

1. Locate the neutral axis for the transformed section.
2. Determine the moment of inertia of the transformed section with respect to the neutral axis.
3. Use the flexure formula to determine the required stresses or moment.

In the design of a reinforced concrete beam to carry a specified moment with a balanced design (a balanced design means that the allowable stresses in the two materials are reached simultaneously), the following four equations can be written in terms of four unknown properties of the cross section and solved simultaneously.

1. The flexure formula for the allowable stress in the concrete.
2. The flexure formula for the allowable stress in the steel.
3. The moment equation for the location of the neutral axis.
4. The equation for the moment of inertia of the cross section with respect to the neutral axis.

The procedure is illustrated in the following examples.

EXAMPLE 4-13

A simple reinforced concrete beam carries a uniformly distributed load of 1500 lb per ft on a span of 16 ft. The beam has a rectangular cross section 12 in. wide by 21 in. deep, and 2 sq. in. of steel reinforcing rods are placed with their centers 3 in. from the bottom of the beam. The moduli of elasticity for the concrete and steel are 2500 ksi and 30,000 ksi, respectively. Determine the maximum fiber stress in the concrete and the average fiber stress in the steel.

Solution

The ratio of the moduli of elasticity is

$$n = 30{,}000/2500 = 12$$

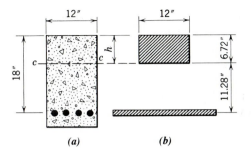

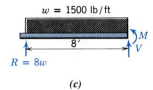

(a) (b) (c) FIG. 4-31

which means that 12(2) or 24 sq. in. of hypothetical concrete that resists tension is required for the transformed section. The actual and transformed sections are shown in Figs. 4-31a and b. The principle of moments with respect to the centroidal axis for the transformed section gives

$$12h(h/2) = 24(18 - h)$$

from which

$$h = 6.72 \text{ in.} \quad \text{and} \quad 18 - h = 11.28 \text{ in.}$$

The moment of inertia of the transformed section with respect to the neutral axis is

$$I = 24(11.28)^2 + (12)(6.73)^3/3 = 4270 \text{ in.}^4$$

For a simple beam with a uniformly distributed load, the maximum moment occurs at the center of the span and, from the free-body diagram of Fig. 4-31c, is

$$M = 8w(8) - 8w(4) = +32w \text{ ft-lb}$$

The flexure formula applied to the transformed section gives the maximum stress in the concrete as

$$\sigma = \frac{Mh}{I} = \frac{32(1500)(12)\ (6.72)}{4270} = \underline{907 \text{ psi } C} \qquad \text{Ans.}$$

Since the stresses in the transformed section have a linear distribution, the average stress in the steel is

$$\sigma = \frac{(18 - 6.72)}{6.72}(970)n = \frac{11.28}{6.72}(907)(12) = \underline{18,260 \text{ psi } T} \quad \text{Ans.}$$

EXAMPLE 4-14

Determine the distance from the top of the beam to the center of the steel and the area of steel required for balanced design of a reinforced concrete beam that is to be 10 in. wide and that must resist a bending moment of 600 in.-kips. The working stresses in the concrete and steel are 1000 psi and 18,000 psi, respectively. The moduli of elasticity for the concrete and steel are 2500 ksi and 30,000 ksi, respectively.

Solution

The ratio of the moduli of elasticity is

$$n = 30,000/2500 = 12$$

The actual and transformed sections are shown in Figs. 4-32a and b. The four unknown quantities are d, h, A, and I for the transformed section where A is the area of the steel. The four available equations involving these four unknowns are as follows.

1. The flexure formula for the maximum stress in the concrete is

$$1000 = 600,000h/I$$

2. The flexure formula modified for the average stress in the steel is

$$18,000 = 12(600,000)(d - h)/I$$

3. The principle of moments with respect to the neutral axis gives

$$10h(h/2) = 12A(d - h)$$

4. The moment of inertia with respect to the neutral axis is

$$I = 10h^3/3 + 12A(d - h)^2$$

Simultaneous solution of these four equations yields

$h = 7.44$ in. $d = 18.61$ in.   and $A = 2.07$ sq. in.   Ans.

The design of a reinforced concrete beam requires a consideration of other factors, such as the bond (shearing) stresses between the concrete and reinforcing steel, the diagonal tensile stresses that may

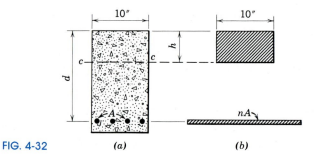

FIG. 4-32   (a)   (b)

be developed, and the amount of concrete that is needed beyond the reinforcing bars. Discussion of such topics can be found in textbooks devoted to reinforced concrete design.

PROBLEMS

4-166* A simply supported reinforced concrete beam carries a uniformly distributed load of 12 kN/m on a span of 4 m. The beam has a rectangular cross section 250 mm wide by 450 mm deep and 825 mm^2 of steel reinforcing rods are placed 60 mm from the bottom of the beam. The moduli of elasticity of the concrete and steel are 15 GPa and 200 GPa, respectively. Determine the average tensile stress in the steel and the maximum compressive stress in the concrete at the section of maximum moment.

4-167 A simply supported reinforced concrete beam is 8 in. wide and has a span of 16 ft. The tension reinforcement, which is located 16 in. below the top surface of the beam, consists of three $\frac{7}{8}$-in.-diameter steel bars. The moduli of elasticity of the concrete and steel are 2400 ksi and 30,000 ksi, respectively. If the working stresses are 1000 psi in the concrete and 16,000 psi in the steel, determine the maximum load per foot of length that can be uniformly distributed over the full length of the beam.

4-168* A simply supported reinforced concrete beam 200 mm wide with a depth to center of steel of 300 mm has a span of 4 m. The tension reinforcement consists of three 16-mm-diameter steel bars. The ratio of the moduli of elasticity is 12. If the working stresses are 6.5 MPa in the concrete and 120 MPa in the steel, determine the maximum load per meter of length that can be uniformly distributed over the middle half of the beam.

A steel-reinforced concrete beam of balanced design is needed to support a uniformly distributed load of 1200 lb/ft on a simply supported span of 20 ft. The width of the beam must be 10 in. and the allowable stresses are 800 psi in the concrete and 16,000 psi in the steel. The moduli of elasticity of the concrete and steel are 2400 ksi and 30,000 ksi, respectively. Determine

   (a) the required cross-sectional area for the steel rods.
   (b) the depth from the top surface of the beam to the center of the steel rods.

4-170* A steel-reinforced concrete beam of balanced design has a width of 300 mm and a depth to center of steel of 500 mm. Use 16.5 GPa and 198 GPa for the moduli of elasticity of the concrete and steel, respectively. If the allowable stresses are 7 MPa in the concrete and 125 MPa in the steel, determine

(a) the required cross-sectional area for the steel rods.

(b) the maximum moment that can be resisted by the beam.

4-171* The effective cross section for a steel-reinforced concrete beam is shown in Fig. P4-171. For allowable stresses of 1200 psi in the concrete and 18,000 psi in the steel, determine the maximum permissible bending moment that the section can resist. Use 3000 ksi and 30,000 ksi for the moduli of elasticity of the concrete and steel, respectively.

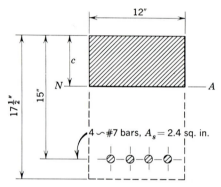

FIG. P4-171

COMPUTER PROBLEMS

Note: The following problems have been designed to be solved with a programmable calculator, microcomputer, or mainframe computer. Appendix F contains a description of a few numerical methods, together with a few simple programs in BASIC and FORTRAN, that can be modified for use in the solution of these problems.

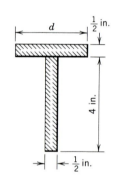

FIG. PC4-1

C4-1 A steel beam with a ½-by-4-in. rectangular cross section will be reinforced by attaching a ½-in.-thick steel plate to the top surface to form the T-beam shown in Fig. PC4-1. If the yield strength of the steel is 40 ksi, plot the percent increase in moment carrying capability ΔM of the reinforced beam compared to the unreinforced beam as a function of the width d (½ in. $\leq d \leq$ 10 in.) of the reinforcing plate.

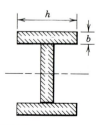

FIG. PC4-2

C4-2 Three identical rectangular steel plates are used to fabricate an I-beam with the cross section shown in Fig. PC4-2. The cross-sectional area of the beam is 6000 mm². If the thickness b must be larger than 3 mm and the minimum depth d ($d = h + 2b$) is 80 mm, prepare a design curve that shows the moment carrying capability M about the x–x axis of the beam as a function of the depth d (80 mm $\leq d \leq$ 673 mm) of the beam. The maximum fiber stress in the beam cannot exceed 150 MPa.

C4-3 A beam with a hollow circular cross section is being designed to support a maximum moment of 100 in.-kips. If the maximum fiber stress in the beam must not exceed 40 ksi and the maximum diameter of the beam must not exceed 5 in., prepare a design curve that shows the acceptable values of the inside diameter d_i as a function of the outside diameter d_o (0 in. $\leqslant d_o \leqslant$ 5 in.).

C4-4 A beam with a solid rectangular cross section is being designed to support a maximum moment of 4 kN-m. If the maximum fiber stress in the beam must not exceed 50 MPa and the width of the beam must not exceed 75 mm, prepare a design curve that shows the acceptable values of depth h as a function of width b (0 mm $\leqslant b \leqslant$ 75 mm).

C4-5 An S4 $\times$ 9.5 structural steel beam (see Appendix C) is carrying a moment M that produces a maximum fiber stress of 20 ksi. The moment M must be increased by 75 percent, but the maximum fiber stress must not be increased. In order to strengthen the beam, rectangular steel plates are to be attached to the top and bottom flanges. If the thickness t of the plates must not exceed $\frac{1}{2}$ in., prepare a design curve that shows the acceptable values of plate width b as a function of plate thickness t (0 in. $\leqslant t \leqslant \frac{1}{2}$ in.).

C4-6 A beam with a hollow square cross section is being designed to support a maximum moment of 10 kN-m. If the maximum fiber stress in the beam must not exceed 100 MPa and the outside dimension a_o must not exceed 120 mm, prepare a design curve that shows the acceptable values of inside dimension a_i as a function of outside dimension a_o (0 mm $\leqslant a_o \leqslant$ 120 mm).

C4-7 Write a program to compute and print out (or compute and plot) the shear force V and bending moment M as functions of position x for a beam loaded and supported as shown in Fig. PC4-7.

C4-8 The supports for the beam shown in Fig. PC4-8 are symmetrically located. The distance d from the supports to the ends of the beam is adjustable.

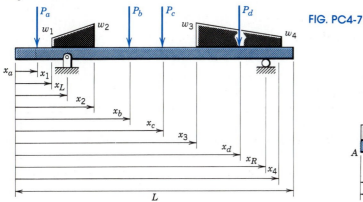

FIG. PC4-7

FIG. PC4-8

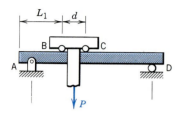

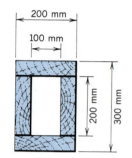

(a) Compute and plot the maximum bending moment M_{AB} in segment AB of the beam as a function of distance d (0 m $\leq$ d $\leq$ 2 m).

(b) Compute and plot the maximum bending moment M_{BC} in segment BC of the beam as a function of distance d (0 m $\leq$ d $\leq$ 2 m).

(c) Determine the position d_s of the supports that will give the smallest maximum fiber stress in the beam.

(d) If the maximum allowable fiber stress is 150 MPa, determine the lightest structural steel American standard beam (see Appendix C) that satisfies part (c).

FIG. PC4-9

C4-9 The load P shown in Fig. PC4-9 is supported by a carriage that distributes the load equally between rollers at B and C. The carriage rolls slowly along beam AD, which is an S3 $\times$ 5.7 structural steel section (see Appendix C). If P = 1000 lb and L = 200 in., determine the maximum fiber stress in the beam and its location when

(a) the wheel spacing d equals 25 in.

(b) the wheel spacing d equals 50 in.

C4-10 The transverse shear V at a certain section of a timber beam is 18 kN. If the beam has the cross section shown in Fig. PC4-10, compute and plot the vertical shearing stress τ as a function of distance y (-150 mm $\leq y \leq$ 150 mm) from the neutral axis.

FIG. PC4-10

C4-11 A timber beam is simply supported and carries a uniformly distributed load w of 360 lb per ft over its entire 18-ft span. If the beam has the cross section shown in Fig. PC4-11, compute and plot the vertical shearing stress τ as a function of distance y from the neutral axis, for a cross section 2 ft from the left end of the beam.

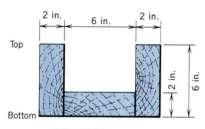

FIG. PC4-11

C4-12 The beam shown in Fig. PC4-12a is fabricated by gluing two pieces of timber together to form the cross section shown in Fig. PC4-12b. Compute and plot the vertical shearing stress τ as a function of distance y from the neutral axis for a cross section 0.5 m from the left end of the beam.

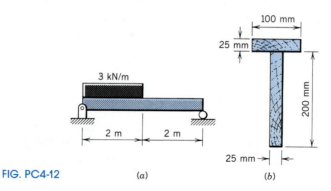

FIG. PC4-12 (a) (b)

CHAPTER 5

FLEXURAL LOADING: DEFLECTIONS

5-1

INTRODUCTION

The important relations between applied load and stress (fiber and shear) developed in a beam were presented in Chapter 4. However, a beam design is frequently not complete until the amount of deflection has been determined for the specified load. Failure to control beam deflections within proper limits in building construction is frequently reflected by the development of cracks in plastered walls and ceilings. Beams in many machines must deflect just the right amount for gears or other parts to make proper contact. In innumerable instances the requirements for a beam involve a given load-carrying capacity with a specified maximum deflection.

The deflection of a beam depends on the stiffness of the material and the dimensions of the beam as well as on the applied loads and supports. Three common methods for calculating beam deflections due to fiber stresses are presented here: (1) the double integration method, (2) the area-moment method, and (3) the superposition method. The material is presented so that methods 1 and 2 can be studied independently of each other. Deflections due to shearing stresses are also discussed.

5-2

THE RADIUS OF CURVATURE

When a straight beam is loaded and the action is elastic, *the longitudinal centroidal axis of the beam becomes a curve defined as the elastic curve*. In regions of constant bending moment, the elastic curve is an arc of a circle of radius ρ as indicated in Fig. 5-1, in which the portion *AB* of a beam is bent only with couples. Therefore, the plane sections *A* and *B* remain plane (see Section 4-2), and the deformation (elongation and compression) of the fibers is proportional to the distance from the neutral surface, which is unchanged in length. From Fig. 5-1,

$$\theta = \frac{L}{\rho} = \frac{L + \delta}{\rho + c}$$

from which

$$\frac{c}{\rho} = \frac{\delta}{L} = \epsilon = \frac{\sigma}{E} = \frac{Mc}{EI} \qquad (5\text{-}1)$$

Therefore,

$$\frac{1}{\rho} = \frac{M}{EI} \qquad (5\text{-}2)$$

which relates the radius of curvature of the neutral surface of the beam to the bending moment M, the stiffness of the material E, and the moment of inertia of the cross-sectional area I.

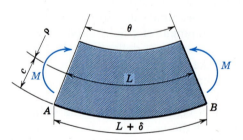

FIG. 5-1

EXAMPLE 5-1

The cantilever beam shown in Fig. 5-2a is fabricated from two 25-by-100-mm aluminum alloy ($E = 73$ GPa) bars as shown in Fig. 5-2b. Determine

(a) the maximum fiber stress in the beam.
(b) the radius of curvature of the beam.
(c) the deflection at the right end of the beam.

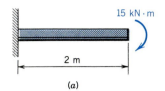

(a)

Solution

The neutral surface is located at a distance y_c from the bottom of the beam. Thus

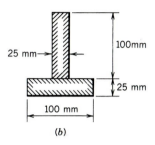

(b)

$$y_c = \frac{100(25)(12.5) + 100(25)(75)}{2(25)(100)} = 43.75 \text{ mm}$$

The moment of inertia I about the neutral axis of the beam is

$$I = \frac{25(81.25)^3}{3} + \frac{100(43.75)^3}{3} - \frac{75(18.75)^3}{3} = 7.096(10^6) \text{ mm}^4$$

(a) The maximum fiber stress σ_{max} is given by Eq. 4-8 as

$$\sigma_{max} = \frac{Mc}{I} = \frac{-15(10^3)(-81.25)(10^{-3})}{7.096(10^{-6})}$$

$$= 171.8(10^6) \text{ N/m}^2 = \underline{171.8 \text{ MPa } T} \qquad \text{Ans.}$$

(b) The radius of curvature of the beam is given by Eq. 5-2 as

$$\rho = \frac{EI}{M} = \frac{73(10^9)(7.096)(10^{-6})}{15(10^3)} = \underline{34.53 \text{ m}} \qquad \text{Ans.}$$

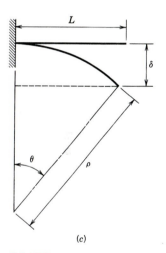

(c)

FIG. 5-2

(c) The angle θ (see Fig. 5-2c) is

$$\theta = \frac{L}{\rho} = \frac{2}{34.53} = 0.05792 \text{ rad} = 3.319°$$

The deflection δ at the right end of the beam (see Fig. 5-2c) is

$$\delta = \rho - \rho \cos \theta = 34.53 - 34.53 \cos 3.319°$$

$$= 57.9(10^{-3}) \text{ m} = \underline{57.9 \text{ mm}} \qquad \text{Ans.}$$

PROBLEMS

5-1* During fabrication of a laminated timber arch, one of the 10-in.-wide by 1-in.-thick Douglas fir (E = 1900 ksi) planks is bent to a radius of 12 ft. Determine the maximum flexural stress developed in the plank.

5-2* A high-strength steel (E = 200 GPa) tube having an outside diameter of 80 mm and a wall thickness of 3 mm is bent into a circular curve of 24-m radius. Determine the maximum fiber stress in the tube.

5-3 Steel (E = 30,000 ksi) wire with a diameter of 0.010 in. will be wound around a 10-in.-diameter tube. If a factor of safety of 2 with respect to failure by slip is specified, determine the minimum permissible value for the yield strength of the steel wire.

5-4 The boards for a concrete form are to be bent to a circular curve of 5-m radius. What maximum thickness can be used if the stress is not to exceed 15 MPa? The modulus of elasticity for the wood is 10 GPa.

5-5* A strip of hardened steel (E = 30,000 ksi) with a rectangular cross section $\frac{3}{4}$ in. wide by $\frac{1}{64}$ in. thick is bent to form a circular ring. If the maximum flexural stress in the strip must not to exceed 40,000 psi, determine the minimum allowable diameter for the ring.

5-6* Determine the minimum diameter required for a bandsaw pulley if the maximum fiber stress due to bending in a 0.75-mm-thick steel (E = 200 GPa) bandsaw blade is not to exceed 250 MPa.

5-7 A 4-in.-wide by $\frac{1}{2}$-in.-thick aluminum alloy (E = 10,600 ksi) bar must be bent into a circular arc of 10-ft radius. Determine

(a) the moment required to bend the bar.

(b) the maximum fiber stress produced in the bar.

5-8 Steel (E = 190 GPa) wire with a diameter of 3 mm is being wrapped around a circular wood spool. If the maximum fiber stress in the wire must not exceed 1100 MPa, determine the minimum allowable diameter for the spool.

5-9* For the W6 × 25 steel (E = 30,000 ksi) beam shown in Fig. P5-9, determine

(a) the radius of curvature at any point between the supports.

(b) the deflection at a section midway between the supports.

(c) the maximum tensile fiber stress in the beam.

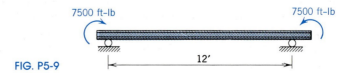

FIG. P5-9

5-10* For the WT229 × 57 steel (E = 200 GPa) beam shown in Fig. P5-10, determine

(a) the radius of curvature at any point between the supports.

(b) the deflection at a section midway between the supports.

(c) the maximum tensile fiber stress in the beam.

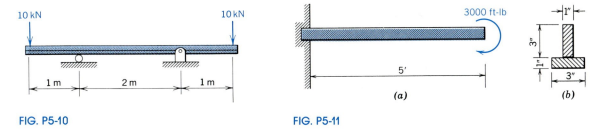

FIG. P5-10   FIG. P5-11

5-11 The cantilever beam shown in Fig. P5-11a is fabricated from two 1-by-3-in. steel (E = 30,000 ksi) bars, as shown in Fig. P5-11b. Determine

(a) the radius of curvature of the beam.

(b) the deflection at the right end of the beam.

(c) the deflection 3 ft from the support.

5-12 The cantilever beam shown in Fig. P5-12a is fabricated from three 30-by-120-mm aluminum (E = 70 GPa) bars as shown in Fig. P5-12b. Determine

(a) the maximum fiber stress in the beam.

(b) the radius of curvature of the beam.

(c) the deflection at the left end of the beam.

(d) the deflection 750 mm from the support.

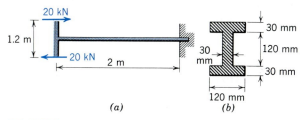

FIG. P5-12

5-13* A 1-in.-thick board is simply supported at the ends of a span of 10 ft. Loads are applied such that the elastic curve between supports is an arc of a circle. The deflection at the midpoint of the span is 5 in. Determine the maximum fiber stress in the wood. Use 1200 ksi for the modulus of elasticity of the wood.

5-14* An aluminum alloy (E = 73 GPa) cantilever beam with a rectangular cross section has a moment of 600 N·m applied at the free end. If the radius of curvature of the beam must be 10 m when the maximum fiber stress is 140 MPa, determine

(a) the required moment of inertia for the cross section.

(b) the required width and thickness for the beam.

5-3

THE DIFFERENTIAL EQUATION OF THE ELASTIC CURVE

The expression for the curvature of the elastic curve derived in Section 5-2 is useful only when the bending moment is constant for the interval of the beam involved. For most beams the bending moment is a function of the position along the beam and a more general expression is required.

The curvature from the calculus (see any standard calculus textbook) is

$$\frac{1}{\rho} = \frac{d^2y/dx^2}{[1 + (dy/dx)^2]^{3/2}}$$

For actual beams the slope dy/dx is very small, and its square can be neglected in comparison to unity. With this approximation,

$$\frac{1}{\rho} = \frac{d^2y}{dx^2}$$

and Eq. 5-2 becomes

$$EI \frac{d^2y}{dx^2} = M_x \tag{5-3}$$

which is the differential equation for the elastic curve of a beam where the subscript x on M is used as a reminder that M is a function of x.

The differential equation of the elastic curve can also be obtained from the geometry of the bent beam as shown in Fig. 5-3, where it is evident that $dy/dx = \tan\theta = \theta$ (approximately) for small angles and that $d^2y/dx^2 = d\theta/dx$. Again from Fig. 5-3,

$$d\theta = \frac{dL}{\rho} = \frac{dx}{\rho}$$

(approximately) for small angles. Therefore,

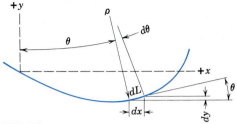

FIG. 5-3

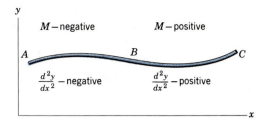

FIG. 5-4

$$\frac{d^2y}{dx^2} = \frac{d\theta}{dx} = \frac{1}{\rho} = \frac{M_x}{EI}$$

or

$$EI\frac{d^2y}{dx^2} = M_x \qquad\qquad (5\text{-}3)$$

The sign convention for bending moments established in Section 4-5 will be used for Eq. 5-3. Both E and I are always positive; therefore, the signs of the bending moment and the second derivative must be consistent. With the coordinate axes selected as shown in Fig. 5-4, the slope changes from positive to negative in the interval from A to B; therefore, the second derivative is negative, which agrees with the sign convention of Section 4-5. For the interval BC, both d^2y/dx^2 and M_x are seen to be positive.

Careful study of Fig. 5-4 reveals that the signs of the bending moment and the second derivative are also consistent when the origin is selected at the right with x positive to the left and y positive upward. Unfortunately, the signs are inconsistent when y is positive downward. *In this book, for horizontal beams, y will always be selected positive upward for beam deflection problems.*

Before proceeding with the solution of Eq. 5-3, it is desirable to correlate the successive derivatives of the deflection y of the elastic curve with the physical quantities that they represent in beam action. They are

$$\text{Deflection} = y$$

$$\text{Slope} = \frac{dy}{dx}$$

$$\text{Moment} = EI\frac{d^2y}{dx^2} \quad (\text{from Eq. 5-3})$$

$$\text{Shear} = \frac{dM}{dx} \quad (\text{from Eq. 4-11}) = EI\frac{d^3y}{dx^3} \ (\text{for } EI \text{ constant})$$

$$\text{Load} = \frac{dV}{dx} \quad (\text{from Eq. 4-9}) = EI\frac{d^4y}{dx^4} \ (\text{for } EI \text{ constant})$$

where the signs are as given in Sections 4-5 and 4-6.

In Section 4-7 a method based on these differential relations was presented for starting from the load diagram and drawing first the shear diagram and then the moment diagram. This method can readily be extended to the construction of the slope diagram from the relation

$$M = EI \frac{d\theta}{dx}$$

from which

$$\int_{\theta_A}^{\theta_B} d\theta = \int_{x_A}^{x_B} \frac{M}{EI} dx$$

This relation shows that except for a factor EI, the area under the moment diagram between any two points along the beam gives the change in slope between the same two points. Likewise, the area under the slope diagram between two points along a beam gives the change in deflection between these points. These relations have been used to construct the complete series of diagrams shown in Fig. 5-5 for a simply supported beam with a concentrated load at the center of the span. The geometry of the beam was used to locate the points of zero slope and deflection, required as starting points for the construction. More commonly used methods for calculating beam deflections will be developed in succeeding sections.

Use of the successive derivatives of the deflection y of the elastic curve for a beam to obtain the various physical quantities (θ, M, V, and w) involved in beam action is illustrated in Example 5-2.

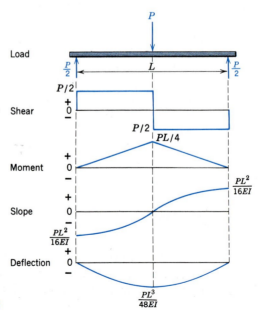

FIG. 5-5

EXAMPLE 5-2

For a beam of uniform cross section and length L, the equation of the elastic curve between sections at $x = 0$ and $x = L$ is

$$y = \left(\frac{k}{24EI}\right)(-x^4 + 2Lx^3 - L^3x)$$

Determine the type of load that the beam carries and the end conditions.

Solution

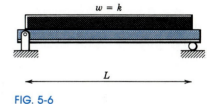

FIG. 5-6

The deflection:

$$y = \left(\frac{k}{24EI}\right)(-x^4 + 2Lx^3 - L^3x)$$

The slope:

$$\theta = \frac{dy}{dx} = \left(\frac{k}{24EI}\right)(-4x^3 + 6Lx^2 - L^3)$$

The moment:

$$M = EI\left(\frac{d^2y}{dx^2}\right) = \left(\frac{k}{24}\right)(-12x^2 + 12Lx)$$

$$= \left(\frac{k}{2}\right)(-x^2 + Lx)$$

The shear:

$$V = \frac{dM}{dx} = EI\left(\frac{d^3y}{dx^3}\right) = \left(\frac{k}{24}\right)(-24x + 12L)$$

$$= k(-x + L/2)$$

The load:

$$w = \frac{dV}{dx} = EI\left(\frac{d^4y}{dx^4}\right) = -k$$

End conditions:

At $x = 0$: $y = 0$, $\theta = -\dfrac{kL^3}{24EI}$, $M = 0$, and $V = +\dfrac{kL}{2}$

At $x = L$: $y = 0$, $\theta = +\dfrac{kL^3}{24EI}$, $M = 0$, and $V = -\dfrac{kL}{2}$

Support reactions:

$$R_0 = \frac{kL}{2}$$

$$R_L = \frac{kL}{2}$$

The beam is simply supported and carries a uniformly distributed load $w = -k$ for the full length of the beam, as shown in Fig. 5-6.

Before proceeding with specific methods for calculating beam deflections, it is advisable to consider the assumptions used in the development of the basic relation Eq. 5-3. All of the limitations that apply to the flexure formula apply to the calculation of deflections because the flexure formula was used in the derivation of Eq. 5-3. It is further assumed that

1. the square of the slope of the beam is negligible compared to unity.
2. the beam deflection due to shearing stresses is negligible (a plane section is assumed to remain plane). See Section 5-11 for further information.
3. the values of E and I remain constant for any interval along the beam. In case either of them varies and can be expressed as a function of the distance x along the beam, a solution of Eq. 5-3 that takes this variation into account may be possible.

PROBLEMS

Note: In the following problems, the equation of the elastic curve is given for a beam with a uniform cross section. The equation is valid for the full length L of the beam. Determine the type of load that the beam carries and the end conditions. Prepare a sketch showing the beam, loads, and supports.

5-15 $y = \left(\dfrac{k}{2EI}\right)(-x^2)$ **5-16** $y = \left(\dfrac{k}{6EIL}\right)(-x^3 + L^2 x)$

5-17 $y = \left(\dfrac{k}{6EI}\right)(x^3 - 3Lx^2)$ **5-18** $y = \left(\dfrac{k}{6EI}\right)(-x^3 + 3L^2 x - 2L^3)$

5-19 $y = \left(\dfrac{k}{24EI}\right)(-x^4 + 4L^3 x - 3L^4)$

5-20 $y = \left(\dfrac{k}{24EI}\right)(-x^4 + 4Lx^3 - 6L^2 x^2)$

5-21 $y = \left(\dfrac{k}{360EIL}\right)(-3x^5 + 10L^2 x^3 - 7L^4 x)$

5-22 $y = \left(\dfrac{k}{120EIL}\right)(-x^5 + 5L^4 x - 4L^5)$

5-23 $y = \left(\dfrac{k}{360EIL}\right)(3x^5 - 15Lx^4 + 20L^2 x^3 - 8L^4 x)$

5-24 $y = \left(\dfrac{k}{120EIL}\right)(-x^5 + 10L^2x^3 - 20L^3x^2)$

5-25 $y = \left(\dfrac{k}{24EI}\right)(-x^4 + 2Lx^3 - L^2x^2)$

5-26 $y = \left(\dfrac{k}{48EI}\right)(-2x^4 + 3Lx^3 - L^3x)$

5-4

DEFLECTIONS BY INTEGRATION OF A MOMENT EQUATION

Whenever the assumptions of the previous section are essentially correct and the bending moment can be readily expressed as an integrable function of x, Eq. 5-3 can be solved for the deflection y of the elastic curve of a beam at any point x along the beam. The constants of integration can be evaluated from the applicable boundary or matching conditions.

A boundary condition is defined as a known set of values for x and y, or x and dy/dx, at a specific location along the beam. One boundary condition (one set of values) can be used to determine one and only one constant of integration.

Many beams are subjected to abrupt changes in loading along the beam, such as concentrated loads, reactions, or even distinct changes in the amount of a uniformly distributed load. Since the expressions for the bending moment on the left and right of any abrupt change in load are different functions of x, it is impossible to write a single equation for the bending moment in terms of ordinary algebraic functions that is valid for the entire length of the beam. This can be resolved by writing separate bending moment equations for each interval of the beam. Although the intervals are bounded by abrupt changes in load, the beam is continuous at such locations and, consequently, the slope and the deflection at the junction of adjacent intervals must match. *A matching condition is defined as the equality of slope or deflection, as determined at the junction of two intervals from the elastic curve equations for both intervals.* One matching condition (for example, at x equals $L/3$, y from the left equation equals y from the right equation) can be used to determine one and only one constant of integration.

The procedure for obtaining beam deflections when matching conditions are required is lengthy and tedious. A method is presented in the next section in which *singularity functions* are used to write a single equation for the bending moment that is valid for the entire length of the beam; this eliminates the need for matching conditions and, accordingly, reduces the labor involved.

Calculating the deflection of a beam by the double integration method involves four definite steps, and the following sequence for these steps is strongly recommended.

1. Select the interval or intervals of the beam to be used; next, place a set of coordinate axes on the beam *with the origin at one end of an interval* and then indicate the range of values of x in each interval. For example, two adjacent intervals might be

$$0 \leq x \leq L \qquad \text{and} \qquad L \leq x \leq 3L$$

2. List the available boundary and matching (where two or more adjacent intervals are used) conditions for each interval selected. Remember that two conditions are required to evaluate the two constants of integration for each interval used.
3. Express the bending moment as a function of x, using the sign convention established in Section 4-5, for each interval selected and equate it to $EI\, d^2y/dx^2$.
4. Solve the differential equation or equations from item 3 and evaluate all constants of integration. Check the resulting equations for dimensional homogeneity. Calculate the deflection at specific points when required.

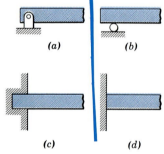

(a)

(b)

(c)

(d)

FIG. 5-7

A roller or pin at any point in a beam (see Figs. 5-7a and b) represents a simple support at which the beam cannot deflect (unless otherwise stated in the problem) but can rotate. At a fixed end, as represented by Figs. 5-7c and d, the beam can neither deflect nor rotate unless otherwise stated.

The following examples illustrate the use of the double integration method for calculating beam deflections.

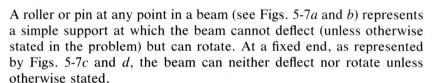

EXAMPLE 5-2

For the beam loaded and supported as shown in Fig. 5-8a, locate the point of maximum deflection between the supports.

Solution

From a free-body diagram of the beam and the equations of equilibrium, the left reaction is found to be $7wL/12$ upward. As indicated in Fig. 5-8b, the origin of coordinates is selected at the left support, and the interval to be used is $0 \leq x \leq L$.

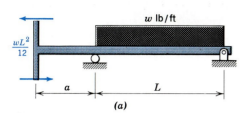

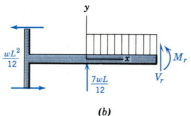

(a)   (b)   FIG. 5-8

The two required boundary conditions are $y = 0$ when $x = 0$ and $y = 0$ when $x = L$. From the free-body diagram of a portion of the beam as shown in Fig. 5-8b, the expression for the bending moment at any point in the selected interval is seen to be

$$EI \frac{d^2y}{dx^2} = M_x = \frac{7wL}{12} x - \frac{wL^2}{12} - wx \left(\frac{x}{2} \right)$$

Successive integration gives

$$EI \frac{dy}{dx} = \frac{7wL}{24} x^2 - \frac{wL^2}{12} x - \frac{w}{6} x^3 + C_1$$

and

$$EIy = \frac{7wL}{72} x^3 - \frac{wL^2}{24} x^2 - \frac{w}{24} x^4 + C_1 x + C_2$$

Substitution of the boundary condition $y = 0$ when $x = 0$ yields $C_2 = 0$. Substitution of the remaining boundary condition $y = 0$ when $x = L$ gives $C_1 = -wL^3/72$, and therefore the elastic curve equation is

$$EIy = \frac{7wL}{72} x^3 - \frac{wL^2}{24} x^2 - \frac{w}{24} x^4 - \frac{wL^3}{72} x$$

The maximum deflection occurs where the slope dy/dx is zero or

$$0 = \frac{7wL}{24} x^2 - \frac{wL^2}{12} x - \frac{w}{6} x^3 - \frac{wL^3}{72}$$

The solution of this cubic equation in x gives the point of maximum deflection at

$$x = 0.541\ L \text{ to the right of the left support} \qquad \text{Ans.}$$

Although not requested here, the amount of the maximum deflection can readily be obtained by substituting $0.541L$ for x in the elastic curve equation. The result is

$$y = -\frac{7.88wL^4}{10^3 EI} = \frac{7.88wL^4}{10^3 EI} \text{ downward}$$

EXAMPLE 5-3

For the beam loaded and supported as shown in Fig. 5-9*a*, determine the deflection of the right end.

Solution

From the free-body diagram of Fig. 5-9*b* the equations of equilibrium give the shear at the wall as $wL/3$ upward and the moment at the wall as $5wL^2/18$ counterclockwise. With the sign conventions established in Section 4-5, these results are

$$V_W = +\frac{wL}{3} \qquad \text{and} \qquad M_W = -\frac{5wL^2}{18}$$

In Fig. 5-9*b* there are two intervals to be considered, namely, the loaded and the unloaded portions of the beam. The loaded portion of the beam must be used because it contains the point where the deflection is required. However, a quick check reveals the absence of boundary conditions in this interval. It therefore becomes necessary to use both intervals as well as matching and boundary conditions; hence, the origin of coordinates is selected, as shown. For clarity, the two intervals are written side by side with available boundary conditions, matching conditions, and equations listed under the appropriate interval. The intervals are

$$0 \le x \le 2L/3 \qquad\qquad 2L/3 \le x \le L$$

The available boundary conditions are

$$\frac{dy}{dx} = 0 \quad \text{when} \quad x = 0$$

and

$$y = 0 \quad \text{when} \quad x = 0$$

The available matching conditions are

$$\frac{dy}{dx} \text{ from the left equation} \quad = \frac{dy}{dx} \text{ from the right equation}$$
$$\text{when} \quad x = 2L/3$$

and

$$y \text{ from the left equation} \quad = y \text{ from the right equation}$$
$$\text{when} \quad x = 2L/3$$

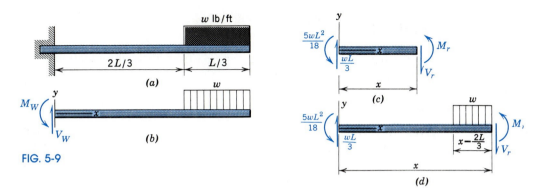

FIG. 5-9

Four conditions (two boundary and two matching) are sufficient for the evaluation of the four constants of integration (two in each of the two elastic curve differential equations); therefore, the problem can be solved in this manner.

From the free-body diagram of an unloaded portion of the beam in Fig. 5-9c, the expression for the bending moment at any point in the left interval is seen to be

$$EI \frac{d^2y}{dx^2} = M_x = \frac{wL}{3} x - \frac{5wL^2}{18}$$

From the free-body diagram of Fig. 5-9d, where the beam is cut in the loaded interval, the bending moment at any point in the right interval is found to be

$$EI \frac{d^2y}{dx^2} = M_x = \frac{wL}{3} x - \frac{5wL^2}{18}$$
$$- w \left(x - \frac{2L}{3} \right) \left(\frac{x - 2L/3}{2} \right)$$

Integration of each differential equation gives

$$EI \frac{dy}{dx} = \frac{wL}{6} x^2 - \frac{5wL^2}{18} x + C_1 \qquad EI \frac{dy}{dx} = \frac{wL}{6} x^2 - \frac{5wL^2}{18} x$$
$$- \frac{w}{6} \left(x - \frac{2L}{3} \right)^3 + C_3$$

Substitution of the boundary condition

$$\frac{dy}{dx} = 0 \quad \text{when} \quad x = 0$$

gives

$$C_1 = 0$$

Substitution of the matching condition

$$\frac{dy}{dx} \text{ from the left equation} \quad = \frac{dy}{dx} \text{ from the right equation}$$

$$\text{when} \quad x = \frac{2L}{3}$$

gives

$$C_1 = C_3 = 0$$

Integration of the resulting differential equations gives

$$EIy = \frac{wL}{18} x^3 - \frac{5wL^2}{36} x^2 + C_2 \quad \bigg| \quad EIy = \frac{wL}{18} x^3 - \frac{5wL^2}{36} x^2$$

$$- \frac{w}{24} \left(x - \frac{2L}{3} \right)^4 + C_4$$

Use of the remaining boundary condition yields $C_2 = 0$. Use of the final matching condition yields $C_2 = C_4 = 0$.

The deflection of the right end of the beam can now be obtained from the elastic curve equation for the right interval by replacing x by its value L. The result is

$$EIy = \frac{wL}{18} L^3 - \frac{5wL^2}{36} L^2 - \frac{w}{24} \left(L - \frac{2L}{3} \right)^4$$

from which

$$y = - \frac{163}{1944} \frac{wL^4}{EI} = \frac{163}{1944} \frac{wL^4}{EI} \text{ downward} \qquad \text{Ans.}$$

EXAMPLE 5-4

Determine the deflection at the left end of the cantilever beam with variable width shown in Fig. 5-10.

Solution

Since the width w of the beam varies linearly with respect to position x along the length of the beam, the moment of inertia I of the cross section also varies linearly with respect to position x. Thus, at position x,

$$I = \frac{wh^3}{12} = \frac{(bx/L)h^3}{12} = \left(\frac{bh^3}{12}\right)\left(\frac{x}{L}\right) = I_L\left(\frac{x}{L}\right)$$

where $I_L = bh^3/12$ is the moment of inertia of the cross section of the beam at the support. This variation in moment of inertia I must be included in the integration process used to determine the equation of the elastic curve for the beam.

The equation of the elastic curve for the beam is obtained by successive integration from Eq. 5-3. Thus

$$E\left(\frac{d^2y}{dx^2}\right) = \frac{M}{I} = -\frac{Px}{I} = -\frac{Px}{I_L(x/L)} = -\frac{PL}{I_L}$$

Successive integration gives

$$EI_L\left(\frac{dy}{dx}\right) = -PLx + C_1$$

$$EI_Ly = -\frac{PLx^2}{2} + C_1x + C_2$$

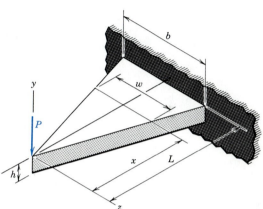

From the boundary conditions:

$$\frac{dy}{dx} = 0 \quad \text{when} \quad x = L, \qquad C_1 = PL^2$$

$$y = 0 \quad \text{when} \quad x = L, \qquad C_2 = -\frac{PL^3}{2}$$

FIG. 5-10

Thus, the equation of the elastic curve is

$$y = -\left(\frac{PL}{EI_L}\right)\left(\frac{x^2}{2} - Lx + \frac{L^2}{2}\right)$$

At the left end of the beam ($x = 0$), the deflection is

$$y = -\frac{PL^3}{2EI_L} = \frac{6PL^3}{Ebh^3} \quad \underline{\text{downward}} \qquad\qquad \text{Ans.}$$

It is also interesting to note how the maximum fiber stress varies along the length of the beam. From Eq. 4-8,

$$\sigma_{max} = \frac{Mc}{I} = \frac{Px\left(\frac{h}{2}\right)}{\left(\frac{bh^3}{12}\right)\left(\frac{x}{L}\right)} = \frac{6PL}{bh^2}$$

The maximum fiber stress does not depend on position x but is constant along the entire length of the beam. This type of beam is frequently referred to as a constant-stress beam.

PROBLEMS

5-27* A beam is loaded and supported as shown in Fig. P5-27.

(a) Derive the equation of the elastic curve in terms of P, L, x, E, and I. Use the designated axes.

(b) Determine the slope at the right end of the beam.

(c) Determine the deflection at the right end of the beam.

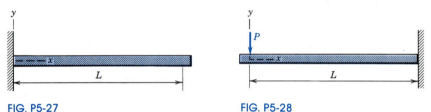

FIG. P5-27   FIG. P5-28

5-28* A beam is loaded and supported as shown in Fig. P5-28.

(a) Derive the equation of the elastic curve in terms of P, L, x, E, and I. Uses the designated axes.

(b) Determine the slope at the left end of the beam.

(c) Determine the deflection at the left end of the beam.

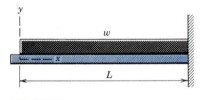

FIG. P5-29

5-29 A beam is loaded and supported as shown in Fig. P5-29.

(a) Derive the equation of the elastic curve in terms of w, L, x, E, and I. Use the designated axes.

(b) Determine the slope at the left end of the beam.

(c) Determine the deflection at the left end of the beam.

5-30 A beam is loaded and supported as shown in Fig. P5-30.

(a) Derive the equation of the elastic curve in terms of w, L, x, E, and I. Use the designated axes.

(b) Determine the slope at the right end of the beam.

(c) Determine the deflection at the right end of the beam.

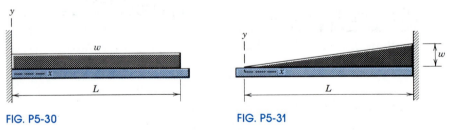

FIG. P5-30   FIG. P5-31

5-31* A beam is loaded and supported as shown in Fig. P5-31.

(a) Derive the equation of the elastic curve in terms of w, L, x, E, and I. Use the designated axes.

(b) Determine the slope at the left end of the beam.

(c) Determine the deflection at the left end of the beam.

5-32* A beam is loaded and supported as shown in Fig. P5-32.

   (a) Derive the equation of the elastic curve in terms of w, L, x, E, and I. Use the designated axes.

   (b) Determine the slope at the right end of the beam.

   (c) Determine the deflection at the right end of the beam.

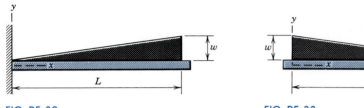

FIG. P5-32   FIG. P5-33

5-33 A beam is loaded and supported as shown in Fig. P5-33.

   (a) Derive the equation of the elastic curve in terms of w, L, x, E, and I. Use the designated axes.

   (b) Determine the slope at the left end of the beam.

   (c) Determine the deflection at the left end of the beam.

5-34 A beam is loaded and supported as shown in Fig. P5-34.

   (a) Derive the equation of the elastic curve in terms of w, L, x, E, and I. Use the designated axes.

   (b) Determine the slope at the right end of the beam.

   (c) Determine the deflection at the right end of the beam.

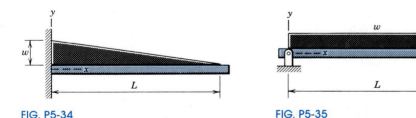

FIG. P5-34   FIG. P5-35

5-35* A beam is loaded and supported as shown in Fig. P5-35.

   (a) Derive the equation of the elastic curve in terms of w, L, x, E, and I. Use the designated axes.

   (b) Determine the slope at the left end of the beam.

   (c) Determine the deflection midway between the supports.

5-36* A beam is loaded and supported as shown in Fig. P5-36.

(a) Derive the equation of the elastic curve in terms of w, L, x, E, and I. Use the designated axes.

(b) Determine the slope at the right end of the beam.

(c) Determine the deflection midway between the supports.

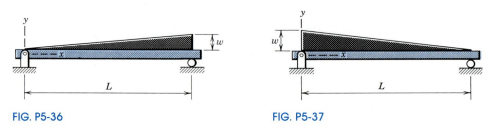

FIG. P5-36   FIG. P5-37

5-37 A beam is loaded and supported as shown in Fig. P5-37.

(a) Derive the equation of the elastic curve in terms of w, L, x, E, and I. Use the designated axes.

(b) Determine the slope at the right end of the beam.

(c) Determine the deflection midway between the supports.

5-38 A beam is loaded and supported as shown in Fig. P5-38.

(a) Derive the equation of the elastic curve in terms of P, L, x, E, and I. Use the designated axes.

(b) Determine the slope at the right end of the beam.

(c) Determine the deflection midway between the supports.

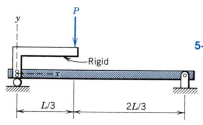

FIG. P5-38

5-39* The beam shown in Fig. P5-39 is a W10 × 30 structural steel ($E = 29{,}000$ ksi) wide-flange section (see Appendix C).

(a) Derive the equation of the elastic curve in terms of w, L, x, E, and I. Use the designated axes.

(b) Determine the deflection at the left end of the beam if $w = 2000$ lb/ft and $L = 10$ ft.

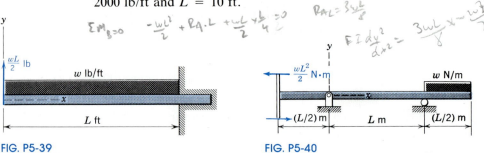

FIG. P5-39   FIG. P5-40

5-40* The beam shown in Fig. P5-40 is a W203 × 60 structural steel ($E = 200$ GPa) wide-flange section (see Appendix C).

(a) Derive the equation of the elastic curve for the region of the beam between the supports in terms of w, L, x, E, and I. Use the designated axes.

(b) Determine the deflection midway between the supports if $w = 3.5$ kN/m and $L = 5$ m.

5-41 The beam shown in Fig. P5-41 is a WT8 × 25 structural steel ($E = 29,000$ ksi) T section (see Appendix C).

(a) Derive the equation of the elastic curve for the region of the beam between the supports in terms of w, L, x, E, and I. Use the designated axes.

(b) Determine the deflection midway between the supports if $w = 400$ lb/ft and $L = 12$ ft.

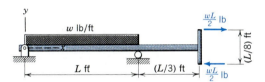

FIG. P5-41

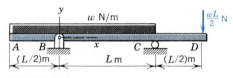

FIG. P5-42

5-42 The beam shown in Fig. P5-42 is a WT229 × 45 structural steel ($E = 200$ GPa) T section (see Appendix C).

(a) Derive the equation of the elastic curve for the region of the beam between the supports in terms of w, L, x, E, and I. Use the designated axes.

(b) Determine the deflection midway between the supports if $w = 2.75$ kN/m and $L = 4$ m.

5-43* A timber beam 6 in. wide by 12 in. deep is loaded and supported as shown in Fig. P5-43. The modulus of elasticity of the timber is 1500 ksi. A pointer is attached to the right end of the beam. Determine

(a) the deflection of the right end of the pointer.

(b) the maximum deflection of the beam.

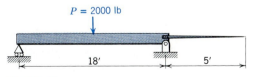

FIG. P5-43

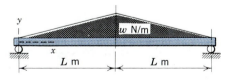

FIG. P5-44

5-44* A timber beam with a rectangular cross section 150 mm wide by 200 mm deep is loaded and supported as shown in Fig. P5-44. Determine

(a) the equation of the elastic curve for the interval $0 \leqslant x \leqslant L$ in terms of w, L, x, E, and I.

(b) the deflection at the middle of the beam if $w = 5000$ N/m, $L = 3$ m, and $E = 10$ GPa.

5-45 The beam AB shown in Fig. P5-45 is the flexural member of a scale that is used to obtain the weight W of food in a microwave oven. Determine

(a) the equation of the elastic curve for beam AB in terms of W, L, x, E, and I.

(b) the deflection at point C when $W = 20$ lb, $L = 2$ in., and $EI = 100$ lb-in.^{2}.

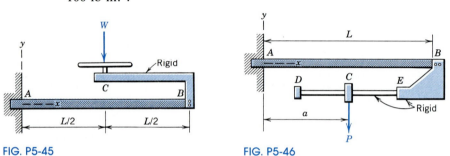

FIG. P5-45   FIG. P5-46

5-46 A beam AB is loaded and supported as shown in Fig. P5-46. The load P is applied through a collar that can be positioned on the load bar at any location in the interval $L/4 \leqslant a \leqslant 3L/4$. Determine

(a) the equation of the elastic curve for beam AB in terms of P, L, x, a, E, and I.

(b) the location of load P for maximum deflection at end B.

(c) the location of load P for zero deflection at end B.

5-47* A beam is loaded and supported as shown in Fig. P5-47. Determine

(a) the slope at the left end of the beam in terms of P, L, E, and I.

(b) the maximum deflection between the supports in terms of P, L, E, and I.

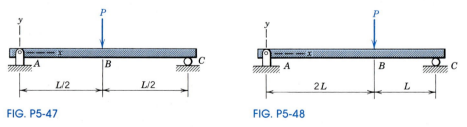

FIG. P5-47   FIG. P5-48

5-48* A beam is loaded and supported as shown in Fig. P5-48. Determine

(a) the slope at the left end of the beam in terms of P, L, E, and I.

(b) the deflection midway between the supports in terms of P, L, E, and I.

(c) the maximum deflection between the supports in terms of P, L, E, and I.

5-49 A beam is loaded and supported as shown in Fig. P5-49. Determine

(a) the slope at the left end of the beam in terms of P, L, E, and I.

(b) the maximum deflection between the supports in terms of P, L, E, and I.

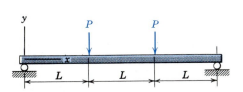

FIG. P5-49

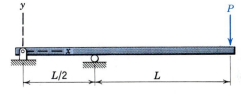

FIG. P5-50

5-50 A beam is loaded and supported as shown in Fig. P5-50. Determine

(a) the slope at the left end of the beam in terms of P, L, E, and I.

(b) the maximum deflection between the supports in terms of P, L, E, and I.

(c) the deflection at the right end of the beam in terms of P, L, E, and I.

5-51 Select the lightest structural steel ($E = 29,000$ ksi) wide-flange or American standard beam (see Appendix C) that can be used for the beam shown in Fig. P5-51 if the maximum fiber stress must not exceed 10 ksi and if the maximum deflection must not exceed 0.200 in. when $L = 8$ ft and $w = 2000$ lb/ft.

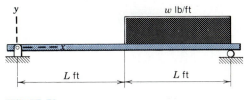

FIG. P5-51

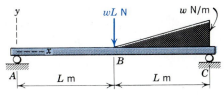

FIG. P5-52

5-52 Select the lightest structural steel ($E = 200$ GPa) wide-flange or American standard beam (see Appendix C) that can be used for the beam shown in Fig. P5-52 if the maximum fiber stress must not exceed 75 MPa and if the maximum deflection must not exceed 8 mm when $L = 3$ m and $w = 15$ kN/m.

5-53 Select the lightest pair of structural steel ($E = 29,000$ ksi) channels (see Appendix C) that can be used for the beam shown in Fig. P5-53 if the maximum fiber stress must not exceed 12 ksi and if the deflection at the left end of the beam must not exceed 0.500 in. when $L = 8$ ft and $w = 1000$ lb/ft.

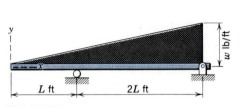

FIG. P5-53

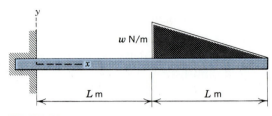

FIG. P5-54

5-54 Select the lightest pair of structural steel ($E = 200$ GPa) channels (see Appendix C) that can be used for the beam shown in Fig. P5-54 if the maximum fiber stress must not exceed 150 MPa and if the deflection at the right end of the beam must not exceed 20 mm when $L = 3$ m and $w = 3.5$ kN/m.

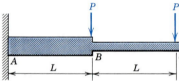

FIG. P5-55

5-55* The cantilever beam ABC shown in Fig. P5-55 has a moment of inertia $2I$ in the interval AB and a moment of inertia I in the interval BC. Determine

(a) the deflection at section B in terms of P, L, E, and I.

(b) the deflection at section C in terms of P, L, E, and I.

5-56* The simply supported beam $ABCD$ shown in Fig. P5-56 has a moment of inertia $2I$ in the center section BC and a moment of inertia I in the other two sections near the supports. Determine

(a) the deflection at section B in terms of P, L, E, and I.

(b) the maximum deflection in the beam in terms of P, L, E, and I.

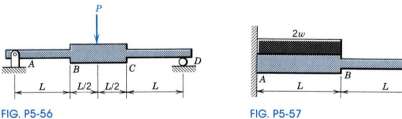

FIG. P5-56

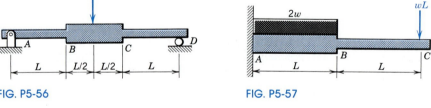

FIG. P5-57

5-57 The cantilever beam ABC shown in Fig. P5-57 has a moment of inertia $4I$ in the interval AB and a moment of inertia I in the interval BC. Determine

(a) the deflection at section B in terms of w, L, E, and I.

(b) the deflection at section C in terms of w, L, E, and I.

5-58 The simply supported beam $ABCD$ shown in Fig. P5-58 has a moment of inertia $2I$ in the center section BC and a moment of inertia I in the other two sections near the supports. Determine

(a) the deflection at section B in terms of P, L, E, and I.

(b) the maximum deflection in the beam in terms of P, L, E, and I.

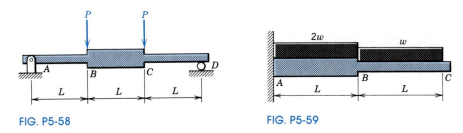

FIG. P5-58   FIG. P5-59

5-59* The cantilever beam ABC shown in Fig. P5-59 has a moment of inertia $4I$ in the interval AB and a moment of inertia I in the interval BC. Determine

(a) the deflection at section B in terms of w, L, E, and I.

(b) the deflection at section C in terms of w, L, E, and I.

5-60* The simply supported beam $ABCD$ shown in Fig. P5-60 has a moment of inertia $3I$ in the center section BC and a moment of inertia I in the other two sections near the supports. Determine

(a) the deflection at section B in terms of w, L, E, and I.

(b) the maximum deflection in the beam in terms of w, L, E, and I.

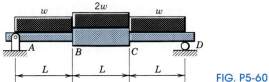

FIG. P5-60

5-5

DEFLECTIONS BY INTEGRATION OF SHEAR-FORCE OR LOAD EQUATIONS

In Section 5-4 the equation of the elastic curve was obtained by integrating the differential equation

$$EI \left(\frac{d^2y}{dx^2} \right) = M_x \qquad (5\text{-}3)$$

and applying the appropriate boundary conditions to evaluate the two constants of integration. In a similar manner, the equation of the elastic curve can be obtained from load and shear-force equations. The differential equations that relate deflection y to load w_x or shear-force V_x are obtained by substituting Eq. 4-11 or 4-9, respectively, into Eq. 5-3. Thus

$$EI\left(\frac{d^3y}{dx^3}\right) = V_x \tag{5-4}$$

$$EI\left(\frac{d^4y}{dx^4}\right) = w_x \tag{5-5}$$

When Eqs. 5-4 or 5-5 are used to obtain the equation of the elastic curve, either three or four integrations will be required instead of the two integrations required with Eq. 5-3. These additional integrations will introduce additional constants of integration. The boundary conditions, however, now include conditions on the shear forces and bending moments, in addition to the conditions on slopes and deflections. Use of a particular differential equation is usually made on the basis of mathematical convenience or personal preference. In those instances when the expression for the load is easier to write than the expression for the moment, Eq. 5-5 would be preferred over Eq. 5-3. The following example illustrates the use of Eq. 5-5 for calculating beam deflections.

EXAMPLE 5-5

A beam is loaded and supported as shown in Fig. 5-11. Determine
   (a) the equation of the elastic curve in terms of w, L, x, E, and I.
   (b) the deflection at the right end of the beam.
   (c) the support reactions V_A and M_A at the left end of the beam.

Solution

Since the equation for the load distribution is given and the moment equation is not easy to write, Eq. 5-5 will be used to determine the deflections.
   (a) In Section 4-6 (see Fig. 4-18), the upward direction was considered positive for a distributed load w; therefore, Eq. 5-5 is written as

$$EI\left(\frac{d^4y}{dx^4}\right) = w_x = -w\cos\left(\frac{\pi x}{2L}\right)$$

Successive integration gives

$$EI\left(\frac{d^3y}{dx^3}\right) = V_x = -\left(\frac{2wL}{\pi}\right)\sin\left(\frac{\pi x}{2L}\right) + C_1$$

$$EI\left(\frac{d^2y}{dx^2}\right) = M_x = \left(\frac{4wL^2}{\pi^2}\right)\cos\left(\frac{\pi x}{2L}\right) + C_1x + C_2$$

$$EI\left(\frac{dy}{dx}\right) = EI\,\theta = \left(\frac{8wL^3}{\pi^3}\right)\sin\left(\frac{\pi x}{2L}\right) + C_1\frac{x^2}{2} + C_2x + C_3$$

$$EIy = EI\delta = -\left(\frac{16wL^4}{\pi^4}\right)\cos\left(\frac{\pi x}{2L}\right) + C_1\frac{x^3}{6} + C_2\frac{x^2}{2} + C_3x + C_4$$

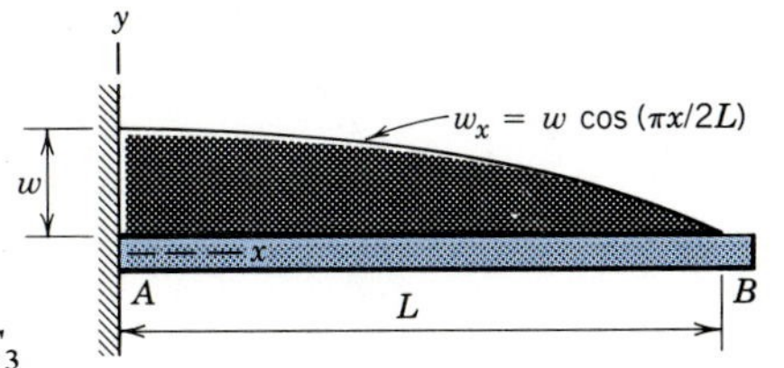

$w_x = w\cos(\pi x/2L)$

FIG. 5-11

The four constants of integration are determined by applying the boundary conditions. Thus

At $x = 0$, $y = 0$;   therefore, $C_4 = \dfrac{16wL^4}{\pi^4}$

At $x = 0$, $\dfrac{dy}{dx} = 0$;   therefore, $C_3 = 0$

At $x = L$, $V = 0$;   therefore, $C_1 = \dfrac{2wL}{\pi}$

At $x = L$, $M = 0$;   therefore, $C_2 = -\dfrac{2wL^2}{\pi}$

Thus,

$$y = -\frac{w}{3\pi^4 EI}\left[48L^4\cos\left(\frac{\pi x}{2L}\right) - \pi^3 Lx^3 + 3\pi^3 L^2 x^2 - 48L^4\right]$$

Ans.

(b) The deflection at the right end of the beam is

$$\delta_B = y_{x=L} = -\frac{w}{3\pi^4 EI}(-\pi^3 L^4 + 3\pi^3 L^4 - 48L^4)$$

$$= -\frac{(2\pi^3 - 48)wL^4}{(3\pi^4 EI)} = -0.04795wL^4/EI \quad \text{Ans.}$$

(c) The shear force V and bending moment M at any distance x from the support are

$$V_x = \frac{2wL}{\pi}\left[1 - \sin\left(\frac{\pi x}{2L}\right)\right]$$

$$M_x = \frac{2wL}{\pi^2}\left[2L\cos\left(\frac{\pi x}{2L}\right) + \pi x - \pi L\right]$$

Thus, the support reactions at the left end of the beam are

$$V_A = V_{x=0} = \frac{2wL}{\pi} \qquad\qquad \text{Ans.}$$

$$M_A = M_{x=0} = -\frac{2(\pi - 2)wL^2}{\pi^2} \qquad \text{Ans.}$$

PROBLEMS

5-61* A beam is loaded and supported as shown in Fig. P5-61. Determine

  (a) the equation of the elastic curve in terms of w, L, x, E, and I.
  (b) the deflection at the left end of the beam.
  (c) the support reactions V_B and M_B.

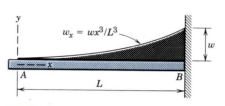

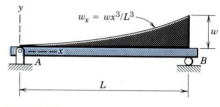

FIG. P5-61

FIG. P5-62

5-62* A beam is loaded and supported as shown in Fig. P5-62. Determine

  (a) the equation of the elastic curve in terms of w, L, x, E, and I.
  (b) the deflection midway between the supports.
  (c) the maximum deflection in the beam. $\quad$ at $\frac{dy}{dx} = 0$
  (d) the support reactions R_A and R_B.

5-63 A beam is loaded and supported as shown in Fig. P5-63. Determine

  (a) the equation of the elastic curve in terms of w, L, x, E, and I.
  (b) the deflection at the left end of the beam.
  (c) the support reactions V_B and M_B.

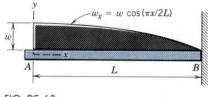

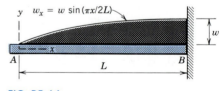

FIG. P5-63

FIG. P5-64

5-64 A beam is loaded and supported as shown in Fig. P5-64. Determine

  (a) the equation of the elastic curve in terms of w, L, x, E, and I.
  (b) the deflection at the left end of the beam.
  (c) the slope at the left end of the beam.
  (d) the support reactions V_B and M_B.

5-65 A beam is loaded and supported as shown in Fig. P5-65. Determine

  (a) the equation of the elastic curve in terms of w, L, x, E, and I.

(b) the deflection midway between the supports.

(c) the slope at the left end of the beam.

(d) the support reactions R_A and R_B.

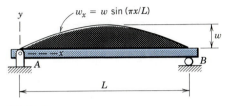

FIG. P-65

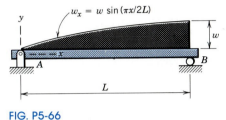

FIG. P5-66

5-66 A beam is loaded and supported as shown in Fig. P5-66. Determine

    (a) the equation of the elastic curve in terms of w, L, x, E, and I.

    (b) the deflection midway between the supports.

    (c) the slope at the left end of the beam.

    (d) the support reactions R_A and R_B.

5-6

SINGULARITY FUNCTIONS FOR BEAM DEFLECTIONS

The double integration method of Section 5-4 becomes extremely tedious and time-consuming when several intervals and several sets of matching conditions are required. The labor involved in solving problems of this type can be diminished by making use of *singularity functions* following the method originally developed in 1862 by the German mathematician A. Clebsch (1833–1872).[1]

Singularity functions are closely related to the unit step function used by the British physicist O. Heaviside (1850–1925) to analyze the transient response of electrical circuits. Singularity functions will be used here for writing one bending moment equation that applies in all intervals along a beam, thus eliminating the need for matching conditions.

A singularity function of x is written as $\langle x - x_0 \rangle^n$, where n is any *integer* (positive or negative) including zero, and x_0 is a constant equal to the value of x at the initial boundary of a specific interval along a beam. Selected properties of singularity functions required for beam-deflection problems are listed here for emphasis and ready reference:

---

[1] For a rather complete history on the Celbsch method and the numerous extensions thereof, see "Clebsch's Method for Beam Deflections," Walter D. Pilkey, *J. Eng. Educ.*, Jan. 1964, p. 170.

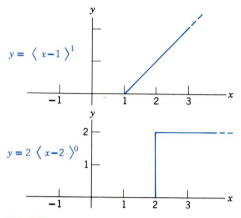

FIG. 5-12

$$\langle x - x_0 \rangle^n = \begin{cases} (x - x_0)^n & \text{when } n > 0 \text{ and } x \geqslant x_0 \\ 0 & \text{when } n > 0 \text{ and } x < x_0 \end{cases}$$

$$\langle x - x_0 \rangle^0 = \begin{cases} 1 & \text{when } x \geqslant x_0 \\ 0 & \text{when } x < x_0 \end{cases}$$

$$\int \langle x - x_0 \rangle^n \, dx = \frac{1}{n+1} \langle x - x_0 \rangle^{n+1} + C \qquad \text{when } n \geqslant 0$$

$$\frac{d}{dx} \langle x - x_0 \rangle^n = n \langle x - x_0 \rangle^{n-1} \qquad \text{when } n \geqslant 1$$

Several examples of singularity functions are shown in Fig. 5-12.

By making use of these properties of singularity functions, one is able to write a single equation for the bending moment for a beam and obtain the correct value of the moment in any interval along the beam. To amplify this statement, consider the beam of Fig. 5-13. By the methods of Sections 4-5 and 5-4, the moment equations at the four designated sections are

$$\begin{array}{ll} M_1 = R_L x & 0 < x < x_1 \\[4pt] M_2 = R_L x - P(x - x_1) & x_1 < x < x_2 \\[4pt] M_3 = R_L x - P(x - x_1) + M_A & x_2 < x < x_3 \\[4pt] M_4 = R_L x - P(x - x_1) + M_A - \left(\dfrac{w}{2}\right)(x - x_3)^2 & x_3 < x < L \end{array}$$

These four moment equations can be combined into a single equation by means of singularity functions to give

$$M_x = R_L x - P\langle x - x_1 \rangle^1 + M_A \langle x - x_2 \rangle^0 - \frac{w}{2} \langle x - x_3 \rangle^2 \quad 0 \leqslant x \leqslant L$$

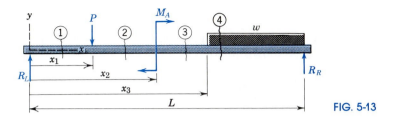

FIG. 5-13

Distributed beam loadings that are sectionally continuous (the distributed load cannot be represented by a single function of x for all values of x) require special consideration. Moment equations for such loads can be readily obtained by superposition, as illustrated in the following example. Further applications of singularity functions to the determination of beam deflections are given in Examples 5-7 and 5-8.

EXAMPLE 5-6

Use singularity functions to write a single equation for the bending moment at any section of the beam shown in Fig. 5-14a.

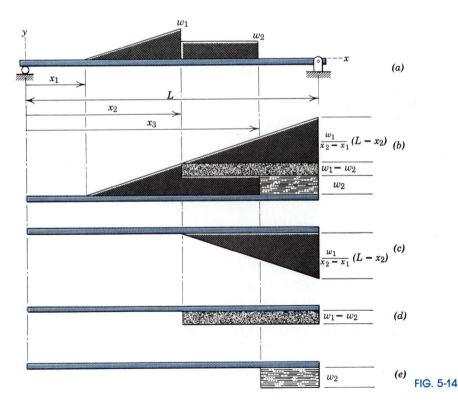

FIG. 5-14

Solution

The loading on the beam of Fig. 5-14a can be considered a combination of the loadings shown in Fig. 5-14b, c, d, and e, where downward-acting loads are shown on top of the beam and upward-acting loads on the bottom. The moment of the linearly varying load can be obtained from the geometry of the load diagram. The moment equation is

$$M_x = R_L x - \frac{w_1}{6(x_2 - x_1)} \langle x - x_1 \rangle^3 + \frac{w_1}{6(x_2 - x_1)} \langle x - x_2 \rangle^3$$

$$+ \frac{w_1 - w_2}{2} \langle x - x_2 \rangle^2 + \frac{w^2}{2} \langle x - x_3 \rangle^2 \quad \text{Ans.}$$

EXAMPLE 5-7

A beam is loaded and supported as shown in Fig. 5-9a. Use singularity functions to determine the deflection at the right end of the beam.

Solution

The boundary conditions, in the interval $0 \langle x \rangle L$, are $dy/dx = 0$ when $x = 0$ and $y = 0$ when $x = 0$ with the axes as shown in Fig. 5-9b. From the free-body diagram of Fig. 5-9d, the expression for the bending moment is

$$EI \frac{d^2y}{dx^2} = -\frac{5wL^2}{18} + \frac{wLx}{3} - \frac{w}{2} \left\langle x - \frac{2L}{3} \right\rangle^2$$

The first integration gives

$$EI \frac{dy}{dx} = -\frac{5wL^2 x}{18} + \frac{wLx^2}{6} - \frac{w}{6} \left\langle x - \frac{2L}{3} \right\rangle^3 + C_1$$

Substituting the boundary condition $dy/dx = 0$ when $x = 0$ and noting that for $x < 2L/3$ the term in the brackets is zero gives $C_1 = 0$. Integrating again gives

$$EIy = -\frac{5wL^2}{36} x^2 + \frac{wL}{18} x^3 - \frac{w}{24} \left\langle x - \frac{2L}{3} \right\rangle^4 + C_2$$

The boundary condition $y = 0$ when $x = 0$ gives $C_2 = 0$. The deflection at the right end is obtained by substituting $x = L$ in the above elastic curve equation. The result is

$$y = -\frac{163}{1944} \frac{wL^4}{EI} = \frac{163}{1944} \frac{wL^4}{EI} \text{ downward} \quad \text{Ans.}$$

The result agrees with that of Example 5-3.

EXAMPLE 5-8

A beam is loaded and supported as shown in Fig. 5-15a. Use singularity functions to determine the deflection at the left end of the beam.

Solution

The equations of equilibrium, applied to the free-body diagram of Fig. 5-15b, give the reaction R_L as $3wL/8$ upward. By the use of singularity functions and the superposed loads of Fig. 5-15c, the bending moment equation becomes

$$EI\frac{d^2y}{dx^2} = -\frac{w}{2}(x+L)^2 + \frac{w}{2}\left\langle x+\frac{L}{2}\right\rangle^2 + \frac{3wL}{8}\langle x\rangle^1 + \frac{wL^2}{2}\left\langle x-\frac{L}{2}\right\rangle^0$$

The boundary conditions are $y = 0$ when $x = 0$, and $y = 0$ when $x = L$. Two integrations of the moment equation give

$$EI\frac{dy}{dx} = -\frac{w}{6}(x+L)^3 + \frac{w}{6}\left\langle x+\frac{L}{2}\right\rangle^3 + \frac{3wL}{16}\langle x\rangle^2 + \frac{wL^2}{2}\left\langle x-\frac{L}{2}\right\rangle^1 + C_1$$

and

$$EIy = -\frac{w}{24}(x+L)^4 + \frac{w}{24}\left\langle x+\frac{L}{2}\right\rangle^4 + \frac{wL}{16}\langle x\rangle^3 + \frac{wL^2}{4}\left\langle x-\frac{L}{2}\right\rangle^2 + C_1x + C_2$$

The first boundary condition, $y = 0$ when $x = 0$, gives

$$0 = -\frac{wL^4}{24} + \frac{wL^4}{384} + 0 + 0 + 0 + C_2$$

from which

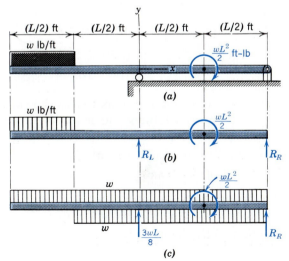

(a)

(b)

(c)   FIG. 5-15

$$C_2 = + \frac{5}{128} wL^4$$

The second boundary condition, $y = 0$ when $x = L$, gives

$$0 = -\frac{16wL^4}{24} + \frac{81wL^4}{384} + \frac{wL^4}{16} + \frac{wL^4}{16} + C_1 L + \frac{5wL^4}{128}$$

from which

$$C_1 = +\frac{7}{24} wL^3$$

The deflection at $x = -L$ is found, by substitution, to be

$$y = -\frac{97}{384} \frac{wL^4}{EI} = \frac{97}{384} \frac{wL^4}{EI} \quad \text{downward} \qquad \text{Ans.}$$

PROBLEMS

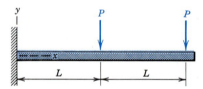

FIG. P5-67

5-67* Use singularity functions to determine the deflection, in terms of P, L, E, and I, at the right end of the cantilever beam shown in Fig. P5-67.

5-68* Use singularity functions to determine the deflection, in terms of P, L, E, and I, at the right end of the cantilever beam shown in Fig. P5-68.

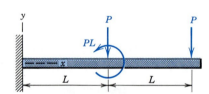

FIG. P5-68   FIG. P5-69

5-69 A cantilever beam is loaded and supported as shown in Fig. P5-69. Use singularity functions to determine the deflection, in terms of P, L, E, and I,

(a) at a distance $x = L$ from the support.

(b) at the right end of the beam.

5-70 A beam is loaded and supported as shown in Fig. P5-70. Use singularity functions to determine, in terms of P, L, E, and I,

(a) the deflection under the load P.

(b) the deflection at the middle of the span.

(c) the maximum deflection in the beam.

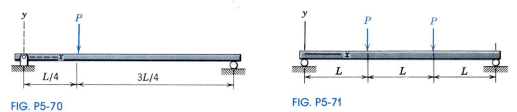

FIG. P5-70

FIG. P5-71

5-71* A beam is loaded and supported as shown in Fig. P5-71. Use singularity functions to determine the deflection, in terms of P, L, E, and I,

(a) at a distance $x = L$ from the left support.

(b) at the middle of the span.

5-72* A beam is loaded and supported as shown in Fig. P5-72. Use singularity functions to determine the deflection, in terms of P, L, E, and I,

(a) at the right end of the beam.

(b) at a section midway between the supports.

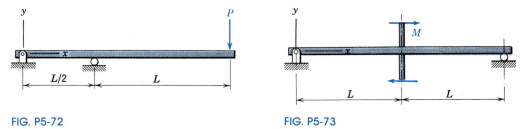

FIG. P5-72

FIG. P5-73

5-73 A beam is loaded and supported as shown in Fig. P5-73. Use singularity functions to determine, in terms of M, L, E, and I,

(a) the deflection at middle of the span.

(b) the maximum deflection in the beam.

5-74 Use singularity functions to determine the deflection, in terms of w, L, E, and I, at the left end of the cantilever beam shown in Fig. P5-74.

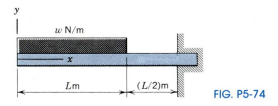

FIG. P5-74

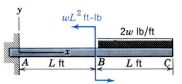

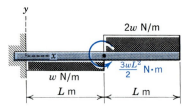

FIG. P5-75

FIG. P5-76

5-75* A cantilever beam is loaded and supported as shown in Fig. P5-75. Use singularity functions to determine the deflection, in terms of w, L, E, and I,

(a) at a distance $x = L$ from the support.

(b) at the right end of the beam.

5-76* A cantilever beam is loaded and supported as shown in Fig. P5-76. Use singularity functions to determine the deflection, in terms of w, L, E, and I,

(a) at a distance $x = L$ from the support.

(b) at the right end of the beam.

5-77 A beam is loaded and supported as shown in Fig. P5-77. Use singularity functions to determine the deflection, in terms of w, L, E, and I, at a section midway between the supports.

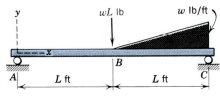

FIG. P5-77

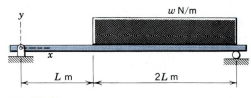

FIG. P5-78

5-78 A beam is loaded and supported as shown in Fig. P5-78. Use singularity functions to determine the deflection, in terms of w, L, E, and I,

(a) at the left end of the distributed load.

(b) at a section midway between the supports.

5-79* A beam is loaded and supported as shown in Fig. P5-79. Use singularity functions to determine the deflection, in terms of w, L, E, and I,

(a) at the left end of the distributed load.

(b) at a section midway between the supports.

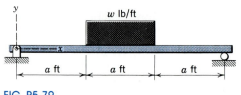

FIG. P5-79

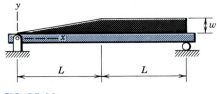

FIG. P5-80

5-80* A beam is loaded and supported as shown in Fig. P5-80. Use singularity functions to determine, in terms of w, L, E, and I,

(a) the deflection at the middle of the span.

(b) the maximum deflection in the beam.

5-81 A beam is loaded and supported as shown in Fig. P5-81. Use singularity functions to determine, in terms of w, L, E, and I,

(a) the deflection at the middle of the span.

(b) the maximum deflection in the beam.

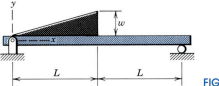

FIG. P5-81

5-7

THE DEVELOPMENT OF THE AREA-MOMENT METHOD

The area-moment method of determining the deflection at any specified point along a beam is a semigraphic method utilizing the relations between successive derivatives of the deflection y and the moment diagram. For problems involving several changes in loading, the area-moment method is usually much faster than the double integration method; consequently, it is widely used in practice. The method is based on two propositions or theorems derived as follows.

From Fig. 5-3 it is seen that for any arc length dL along the elastic curve of a loaded beam

$$d\theta = \frac{dL}{\rho}$$

and from Eq. 5-2 this becomes

$$d\theta = \frac{M}{EI} dL$$

For actual beams the curvature is so small that dL can be replaced by the distance dx along the unloaded beam, and the relation becomes

$$d\theta = \frac{M}{EI} dx$$

Reference to Fig. 5-16 (which consists of a loaded beam, its moment diagram, and a sketch, greatly exaggerated, of the elastic curve of the deflected beam) shows that $d\theta$ is also the angle between tangents to the elastic curve at two points a distance dL apart. When any two points A and B along the elastic curve are selected as shown in Fig.

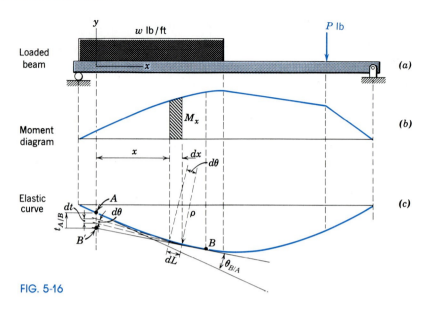

Loaded beam

Moment diagram

Elastic curve

FIG. 5-16

5-16*c* and the relation for $d\theta$ is integrated between these limits, the result is

$$\int_{\theta_A}^{\theta_B} d\theta = \theta_B - \theta_A = \int_{x_A}^{x_B} \frac{M}{EI} \, dx$$

The integral on the left, designated $\theta_{B/A}$, is the angle between tangents to the elastic curve at A and B. The theorem can be written as

$$\theta_{B/A} = \int_{x_A}^{x_B} \frac{M}{EI} \, dx \tag{5-6}$$

From Fig. 5-16*b*, the integral is clearly equal to $1/(EI)$ times the area under the moment diagram when E and I are constants (as they are in many beams). In case either E or I vary along the beam, it is advisable to construct an $M/(EI)$ diagram instead of a moment diagram. The first area-moment theorem can be expressed in words as follows: *The angle between the tangents to the elastic curve of a loaded beam at any two points A and B along the elastic curve equals the area under the M/(EI) diagram of the beam between ordinates at A and B.*

The correct interpretation of numerical results frequently depends on a reliable sign convention, and the following convention is recommended. Areas under positive bending moments are considered positive. A positive area means that the angle is positive or counterclockwise when measured from the tangent to the point on the left to the other tangent, as shown in Fig. 5-17.

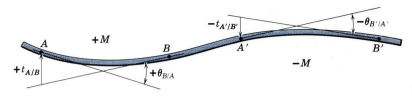

FIG. 5-17

In order to obtain the second area-moment theorem, let B' in Fig. 5-16c be the intersection of the tangent to the elastic curve at B with a vertical from point A on the elastic curve. Since the deflections of a beam are assumed to be small, it can be seen from Fig. 5-16c that

$$dt = x\, d\theta = \frac{M}{EI} x\, dx$$

Integration gives

$$\int_0^{t_{A/B}} dt = t_{A/B} = \int_{x_A}^{x_B} \frac{M}{EI} x\, dx \qquad (5\text{-}7)$$

A study of Fig. 5-16c shows that the left integral is the vertical (actually perpendicular to the beam) distance $t_{A/B}$ of any point A on the elastic curve from a tangent drawn to any other point B on the elastic curve. This distance is frequently called the *tangential deviation* to distinguish it from the beam deflection y. The integral on the right, as seen from Fig. 5-16b, is the first moment, with respect to an axis through A parallel to the y axis, of the area under the $M/(EI)$ diagram between ordinates at A and B.

The second area-moment theorem (Eq. 5-7) can be stated in words as follows: *The vertical distance (tangential deviation) of any point A on the elastic curve of a beam from a tangent drawn at any other point B on the elastic curve equals the first moment, with respect to an axis at A, of the area under the $M/(EI)$ diagram between ordinates at A and B.*

A sign convention is required for those instances in which the shape of the elastic curve is not immediately evident from the given loads. The following sign convention is recommended. The first moment of the area under a positive $M/(EI)$ diagram is considered positive, and from the second area-moment theorem a positive first moment gives a positive tangential deviation. A positive tangential deviation means that point A on the beam (the moment center) is above the tangent drawn from the other point B. Examples of the sign conventions for both the first and second area-moment theorems are shown in Fig. 5-17.

The effectiveness of the area-moment method for determining beam deflections depends, to a considerable extent, on the ease of calculating areas and first moments of areas under the $M/(EI)$ diagrams. Naturally, these values can be obtained by integration of the general expressions from Eqs. 5-6 and 5-7. In general, however, this procedure offers little or no advantage over the double integration method of Sections 5-4 and 5-6. The method of drawing a separate moment diagram for each load and reaction on a beam, as explained in the next section, greatly simplifies the finding of areas and moments of areas because in this way nearly all areas are triangular, rectangular, or parabolic.

5-8
MOMENT DIAGRAMS BY PARTS

In Chapter 4 a composite or complete moment diagram was drawn that included the contribution or effect of all loads and reactions acting on a beam. In the application of the area-moment method for obtaining deflections, it is sometimes easier and more practical to draw a series of diagrams showing the moment due to each load or reaction on a separate sketch. For this purpose, each moment diagram is drawn from the load to an arbitrarily selected section called the reference section. It is preferable to select the reference section at one end of the beam or at one end of a uniformly distributed load. With a reference section at one end, start with the load or reaction at the opposite end of the beam and draw the bending moment diagram immediately below the beam as if no other loads or reactions were acting on the beam except at the reference section. Proceed toward the right (or left), drawing the moment diagram for each load and reaction below the previous one. The composite moment diagram can be obtained, when necessary, by superposition (addition) of the component diagrams. The actual or total moment on the beam at any section is the algebraic sum of the moments from the separate diagrams, except at an interior reference section where the sum of the moments on either side of the reference section gives the total moment. The following example illustrates the procedure for drawing moment diagrams by parts.

EXAMPLE 5-9

A beam is loaded and supported as shown in Fig. 5-18a. Construct the moment diagram by parts with the reference section at
  (a) the right end of the beam.
  (b) the left end of the distributed load.
  (c) the left end of the beam.

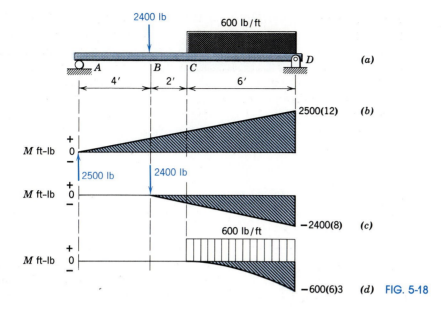

2400 lb

600 lb/ft

A *B* *C* *D* (a)

4'   2'   6'

2500(12) (b)

M ft-lb 0   + −

2500 lb   2400 lb

M ft-lb 0   + −

−2400(8) (c)

600 lb/ft

M ft-lb 0   + −

−600(6)3 (d) FIG. 5-18

Solution

From a free-body diagram of the beam, the equations of equilibrium give the left reaction as 2500 lb upward and the right reaction as 3500 lb upward. The reference section can be selected anywhere along the beam, and each different reference section yields a different series of component moment diagrams. Three different solutions of the example are given to emphasize the importance of a careful selection of the reference section.

(a) The reference section is selected at the right end of the beam. The bending moment due to the 2500-lb left reaction is positive and equal to 2500 times the distance from the left end, as shown in Fig. 5-18*b*. The bending moment due to the 2400-lb concentrated load is negative and equal to 2400 times the distance from the load, as shown in Fig. 5-18*c*. The bending moment due to the distributed load is negative and equal to 300 times the square of the distance from *C* (the left edge of the uniform load), as shown in Fig. 5-18*d*. Since the diagrams were drawn from left to right, the right reaction makes no contribution to the moment. It is in the reference section.

(b) The reference section is selected at *C*, the left edge of the distributed load. The bending moment due to the 3500-lb right reaction is positive, varies directly as the distance from the right end, and is shown in Fig. 5-19*b*. The bending moment due to the distributed load is negative, equal to 300 times the square of the

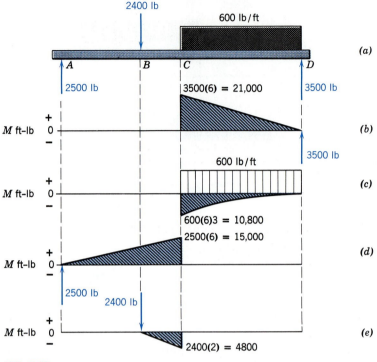

FIG. 5-19

distance from the right end, as shown in Fig. 5-19c. Likewise, the moments due to the left reaction and the concentrated load are shown in Figs. 5-19d and e. The total bending moment at the reference section C is the sum of the individual moments just to the left or right of the section—namely, +10,200 ft-lb.

(c) The reference section is selected at the left end. The bending moment due to the 3500-lb right reaction is shown in Fig. 5-20b. The bending moment due to the distributed load is shown in Fig. 5-20c, and the area is subdivided into a triangle, a rectangle, and a parabolic area. This shows the disadvantage of selecting the reference section away from one end of a distributed load. The bending moment due to the concentrated load is shown in Fig. 5-20d.

A study of the three solutions reveals that the amount of numerical work involved in finding areas and first moments is greatly altered by the choice of reference section to be used. In solution (a) there are three component areas, whereas solution (b) involves four component areas, and solution (c) is the least desirable with five component areas, three of which are due to the distributed load.

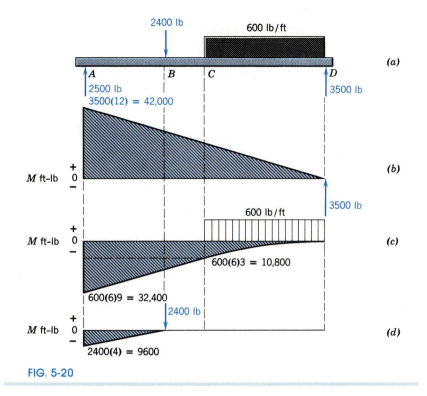

FIG. 5-20

PROBLEMS

5-82 Construct the moment diagram by parts for the beam of Fig. P5-82.

(a) Use a reference section at the right support.

(b) Use a reference section at the middle of the span.

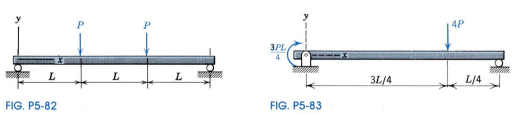

FIG. P5-82

FIG. P5-83

5-83 Construct the moment diagram by parts for the beam of Fig. P5-83.

(a) Use a reference section at the right support.

(b) Use a reference section at the left support.

5-84 Construct the moment diagram by parts for the beam of Fig. P5-84.
(a) Use a reference section at the right support.
(b) Use a reference section at the middle of the span.

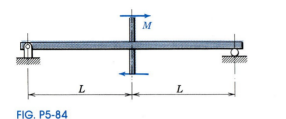

FIG. P5-84

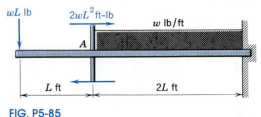

FIG. P5-85

5-85 Construct the moment diagram by parts for the beam of Fig. P5-85.
(a) Use a reference section at the support.
(b) Use a reference section at the left end of the distributed load.

5-86 Construct the moment diagram by parts for the beam of Fig. P5-86.
(a) Use a reference section at the support.
(b) Use a reference section at the right end of the distributed load.

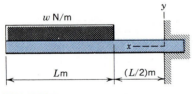

FIG. P5-86

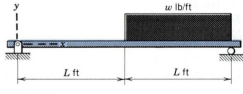

FIG. P5-87

5-87 Construct the moment diagram by parts for the beam of Fig. P5-87.
(a) Use a reference section at the right support.
(b) Use a reference section at the left end of the distributed load.

5-88 Construct the moment diagram by parts for the beam of Fig. P5-88.
(a) Use a reference section at the support.
(b) Use a reference section at the left end of the distributed load.

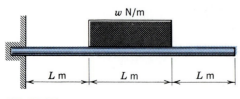

FIG. P5-88

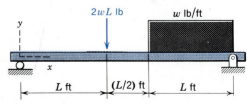

FIG. P5-89

5-89 Construct the moment diagram by parts for the beam of Fig. P5-89.

(a) Use a reference section at the right support.

(b) Use a reference section at the left end of the distributed load.

5-90 Construct the moment diagram by parts for the beam of Fig. P5-90.

(a) Use a reference section at the support.

(b) Use a reference section at the middle of the span.

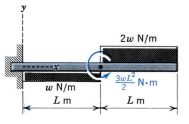

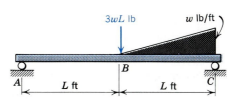

FIG. P5-90   FIG. P5-91

5-91 Construct the moment diagram by parts for the beam of Fig. P5-91. Use a reference section at the right support.

5-92 Construct the moment diagram by parts for the beam of Fig. P4-54. Use a reference section at the left support.

5-93 Construct the moment diagram by parts for the beam of Fig. P4-55. Use a reference section at the right end of the distributed load.

5-94 Construct the moment diagram by parts for the beam of Fig. P4-58. Use a reference section at the left support.

5-95 Construct the moment diagram by parts for the beam of Fig. P4-59. Use a reference section at the right support.

5-96 Construct the moment diagram by parts for the beam of Fig. P4-60. Use a reference section at the left support.

5-97 Construct the moment diagram by parts for the beam of Fig. P4-61. Use a reference section at the left end of the distributed load.

5-98 Construct the moment diagram by parts for the beam of Fig. P4-80. Use a reference section midway between the supports.

5-99 Construct the moment diagram by parts for the beam of Fig. P5-80.

(a) Use a reference section at the middle of the span.

(b) Use a reference section at the right support.

5-100 Construct the moment diagram by parts for the beam of Fig. P5-81.

(a) Use a reference section at the middle of the span.

(b) Use a reference section at the right support.

5-9

DEFLECTIONS BY AREA MOMENTS

The deflection of a specific point along the elastic curve of a beam can normally be determined by one or more applications of the second area-moment theorem. The amount of time and effort required for a given problem depends to a considerable extent on the judgment used in selecting the moment center for the second theorem and in drawing the moment diagram by parts. The ability to construct the correct moment diagram and sketch the elastic curve is essential to the application of the method.

A few general suggestions are presented here to facilitate the application of the method.

1. It is frequently easier to use a composite moment diagram instead of a moment diagram by parts for those cases in which the beam carries only one load.
2. It will be found that selection of the reference section of a cantilever beam at the fixed end for construction of the moment diagram by parts usually gives the minimum number of areas.
3. For overhanging beams it is advisable to avoid taking a tangent to the elastic curve at the overhanging end.
4. It is sometimes advantageous to solve the problem with symbols and substitute numerical values in the last step.

The list of steps given below for the solution of deflection problems by the area-moment method may prove helpful. There is no point in memorizing these steps, as they are not a substitute for an understanding of the principles involved. They are:

1. Sketch the loaded beam, the moment [or $M/(EI)$] diagrams (either by parts or composite diagram, depending on the complexity of the problem), and the elastic curve (if the shape is not known, it may be assumed).
2. Visualize which tangent lines may be most helpful and draw such lines on the elastic curve. When a point of zero slope is known, either from symmetry or from supports, a tangent drawn at this point will frequently be useful.
3. By application of the second area-moment theorem, determine the tangential deviation at the point where the beam deflection is desired and at any other points required.
4. From geometry, determine the perpendicular distance from the unloaded beam to the tangent line at the point where the beam deflection is desired, and, using the results of step 3, solve for the required deflection.

For essential data on properties of curves, see Appendix B. Examples 5-10, 5-11, 5-12, which follow, illustrate the method.

EXAMPLE 5-10

Use the area-moment method to determine the deflection of the free end of the cantilever beam shown in Fig. 5-21a, in terms of w, L, E, and I.

Solution

Since E and I are constant, a bending moment diagram is used instead of an $M/(EI)$ diagram. The moment diagram is shown in Fig. 5-2b, and the area under it has been divided into rectangular, triangular, and parabolic parts for ease in calculating areas and moments. The elastic curve is shown in Fig. 5-21c with the deflections greatly exaggerated. Points A and B are selected at the ends of the beam because the beam has a horizontal tangent at B and the deflection at A is required. The vertical distance to A from the tangent at B ($t_{A/B}$ in Fig. 5-21c) equals the deflection of the free end of the beam y_A. For this reason, the second area-moment theorem can be used directly to obtain the required deflection. The area of each of the three portions of the area under the moment diagram is shown in Fig. 5-21b along with the distance from the centroid of each part to the moment axis at the free end A. The second area-moment theorem gives

$$EIt_{A/B} = -\frac{wL^3}{6}\left(\frac{5L}{4}\right) - \frac{wL^3}{2}(2L) - \frac{wL^3}{2}\left(\frac{13L}{2}\right)$$

from which

$$y_A = t_{A/B} = -\frac{55wL^4}{24EI} = \frac{55wL^4}{24EI} \quad \text{downward} \qquad \text{Ans.}$$

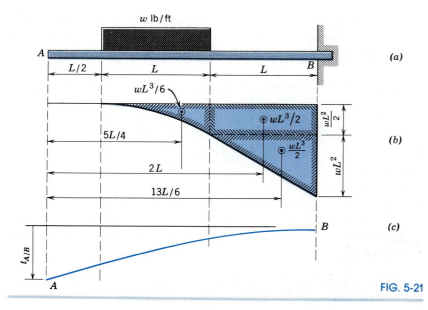

FIG. 5-21

EXAMPLE 5-11

A beam is loaded and supported as shown in Fig. 5-22a. Use the area-moment method to determine the deflection at the left end of the beam in terms of P, L, E, and I.

Solution

The equations of equilibrium yield the left reaction equal to $4P/3$ upward and the right reaction equal to $P/3$ downward. Since E and I are constant, a moment diagram (instead of an $M/(EI)$ diagram) is drawn, and, with only one applied load, a composite moment diagram as shown in Fig. 5-22b is used.

The elastic curve with a tangent drawn at C is shown in Fig. 5-22c. The tangent line could have been drawn at B for an equally satisfactory solution. The second area-moment theorem gives the tangential deviation at B as

$$EIt_{B/C} = -\frac{3PL^2}{2}(L) = -\frac{3PL^3}{2}$$

where the negative sign indicates that the moment center B is below the tangent drawn from C. Another application of the second area-moment theorem gives the tangential deviation at A as

$$EIt_{A/C} = -\frac{3PL^2}{2}(2L) - \frac{PL^2}{2}\left(\frac{2L}{3}\right) = -\frac{10PL^3}{3}$$

where the negative sign indicates that A is below the tangent from C. Finally,

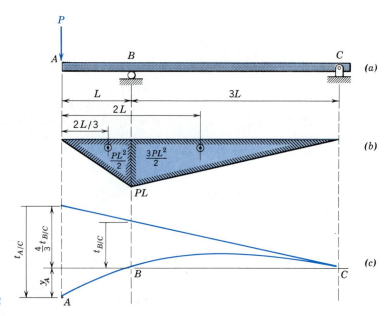

FIG. 5-22

$$EIy_A = EI\left(t_{A/C} - \frac{4}{3}t_{B/C}\right) = -\frac{10PL^3}{3} - \frac{4}{3}\left(-\frac{3PL^3}{2}\right) = -\frac{4}{3}PL^3$$

and

$$y_A = \frac{4}{3}\frac{PL^3}{EI} \quad \text{downward} \qquad \text{Ans.}$$

EXAMPLE 5-12

A stepped cantilever beam is loaded as shown in Fig. 5-23a. Use the area-moment method to determine the deflection at the free end in terms of P, L, E, and I.

Solution

Since E is constant but I is not, an M/I diagram will be used to solve the problem. The moment diagram is shown in Fig. 5-23b. The M/I diagram is shown in Fig. 5-23c. The M/I diagram has been divided into rectangular and triangular parts for ease in calculating moments and areas. The individual areas and the centroidal distances of each from point C are shown on Fig. 5-23c. A tangent to the elastic curve at point A is horizontal; therefore, the tangential deviation $t_{C/A}$, shown in Fig. 5-23d, is the required deflection. The second area-moment theorem gives

$$Et_{C/A} = -\frac{PL^2}{2I}\frac{2L}{3} - \frac{PL^2}{2I}\frac{3L}{2} - \frac{PL^2}{4I}\frac{5L}{3}$$

from which

$$y_C = t_{C/A} = -\frac{3PL^3}{2EI} = \frac{3PL^3}{2EI} \quad \text{downward} \qquad \text{Ans.}$$

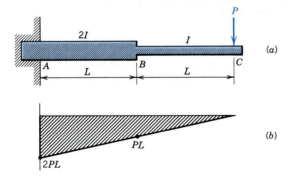

(a)

(b)

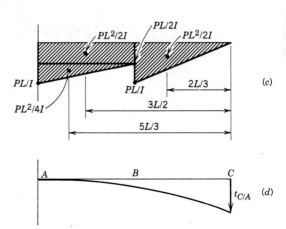

(c)

(d)

FIG. 5-23

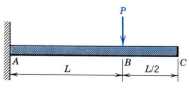

FIG. P5-101

PROBLEMS

5-101* A cantilever beam is loaded as shown in Fig. P5-101. Use area moments to determine the deflection, in terms of P, L, E, and I,

(a) at the point of application of the load P.

(b) at the free end of the beam.

5-102* A cantilever beam is loaded as shown in Fig. P5-102. Use area moments to determine the deflection, in terms of w, L, E, and I,

(a) at the left end of the distributed load w.

(b) at the free end of the beam.

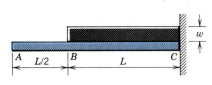

FIG. P5-102

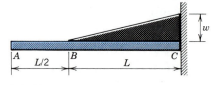

FIG. P5-103

5-103* A cantilever beam is loaded as shown in Fig. P5-103. Use area moments to determine the deflection, in terms of w, L, E, and I,

(a) at the left end of the distributed load w.

(b) at the free end of the beam.

5-104* A cantilever beam is loaded as shown in Fig. P5-104. Use area moments to determine the deflection, in terms of w, L, E, and I,

(a) at the right end of the distributed load w.

(b) at the free end of the beam.

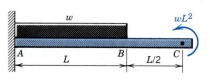

FIG. P5-104

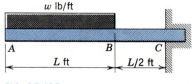

FIG. P5-105

5-105 A cantilever beam is loaded as shown in Fig. P5-105. Use area moments to determine, in terms of w, L, E, and I,

(a) the slope at the left end of the beam.

(b) the deflection at the left end of the beam.

5-106 A cantilever beam is loaded as shown in Fig. P5-106. Use area moments to determine, in terms of w, L, E, and I,

(a) the slope at the right end of the beam.

(b) the deflection at the right end of the beam.

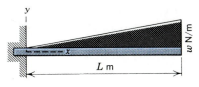

FIG. P5-106

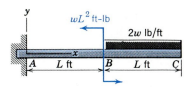

FIG. P5-107

5-107* A cantilever beam is loaded as shown in Fig. P5-107. Use area moments to determine the deflection, in terms of w, L, E, and I,

    (a) at the left end of the distributed load $2w$.

    (b) at the free end of the beam.

5-108* A cantilever beam is loaded as shown in Fig. P5-108. Use area moments to determine the deflection, in terms of w, L, E, and I,

    (a) at the right end of the distributed load w.

    (b) at the free end of the beam.

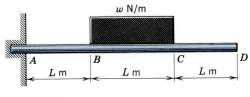

FIG. P5-108

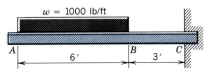

FIG. P5-109

5-109 Select the lightest structural steel ($E = 29{,}000$ ksi) T-section (see Appendix C) that can be used for the cantilever beam shown in Fig. P5-109 if the working fiber stress must be limited to 18 ksi and the deflection at the free end of the beam must be limited to 0.500 in. Use the area-moment method to determine the deflection.

5-110 Select the lightest structural steel ($E = 200$ GPa) wide-flange or American standard section (see Appendix C) that can be used for the cantilever beam shown in Fig. P5-110 if the working fiber stress must be limited to 140 MPa and the deflection at the free end of the beam must be limited to 30 mm. Use the area-moment method to determine the deflection.

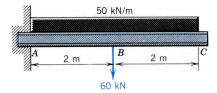

FIG. P5-110

5-111* A beam is loaded and supported as shown in Fig. P5-111. Use area moments to determine the deflection, in terms of P, L, E, and I,

(a) at the point of application of the load P.

(b) at a point midway between the supports.

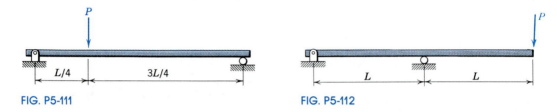

FIG. P5-111   FIG. P5-112

5-112* A beam is loaded and supported as shown in Fig. P5-112. Use area moments to determine the deflection, in terms of P, L, E, and I,

(a) at the point of application of the load P.

(b) at a point midway between the supports.

5-113 A beam is loaded and supported as shown in Fig. P5-113. Use area moments to determine the deflection, in terms of P, L, E, and I,

(a) at the point of application of the load $2P$.

(b) at the point of application of the load P.

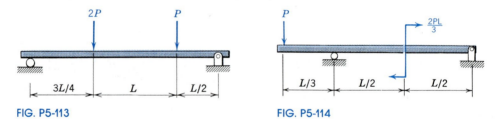

FIG. P5-113   FIG. P5-114

5-114 A beam is loaded and supported as shown in Fig. P5-114. Use area moments to determine the deflection, in terms of P, L, E, and I,

(a) at the left end of the beam.

(b) at a point midway between the supports.

5-115* A beam is loaded and supported as shown in Fig. P5-115. Use area moments to determine the deflection, in terms of w, L, E, and I, at a point midway between the supports.

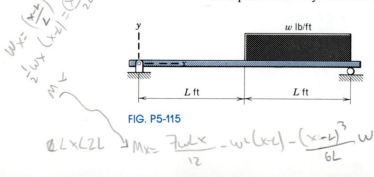

FIG. P5-115   FIG. P5-116

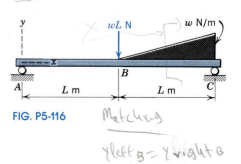

5-116* A beam is loaded and supported as shown in Fig. P5-116. Use area moments to determine the deflection, in terms of w, L, E, and I, at a point midway between the supports.

5-117 A beam is loaded and supported as shown in Fig. P5-117. Use area moments to determine the deflection, in terms of w, L, E, and I,

    (a) at the left end of the beam.

    (b) at the right end of the beam.

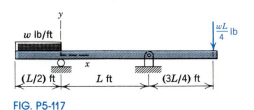

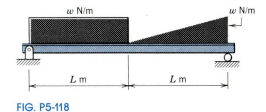

FIG. P5-117   FIG. P5-118

5-118 A beam is loaded and supported as shown in Fig. P5-118. Use area moments to determine the deflection, in terms of w, L, E, and I, at a point midway between the supports.

5-119* A beam is loaded and supported as shown in Fig. P5-119. Use area moments to determine the midspan deflection if $w = 2500$ lb/ft, $a = 4$ ft, $I = 148$ in.⁴, and $E = 29{,}000$ ksi.

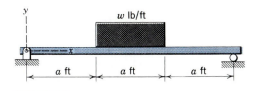

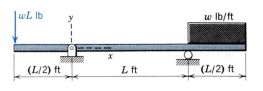

FIG. P5-119   FIG. P5-120

5-120* A beam is loaded and supported as shown in Fig. P5-120. Determine the deflection at the left end of the beam if $w = 45$ kN/m, $L = 4$ m, $I = 150(10^6)$ mm⁴, and $E = 200$ GPa. Use the area-moment method.

5-121 Select the lightest structural steel ($E = 29{,}000$ ksi) wide-flange or American standard section (see Appendix C) that can be used for the beam shown in Fig. P5-121 ($w = 1000$ lb/ft and $L = 6$ ft) if the working fiber stress must be limited to 12 ksi and the deflection at the middle of the span must be limited to 0.250 in. Use the area-moment method to determine the deflection.

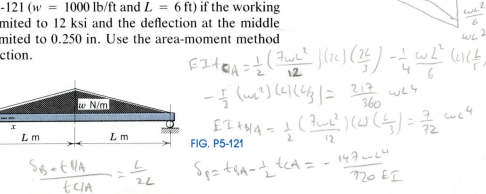

FIG. P5-121

$$EIt_{C/A} = \frac{1}{2}\left(\frac{7wL^2}{12}\right)(2L)\left(\frac{2L}{3}\right) - \frac{1}{4}\frac{wL^2}{6}(L)\left(\frac{L}{5}\right)$$

$$-\frac{1}{2}(wL^2)(L)\left(\frac{L}{3}\right) = \frac{217}{360}wL^4$$

$$EIt_{B/A} = \frac{1}{2}\left(\frac{7wL^2}{12}\right)(L)\left(\frac{L}{3}\right) = \frac{7}{72}wL^4$$

$$\frac{\delta_B + t_{B/A}}{t_{C/A}} = \frac{L}{2L}$$

$$\delta_B = t_{B/A} - \frac{1}{2}t_{C/A} = -\frac{147wL^4}{720EI}$$

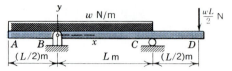

FIG. P5-122

5-122 Select the lightest structural steel (E = 200 GPa) wide-flange or American standard section (see Appendix C) that can be used for the beam shown in Fig. P5-122 (w = 30 kN/m and L = 3.50 m) if the working fiber stress must be limited to 80 MPa and the deflection at the right end of the beam must be limited to 6.5 mm. Use the area-moment method to determine the deflection.

5-123* The cantilever beam ABC shown in Fig. P5-55 has a moment of inertia $3I$ in the interval AB and a moment of inertia I in the interval BC. Use area moments to determine the deflection at the free end of the beam in terms of P, L, E, and I.

5-124* The simply supported beam $ABCD$ shown in Fig. P5-56 has a moment of inertia $3I$ in the center section BC and a moment of inertia I in the other two sections near the supports. Use area moments to determine the deflection at the middle of the span in terms of P, L, E, and I.

5-125 The cantilever beam ABC shown in Fig. P5-57 has a moment of inertia $2I$ in the interval AB and a moment of inertia I in the interval BC. Use area moments to determine the deflection at the free end of the beam in terms of w, L, E, and I.

5-126 The simply supported beam $ABCD$ shown in Fig. P5-58 has a moment of inertia $3I$ in the center section BC and a moment of inertia I in the other two sections near the supports. Use area moments to determine the deflection at the middle of the span in terms of P, L, E, and I.

5-127* The cantilever beam ABC shown in Fig. P5-59 has a moment of inertia $3I$ in the interval AB and a moment of inertia I in the interval BC. Use area moments to determine the deflection at the free end of the beam in terms of w, L, E, and I.

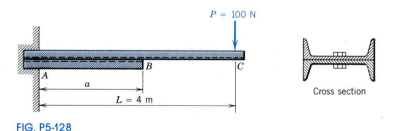

FIG. P5-128

5-128 A C76 × 7 steel ($E = 200$ GPa) channel (see Appendix C) 4 m long is used as a cantilever beam to support a 100-N load as shown in Fig. P5-128. The web of the channel is horizontal. Use area moments to determine the minimum length of C76 × 7 channel that must be bolted back-to-back to the original beam, as shown in Fig. P5-128, in order to limit the tip deflection of the 4-m channel to 50 mm.

5-10

MAXIMUM DEFLECTIONS BY AREA MOMENTS

In most cantilever beams and in symmetrically loaded simply supported beams, the point of maximum deflection can be determined by inspection. In such instances the procedure outlined in Section 5-9 is usually adequate for finding the maximum deflection. When the location of the point of maximum deflection is unknown, it is necessary to use both the first and second area-moment theorems for determining the maximum deflection.

Figure 5-24 shows a typical situation in which the maximum deflection, at point B on the elastic curve, is required. The deflection y_B, or the tangential deviation $t_{A/B}$, is wanted, but the distance x along the beam is unknown and must be determined before the deflection can be obtained. Once $t_{C/A}$ is determined from the second area-moment theorem, $t_{C/A}/L$ gives $\tan \theta_A$ that is approximately equal to θ_A. The angle θ_A can also be determined, in terms of x, from the first area-moment theorem because the tangent to the elastic curve at B is horizontal and $\theta_A = \theta_{A/B}$. When these two expressions for θ_A are equated, the distance x is found. Finally, $t_{A/B}$ is determined from the second area-moment theorem. The following example illustrates the procedure for a specific case.

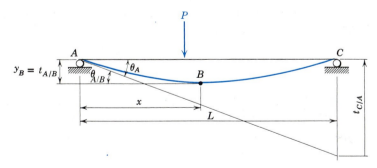

FIG. 5-24

EXAMPLE 5-13

A beam is loaded and supported as shown in Fig. 5-25a. Determine the maximum deflection in terms of w, L, E, and I.

Solution

The equations of equilibrium give the left reaction as $wL/2$ upward and the right reaction as wL upward. The moment diagram by parts is shown in Figs. 5-25b, c, and d. The elastic curve, very much exaggerated, of the loaded beam is shown in Fig. 5-25e with tangents drawn at A and B. The second area-moment theorem, with the moment axis at C, gives the vertical distance to C from the tangent at A as

$$EIt_{C/A} = L\left(\frac{9wL^3}{4}\right) - \frac{2L}{3}(wL^3) - \frac{L}{4}\left(\frac{wL^3}{6}\right) = \frac{37}{24}wL^4$$

Since the deflections are small,

$$\theta_A = \tan\theta_A = \frac{t_{C/A}}{3L} = \frac{37wL^3}{72EI}$$

Because the tangent at B is horizontal, the first area-moment theorem applied to the segment AB gives

$$\theta_A = \theta_{A/B} = \frac{1}{EI}\left[\frac{wL}{2}x\left(\frac{x}{2}\right) - \frac{wL}{2}(x-L)\left(\frac{x-L}{2}\right)\right]$$
$$= \frac{wL}{4EI}(2Lx - L^2)$$

Equating the two expressions for θ_A gives

$$\frac{37wL^3}{72EI} = \frac{wL}{4EI}(2Lx - L^2)$$

from which

$$x = 55L/36$$

Application of the second theorem to the segment AB, with the moment axis at A, gives the vertical distance to A from the horizontal tangent at B as

$$|y_B| = t_{A/B} = \frac{1}{EI}\left\{\left(\frac{2x}{3}\right)\left(\frac{wLx}{2}\right)\left(\frac{x}{2}\right)\right.$$
$$\left. - \left[L + \frac{2}{3}(x-L)\right]\frac{wL}{2}\frac{(x-L)^2}{2}\right\}$$
$$= \frac{wL^4}{2EI}\left[\frac{1}{3}\left(\frac{55}{36}\right)^3 - \frac{1}{2}\left(\frac{19}{36}\right)^2 - \frac{1}{3}\left(\frac{19}{36}\right)^3\right]$$

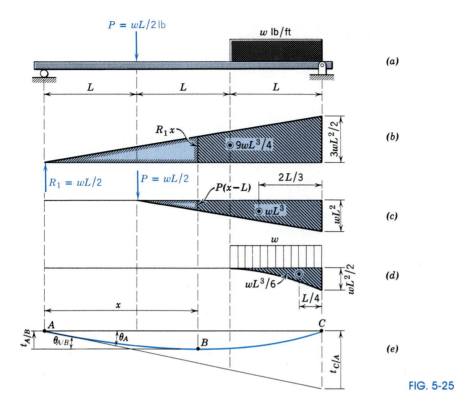

FIG. 5-25

where the positive result means that A is above the tangent line. There-fore,

$$y_B = 0.500 \frac{wL^4}{EI} \quad \text{downward} \qquad \text{Ans.}$$

PROBLEMS

5-129* A beam is loaded and supported as shown in Fig. P5-129. Use area moments to determine the maximum deflection between the sup-ports in terms of P, L, E, and I.

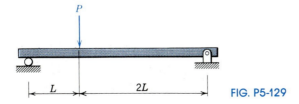

FIG. P5-129

5-130* A beam is loaded and supported as shown in Fig. P5-130. Use area moments to determine the maximum deflection between the supports in terms of P, L, E, and I.

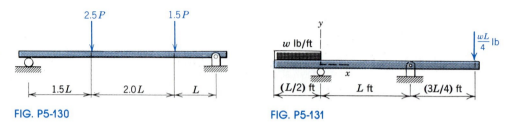

FIG. P5-130

FIG. P5-131

5-131 A beam is loaded and supported as shown in Fig. P5-131. Use area moments to determine the maximum deflection between the supports in terms of w, L, E, and I.

5-132 A beam is loaded and supported as shown in Fig. P5-132. Use area moments to determine the maximum deflection between the supports in terms of w, L, E, and I.

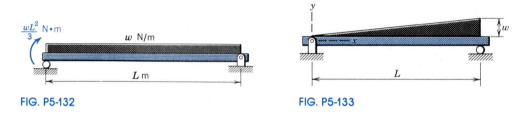

FIG. P5-132

FIG. P5-133

5-133* A beam is loaded and supported as shown in Fig. P5-133. Use area moments to determine the maximum deflection between the supports if $w = 2500$ lb/ft, $L = 12$ ft, $I = 120$ in.⁴, and $E = 29,000$ ksi.

5-134* A beam is loaded and supported as shown in Fig. P5-134. Use area moments to determine the maximum deflection between the supports if $w = 50$ kN/m, $L = 4$ m, $I = 50(10^6)$ mm⁴, and $E = 200$ GPa.

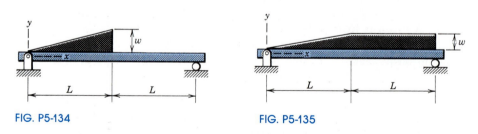

FIG. P5-134

FIG. P5-135

5-135 A beam is loaded and supported as shown in Fig. P5-135. Use area moments to determine the maximum deflection between the supports if $w = 3000$ lb/ft, $L = 15$ ft, $I = 250$ in.⁴, and $E = 29,000$ ksi.

5-136 A beam is loaded and supported as shown in Fig. P5-136. Use area moments to determine the maximum deflection between the supports if $w = 50 \text{ kN/m}$, $L = 2 \text{ m}$, $I = 300(10^6) \text{ mm}^4$, and $E = 200 \text{ GPa}$.

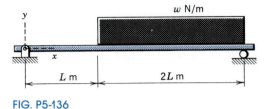

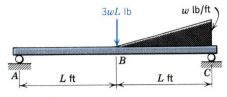

FIG. P5-136

FIG. P5-137

5-137* Select the lightest structural steel ($E = 29,000$ ksi) wide-flange or American standard section (see Appendix C) that can be used for the beam shown in Fig. P5-137 ($w = 2000$ lb/ft and $L = 8$ ft) if the working fiber stress must be limited to 15 ksi and the maximum deflection between the supports must be limited to 0.375 in. Use the area-moment method to determine the maximum deflection.

5-138* Select the lightest structural steel ($E = 200$ GPa) T-section (see Appendix C) that can be used for the beam shown in Fig. P5-138 ($w = 2.5$ kN/m and $L = 3.50$ m) if the working fiber stress must be limited to 100 MPa and the maximum deflection between the supports must be limited to 15 mm. Use the area-moment method to determine the maximum deflection.

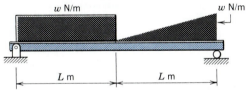

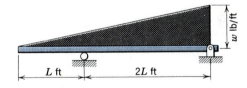

FIG. P5-138

FIG. P5-139

5-139 Select the lightest structural steel ($E = 29,000$ ksi) wide-flange or American standard section (see Appendix C) that can be used for the beam shown in Fig. P5-139 ($w = 2000$ lb/ft and $L = 8$ ft) if the working fiber stress must be limited to 12 ksi and the maximum deflection between the supports must be limited to 0.250 in. Use the area-moment method to determine the maximum deflection.

5-140 Select the lightest structural steel ($E = 200$ GPa) wide-flange or American standard section (see Appendix C) that can be used for the beam shown in Fig. P5-140 ($w = 30$ kN/m) if the working fiber stress must be limited to 75 MPa and the maximum deflection between the supports must be limited to 7.5 mm. Use the area-moment method to determine the maximum deflection.

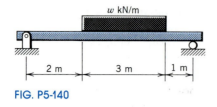

FIG. P5-140

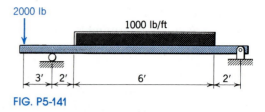

FIG. P5-141

5-141* A structural steel beam (E = 29,000 ksi and I = 36.7 in.⁴) is loaded and supported as shown in Fig. P5-141. Use area moments to determine

(a) the maximum deflection between the supports.

(b) the difference between the maximum deflection and the midspan deflection.

5-142 A structural steel beam [E = 200 GPa and I = 51.6(10^6) mm⁴] is loaded and supported as shown in Fig. P5-142. Use area moments to determine

(a) the maximum dcflection between the supports.

(b) the difference between the maximum deflection and the midspan deflection.

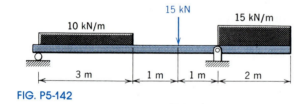

FIG. P5-142

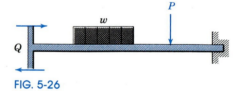

FIG. 5-26

5-11

DEFLECTIONS BY SUPERPOSITION

As explained in Section 1-15, the concept or method of superposition consists of finding the resultant effect of several loads acting on a member simultaneously as the sum of the contributions from each of the loads applied individually. The results for the separate loads are frequently available from previous work or readily determined by a previous method. In such instances the superposition method becomes a powerful concept or tool for finding stresses, deflections, and the like. The method is applicable in all cases in which a linear relation exists between the stresses or deflections and the applied loads.

To show that beam deflections can be accurately determined by superposition, consider the cantilever beam of Fig. 5-26 with loads Q, w, and P. In order to determine the deflection at any point of this beam by the double integration method, it is necessary to express the bending moment in terms of the applied loads. For each interval along the beam, the value of M is the algebraic sum of the moments due to the separate loads. After two successive integrations, the solution for the deflection at any point will still be the algebraic sum of the contributions from each applied load. Furthermore, for any given value of x, the relation between applied load and resulting deflection will be linear. It is evident, therefore, that the deflection of a beam is the sum of the deflections produced by the individual loads. Once the deflections produced by a few typical individual loads have been determined by one of the methods already presented, the superposition method provides a means of rapidly solving a wide variety of more complicated problems by various combinations of known results. As more data becomes available, a wider range of problems can be solved by superposition.

The data in Appendix D are provided for use in mastering the superposition method. No attempt is made to give a large number of results because such data are readily available in various handbooks. The data given and the illustrative examples are for the purpose of making the concept and methods clear.

EXAMPLE 5-14

Use the method of superposition to determine the deflection at midspan for the steel beam ($E = 30,000$ ksi and $I = 180$ in.^{4}) shown in Fig. 5-27a.

Solution

The given loading is equivalent to the two loads, parts 1 and 2, shown in Fig. 5-27b. For part 1, the deflection at the center of the span is given as case 7 (Table D-2) of Appendix D. It is

$$(y_C)_1 = -\frac{5wL^4}{384EI} = -\frac{5(600)(16^4)(12^3)}{384(30)(10^6)(180)} = -0.1638 \text{ in.}$$

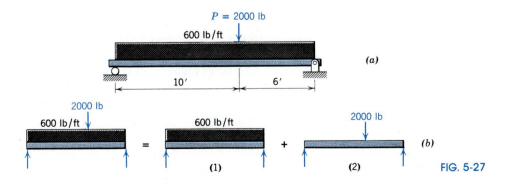

FIG. 5-27

For part 2, the deflection at the center is given as a part of case 5 (Table D-2) of Appendix D. It is

$$(y_C)_2 = -\frac{Pb(3L^2 - 4b^2)}{48EI} = -\frac{2000(6)[3(16^2) - 4(6^2)](12^3)}{48(30)(10^6)(180)}$$

$$= -0.499 \text{ in.}$$

The algebraic sum of the deflections for midspan is

$$y_C = -0.1638 + (-0.0499) = -0.2137 \text{ in.} = \underline{0.214 \text{ in. downward}}$$

Ans.

EXAMPLE 5-15

A beam is loaded and supported as shown in Fig. 5-28a. Use the method of superposition to determine the maximum deflection when $E = 12,000$ ksi and $I = 81$ in.^{4}.

Solution

From symmetry and $\Sigma F_y = 0$, the reactions are each 4500 lb upward. Because of the symmetrical loading, the slope of the beam is zero at the center of the span, and the right (or left) half of the beam can be considered a cantilever beam with two loads. As shown in Fig. 5-28b, the cantilever with two loads can be replaced by two beams (designated 1 and 2), each carrying one of the two loads.

The elastic curve (exaggerated) for part 1, as shown in Fig. 5-28c, gives the deflection at the right end as $y_{4.5} + y_3$, where $y_{4.5}$ is the deflection at the end of the uniformly distributed load and y_3 is the additional deflection of the unloaded 3 ft. From case 2 of Appendix D,

$$y_{4.5} = -\frac{wL^4}{8EI} = -\frac{1000(4.5^4)(12^3)}{8(12)(10^6)(81)} = -\frac{9^3}{8(10^3)} = -0.0911 \text{ in.}$$

Note that in determining deflection y, w lb/ft times L ft gives the applied load in pounds. The remaining L^3 must be expressed in inches if E and I are in inches and the result is to be in inches.

Similarly,

$$\theta_{4.5} = -\frac{wL^3}{6EI} = -\frac{1000(4.5^3)(12^2)}{6(12)(10^6)(81)} = -\frac{9}{4(10^3)} = -0.00225 \text{ rad}$$

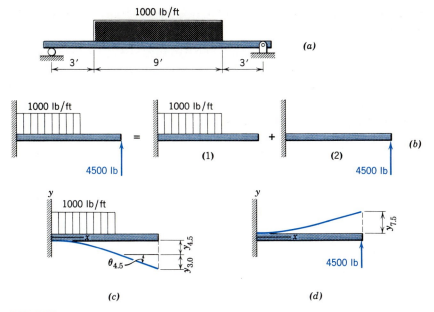

FIG. 5-28

from which

$$y_{3.0} = 3(12)(-0.00225) = -0.0810 \text{ in.}$$

Consequently, the total deflection of the right end is 0.1721 in. downward.

The elastic curve (exaggerated) for part 2 is shown in Fig. 5-28d. From case 1 (Table D-2) of Appendix D,

$$y_{7.5} = + \frac{PL^3}{3EI} = + \frac{5400(7.5^3)(12^3)}{3(12)(10^6)(81)} = 1.125 \text{ in. upward}$$

The algebraic sum of the deflections for parts 1 and 2 is +0.953 in., which means that the right end of the beam is 0.953 in. above the center. Obviously, the right end does not move, and the maximum deflection is at the center and is

$$y_{\text{max}} = \underline{0.953 \text{ in. downward}} \qquad \text{Ans.}$$

EXAMPLE 5-16

A beam is loaded and supported as shown in Fig. 5-29a. Use the method of superposition to determine the deflection, in terms of w, L, E, and I,

    (a) at a point midway between the supports.

    (b) at the right end of the beam.

Solution

(a) The deflection at a point midway between the supports is determined by using the beam shown in Fig. 5-29b. The effects of the loaded overhang on span AC of the beam can be represented by a shear force $V = wL$ and a moment $M = wL^2/2$. Since the shear force V does not contribute to the deflection at any point in span AC of the beam, the deflection at the middle of the span, as shown in Figs. 5-29c and d, can be expressed as

$$y_B = y_w + y_M$$

The deflections y_w and y_M are listed as cases 7 and 8 (Table D-2), respectively, of Appendix D. Thus,

$$y_B = -\frac{5w(2L)^4}{384EI} + \frac{(wL^2/2)(2L)^2}{16EI}$$

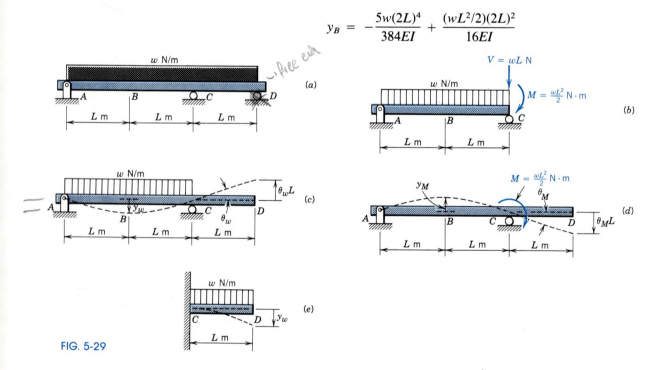

FIG. 5-29

$$= -\frac{wL^4}{12EI} = \frac{wL^4}{12EI} \quad \text{downward} \qquad \text{Ans.}$$

(b) The deflection at the right end of the beam is produced by the combined effects of the distributed load on the overhang and the rotation of the cross section of the beam at support C as shown in Figs. 5-29c, d, and e. Thus,

$$y_D = \theta_w L + \theta_M L + y_w$$

The angles θ_w and θ_M and the deflection y_w are listed as cases 7, 8, and 2 (Table D-2), respectively, of Appendix D. Thus,

$$y_D = \frac{w(2L)^3(L)}{24EI} - \frac{(wL^2/2)(2L)(L)}{3EI} - \frac{wL^4}{8EI}$$

$$= -\frac{wL^4}{8EI} = \frac{wL^4}{8EI} \quad \text{downward} \qquad \text{Ans.}$$

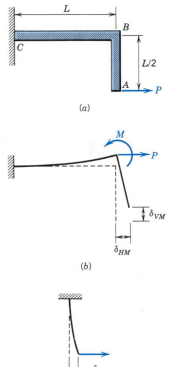

(a)

(b)

(c)

FIG. 5-30

EXAMPLE 5-17

An S102 × 14 American standard section is used to fabricate the simple frame shown in Fig. 5-30a. Use the method of superposition to determine the horizontal and vertical components of the deflection at the point of application of the load P if $E = 200$ GPa, $P = 5$ kN, and $L = 2$ m.

Solution

The frame can be separated for analysis into the two parts shown in Figs. 5-30b and c. The horizontal part BC of the frame acts as a cantilever beam that is subjected to an axial load P and a bending moment $M = PL/2$, as shown in Fig. 5-30b. The vertical part AB rotates as a rigid body, as shown in Fig. 5-30b and, in addition, acts as a cantilever beam that is subjected to a transverse load P at the free end, as shown in Fig. 5-30c. In the analysis of simple frames it is usually assumed that angles, such as the angle at B, do not change and that deformations due to axial loads are negligible in comparison to deformations due to bending. As a result, the vertical component of the deflection at point A is equal to the vertical deflection of beam BC at point B and the horizontal component of the deflection at point A is sum of the deflections resulting from the rotation of beam AB and the deflection of beam AB due to load P. From cases 1 and 4 (Table D-2) in Appendix D,

$$\delta_H = \delta_{HM} + \delta_{HP} = \theta_B(L/2) + \delta_{HP}$$

$$= \frac{(PL/2)(L)}{EI}(L/2) + \frac{P(L/2)^3}{3EI}$$

$$= \frac{7PL^3}{24EI}$$

$$\delta_V = \delta_{VM} = \frac{(PL/2)(L^2)}{2EI}$$

$$= \frac{PL^3}{4EI}$$

The properties of an S102 $\times$ 14 American standard section, as obtained from Appendix C, are $A = 1800$ mm^2, $I = 2.83(10^6)$ mm^4, and $S = 55.6(10^3)$ mm^3. Substituting the appropriate values for P, L, E, and I into the equations for δ_H and δ_V yields

$$\delta_H = \frac{7(5)(10^3)(2^3)}{24(200)(10^9)(2.83)(10^{-6})}$$

$$= 20.61(10^{-3}) \text{ m} = \underline{20.6 \text{ mm}} \qquad \text{Ans.}$$

$$\delta_V = \frac{5(10^3)(2^3)}{4(200)(10^9)(2.83)(10^{-6})}$$

$$= 17.67(10^{-3}) \text{ m} = \underline{17.67 \text{ mm}} \qquad \text{Ans.}$$

The extension of beam BC resulting from the axial load P is

$$\delta_P = \frac{PL}{AE} = \frac{5(10^3)(2)}{1800(10^{-6})(200)(10^9)}$$

$$= 0.0278(10^{-3}) \text{ m} = 0.0278 \text{ mm}$$

Clearly, this extension is negligible in comparison to the bending deflections.

PROBLEMS

5-143* Use the method of superposition to determine the deflection at the free end of the cantilever beam shown in Fig. P5-143 when $w = 1000$ lb/ft, $L = 8$ ft, $I = 75$ in.^{4}, and $E = 29,000$ ksi.

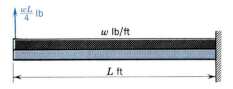

FIG. P5-143

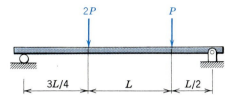

FIG. P5-144

5-144* Use the method of superposition to determine the deflection midway between the supports of the beam shown in Fig. P5-144 when $P = 13.5$ kN, $L = 3$ m, $I = 80(10^6)$ mm^4, and $E = 200$ GPa.

5-145* Use the method of superposition to determine the deflection at the free end of the cantilever beam shown in Fig. P5-145 in terms of P, L, E, and I.

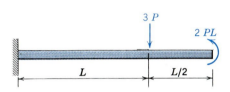

FIG. P5-145

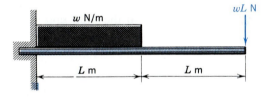

FIG. P5-146

5-146* Use the method of superposition to determine the deflection at the free end of the cantilever beam shown in Fig. P5-146 in terms of w, L, E, and I.

5-147 Use the method of superposition to determine the deflection at the free end of the cantilever beam shown in Fig. P5-147 in terms of w, L, E, and I.

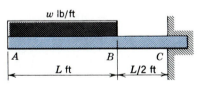

FIG. P5-147

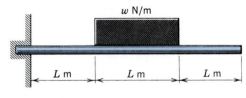

FIG. P5-148

5-148 Use the method of superposition to determine the deflection at the free end of the cantilever beam shown in Fig. P5-148 in terms of w, L, E, and I.

5-149* A beam is loaded and supported as shown in Fig. P5-149. Use the method of superposition to determine the deflection midway between the supports in terms of w, L, E, and I.

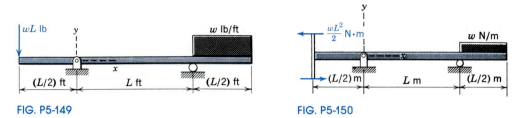

FIG. P5-149   FIG. P5-150

5-150* A beam is loaded and supported as shown in Fig. P5-150. Use the method of superposition to determine the deflection midway between the supports in terms of w, L, E, and I.

5-151 A beam is loaded and supported as shown in Fig. P5-151. Use the method of superposition to determine the deflection, in terms of w, L, E, and I,

(a) at a point midway between the supports.

(b) at the right end of the beam.

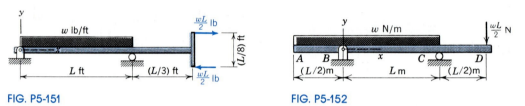

FIG. P5-151   FIG. P5-152

5-152 A beam is loaded and supported as shown in Fig. P5-152. Use the method of superposition to determine the deflection, in terms of w, L, E, and I,

(a) at a point midway between the supports.

(b) at the right end of the beam.

5-153 Select the lightest structural steel T-section (see Appendix C) that can be used for the beam shown in Fig. P5-153 if $w = 500$ lb/ft, $L = 4$ ft, and $E = 29,000$ ksi. The working fiber stress must be limited to 12,500 psi and the deflection at the left end of the beam must not exceed 0.175 in. Use the method of superposition to determine the deflection.

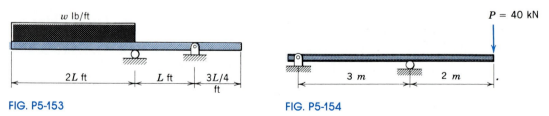

FIG. P5-153   FIG. P5-154

5-154 Select the lightest structural steel ($E = 200$ GPa) wide-flange or American standard section (see Appendix C) that can be used for the beam shown in Fig. P5-154 if the working fiber stress must be limited to 85 MPa and the deflection at the right end of the beam must not exceed 8 mm. Use the method of superposition to determine the deflection.

5-155* Use the method of superposition to determine the deflection at the free end of the cantilever beam shown in Fig. P5-155 when $w = 500$ lb/ft, $L = 6$ ft, $I = 305$ in.^{4}, and $E = 29,000$ ksi.

5-156* Use the method of superposition to determine the deflection at the free end of the cantilever beam shown in Fig. P5-156 when $w = 7.5$ kN/m, $L = 3$ m, $I = 180(10^6)$ mm^4, and $E = 200$ GPa.

5-157 The cantilever beam ABC shown in Fig. P5-157 has a moment of inertia $3I$ in the interval AB and a moment of inertia I in the interval BC. Use the method of superposition to determine the deflection at the free end of the beam in terms of w, L, E, and I.

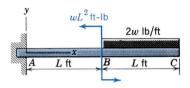

FIG. P5-155

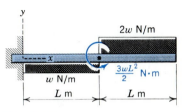

FIG. P5-156

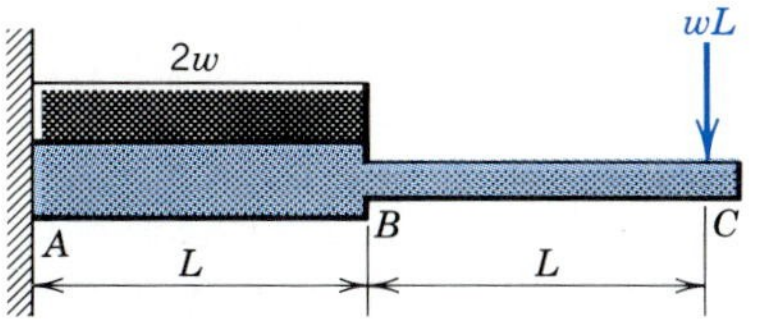

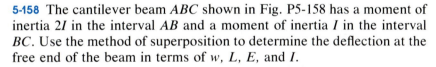

FIG. P5-157

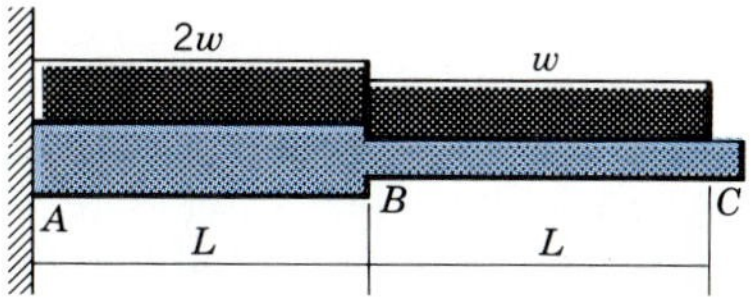

FIG. P5-158

5-158 The cantilever beam ABC shown in Fig. P5-158 has a moment of inertia $2I$ in the interval AB and a moment of inertia I in the interval BC. Use the method of superposition to determine the deflection at the free end of the beam in terms of w, L, E, and I.

5-159 The simply supported beam $ABCD$ shown in Fig. P5-159 has a moment of inertia $2I$ in the center section CB and a moment of inertia I in the other two sections near the supports. Use the method of superposition to determine the deflection at the middle of the span in terms of P, L, E, and I.

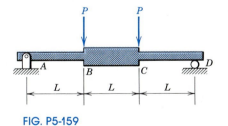

FIG. P5-159

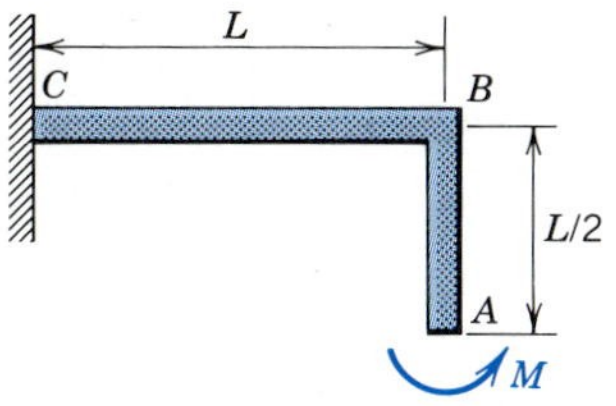

FIG. P5-160

5-160* A simple frame with constant EI is loaded and supported as shown in Fig. P5-160. Use the method of superposition to determine, in terms of M, L, E, and I,

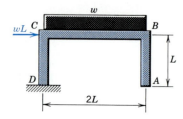

FIG. P5-161

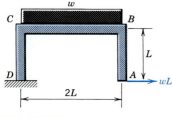

FIG. P5-162

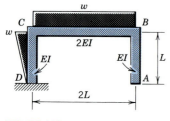

FIG. P5-163

(a) the vertical component of the deflection at point A.

(b) the horizontal component of the deflection at point A.

5-161* A simple frame with constant EI is loaded and supported as shown in Fig. P5-161. Use the method of superposition to determine, in terms of w, L, E, and I,

(a) the vertical component of the deflection at point A.

(b) the horizontal component of the deflection at point A.

5-162 A simple frame with constant EI is loaded and supported as shown in Fig. P5-162. Use the method of superposition to determine, in terms of w, L, E, and I,

(a) the vertical component of the deflection at point A.

(b) the horizontal component of the deflection at point A.

5-163 A simple frame with constant EI is loaded and supported as shown in Fig. P5-163. Use the method of superposition to determine, in terms of w, L, E, and I,

(a) the vertical component of the deflection at point B.

(b) the horizontal component of the deflection at point B.

5-164 A simple frame with constant EI is loaded and supported as shown in Fig. P5-163. Use the method of superposition to determine, in terms of w, L, E, and I,

(a) the vertical component of the deflection at point A.

(b) the horizontal component of the deflection at point A.

5-12

DEFLECTIONS DUE TO SHEARING STRESS

As mentioned in Section 5-3, the beam deflections calculated so far neglect the deflection produced by the shearing stresses in the beam. For short, heavily loaded beams this deflection can be significant, and an approximate method for evaluating such deflections will now be developed. The deflection of the neutral surface dy due to shearing stresses in the interval dx along the beam of Fig. 5-31 is

$$dy = \gamma\, dx = \frac{\tau}{G}\, dx = \frac{VQ}{GIt}\, dx$$

from which, since the shear in Fig. 5-31 is negative,

$$\frac{GIt}{Q}\frac{dy}{dx} = -V \tag{5-8}$$

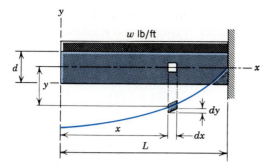

FIG. 5-31

Since the vertical shearing stress varies from top to bottom of a beam, the deflection due to shear is not uniform. This nonuniform deflection due to shear is reflected in the slight warping of the sections of a beam. Equation 5-8 gives values too high because the maximum shearing stress (at the neutral surface) is used and also because the rotation of the differential shear element is ignored.

In order to obtain an idea of the relative amount of beam deflection due to shearing stress, consider a rectangular cross section for the beam of Fig. 5-31 and use the maximum stress for which

$$\frac{Q}{It} = \frac{3}{2A}$$

where A is the cross-sectional area of the beam. The expression for dy becomes

$$dy = \frac{3wxdx}{2AG}$$

where V is replaced by its value $-wx$. Integration along the entire beam gives the change in y due to shear as

$$\Delta y = \frac{3w}{2AG} \int_0^L x\, dx = \frac{3wL^2}{4AG}$$

which equals the deflection at the left end. For this same beam the magnitude of the deflection at the left end due to flexural stresses is

$$\frac{wL^4}{8EI} = \frac{3wL^4}{2EAd^2}$$

and the magnitude of the total deflection at the left end becomes

$$y = \frac{3wL^4}{2EAd^2} + \frac{3wL^2}{4AG} = \frac{3wL^2}{4AG}\left(\frac{2L^2G}{d^2E} + 1\right) \qquad (a)$$

Equation a indicates that the ratio of the two deflections increases as the square of L/d, which means that the deflection due to shear is of importance only in the case of very short, deep beams.

EXAMPLE 5-18

A structural steel ($E = 29,000$ ksi and $G = 11,000$ ksi) cantilever beam with a rectangular cross section 2 in. wide by 4 in. deep supports a concentrated load of 1000 lb at the end of a 3-ft span. Determine the percent increase in deflection at the free end of the beam resulting from the shearing stresses.

Solution

The deflection dy of the neutral surface due to shearing stresses in an interval dx along the beam is given by Eq. 5-8 as

$$dy = \gamma \, dx = \frac{\tau}{G} \, dx = \frac{VQ}{ItG} \, dx$$

For a beam with a rectangular cross section, $I = bh^3/12$ and $Q = bh^2/8$. Also, $V = -P$ for a cantilever beam that supports a concentrated load P at the free end. Thus,

$$dy = -\frac{3P}{2bhG} \, dx$$

The deflection y_s at the free end of the beam due to shearing stresses is

$$y_s = \int_0^L dy = -\frac{3PL}{2bhG}$$

$$= -\frac{3(1000)(3)(12)}{2(2)(4)(11)(10^6)}$$

$$= -0.0006136 \text{ in.}$$

The deflection y_f at the free end of the beam due to flexure is

$$y_f = -\frac{PL^3}{3EI} = -\frac{4PL^3}{Ebh^3}$$

$$= -\frac{4(1000)(3^3)(12^3)}{29(10^6)(2)(4^3)}$$

$$= -0.05028 \text{ in.}$$

$$\text{Increase} = \frac{0.0006136}{0.05028} \, (100) = \underline{1.220\%} \qquad \text{Ans.}$$

PROBLEMS

5-165* A structural steel ($E = 29,000$ ksi and $G = 11,000$ ksi) cantilever beam with a 4-in.-diameter circular cross section supports a concentrated load of 1200 lb at the end of a 4-ft span. Determine the percent increase in deflection at the free end of the beam resulting from the shearing stresses.

5-166* An aluminum alloy ($E = 73$ GPa and $G = 28$ GPa) cantilever beam with a rectangular cross section 50 mm wide by 100 mm deep supports a uniformly distributed load of 5.0 kN/m over a 1.5-m span. Determine the percent increase in deflection at the free end of the beam resulting from the shearing stresses.

5-167 A structural steel ($E = 29,000$ ksi and $G = 11,000$ ksi) beam with a hollow rectangular cross section 3 in. wide by 5 in. deep is made from $\frac{1}{2}$-in.-thick plate. The beam is simply supported and carries a concentrated load of 4000 lb at the center of an 8-ft span. Determine the percent increase in deflection at the center of the span resulting from the shearing stresses.

5-168 A W203 × 60 structural steel ($E = 200$ GPa and $G = 76$ GPa) wide-flange section is used as a simply supported beam to support a distributed load of 20 kN/m over a 4-m span. Determine the percent increase in deflection at the center of the span resulting from the shearing stresses.

COMPUTER PROBLEMS

Note: The following problems have been designed to be solved with a programmable calculator, microcomputer, or mainframe computer. Appendix F contains a description of a few numerical methods, together with a few simple programs in BASIC and FORTRAN, that can be modified for use in the solution of these problems.

C5-1 A 160-lb diver walks slowly onto a diving board. The diving board is a wood ($E = 1800$ ksi) plank 10 ft long, 18 in. wide, and 2 in. thick, and is modeled as the cantilever beam shown in Fig. PC5-1. For the diver at positions $a = nL/5$ ($n = 1, 2, \ldots, 5$), compute and plot the deflection curve for the diving board (plot y as a function of x for $0 \leqslant x \leqslant L$).

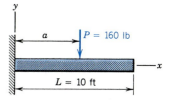

FIG. PC5-1

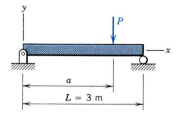

FIG. PC5-2

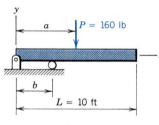

FIG. PC5-3

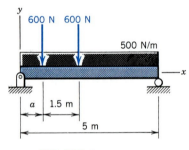

FIG. PC5-4

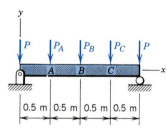

FIG. PC5-6

C5-2 The gangplank between a fishing boat and a dock consists of a wood (E = 12 GPa) plank 3 m long, 300 mm wide, and 50 mm thick. If the plank is modeled as the simply supported beam shown in Fig. PC5-2, compute and plot the deflection curve for the gangplank (plot y as a function of x for $0 \leqslant x \leqslant L$) as an 80-kg man walks across the plank. Plot curves for the man at positions $a = nL/5$ (n = 1, 2 . . . , 5).

C5-3 Repeat Problem C5-1 if the diving board is modeled as a simply supported beam with a large overhang (b = 3 ft) as shown in Fig. PC5-3.

C5-4 A bridge over a small stream on a golf course consists of a wood deck (weighing 1000 newtons per meter) laid over two wood (E = 12 GPa) beams (5 m long, 100 mm wide, and h mm high). It is desired that the bridge support the weight of a loaded golf cart (weighing 2400 N) with a maximum deflection of no more than 50 mm. Model the beams and loading as shown in Fig. PC5-4.

(a) Determine the minimum depth h of the beams that will support the given weight.

(b) Compute and plot the deflection curve for the beam (plot y as a function of x for $0 \leqslant x \leqslant L$) for the cart at positions a = 1 m, 2 m, and 3 m.

C5-5 The diver in Problem C5-3 stands at the right end of the diving board ($a = L$). Plot the deflection curve for the diving board (plot y as a function of x for $0 \leqslant x \leqslant L$) for the right support at positions b = 6 in., 12 in., 24 in., 36 in., and 60 in.

C5-6 A Cub Scout troop is marching across a small footbridge consisting of a wood (E = 12 GPa) plank 2 m long, 300 mm wide, and 40 mm thick. The scouts are separated by 0.5 m and are modeled as static, concentrated loads (40 kg each) on a simply supported beam, as in Fig. PC5-6. Plot the deflection curve of the bridge (plot y as a function of x for $0 \leqslant x \leqslant L$) as the troop marches across the bridge. (Initially, only one scout is on the bridge—at position A. Then two scouts are on the bridge—at positions A and B. The troop continues to march across until finally only a single scout remains on the bridge—at position C).

CHAPTER 6

STATICALLY INDETERMINATE BEAMS

6-1

INTRODUCTION

As explained in Section 4-1, a beam, subjected only to transverse loads, with more than two reaction components is statically indeterminate because the equations of equilibrium are not sufficient to determine all the reactions. In such cases the geometry of the deformation of the loaded beam is utilized to obtain the additional relations needed for an evaluation of the reactions (or other unknown forces). For problems involving elastic action, each additional constraint on a beam provides additional information concerning slopes or deflections. Such information, when utilized with appropriate slope or deflection equations, yields expressions that supplement the independent equations of equilibrium.

Any of the methods for determining beam deflections outlined in Chapter 5 can be utilized to provide the additional equations needed to solve for the unknown reactions or loads in a statically indeterminate beam, provided the action is linearly elastic. Once the reactions and loads are known, the methods of Chapters 4 and 5 will give the required stresses and deflections.

6-2

THE INTEGRATION METHOD

For statically determinate beams, known slopes and deflections were used to obtain boundary and matching conditions, from which the constants of integration in the elastic curve equation could be evaluated. For statically indeterminate beams, the procedure is identical. However, the moment equations will contain reactions or loads that cannot be evaluated from the available equations of equilibrium, and one additional boundary condition is needed for the evaluation of each such unknown. For example, if a beam is subjected to a force system for which there are two independent equilibrium equations and if there are four unknown reactions or loads on the beam, two boundary or matching conditions are needed in addition to those necessary for the determination of the constants of integration. These extra boundary conditions, when substituted in the appropriate elastic curve equations (slope or deflection), will yield the necessary additional equations. The following example illustrates the method.

EXAMPLE 6-1

A beam is loaded and supported as shown in Fig. 6-1a. Determine the reactions in terms of w and L.

Solution

From the free-body diagram of Fig. 6-1b it is seen that there are three unknown reaction components (M, V, and R) and that only two independent equations of equilibrium are available. The additional unknown requires the use of the elastic curve equation, for which one extra boundary condition is required in addition to the two required for the constants of integration. Because three boundary conditions are available in the interval between the supports, only one elastic curve equation need be written. The origin of coordinates is arbitrarily placed at the wall, and, for the interval $0 \leqslant x \leqslant L$, the boundary conditions are: when $x = 0$, $dy/dx = 0$; when $x = 0$, $y = 0$; and when $x = L$, $y = 0$. From Fig. 6-1c the bending moment equation is

$$EI \frac{d^2y}{dx^2} = M_x = Vx + M - \frac{wx^2}{2}$$

Integration gives

$$EI \frac{dy}{dx} = \frac{Vx^2}{2} + Mx - \frac{wx^3}{6} + C_1$$

The first boundary condition gives $C_1 = 0$. A second integration yields

$$EIy = \frac{Vx^3}{6} + \frac{Mx^2}{2} - \frac{wx^4}{24} + C_2$$

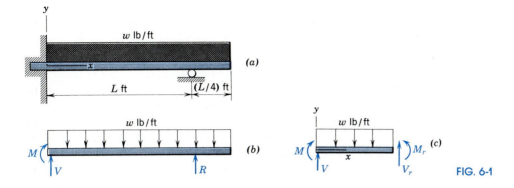

FIG. 6-1

The second boundary condition gives $C_2 = 0$, and the last boundary condition gives

$$0 = \frac{VL^3}{6} + \frac{ML^2}{2} - \frac{wL^4}{24}$$

which reduces to

$$4VL + 12M = wL^2$$

The equation of equilibrium $\Sigma M_R = 0$ for the free-body diagram of Fig. 6-1b yields

$$VL + M = w\left(\frac{5L}{4}\right)\left(\frac{5L}{8} - \frac{L}{4}\right) = \frac{15wL^2}{32}$$

The simultaneous solution of these equations gives

$$M = -\frac{7wL^2}{64} = \frac{7wL^2}{64} \text{ ft-lb counterclockwise} \qquad \text{Ans.}$$

and

$$V = \frac{37wL}{64} \text{ lb upward} \qquad \text{Ans.}$$

Finally, the equation $\Sigma F_y = 0$ for Fig. 6-1b gives

$$R + V = 5wL/4$$

from which

$$R = \frac{43wL}{64} \text{ lb as shown} \qquad \text{Ans.}$$

An alternate solution would be to place the origin of coordinates at the right support and write the moment equation for the interval $0 \leqslant x \leqslant L$. This equation would involve only one unknown, the reaction R; upon integration and evaluation of the constants, the third boundary condition would directly yield the value of R. The two independent equilibrium equations could then be used to evaluate M and V.

PROBLEMS

Note: Use the integration method to solve the following problems.

6-1* When the moment M is applied to the left end of the cantilever beam shown in Fig. P6-1, the slope at the left end of the beam is zero. Determine

    (a) the magnitude of the moment M in terms of P and L.

    (b) the deflection at the left end of the beam in terms of P, L, E, and I.

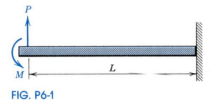

FIG. P6-1

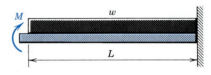

FIG. P6-2

6-2* A beam is loaded and supported as shown in Fig. P6-2. Determine the magnitude of the moment M, in terms of w and L, required to

    (a) make the slope at the left end of the beam zero.

    (b) make the deflection at the left end of the beam zero.

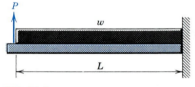

FIG. P6-3

6-3 When the load P is applied to the left end of the cantilever beam shown in Fig. P6-3, the deflection at the left end of the beam is zero. Determine

    (a) the magnitude of the load P in terms of w and L.

    (b) the slope at the left end of the beam in terms of w, L, E, and I.

6-4 A beam is loaded and supported as shown in Fig. P6-4. Determine the magnitude of the load P, in terms of w and L, required to

    (a) make the slope at the left end of the beam zero.

    (b) make the deflection at the left end of the beam zero.

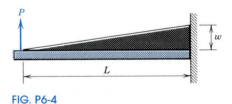

FIG. P6-4

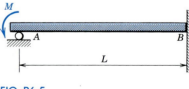

FIG. P6-5

6-5* A beam is loaded and supported as shown in Fig. P6-5. Determine

    (a) the reactions at supports A and B in terms of M and L.

(b) the deflection at the middle of the span in terms of M, L, E, and I.

6-6* A beam is loaded and supported as shown in Fig. P6-6. Determine
(a) the reactions at supports A and B in terms of w and L.
(b) the deflection at the middle of the span in terms of w, L, E, and I.

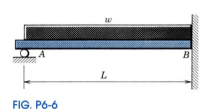

FIG. P6-6

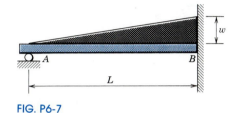

FIG. P6-7

6-7 A beam is loaded and supported as shown in Fig. P6-7. Determine
(a) the reactions at supports A and B in terms of w and L.
(b) the maximum deflection in terms of w, L, E, and I.

6-8 A beam is loaded and supported as shown in Fig. P6-8. Determine
(a) the reaction at supports A and B in terms of P and L.
(b) the deflection at the middle of the span in terms of P, L, E, and I.

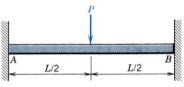

FIG. P6-8

6-9* A beam is loaded and supported as shown in Fig. P6-9. Determine
(a) the reactions at supports A and B in terms of w and L.
(b) the maximum deflection in terms of w, L, E, and I.

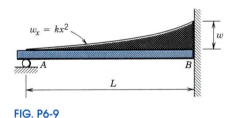

FIG. P6-9

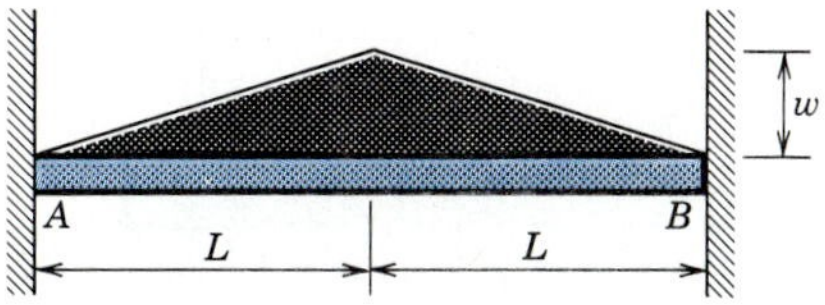

FIG. P6-10

6-10 A beam is loaded and supported as shown in Fig. P6-10. Determine
(a) the reactions at supports A and B in terms of w and L.
(b) the deflection at the middle of the span in terms of w, L, E, and I.
(c) the moment at the middle of the span in terms of w and L.

6-11* The right end of the beam shown in Fig. P6-11 is only partially fixed and under action of the load the slope becomes $wL^3/(96EI)$ upward to the right. Determine

(a) the reactions at supports A and B in terms of w and L.

(b) the maximum deflection in terms of w, L, E, and I.

(c) the maximum bending moment in the beam.

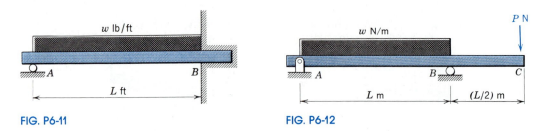

FIG. P6-11   FIG. P6-12

6-12* When the load P is applied to the right end of the beam shown in Fig. P6-12, the slope of the beam over the right support becomes $wL^3/(72EI)$ upward to the right. Determine

(a) the magnitude of the force P in terms of w and L.

(b) the reactions at supports A and B in terms of w and L.

(c) the maximum deflection between the supports in terms of w, L, E, and I.

6-13 When the moment M is applied to the left end of the beam shown in Fig. P6-13, the slope of the beam over the left support becomes $wL^3/(36EI)$ upward to the right. Determine

(a) the magnitude of the moment M in terms of w and L.

(b) the reactions at supports A and B in terms of w and L.

(c) the deflection midway between the supports in terms of w, L, E, and I.

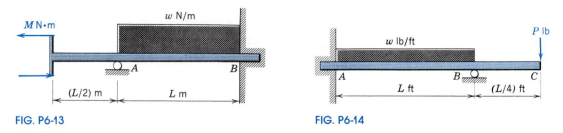

FIG. P6-13   FIG. P6-14

6-14 When the load P is applied to the right end of the beam shown in Fig. P6-14, the slope of the beam over the right support becomes $wL^3/(72EI)$ downward to the right. Determine

(a) the magnitude of the force P in terms of w and L.

(b) the reactions at supports A and B in terms of w and L.

(c) the deflection midway between the supports in terms of w, L, E, and I.

6-15* A beam is loaded and supported as shown in Fig. P6-15. Determine

(a) the reactions at supports A and B in terms of M and L.

(b) the deflection at the middle of the span in terms of M, L, E, and I.

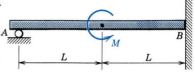

FIG. P6-15

6-16* A beam is loaded and supported as shown in Fig. P6-16. Determine

(a) the reactions at supports A and B in terms of P and L.

(b) the deflection at the middle of the span in terms of P, L, E, and I.

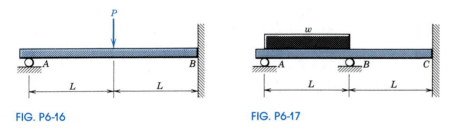

FIG. P6-16

FIG. P6-17

6-17 A beam is loaded and supported as shown in Fig. P6-17. Determine

(a) the reactions at supports A, B, and C in terms of w and L.

(b) the moment over the middle support in terms of w and L.

(c) the deflection at the middle of span AB in terms of w, L, E, and I.

6-18 A beam is loaded and supported as shown in Fig. P6-18. Determine

(a) the reactions at supports A, B, and C in terms of w and L.

(b) the moment over the middle support in terms of w and L.

(c) the deflection at the middle of span BC in terms of w, L, E, and I.

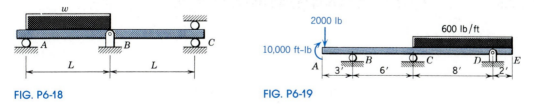

FIG. P6-18

FIG. P6-19

6-19* A beam is loaded and supported as shown in Fig. P6-19. Determine

(a) the reactions at supports B, C, and D.

(b) the bending moment over the center support.

(c) the maximum bending moment in the beam.

6-20* A beam is loaded and supported as shown in Fig. P6-20. Determine

(a) the reactions at supports A and D.

(b) the deflection at C if $E = 200$ GPa and $I = 350(10^6)$ mm⁴.

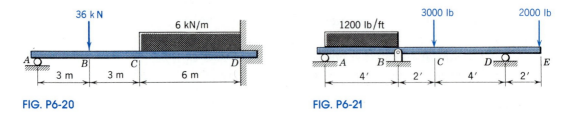

FIG. P6-20   FIG. P6-21

6-21 A beam is loaded and supported as shown in Fig. P6-21. Determine

(a) the reactions at supports A, B, and D.

(b) the bending moment over the center support.

(c) the maximum bending moment in the beam.

6-22 A beam is loaded and supported as shown in Fig. P6-22. Determine

(a) the reactions at supports A and B.

(b) the deflection midway between the supports if $E = 200$ GPa and $I = 150(10^6)$ mm⁴.

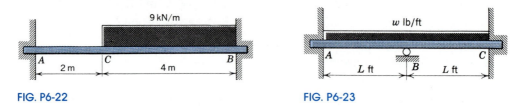

FIG. P6-22   FIG. P6-23

6-23* A beam is loaded and supported as shown in Fig. P6-23. Determine

(a) the reactions at supports A, B, and C.

(b) the bending moment over the center support.

6-24 A beam is loaded and supported as shown in Fig. P6-24. Determine

(a) the reactions at supports A, B, and C.

(b) the bending moment over the center support.

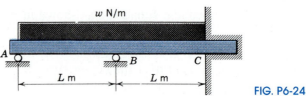

FIG. P6-24

(c) the maximum bending moment in the beam.

6-25* A steel (E = 30,000 ksi) beam 30 ft long is simply supported at the ends and at the center. When a uniformly distributed load of 2000 lb/ft is applied over the entire length, the center settles 0.40 in. The moment of inertia of the cross section with respect to the neutral axis is 600 in.^{4}. Determine the reaction at the center support.

6-26* A steel (E = 200 GPa) beam with a span of 2.5 m is fixed at both ends. When a uniformly distributed load of 120 N/m is applied to the beam, the right support settles 2.5 mm. There is no rotation at either support. The moment of inertia of the cross section with respect to the neutral axis is 20(10^6) mm^4. Determine all of the reactions.

6-27 The beam shown in Fig. P6-27 is simply supported at the left end and framed into a column at the right end. When the load is applied, the ends of the beam remain at the same level, but the right end rotates (due to loading of the adjacent span) clockwise until the slope of the elastic curve is $wL^3/(48EI)$ downward to the right. Determine the left reaction in terms of w and L.

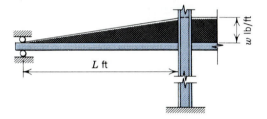

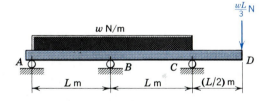

FIG. P6-27   FIG. P6-28

6-28 A beam is loaded and supported as shown in Fig. P6-28. When the loads are applied, the center support settles an amount equal to $wL^4/(24EI)$. Determine all of the reactions in terms of w and L.

6-29 A beam of span L is fixed at the left support and restrained at the right support. When a uniformly distributed load of w lb/ft is applied over the entire length, the right support settles an amount equal to $wL^4/(24EI)$ and rotates (as a result of loading the adjacent span) until the slope of the elastic curve is $wL^3/(72EI)$ downward to the right. Determine the reaction components at the left support in terms of w and L.

6-30 A structural steel (E = 200 GPa) S152 × 26 American standard section (see Appendix C) is welded into place at its left end. After the welding is completed it is observed that the beam has a slope of 0.01 rad downward to the right. The right end of the beam is then jacked into position at the same elevation as the left end at zero slope. Determine the maximum fiber stress introduced into the 3-m-long beam by this positioning procedure.

6-3

THE AREA-MOMENT METHOD

For statically determinate beams, points with known slopes or deflections were used as tangent points and moment centers for the application of the area-moment theorems. For statically indeterminate beams, each additional constraint provides information regarding a slope or deflection, which makes it possible to write an area-moment equation to supplement the equilibrium equations. The following example illustrates the application of the concepts.

EXAMPLE 6-2

A beam is loaded and supported as shown in Fig. 6-2a. Determine the reactions at supports A and B in terms of w and L.

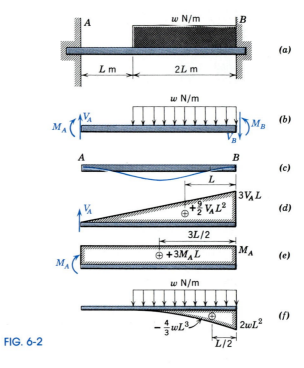

FIG. 6-2

Solution

There are four unknown reactions (V_A, M_A, V_B, and M_B) on the free-body diagram of Fig. 6-2b, and there are only two independent equations of equilibrium. Therefore, two area-moment relations are necessary to supplement the equations of equilibrium.

Inspection of the elastic curve (very much exaggerated) of Fig. 6-2c reveals that the angle between tangents drawn at A and B is zero. Also, the distance from a tangent drawn at A to point B is zero.

These geometric statements can be expressed mathematically by the area-moment theorems. Thus, since the angle between the tangents at A and B is zero, the area under the moment diagram between A and B must be zero. Also, since the distance from the tangent drawn at A to point B is zero, the moment of the area under the $M/(EI)$ diagram between A and B, with respect to an axis at B, is zero. (In this particular problem the moment axis could also have been selected at A, since the deflection of A from a tangent drawn at B is also zero.) These two area-moment relations give

$$EI\theta_{A/B} = 0 = \frac{9}{2} V_A L^2 + 3M_A L - \frac{4}{3} wL^3$$

from which

$$27V_A L + 18M_A = 8wL^2$$

and

$$EIt_{B/A} = 0 = \frac{9}{2} V_A L^2(L) + 3M_A L \left(\frac{3L}{2}\right) - \frac{4wL^3}{3}\left(\frac{L}{2}\right)$$

from which

$$27V_A L + 27M_A = 4wL^2$$

The simultaneous solution of these equations yields

$$M_A = -\frac{4wL^2}{9} = \underline{\frac{4wL^2}{9} \text{ N·m counterclockwise}} \qquad \text{Ans.}$$

and

$$V_A = \underline{\frac{16wL}{27} \text{ N upward}} \qquad \text{Ans.}$$

With two reactions known, the equations of equilibrium give the other two as

$$M_B = -\frac{2wL^2}{3} = \underline{\frac{2wL^2}{3} \text{ N·m clockwise}} \qquad \text{Ans.}$$

and

$$V_B = -\frac{38wL}{27} = \underline{\frac{38wL}{27} \text{ N upward}} \qquad \text{Ans.}$$

PROBLEMS

Note: Use the area-moment method to solve the following problems.

6-31* When the moment *M* is applied to the left end of the cantilever beam shown in Fig. P6-31, the deflection at the left end of the beam is zero. Determine the magnitude of the moment *M* in terms of *w* and *L*.

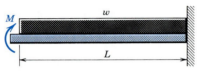

FIG. P6-31

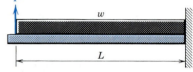

FIG. P6-32

6-32* A beam is loaded and supported as shown in Fig. P6-32. Determine the magnitude of the load *P*, in terms of *w* and *L*, required to make the slope at the left end of the beam zero.

6-33 A beam is loaded and supported as shown in Fig. P6-33. Determine the reactions at supports *A* and *B* in terms of *P* and *L*.

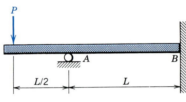

FIG. P6-33

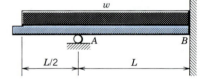

FIG. P6-34

6-34 A beam is loaded and supported as shown in Fig. P6-34. Determine the reactions at supports *A* and *B* in terms of *w* and *L*.

6-35* A beam is loaded and supported as shown in Fig. P6-35. Determine

    (a) the reactions at supports *A* and *B* in terms of *P* and *L*.

    (b) the deflection at the middle of the span in terms of *P*, *L*, *E*, and *I*.

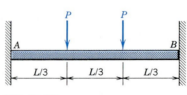

FIG. P6-35

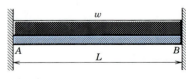

FIG. P6-36

6-36* A beam is loaded and supported as shown in Fig. P6-36. Determine

  (a) the reactions at supports A and B in terms of w and L.
  (b) the deflection at the middle of the span in terms of w, L, E, and I.

6-37 A beam is loaded and supported as shown in Fig. P6-37. Determine

  (a) the reactions at supports A and B in terms of w and L.
  (b) the deflection at the middle of the span in terms of w, L, E, and I.

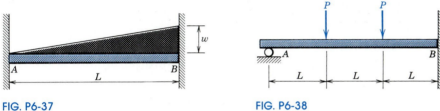

FIG. P6-37   FIG. P6-38

6-38 A beam is loaded and supported as shown in Fig. P6-38. Determine

  (a) the reactions at supports A and B in terms of P and L.
  (b) the deflection at the middle of the span in terms of P, L, E, and I.

6-39* A beam is loaded and supported as shown in Fig. P6-39. Determine

  (a) the reactions at supports A and C in terms of P and L.
  (b) the deflection at the middle of span AC in terms of P, L, E, and I.

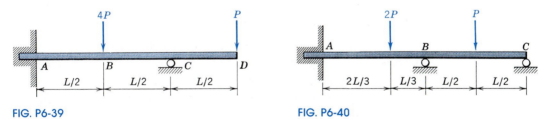

FIG. P6-39   FIG. P6-40

6-40 A beam is loaded and supported as shown in Fig. P6-40. Determine

  (a) the reactions at supports A, B, and C in terms of P and L.
  (b) the deflection at the middle of span BC in terms of P, L, E, and I.

6-41* A beam is loaded and supported as shown in Fig. P6-41. Determine

(a) the reactions at supports A and C in terms of w and L.

(b) the deflection at the middle of the span in terms of w, L, E, and I.

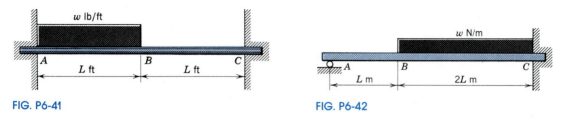

FIG. P6-41

FIG. P6-42

6-42* A beam is loaded and supported as shown in Fig. P6-42. Determine

(a) the reactions at supports A and C in terms of w and L.

(b) the deflection at the left end of the distributed load in terms of w, L, E, and I.

6-43 A beam is loaded and supported as shown in Fig. P6-43. Determine

(a) the reactions at supports A and D in terms of w and L.

(b) the deflection at the left end of the distributed load in terms of w, L, E, and I.

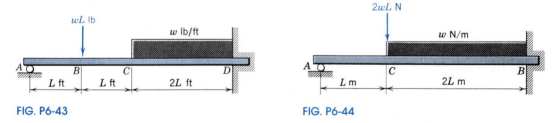

FIG. P6-43

FIG. P6-44

6-44 A beam is loaded and supported as shown in Fig. P6-44. Determine

(a) the reactions at supports A and B in terms of w and L.

(b) the deflection at the left end of the distributed load in terms of w, L, E, and I.

6-45* A beam is loaded and supported as shown in Fig. P6-45. Determine

(a) the reactions at supports A, B, and C in terms of P and L.

(b) the slope over the middle support in terms of P, L, E, and I.

(c) the deflection under the concentrated load P in terms of P, L, E, and I.

6-46* A beam is loaded and supported as shown in Fig. P6-46. Determine

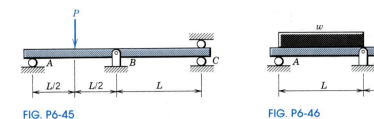

FIG. P6-45   FIG. P6-46

(a) the reactions at supports A, B, and C in terms of w and L.
(b) the slope over the middle support in terms of w, L, E, and I.
(c) the deflection under the concentrated load wL in terms of w, L, E, and I.

6-47 A beam is loaded and supported as shown in Fig. P6-47. Determine
(a) the reactions at supports A, B, and C in terms of w and L.
(b) the slope over the middle support in terms of w, L, E, and I.
(c) the deflection at the right end of the beam in terms of w, L, E, and I.

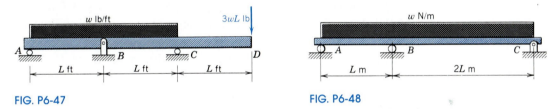

FIG. P6-47   FIG. P6-48

6-48 A beam is loaded and supported as shown in Fig. P6-48. Determine
(a) the reactions at supports A, B, and C in terms of w and L.
(b) the slope over the middle support in terms of w, L, E, and I.
(c) the deflection at the middle of span BC in terms of w, L, E, and I.

6-49* A beam is loaded and supported as shown in Fig. P6-49. Determine
(a) the reactions at supports A, B, and C in terms of P and L.
(b) the slope over the middle support in terms of P, L, E, and I.
(c) the deflection under the concentrated load P in terms of P, L, E, and I.

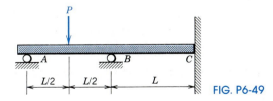

FIG. P6-49

6-50 A beam is loaded and supported as shown in Fig. P6-50. Determine
(a) the reactions at supports A, B, and C in terms of w and L.
(b) the slope over the middle support in terms of w, L, E, and I.
(c) the deflection at the middle of span BC in terms of w, L, E, and I.

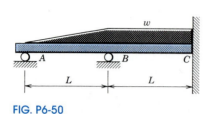

FIG. P6-50

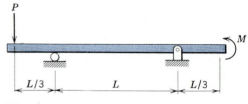

FIG. P6-51

6-51* A beam is loaded and supported as shown in Fig. P6-51. When the load P and the moment M are applied, the slope of the elastic curve at the left support is $PL^2/(24EI)$ downward to the right. Determine
(a) the value of M in terms of P and L.
(b) the reactions at the left and right supports in terms of P and L.
(c) the slope of the elastic curve at the right support in terms of P, L, E, and I.
(d) the deflection at the left end of the beam in terms of P, L, E, and I.

6-52 A beam is loaded and supported as shown in Fig. P6-52. When the moments M and PL are applied, the slope of the elastic curve over the center support is $PL^2/(12EI)$ downward to the right. Determine
(a) the value of M in terms of P and L.
(b) the reactions at supports A, B, and C in terms of P and L.
(c) the slope of the elastic curve at the right support in terms of P, L, E, and I.
(d) the deflection midway between supports A and B in terms of P, L, E, and I.

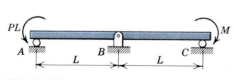

FIG. P6-52

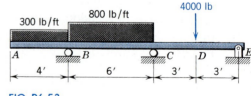

FIG. P6-53

6-53* A structural steel ($E = 29,000$ ksi) S6 × 12.5 American standard beam (see Appendix C) is loaded and supported as shown in Fig. P6-53. Determine

(a) the reactions at supports B, C, and E.

(b) the slope over the middle support.

(c) the deflection at the left end of the beam.

6-54 A structural steel ($E = 200$ GPa) W203 × 22 wide-flange beam (see Appendix C) is loaded and supported as shown in Fig. P6-54. Determine

(a) the reactions at supports A, B, and C.

(b) the slope over the middle support.

(c) the deflection at the middle of span BC.

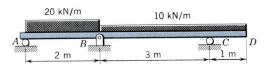

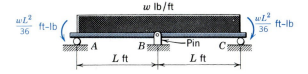

FIG. P6-54   FIG. P6-55

6-55* A beam is loaded and supported as shown in Fig. P6-55. When the loads are applied, the center support settles an amount equal to $wL^4/(16EI)$. Determine the reactions at supports A, B, and C in terms of w and L.

6-56* A beam is loaded and supported as shown in Fig. P6-56. When the loads are applied, the right support settles an amount equal to $wL^4/(8EI)$. Determine the reactions at supports A, B, and C in terms of w and L.

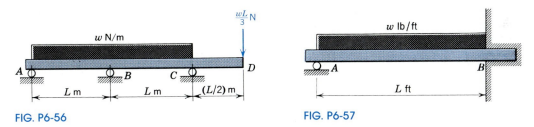

FIG. P6-56   FIG. P6-57

6-57* The beam of span L shown in Fig. P6-57 is simply supported at the left end and restrained at the right end. When a uniformly distributed load of w lb/ft is applied over the entire length, the right support settles an amount equal to $wL^4/(24EI)$ and rotates until the slope of the elastic curve is $wL^3/(48EI)$ upward to the right. Determine the reactions at supports A and B in terms of w and L.

6-58 The right end of the beam shown in Fig. P6-58 is only partially fixed and under action of the load the slope becomes $wL^3/(96EI)$ upward to the right. Determine the reactions at supports A, B, and C in terms of w and L.

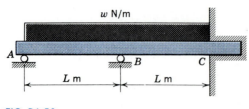

FIG. P6-58

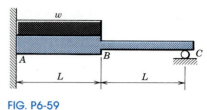

FIG. P6-59

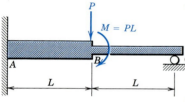

FIG. P6-60

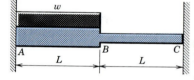

FIG. P6-61

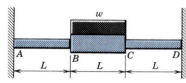

FIG. P6-62

6-59* A beam is loaded and supported as shown in Fig. P6-59. The beam has a moment of inertia $2I$ in the interval AB and a moment of inertia I in the interval BC. Determine

(a) the reactions at supports A and C in terms of w and L.

(b) the deflection at section B in terms of w, L, E, and I.

6-60* A beam is loaded and supported as shown in Fig. P6-60. The beam has a moment of inertia $3I$ in the interval AB and a moment of inertia I in the interval BC. Determine

(a) the reactions at supports A and C in terms of P and I.

(b) the deflection at section B in terms of P, L, E, and I.

6-61 A beam is loaded and supported as shown in Fig. P6-61. The beam has a moment of inertia $2I$ in the interval AB and a moment of inertia I in the interval BC. Determine

(a) the reactions at supports A and C in terms of w and L.

(b) the deflection at section B in terms of w, L, E, and I.

6-62 A beam is loaded and supported as shown in Fig. P6-62. The beam has a moment of inertia $2I$ in the center interval BC and a moment of inertia I in the other two intervals near the supports. Determine the deflection midway between the supports in terms of w, L, E, and I.

6-4

THE SUPERPOSITION METHOD

The concept (discussed in Section 5-11) that a slope or deflection due to several loads is the algebraic sum of the slopes or deflections due to each of the loads acting individually is frequently used to provide the deformation equations needed to supplement the equilibrium equations in the solution of statically indeterminate beam problems. The general method of attack is outlined in Section 1-15. To provide the necessary deformation equations, selected restraints are removed and replaced by unknown loads (forces and couples); the deformation diagrams corresponding to individual loads (both known and unknown) are sketched; and the component deflections or slopes are summed to produce the known configuration. The following examples illustrate the use of superposition for this purpose.

EXAMPLE 6-3

A steel ($E = 30,000$ ksi) beam 20 ft long is simply supported at the ends and at the midpoint. Under a load of 400 lb/ft, uniformly distributed over the entire beam, the center support is 0.12 in. above the end supports. Determine the reactions. The moment of inertia of the cross section with respect to the neutral axis is 100 in.[4].

Solution

With three unknown reactions and only two equations of equilibrium available, the beam is statically indeterminate. Replace the center support with an unknown applied load. The resulting simply supported beam is equivalent to two beams with individual loads as shown in Fig. 6-3. The resulting deflection at the midpoint of the beam is the upward deflection y_R due to R_C, minus the downward deflection y_w due to the uniform load; that is,

$$y = y_R - y_w = 0.12 \text{ in.} \tag{a}$$

The deflection y_w can be obtained from Appendix D, case 7 (Table D-2), and is

$$y_w = \frac{5wL^4}{384EI} = \frac{5[400(20)][20(12)]^3}{384(30)(10^6)(100)} = 0.48 \text{ in. downward}$$

The deflection y_R can be obtained in terms of R_C from Appendix D, case 6 (Table D-2), and is

$$y_R = \frac{R_C L^3}{48EI} = \frac{R_C[20(12)]^3}{48(30)(10^6)(100)} = \frac{96R_C}{10^6} \text{ in. upward}$$

When these values are substituted in Eq. a, the result is

$$\frac{96R_C}{10^6} - 0.48 = 0.12$$

from which

$$R_C = 6250 \text{ lb upward} \qquad\qquad \text{Ans.}$$

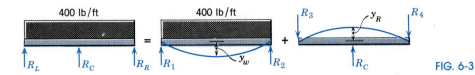

FIG. 6-3

The equilibrium equation $\Sigma F_y = 0$ and symmetry give

$$R_L = R_R = \underline{875 \text{ lb upward}} \qquad \text{Ans.}$$

Note that in determining y_w, w lb/ft $\times$ L ft gives the applied load in pounds. The remaining L^3 must be expressed in inches if E and I are in inches and the result is to be in inches. The arithmetic will frequently be simplified if the expressions for the deflections are substituted in the deflection equation in symbol form. In this example Eq. a becomes

$$\frac{R_C L^3}{48EI} - \frac{5wL^4}{384EI} = 0.12$$

which reduces (when multiplied by $48EI/L^3$) to

$$R_C - \frac{5wL}{8} = \frac{0.12(48EI)}{L^3}$$

or

$$R_C = \frac{5(400)(20)}{8} + \frac{0.12(48)(30)(10^6)(100)}{[20(12)]^3} = 6250 \text{ lb upward}$$

EXAMPLE 6-4

A beam is loaded and supported as shown in Fig. 6-4a. Determine the reactions at supports A and B in terms of P, L, and a.

Solution

There are four unknown reactions (a shear and moment at each end), and only two equations of equilibrium are available; therefore, the beam is statically indeterminate, and two deformation equations are necessary. Replace the constraint at the right end with an unknown force and couple. The resulting cantilever beam is equivalent to three beams with individual loads, as shown in Fig. 6-4b. Note that the unknown shear and moment at the right end are both shown as positive values so that the algebraic sign of the result will be correct. From the geometry of the constrained beam, the resultant slope and the resultant deflection at the right end are both zero. The slope and deflection at the end of each of the three replacement beams can be obtained from the expressions in Appendix D. Thus, the first beam with load P (see case 1 of Table D-2) has a constant slope from P to the end of the beam, which is

$$\theta_P = -\frac{Pa^2}{2EI}$$

The deflection y_P at the end is made up of two parts: y_1 for a beam of length a, and y_2 the added deflection of the tangent segment (straight line) from P to the end of the beam. This deflection is

$$y_P = y_1 + y_2 = -\frac{Pa^3}{3EI} + (L - a)\theta_P$$

$$= -\frac{Pa^3}{3EI} + (L - a)\left(-\frac{Pa^2}{2EI}\right) = \frac{Pa^3}{6EI} - \frac{Pa^2L}{2EI}$$

The slope and deflection at the end of the beam due to the shear V_R are (also from case 1 of Table D-2)

$$\theta_V = -\frac{V_R L^2}{2EI} \quad \text{and} \quad y_V = -\frac{V_R L^3}{3EI}$$

Finally, the slope and deflection at the right end of the beam due to M_R are (see case 4 of Table D-2)

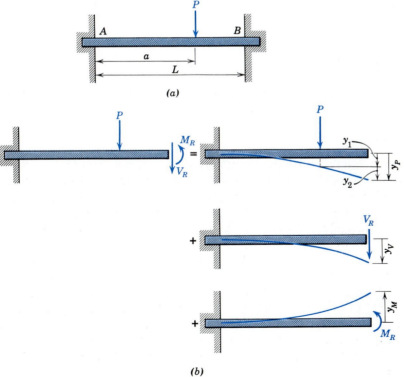

(a)

(b)

FIG. 6-4

$$\theta_M = \frac{M_R L}{EI} \quad \text{and} \quad y_M = \frac{M_R L^2}{2EI}$$

Since the resultant slope is zero,

$$\theta_P + \theta_V + \theta_M = -\frac{Pa^2}{2EI} - \frac{V_R L^2}{2EI} + \frac{M_R L}{EI} = 0$$

Similarly,

$$y_P + y_V + y_M = \frac{Pa^3}{6EI} - \frac{Pa^2 L}{2EI} - \frac{V_R L^3}{3EI} + \frac{M_R L^2}{2EI} = 0$$

The simultaneous solution of these two equations yields

$$M_R = -\frac{Pa^2(L - a)}{L^2} \quad \text{and} \quad V_R = -\frac{Pa^2(3L - 2a)}{L^3} \quad \text{Ans.}$$

When the equations of equilibrium are applied to a free-body diagram of the beam, the shear and moment at the left end are found to be

$$M_L = -\frac{Pa(L - a)^2}{L^2} \quad \text{and} \quad V_L = +\frac{P(L^3 - 3a^2 L + 2a^3)}{L^3} \quad \text{Ans.}$$

PROBLEMS

> *Note:* Use the method of superposition to solve the following problems.

6-63* A beam is loaded and supported as shown in Fig. P6-63. Determine the magnitude of the moment M, in terms of w and L, required to make

(a) the slope at the left end of the beam zero.

(b) the deflection at the left end of the beam zero.

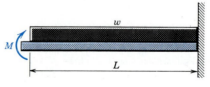

FIG. P6-63

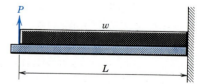

FIG. P6-64

6-64* A beam is loaded and supported as shown in Fig. P6-64. Determine the magnitude of the load P, in terms of w and L, required to make

(a) the slope at the left end of the beam zero.

(b) the deflection at the left end of the beam zero.

6-65* A beam is loaded and supported as shown in Fig. P6-65. Determine

(a) the reactions at supports A, B, and C in terms of w and L.

(b) the deflection at the middle of span AB in terms of w, L, E, and I.

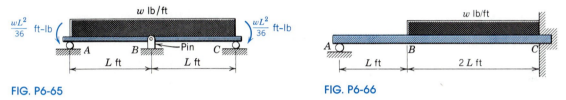

FIG. P6-65   FIG. P6-66

6-66* A beam is loaded and supported as shown in Fig. P6-66. Determine

(a) the reactions at supports A and C in terms of w and L.

(b) the deflection at the left end of the distributed load in terms of w, L, E, and I.

6-67 A beam is loaded and supported as shown in Fig. P6-67. Determine

(a) the reactions at supports A and D in terms of w and L.

(b) the deflection at the left end of the distributed load in terms of w, L, E, and I.

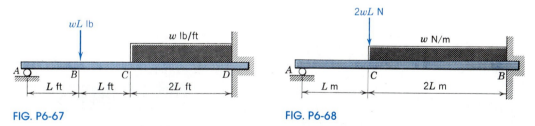

FIG. P6-67   FIG. P6-68

6-68 A beam is loaded and supported as shown in Fig. P6-68. Determine

(a) the reactions at supports A and B in terms of w and L.

(b) the deflection at the left end of the distributed load in terms of w, L, E, and I.

6-69* A beam is loaded and supported as shown in Fig. P6-69. Determine the magnitude of the moment M, in terms of P and L, required to make

(a) the slope at the right end of the beam zero.

(b) the deflection at the right end of the beam zero.

6-70 A beam is loaded and supported as shown in Fig. P6-69. Determine the magnitude of the moment M, in terms of P and L, required to make

(a) the slope at the left end of the beam zero.

(b) the deflection at the left end of the beam zero.

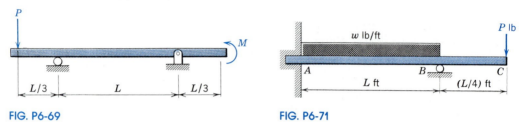

FIG. P6-69   FIG. P6-71

6-71* A beam is loaded and supported as shown in Fig. P6-71. When the load P is applied, the slope at the right end of the beam is zero. Determine

(a) the value of P in terms of w and L.

(b) the reactions at supports A and B in terms of w and L.

6-72 A beam is loaded and supported as shown in Fig. P6-71. When the load P is applied, the deflection at the right end of the beam is zero. Determine

(a) the value of P in terms of w and L.

(b) the reactions at supports A and B in terms of w and L.

6-73* A beam is loaded and supported as shown in Fig. P6-73. Determine

(a) the reactions at supports A and C in terms of w and L.

(b) the deflection at the right end of the distributed load in terms of w, L, E, and I.

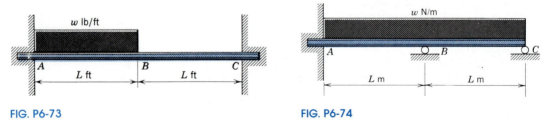

FIG. P6-73   FIG. P6-74

6-74* A beam is loaded and supported as shown in Fig. P6-74. Determine

(a) the reactions at supports A, B, and C in terms of w and L.

(b) the slope of the beam over support C in terms of w, L, E, and I.

6-75 A beam is loaded and supported as shown in Fig. P6-75. Determine

(a) the reactions at supports A and B in terms of w and L.

(b) the deflection at the middle of span AB in terms of w, L, E, and I.

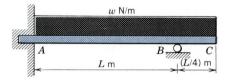

FIG. P6-75

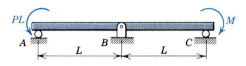

FIG. P6-76

6-76 A beam is loaded and supported as shown in Fig. P6-76. Determine the magnitude of the moment M, in terms of P and L, required to make

(a) the slope at the left end of the beam zero.

(b) the slope at the right end of the beam zero.

6-77* A structural steel ($E = 29,000$ ksi) W5 × 16 wide-flange section (see Appendix C) is loaded and supported as shown in Fig. P6-77. When the uniformly distributed load is applied, the support at B settles 0.125 in. Determine

(a) the reactions at supports A, B, and C.

(b) the maximum fiber stress produced in the beam.

6-78 A structural steel ($E = 29,000$ ksi) W5 × 16 wide-flange section (see Appendix C) is loaded and supported as shown in Fig. P6-77. When the uniformly distributed load is applied, the support at C settles 0.125 in. Determine

(a) the reactions at supports A, B, and C.

(b) the maximum fiber stress produced in the beam.

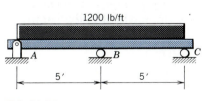

FIG. P6-77

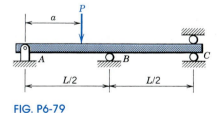

FIG. P6-79

6-79* A beam is simply supported at the ends and at the center of the span as shown in Fig. P6-79. The concentrated load P moves slowly across the beam. Determine

(a) the reactions at supports A, B, and C in terms of P, L, and a.

(b) the maximum moment produced in the beam in terms of P and L.

6-80 Two parallel cold-rolled stainless-steel ($E = 190$ GPa) bars will be used to support a concentrated load P of 50 kN at the end of a 300-mm span. The bars will be fixed to a rigid wall at the left end and securely attached to a rigid block at the right end, as shown in Fig. P6-80. If the maximum deflection must be limited to 2 mm and the maximum fiber stress must be limited to 300 MPa, determine the required thickness t and width b for the bars.

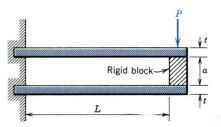

FIG. P6-80

6-81* A 6-in.-wide by 10-in.-deep timber ($E = 1200$ ksi) beam 20 ft long is simply supported at the ends and at the middle of the span. Under a uniformly distributed load of 500 lb/ft, the middle support settles $\frac{3}{4}$ in. relative to the end supports. Determine the reactions at the center and ends of the beam.

6-82* A timber ($E = 10$ GPa) beam 100 mm wide by 250 mm deep is 4 m long. The beam is fixed at the left end, simply supported at the right end, and subjected to a uniformly distributed load of w N/m. If the right end settles an amount equal to $wL^4/(48EI)$ when the load w is applied, determine the maximum total load that can be carried by the beam if the maximum allowable fiber stress is 10 MPa.

6-83 The 4-in.-wide by 6-in.-deep timber ($E = 1200$ ksi) beam shown in Fig. P6-83 is fixed at the left end and supported at the right end with a tie rod that has a cross-sectional area of 0.125 in.^{2}. Determine the tension in the tie rod if it is unstressed before the load is applied to the timber beam and

(a) the tie rod is made of steel ($E = 30,000$ ksi).

(b) the tie rod is made of aluminum alloy ($E = 10,000$ ksi).

6-84 Solve Problem 6-83 if there is a 0.30-in. gap between the bottom of the beam and the bearing plate attached to the tie rod before the load is applied to the beam.

6-85* A steel beam [$E = 200$ GPa and $I = 50(10^6)$ mm^4] is loaded and supported as shown in Fig. P6-85. The post BD is a 150×150-mm

FIG. P6-83

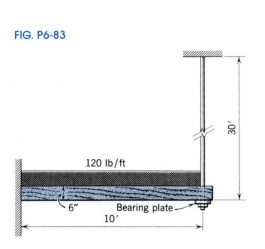

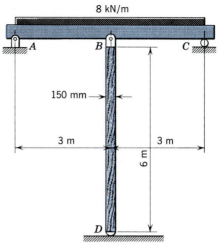

FIG. P6-85

timber ($E = 10$ GPa) that is braced to prevent buckling. Determine the load carried by the post if it is unstressed before the 8-kN/m distributed load is applied to the beam.

6-86* Solve Problem 6-85 if there is a 5-mm gap between the bottom of the beam and the bearing plate at the top of the timber post before the load is applied to the beam.

6-87 A timber ($E = 1800$ ksi) beam is loaded and supported as shown in Fig. P6-87. The tension in the $\frac{1}{2}$-in.-diameter tie rod BD is zero before the load is applied to the beam. Determine the reaction at B if

(a) the tie rod is made of steel ($E = 30{,}000$ ksi).

(b) the tie rod is made of aluminum alloy ($E = 10{,}000$ ksi).

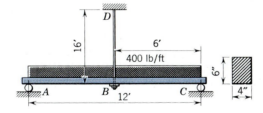

FIG. P6-87

6-88* The steel ($E = 200$ GPa) beam AB of Fig. P6-88 is fixed at ends A and B and supported at the center by the pin-connected timber ($E = 10$ GPa) struts CD and CE. The cross-sectional area of each strut is 6400 mm^2 and the moment of inertia for the beam is $25(10^6)$ mm^4. Determine the force in each strut after the 6-kN/m distributed load is applied to the beam. The constants in the expression for maximum deflection of a beam with fixed ends are 1/384 for a uniformly distributed load and 1/192 for a concentrated load at the center.

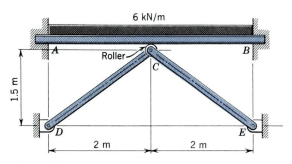

FIG. P6-88

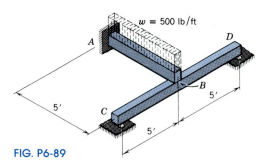

FIG. P6-89

6-89* A uniformly distributed load of 500 lb/ft is supported by two beams, as shown in Fig. P6-89. Before the load is applied, the beams are in contact, but the reaction at B is zero. The moment of inertia of the cross section with respect to the neutral axis is 10 in.^{4} for beam AB and 20 in.^{4} for beam CD. The modulus of elasticity is 1200 ksi for both beams. Determine the reaction at B when the 500 lb/ft load is applied to beam AB.

6-90* A simple frame with constant EI is loaded and supported as shown in Fig. P6-90. Determine

(a) the reaction at support A.

(b) the deflection at support A in terms of P, L, E, and I.

6-91* A simple frame with constant EI is loaded and supported as shown in Fig. P6-91. Determine

(a) the horizontal component of the reaction at support A.

(b) the vertical component of the reaction at support A.

6-92 The simple frame ($L = 5$ m) shown in Fig. P6-92 was fabricated using structural steel ($E = 200$ GPa) S102 × 14 American standard sections (see Appendix C). During construction the foundation at support A moved 50 mm to the right. Determine the maximum fiber stress produced in the frame.

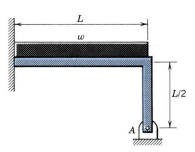

FIG. P6-90

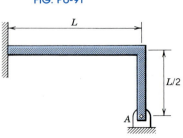

FIG. P6-91

FIG. P6-92

COMPUTER PROBLEMS

Note: The following problems have been designed to be solved with a programmable calculator, microcomputer, or mainframe computer. Appendix F contains a description of a few numerical methods, together with a few simple programs in BASIC and FORTRAN, that can be modified for use in the solution of these problems.

C6-1 A beam carries a uniformly distributed load w and is supported as shown in Fig. PC6-1a. Write a program that will

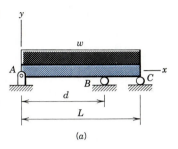

(a) compute the deflection $y_1(x)$ of the beam due to the load w by ignoring the support at B (see Fig. PC6-1b).

(b) compute the deflection $y_2(x)$ of the beam due to a unit concentrated load P_B ($P_B = 1$) at B by ignoring the support at B (see Fig. PC6-1c).

(c) compute

$$y(x) = y_1(x) - \frac{y_1(d)}{y_2(d)} y_2(x)$$

which is the deflection of the original beam (see Fig. PC6-1a).

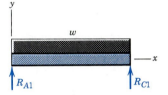

(d) compute

$$R_A = R_{A1} - \frac{y_1(d)}{y_2(d)} R_{A2}$$

$$R_B = \frac{y_1(d)}{y_2(d)}$$

$$R_C = R_{C1} - \frac{y_1(d)}{y_2(d)} R_{C2}$$

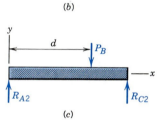

which are the three support reactions.

C6-2 The beam of Problem C6-1 has a length of 4.5 m and is constructed of wood ($EI = 60$ kN-m²). If $w = 7.5$ kN/m, compute and plot

FIG. PC6-1

(a) the deflection curve for the beam (plot y as a function of x; $0 \le x \le L$) for distance $d = 1$ m, 2 m, and 3 m.

(b) the bending moment distribution along the beam (plot $M(x)$ as a function of x for $0 \le x \le L$) for a distance $d = 1$ m, 2 m, and 3 m.

C6-3 A 15-ft-long W4 × 13 (see Appendix C) structural steel ($E = 29,000$ ksi) section is used for a beam that is subjected to a uniformly distributed load w of 500 lb/ft as shown in Fig. PC6-1a. Compute and plot the maximum tensile and compressive fiber stresses in the beam as a function of the location d of the middle support.

C6-4 A 4.5-m-long WT178 × 51 (see Appendix C) structural steel ($E = 200$ GPa) section is used for a beam that supports a uniformly distributed load w of 7 kN/m as shown in Fig. PC6-1a. The flange is at the top of the beam. Compute and plot the maximum tensile and compressive fiber stresses in the beam as a function of the location d of the middle support.

C6-5 A concentrated load P of 4000 lb moves slowly across the beam shown in Fig. PC6-5. The beam is a structural steel ($E = 29,000$ ksi)

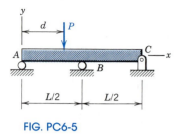

FIG. PC6-5

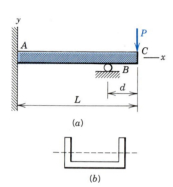

FIG. PC6-9

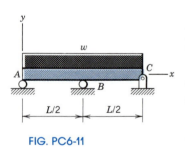

FIG. PC6-11

WT5 × 30 section (see Appendix C) and is 100 in. long. The flange is located at the bottom of the beam. Compute and plot the maximum and minimum bending moments as a function of the location d ($0 \leq d \leq L/2$) of the load.

C6-6 A concentrated load P of 15 kN moves slowly across the beam shown in Fig. PC6-5. The beam is a structural steel ($E = 200$ GPa) WT127 × 45 (see Appendix C) section and is 2.5 m long. The flange is located on the bottom of the beam. Compute and plot the magnitude of the support reaction at B as a function of the location d of the load.

C6-7 For the conditions given in Problem C6-5, compute and plot the maximum tensile fiber stress in the beam as a function of the position d of the concentrated load P.

C6-8 For the conditions given in Problem C6-6, compute and plot the elastic curve for load locations $d = L/4$, $L/3$, and $3L/5$.

C6-9 A 6-ft-long C12 × 30 (see Appendix C) structural steel ($E = 29,000$ ksi) section is used for a beam that supports a concentrated load P of 1 kip as shown in Fig. PC6-9. Compute and plot the deflection y_C at the right end of the beam as a function of the location d ($0.1 \leq d/L \leq 0.9$) of the right support.

C6-10 The 4-m-long beam shown in Fig. PC6-9a must support a concentrated load P of 50 kN. The maximum fiber stress must be limited to 125 MPa for the structural steel ($E = 200$ GPa) wide-flange section being considered for the beam. Compute and plot the minimum section modulus S required for the beam as a function of the location d ($0.1 \leq d/L \leq 0.9$) of the right support.

C6-11 A 20-ft-long S18 × 70 (see Appendix C) structural steel ($E = 29,000$ ksi) section will be used for the two-span beam shown in Fig. PC6-11. The beam supports a uniformly distributed load w of 5500 lb/ft. The support at B settles with time until it provides no resistance to deflection of the beam. Compute and plot the maximum fiber stress in the beam as a function of the vertical settlement at B.

C6-12 A 6-m-long S457 × 104 (see Appendix C) structural steel ($E = 200$ GPa) section will be used for the two-span beam shown in Fig. PC6-11. The beam supports a uniformly distributed load w of 80 kN/m. The support at A settles with time until it provides no resistance to deflection of the beam. Compute and plot the maximum fiber stress in the beam as a function of the vertical settlement at A.

CHAPTER 7

STRESS AND STRAIN TRANSFORMATION EQUATIONS

7-1

INTRODUCTION

In previous chapters, formulas were developed for determining normal and shearing stresses on specific planes in axially loaded bars, thin-walled spherical and cylindrical pressure vessels, circular shafts, and beams. For example, the normal stress at a point on a transverse cross section of a beam can be determined by using the flexure formula (Eq. 4-7), while the shearing stress at the same point on the cross section can be determined by using the shearing stress formula (Eq. 4-13). In this chapter, methods will be developed for determining the normal and shearing stresses on other planes through the point of interest.

For the special case of an axially loaded bar, expressions were developed in Section 1-6 for determining the normal (Eq. 1-7) and shearing (Eq. 1-8) stresses on inclined planes through the bar. This analysis indicated that maximum normal stresses occur on transverse planes and that maximum shearing stresses occur on planes inclined at 45° to the axis of the bar. Similarly, for the case of pure torsion in a circular shaft, maximum shearing stresses (Eq. 3-6) occur on transverse planes and maximum tensile and compressive stresses (Eq. 3-7) occur on planes inclined at 45° to the axis of the shaft. For the case of a beam subjected to an arbitrary type of transverse loading, both normal (Eq. 4-7) and shearing (Eq. 4-13) stresses develop on a transverse cross section. The stresses on a specified inclined plane can be determined by using the free-body diagram method of Section 1.8; however, this method is not suitable for the determination of maximum normal and maximum shearing stresses that are often required. A more general approach for determining stresses on inclined planes will be developed to handle this type of analysis.

The state of stress at a general point in an arbitrarily loaded body was described in Section 1-7. This state of stress can be represented by six stress components, as shown in Fig. 1-22. Three of these components σ_x, σ_y, and σ_z represent the normal stresses that develop on the faces of a small cubic element of material centered at the point of interest and having faces of the element perpendicular to the coordinate axes. The other three components τ_{xy}, τ_{yz}, and τ_{zx} represent the shearing stresses on the faces of the same element. Recall that $\tau_{xy} = \tau_{yx}$, $\tau_{yz} = \tau_{zy}$, and $\tau_{zx} = \tau_{xz}$. The notation and sign convention for stresses was discussed in Section 1-7. The state of stress shown in Fig. 1-22 can be represented by a different set of components if the coordinate axes are rotated. Stress transformation equations are used to determine the stress components on the faces of the rotated element, assuming that the stresses are known for the initial reference position.

Two-dimensional or plane stress was first introduced in Section 1-8. For this case, two parallel faces of the cubic element are free of stress. If the z axis is chosen perpendicular to these two faces, then $\sigma_z = \tau_{zx} = \tau_{zy} = 0$. The remaining stress components are σ_x, σ_y, and τ_{xy} as shown in three-dimensional representation in Fig. 7-1a and in two-dimensional representation in Fig. 7-1b. The two-dimensional or plane-stress state of stress is very important in engineering design, since it occurs in a thin plate subjected to forces acting in the midplane of the plate and on the free surfaces of all structural elements or machine components. The stress transformation equations for plane stress will be developed first. The general state of stress is treated in Section 7-10.

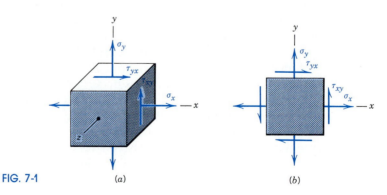

FIG. 7-1   (a)   (b)

7-2

STRESS TRANSFORMATION EQUATIONS FOR PLANE STRESS

As mentioned in Section 1-3, a thorough understanding of the physical significance of stress is fundamental in mechanics of materials. The concept of stress (internal force per unit area) at a point in a body on a plane through the point must be kept constantly in mind. The fact that the stresses are, in general, different for different planes through any one point has been emphasized previously.

Three important facts about stresses were developed in Section 1-6:

1. The maximum normal stress for a centric tensile or compressive load in one direction is in the direction of the load on a plane normal to the load, and its magnitude is $\sigma = P/A$.

2. The maximum shearing stress for a centric tensile or compressive load in one direction is on a plane at an angle of 45° to the load, and its magnitude is $\tau = \sigma/2 = P/(2A)$.

3. For all types of loads, a shearing stress at any one point on any plane is accompanied by a shearing stress of equal magnitude on an orthogonal plane through the point.

Equations relating the desired normal and shearing stress σ_n and τ_{nt} on an arbitrary plane through a point oriented at an angle θ with respect to a reference x axis and the known stresses σ_x, σ_y, and $\tau_{xy} = \tau_{yx}$ on the reference planes can be developed using the free-body diagram method of Section 1-8. Consider the plane-stress situation indicated in Fig. 7-2a, where the dotted line A–A represents the trace of any plane through the point (all planes are perpendicular to the plane of zero stress–the plane of the paper). In the following derivation, a counterclockwise angle θ is positive.

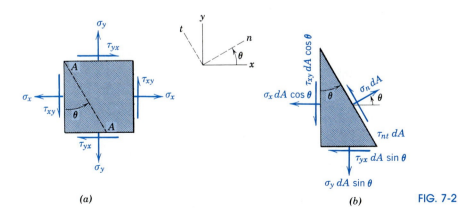

(a)   (b)   FIG. 7-2

Figure 7-2*b* is a free-body diagram of a wedge-shaped element in which the areas of the faces are *dA* for the inclined face (plane *A*–*A*), *dA* cos *θ* for the vertical face, and *dA* sin *θ* for the horizontal face. The force equation of equilibrium in the *n* direction gives

$$\sigma_n dA - \tau_{yx}(dA \sin \theta)\cos \theta - \tau_{xy}(dA \cos \theta)\sin \theta - \sigma_x(dA \cos \theta)\cos \theta$$
$$- \sigma_y(dA \sin \theta)\sin \theta = 0$$

from which, since $\tau_{yx} = \tau_{xy}$,

$$\sigma_n = \sigma_x \cos^2 \theta + \sigma_y \sin^2 \theta + 2\tau_{xy} \sin \theta \cos \theta \qquad (7\text{-}1a)$$

or, in terms of the double angle,

$$\sigma_n = \frac{\sigma_x(1 + \cos 2\theta)}{2} + \frac{\sigma_y(1 - \cos 2\theta)}{2} + \frac{2\tau_{xy}(\sin 2\theta)}{2} \qquad (7\text{-}1b)$$
$$= \frac{\sigma_x + \sigma_y}{2} + \frac{\sigma_x - \sigma_y}{2} \cos 2\theta + \tau_{xy} \sin 2\theta$$

For the free-body diagram of Fig. 7-2*b*, the force equation of equilibrium in the *t* direction gives

$$\tau_{nt} dA - \tau_{xy}(dA \cos \theta)\cos \theta + \tau_{yx}(dA \sin \theta)\sin \theta + \sigma_x(dA \cos \theta)\sin \theta$$
$$- \sigma_y(dA \sin \theta)\cos \theta = 0$$

from which

$$\tau_{nt} = -(\sigma_x - \sigma_y)\sin \theta \cos \theta + \tau_{xy}(\cos^2 \theta - \sin^2 \theta) \qquad (7\text{-}2a)$$

or, in terms of the double angle,

$$\tau_{nt} = -\frac{\sigma_x - \sigma_y}{2} \sin 2\theta + \tau_{xy} \cos 2\theta \qquad (7\text{-}2b)$$

Equations 7-1 and 7-2 provide a means for determining normal and shearing stresses on any plane whose outward normal is perpendicular to the *z* axis and is oriented at an angle *θ* with respect to the reference *x* axis. When these equations are used, the sign conventions used in their development must be rigorously followed; otherwise, erroneous results will be obtained. These sign conventions can be summarized as follows:

1. Tensile normal stresses are positive; compressive normal stresses are negative. All of the normal stresses shown in Fig. 7-2 are positive.

2. A shearing stress is positive if it points in the positive direction of the coordinate axis of the second subscript when it is acting on a surface whose outward normal is in a positive direction. Similarly, if the outward normal of the surface is in a negative direction, then a positive shearing stress points in the negative direction of the coordinate axis of the second subscript. All of the shearing stresses shown on Fig. 7-2 are positive. Shearing stresses pointing in the opposite directions would be negative.

3. An angle measured counterclockwise from the reference x axis is positive. Conversely, angles measured clockwise from the reference x axis are negative.

4. The (n, t, z) axes have the same order as the (x, y, z) axes. Both sets of axes form a right-hand coordinate system.

EXAMPLE 7-1

At a point in a structural member subjected to plane stress there are normal and shearing stresses on horizontal and vertical planes through the point, as shown in Fig. 7-3a. Use the stress transformation equations to determine

(a) the normal and shearing stresses on plane AB.

(b) the normal and shearing stresses on plane CD, which is perpendicular to plane AB.

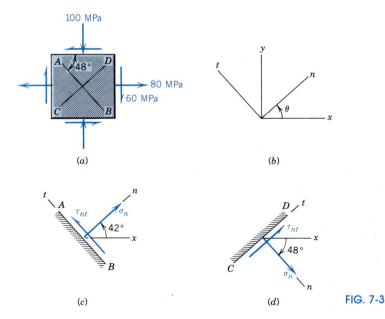

FIG. 7-3

Solution

First define the x–y and n–t (if needed) directions if they have not been specified.[1] On the basis of the axes shown in Fig. 7-3b and the established sign conventions, σ_x is positive, whereas σ_y and τ_{xy} are negative. Thus, the given values for use in Eqs. 7-1 and 7-2 are

$$\sigma_x = +80 \text{ MPa} \qquad \sigma_y = -100 \text{ MPa} \qquad \text{and} \qquad \tau_{xy} = -60 \text{ MPa}$$

(a) The angle θ for plane AB is $+42°$, as shown in Fig. 7-3c. Thus, from Eqs. 7-1 and 7-2

$$\sigma_n = 80 \cos^2(+42°) + (-100) \sin^2(+42°)$$
$$+ 2(-60) \sin(+42°) \cos(+42°)$$
$$= -60.3 \text{ MPa} = \underline{60.3 \text{ MPa } C} \qquad \text{Ans.}$$

$$\tau_{nt} = -(80 + 100) \sin(+42°) \cos(+42°)$$
$$+ (-60)[\cos^2(+42°) - \sin^2(+42°)]$$
$$= \underline{-95.8 \text{ MPa}} \qquad \text{Ans.}$$

(b) The angle θ for plane CD is $-48°$ (or $+132°$), as shown in Fig. 7-3d. Thus, from Eqs. 7-1 and 7-2

$$\sigma_n = 80 \cos^2(-48°) + (-100) \sin^2(-48°)$$
$$+ 2(-60) \sin(-48°) \cos(-48°)$$
$$= +40.3 \text{ MPa} = \underline{40.3 \text{ MPa } T} \qquad \text{Ans.}$$

$$\tau_{nt} = -(80 + 100) \sin(-48°) \cos(-48°)$$
$$+ (-60)[\cos^2(-48°) - \sin^2(-48°)]$$
$$= \underline{+95.8 \text{ MPa}} \qquad \text{Ans.}$$

Since planes AB and CD are orthogonal, the shearing stresses on the two planes must be equal in magnitude to satisfy Eq. 1-11. Also, one of the stresses must tend to produce a clockwise rotation of the element, while the other tends to produce a counterclockwise rotation. With the coordinate systems shown in Figs. 7-3c and 7-3d, this means that the shearing stresses calculated using Eq. 7-2 will have opposite signs on the two planes.

---

[1] Coordinate systems and axes are tools that may be used to advantage by the analyst. Machine components and structural elements do not come inscribed with x–y axes; therefore, the analyst is free to select directions that are convenient for his work.

PROBLEMS

Note: Use the stress transformation equations for plane stress (Eqs. 7-1 and 7-2) to solve the following problems.

7-1* The stresses shown in Fig. P7-1 act at a point on the free surface of a stressed body. Determine the normal and shearing stresses at this point on the inclined plane *AB* shown in the figure.

7-2* The stresses shown in Fig. P7-2 act at a point on the free surface of a stressed body. Determine the normal and shearing stresses at this point on the inclined plane *AB* shown in the figure.

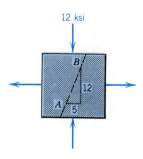

FIG. P7-1

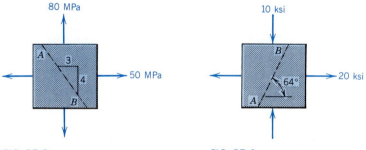

FIG. P7-2   FIG. P7-3

7-3 The stresses shown in Fig. P7-3 act at a point on the free surface of a stressed body. Determine the normal and shearing stresses at this point on the inclined plane *AB* shown in the figure.

7-4 The stresses shown in Fig. P7-4 act at a point on the free surface of a stressed body. Determine the normal and shearing stresses at this point on the inclined plane *AB* shown in the figure.

7-5* The stresses shown in Fig. P7-5 act at a point on the free surface of a stressed body. Determine the normal and shearing stresses at this point on the inclined plane *AB* shown in the figure.

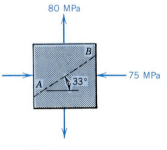

FIG. P7-4

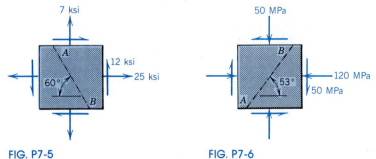

FIG. P7-5   FIG. P7-6

7-6* The stresses shown in Fig. P7-6 act at a point on the free surface

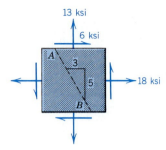

FIG. P7-7

of a stressed body. Determine the normal and shearing stresses at this point on the inclined plane *AB* shown in the figure.

7-7 The stresses shown in Fig. P7-7 act at a point on the free surface of a stressed body. Determine the normal and shearing stresses at this point on the inclined plane *AB* shown in the figure.

7-8 The stresses shown in Fig. P7-8 act at a point on the free surface of a stressed body. Determine the normal and shearing stresses at this point on the inclined plane *AB* shown in the figure.

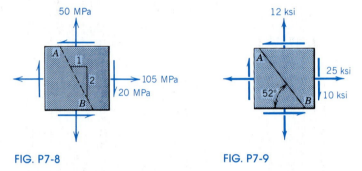

FIG. P7-8   FIG. P7-9

7-9* The stresses shown in Fig. P7-9 act at a point on the free surface of a stressed body. Determine the normal and shearing stresses at this point on the inclined plane *AB* shown in the figure.

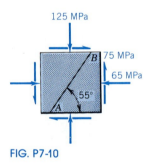

FIG. P7-10

7-10* The stresses shown in Fig. P7-10 act at a point on the free surface of a stressed body. Determine the normal and shearing stresses at this point on the inclined plane *AB* shown in the figure.

7-11 The stresses shown in Fig. P7-11*a* act at a point on the free surface of a stressed body. Determine the normal and shearing stresses at this point on the inclined plane *AB* shown in the figure.

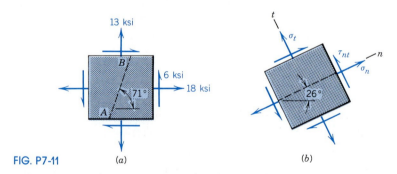

FIG. P7-11   (*a*)   (*b*)

7-12 The stresses shown in Fig. P7-12*a* act at a point on the free surface of a stressed body. Determine the normal and shearing stresses at this point on the inclined plane *AB* shown in the figure.

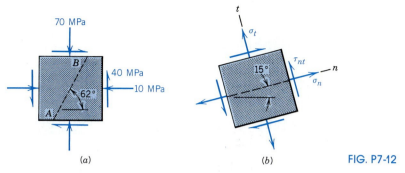

(a)   (b)   FIG. P7-12

7-13* The stresses shown in Fig. P7-11a act at a point on the free surface of a stressed body. Determine the normal stresses σ_n and σ_t and the shearing stress τ_{nt} at this point if they act on the rotated stress element shown in Fig. P7-11b.

7-14* The stresses shown in Fig. P7-12a act at a point on the free surface of a stressed body. Determine the normal stresses σ_n and σ_t and the shearing stress τ_{nt} at this point if they act on the rotated stress element shown in Fig. P7-12b.

7-15 The stresses shown in Fig. P7-15 act at a point on the free surface of a machine component. Determine the normal stresses σ_n and σ_t and the shearing stress τ_{nt} at the point.

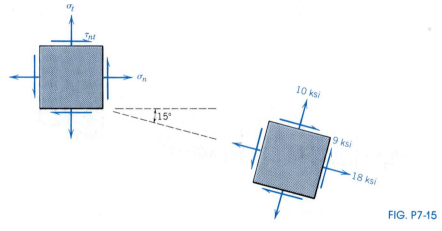

FIG. P7-15

7-16 The stresses shown in Fig. P7-16 act at a point on the free surface of a structural member. Determine the normal stresses σ_n and σ_t and the shearing stress τ_{nt} at the point.

7-17* The stresses shown in Fig. P7-17 act at a point on the free surface of a steel ($E = 30,000$ ksi and $\nu = 0.30$) machine component. Determine the normal strain ϵ_n that would be indicated by the strain gage shown in the figure.

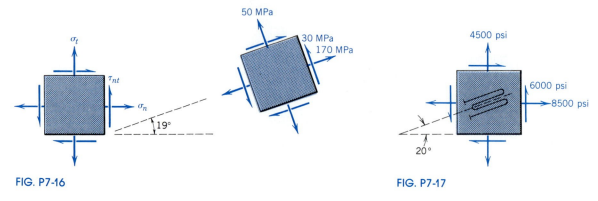

FIG. P7-16

FIG. P7-17

FIG. P7-18

7-18 The stresses shown in Fig. P7-18 act at a point on the free surface of an aluminum alloy ($E = 70$ GPa and $\nu = 0.33$) machine component. Determine the normal strain ϵ_n that would be indicated by the strain gage shown in the figure.

7-3

PRINCIPAL STRESSES AND MAXIMUM SHEARING STRESS

The transformation equations for plane stress (Eqs. 7-1 and 7-2) provide a means for determining the normal stress σ_n and the shearing stress τ_{nt} on different planes through a point in a stressed body. As an example, consider the state of stress at a point on the free surface of a machine component or structural member shown in Fig. 7-4a. As the element is rotated through an angle θ about an axis perpendicular to the free surface, the normal stress σ_n and the shearing stress τ_{nt} on the different planes vary continuously as shown in Fig. 7-4b. For design purposes, the critical stresses at the point are usually the maximum tensile stress and the maximum shearing stress.

For a centric load in one direction and for a shaft in pure torsion, the planes on which maximum normal stresses and maximum shearing stresses act are known from the developments of Section 1-6 and 3-4, respectively. For more complicated forms of loading, these stresses can be determined by plotting curves similar to those shown in Fig. 7-4b for each different state of stress encountered, but this process is time-consuming and inefficient. Therefore, more general methods for finding the critical stresses have been developed.

The transformation equations for plane stress developed in Section 7-2 are:

For normal stress σ_n:

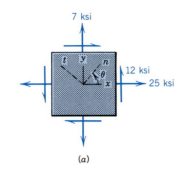

(a)

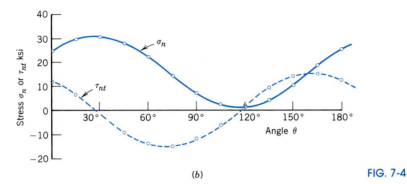

(b)

FIG. 7-4

$$\sigma_n = \sigma_x \cos^2 \theta + \sigma_y \sin^2 \theta + 2\tau_{xy} \sin \theta \cos \theta \qquad (7\text{-}1a)$$

or, in terms of the double angle,

$$\sigma_n = \frac{\sigma_x + \sigma_y}{2} + \frac{\sigma_x - \sigma_y}{2} \cos 2\theta + \tau_{xy} \sin 2\theta \qquad (7\text{-}1b)$$

For shear stress τ_{nt}:

$$\tau_{nt} = -(\sigma_x - \sigma_y)\sin \theta \cos \theta + \tau_{xy}(\cos^2 \theta - \sin^2 \theta) \qquad (7\text{-}2a)$$

or, in terms of the double angle,

$$\tau_{nt} = -\frac{\sigma_x - \sigma_y}{2} \sin 2\theta + \tau_{xy} \cos 2\theta \qquad (7\text{-}2b)$$

Maximum and minimum values of σ_n will occur at the values of θ for which $d\sigma_n/d\theta$ is equal to zero. Differentiation of σ_n with respect to θ yields

$$\frac{d\sigma_n}{d\theta} = -(\sigma_x - \sigma_y)\sin 2\theta + 2\tau_{xy} \cos 2\theta \qquad (a)$$

Setting Eq. *a* equal to zero and solving gives

$$\tan 2\theta_p = \frac{\tau_{xy}}{(\sigma_x - \sigma_y)/2} \tag{7-3}$$

Note that the expression for $d\sigma_n/d\theta$ from Eq. a is numerically twice the value of the expression for τ_{nt} from Eq. 7-2b. Consequently, the shearing stress is zero on planes experiencing maximum and minimum values of normal stress. Planes free of shear stress are known as *principal planes*. The normal stresses occurring on principal planes are known as *principal stresses*. The values of θ_p from Eq. 7-3 give the orientation of two principal planes. A third principal plane for the plane-stress state has an outward normal in the z direction. For a given set of values of σ_x, σ_y, and τ_{xy}, there are two values of $2\theta_p$ differing by 180° and, consequently, two values of θ_p 90° apart. This proves that the principal planes are normal to each other.

When τ_{xy} and $(\sigma_x - \sigma_y)$ have the same sign, $\tan 2\theta_p$ is positive and one value of $2\theta_p$ is between 0° and 90°, with the other value 180° greater, as shown in Fig. 7-5. Consequently, one value of θ_p is between 0° and 45°, and the other one is 90° greater. In the first case, both $\sin 2\theta_p$ and $\cos 2\theta_p$ are positive, and in the second case both are negative. When these functions of $2\theta_p$ are substituted into Eq. 7-1b, two in-plane principal stresses σ_{p1} and σ_{p2} are found to be

$$\sigma_{p1,\ p2} = \frac{\sigma_x + \sigma_y}{2} + \frac{\sigma_x - \sigma_y}{2}\left[\pm \frac{(\sigma_x - \sigma_y)/2}{\sqrt{\left(\dfrac{\sigma_x - \sigma_y}{2}\right)^2 + \tau_{xy}^2}} \right]$$

$$+ \ \tau_{xy}\left[\pm \frac{\tau_{xy}}{\sqrt{\left(\dfrac{\sigma_x - \sigma_y}{2}\right)^2 + \tau_{xy}^2}} \right]$$

which reduces to

$$\sigma_{p1,\ p2} = \frac{\sigma_x + \sigma_y}{2} \pm \sqrt{\left(\frac{\sigma_x - \sigma_y}{2}\right)^2 + \tau_{xy}^2} \tag{7-4}$$

Equation 7-4 gives the two principal stresses in the xy plane, and the third one is $\sigma_{p3} = \sigma_z = 0$. Equation 7-3 gives the angles θ_p and $\theta_p + 90°$ between the x (or y) plane and the mutually perpendicular planes on which the principal stresses act. When $\tan 2\theta_p$ is positive, θ_p is positive, and the rotation is counterclockwise from the x and y planes to the planes on which the two principal stresses act. When $\tan 2\theta_p$ is negative, the rotation is clockwise. Note that one value of θ_p will always be between positive and negative 45° (inclusive) with the other value 90° greater. The numerically greater principal stress will act on the plane that makes an angle of 45° or less with the plane of the numerically larger of the two given normal stresses. This statement

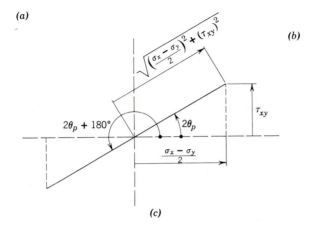

FIG. 7-5

can be confirmed in any case by substitution of the value of θ_p from Eq. 7-3 into Eq. 7-1b to obtain the corresponding principal stress.

Note that if one or both of the principal stresses from Eq. 7-4 is negative, the algebraic maximum stress can have a smaller absolute value than the "minimum" stress.

The *maximum in-plane shearing stress* τ_p occurs on planes located by values of θ where $d\tau_{nt}/d\theta$ is equal to zero. Differentiation of Eq. 7-2b yields

$$\frac{d\tau_{nt}}{d\theta} = -(\sigma_x - \sigma_y)\cos 2\theta - 2\tau_{xy} \sin 2\theta$$

When $d\tau_{nt}/d\theta$ is equated to zero, the value of θ_τ is given by the expression

$$\tan 2\theta_\tau = -\frac{(\sigma_x - \sigma_y)/2}{\tau_{xy}} \qquad (b)$$

Comparison of Eqs. b and 7-3 reveals that the two tangents are negative reciprocals. Therefore, the two angles $2\theta_p$ and $2\theta_\tau$ differ by 90°, and θ_p and θ_τ are 45° apart. This means that the planes on which the maximum in-plane shearing stresses occur are 45° from the principal planes. The maximum in-plane shearing stresses are found by substituting values of angle functions obtained from Eq. b in Eq. 7-2b. The results are

$$\tau_p = -\frac{\sigma_x - \sigma_y}{2}\left[\pm \frac{(\sigma_x - \sigma_y)/2}{\sqrt{\left(\dfrac{\sigma_x - \sigma_y}{2}\right)^2 + \tau_{xy}^2}}\right]$$

$$+ \tau_{xy}\left[\pm \frac{\tau_{xy}}{\sqrt{\left(\dfrac{\sigma_x - \sigma_y}{2}\right)^2 + \tau_{xy}^2}}\right]$$

which reduces to

$$\tau_p = \pm \sqrt{\left(\frac{\sigma_x - \sigma_y}{2}\right)^2 + \tau_{xy}^2} \qquad (7\text{-}5)$$

Equation 7-5 has the same magnitude as the second term of Eq. 7-4.

A useful relation between the principal stresses and the maximum in-plane shearing stress is obtained from Eqs. 7-4 and 7-5 by subtracting the values for the two in-plane principal stresses and substituting the value of the radical from Eq. 7-5. The result is

$$\tau_p = (\sigma_{p1} - \sigma_{p2})/2 \qquad (7\text{-}6)$$

or, in words, the maximum value of τ_{nt} is equal in magnitude to one-half the difference between the two in-plane principal stresses.

In general, when stresses act in three directions it can be shown (see Section 7-10) that there are three orthogonal planes on which the shearing stress is zero. These planes are known as the principal planes, and the stresses acting on them (the principal stresses) will have three values: one maximum, one minimum, and a third stress between the other two. The maximum shearing stress, τ_{max} on any plane that could be passed through the point, is one-half the difference between the maximum and minimum principal stresses and acts on planes that bisect the angles between the planes of the maximum and minimum normal stresses:

$$\tau_{max} = \frac{\sigma_{max} - \sigma_{min}}{2} \qquad (7\text{-}7)$$

When a state of plane stress exists, one of the principal stresses is zero. If the values of σ_{p1} and σ_{p2} from Eq. 7-4 have the same sign, then the third principal stress, σ_{p3} equals zero, will be either the maximum or the minimum normal stress. Thus, the maximum shearing stress may be $(\sigma_{p1} - \sigma_{p2})/2$, $(\sigma_{p1} - 0)/2$, or $(\sigma_{p2} - 0)/2$, depending on the relative magnitudes and signs of the principal stresses. These three possibilities are illustrated in Fig. 7-6, in which one of the two orthogonal planes on which the maximum shearing stress acts is hatched for each example.

The direction of the maximum shearing stress can be determined by drawing a wedge-shaped block with two sides parallel to the planes having the maximum and minimum principal stresses, and with the third side at an angle of 45° with the other two sides. The direction of the maximum shearing stress must oppose the larger of the two principal stresses.

Another useful relation between the principal stresses and the normal stresses on the orthogonal planes, shown in Fig. 7-7, is obtained by

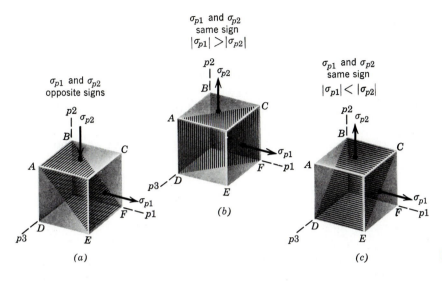

FIG. 7-6

adding the values for the two principal stresses as given by Eq. 7-4. The result is

$$\sigma_{p1} + \sigma_{p2} = \sigma_x + \sigma_y \qquad (7\text{-}8)$$

or, in words, *for plane stress, the sum of the normal stresses on any two orthogonal planes through a point in a body is a constant or invariant.*

In the preceding discussion, "maximum" and "minimum" stresses were considered as algebraic quantities, and it has already been pointed out that the minimum algebraic stress may have a larger magnitude than the maximum stress. However, in the application to engineering problems (which includes the problems in this book), the term *maximum* will always refer to the largest absolute value (largest magnitude).

The application of the formulas and procedures developed in this section are illustrated by the following examples.

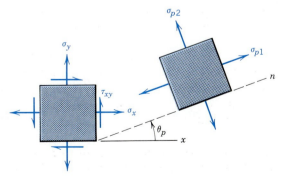

FIG. 7-7

EXAMPLE 7-2

At a point in a structural member subjected to plane stress there are normal and shearing stresses on horizontal and vertical planes through the point, as shown in Fig. 7-8a.

   (a) Determine the principal stresses and the maximum shearing stress at the point.

   (b) Locate the planes on which these stresses act and show the stresses on a complete sketch.

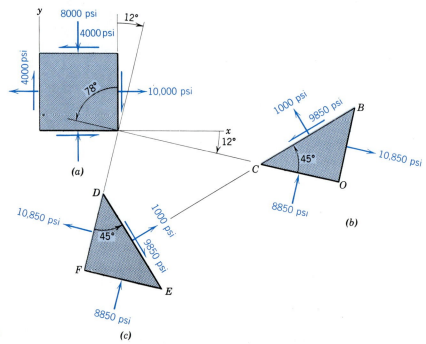

FIG. 7-8

Solution

   (a) On the basis of the axes shown in Fig. 7-8a and the established sign conventions, σ_x is positive, whereas σ_y and τ_{xy} are negative. For use in Eqs. 7-3 and 7-4, the given values are

$$\sigma_x = +10{,}000 \text{ psi}, \qquad \sigma_y = -8000 \text{ psi}, \qquad \text{and}$$

$$\tau_{xy} = -4000 \text{ psi}$$

When these values are substituted in Eq. 7-4, the principal stresses are found to be

$$\sigma_p = \frac{10,000 - 8000}{2} \pm \sqrt{\left(\frac{10,000 + 8000}{2}\right)^2 + (-4000)^2}$$

$$= 1000 \pm 9850$$

$$\sigma_{p1} = +10,850 \, \text{psi} = 10,850 \, \text{psi} \, T \qquad \text{Ans.}$$

$$\sigma_{p2} = -8850 \, \text{psi} = 8850 \, \text{psi} \, C \qquad \text{Ans.}$$

$$\sigma_{p3} = \sigma_z = 0 \qquad \text{Ans.}$$

Since σ_{p1} and σ_{p2} are of opposite sign, the maximum shearing stress is

$$\tau_{max} = \frac{10,850 - (-8850)}{2} = 9850 \, \text{psi} \qquad \text{Ans.}$$

(b) When the given data are substituted in Eq. 7-3, the results are

$$2 c\gamma / \sigma x - \sigma y$$

$$\tan 2\theta_p = -4000/9000 = -0.4444$$

from which

$$2\theta_p = -23.96°$$

and

$$\theta_p = -11.98° \text{ or } 12° \quad = 90 - 12 = 78°$$

The required sketch is shown in Fig. 7-8b or c.

From Eq. 7-8, the sum of the normal stresses on any two orthogonal planes is a constant for plane stress. Planes CB and DE of Fig. 7-8 are orthogonal, and the normal stresses on them are obviously equal. Therefore,

$$2\sigma_n = \sigma_x + \sigma_y = 10,000 - 8000 = 2000$$

and

$$\sigma_n = 1000 \, \text{psi on the planes of maximum shear}$$

EXAMPLE 7-3

At a point in a structural member subjected to plane stress there are normal and shearing stresses on horizontal and vertical planes through the point, as shown in Fig. 7-9a.

(a) Determine the principal stresses and the maximum shearing stress at the point.

(b) Locate the planes on which these stresses act and show the stresses on a complete sketch.

Solution

(a) On the basis of the axes shown in Fig. 7-9a and the established sign conventions, σ_x is positive, σ_y is positive, and τ_{xy} is negative. For use in Eqs. 7-3 and 7-4, the given values are

$$\sigma_x = +100 \text{ MPa}, \qquad \sigma_y = +80 \text{ MPa}, \qquad \text{and}$$
$$\tau_{xy} = -40 \text{ MPa}$$

When these values are substituted in Eq. 7-4, the principal stresses are found to be

$$\sigma_p = \frac{100 + 80}{2} \pm \sqrt{\left(\frac{100 - 80}{2}\right)^2 + (-40)^2}$$

$$= 90 \pm 41.23$$

$$\sigma_{p1} = \underline{131.2 \text{ MPa } T} \qquad\qquad \text{Ans.}$$

$$\sigma_{p2} = \underline{48.8 \text{ MPa } T} \qquad\qquad \text{Ans.}$$

$$\sigma_{p3} = \sigma_z = \underline{0} \qquad\qquad\qquad \text{Ans.}$$

Since σ_{p1} and σ_{p2} have the same sign, the maximum shearing stress is

$$\tau_{max} = (131.2 - 0)/2 = \underline{65.6 \text{ MPa}} \qquad \text{Ans.}$$

(b) When the given data are substituted in Eq. 7-3, the results are

$$\tan 2\theta_p = -40/10 = -4$$

from which

$$2\theta_p = -75.97°$$

and

$$\theta_p = -37.985° \text{ or } 38° \circlearrowleft$$

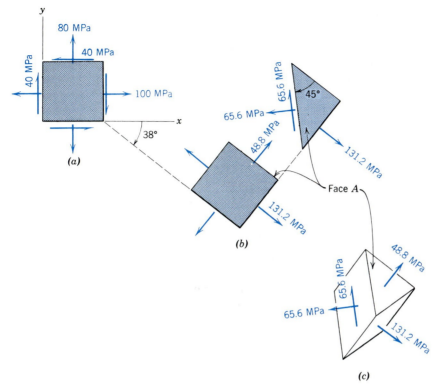

FIG. 7-9

The maximum shearing stress occurs on the plane making an angle of 45° with the planes of maximum and minimum normal stress—in this case, 131.2 MPa and zero. The complete sketch is given in Fig. 7-9b, where the upper (wedge-shaped) block is the orthographic projection of the lower block. The three-dimensional Fig. 7-9c is presented as an aid to the visualization of Fig. 7-9b.

PROBLEMS

7-19* At a point in a structural member subjected to plane stress there are normal and shearing stresses on horizontal and vertical planes through the point, as shown in Fig. P7-19.

(a) Determine the principal stresses and the maximum shearing stress at the point.

(b) Locate the planes on which these stresses act and show the stresses on a complete sketch.

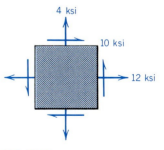

FIG. P7-19

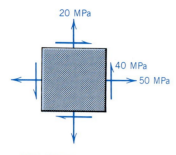

FIG. P7-20

7-20* At a point in a structural member subjected to plane stress there are normal and shearing stresses on horizontal and vertical planes through the point, as shown in Fig. P7-20.

(a) Determine the principal stresses and the maximum shearing stress at the point.

(b) Locate the planes on which these stresses act and show the stresses on a complete sketch.

7-21* At a point in a structural member subjected to plane stress there are normal and shearing stresses on horizontal and vertical planes through the point, as shown in Fig. P7-21. Determine and show on a sketch the principal and maximum shearing stresses at the point.

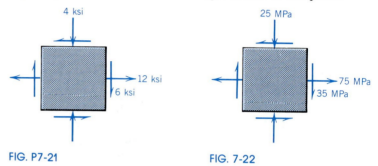

FIG. P7-21   FIG. 7-22

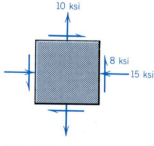

FIG. P7-23

7-22* At a point in a structural member subjected to plane stress there are normal and shearing stresses on horizontal and vertical planes through the point, as shown in Fig. P7-22. Determine and show on a sketch the principal and maximum shearing stresses at the point.

7-23 At a point in a structural member subjected to plane stress there are normal and shearing stresses on horizontal and vertical planes through the point, as shown in Fig. P7-23. Determine and show on a sketch the principal and maximum shearing stresses at the point.

7-24 At a point in a structural member subjected to plane stress there are normal and shearing stresses on horizontal and vertical planes through the point, as shown in Fig. P7-24. Determine and show on a sketch the principal and maximum shearing stresses at the point.

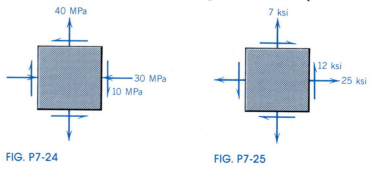

FIG. P7-24   FIG. P7-25

7-25* At a point in a structural member subjected to plane stress there are normal and shearing stresses on horizontal and vertical planes through the point, as shown in Fig. P7-25. Determine and show on a sketch the principal and maximum shearing stresses at the point.

7-26* At a point in a structural member subjected to plane stress there are normal and shearing stresses on horizontal and vertical planes through the point, as shown in Fig. P7-26. Determine and show on a sketch the principal and maximum shearing stresses at the point.

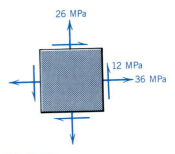

FIG. P7-26

7-27 At a point in a structural member subjected to plane stress there are normal and shearing stresses on horizontal and vertical planes through the point, as shown in Fig. P7-27. Determine and show on a sketch the principal and maximum shearing stresses at the point.

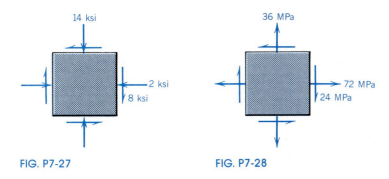

FIG. P7-27   FIG. P7-28   FIG. P7-29

7-28 At a point in a structural member subjected to plane stress there are normal and shearing stresses on horizontal and vertical planes through the point, as shown in Fig. P7-28. Determine and show on a sketch the principal and maximum shearing stresses at the point.

7-29 At a point in a structural member subjected to plane stress there are normal and shearing stresses on horizontal and vertical planes through the point, as shown in Fig. P7-29. Determine and show on a sketch the principal and maximum shearing stresses at the point.

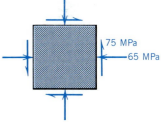

FIG. P7-30

7-30 At a point in a structural member subjected to plane stress there are normal and shearing stresses on horizontal and vertical planes through the point, as shown in Fig. P7-30. Determine and show on a sketch the principal and maximum shearing stresses at the point.

7-31* The stresses shown in Fig. P7-31 act at a point on the free surface of a stressed body. The principal stresses at the point are 20 ksi C and 12 ksi T. Determine the unknown normal stresses on the horizontal and vertical planes and the angle θ_p between the x axis and the maximum tensile stress at the point.

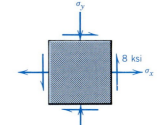

FIG. P7-31

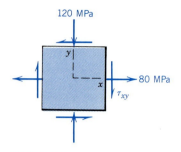

FIG. P7-32

7-32* The stresses shown in Fig. P7-32 act at a point on the free surface of a stressed body. The magnitude of the maximum shearing stress at the point is 125 MPa. Determine the principal stresses, the magnitude of the unknown shearing stress on the vertical plane, and the angle θ_p between the x axis and the maximum tensile stress at the point.

7-33 The principal compressive stress on a vertical plane through a point in a wooden block is equal to four times the principal compressive stress on a horizontal plane. The plane of the grain is 30° clockwise from the vertical plane. If the normal and shearing stresses on the plane of the grain must not exceed 300 psi C and 125 psi shear, determine the maximum allowable compressive stress on the horizontal plane.

7-34 At a point on the free surface of a stressed body, a normal stress of 64 MPa C and an unknown positive shearing stress exist on a horizontal plane. One principal stress at the point is 8 MPa C. The maximum shearing stress at the point has a magnitude of 95 MPa. Determine the unknown stresses on the horizontal and vertical planes and the unknown principal stress at the point.

7-35* At a point on the free surface of a stressed body, the normal stresses are 20 ksi T on a vertical plane and 30 ksi C on a horizontal plane. An unknown negative shearing stress acts on the vertical plane. The maximum shearing stress at the point has a magnitude of 32 ksi. Determine the principal stresses and the shearing stress on the vertical plane at the point.

7-36* At a point on the free surface of a stressed body, a normal stress of 75 MPa T and an unknown negative shearing stress exist on a horizontal plane. One principal stress at the point is 200 MPa T. The maximum in-plane shearing stress at the point has a magnitude of 85 MPa. Determine the unknown stresses on the vertical plane, the unknown principal stress, and the maximum shearing stress at the point.

7-4

MOHR'S CIRCLE FOR PLANE STRESS

The equations for finding the principal stresses and the maximum shearing stress at a point in a stressed member were developed in Section 7-3. The German engineer Otto Mohr (1835–1918) developed a useful pictorial or graphic interpretation of these equations.[2] This method,

---

[2]The particular case for plane stress was presented by the German engineer K. Culmann (1821–1881) about sixteen years before Mohr's paper.

commonly called Mohr's circle,[3] involves the construction of a circle in such a manner that the coordinates of each point on the circle represent the normal and shearing stresses on one plane through the stressed point, and the angular position of the radius to the point gives the orientation of the plane. The proof that normal and shearing components of stress on an arbitrary plane through a point can be represented as points on a circle follows from the general expressions for normal and shearing stresses derived in Section 7-3. Recall Eqs. 7-1*b* and 7-2*b* and rewrite in the following form:

$$\sigma_n - \frac{\sigma_x + \sigma_y}{2} = \frac{\sigma_x - \sigma_y}{2} \cos 2\theta + \tau_{xy} \sin 2\theta$$

$$\tau_{nt} = -\frac{\sigma_x - \sigma_y}{2} \sin 2\theta + \tau_{xy} \cos 2\theta$$

Squaring both equations, adding, and simplifying yield

$$\left(\sigma_n - \frac{\sigma_x + \sigma_y}{2}\right)^2 + \tau_{nt}^2 = \left(\frac{\sigma_x - \sigma_y}{2}\right)^2 + \tau_{xy}^2$$

This is the equation of a circle in terms of the variables σ_n and τ_{nt}. The circle is centered on the σ axis at a distance $(\sigma_x + \sigma_y)/2$ from the τ axis, and the radius of the circle is given by

$$R = \sqrt{\left(\frac{\sigma_x - \sigma_y}{2}\right)^2 + \tau_{xy}^2}$$

Normal stresses are plotted as horizontal coordinates, with tensile stresses (positive) plotted to the right of the origin and compressive stresses (negative) plotted to the left. Shearing stresses are plotted as vertical coordinates, with those tending to produce a clockwise rotation of the stress element plotted above the σ axis, and those tending to produce a counterclockwise rotation of the stress element plotted below the σ axis. The method for interpreting the sign to be associated with a particular shear stress value obtained from a Mohr's circle analysis will be illustrated in the discussion and example problems that follow.

The Mohr's circle for any point subjected to plane stress can be drawn when the stresses on two mutually perpendicular planes through the point are known. Consider, for example, the stress situation of Fig. 7-2*a* with σ_x greater than σ_y and plot in Fig. 7-10 the points representing

---

[3] Mohr's circle is also used as a pictorial interpretation of the formulas for maximum and minimum second moments of areas, maximum products of inertia, plane strain, three-dimensional stress, and maximum and minimum moments of inertia of mass.

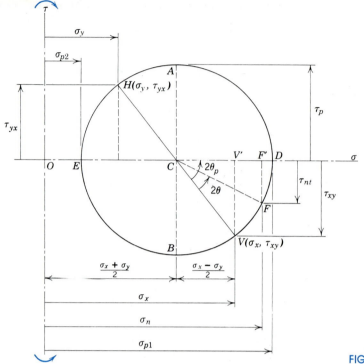

FIG. 7-10

the given stresses. The coordinates on point V of Fig. 7-10 are the stresses on the vertical plane through the stressed point of Fig. 7-2a, and point H is determined by the stresses on the horizontal plane through the point. Line VH is the diameter of Mohr's circle for the stresses at the point. Because τ_{yx} is equal to τ_{xy}, point C, the center of the circle, is on the σ axis. Line CV on Mohr's circle represents the plane (the vertical plane of Fig. 7-2a) through the stressed point from which the angle θ is measured. The coordinates of each point on the circle represent σ_n and τ_{nt} for the particular plane through the stressed point, the abscissa representing σ_n and the ordinate representing τ_{nt}. To demonstrate this statement, draw any radius CF in Fig. 7-10 at an angle 2θ counterclockwise from radius CV. From the figure, it is apparent that

$$OF' = OC + CF \cos(2\theta_p - 2\theta)$$

and since CF equals CV, the above equations reduces to

$$OF' = OC + CV \cos 2\theta_p \cos 2\theta + CV \sin 2\theta_p \sin 2\theta$$

Referring to Fig. 7-10, note that

$$CV \cos 2\theta_p = CV' = (\sigma_x - \sigma_y)/2$$
$$CV \sin 2\theta_p = VV' = \tau_{xy}$$

and

$$OC = (\sigma_x + \sigma_y)/2 = \sigma_{avg}$$

Therefore,

$$OF' = OC + CV' \cos 2\theta + VV' \sin 2\theta$$
$$= \frac{\sigma_x + \sigma_y}{2} + \frac{\sigma_x - \sigma_y}{2} \cos 2\theta + \tau_{xy} \sin 2\theta$$

This expression is identical to Eq. 7-1b of Section 7-3. Therefore, OF' is equal to σ_n. In a similar manner,

$$F'F = CF \sin(2\theta_p - 2\theta)$$
$$= CV \sin 2\theta_p \cos 2\theta - CV \cos 2\theta_p \sin 2\theta$$
$$= V'V \cos 2\theta - CV' \sin 2\theta$$
$$= \tau_{xy} \cos 2\theta - \frac{\sigma_x - \sigma_y}{2} \sin 2\theta = \tau_{nt}$$

which is Eq. 7-2b of Section 7-3.

Since the horizontal coordinate of each point on the circle represents a particular value of σ_n, the maximum normal stress is represented by OD, and its value is

$$\sigma_{p1} = OD = OC + CD = OC + CV$$
$$= \frac{\sigma_x + \sigma_y}{2} + \sqrt{\left(\frac{\sigma_x - \sigma_y}{2}\right)^2 + \tau_{xy}^2}$$

which agrees with Eq. 7-4.

Likewise, the vertical coordinate of each point on the circle represents a particular value of τ_{nt} (called the in-plane shearing stress), which means that the maximum in-plane shearing stresses are represented by CA and CB, and their value is

$$\tau_p = CA = CV = \sqrt{\left(\frac{\sigma_x - \sigma_y}{2}\right)^2 + \tau_{xy}^2}$$

which agrees with Eq. 7-5. As noted in Section 7-3, if the two nonzero principal stresses have the same sign, the maximum shearing stress at the point will not be in the plane of the applied stresses.

The angle $2\theta_p$ from CV to CD is counterclockwise or positive, and its tangent is

$$\tan 2\theta_p = \frac{\tau_{xy}}{(\sigma_x - \sigma_y)/2}$$

which is Eq. 7-3. From the derivation of Eq. 7-4, the angle between the vertical plane and one of the principal planes was θ_p. In obtaining the same equation from Mohr's circle, the angle between the radii representing these same two planes is $2\theta_p$. In other words, all angles in Mohr's circle are twice the corresponding angles for the actual stressed body. The angle from the vertical plane to the horizontal plane in Fig. 7-2a is 90°, but in Fig. 7-10 the angle between line CV (which represents the vertical plane) and line CH (which represents the horizontal plane) in Mohr's circle is 180°.

The results obtained from Mohr's circle have been shown to be identical with the equations derived from the free-body diagram of Section 7-3. Consequently, Mohr's circle provides an extremely useful aid in the solution for, and visualization of, the stresses on various planes through a point in a stressed body in terms of the stresses on two mutually perpendicular planes through the point. Although Mohr's circle can be drawn to scale and used to obtain values of stresses and angles by direct measurements on the figure, it is probably more useful as a pictorial aid to the analyst who is performing analytical determinations of stresses and their directions at the point.

When the state of stress at a point is specified by means of a sketch of a small element, the procedure for drawing and using Mohr's circle to obtain specific stress information can be briefly summarized as follows:

1. Choose a set of x–y reference axes.
2. Identify the stresses σ_x, σ_y, and $\tau_{xy} = \tau_{yx}$ and list them with the proper sign.
3. Draw a set of σ–τ coordinate axes with σ and τ positive to the right and upward, respectively.
4. Plot the point $(\sigma_x, -\tau_{xy})$ and label it point V (vertical plane).
5. Plot the point (σ_y, τ_{yx}) and label it point H (horizontal plane).
6. Draw a line between V and H. This establishes the center C and the radius R of Mohr's circle.
7. Draw the circle.
8. An extension of the radius between C and V can be identified as the x axis or the reference line for angle measurements (i.e., $\theta = 0°$).

By plotting points V and H as $(\sigma_x, -\tau_{xy})$ and (σ_y, τ_{yx}), respectively, shear stresses that tend to rotate the stress element clockwise will plot above the σ axis, while those tending to rotate the element counterclockwise will plot below the σ axis. The use of a negative sign with one of the shearing stresses (τ_{xy} or τ_{yx}) is required for plotting purposes, since for a given state of stress, the shearing stresses ($\tau_{xy} = \tau_{yx}$) have only one sign (both are positive or both are negative). The use of the negative sign at point V on Mohr's circle brings the direction of angular measurements 2θ on Mohr's circle into agreement with the direction of angular measurements θ on the stress element.

Once the circle has been drawn, the normal and shearing stresses on an arbitrary inclined plane AB having an outward normal n that is oriented at an angle θ with respect to the reference x axis can be obtained from the coordinates at a point F (see Fig. 7-10) on the circle that is located at angular position 2θ from the reference axis through point V. The coordinates of point F must be interpreted as stresses σ_n and $-\tau_{nt}$. Other points on Mohr's circle that provide stresses of interest are

1. Point D that provides the principal stress σ_{p1}.
2. Point E that provides the principal stress σ_{p2}.
3. Point A that provides the maximum in-plane shearing stress $-\tau_p$ and the accompanying normal stress σ_{avg} that acts on the plane.

A negative sign must be used when interpreting shearing stresses τ_{nt} and τ_p obtained from the circle since a shearing stress tending to produce a clockwise rotation of the stress element is a negative shearing stress when a right-hand n–t coordinate system is used.

Problems of the type presented in Section 7-2 as well as those of Section 7-3 can readily be solved by this semigraphic method, as will be illustrated in the following examples.

EXAMPLE 7-4

At a point in a structural member subjected to plane stress there are normal and shearing stresses on horizontal and vertical planes through the point, as shown in Fig. 7-11a. Determine and show the following on a sketch.

(a) The principal and maximum shearing stresses at the point.
(b) The normal and shearing stresses on plane A–A through the point.

Solution

Mohr's circle is constructed from the given data as follows: on a set of coordinate axes (Fig. 7-11b) plot point V (representing the stresses on the vertical plane) at $(8, -4)$ because the stresses on the vertical

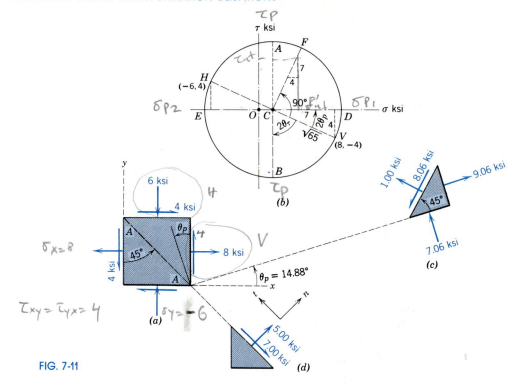

FIG. 7-11

plane are 8 ksi T and 4 ksi (counterclockwise) shear. Likewise, point H (representing the stresses on the horizontal plane) has the coordinates $(-6, 4)$. Draw line HV, which is a diameter of Mohr's circle, and note that the center of the circle is at $(1, 0)$. The radius of the circle is

$$CV = \sqrt{7^2 + 4^2} = 8.06 \text{ ksi}$$

(a) The principal stresses and the maximum shearing stress at the point are

$$\sigma_{p1} = OD = 1 + 8.06 = +9.06 = \underline{9.06 \text{ ksi } T} \qquad \text{Ans.}$$

$$\sigma_{p2} = OE = 1 - 8.06 = -7.06 = \underline{7.06 \text{ ksi } C} \qquad \text{Ans.}$$

$$\sigma_{p3} = \sigma_z = \underline{0} \qquad \text{Ans.}$$

Since σ_{p1} and σ_{p2} have opposite signs, the maximum shearing stress is

$$\tau_p = \tau_{\max} = CA = CB = \underline{8.06 \text{ ksi}} \qquad \text{Ans.}$$

The principal planes are represented by lines CD and CE, where

$$\tan 2\theta_p = 4/7 = 0.5714$$

which gives

$$2\theta_p = 29.75°$$

Since the angle $2\theta_p$ is counterclockwise, the principal planes are counterclockwise from the vertical and horizontal planes of the stress block, as shown in Fig. 7-11c. To determine which principal stress acts on which plane, note that as the radius of the circle rotates counterclockwise, the end of the radius CV moves from V to D, indicating that as the initially vertical plane rotates through 14.88°, the stresses change to 9.06 ksi T and zero. Note also that the end H or radius CH moves to E, indicating that as the initially horizontal plane rotates through 14.88°, the stresses change to 7.06 ksi C and zero. The required sketch is Fig. 7-11c.

(b) Plane A–A is 45° counterclockwise from the vertical plane; therefore, the corresponding radius of Mohr's circle is 90°, 2θ, counterclockwise from the line CV, and is shown as CF on Fig. 7-11b. The coordinates of the point F are seen to be (5, 7), which means that the stresses on plane A–A are 5.00 ksi T and 7.00 ksi shear in a direction to produce a clockwise couple on a stress element having plane A–A as one of its faces. This shear stress would be defined as a negative shear stress since to produce a clockwise couple, it must be directed in the negative t direction associated with plane A–A (see Fig. 7-11d). Note that Fig. 7-11d does not show all the stresses acting on the element.

$$\tau_{nt} = FF'$$
$$= CF \sin(90 - 2\theta_p)$$
$$= 8.06 \sin(90 - 29.75)$$
$$= 7 \text{ ksi}$$

$$\sigma_n = CF \cos(90 - 2\theta_p) + C$$
$$= 8.06 \cos(90 - 2\theta_p) + 1$$
$$= 5 \text{ ksi}$$

EXAMPLE 7-5

At a point in a structural member subjected to plane stress there are normal and shearing stresses on horizontal and vertical planes through the point, as shown in Fig. 7-12a. Determine, and show on a sketch, the principal stresses and the maximum shearing stress at the point.

Solution

Mohr's circle (Fig. 7-12b) is constructed from the given data by plotting point V (representing the stresses on the vertical plane) at (72, 24) and point H (representing the stresses on the horizontal plane) at (36, −24). Line HV drawn between the two points is a diameter of Mohr's circle. The circle is centered at point C (54, 0) and has a radius CV equal to 30 MPa. The principal stresses at the point are

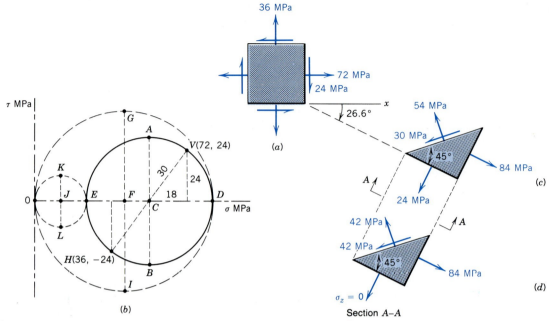

FIG. 7-12

$$\sigma_{p1} = OD = 54 + 30 = \underline{84 \text{ MPa T}} \qquad \text{Ans.}$$

$$\sigma_{p2} = OE = 54 - 30 = \underline{24 \text{ MPa T}} \qquad \text{Ans.}$$

$$\sigma_{p3} = \sigma_z = \underline{0} \qquad \text{Ans.}$$

The principal planes are represented by lines CD and CE, where

$$\tan 2\theta_p = \frac{-24}{28} = -1.3333$$

which gives

$$2\theta_p = -53.13° = 53.13°$$
$$\theta_p = -26.57° = 26.6°\curvearrowright$$

Since θ_{p1} and σ_{p2} have the same sign, the maximum in-plane shearing stress τ_p is not the maximum shearing stress τ_{max} at the point. The maximum shearing stress at the point can be represented on Mohr's circle by drawing an additional circle (shown dashed in Fig. 7-12b) that has line OD as a diameter. This circle is centered at point F (42, 0) and has a radius FG equal to 42 MPa. This circle represents the combinations of normal and shearing stresses existing on planes obtained by rotating the element about the principal axis associated with the principal stress σ_{p2}. A third Mohr's circle (shown dotted in Fig. 7-12b) has line OE as a diameter. This circle is centered at point J (12, 0) and has a radius JK equal to 12 MPa. This circle represents the combinations of normal and shearing stresses existing on planes obtained by rotating the element about the principal axis associated with the principal stress σ_{p1}. Thus

$$\tau_p = CA = 30 \text{ MPa}$$

$$\tau_{\max} = FG = \underline{42 \text{ MPa}} \qquad \text{Ans.}$$

The principal stresses σ_{p1}, σ_{p2}, and $\sigma_z = \sigma_{p3} = 0$, the maximum in-plane shearing stress τ_p, and the maximum shearing stress $\tau_{\max}$ at the point are all shown on Figs. 7-12c and 7-12d.

PROBLEMS

Note: The bodies in the following problems are assumed to be in a state of plane stress. For interpretation of signs on shearing stresses, the horizontal (x) and vertical (y) directions are positive to the right and upward, respectively.

7-37* The stresses shown in Fig. P7-37 act at a point on the free surface of a stressed body. Use Mohr's circle to determine the normal and shearing stresses at this point on the inclined plane *AB* shown in the figure.

7-38* The stresses shown in Fig. P7-38 act at a point on the free surface of a stressed body. Use Mohr's circle to determine the normal and shearing stresses at this point on the inclined plane *AB* shown in the figure.

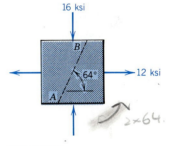

FIG. P7-37

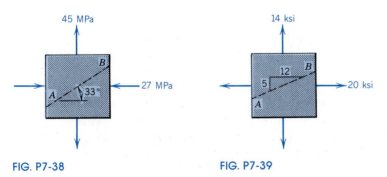

FIG. P7-38

FIG. P7-39

7-39 The stresses shown in Fig. P7-39 act at a point on the free surface of a stressed body. Use Mohr's circle to determine the normal and shearing stresses at this point on the inclined plane *AB* shown in the figure.

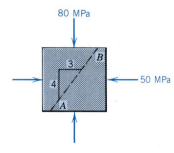

FIG. P7-40

7-40 The stresses shown in Fig. P7-40 act at a point on the free surface of a stressed body. Use Mohr's circle to determine the normal and shearing stresses at this point on the inclined plane *AB* shown in the figure.

7-41* At a point in a structural member subjected to plane stress there are normal and shearing stresses on horizontal and vertical planes through the point, as shown in Fig. P7-41. Use Mohr's circle to determine

(a) the principal stresses and the maximum shearing stress at the point.

(b) the normal and shearing stresses on the inclined plane *AB* shown in the figure.

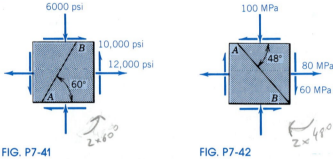

FIG. P7-41   FIG. P7-42

7-42* At a point in a structural member subjected to plane stress there are normal and shearing stresses on horizontal and vertical planes through the point, as shown in Fig. P7-42. Use Mohr's circle to determine

(a) the principal stresses and the maximum shearing stress at the point.

(b) the normal and shearing stresses on the inclined plane *AB* shown in the figure.

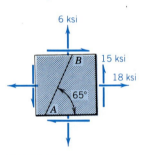

FIG. P7-43

7-43 At a point in a structural member subjected to plane stress there are normal and shearing stresses on horizontal and vertical planes through the point, as shown in Fig. P7-43. Use Mohr's circle to determine

(a) the principal stresses and the maximum shearing stress at the point.

(b) the normal and shearing stresses on the inclined plane *AB* shown in the figure.

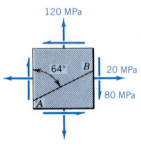

FIG. P7-44

7-44 At a point in a structural member subjected to plane stress there are normal and shearing stresses on horizontal and vertical planes through the point, as shown in Fig. P7-44. Use Mohr's circle to determine

(a) the principal stresses and the maximum shearing stress at the point.

(b) the normal and shearing stresses on the inclined plane AB shown in the figure.

7-45* At a point in a structural member subjected to plane stress there are normal and shearing stresses on horizontal and vertical planes through the point, as shown in Fig. P7-45. Use Mohr's circle to determine

(a) the principal stresses and the maximum shearing stress at the point.

(b) the normal and shearing stresses on the inclined plane AB shown in the figure.

7-46* At a point in a structural member subjected to plane stress there are normal and shearing stresses on horizontal and vertical planes through the point, as shown in Fig. P7-46. Use Mohr's circle to determine

(a) the principal stresses and the maximum shearing stress at the point.

(b) the normal and shearing stresses on the inclined plane AB shown in the figure.

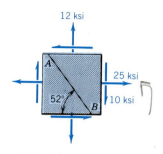

FIG. P7-45

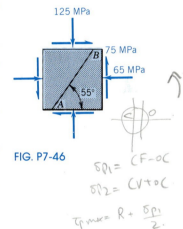

FIG. P7-46

τ_{max}

$\sigma_{p1} = (F - o($

$\sigma_{p2} = (V + o($

$\tau_{pmax} = R + \dfrac{\sigma_{p1}}{2}$

7-47 At a point in a stressed body, the nonzero principal stresses are oriented as shown in Fig. P7-47. Use Mohr's circle to determine

(a) the stresses on plane $a–a$.

(b) the stresses on horizontal and vertical planes at the point.

(c) the maximum shearing stress at the point.

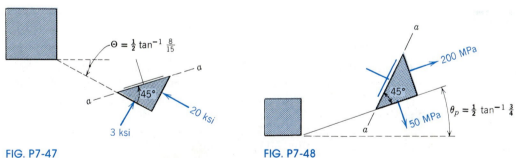

FIG. P7-47   FIG. P7-48

7-48 At a point in a stressed body, the nonzero principal stresses are oriented as shown in Fig. P7-48. Use Mohr's circle to determine

(a) the stresses on plane $a–a$.

(b) the stresses on horizontal and vertical planes at the point.

(c) the maximum shearing stress at the point.

7-49* At a point on the free surface of a stressed body, the normal stresses are 18 ksi T on a vertical plane and 9 ksi T on a horizontal plane. A negative shearing stress of 6 ksi acts on the vertical plane at the point. Use Mohr's circle to determine

(a) the principal and maximum shearing stresses at the point.

(b) the stresses on a plane 30° clockwise from the vertical plane.

7-50* At a point on the free surface of a stressed body, the normal stresses are 54 MPa T on a vertical plane and 27 MPa T on a horizontal plane. A positive shearing stress of 18 MPa acts on the vertical plane at the point. Use Mohr's circle to determine

(a) the principal and maximum shearing stresses at the point.

(b) the stresses on a plane 40° counterclockwise from the vertical plane.

7-51 Use Mohr's circle to solve Problem 7-15.

7-52 Use Mohr's circle to solve Problem 7-16.

7-53* Use Mohr's circle to solve Problem 7-13.

7-54* Use Mohr's circle to solve Problem 7-14.

7-55 Use Mohr's circle to solve Problem 7-33.

7-56 Use Mohr's circle to solve Problem 7-34.

7-5

STRAIN TRANSFORMATION EQUATIONS FOR PLANE STRAIN

The method of relating the components of strain associated with a Cartesian coordinate system to the normal and shearing strains associated with other orthogonal directions will be illustrated by considering the *two-dimensional* or *plane strain* case. If the xy plane is taken as the reference plane, then for conditions of plane strain $\epsilon_z = \gamma_{zx} = \gamma_{zy} = 0$. Consider Fig. 7-13, in which the shaded rectangle represents the small unstrained element of material having the configuration of a rectangular parallelepiped. When the body is subjected to a system of loads, the element assumes the shape indicated by the dashed lines in Fig. 7-13. The dimensions of the deformed element are given in terms of strains obtained from Eq. 1-18.

An expression for normal strain in the n direction, ϵ_n, can be obtained by applying the law of cosines to the triangle $OC'B'$ shown in Fig. 7-13. Thus,

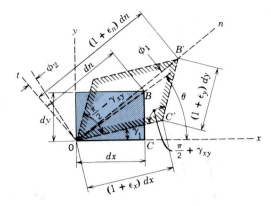

FIG. 7-13

$$(OB')^2 = (OC')^2 + (C'B')^2 - 2(OC')(C'B') \cos \left(\frac{\pi}{2} + \gamma_{xy} \right)$$

or in terms of strains

$$[(1 + \epsilon_n)dn]^2 = [(1 + \epsilon_x)dx]^2 + [(1 + \epsilon_y)dy]^2$$
$$- 2[(1 + \epsilon_x)dx][(1 + \epsilon_y)dy][- \sin \gamma_{xy}] \qquad (a)$$

Substituting $dx = dn \cos \theta$ and $dy = dn \sin \theta$ into Eq. a yields

$$(1 + \epsilon_n)^2(dn)^2 = (1 + \epsilon_x)^2(dn)^2(\cos^2 \theta) + (1 + \epsilon_y)^2(dn)^2(\sin^2 \theta)$$
$$+ 2(dn)^2(\sin \theta)(\cos \theta)(1 + \epsilon_x)(1 + \epsilon_y)(\sin \gamma_{xy}) \qquad (b)$$

Since the strains are small, it follows that $\epsilon^2 << \epsilon$, $\sin \gamma \approx \gamma$, and so forth; hence, all second-degree terms such as ϵ^2, $\gamma\epsilon$, and the like, can be neglected as Eq. b is expanded to become

$$1 + 2\epsilon_n = (1 + 2\epsilon_x) \cos^2 \theta + (1 + 2\epsilon_y) \sin^2 \theta + 2\gamma_{xy} \sin \theta \cos \theta$$

which reduces to

$$\epsilon_n = \epsilon_x \cos^2 \theta + \epsilon_y \sin^2 \theta + \gamma_{xy} \sin \theta \cos \theta \qquad (7\text{-}9a)$$

or in terms of the double angle

$$\epsilon_n = \frac{\epsilon_x + \epsilon_y}{2} + \frac{\epsilon_x - \epsilon_y}{2} \cos 2\theta + \frac{\gamma_{xy}}{2} \sin 2\theta \qquad (7\text{-}9b)$$

An expression for the shearing strain γ_{nt} can be obtained by applying the law of sines to the triangle $OC'B'$ (shown in Fig. 7-13) to obtain the rotations ϕ_1 and ϕ_2 associated with the n and t directions. Thus,

$$\frac{(1 + \epsilon_n)\, dn}{\sin\left(\dfrac{\pi}{2} + \gamma_{xy}\right)} = \frac{(1 + \epsilon_y)\, dy}{\sin(\theta + \phi_1 - \gamma_1)} \tag{c}$$

Since the strains are small,

$$\frac{(1 + \epsilon_y)\, dy}{(1 + \epsilon_n)\, dn} \approx (1 + \epsilon_y - \epsilon_n) \sin \theta$$

and

$$\frac{\sin(\theta + \phi_1 - \gamma_1)}{\sin\left(\dfrac{\pi}{2} + \gamma_{xy}\right)} \approx \sin \theta + (\phi_1 - \gamma_1) \cos \theta$$

Hence, Eq. c can be reduced to

$$(\epsilon_y - \epsilon_n) \sin \theta = (\phi_1 - \gamma_1) \cos \theta \tag{d}$$

Substituting Eq. 7-9a into Eq. d and solving for ϕ_1 yields

$$\phi_1 = -(\epsilon_x - \epsilon_y) \sin \theta \cos \theta - \gamma_{xy} \sin^2 \theta + \gamma_1 \tag{e}$$

The rotation ϕ_2 associated with the t direction is obtained by substituting $\pi/2 + \theta$ for θ in Eq. e. Thus,

$$\phi_2 = (\epsilon_x - \epsilon_y) \sin \theta \cos \theta - \gamma_{xy} \cos^2 \theta + \gamma_1$$

The shearing strain γ_{nt} is equal to the difference between the rotations ϕ_1 and ϕ_2. Therefore,

$$\gamma_{nt} = -2(\epsilon_x - \epsilon_y) \sin \theta \cos \theta + \gamma_{xy}(\cos^2 \theta - \sin^2 \theta) \tag{7-10a}$$

or in terms of the double angle

$$\gamma_{nt} = -(\epsilon_x - \epsilon_y) \sin 2\theta + \gamma_{xy} \cos 2\theta \tag{7-10b}$$

Equations 7-9 and 7-10 provide a means for determining the normal strain ϵ_n associated with a line oriented in an arbitrary n direction in the x–y plane and the shearing strain γ_{nt} associated with any two orthogonal lines oriented in the n and t directions in the x–y plane when the strains ϵ_x, ϵ_y, and γ_{xy} associated with the coordinate directions are known. When these equations are used, the sign conventions used in their development must be rigorously followed. The sign conventions used are as follows:

1. Tensile strains are positive; compressive strains are negative.
2. Shearing strains that decrease the angle between the two lines at the origin of coordinates are positive.
3. Angles measured counterclockwise from the reference x axis are positive.
4. The (n, t, z) axes have the same order as the (x, y, z) axes. Both sets of axes form a right-hand coordinate system.

PROBLEMS

7-57* The strain components at a point are $\epsilon_x = +800\mu$, $\epsilon_y = -1000\mu$, and $\gamma_{xy} = -600\mu$. Determine the strain components ϵ_n, ϵ_t, and γ_{nt} if the $x-y$ and $n-t$ axes are oriented as shown in Fig. P7-57.

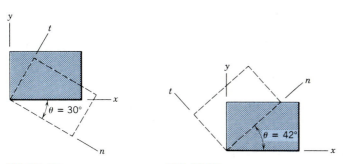

FIG. P7-57   FIG. P7-58

7-58* The strain components at a point are $\epsilon_x = +900\mu$, $\epsilon_y = +650\mu$, and $\gamma_{xy} = +300\mu$. Determine the strain components ϵ_n, ϵ_t, and γ_{nt} if the $x-y$ and $n-t$ axes are oriented as shown in Fig. P7-58.

In Problems 7-59 through 7-64 the strain components ϵ_x, ϵ_y, and γ_{xy} are given for a point in a body subjected to plane strain. Determine the strain components ϵ_n, ϵ_t, and γ_{nt} at the point if the $n-t$ axes are rotated with respect to the $x-y$ axes by the amount and in the direction indicated by the angle θ.

Problem	ϵ_x	ϵ_y	γ_{xy}	θ
7-59	$+800\mu$	-950μ	-800μ	$-42°$
7-60	$+750\mu$	$+360\mu$	-300μ	$+52°$
7-61*	-100μ	-700μ	$+400\mu$	$-28°$
7-62*	$+900\mu$	$+650\mu$	-300μ	$+32°$
7-63	$+720\mu$	-480μ	$+360\mu$	$-30°$
7-64	$+850\mu$	$+450\mu$	$+600\mu$	$+20°$

7-6

PRINCIPAL STRAINS AND MAXIMUM SHEARING STRAIN

The similarity between Eqs. 7-9 and 7-10 for plane strain and Eqs. 7-1 and 7-2 for plane stress in Section 7-2 indicates that all the relationships developed for plane stress can be applied to plane strain by substituting ϵ_x for σ_x, ϵ_y for σ_y, and $\gamma_{xy}/2$ for τ_{xy}. Thus, from Eqs. 7-3, 7-4, and 7-5, expressions are obtained for determining the in-plane principal directions, the in-plane principal strains, and the maximum in-plane shearing strain

$$\tan 2\theta_p = \frac{\gamma_{xy}}{\epsilon_x - \epsilon_y} \tag{7-11}$$

$$\epsilon_{p1}, \epsilon_{p2} = \frac{\epsilon_x + \epsilon_y}{2} \pm \sqrt{\left(\frac{\epsilon_x - \epsilon_y}{2}\right)^2 + \left(\frac{\gamma_{xy}}{2}\right)^2} \tag{7-12}$$

$$\gamma_p = 2\sqrt{\left(\frac{\epsilon_x - \epsilon_y}{2}\right)^2 + \left(\frac{\gamma_{xy}}{2}\right)^2} \tag{7-13}$$

In the previous equations, normal strains that are tensile and shearing strains that decrease the angle between the faces of the element at the origin of coordinates (see Fig. 7-13) are positive.

When a state of plane strain exists, Eq. 7-12 gives the two in-plane principal strains while the third principal strain, ϵ_{p3}, is $\epsilon_z = 0$. An examination of Eqs. 7-12 and 7-13 indicates that the maximum in-plane shearing strain is the difference between the in-plane principal strains, but this may not be the maximum shearing strain at the point. The maximum shearing strain at the point may be $(\epsilon_{p1} - \epsilon_{p2})$, $(\epsilon_{p1} - 0)$, or $(0 - \epsilon_{p2})$, depending on the relative magnitudes and signs of the principal strains. The lines associated with the maximum shearing strain bisect the angles between lines experiencing maximum and minimum normal strains. The three possibilities are illustrated in Fig. 7-14.

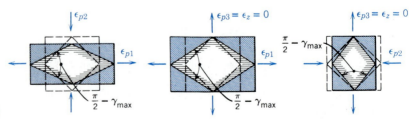

FIG. 7-14

EXAMPLE 7-6

The strain components at a point in a body under a state of plane strain are $\epsilon_x = +1200\mu$, $\epsilon_y = -600\mu$, and $\gamma_{xy} = +900\mu$. Determine the principal strains and the maximum shearing strain at the point. Show the principal strain deformations and the maximum shearing strain distortion on a sketch.

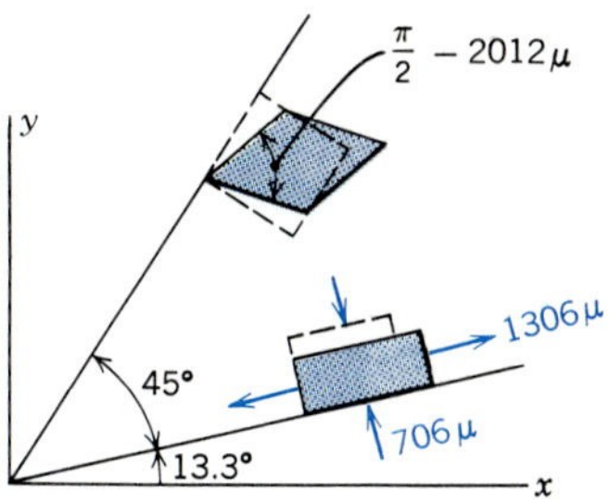

FIG. 7-15

Solution

When the given data are substituted in Eqs. 7-11, 7-12, and 7-13 they yield the in-plane principal strains at the point, their orientations, and the maximum in-plane shearing strain. Thus,

$$\tan 2\theta_p = \frac{900}{1200 + 600}$$

$$(\epsilon_{p1}, \epsilon_{p2})(10^6) = \frac{1200 - 600}{2} \pm \sqrt{\left(\frac{1200 + 600}{2}\right)^2 + \left(\frac{900}{2}\right)^2}$$

$$\gamma_p(10^6) = 2\sqrt{\left(\frac{1200 + 600}{2}\right)^2 + \left(\frac{900}{2}\right)^2}$$

from which

$$\epsilon_{p1} = \underline{+1306\mu}$$

$$\epsilon_{p2} = \underline{-706\mu}$$

$$\epsilon_{p3} = \underline{0} \qquad\qquad \text{Ans.}$$

$$\theta_p = \underline{13.3°}$$

$$\gamma_p = \underline{+2012\mu} = \gamma_{max}$$

The required sketch is shown in Fig. 7-15.

PROBLEMS

In Problems 7-65 through 7-76 the strain components ϵ_x, ϵ_y, and γ_{xy} are given for a point in a body subjected to plane strain. Determine the principal strains and the maximum shearing strain at the point. Show the principal strain deformations and the maximum shearing strain distortion on a sketch.

Problem	ϵ_x	ϵ_y	γ_{xy}
7-65*	$+600\mu$	-200μ	$+500\mu$
7-66*	$+960\mu$	-320μ	-480μ
7-67	$+900\mu$	-300μ	-420μ
7-68	-900μ	$+600\mu$	$+480\mu$
7-69*	-750μ	$+1000\mu$	-250μ
7-70*	$+750\mu$	-410μ	$+360\mu$
7-71	$+720\mu$	$+520\mu$	$+480\mu$
7-72	-540μ	-980μ	-560μ
7-73*	$+864\mu$	$+432\mu$	-288μ
7-74*	$+650\mu$	$+900\mu$	$+300\mu$
7-75	-325μ	-625μ	$+380\mu$
7-76	$+900\mu$	$+650\mu$	$+600\mu$

7-7

MOHR'S CIRCLE FOR PLANE STRAIN

The pictorial or graphic representation of Eqs. 7-1 and 7-2, known as Mohr's circle for stress, can be used with Eqs. 7-9 and 7-10 to yield a Mohr's circle for strain. The equation for the strain circle obtained from the equation for the stress circle by using a change in variables is

$$\left(\epsilon_n - \frac{\epsilon_x + \epsilon_y}{2}\right)^2 + \left(\frac{\gamma_{nt}}{2}\right)^2 = \left(\frac{\epsilon_x - \epsilon_y}{2}\right)^2 + \left(\frac{\gamma_{xy}}{2}\right)^2$$

The variables in this equation are ϵ_n and $\gamma_{nt}/2$. The circle is centered on the ϵ axis at a distance $(\epsilon_x + \epsilon_y)/2$ from the origin and has a radius

$$R = \sqrt{\left(\frac{\epsilon_x - \epsilon_y}{2}\right)^2 + \left(\frac{\gamma_{xy}}{2}\right)^2}$$

Mohr's circle for the strains of Fig. 7-13 (with $\epsilon_x > \epsilon_y$) is given in Fig. 7-16. It is apparent that the sign convention for shearing strain needs to be extended to cover the construction of Mohr's circle. Observe that for positive shearing strain (indicated in Fig. 7-13), the edge of the

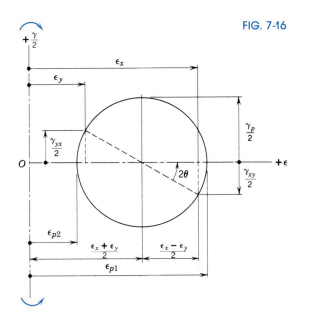

FIG. 7-16

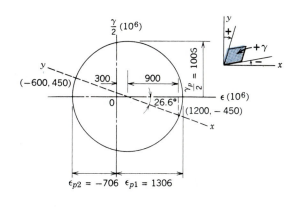

FIG. 7-17

element parallel to the x axis tends to rotate counterclockwise while the edge parallel to the y axis tends to rotate clockwise. For Mohr's circle construction, the clockwise rotation will be designated positive and the counterclockwise designated negative. This is consistent with the sign convention for shearing stresses given in Section 7-4. The Mohr's circle solution for Example 7-6 is shown in Fig. 7-17.

PROBLEMS

In Problems 7-77 through 7-88 certain strain components and angles are given for a point in a body subjected to plane strain. Use Mohr's circle to determine the unknown quantities for each problem and prepare a sketch showing the angle θ_p, the principal strain deformations, and the maximum shearing strain distortions. In some problems there may be more than one possible value of θ_p depending on the sign of γ_{xy}.

Problem	ϵ_x	ϵ_y	γ_{xy}	ϵ_{p1}	ϵ_{p2}	γ_p	γ_{max}	θ_p
✓ 7-77*	500	-300	600	$+600\mu$	-400μ	1000	1000	$+18.43°$
7-78*	—	—	—	$+785\mu$	-945μ	—	—	$+16.85°$
7-79	—	—	—	$+708\mu$	-104μ	—	—	$-34.1°$
7-80	—	—	—	-114μ	-903μ	—	—	$+19.26°$

Problem	ϵ_x	ϵ_y	γ_{xy}	ϵ_{p1}	ϵ_{p2}	γ_p	γ_{max}	θ_p
7-81*	$+950\mu$	-225μ	$+275\mu$	—	—	—	—	—
7-82*	$+900\mu$	-333μ	$+982\mu$	—	—	—	—	—
7-83	-750μ	-390μ	$+900\mu$	—	—	—	—	—
7-84	$+600\mu$	$+480\mu$	-480μ	—	—	—	—	—
7-85*	-680μ	$+320\mu$	—	$+414\mu$	—	—	—	—
7-86*	$+450\mu$	$+150\mu$	—	$+780\mu$	—	—	—	—
7-87	$+360\mu$	$+750\mu$	—	—	$+197\mu$	—	—	—
7-88	-300μ	$+600\mu$	—	—	-522μ	—	—	—

7-8

PRINCIPAL STRAINS ASSOCIATED WITH PLANE STRESS

In most experimental work involving strain measurement, the strains are measured on a free (unstressed) surface of a member where a state of plane stress exists. If the outward normal to the surface is taken as the z direction, then $\sigma_z = \tau_{zx} = \tau_{zy} = 0$. Since this state of stress offers no restraint to out-of-plane deformations, a normal strain ϵ_z develops in addition to the in-plane strains ϵ_x, ϵ_y, and γ_{xy}. The shearing strains γ_{zx} and γ_{zy} remain zero; therefore, the normal strain ϵ_z is a principal strain. In Section 7-6, expressions were developed for the plane strain case relating the in-plane principal strains (ϵ_{p1} and ϵ_{p2}) and their orientations to the in-plane strains ϵ_x, ϵ_y, and γ_{xy}. For the plane-stress case, which involves the normal strain ϵ_z in addition to the in-plane strains, similar expressions are needed.

As an illustration of the effects of an out-of-plane displacement on the deformation (change in length) of a line segment originally located in the xy plane, consider line AB of Fig. 7-18. As a result of the loads imposed on the member, the line AB is displaced and extended into the line $A'B'$. The displacements associated with point A' are u in the x direction, v in the y direction, and w in the z direction. Point B' displaces $u + du$ in the x direction, $v + dv$ in the y direction, and $w + dw$ in the z direction. The deformation δ_{AB} is obtained from the original length of the line and the displacements du, dv, and dw of point B' with respect to A'. Thus,

$$(L + \delta_{AB})^2 = (L + du)^2 + (dv)^2 + (dw)^2$$

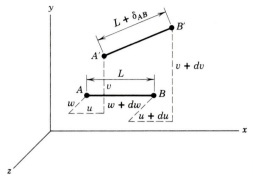

FIG. 7-18

and after squaring both sides, the result is

$$L^2 + 2L\delta_{AB} + \delta_{AB}^2 = L^2 + 2L(du) + (du)^2 + (dv)^2 + (dw)^2$$

If the deformations are small, the second-degree terms can be neglected; hence,

$$\delta_{AB} = du$$

This indicates that the normal strain along AB (δ_{AB} divided by L) is not affected by the presence of the out-of-plane displacements. In fact, none of the in-plane strains are affected; therefore, Eqs. 7-9 and 7-10 and their Mohr's circle representation are valid not only for the plane-strain case but also for the plane-stress case present when strain measurements are made on a free surface.

7-9

STRAIN MEASUREMENT: ROSETTE ANALYSIS

Electrical resistance strain gages have been developed to provide accurate measurements of normal strain. The resistance gage may be a wire gage consisting of a length of 0.001-in.-diameter wire arranged as shown in Fig. 7-19*a* and cemented between two pieces of paper, or it may be an etched foil conductor mounted on epoxy or polyimide backing (see Fig. 7-19*b*). The wire or foil gage is cemented to the material for which the strain is to be determined. As the material is strained, the wires are lengthened or shortened; this changes the electrical resistance of the gage. The change in resistance is measured by means of a Wheatstone bridge, which may be calibrated to directly read strain. Figure 4-1 shows SR-4 wire strain gages mounted on an aluminum

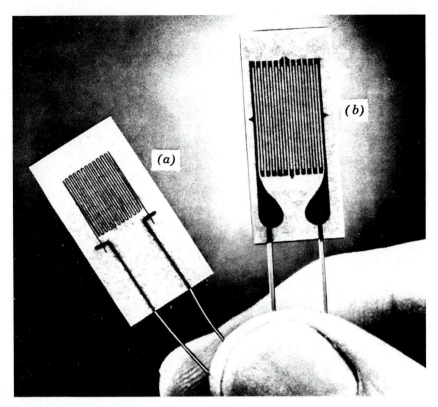

FIG. 7-19

beam. Shearing strains are more difficult to measure directly than normal strains and are often obtained by measuring normal strains in two or three different directions. The shearing strain γ_{xy} can be computed from the normal strain data by using Eq. 7-9a. For example, consider the most general case of three arbitrary normal strain measurements as shown in Fig. 7-20. From Eq. 7-9a

$$\epsilon_a = \epsilon_x \cos^2 \theta_a + \epsilon_y \sin^2 \theta_a + \gamma_{xy} \sin \theta_a \cos \theta_a$$
$$\epsilon_b = \epsilon_x \cos^2 \theta_b + \epsilon_y \sin^2 \theta_b + \gamma_{xy} \sin \theta_b \cos \theta_b$$
$$\epsilon_c = \epsilon_x \cos^2 \theta_c + \epsilon_y \sin^2 \theta_c + \gamma_{xy} \sin \theta_c \cos \theta_c$$

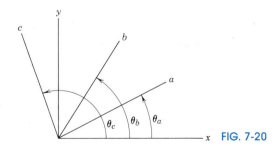

FIG. 7-20

From the measured values of ϵ_a, ϵ_b, and ϵ_c and a knowledge of the gage orientations θ_a, θ_b, and θ_c with respect to the reference x axis, the values of ϵ_x, ϵ_y, and γ_{xy} can be determined by simultaneous solution of the three equations. In practice the angles θ_a, θ_b, and θ_c are selected to simplify the calculations. Multiple-element strain gages used for this type of measurement are known as strain rosettes. Several rosette configurations marketed commercially are shown in Fig. 7-21.

In this book the angles used to identify the normal strain directions of the various elements of a rosette will always be measured counterclockwise from the reference x axis. Once ϵ_x, ϵ_y, and γ_{xy} have been determined, Eqs. 7-11, 7-12, and 7-13 or the corresponding Mohr's circle can be used to determine the in-plane principal strains, their orientations, and the maximum in-plane shearing strain at the point.

The principal strain $\epsilon_z = \epsilon_{p3}$ can be determined from the measured in-plane data. Equations 1-22 and 1-24, which represents Hooke's law for the case of plane stress, are

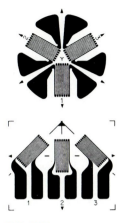

FIG. 7-21

$$\epsilon_x = \frac{1}{E}(\sigma_x - \nu\sigma_y)$$

$$\epsilon_y = \frac{1}{E}(\sigma_y - \nu\sigma_x)$$

$$\epsilon_z = -\frac{\nu}{E}(\sigma_x + \sigma_y) \qquad (1\text{-}22)$$

$$\sigma_x = \frac{E}{1 - \nu^2}(\epsilon_x + \nu\epsilon_y)$$

$$\sigma_y = \frac{E}{1 - \nu^2}(\epsilon_y + \nu\epsilon_x) \qquad (1\text{-}24)$$

Substituting Eq. 1-24 into Eq. 1-22 yields

$$\epsilon_z = -\frac{\nu(\sigma_x + \sigma_y)}{E} = -\frac{\nu}{E}\frac{E}{1 - \nu^2}[(\epsilon_x + \nu\epsilon_y) + (\epsilon_y + \nu\epsilon_x)]$$

$$= -\frac{\nu}{(1 - \nu)(1 + \nu)}[(1 + \nu)\epsilon_x + (1 + \nu)\epsilon_y]$$

$$\epsilon_z = -\frac{\nu}{1 - \nu}(\epsilon_x + \epsilon_y) \qquad (7\text{-}14)$$

This out-of-plane principal strain is important since the maximum shearing strain at the point may be $(\epsilon_{p1} - \epsilon_{p2})$, $(\epsilon_{p1} - \epsilon_{p3})$, or $(\epsilon_{p3} - \epsilon_{p2})$, depending on the relative magnitudes and signs of the principal strains at the point.

The following example illustrates the application of Eqs. 7-9 through 7-14 and Mohr's circle to principal stress and strain and maximum shearing stress and strain determinations under conditions of plane stress.

EXAMPLE 7-7

A strain rosette, composed of three resistance gages making angles of $0°$, $60°$, and $120°$ with the x axis, was mounted on the free surface of a steel machine component ($E = 30,000$ ksi and $\nu = 0.30$). Under load, the following strains were measured:

$$\epsilon_0 = \epsilon_x = +1000\mu \qquad \epsilon_{60} = -650\mu \qquad \epsilon_{120} = +750\mu$$

(a) Determine the principal strains and the maximum shearing strain at the point. Show the directions of the in-plane principal strains on a sketch.

(b) Determine the principal stresses and the maximum shearing stress at the point. Show the principal stresses and the maximum shearing stress on a sketch.

Solution

(a) When the given data are substituted into Eq. 7-9a, the following equations (in terms of μ) are obtained:

$$\epsilon_x = +1000 \text{ (measured)}$$

$$-650 = \frac{1000 + \epsilon_y}{2} + \frac{1000 - \epsilon_y}{2}(\cos 120°) + \frac{\gamma_{xy}}{2}(\sin 120°)$$

$$= \frac{1000 + \epsilon_y}{2} + \frac{1000 - \epsilon_y}{2}\left(-\frac{1}{2}\right) + \frac{\gamma_{xy}}{2}\left(\frac{\sqrt{3}}{2}\right)$$

$$750 = \frac{1000 + \epsilon_y}{2} + \frac{1000 - \epsilon_y}{2}(\cos 240°) + \frac{\gamma_{xy}}{2}(\sin 240°)$$

$$= \frac{1000 + \epsilon_y}{2} + \frac{1000 - \epsilon_y}{2}\left(-\frac{1}{2}\right) + \frac{\gamma_{xy}}{2}\left(-\frac{\sqrt{3}}{2}\right)$$

from which $\epsilon_y = -266.7$ and $\gamma_{xy} = -1616.6$. These values are then used to construct the Mohr's circle of Fig. 7-22. The extensions of the radii to the plotted points ϵ_x, ϵ_y are designated x and y as an aid in determining the principal directions. Since γ_{xy} is negative, the angle between the edges of the element at the origin increases; hence, the edge parallel to the x axis rotates clockwise ($\gamma_{xy}/2$ is plotted positive), and the edge parallel to the y axis rotates counterclockwise ($\gamma_{xy}/2$ is plotted negative). The in-plane principal strains obtained from the circle are $\epsilon_{p1} = 1394\mu$ and $\epsilon_{p2} = -660\mu$. The principal strain perpendicular to the surface, $\epsilon_z = \epsilon_{p3}$, is found from Eq. 7-14 as follows:

$$\epsilon_z = \epsilon_{p3} = -\frac{0.30}{1 - 0.30}(1000\mu - 267\mu) = -314\mu$$

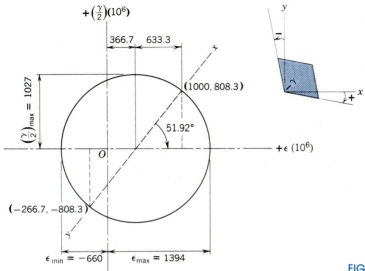

FIG. 7-22

Since ϵ_{p3} is less compressive than the in-plane principal strain ϵ_{p2}, the maximum shearing strain at the point is the maximum in-plane shearing strain γ_p, obtained from the circle, as

$$\gamma_{max} = \gamma_p = \pm(1394\mu + 660\mu) = \pm 2054\mu = \underline{\pm 2050\mu} \quad \text{Ans.}$$

The principal strains are

$$\epsilon_{p1} = \epsilon_{max} = \underline{+1394\mu}$$
$$\epsilon_{p3} = \epsilon_{int} = \underline{-314\mu} \qquad \text{Ans.}$$
$$\epsilon_{p2} = \epsilon_{min} = \underline{-660\mu}$$

as shown in the accompanying sketch.

(b) The principal stresses and the maximum shearing stress at the point are obtained from the strain results of part (a) by using Eqs. 1-24 and 1-28. Thus, from Eq. 1-24,

$$\sigma_{max} = \frac{30(10^6)}{1 - 0.09}[1394 + 0.3(-660)](10^{-6})$$

$$\sigma_{max} = \frac{E}{1-v^2}\left(\epsilon_{max} + v\,\epsilon_{max}\right).$$

$$= +39,400 \text{ psi} = \underline{39.4 \text{ ksi } T} \quad \text{Ans.}$$

$$\sigma_{min} = \frac{30(10^6)}{1 - 0.09}[-660 + 0.3(1394)](10^{-6})$$

$$= -7980 \text{ psi} = \underline{7.98 \text{ ksi } C} \quad \text{Ans.}$$

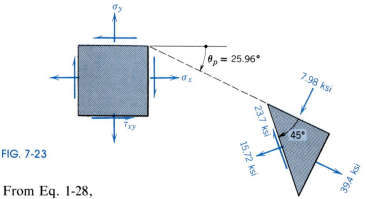

FIG. 7-23

From Eq. 1-28,

$$\tau_{max} = \frac{E}{2(1+v)}\,\gamma_{max}$$

$$\tau_{max} = \frac{30(10^6)}{2(1 + 0.3)}\,(2054)(10^{-6}) = 23,700 \text{ psi} = \underline{23.7 \text{ ksi}}\quad \text{Ans.}$$

Note that the value for τ_{max} checks that given by Eq. 7-7; thus,

$$\tau_{max} = \frac{\sigma_{p1} - \sigma_{p2}}{2} = \frac{39,400 + 7980}{2} = 23,690 \text{ psi} = 23.7 \text{ ksi}$$

The directions of the principal stresses will be the same as those for the principal strains; hence, from Eq. 7-22, the directions are obtained for the sketch of Fig. 7-23.

Alternate Solution

The Cartesian components of strain obtained from the original rosette data were

$$\epsilon_x = 1000\mu \qquad \epsilon_y = -266.7\mu \qquad \gamma_{xy} = -1616.6\mu$$

The Cartesian components of stress at the point can be obtained from these strains by using Eqs. 1-24 and 1-28. Thus, from Eq. 1-24,

$$\sigma_x = \frac{30(10^6)}{1 - 0.09}[1000 + 0.3(-266.7)](10^{-6}) = +30,330 \text{ psi} = 30.3 \text{ ksi T}$$

$$\sigma_y = \frac{30(10^6)}{1 - 0.09}[-266.7 + 0.3(1000)](10^{-6}) = +1098 \text{ psi} = 1.098 \text{ ksi T}$$

and from Eq. 1-28,

$$\tau_{xy} = \frac{30(10^6)}{2(1 + 0.3)}(-1616.6)(10^{-6}) = -18,650 \text{ psi} = -18.65 \text{ ksi}$$

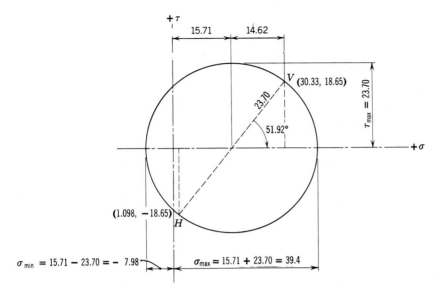

$\sigma_{min} = 15.71 - 23.70 = -7.98$

$\sigma_{max} = 15.71 + 23.70 = 39.4$

FIG. 7-24

The Mohr's circle of Fig. 7-24 is constructed from these data and the following results are obtained:

$$\sigma_{max} = \underline{39.4 \text{ ksi } T} \qquad \text{Ans.}$$
$$\sigma_{min} = \underline{7.98 \text{ ksi } C} \qquad \text{Ans.}$$
$$\tau_{max} = \underline{23.7 \text{ ksi}} \qquad \text{Ans.}$$
$$\theta_p = \underline{25.96°} \qquad \text{Ans.}$$

The complete answer is given in Fig. 7-23.

PROBLEMS

Note: In the following problems, when the material is named but the properties are not given, refer to Table A-1 in Appendix A for the properties. Use Eq. 1-27 to determine Poisson's ratio when data from Table A-1 are used.

7-89* At a point on the free surface of a steel ($E = 30,000$ ksi and $\nu = 0.30$) machine part, the strain rosette shown in Fig. P7-89 was used to obtain the following normal strain data: $\epsilon_a = +750\mu$, $\epsilon_b = -125\mu$, and $\epsilon_c = -250\mu$. Determine

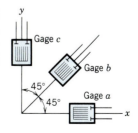

FIG. P7-89

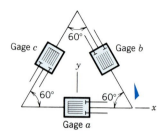

FIG. P7-90

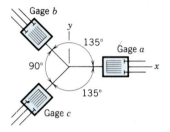

FIG. P7-91

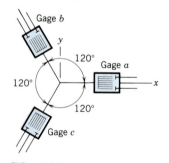

FIG. P7-92

(a) the strain components ϵ_x, ϵ_y, and γ_{xy} at the point.

(b) the principal strains and the maximum shearing strain at the point.

7-90* At a point on the free surface of an aluminum alloy ($E = 73$ GPa and $\nu = 0.33$) machine part, the strain rosette shown in Fig. P7-90 was used to obtain the following normal strain data: $\epsilon_a = +780\mu$, $\epsilon_b = +345\mu$, and $\epsilon_c = -332\mu$. Determine

(a) the strain components ϵ_x, ϵ_y, and γ_{xy} at the point.

(b) the principal strains and the maximum shearing strain at the point.

7-91 At a point on the free surface of an aluminum alloy ($E = 10,600$ ksi and $\nu = 0.33$) machine part, the strain rosette shown in Fig. P7-91 was used to obtain the following normal strain data: $\epsilon_a = +875\mu$, $\epsilon_b = +700\mu$, and $\epsilon_c = -350\mu$. Determine

(a) the strain components ϵ_x, ϵ_y, and γ_{xy} at the point.

(b) the stress components σ_x, σ_y, and τ_{xy} at the point.

(c) the principal strains and the maximum shearing strain at the point.

(d) the principal stresses and the maximum shearing stress at the point.

7-92 At a point on the free surface of a steel ($E = 200$ GPa and $\nu = 0.30$) machine part, the strain rosette shown in Fig. P7-92 was used to obtain the following normal strain data: $\epsilon_a = -555\mu$, $\epsilon_b = +925\mu$, and $\epsilon_c = +740\mu$. Determine

(a) the strain components ϵ_x, ϵ_y, and γ_{xy} at the point.

(b) the stress components σ_x, σ_y, and τ_{xy} at the point.

(c) the principal strains and the maximum shearing strain at the point.

(d) the principal stresses and the maximum shearing stress at the point.

7-93* Use Mohr's circle to determine the strains that would be indicated by the strain gages shown in Fig. P7-89 if the two in-plane principal strains at the point are $\epsilon_{p1} = +1035\mu$ and $\epsilon_{p2} = -115\mu$. The principal strain ϵ_{p1} is located 30° counterclockwise from the x axis.

7-94* Use Mohr's circle to determine the strains that would be indicated by the strain gages shown in Fig. P7-90 if the two in-plane principal strains at the point are $\epsilon_{p1} = +1080\mu$ and $\epsilon_{p2} = +120\mu$. The principal strain ϵ_{p1} is located 30° clockwise from the x axis.

7-95 Use Mohr's circle to determine the strains that would be indicated by the strain gages shown in Fig. P7-91 if the two in-plane principal strains at the point are $\epsilon_{p1} = +1260\mu$ and $\epsilon_{p2} = -140\mu$. The principal strain ϵ_{p1} is located 20° counterclockwise from the x axis.

7-96 Use Mohr's circle to determine the strains that would be indicated by the strain gages shown in Fig. P7-92 if the two in-plane principal strains at the point are $\epsilon_{p1} = +966\mu$ and $\epsilon_{p2} = -414\mu$. The principal strain ϵ_{p1} is located 25° clockwise from the x axis.

7-97* The strain rosette shown in Fig. P7-97 was used to obtain normal strain data at a point on the free surface of a structural steel machine part. The gage readings were: $\epsilon_a = -350\mu$, $\epsilon_b = +1000\mu$, and $\epsilon_c = +550\mu$. Determine

(a) the strain components ϵ_x, ϵ_y, and γ_{xy} at the point.
(b) the stress components σ_x, σ_y, and τ_{xy} at the point.
(c) the principal stresses and the maximum shearing stress at the point. Express the stresses in U.S. Customary units.

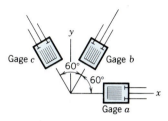

FIG. P7-97

7-98* The strain rosette shown in Fig. P7-98 was used to obtain normal strain data at a point on the free surface of a 2024-T4 aluminum alloy machine part. The gage readings were: $\epsilon_a = +525\mu$, $\epsilon_b = +450\mu$, and $\epsilon_c = +1425\mu$. Determine

(a) the strain components ϵ_x, ϵ_y, and γ_{xy} at the point.
(b) the stress components σ_x, σ_y, and τ_{xy} at the point.
(c) the principal stresses and the maximum shearing stress at the point. Express the stresses in SI units.

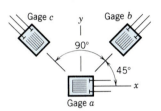

FIG. P7-98

7-99 The strain rosette shown in Fig. P7-99 was used to obtain normal strain data at a point on the free surface of an 18-8 cold-rolled stainless-steel machine part. The gage readings were: $\epsilon_a = +665\mu$, $\epsilon_b = +390\mu$, and $\epsilon_c = +870\mu$. Determine

(a) the strain components ϵ_x, ϵ_y, and γ_{xy} at the point.
(b) the stress components σ_x, σ_y, and τ_{xy} at the point.
(c) the principal stresses and the maximum shearing stress at the point. Express the stresses in U.S. Customary units.

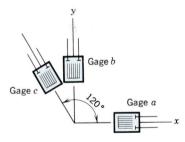

FIG. P7-99

7-100 The strain rosette shown in Fig. P7-100 was used to obtain normal strain data at a point on the free surface of an annealed titanium alloy machine part. The gage readings were: $\epsilon_a = -306\mu$, $\epsilon_b = -456\mu$, and $\epsilon_c = +906\mu$. Determine

(a) the strain components ϵ_x, ϵ_y, and γ_{xy} at the point.
(b) the stress components σ_x, σ_y, and τ_{xy} at the point.
(c) the principal stresses and the maximum shearing stress at the point. Express the stresses in SI units.

7-101* At a point on the free surface of a steel ($E = 30,000$ ksi and $\nu = 0.30$) machine part, the strain rosette shown in Fig. P7-101 was

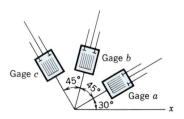

FIG. P7-100

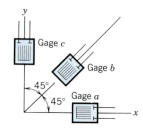

FIG. P7-101

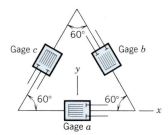

FIG. P7-102

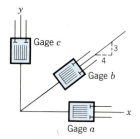

FIG. P7-103

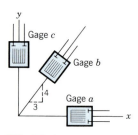

FIG. P7-104

used to obtain the following normal strain data: $\epsilon_a = +650\mu$, $\epsilon_b = +475\mu$, and $\epsilon_c = -250\mu$. Determine

(a) the stress components σ_x, σ_y, and τ_{xy} at the point.

(b) the principal strains and the maximum shearing strain at the point. Prepare a sketch showing all of these strains.

(c) the principal stresses and the maximum shearing stress at the point. Prepare a sketch showing all of these stresses.

7-102* At a point on the free surface of an aluminum alloy ($E = 73$ GPa and $\nu = 0.33$) machine part, the strain rosette shown in Fig. P7-102 was used to obtain the following normal strain data: $\epsilon_a = +875\mu$, $\epsilon_b = +700\mu$, and $\epsilon_c = -650\mu$. Determine

(a) the stress components σ_x, σ_y, and τ_{xy} at the point.

(b) the principal strains and the maximum shearing strain at the point. Prepare a sketch showing all of these strains.

(c) the principal stresses and the maximum shearing stress at the point. Prepare a sketch showing all of these stresses.

7-103 At a point on the free surface of an aluminum alloy ($E = 10,600$ ksi and $\nu = 0.33$) machine part, the strain rosette shown in Fig. P7-103 was used to obtain the following normal strain data: $\epsilon_a = +800\mu$, $\epsilon_b = +950\mu$, and $\epsilon_c = +600\mu$. Determine

(a) the stress components σ_x, σ_y, and τ_{xy} at the point.

(b) the principal strains and the maximum shearing strain at the point. Prepare a sketch showing all of these strains.

(c) the principal stresses and the maximum shearing stress at the point. Prepare a sketch showing all of these stresses.

7-104 At a point on the free surface of a steel ($E = 200$ GPa and $\nu = 0.30$) machine part, the strain rosette shown in Fig. P7-104 was used to obtain the following normal strain data: $\epsilon_a = +875\mu$, $\epsilon_b = +700\mu$, and $\epsilon_c = +350\mu$. Determine

(a) the stress components σ_x, σ_y, and τ_{xy} at the point.

(b) the principal strains and the maximum shearing strain at the point. Prepare a sketch showing all of these strains.

(c) the principal stresses and the maximum shearing stress at the point. Prepare a sketch showing all of these stresses.

7-10

GENERAL STATE OF STRESS AT A POINT

The general (three-dimensional) state of stress at a point was previously introduced in Section 1-7. This state of stress has three normal stress

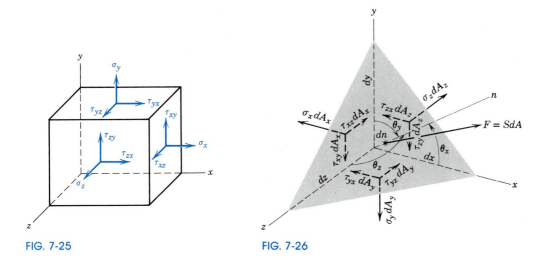

FIG. 7-25   FIG. 7-26

components and six shearing stress components, as illustrated in Fig. 7-25. The shearing stress components shown on Fig. 7-25 are not all independent, however, since moment equilibrium (Eq. 1-11) requires that

$$\tau_{xy} = \tau_{yx}$$
$$\tau_{yz} = \tau_{zy} \qquad (1\text{-}11)$$
$$\tau_{zx} = \tau_{xz}$$

The stresses shown on Fig. 7-25 are all positive according to the sign conventions for normal and shearing stresses outlined in Section 1-7.

Expressions for the stresses on any oblique plane through the point, in terms of stresses on the reference planes, can be developed with the aid of the free-body diagram shown in Fig. 7-26, where the n axis is normal to the oblique (shaded) face. The areas of the faces of the element are dA for the oblique face and $dA \cos \theta_x$, $dA \cos \theta_y$, and $dA \cos \theta_z$ for the x, y, and z faces, respectively.[4] The resultant force F on the oblique face is $S\, dA$, where S is the resultant stress (stress vector— see Section 1-7) on the area. The resultant stress S is related to the normal and shearing stress components on the oblique face by the expression

$$S = \sqrt{\sigma_n^2 + \tau_{nt}^2} \qquad (7\text{-}15)$$

[4]These relationships can be established by considering the volume of the tetrahedron in Fig. 7-26. Thus, $V = (1/3)\, dn\, dA = (1/3)\, dx\, dA_x = (1/3)\, dy\, dA_y = (1/3)\, dz\, dA_z$. But $dn = dx \cos \theta_x = dy \cos \theta_y = dz \cos \theta_z$; therefore, $dA_x = dA \cos \theta_x$, $dA_y = dA \cos \theta_y$, $dA_z = dA \cos \theta_z$.

The forces on the x, y, and z faces are shown as three components, the magnitude of each being the product of the area by the appropriate stress. If we use l, m, and n for $\cos \theta_x$, $\cos \theta_y$, and $\cos \theta_z$, respectively, the three force equations of equilibrium in the x, y, and z directions are

$$F_x = S_x \, dA = \sigma_x \, dA \, l + \tau_{yx} \, dA \, m + \tau_{zx} \, dA \, n$$
$$F_y = S_y \, dA = \sigma_y \, dA \, m + \tau_{zy} \, dA \, n + \tau_{xy} \, dA \, l$$
$$F_z = S_z \, dA = \sigma_z \, dA \, n + \tau_{xz} \, dA \, l + \tau_{yz} \, dA \, m$$

from which the three orthogonal components of the resultant stress are

$$S_x = \sigma_x l + \tau_{yx} m + \tau_{zx} n$$
$$S_y = \tau_{xy} l + \sigma_y m + \tau_{zy} n \qquad (a)$$
$$S_z = \tau_{xz} l + \tau_{yz} m + \sigma_z n$$

The normal component σ_n of the resultant stress S equals $S_x l + S_y m + S_z n$; therefore, from Eq. a, the following equation for the normal stress on any oblique plane through the point is obtained:

$$\sigma_n = \sigma_x l^2 + \sigma_y m^2 + \sigma_z n^2 + 2\tau_{xy} lm + 2\tau_{yz} mn + 2\tau_{zx} nl \quad (7\text{-}16)$$

The shearing stress τ_{nt} on the oblique plane can be obtained from the relation $S^2 = \sigma_n^2 + \tau_{nt}^2$. For a given problem, the values of S and σ_n will be obtained from Eqs. a and 7-16.

A principal plane was previously defined as a plane on which the shearing stress τ_{nt} is zero. The normal stress σ_n on such a plane was defined as a principal stress σ_p. If the oblique plane of Fig. 7-26 is a *principal plane*, then $S = \sigma_p$ and $S_x = \sigma_p l$, $S_y = \sigma_p m$, $S_z = \sigma_p n$. When these components are substituted into Eq. a, the equations can be rewritten to produce the following homogeneous linear equations in l, m, and n:

$$(\sigma_p - \sigma_x)l - \tau_{yx} m - \tau_{zx} n = 0$$
$$(\sigma_p - \sigma_y)m - \tau_{zy} n - \tau_{xy} l = 0 \qquad (b)$$
$$(\sigma_p - \sigma_z)n - \tau_{xz} l - \tau_{yz} m = 0$$

This set of equations has a nontrivial solution only if the determinant of the coefficients of l, m, and n is equal to zero. Thus,

$$\begin{vmatrix} (\sigma_p - \sigma_x) & -\tau_{yx} & -\tau_{zx} \\ -\tau_{xy} & (\sigma_p - \sigma_y) & -\tau_{zy} \\ -\tau_{xz} & -\tau_{yz} & (\sigma_p - \sigma_z) \end{vmatrix} = 0$$

Expansion of the determinant yields the following cubic equation for determining the principal stresses:

$$
\begin{aligned}
\sigma_p^3 &- (\sigma_x + \sigma_y + \sigma_z)\sigma_p^2 \\
&+ (\sigma_x\sigma_y + \sigma_y\sigma_z + \sigma_z\sigma_x - \tau_{xy}^2 - \tau_{yz}^2 - \tau_{zx}^2)\sigma_p \\
&- (\sigma_x\sigma_y\sigma_z - \sigma_x\tau_{yz}^2 - \sigma_y\tau_{zx}^2 - \sigma_z\tau_{xy}^2 + 2\tau_{xy}\tau_{yz}\tau_{zx}) = 0 \quad \text{(7-17)}
\end{aligned}
$$

For given values of σ_x, σ_y, . . . , τ_{zx}, Eq. 7-17 gives three values for the principal stresses σ_{p1}, σ_{p2}, σ_{p3}. By substituting these values for σ_p, in turn, into Eq. *b* and using the relation

$$
l^2 + m^2 + n^2 = 1 \qquad\qquad (c)
$$

three sets of direction cosines may be determined for the normals to the three principal planes. The foregoing discussion verifies the existence of three mutually perpendicular principal planes for the most general state of stress.

In developing equations for maximum and minimum normal stresses, the special case will be considered in which $\tau_{xy} = \tau_{yz} = \tau_{zx} = 0$. No loss in generality is introduced by considering this special case, since it involves only a reorientation of the reference x, y, z axes to coincide with the principal directions. Since the x, y, z planes are now principal planes, the stresses σ_x, σ_y, σ_z become σ_{p1}, σ_{p2}, and σ_{p3}. Solving Eq. *a* for the direction cosines yields

$$
l = S_x/\sigma_{p1} \qquad m = S_y/\sigma_{p2} \qquad n = S_z/\sigma_{p3}
$$

By substituting these values into Eq. *c*, the following equation is obtained:

$$
\frac{S_x^2}{\sigma_{p1}^2} + \frac{S_y^2}{\sigma_{p2}^2} + \frac{S_z^2}{\sigma_{p3}^2} = 1 \qquad\qquad (d)
$$

The plot of Eq. *d* is the ellipsoid shown in Fig. 7-27. It can be observed that the magnitude of σ_n is everywhere less than that of S (since $S^2 = \sigma_n^2 + \tau_{nt}^2$) except at the intercepts, where S is σ_{p1}, σ_{p2}, or σ_{p3}. Therefore, it can be concluded that two of the principal stresses (σ_{p1} and σ_{p3} of Fig. 7-27) are the maximum and minimum normal stresses at the point. The third principal stress is intermediate in value and has no particular significance. The discussion above demonstrates that the set of principal stresses includes the maximum and minimum normal stresses at the point.

Continuing with the special case where the given stresses σ_x, σ_y, and σ_z are principal stresses, we can develop equations for the maxi-

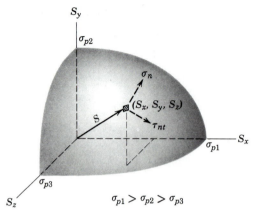

FIG. 7-27

mum shearing stress at the point. The resultant stress on the oblique plane is given by the expression

$$S^2 = S_x^2 + S_y^2 + S_z^2$$

Substitution of values for S_x, S_y, and S_z from Eq. a, with zero shearing stresses, yields the expression

$$S^2 = \sigma_x^2 l^2 + \sigma_y^2 m^2 + \sigma_z^2 n^2 \tag{e}$$

Also, from Eq. 7-16,

$$\sigma_n^2 = (\sigma_x l^2 + \sigma_y m^2 + \sigma_z n^2)^2 \tag{f}$$

Since $S^2 = \sigma_n^2 + \tau_{nt}^2$, an expression for the shearing stress τ_{nt} on the oblique plane is obtained from Eqs. e and f as

$$\tau_{nt} = \sqrt{\sigma_x^2 l^2 + \sigma_y^2 m^2 + \sigma_z^2 n^2 - (\sigma_x l^2 + \sigma_y m^2 + \sigma_z n^2)^2} \tag{7-18}$$

The planes on which maximum and minimum shearing stresses occur can be obtained from Eq. 7-18 by differentiating with respect to the direction cosines l, m, and n. One of the direction cosines in Eq. 7-18 (n for example) can be eliminated by solving Eq. c for n^2 and substituting into Eq. 7-18. Thus,

$$\tau_{nt} = \{(\sigma_x^2 - \sigma_z^2)l^2 + (\sigma_y^2 - \sigma_z^2)m^2$$
$$+ \sigma_z^2 - [(\sigma_x - \sigma_z)l^2 + (\sigma_y - \sigma_z)m^2 + \sigma_z]^2\}^{1/2} \tag{g}$$

By taking the partial derivatives of Eq. g, first with respect to l and then with respect to m and equating to zero, the following equations are obtained for determining the direction cosines associated with planes having maximum and minimum shearing stresses:

TABLE 7-1 Direction Cosines for Shearing Stresses

	Minimum			Maximum		
	1	2	3	4	5	6
l	± 1	0	0	$\pm\sqrt{1/2}$	$\pm\sqrt{1/2}$	0
m	0	± 1	0	$\pm\sqrt{1/2}$	0	$\pm\sqrt{1/2}$
n	0	0	± 1	0	$\pm\sqrt{1/2}$	$\pm\sqrt{1/2}$

$$l[\tfrac{1}{2}(\sigma_x - \sigma_z) - (\sigma_x - \sigma_z)l^2 - (\sigma_y - \sigma_z)m^2] = 0 \qquad (h)$$

$$m[\tfrac{1}{2}(\sigma_y - \sigma_z) - (\sigma_x - \sigma_z)l^2 - (\sigma_y - \sigma_z)m^2] = 0 \qquad (i)$$

One solution of these equations is obviously $l = m = 0$. Then from Eq. c, $n = \pm 1$. Solutions different from zero are also possible for this set of equations. Consider first that $m = 0$; then from Eq. h, $l = \pm\sqrt{1/2}$ and from Eq. c, $n = \pm\sqrt{1/2}$. Also if $l = 0$, then from Eq. i, $m = \pm\sqrt{1/2}$ and from Eq. c, $n = \pm\sqrt{1/2}$. Repeating the above procedure by eliminating l and m in turn from Eq. g yields other values for the direction cosines that make the shearing stresses maximum or minimum. All the possible solutions are listed in Table 7-1. In the last line of the table the planes corresponding to the direction cosines in the column above are shown shaded. Note that in each case only one of the two possible planes is shown. Substituting values for the direction cosines from column 4 of Table 7-1 into Eq. 7-18, with σ_x, σ_y, σ_z replaced with σ_{p1}, σ_{p2}, σ_{p3}, yields

$$\tau_{\max} = \sqrt{\tfrac{1}{2}\sigma_{p1}^2 + \tfrac{1}{2}\sigma_{p2}^2 + 0 - (\tfrac{1}{2}\sigma_{p1} + \tfrac{1}{2}\sigma_{p2})^2}$$

$$= \frac{\sigma_{p1} - \sigma_{p2}}{2}$$

Similarly, using the values of the cosines from columns 5 and 6 gives

$$\tau_{\max} = \frac{\sigma_{p1} - \sigma_{p3}}{2} \qquad \text{and} \qquad \tau_{\max} = \frac{\sigma_{p2} - \sigma_{p3}}{2}$$

Of these three possible results, the largest magnitude will be the maximum stress; hence, the expression for the maximum shearing stress is

$$\tau_{\max} = \frac{\sigma_{\max} - \sigma_{\min}}{2} \qquad (7\text{-}19)$$

which verifies the statement in Section 1-7 regarding maximum shearing stress.

PROBLEMS

7-105 Demonstrate that Eq. 7-17 reduces to Eq. 7-4 for the state of plane stress.

7-106* At a point in a stressed body, the known stresses are $\sigma_x = 40$ MPa T, $\sigma_y = 20$ MPa C, $\sigma_z = 20$ MPa T, $\tau_{xy} = +40$ MPa, $\tau_{yz} = 0$, and $\tau_{zx} = +30$ MPa. Determine

(a) the normal and shearing stresses on a plane whose outward normal is oriented at angles of 40°, 75°, and 54° with the x, y, and z axes, respectively.

(b) the principal stresses and the maximum shearing stress at the point.

7-107* At a point in a stressed body, the known stresses are $\sigma_x = 14$ ksi T, $\sigma_y = 12$ ksi T, $\sigma_z = 10$ ksi T, $\tau_{xy} = +4$ ksi, $\tau_{yz} = -4$ ksi, and $\tau_{zx} = 0$. Determine

(a) the normal and shearing stresses on a plane whose outward normal is oriented at angles of 40°, 60°, and 66.2° with the x, y, and z axes, respectively.

(b) the principal stresses and the maximum shearing stress at the point.

7-108 At a point in a stressed body, the known stresses are $\sigma_x = 60$ MPa T, $\sigma_y = 90$ MPa T, $\sigma_z = 60$ MPa T, $\tau_{xy} = +120$ MPa, $\tau_{yz} = +75$ MPa, and $\tau_{zx} = +90$ MPa. Determine

(a) the normal and shearing stresses on a plane whose outward normal is oriented at angles of 60°, 70°, and 37.3° with the x, y, and z axes, respectively.

(b) the principal stresses and the maximum shearing stress at the point.

7-109 At a point in a stressed body, the known stresses are $\sigma_x = 0$, $\sigma_y = 0$, $\sigma_z = 0$, $\tau_{xy} = +6$ ksi, $\tau_{yz} = +10$ ksi, and $\tau_{zx} = +8$ ksi. Determine

(a) the normal and shearing stresses on a plane whose outward normal makes equal angles with the x, y, and z axes.

(b) the principal stresses and the maximum shearing stress at the point.

7-110* At a point in a stressed body, the known stresses are $\sigma_x = 72$ MPa T, $\sigma_y = 32$ MPa C, $\sigma_z = 0$, $\tau_{xy} = +21$ MPa, $\tau_{yz} = 0$, and $\tau_{zx} = +21$ MPa. Determine

(a) the normal and shearing stresses on a plane whose outward normal makes equal angles with the x, y, and z axes.

(b) the principal stresses and the maximum shearing stress at the point.

7-111* At a point in a stressed body, the known stresses are $\sigma_x = 60$ MPa T, $\sigma_y = 50$ MPa C, $\sigma_z = 40$ MPa T, $\tau_{xy} = +40$ MPa, $\tau_{yz} = -50$ MPa, and $\tau_{zx} = +60$ MPa. Determine

(a) the normal and shearing stresses on a plane whose outward normal is oriented at angles of 30°, 80°, and 62° with the x, y, and z axes, respectively.

(b) the principal stresses and the maximum shearing stress at the point.

7-112 At a point in a stressed body, the known stresses are $\sigma_x = 14$ ksi T, $\sigma_y = 12$ ksi T, $\sigma_z = 10$ ksi T, $\tau_{xy} = +8$ ksi, $\tau_{yz} = -10$ ksi, and $\tau_{zx} = +6$ ksi. Determine

(a) the normal and shearing stresses on a plane whose outward normal is oriented at angles of 61.3°, 53.1°, and 50.2° with the x, y, and z axes, respectively.

(b) the principal stresses and the maximum shearing stress at the point.

7-113* At a point in a stressed body, the known stresses are $\sigma_x = 60$ MPa T, $\sigma_y = 40$ MPa T, $\sigma_z = 20$ MPa T, $\tau_{xy} = +40$ MPa, $\tau_{yz} = +20$ MPa, and $\tau_{zx} = +30$ MPa. Determine

(a) the principal stresses and the maximum shearing stress at the point.

(b) the orientation of the plane on which the maximum tensile stress acts.

7-114* At a point in a stressed body, the known stresses are $\sigma_x = 18$ ksi T, $\sigma_y = 12$ ksi T, $\sigma_z = 6$ ksi T, $\tau_{xy} = +12$ ksi, $\tau_{yz} = -6$ ksi, and $\tau_{zx} = +9$ ksi. Determine

(a) the principal stresses and the maximum shearing stress at the point.

(b) the orientation of the plane on which the maximum tensile stress acts.

7-115 At a point in a stressed body, the known stresses are $\sigma_x = 100$ MPa T, $\sigma_y = 100$ MPa C, $\sigma_z = 80$ MPa T, $\tau_{xy} = +50$ MPa, $\tau_{yz} = -70$ MPa, and $\tau_{zx} = -64$ MPa. Determine

(a) the principal stresses and the maximum shearing stress at the point.

(b) the orientation of the plane on which the maximum compressive stress acts.

7-116 At a point in a stressed body, the known stresses are $\sigma_x = 18$ ksi C, $\sigma_y = 15$ ksi C, $\sigma_z = 12$ ksi C, $\tau_{xy} = -15$ ksi, $\tau_{yz} = +12$ ksi, and $\tau_{zx} = -9$ ksi. Determine

(a) the principal stresses and the maximum shearing stress at the point.

(b) the orientation of the plane on which the maximum compressive stress acts.

COMPUTER PROBLEMS

Note: The following problems have been designed to be solved with a programmable calculator, microcomputer, or mainframe computer. Appendix F contains a description of a few numerical methods, together with a few simple programs in BASIC and FORTRAN, that can be modified for use in the solution of these problems.

C7-1 A strain gage rosette is used to obtain the strains ϵ_a, ϵ_b, and ϵ_c at the angles θ_a, θ_b, and θ_c (see Fig. PC7-1). For a state of plane stress, write a program to

(a) compute and print the Cartesian strains, ϵ_x, ϵ_y, and γ_{xy}.

(b) compute and print the principal strains ϵ_1, ϵ_2, and ϵ_3.

(c) compute and print the angles for the in-plane principal strains, θ_{p1} and θ_{p2}.

(d) compute and print the maximum shearing strains, γ_{12}, γ_{23}, and γ_{31};

(e) compute (from ϵ_1, ϵ_2, and ϵ_3) and print the principal stresses, σ_1, σ_2, and σ_3.

(f) compute (from γ_{12}, γ_{23}, and γ_{31}) and print the maximum shearing stresses τ_{12}, τ_{23}, and τ_{31}.

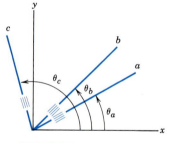

FIG. PC7-1

C7-2 Use the data given below in the program developed in Problem C7-1.

Case	ϵ_a	ϵ_b	ϵ_c	θ_a	θ_b	θ_c	E	ν
1	750 μ	-125 μ	-250 μ	0°	45°	90°	30,000 ksi	0.3
2	780 μ	345 μ	-332 μ	0°	$-60°$	60°	10,000 ksi	1/3
3	875 μ	700 μ	-350 μ	30°	165°	$-105°$	10,600 ksi	0.33
4	-555 μ	925 μ	740 μ	0°	120°	$-120°$	200 GPa	0.3
5	525 μ	450 μ	1425 μ	0°	45°	135°	73 GPa	0.304
6	-306 μ	-456 μ	906 μ	30°	75°	120°	96 GPa	1/3
7	0 μ	1000 μ	0 μ	0°	45°	90°	10,000 ksi	1/3
8	0 μ	1000 μ	0 μ	60°	105°	150°	10,000 ksi	1/3

C7-3 A strain gage rosette is used to obtain the strains ϵ_a, ϵ_b, and ϵ_c at the angles θ_a, θ_b, and θ_c (see Fig. PC7-1). For a state of plane stress, write a program to

(a) compute and print the Cartesian strains, ϵ_x, ϵ_y, and γ_{xy}.

(b) compute and print the Cartesian stresses, σ_x, σ_y, and τ_{xy}.

(c) compute and print the principal stresses, σ_1, σ_2, and σ_3.

(d) compute and print the angles for the in-plane principal stresses θ_{p1} and θ_{p2}.

(e) compute and print the maximum shearing stresses, τ_{12}, τ_{23}, and τ_{31}.

C7-4 Use the data given in Problem C7-2 in the program developed in Problem C7-3.

C7-5 For a general (three-dimensional) state of stress, the principal stresses are the roots of the cubic equation

$$\sigma^3 - a\sigma^2 + b\sigma - c = 0 \qquad (a)$$

where

$$a = \sigma_x + \sigma_y + \sigma_z$$

$$b = \sigma_x\sigma_y + \sigma_y\sigma_z + \sigma_z\sigma_x - \tau_{xy}^2 - \tau_{yz}^2 - \tau_{zx}^2$$

$$c = \sigma_x\sigma_y\sigma_z - \sigma_x\tau_{yz}^2 - \sigma_y\tau_{zx}^2 - \sigma_z\tau_{xy}^2 + 2\,\tau_{xy}\tau_{yz}\tau_{zx}$$

Write a program to compute and print the three principal stresses $\sigma_1 \geq \sigma_2 \geq \sigma_3$ and the maximum shearing stresses τ_{12}, τ_{23}, and τ_{31} by using the following procedure.

(a) First, solve Eq. a for one of the principal stresses, say σ_1, by using the Newton–Raphson method (see Appendix F).

(b) Next, since σ_1 is a solution of Eq. a, the equation can be written

$$(\sigma - \sigma_1)(\sigma^2 + A\sigma + B) = 0 \qquad (b)$$

Multiplying out the terms in Eq. b and comparing them with the terms of Eq. a shows that

$$A = \sigma_1 - a$$

$$B = b + \sigma_1(\sigma_1 - a)$$

(c) Next, solve the quadratic equation part of Eq. b for

$$\sigma_2, \sigma_3 = \frac{-A \pm \sqrt{A^2 - 4B}}{2}$$

(d) Finally, renumber the principal stresses (if necessary) so that $\sigma_1 > \sigma_2 > \sigma_3$ and compute the maximum shearing stresses τ_{12}, τ_{23}, and τ_{31}.

C7-6 Use the program for Problem C7-5 to solve for the principal stresses and the maximum shearing stresses for each of the following states of stress:

Case	σ_x	σ_y	σ_z	τ_{xy}	τ_{yz}	τ_{zx}
1	14 ksi	12 ksi	10 ksi	4 ksi	− 4 ksi	0 ksi
2	40 MPa	− 20 MPa	20 MPa	40 MPa	0 MPa	30 MPa
3	15 ksi	15 ksi	15 ksi	6 ksi	0 ksi	8 ksi
4	60 MPa	90 MPa	60 MPa	120 MPa	75 MPa	90 MPa
5	18 ksi	12 ksi	6 ksi	12 ksi	− 6 ksi	9 ksi
6	72 MPa	− 32 MPa	0 MPa	21 MPa	0 MPa	21 MPa
7	− 18 ksi	− 15 ksi	− 12 ksi	− 15 ksi	12 ksi	− 9 ksi
8	100 MPa	− 100 MPa	80 MPa	50 MPa	− 70 MPa	− 64 MPa

C7-7 For a general (three-dimensional) state of stress, the directions of the principal stresses (direction cosines, l, m, and n) are obtained by solving the system of equations

$$(\sigma_i - \sigma_x)\, l_i \qquad\quad -\ \tau_{xy}\, m_i \qquad\quad -\ \tau_{zx}\, n_i = 0 \qquad (a)$$

$$-\ \tau_{xy}\, l_i + (\sigma_i - \sigma_y)\, m_i \qquad\quad -\ \tau_{yz}\, n_i = 0 \qquad (b)$$

$$-\ \tau_{zx}\, l_i \qquad\quad -\ \tau_{yz}\, m_i + (\sigma_i - \sigma_z)\, n_i = 0 \qquad (c)$$

$$l_i^2 + m_i^2 + n_i^2 = 1 \qquad (d)$$

where σ_i ($i = 1$, 2, or 3) is one of the principal stresses and l_i, m_i, and n_i are the corresponding direction cosines. Write a program to compute and print the three direction cosines, l_i, m_i, and n_i for each of the principal stresses σ_i, $i = 1, 2$, and 3. In writing the program, you should note the following.

(a) When the principal stress, σ_i, used in Eqs. a, b, and c is different than the other two principal stresses (a single root), then Eqs. a, b, and c are not independent—one of the equations is a combination of the other two. Then Eq. d must be used along with the two independent equations to get a solution for the direction cosines.

(b) When the principal stress, σ_i, used in Eqs. a, b, and c is the same as another value of principal stress (a double root), then only one of the three equations is unique. In this case, only Eqs. a and d can be used to solve for the direction cosines. The problem statement is completed by requiring that the unit vectors defined by the three sets of direction cosines be perpendicular.

(c) Although not all three of the direction cosines can be zero for a particular principal stress, σ_i, it is quite possible that one or two of the direction cosines will be zero.

C7-8 Use the program of Problem C7-7 to compute and print the direction cosines for each of the states of stress of Problem C7-6.

CHAPTER 8

COMBINED STATIC LOADINGS

8-1

INTRODUCTION

The stresses and strains produced by three fundamental types of loads (centric, torsional, and flexural) have been analyzed in the preceding chapters. Many machine and structural elements are subjected to a combination of any two or all three of these types of loads, and a procedure for calculating the stresses on a given plane at a point is required. One method is to replace the given force system by an equivalent system of forces and couples, each of which is selected in such a manner that the stresses produced by it can be calculated by methods already developed. The composite effect can be obtained by the principle of superposition, as explained in Section 1-14, provided that the combined stresses do not exceed the proportional limit. Various combinations of loads that can be analyzed in this manner are discussed in the following sections.

8-2

COMBINED AXIAL AND TORSIONAL LOADS

There are numerous situations in which a shaft or other machine member is subjected to both an axial and a torsional load. Examples are the drill rod on a well or other drilling machine and the propeller shaft on a ship. Since radial and circumferential normal stresses are zero, the combination of axial and torsional loads results in a state-of-plane stress at any point in the body. The axial stress is the same for all interior and exterior points, while the torsional stress is maximum at the outer surface; therefore, the investigation for critical stresses would normally be made for points on the surface of the shaft.

The following example illustrates the procedure for combined torsional and axial loads.

EXAMPLE 8-1

A hollow circular shaft of outer diameter 4 in. and inner diameter 2 in. is loaded as shown in Fig. 8-1a.

(a) Determine the principal stresses and the maximum shearing stress at the point (or points) where the stress situation is most severe.

(b) On a sketch show the approximate directions of these stresses.

Solution

The axial stress is constant throughout the shaft and is equal to

$$\sigma = \frac{P}{A} = \frac{24,000\pi}{\pi(2^2 - 1^2)} = 8000 \text{ psi } C$$

The torsional load is highest to the right of the torque T_1, and the torsional stress is maximum on the outer surface of the shaft; therefore, the magnitude of the maximum torsional shearing stress is

$$\tau = \frac{Tc}{J} = \frac{22,500\pi(2)}{\pi(2^4 - 1^4)/2} = 6000 \text{ psi}$$

These stresses are shown acting on mutually perpendicular planes through a point on the surface of the right portion of the shaft in Fig. 8-1b. Mohr's circle for the stresses at any point on the surface of the right portion of the shaft is shown in Fig. 8-1c.

(a) The principal stresses OA and OB from Mohr's circle are

$$\sigma_{p1} = OA = -4.00 + 2\sqrt{13} = \underline{3.21 \text{ ksi } T} \qquad \text{Ans.}$$

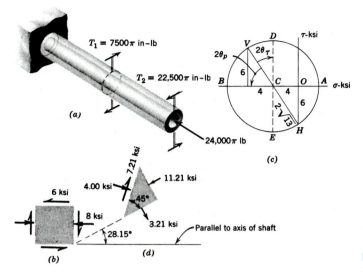

FIG. 8-1

and

$$\sigma_{p2} = OB = -4.00 - 2\sqrt{13} = -11.21 = \underline{11.21 \text{ ksi } C} \quad \text{Ans.}$$

The maximum shearing stress CD or CE from Mohr's circle is

$$\tau_{max} = CD = 2\sqrt{13} = \underline{7.21 \text{ ksi}} \qquad \text{Ans.}$$

(b) From the Mohr's circle,

$$\tan 2\theta_p = 6/4 = 1.5$$

which gives

$$2\theta_p = 56.30° \curvearrowright$$

and

$$\theta_p = 28.15° \curvearrowright$$

The stresses are shown in their proper directions in Fig. 8-1d.

PROBLEMS

8-1* A 4-in.-diameter shaft is subjected to both a torque of 30 in.-kips and an axial tensile load of 50 kips, as shown in Fig. P8-1.

(a) Determine the principal stresses and the maximum shearing stress at point A on the surface of the shaft.

(b) Show the stresses of part (a) and their directions on a sketch.

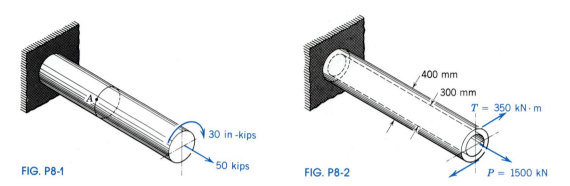

FIG. P8-1

FIG. P8-2

8-2* A hollow shaft with an outside diameter of 400 mm and an inside diameter of 300 mm is subjected to both a torque of 350 kN·m and an axial tensile load of 1500 kN, as shown in Fig. P8-2.

(a) Determine the principal stresses and the maximum shearing stress at a point on the outside surface of the shaft.

(b) Show the stresses of part (a) and their directions on a sketch.

8-3 A 2-in.-diameter shaft is used in an aircraft engine to transmit 360 hp at 1500 rpm to a propeller that develops a thrust of 2800 lb. Determine the principal stresses and the maximum shearing stress produced at any point on the outside surface of the shaft.

8-4 A hollow tube with an outside diameter of 150 mm and an inside diameter of 100 mm is subjected to a torque of 16 kN·m and an axial compressive load of 250 kN. Determine

(a) the principal stresses and the maximum shearing stress on the outside surface of the tube.

(b) the principal stresses and the maximum shearing stress on the inside surface of the tube.

8-5* A 4-in.-diameter shaft must support an axial tensile load of unknown magnitude while it is transmitting a torque of 100 in.-kips. Determine the maximum allowable value for the axial load if the tensile principal stress on the outside surface of the shaft must not to exceed 18,000 psi.

8-6* A 60-mm-diameter shaft must transmit a torque of unknown magnitude while it is supporting an axial tensile load of 150 kN. Determine the maximum allowable value for the torque if the tensile principal stress on the outside surface of the shaft must not exceed 125 MPa.

8-7 A 6-in.-diameter shaft will be used to support the axial load and torques shown in Fig. P8-7.

  (a) Determine the principal stresses and the maximum shearing stress at point A on the surface of the shaft.

  (b) Show the stresses of part (a) and their directions on a sketch.

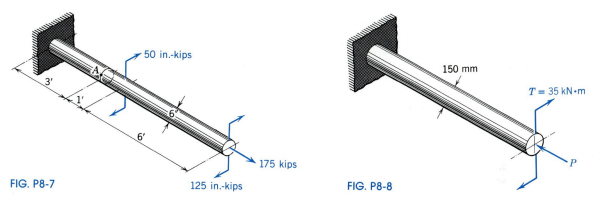

FIG. P8-7

FIG. P8-8

8-8 A steel shaft is loaded and supported as shown in Fig. P8-8. Determine the maximum allowable value for the axial load P if the maximum shearing stress in the shaft is not to exceed 75 MPa and the maximum compressive stress in the shaft is not to exceed 120 MPa.

8-9* A steel shaft is loaded and supported as shown in Fig. P8-9. If the maximum shearing stress in the shaft must not exceed 8 ksi and the maximum tensile stress in the shaft must not exceed 12 ksi, determine the maximum torque T that can be applied to the shaft.

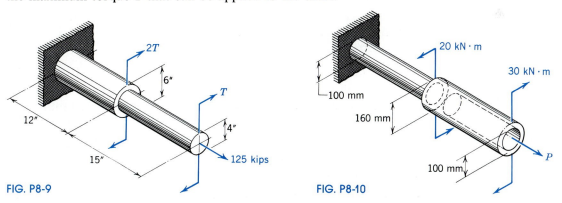

FIG. P8-9

FIG. P8-10

8-10* A steel shaft with the left portion solid and the right portion hollow is loaded as shown in Fig. P8-10. If the maximum shearing stress in the shaft must not exceed 80 MPa and the maximum tensile stress in the shaft must not exceed 140 MPa, determine the maximum axial load P that can be applied to the shaft.

8-11 A shaft that is transmitting 240 hp at 1800 rpm must also support an axial tensile load of 20 kips. If the maximum tensile stress in the shaft is not to exceed 15 ksi, determine the minimum diameter required for the shaft.

8-12 A tube having an inside diameter equal to one-half the outside diameter must transmit a torque of 7.5 kN·m while resisting an axial compressive load of 250 kN. Determine the minimum outside diameter required if the maximum compressive stress in the tube is not to exceed 125 MPa.

8-13* A 1-in.-diameter steel ($E = 30,000$ ksi and $\nu = 0.30$) bar is subjected to a tensile load P and a torque T as shown in Fig. P8-13. Determine the axial load P and the torque T if the strains indicated by gages a and b on the bar are $\epsilon_a = +1084\mu$ and $\epsilon_b = -754\mu$.

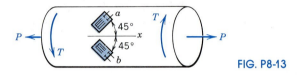

FIG. P8-13

8-14 A 50-mm-diameter steel ($E = 200$ GPa and $\nu = 0.30$) bar is subjected to a tensile load P and a torque T as shown in Fig. P8-13. Determine the axial load P and the torque T if the strains indicated by gages a and b on the bar are $\epsilon_a = +1414\mu$ and $\epsilon_b = -212\mu$.

8-3

COMBINED CROSS SHEAR AND TORSIONAL LOADS

A common example of a cross shear load combined with a torsional load occurs in a closely coiled helical spring. A spring formed by winding a rod around a cylinder in such a manner that the rod forms a helix is called a helical spring. A helical spring is said to be closely coiled when the coils are so close to each other that any one coil can be assumed to lie in a plane essentially perpendicular to the axis of the spring. The discussion in this section will be limited to a spring made of a circular rod of radius r with a helix of mean radius R as shown in Fig. 8-2.

For a closely coiled spring, a cross section of the spring rod perpendicular to the axis of the rod can be assumed to be in the same plane as the axis of the helix (a vertical plane for the cross section shown in Figs. 8-2a through c). The tensile load P applied to the spring is held in equilibrium by the resisting shear V (equal to P) and the

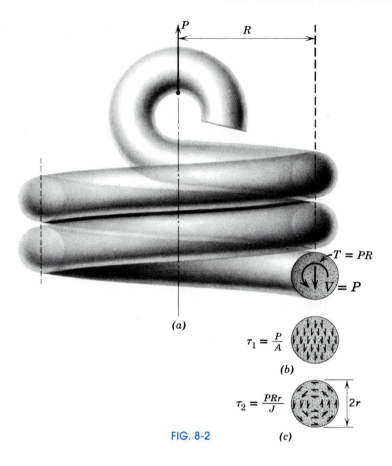

$\tau_1 = \dfrac{P}{A}$

(b)

$\tau_2 = \dfrac{PRr}{J}$

FIG. 8-2   (c)

resisting torque T (equal to PR) as shown in Fig. 8-2a. The cross (frequently called the direct) shearing stress due to the centric load $(V = P)$ is

$$\tau_1 = \frac{P}{A} = \frac{P}{\pi r^2}$$

and is usually considered to be uniformly distributed over the cross section of the rod as shown in Fig. 8-2b. The torsional shearing stress due to the torque $(T = PR)$ is

$$\tau_2 = \frac{Tc}{J} = \frac{PRr}{\pi r^4/2}$$

and is maximum at the outer surface of the rod and has the same direction as the direct shearing stress at the left edge of the cross section as shown in Fig. 8-2c. Therefore, the critical shearing stress occurs at the inside surface of the coil and is the sum of the direct and torsional shearing stresses; that is,

$$\tau_{max} = \frac{P}{A} + \frac{Tr}{J} \tag{8-1}$$

Since the mean radius of the helix, R, is usually large compared to the radius of the rod, r, the direct shearing stress is normally small compared to the torsional shearing stress.

Equation 8-1 must be considered a statement of basic concept (the addition of stresses acting on a common plane) rather than an adequate equation for stress determination in any helical spring, for the following reasons: (1) The cross shearing stress is not uniformly distributed but is a maximum on the transverse diameter of the rod. (2) The torsion formula, which was derived for a straight bar, gives only an approximate value for the torsional stress because the longitudinal elements of the rod in a helical spring have different lengths, those on the inside being shorter than those on the outside of the coil. Since γ is equal to $r\theta/L$, and τ_2 is equal to $G\gamma$, it is observed that the torsional stress will be highest on the inside of the helix where L has the least value.

A more accurate analysis (supported by experimental evidence) of helical springs[1] leads to the Wahl formula

$$\tau_{max} = \frac{2PR}{\pi r^3} \left(\frac{4m - 1}{4m - 4} + \frac{0.615}{m} \right) \tag{8-2}$$

where m is R/r. For a ratio of R/r of 8, this expression gives a maximum stress about 10 percent greater than that from Eq. 8-1.

The deflection of a closely coiled helical spring results primarily from the torsional shearing stress ($\tau_2 = Tr/J$) in the circular rod. The deflection resulting from the direct or cross shearing stress ($\tau_1 = V/A$) is very small; therefore, it can be neglected without introducing significant error into the deflection expression. Since the spring deflection δ results primarily from torsional effects, the increment $d\delta$ contributed by a small increment of length dL of the rod (which twists through an angle $d\theta$) is given by Eq. 3-8b

$$d\delta = R \, d\theta = R \, \frac{T \, dL}{GJ}$$

Integrating over the total length of the rod yields

$$\delta = R\theta = \frac{RTL}{GJ} \tag{8-3a}$$

[1] *Mechanical Springs*, A. M. Wahl, Penton Publishing Co., Cleveland, Ohio, 1944.

or since $T = PR$ and $L = 2\pi Rn$, where n is the number of effective coils in the spring,

$$\delta = \frac{R(PR)(2\pi Rn)}{G(\pi/2)(r^4)} = \frac{4PR^3n}{Gr^4} \tag{8-3b}$$

The relationship between the deflection of a spring and the force applied (so long as this relationship is linear) is conveniently expressed in terms of the modulus of the spring. The modulus k of a spring is the force required to stretch or compress the spring a unit distance and in equation form is

$$k = \frac{Gr^4}{4R^3n} = \frac{P}{\delta} \tag{8-4}$$

PROBLEMS

8-15* A helical spring with a mean diameter of 3 in. is made from a $\frac{7}{8}$-in.-diameter alloy-steel rod. The spring will be subjected to a load of 1500 lb. Determine

(a) the maximum shearing stress (use Eq. 8-1) in the spring.

(b) the percent difference between the results of part (a) and the more accurate Wahl formula.

8-16* A helical spring with a mean diameter of 60 mm is made from a 15-mm-diameter alloy-steel rod. The spring will be subjected to a load of 5 kN. Determine

(a) the maximum shearing stress (use Eq. 8-1) in the spring.

(b) the percent difference between the results of part (a) and the more accurate Wahl formula.

8-17 A helical spring with a mean diameter of 5 in. is made from a 1-in.-diameter brass ($G = 5000$ ksi) rod. The spring has 10 turns and is subjected to a load of 1000 lb. Determine

(a) the maximum stress (use Eq. 8-1) in the spring.

(b) the deflection of the spring.

8-18 A helical spring with a mean diameter of 200 mm is made from a 25-mm-diameter alloy-steel ($G = 80$ GPa) rod. The spring has 8 turns and is subjected to a load of 6 kN. Determine

(a) the maximum shearing stress (use Eq. 8-1) in the spring.

(b) the deflection of the spring.

8-19* Design a helical spring made of $\frac{1}{2}$-in. diameter SAE 4340 heat-treated steel ($G = 11,000$ ksi) rod that will deflect at least 3 in. and have a maximum shearing stress not more than 50 ksi when supporting a 500-lb load. Use Eq. 8-1.

8-20* Design a helical spring made of 25-mm-diameter alloy-steel ($G = 80$ GPa) rod that will deflect at least 80 mm and have a maximum shearing stress not more than 200 MPa when supporting a 5-kN load. Use Eq. 8-1.

8-21 A helical spring with an inside diameter of 5 in. is made from a 1-in.-diameter alloy-steel ($G = 11,000$ ksi) rod. If the spring has 10 turns, determine the modulus (spring constant) of the spring.

8-22 A helical spring with an outside diameter of 150 mm is made from a 20-mm-diameter alloy-steel ($G = 76$ GPa) rod. If the spring has 12 turns, determine the modulus (spring constant) of the spring.

8-23* A helical spring with an outside diameter of 2 in. and a modulus (spring constant) of 200 lb/in. is being designed for an industrial application. The spring will be made from an alloy-steel ($G = 11,000$ ksi) rod and will have 10 turns. Determine the rod diameter required for the spring.

8-24 A helical spring with an inside diameter of 75 mm and a modulus (spring constant) of 1000 N/m is being designed for an industrial application. The spring will be made from an alloy-steel ($G = 76$ GPa) rod and will have 12 turns. Determine the rod diameter required for the spring.

8-4
COMBINED CENTRIC AND FLEXURAL LOADS

A common example of a centric load combined with a bending load is a structural or machine part with a force acting parallel to but not along the axis of the member; for example, the vertical portion AB of the loading hook of Fig. 8-3. Such an eccentric load can be replaced by a centric load (a force acting along the centroidal axis of the part AB) and one or more couples. The stresses resulting from the axial load were discussed in Chapter 2. Procedures developed in Chapter 4 can be applied to evaluate the stresses produced by the couples.

The normal stress at any point on a transverse plane through AB is the algebraic sum of the normal stresses due to the centric load and that due to the bending load (the couples). Since the transverse shear is zero, this stress will be a principal stress. The procedure is illustrated in the following example.

FIG. 8.3 Five-ton Littell coil-loading hook. (*Courtesy. Steel Founders Society of America.*)

EXAMPLE 8-2

The cast-iron frame of a small press is shaped as shown in Fig. 8-4a. Figure 8-4b represents the cross section a–a, and the moment of inertia of the area is 1000/3 in.^{4} with respect to the centroidal axis c–c. For a load P of 16 kips, and assuming linearly elastic action, determine

(a) The normal stress distribution on section a–a.

(b) The principal stresses for the critical points on section a–a.

Solution

(a) When the free-body diagram of the top portion of the frame was drawn, two equal opposite collinear forces P through the centroid of the section a–a were added as shown in Fig. 8-4c. The force P_1 is a centric load equal to the resisting force T on section a–a. The other two forces P and P_2 constitute an 18 P-in.-kip couple equal to the resisting moment M on section a–a.

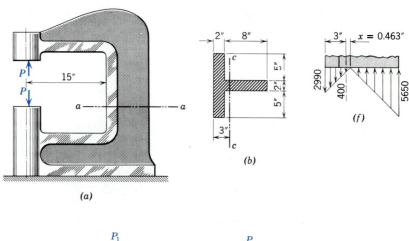

(a)

(b)

(f)

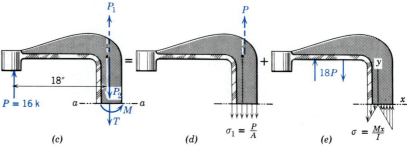

FIG. 8-4

(c)

(d)

(e)

The axial stress uniformly distributed over section a–a, as shown in Fig. 8-4d, is

$$\sigma_1 = \frac{P}{A} = \frac{16,000}{40} = 400 \text{ psi } T$$

The maximum tensile fiber stress occurs at the left edge of the section a–a and is

$$\sigma_2 = \frac{Mc}{I} = \frac{16,000(18)(3)}{1000/3} = 2592 \text{ psi } T$$

The maximum compressive fiber stress occurs at the right edge of the section a–a and is

$$\sigma_3 = \frac{Mc}{I} = \frac{16,000(18)(7)}{1000/3} = 6048 \text{ psi } C$$

The distribution of fiber stresses is shown in Fig. 8-4e.

When the principle of superposition is applied, the total normal stress at any point on section $a–a$ is the algebraic sum of the component stresses at the point. When the stress distributions of Figs. 8-4d and e are added, the resultant stress distribution for section $a–a$ is as shown in Fig. 8-4f, which is the answer for part (a).

(b) With no shearing stresses on section $a–a$, the normal stresses are principal stresses, and the critical points, as observed from the stress distribution of Fig. 8-4f, are the left and right edges of the section. The principal stresses for the right edge are

$$\sigma_p = 5650 \text{ psi } C \quad \text{and} \quad 0 \qquad \text{Ans.}$$

The principal stresses for the left edge are

$$\sigma_p = 2990 \text{ psi } T \quad \text{and} \quad 0 \qquad \text{Ans.}$$

Although not requested in this example, the location of the neutral axis (line of zero stress) can be determined as the place where the compressive fiber stress is 400 psi because this stress will just balance the axial tensile stress of 400 psi. The distance x on Fig. 8-4f can be obtained by writing

$$400 = \frac{16,000(18)x}{1000/3}$$

from which

$$x = 0.463 \text{ in.}$$

to the right of the centroidal axis of the cross section.

EXAMPLE 8-3

A gray cast-iron compression member is subjected to a vertical load P, as shown in Fig. 8-5a. The allowable normal stresses on plane $ABCD$ are 140 MPa C and 35 MPa T. Determine the maximum allowable load P.

Solution

The specified stresses indicate elastic action. When two vertical forces P (labeled P_1 and P_2 for clarity) of opposite sense are introduced at

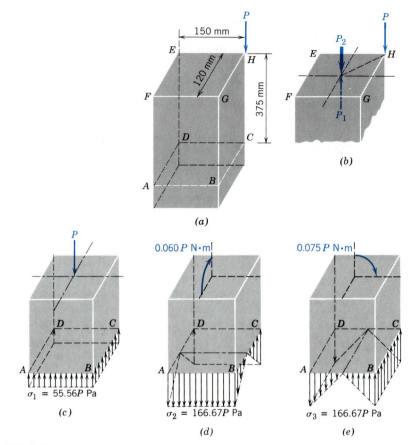

FIG. 8-5

the center of the top section, *EFGH,* the eccentric force *P* is replaced by a centric force P_2 and a couple PP_1, as shown in Fig. 8-5*b.*

The stresses produced by the centric load are easily calculated; however, the couple does not act in a plane of symmetry of the member, and the flexural stresses produced are not readily computed for this orientation. This difficulty can be eliminated by replacing the resultant couple PP_1 by its components in or parallel to planes of symmetry of the compression member. The magnitudes of the components are $60P$ N·mm = $0.060P$ N·m and $75P$ N·mm = $0.075P$ N·m. The sense of each of these couples is shown in Figs. 8-5*d* and *e.*

The centric load P_2 produces a uniformly distributed compressive stress on plane *ABCD.* The magnitude of this stress is

$$\sigma_1 = \frac{P}{A} = \frac{P}{0.150(0.120)} = 55.56P \text{ N/m}^2 = 55.56P \text{ Pa}$$

The $0.060P$ component of the couple PP_1 produces a compressive stress along edge CD of plane $ABCD$ and an equal tensile stress along edge AB. These stresses are

$$\sigma_2 = \frac{Mc}{I} = \frac{0.060P(0.060)}{0.150(0.120)^3/12} = 166.67P \text{ N/m}^2 = 166.67P \text{ Pa}$$

The $0.075P$ component of the couple PP_1 produces a compressive stress along edge BC of plane $ABCD$ and an equal tensile stress along edge AD. These stresses are

$$\sigma_3 = \frac{Mc}{I} = \frac{0.075P(0.075)}{0.120(0.150)^3/12} = 166.67P \text{ N/m}^2 = 166.67P \text{ Pa}$$

The component stresses are shown in the three stress distribution diagrams of Figs. 8-5c, d, and e. From the diagrams, one observes that the maximum compressive stress occurs at C and is

$$\sigma_C = 55.56P + 166.67P + 166.67P = 388.9P \text{ Pa}$$

and that the maximum tensile stress occurs at A and is

$$\sigma_T = 166.67P + 166.67P - 55.56P = 277.8P \text{ Pa}$$

When the allowable stresses are substituted in the above expressions, the values of P become

$$P_C = \frac{140(10^6)}{388.9} = 360(10^3) \text{ N} = 360 \text{ kN}$$

and

$$P_T = \frac{35(10^6)}{277.8} = 126(10^3) \text{ N} = 126 \text{ kN}$$

Since P_T is less than P_C, the maximum allowable load is

$$P = \underline{126 \text{ kN}} \qquad\qquad \text{Ans.}$$

PROBLEMS

> *Note:* In the following problems, the effect of deflection on the eccentricity is to be neglected and elastic action is to be assumed, unless otherwise noted.

8-25* A solid circular bar with a diameter of 4 in. is used to support a tensile load of 10 kips. The line of action of the load is located 1.75 in. from the axis of the bar. Determine the maximum tensile and compressive stresses produced in the bar.

8-26* A hollow circular tube with an outside diameter of 100 mm and an inside diameter of 75 mm is used to support a compressive load of 50 kN. The line of action of the load is located 40 mm from the axis of the tube. Determine the maximum tensile and compressive stresses produced in the tube.

8-27 The T-section shown in Fig. P8-27 is used as a short post to support a compressive load of 150 kips. The load is applied on thc centerline of the stem at a distance $e = 2$ in. from the centroid of the cross section. Determine the normal stresses at surfaces A and B on a transverse plane near the base of the post.

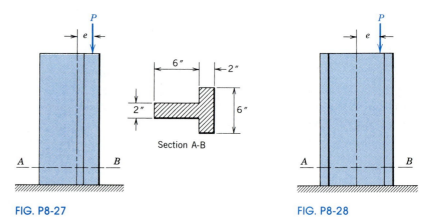

Section A-B

FIG. P8-27   FIG. P8-28

8-28* A W305 × 143 wide-flange section (see Appendix C) is used as a short post to support a load $P = 400$ kN, as shown in Fig. P8-28. The load is applied on the centerline of the web at a distance $e = 100$ mm from the centroid of the cross section. Determine the normal stresses at surfaces A and B on a transverse plane near the base of the post.

8-29* A structural member with a rectangular cross section supports a 15-kip load, as shown in Fig. P8-29. Determine the distribution of normal stress on section AB of the member.

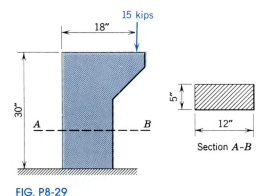

FIG. P8-29

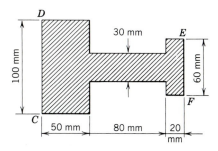

FIG. P8-30

8-30* The cross section of the straight vertical portion of the Littel coil-loading hook illustrated in Fig. P8-3 is shown in Fig. P8-30. The horizontal distance from the line of action of the applied load to the inside face CD of the cross section is 600 mm. Determine the maximum tensile and compressive stresses on section $CDEF$ for a 40-kN load.

8-31 A steel bar with the square cross section shown in Fig. P8-31 will be used to transmit a tensile force P. The maximum tensile stress in the bar must be limited to 10,000 psi. Determine the maximum allowable value of the force P if the line of action of the force

(a) passes through point A of the cross section.

(b) passes through point B of the cross section.

8-32 A short hollow post with the rectangular cross section shown in Fig. P8-32 will be used to support a compressive force P. The maximum compressive stress in the post must be limited to 70 MPa. Determine the maximum allowable value of the load P if the line of action of the force

(a) passes through point A of the cross section.

(b) passes through point B of the cross section.

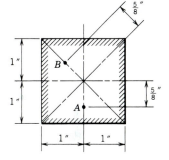

FIG. P8-31

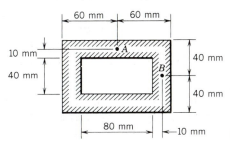

FIG. P8-32

8-33* A short post supports a vertical force P = 9600 lb and a horizontal force H = 800 lb, as shown in Fig. P8-33. Determine the vertical normal stresses at corners A, B, C, and D of the post.

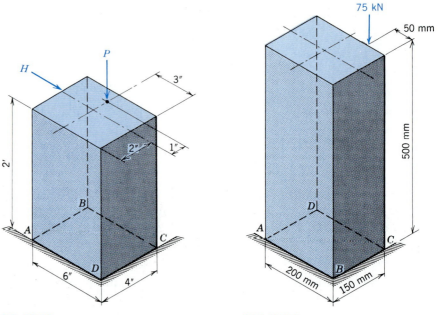

FIG. P8-33   FIG. P8-34

8-34* Determine the vertical normal stresses at points A, B, C, and D of the rectangular post shown in Fig. P8-34.

8-35 The straight portion AB of the cast-steel machine part shown in Fig. P8-35 has a hollow rectangular cross section with outside dimen-

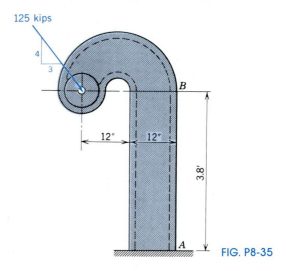

FIG. P8-35

sions 12 in. by 10 in. and walls 2 in. thick. Determine the magnitudes and locations of the maximum tensile and compressive normal stresses on a transverse plane within portion *AB*.

8-36 Determine the magnitudes and locations of the maximum tensile and compressive normal stresses on a transverse plane in the straight portion of the structure shown in Fig. P8-36. The member is braced perpendicular to the plane of symmetry.

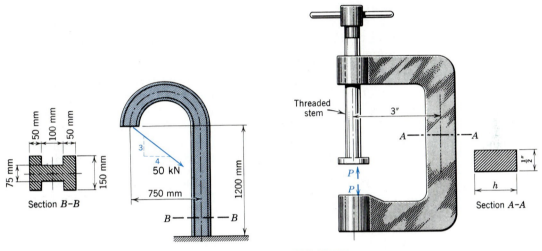

Section *B-B*

FIG. P8-36

FIG. P8-37

8-37* The estimated maximum total force *P* to be exerted on the C-clamp shown in Fig. P8-37 is 450 lb. If the normal stress on section *A–A* is not to exceed 16,000 psi, determine the minimum allowable value for the dimension *h* of the cross section.

8-38* The cross section of the steel ($E = 200$ GPa) beam shown in Fig. P8-38 is a rectangle 100 mm wide by 150 mm deep. Determine the maximum normal stress on a vertical section at the wall

(a) by neglecting the deflection of the member.

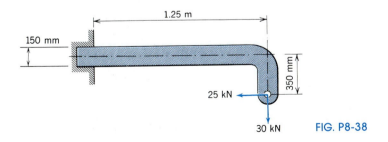

FIG. P8-38

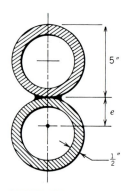

FIG. P8-40

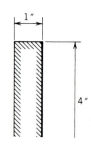

FIG. P8-41

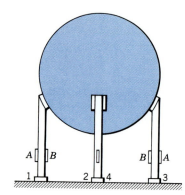

FIG. P8-42

8-39 A solid circular bar will be used to support a compressive load having a line of action parallel to the axis of the bar. Determine the maximum eccentricity permitted if

(a) the tensile stress on a cross section must equal one-half the compressive stress.

(b) the tensile stress on a cross section must be zero.

8-40 A short post is constructed by welding two sections of pipe together to form the cross section shown in Fig. P8-40. The post will be used to support a compressive load P. At what distance e from the centroid of the cross section should the load be placed to make the maximum tensile stress in the post equal to zero.

8-41* A steel pipe with an outside diameter of 2 in. and an inside diameter of 1 in. is welded to a 4-in.-by-1-in. rectangular steel bar to form the tension member shown in Fig. P8-41. A 30-kip load with a line of action coincident with the axis of the pipe is carried by the member. If a factor of 4 with respect to failure by slip is specified, determine the minimum elastic strength required for the steel.

8-42* Eight strain gages are used to monitor the weight of the fluid in the storage tank shown in Fig. P8-42. Two gages (A and B) are mounted on each of the four identical steel ($E = 29,000$ ksi) columns to measure normal strains in the vertical direction. Gages A and B on columns 1 and 2 indicate strains of -250μ and -150μ, respectively. Gages A and B on columns 3 and 4 indicate strains of -200μ and -180μ, respectively. The cross-sectional area of each column is 4 in.^{2}. Determine the weight of the fluid in the tank.

8-43 Four strain gages are mounted at 90° intervals around the circumference of a 100-mm-diameter steel ($E = 210$ GPa and $\nu = 0.30$) bar, as shown in Fig. P8-43. As a result of axial and flexural loadings the four gages indicate longitudinal strains of $\epsilon_1 = -200\mu$, $\epsilon_2 = +820\mu$, $\epsilon_3 = +600\mu$, and $\epsilon_4 = -420\mu$. Determine the axial load P and the two moments M_y and M_z.

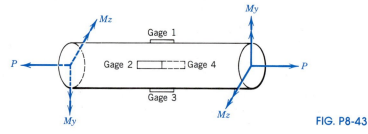

FIG. P8-43

8-44 The output from a strain gage located on the bottom surface of the hat section shown in Fig. P8-44 will be used to indicate the magnitude of the load P applied to the section. The hat section is made of aluminum alloy (E = 10,000 ksi and $v = \frac{1}{3}$) and is 1 in. wide. When the maximum load P = 100 lb is applied to the section, the strain gage should read ϵ = 1000μ. Plot a curve showing the combinations of thickness t and height h that will satisfy the specification. Limit the range of h from 0 to 2 in.

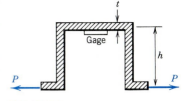

FIG. P8-44

8-45* Three strain gages are mounted on a 1-in.-diameter aluminum alloy (E = 10,000 ksi and $v = \frac{1}{3}$) rod as shown in Fig. P8-45. When loads P and Q are applied to the bar they produce longitudinal strains ϵ_A = 550μ, ϵ_B = 400μ, and ϵ_C = $-$300μ. Determine the magnitudes of loads P and Q and the location x of load P.

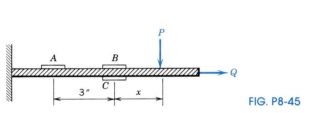

FIG. P8-45

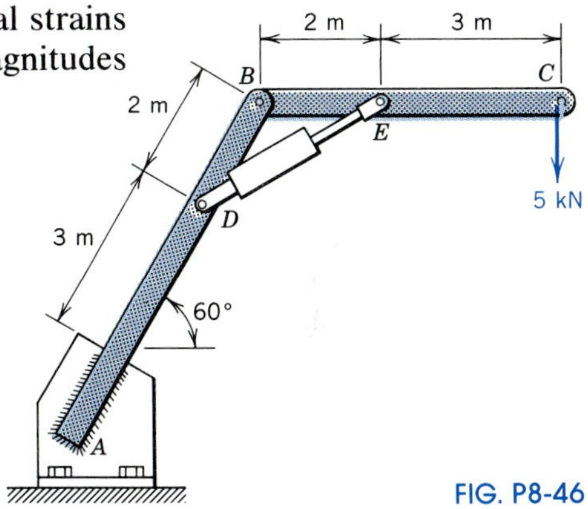

8-46 Members AB and BC of the engine hoist shown in Fig. P8-46 are hollow square 125-mm-by-125-mm structural steel sections that have a wall thickness of 10 mm. Determine the factor of safety with respect to failure by slip for

(a) member AB.   (b) member BC.

FIG. P8-46

8-5

PRINCIPAL STRESSES IN FLEXURAL MEMBERS

Methods for finding the fiber stress at any point in a beam were presented in Sections 4-3 and 4-4. Procedures for locating the critical sections of a beam (maximum M and V) were developed in Sections 4-6 and 4-7. Methods for determining the transverse and longitudinal shearing stresses at any point in a beam were presented in Sections 4-8 and 4-9. However, the discussion of stresses in beams is incomplete without careful consideration of the principal and maximum shearing stresses at carefully selected points on the sections of maximum shear and bending moment.

The fiber stress is maximum at the top or bottom edge of a section, and the transverse and longitudinal shearing stresses are zero at these points. Consequently, the maximum tensile and compressive fiber stresses on a given section are also the principal stresses at these points, and the corresponding maximum shearing stress is equal to one-half the fiber stress [$\tau_{\max} = (\sigma_p - 0)/2$]. The longitudinal and transverse shearing stresses are normally maximum for a given section at the neutral axis, where the fiber stress is zero. Thus, the evidence presented so far is limited to the extremes in a cross section. One might well wonder what the stress situation might be between these points. Unfortunately, the magnitude of the principal stresses throughout a cross section cannot be expressed for all sections as a simple function of position. However, in order to provide some insight into the nature of the problem, two typical sections will be discussed in the following paragraphs.

For a cantilever beam of rectangular cross section subjected to a concentrated load, the theoretical principal stress variation is indicated for two sections in Fig. 8-6. One observes from this figure that, at a distance from the load of one-fourth the depth of the beam, the maximum normal stress does not occur at the surface. However, at a distance of one-half the depth, the maximum fiber stress is the maximum normal stress. For either of these sections, Saint-Venant's principle (see Section 2-6) indicates that the flexure (and transverse shearing stress) formula is inapplicable. Since such a small increase in bending moment is required to overcome the effect of the transverse shearing stress, the conclusion may be drawn that, for a rectangular cross section, in regions where the flexure formula applies, the maximum fiber stress is the maximum normal stress. Although for the rectangular cross section the maximum shearing stress will usually be one-half the maximum normal stress (at a surface of the beam), for materials having a longitudinal plane of weakness (for example, the usual timber beam),

FIG. 8-6

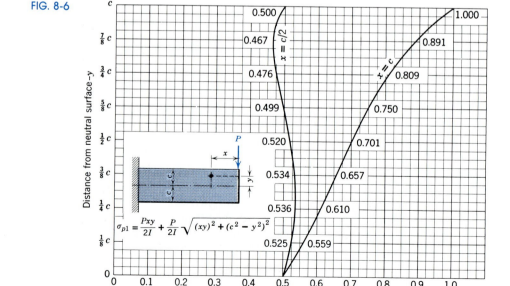

$$\sigma_{p1} = \frac{Pxy}{2I} + \frac{P}{2I}\sqrt{(xy)^2 + (c^2 - y^2)^2}$$

Principal stress in terms of Pc^2/I

the longitudinal shearing stress may frequently be the significant stress—hence, the emphasis on this stress.

The other section to be discussed is the deep, wide-flange section subjected to a combination of large shear and large bending moment. For this combination, the high fiber and transverse shearing stresses occurring simultaneously at the junction of flange and web sometimes yields a principal stress greater than the fiber stress at the surface of the flange (see Example 8-4). In general, at any point in a beam, a combination of large M, V, Q, and y and a small t should suggest a check on the principal stresses at such a point. Otherwise, the maximum flexural (fiber) stress will very likely be the maximum normal stress, and the maximum shearing stress will probably occur at the same point.

A knowledge of the directions of the principal stresses may aid in the prediction of the directions of cracks in a brittle material (concrete, for example) and thus may aid in the design of reinforcement to carry the tensile stresses. Curves drawn with their tangents at each point in the directions of the principal stresses are called *stress trajectories*. Since there are, in general, two nonzero principal stresses at each point (plane stress), there are two stress trajectories passing through each point. These curves will be perpendicular, since the principal stresses are orthogonal; one set of curves will represent the maximum stresses, whereas the other represents the minimum stresses. The trajectories for a simply supported rectangular beam carrying a concentrated load at the midpoint are shown in Fig. 8-7, with dashed lines representing the directions of the compressive stresses and solid lines showing tensile stress directions. In the vicinities of the load and reactions there are stress concentrations, and the trajectories become much more complicated. Figure 8-7 neglects all stress concentrations.

In order to determine the principal stresses and the maximum shearing stresses at a particular point in a beam, it is necessary to calculate the fiber stresses and transverse (or longitudinal) shearing stresses at the point. With the stresses on orthogonal planes through the point known, the methods of Sections 7-3 or 7-4 can be used to calculate the maximum stresses at the point. The following example illustrates the procedure and provides an example in which the principal stresses at some interior point are greater than the maximum fiber stresses.

FIG. 8-7

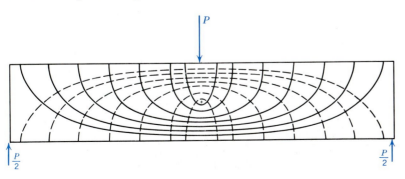

EXAMPLE 8-4

A cantilever beam [$I = 1243(10^6)$ mm^4, $S = 4079(10^3)$ mm^3], supported at the left end, carries a uniformly distributed load of 160 kN/m on a span of 2.5 m. Determine the maximum normal and shearing stresses in the beam.

Solution

For a cantilever beam with a uniformly distriubted load, both the maximum bending moment and the maximum transverse shear occur on the section at the support. In this case they are

$$M = \frac{wL^2}{2} = \frac{160(2.5^2)}{2} = 500 \text{ kN·m}$$

and

$$V = wL = 160(2.5) = 400 \text{ kN}$$

The upper half of the cross section of the beam at the wall is shown in Fig. 8-8a. The distribution of fiber stress for this half of the section is shown in Fig. 8-8b, and the distribution of the average transverse shearing stress is as shown in Fig. 8-8c. The vertical stresses σ_y due to the pressure of the load on the top of the beam are considered negligible.

Values will be calculated for three points—namely, at the neutral axis, in the web at the junction of the web and the top flange, and at the top surface. Both the fillets and the stress concentrations at the junction of the web and flange will be neglected. At the neutral axis, the fiber stress is zero; however,

$$Q = 304.8(19.7)(295) + 285.1(11.9)(142.6) = 2.255(10^6) \text{ mm}^3$$

and

$$\tau = \frac{VQ}{It} = \frac{400(10^3)(2.255)(10^{-3})}{1243(10^{-6})(0.0119)} = 61.0(10^6) \text{ N/m}^2 = 61.0 \text{ MPa}$$

In the web at the junction with the top flange, the fiber stress is

$$\sigma = \frac{My}{I} = \frac{500(10^3)(0.2851)}{1243(10^{-6})} = 114.7(10^6) \text{ N/m}^2 = 114.7 \text{ MPa } T$$

and since

$$Q = 304.8(19.7)(295) = 1.771(10^6) \text{ mm}^3$$

the transverse shearing stress is

$$\tau = \frac{VQ}{It} = \frac{400(10^3)(1.771)(10^{-3})}{1243(10^{-6})(0.0119)} = 47.9(10^6) \text{ N/m}^2 = 47.9 \text{ MPa}$$

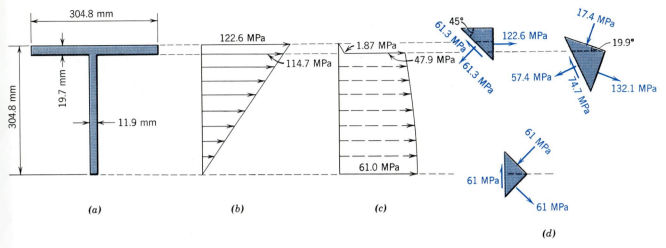

(a)   (b)   (c)   (d)

At the top surface, the transverse shearing stress is zero, and the fiber stress is

FIG. 8-8

$$\sigma = \frac{My}{I} = \frac{M}{S} = \frac{500(10^3)}{4079(10^{-6})} = 122.6(10^6) \text{ N/m}^2 = 122.6 \text{ MPa } T$$

The principal stresses and maximum shearing stresses for each of the three selected points are shown in Fig. 8-8d. The calculations, from the equations of Section 7-3, for the point at the junction of web and flange are

$$\sigma_p = \frac{114.7}{2} \pm \sqrt{\left(\frac{114.7}{2}\right)^2 + 47.9^2}$$
$$= 57.35 \pm 74.72$$

which gives

$$\sigma_{p1} = 132.1 \text{ MPa } T \qquad \sigma_{p2} = 17.4 \text{ MPa } C \qquad \text{and} \qquad \tau_{max} = 74.7 \text{ MPa}$$

The angle is not particularly important in this case, but it can be obtained from the equation

$$\tan 2\theta_p = -\frac{47.90}{57.35} = -0.835$$

The result is

$$2\theta_p = -39.9°$$

or

$$\theta_p = 19.9° \curvearrowleft$$

The other stresses of Fig. 8-8d are obtained by inspection. It should be noted that the maximum tensile stress of 132.1 MPa is 7.75 percent above the maximum tensile fiber stress of 122.6 MPa and that the maximum shearing stress of 74.7 MPa is 22.5 percent above the maximum transverse shearing stress of 61.0 MPa.

PROBLEMS

> *Note:* In the following problems neglect both the fillets and the stress concentrations at the junction of the web and flange. The action is elastic in all problems.

8-47* The cantilever beam shown in Fig. P8-47a has the cross section shown in Fig. P8-47b. A concentrated load $P = 50$ kips is applied at the free end of the beam. Determine, and show on a sketch, the principal stresses and the maximum shearing stress at point A, which is in the web of the beam just below the top flange.

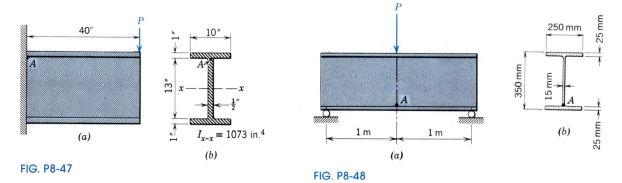

FIG. P8-47

FIG. P8-48

8-48* The simply supported beam shown in Fig. P8-48a has the cross section shown in Fig. P8-48b. A concentrated load $P = 500$ kN is applied at the middle of the span. Determine, and show on a sketch, the principal stresses and the maximum shearing stress at point A, which is in the web of the beam just above the bottom flange on a section just to the left of the load.

8-49 The timber beam shown in Fig. P8-49a has the cross section shown in Fig. P8-49b. If the allowable stresses are 200 psi shear and 400 psi compression at point A, which is 2 ft from the left end of the beam and 2 in. below the top surface of the beam, determine the maximum allowable load P.

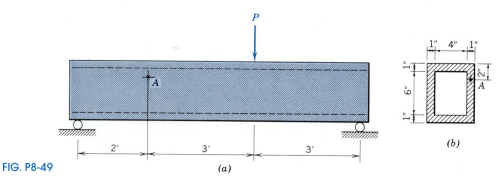

FIG. P8-49

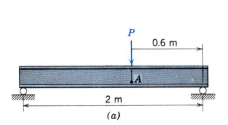

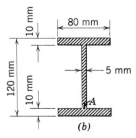

FIG. P8-50

8-50 The steel beam shown in Fig. P8-50a has the cross section shown in Fig. P8-50b. If the allowable stresses are 50 MPa shear and 100 MPa tension at point A, which is on a section just to the right of the load P and in the web of the beam just above the bottom flange, determine the maximum allowable load P.

8-51* A W18 × 60 simply supported beam (see Appendix C) with a span of 20 ft carries a concentrated load of 36 kips at the middle of the span. Determine the maximum principal and shearing stresses in the beam.

8-52* A W838 × 193 cantilever beam (see Appendix C) with a span of 3.5 m carries a uniformly distributed load of 225 kN/m. Determine the maximum principal and shearing stresses in the beam.

8-53 A W24 × 62 simply supported beam (see Appendix C) with a span of 18 ft carries a concentrated load P at the middle of the span. If the maximum principal and shearing stresses in the beam must not exceed 18 ksi and 10 ksi, respectively, determine the maximum allowable load P.

8-54 A W305 × 97 cantilever beam (see Appendix C) with a span of 3 m carries a concentrated load P at the free end of the beam. If the maximum principal and shearing stresses in the beam must not exceed 125 MPa and 75 MPa, respectively, determine the maximum allowable load P.

8-55* The beam shown in Fig. P8-55 has a 3-by-8-in. rectangular cross section and is loaded in a plane of symmetry. Determine and show on a sketch the principal and maximum shearing stresses at point A.

8-56 The machine element shown in Fig. P8-56 is loaded in a plane of symmetry. Determine and show on a sketch the principal and maximum shearing stresses at point A.

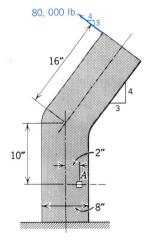

FIG. P8-55

FIG. P8-56

8-6

COMBINED CENTRIC, TORSIONAL, AND FLEXURAL LOADS

In numerous industrial situations, machine members are subjected to all three general types of loads studied so far. When a flexural load is combined with torsional and centric loads, it is frequently difficult to locate the point or points where the most severe stresses occur. The following review or summary statements may be helpful in this regard.

The longitudinal and transverse shearing stresses are maximum at the centroidal axis of the section where V is maximum, except for a few special shapes with increased width at the neutral axis. The flexural stress is maximum at the greatest distance from the centroidal axis on the section where M is maximum. The torsional shearing stress is maximum at the surface of a shaft at the section where T is maximum. With these facts in mind, it is normally possible to locate one or more possible points of high stress. Even so, it may be necessary to calculate the various stresses at more than one point of a member before locating the most severely stressed point. For stresses below the proportional limit of the material, the superposition method can be used to combine stresses on any given plane at any specific point of a loaded member.

After the stresses on a pair of mutually perpendicular planes at a specific point are determined, the methods of Sections 7-3 and 7-4 can be used to find the principal stresses and the maximum shearing stress at the point.

The following example illustrates the procedure for the solution of elastic combined stress problems.

EXAMPLE 8-5

The 100-mm-diameter shaft of Fig. 8-9a is subjected to the loads shown. Determine the maximum permissible value of P for allowable stresses of 120 MPa T and 70 MPa shear.

Solution

When two equal opposite forces P are introduced along line EF as in Fig. 8-9b, it is seen that the shaft is subjected to a couple of $0.6P$ N·m clockwise looking toward the wall and a concentrated load P N acting at E. In other words, the shaft acts as a cantilever beam with a concentrated load P N at the right end, a torsion member subjected to a $0.6P$-N·m torque, and an axially loaded member with a load of $30P$ N.

The couple produces a torsional shearing stress at all surface points of the shaft of

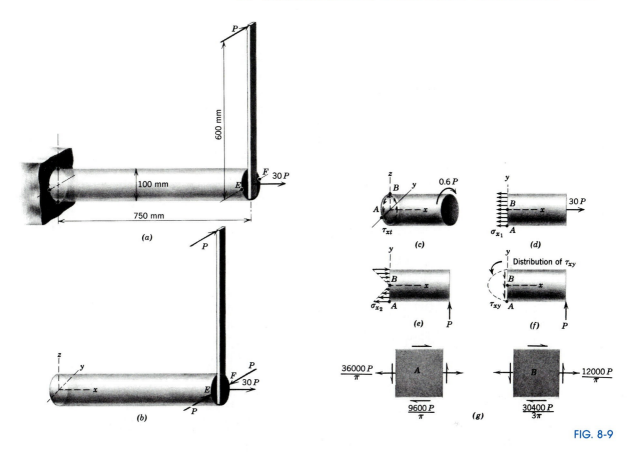

FIG. 8-9

$$\tau_{xt} = \frac{Tc}{J} = \frac{0.6P(0.050)}{(\pi/2)(0.050)^4} = \frac{9600P}{\pi}$$

as shown in Fig. 8-9c.

The axial tensile stress at all points of the shaft is

$$\sigma_{x1} = \frac{P}{A} = \frac{30P}{\pi(0.050)^2} = \frac{12000P}{\pi}$$

as shown in Fig. 8-9d.

The cantilever beam load P produces a maximum tensile fiber stress at point A and an equal compressive stress on the opposite side of the shaft. The axial stress is tensile, and therefore the compressive flexural stress is not critical. The tensile fiber stress at A is (see Fig. 8-9e)

$$\sigma_{x2} = \frac{Mc}{I} = \frac{0.75P(0.050)}{(\pi/4)(0.050)^4} = \frac{24000P}{\pi}$$

The transverse shearing stress is equal to the longitudinal shearing stress, which is maximum on the xz plane, including such surface points as B and C where the torsional shearing stress is also maximum. The maximum shearing stress is (see Fig. 8-9f)

$$\tau_{xy} = \frac{VQ}{It} = \frac{P\left[\dfrac{\pi(0.050)^2}{2}\right]\left[\dfrac{4(0.050)}{3\pi}\right]}{(\pi/4)(0.050)^4(0.100)} = \frac{1600P}{3\pi}$$

The torsional and transverse shearing stresses are in the same direction along the top of the shaft and in opposite directions along the bottom of the shaft. Therefore, any critical situation due to these two stresses will develop along the top of the shaft at some point such as B.

All four stresses are maximum at some point or points in the section of the shaft at the wall, since the fiber stress is not maximum at any other section of the shaft. Therefore, a complete analysis for the member can be made on the section at the wall. A complete analysis in cluding interior points of this section is beyond the scope of this book. However, a fairly good solution can be obtained by investigating the combined stresses at two surface points A and B. At A the fiber stress, axial stress, and torsional shearing stress are all maximum, whereas the transverse shearing stress is zero. At B the axial stress, torsional shearing stress, and transverse shearing stress are all maximum, whereas the fiber stress is zero. The fiber stress and transverse shearing stress are never maximum at the same point. Furthermore, the torsional shearing stress is never maximum at an interior point. Therefore, the chance of finding a more severe stress at an interior point is unlikely.

The stresses on orthogonal planes through A and B are as shown in Fig. 8-9g. When these values are substituted in Eq. 7-4 of Section 7-3 and the resulting expressions compared, it is evident that the combination of stresses of A is more severe than that at B. The maximum stresses at A are

$$\sigma_A = \frac{18000P}{\pi} + \sqrt{\left(\frac{18000P}{\pi}\right)^2 + \left(\frac{9600P}{\pi}\right)^2} = \frac{38400P}{\pi}$$

and

$$\tau_A = \sqrt{\left(\frac{18000P}{\pi}\right)^2 + \left(\frac{9600P}{\pi}\right)^2} = \frac{20400P}{\pi}$$

The allowable values of P from the allowable stresses and these relations are

$$120(10^6) = 38400P_\sigma/\pi$$

or

$$P_\sigma = 9.817(10^3) \text{ N} = 9.82 \text{ kN}$$

and

$$70(10^6) = 20400 \, P_\tau/\pi$$

or

$$P_\tau = 10.780(10^3) \text{ N} = 10.78 \text{ kN}$$

The smaller value controls and the maximum permissible value is

$$P = 9.82 \text{ kN} \qquad\qquad \text{Ans.}$$

PROBLEMS

> *Note:* In the following problems, stress concentrations and the effect of beam deflection on the eccentricity are to be neglected unless otherwise noted. Assume elastic action.

8-57* A 2-in.-diameter steel rod is loaded and supported as shown in Fig. P8-57. Determine, and show on a sketch, the principal stresses and the maximum shearing stress at the top surface of the rod adjacent to the support.

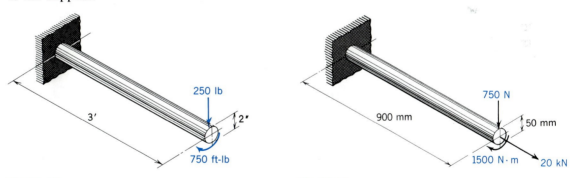

FIG. P8-57   FIG. P8-58

8-58* A 50-mm-diameter steel rod is loaded and supported as shown in Fig. P8-58. Determine, and show on a sketch, the principal stresses and the maximum shearing stress at the top surface of the rod adjacent to the support.

8-59 A 4-in.-diameter steel rod is loaded and supported as shown in Fig. P8-59. Determine the principal stresses and the maximum shearing stress

(a) at point *A* on a section adjacent to the support.

(b) at point *B* on a section adjacent to the support.

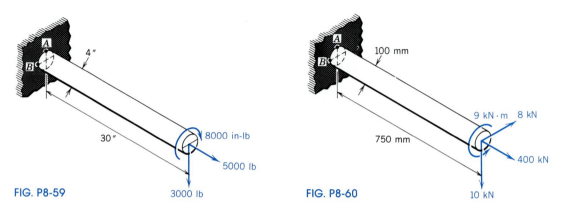

FIG. P8-59

FIG. P8-60

8-60 A 100-mm-diameter steel rod is loaded and supported as shown in Fig. P8-60. Determine the principal stresses and the maximum shearing stress

(a) at point *A* on a section adjacent to the support.

(b) at point *B* on a section adjacent to the support.

8-61* A 2-in.-diameter shaft is loaded and supported as shown in Fig. P8-61. Determine the principal stresses and the maximum shearing stress at point *A* on the surface of the shaft.

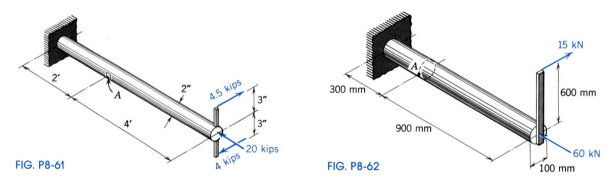

FIG. P8-61

FIG. P8-62

8-62* A 100-mm-diameter shaft is loaded and supported as shown in Fig. P8-62. Determine the principal stresses and the maximum shearing stress at point *A* on the surface of the shaft.

8-63 A solid circular steel shaft is loaded and supported as shown in Fig. P8-63. Because of stress concentrations at point *A* near the support, the maximum normal and shearing stresses at the point must be limited to 10 ksi and 6 ksi, respectively. Determine the minimum permissible diameter that can be used for the shaft.

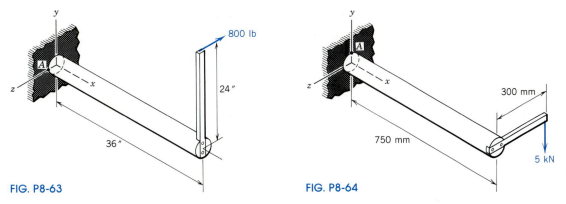

FIG. P8-63

FIG. P8-64

8-64 A solid circular steel shaft is loaded and supported as shown in Fig. P8-64. Because of stress concentrations at point A near the support, the maximum normal and shearing stresses at the point must be limited to 75 MPa and 40 MPa, respectively. Determine the minimum permissible diameter that can be used for the shaft.

8-65* A solid circular shaft is loaded and supported as shown in Fig. P8-65. If the maximum tensile and shearing stresses at point A on the surface of the shaft are not to exceed 18 ksi and 10 ksi, respectively, determine the minimum permissible diameter for the shaft.

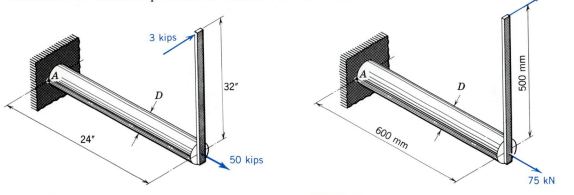

FIG. P8-65

FIG. P8-66

8-66 A solid circular shaft is loaded and supported as shown in Fig. P8-66. if the minimum tensile and shearing stresses at point A on the surface of the shaft are not to exceed 140 MPa and 85 MPa, respectively, determine the minimum permissible diameter for the shaft.

8-67 A steel shaft 4 in. in diameter is supported in flexible bearings at its ends. Two pulleys, each 2 ft in diameter, are keyed to the shaft. The pulleys carry belts as shown in Fig. P8-67. Determine the principal stresses and the maximum shearing stress at point A on the surface of the shaft.

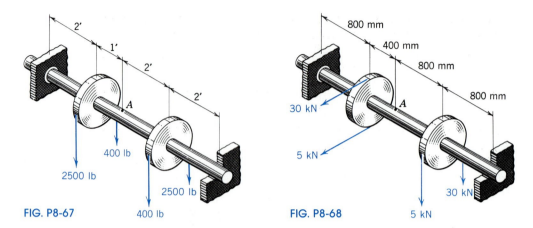

FIG. P8-67   FIG. P8-68

8-68* A steel shaft 120 mm in diameter is supported in flexible bearings at its ends. Two pulleys, each 500 mm in diameter, are keyed to the shaft. The pulleys carry belts as shown in Fig. P8-68. Determine the principal stresses and the maximum shearing stress at point A on the surface of the shaft.

8-69* The thin-walled cylindrical pressure vessel shown in Fig. P8-69 has an inside diameter of 24 in. and a wall thickness of $\frac{1}{2}$ in. The vessel is subjected to an internal pressure of 250 psi. In addition, an axial load of 75 kips and a torque of 150 ft-kips is applied to the vessel through rigid plates on the ends, as shown in Fig. P8-69. Determine the maximum normal and shearing stresses at a point on the outside surface of the vessel.

8-70 A thin-walled cylindrical pressure tank is fabricated by butt-welding 15-mm plate with a spiral seam as shown in Fig. P8-70. The pressure

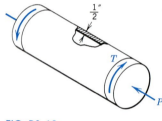

FIG. P8-69

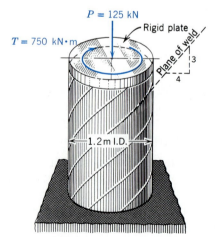

FIG. P8-70

in the tank is 2500 kPa. Additional loads are applied to the cylinder through a rigid end plate as shown in Fig. P8-70. Determine

(a) the normal and shearing stresses on the plane of the weld at the outside surface of the tank.

(b) the principal stresses and the maximum shearing stress at a point on the inside surface of the tank.

8-71* A 4-in.-diameter solid circular steel shaft is loaded and supported as shown in Fig. P8-71. If the maximum normal and shearing stresses at point A must be limited to 7500 psi T and 5000 psi, respectively, determine the maximum permissible value for the transverse load V.

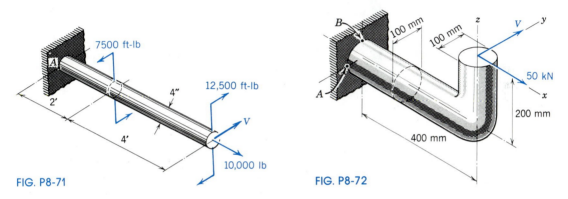

FIG. P8-71

FIG. P8-72

8-72* A 100-mm-diameter solid circular steel shaft is loaded and supported as shown in Fig. P8-72. If the maximum normal and shearing stresses at point A and B must be limited to 120 MPa T and 60 MPa, respectively, determine the maximum permissible value for the transverse load V.

8-73 A hollow circular shaft has an outside diameter of 12 in. and an inside diameter of 4 in., as shown in Fig. P8-73. The shaft is in equi-

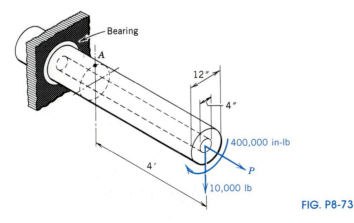

FIG. P8-73

librium when the indicated loads and torque are applied. If the maximum normal and shearing stresses at point A must be limited to 5000 psi T and 3000 psi, respectively, determine the maximum permissible value for the axial load P.

8-74 The motor shown in Fig. P8-74 supplies a torque of 400 N·m to the 100-mm-diameter drive gear. Determine the minimum diameter required for the shaft between the motor and the drive gear if the maximum tensile stress in the shaft must be limited to 75 MPa.

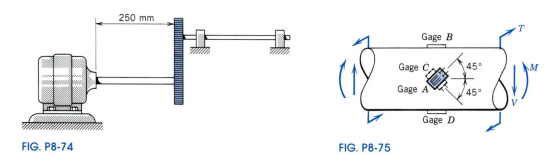

FIG. P8-74   FIG. P8-75

8-75 Four strain gages are mounted at 90° intervals around the circumference of a 100-mm-diameter steel ($E = 200$ GPa and $\nu = 0.30$) shaft, as shown in Fig. P8-75. All of the gages are oriented at an angle of 45° with respect to a line on the surface of the shaft that is parallel to the axis of the shaft. Determine the torque T, shear V, and moment M at the cross section where the gages are located if $\epsilon_A = 450\mu$, $\epsilon_B = 325\mu$, $\epsilon_C = 550\mu$, and $\epsilon_D = 675\mu$.

8-7
STRESS CONCENTRATIONS UNDER COMBINED LOADINGS

In Section 2-6 it was shown that the introduction of a circular hole or other geometric discontinuity into an axially loaded member can cause a significant increase in the magnitude of the stress (stress concentration) in the immediate vicinity of the discontinuity. This is also the case for circular shafts under torsional forms of loading (see Section 3-8) and for beams under flexural forms of loading (see Section 4-10). When a combined form of loading is applied to any of these simple members, the principle of superposition can be used to evaluate the stress concentration associated with a few of the common discontinuities. The procedure is illustrated in the following example.

EXAMPLE 8-6

A small hole has been drilled normal to the surface of a circular shaft in a region where a pure torque T is being transmitted by the shaft as shown in Fig. 8-10a. Determine

    (a) the stress distribution on the boundary of the hole.

    (b) the stress-concentration factor for this type of discontinuity.

Solution

The stress distribution produced by a torsional form of loading in a circular shaft is given by Eq. 3-5. The stresses are maximum at the surface of the shaft and can be represented as shown in Fig. 8-10b. The state of stress at the surface of the shaft can also be represented by the principal stresses $\sigma_{p1} = -\sigma_{p2} = Tc/J$ as shown in Fig. 8-10c.

    (a) The stress distribution σ_θ on the boundary of a small circular hole in a uniform unidirectional stress field σ was discussed in Section 2-6 and can be expressed as

$$\sigma_\theta = \sigma\,(1 + 2\cos 2\theta)$$

The stress distribution σ_θ on the boundary of the small circular hole in the shaft can be obtained by superimposing the stresses resulting from principal stresses σ_{p1} and σ_{p2}. The directions associated with each of these principal stresses with respect to an arbitrary point A on the boundary of the hole are shown in Figs. 8-10d and 8-10e. Thus

$$\sigma_\theta = \sigma_{p1}(1 + 2\cos 2\theta_1) + \sigma_{p2}(1 + 2\cos 2\theta_2)$$
$$= +\frac{Tc}{J}\left[1 + 2\cos 2\left(\alpha + \frac{\pi}{4}\right)\right]$$
$$\quad\; -\frac{Tc}{J}\left[1 + 2\cos 2\left(\alpha - \frac{\pi}{4}\right)\right]$$
$$= \underline{-4\frac{Tc}{J}\sin 2\,\alpha} \qquad\qquad\qquad\text{Ans.}$$

    (b) The maximum tensile stress on the boundary of the hole occurs when $\alpha = -\pi/4$ or $3\pi/4$. At these points,

$$\sigma_\theta = 4\left(\frac{Tc}{J}\right)$$

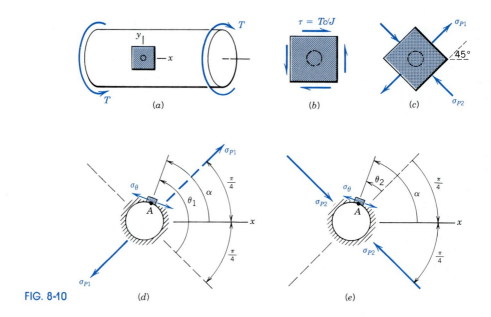

FIG. 8-10

The maximum compressive stress on the boundary of the hole occurs when $\alpha = \pi/4$ or $5\pi/4$. At these points,

$$\sigma_\theta = -4\left(\frac{Tc}{J}\right)$$

Thus, the stress-concentration factor K_t for a small hole drilled normal to the surface of a circular shaft in a region where a pure torque T is being transmitted by the shaft is

$$K_t = \underline{4} \qquad \text{Ans.}$$

Note that this stress-concentration factor is based on the shearing stress $\tau = Tc/J$ at the surface of the shaft.

PROBLEMS

Note: In the following problems, all of the stresses are below the proportional limits of the materials.

A 50-mm-diameter solid circular steel shaft has a small hole drilled normal to the surface as shown in Fig. P8-76. Determine the maximum normal stress at the boundary of the hole if $P = 6.5$ kN, $T = 1000$ N·m, and $M = 650$ N·m.

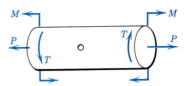

FIG. P8-76

8-77* A 2-in.-diameter solid circular steel shaft has a small hole drilled normal to the surface as shown in Fig. P8-76. Determine the maximum normal stress at the boundary of the hole if $P = 1500$ lb, $T = 800$ in.-lb, and $M = 500$ in.-lb.

8-78 A 50-mm-diameter solid circular steel shaft has a small hole drilled normal to the surface as shown in Fig. P8-78. Determine the maximum normal stress at the boundary of the hole if $P = 7$ kN, $T = 100$ N·m, and $M = 60$ N·m.

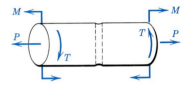

FIG. P8-78

8-79 A hollow circular steel shaft with an outside diameter of 3 in. and an inside diameter of 2 in. has a small hole drilled normal to the surface as shown in Fig. P8-79. If $T = 15$ in.-kips, and $M = 10$ in.-kips, determine

(a) the maximum normal stress at point A on the inside surface of the shaft.

(b) the maximum normal stress at point B at the boundary of the hole.

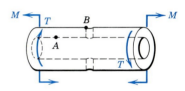

FIG. P8-79

8-80* A cantilever beam with a rectangular cross section 6 in. wide by 4 in. deep has a small hole drilled normal to the top and bottom surfaces of the beam as shown in Fig. P8-80. Determine the maximum normal and shearing stresses produced at the boundary of the hole on surface $a–a$ by a load $P = 30$ kips.

8-81 A cantilever beam with a rectangular cross section 6 in. wide by 4 in. deep has a small hole drilled normal to the top and bottom surfaces of the beam as shown in Fig. P8-80. Determine the maximum normal and shearing stresses produced at the boundary of the hole on surface $b–b$ by a load $P = 30$ kips.

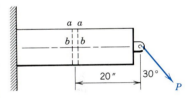

FIG. P8-80

8-82* A cantilever beam with a 50-mm by 50-mm cross section has a small hole drilled normal to the side of the beam as shown in Fig. P8-82. Determine the maximum normal and shearing stresses produced at the boundary of the hole by a load $P = 13$ kN.

8-83* A W8 × 31 wide-flange section (see Appendix C) is used as a cantilever beam. A small hole is drilled normal to the web of the beam as shown in Fig. P8-83. Determine the maximum normal stress produced at the edge of the hole by a load $P = 10$ kips if $y = 0$.

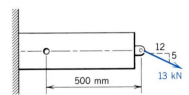

FIG. P8-82

8-84 A W8 × 31 wide-flange section (see Apendix C) is used as a cantilever beam. A small hole is drilled normal to the web of the beam as shown in Fig. P8-83. Determine the maximum normal stress produced at the edge of the hole by a load $P = 10$ kips if $y = 3$ in.

FIG. P8-83

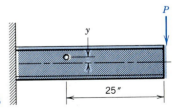

8-8

THEORIES OF FAILURE

A tension test of an axially loaded member is easy to conduct, and the results, for many types of materials, are well known. When such a member fails, the failure occurs at a specific principal (axial) stress, a definite axial strain, a maximum shearing stress of one-half the axial stress, and a specific amount of strain energy per unit volume of stressed material. Since all of these limits are reached simultaneously for an axial load, it makes no difference which criterion (stress, strain, or energy) is used for predicting failure in another axially loaded member of the same material.

For an element subjected to biaxial or triaxial loading, however, the situation is more complicated because the limits of normal stress, normal strain, shearing stress, and strain energy existing at failure for an axial load are not all reached simultaneously. In other words, the cause of failure, in general, is unknown. In such cases, it becomes important to determine the best criterion for predicting failure, because test results are difficult to obtain and the combinations of loads are endless. Several theories have been proposed for predicting failure of various types of material subjected to many combinations of loads. Unfortunately, none of the theories agree with test data for all types of materials and combinations of loads. Several of the more common theories of failure are presented and briefly explained in the following paragraphs.

1. Maximum-Normal-Stress Theory[2]　The maximum-normal-stress theory predicts failure of a specimen subjected to any combination of loads when the maximum normal stress at any point in the specimen reaches the axial failure stress as determined by an axial tensile or compressive test of the same material.

The maximum-normal-stress theory is presented graphically in Fig. 8-11b for an element subjected to biaxial principal stresses in the $p1$ and $p2$ directions, as shown in Fig. 8-11a. The limiting stress σ_f is the failure stress for this material when loaded axially. Any combination of biaxial principal stresses σ_{p1} and σ_{p2} represented by a point inside the square of Fig. 8-11b is safe according to this theory, whereas any combination of stresses represented by a point outside of the square will cause failure of the element on the basis of this theory.

[2]Often called Rankine's theory after W. J. M. Rankine (1820–1872), an eminent engineering educator in Scotland.

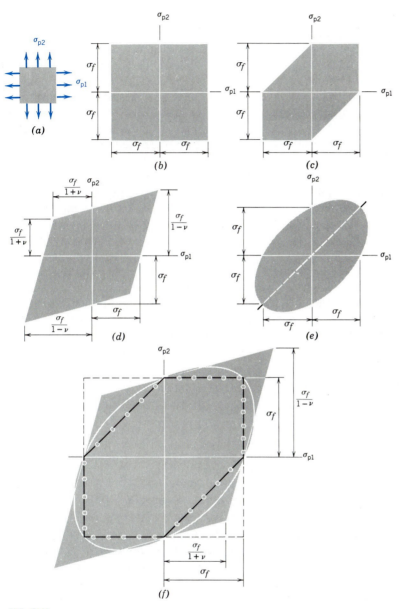

FIG. 8-11

2. Maximum-Shearing-Stress Theory[3]   The maximum-shearing-stress theory predicts failure of a specimen subjected to any combination of loads when the maximum shearing stress at any point in the specimen reaches the failure stress τ_f equal to $\sigma_f/2$, as determined by an axial tensile or compressive test of the same material. For ductile materials the shearing elastic limit, as determined from a torsion test (pure shear), is greater than one-half the tensile elastic limit (with an average value of τ_f about $0.57\sigma_f$). This means that the maximum shearing stress theory errs on the conservative side by being based on the limit obtained from an axial test.

The maximum-shearing-stress theory is presented graphically in Fig. 8-11c for an element subjected to biaxial (σ_z is equal to zero) principal stresses, as shown in Fig. 8-11a. In the first and third quadrants, σ_{p1} and σ_{p2} have the same sign, and the maximum shearing stress is half the numerically larger principal stress σ_{p1} or σ_{p2}, as explained in Section 7-3. In the second and fourth quadrants, where σ_{p1} and σ_{p2} are of an opposite sign, the maximum shearing stress is half the arithmetical sum of the two principal stresses. In the fourth quadrant, the equation of the boundary, or limit stress, line is

$$\sigma_{p1} - \sigma_{p2} = \sigma_f$$

and in the second quadrant the relation is

$$\sigma_{p1} - \sigma_{p2} = -\sigma_f$$

3. Maximum-Normal-Strain Theory[4]   The maximum-normal-strain theory predicts failure of a specimen subjected to any combination of loads when the maximum normal strain at any point in the specimen reaches the failure strain σ_f/E at the proportional limit, as determined by an axial tensile or compressive test of the same material.

The maximum-normal-strain theory is presented graphically in Fig. 8-11d for an element subjected to biaxial principal stresses (not greater than the proportional limit), as shown in Fig. 8-11a. The limiting strain in the positive $p1$ direction is

$$\epsilon_{p1} = \frac{\sigma_f}{E} = \frac{\sigma_{p1} - \nu\sigma_{p2}}{E}$$

---

[3] Sometimes called Coulomb's theory because it was originally stated by him in 1773. More frequently called Guest's theory or law because of the work of J. J. Guest in England in 1900.

[4] Often called Saint-Venant's theory because of the work of Barre de Saint-Venant, a great French mathematician and elastician.

from which

$$\sigma_{p2} = \frac{1}{\nu}(\sigma_{p1} - \sigma_f)$$

which is observed to be a straight line through the point $(\sigma_f, 0)$ with a slope of $1/\nu$. The limiting strain in the positive $p2$ direction is

$$\epsilon_{p2} = \frac{\sigma_f}{E} = \frac{\sigma_{p2} - \nu\sigma_{p1}}{E}$$

from which

$$\sigma_{p2} = \nu\sigma_{p1} + \sigma_f$$

which is observed to be a straight line through the point $(0, \sigma_f)$ with a slope equal to ν.

4. Maximum-Strain-Energy Theory   The maximum-strain-energy theory predicts failure of a specimen subjected to any combination of loads when the strain energy per unit volume of any portion of the stressed member reaches the failure value of the strain energy per unit volume as determined by an axial tensile or compressive test of the same material.

The concept of strain energy is illustrated in Fig. 8-12. A bar of uniform cross section subjected to a slowly applied axial load P is shown in Fig. 8-12a. A load-deformation diagram for the bar is shown in Fig. 8-12b. The work done in elongating the bar an amount δ_2 is

$$W_k = \int_0^{\delta_2} P \, d\delta \qquad\qquad (a)$$

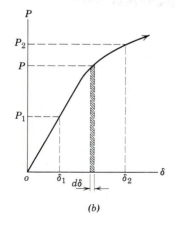
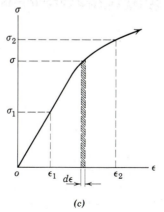

(a) (b) (c) FIG. 8-12

where P is some function of δ. The work done on the bar must equal the change in energy of the material,[5] and this energy change, because it involves the strained configuration of the material, is termed *strain energy U*. If δ is expressed in terms of axial strain ($\delta = L\epsilon$) and P in terms of axial stress ($P = A\sigma$), Eq. *a* becomes

$$W_k = U = \int_0^{\epsilon_2} (\sigma)(A)(L)d\epsilon = AL \int_0^{\epsilon_2} \sigma \, d\epsilon \qquad (b)$$

where σ is a function of ϵ (see Fig. 8-12c). If Hooke's law applies,

$$\epsilon = \sigma/E \qquad d\epsilon = d\sigma/E$$

and Eq. *b* becomes

$$U = \left(\frac{AL}{E}\right) \int_0^{\sigma_2} \sigma \, d\sigma$$

or

$$U = AL \left(\frac{\sigma_2^2}{2E}\right) \qquad (c)$$

Equation *c* gives the *elastic strain energy* (which is, in general, recoverable[6]) for axial loading of a material obeying Hooke's law. The quantity in parentheses, $\sigma_2^2/(2E)$, is the elastic strain energy u in tension or compression per unit volume, or strain energy intensity, for a particular value of stress σ below the proportional limit of the material. Thus

$$u = \frac{\sigma^2}{2E} = \frac{\sigma\epsilon}{2} \qquad (8\text{-}5)$$

For shear loading, the expression would be identical except that σ would be replaced by τ, ϵ by γ, and E by G.

The concept of elastic strain energy can be extended to include biaxial and triaxial loadings by writing the expression for strain energy intensity u as $\sigma\epsilon/2$ and adding the energies due to each of the stresses. Since energy is a positive scalar quantity, the addition is the arithmetic sum of the energies. For a system of triaxial principal stresses σ_{p1}, σ_{p2}, σ_{p3}, the total elastic strain energy intensity is

$$u = (1/2)\sigma_{p1}\epsilon_{p1} + (1/2)\sigma_{p2}\epsilon_{p2} + (1/2)\sigma_{p3}\epsilon_{p3} \qquad (d)$$

When the expressions for strains in terms of stresses from Eq. 1-23 of Section 1-14 are substituted into Eq. *d*, the result is

---

[5]Known as *Clapeyron's theorem*, after the French engineer B. P. E. Clapeyron (1799–1864).

[6]Elastic hysteresis is neglected here as an unnecessary complication.

$$u = \frac{1}{2E} \{ \sigma_{p1}[\sigma_{p1} - \nu(\sigma_{p2} + \sigma_{p3})] + \sigma_{p2}[\sigma_{p2} - \nu(\sigma_{p3} + \sigma_{p1})]$$

$$+ \sigma_{p3}[\sigma_{p3} - \nu(\sigma_{p1} + \sigma_{p2})]\}$$

from which

$$u = \frac{1}{2E} [\sigma_{p1}^2 + \sigma_{p2}^2 + \sigma_{p3}^2 - 2\nu(\sigma_{p1}\sigma_{p2} + \sigma_{p2}\sigma_{p3} + \sigma_{p3}\sigma_{p1})] \quad (8\text{-}6)$$

The maximum-strain-energy theory of failure assumes that inelastic action will occur under a biaxial or triaxial state of stress whenever the elastic strain energy given by Eq. 8-6 exceeds the limiting value obtained from a tensile test. Thus, from Eqs. 8-5 and 8-6,

$$\frac{\sigma_f^2}{2E} = \frac{1}{2E} [\sigma_{p1}^2 + \sigma_{p2}^2 + \sigma_{p3}^2$$

$$- 2\nu(\sigma_{p1}\sigma_{p2} + \sigma_{p2}\sigma_{p3} + \sigma_{p3}\sigma_{p1})]$$

or

$$\sigma_f^2 = \sigma_{p1}^2 + \sigma_{p2}^2 + \sigma_{p3}^2 - 2\nu(\sigma_{p1}\sigma_{p2} + \sigma_{p2}\sigma_{p3} + \sigma_{p3}\sigma_{p1})$$

Since the maximum-strain-energy theory is not verified by experiment, it has been largely replaced by the maximum-distortion-energy theory.

5. Maximum-Distortion-Energy Theory[7]   This theory differs from the maximum-strain-energy theory in that the portion of the strain energy producing volume change is considered ineffective in causing failure by yielding. Supporting evidence comes from experiments showing that homogeneous materials can withstand very high hydrostatic stresses without yielding. Therefore, only the portion of the strain energy producing a change of shape is assumed to be responsible for the failure of the material by inelastic action. The strain energy of distortion is most readily computed by determining the total strain energy of the stressed material and subtracting the strain energy associated with the volume change.

The strain energy can be resolved into two components u_v and u_d, resulting from a volume change and a distortion, respectively, by considering the principal stresses to be made up of two sets of stresses as indicated in Figs. 8-13a, b, and c. The state of stress in Fig. 8-13c will

[7]Frequently called the Huber–Hencky–von Mises theory, because it was proposed by M. T. Huber of Poland in 1904 and independently by R. von Mises of Germany in 1913. The theory was further developed by H. Hencky and von Mises in Germany and the United States.

result in distortion only (no volume change) if the sum of the three normal strains is zero. That is,

$$E(\epsilon_{p1} + \epsilon_{p2} + \epsilon_{p3})_d = [(\sigma_{p1} - p) - \nu(\sigma_{p2} + \sigma_{p3} - 2p)]$$
$$+ [(\sigma_{p2} - p) - \nu(\sigma_{p3} + \sigma_{p1} - 2p)]$$
$$+ [(\sigma_{p3} - p) - \nu(\sigma_{p1} + \sigma_{p2} - 2p)] = 0$$

which reduces to

$$(1 - 2\nu)(\sigma_{p1} + \sigma_{p2} + \sigma_{p3} - 3p) = 0$$

Therefore,

$$p = \frac{1}{3}(\sigma_{p1} + \sigma_{p2} + \sigma_{p3})$$

The three nominal strains due to p are, from Eq. 1-23,

$$\epsilon_v = \frac{1}{E}(1 - 2\nu)p$$

and the energy resulting from the hydrostatic stress (the volume change) is

$$u_v = 3\left(\frac{p\epsilon_v}{2}\right) = \frac{3}{2}\frac{1 - 2\nu}{E}p^2$$
$$= \frac{1 - 2\nu}{6E}(\sigma_{p1} + \sigma_{p2} + \sigma_{p3})^2$$

The energy resulting from the distortion (change of shape) is

$$u_d = u - u_v$$
$$= \frac{1}{6E}[3(\sigma_{p1}^2 + \sigma_{p2}^2 + \sigma_{p3}^2) - 6\nu(\sigma_{p1}\sigma_{p2} + \sigma_{p2}\sigma_{p3} + \sigma_{p3}\sigma_{p1})$$
$$- (1 - 2\nu)(\sigma_{p1} + \sigma_{p2} + \sigma_{p3})^2]$$

When the third term in the brackets is expanded, the expression can be rearranged to give

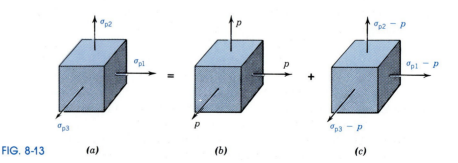

FIG. 8-13   (a)   (b)   (c)

$$u_d = \frac{1 + \nu}{6E} \left[(\sigma_{p1}^2 - 2\sigma_{p1}\sigma_{p2} + \sigma_{p2}^2) + (\sigma_{p2}^2 - 2\sigma_{p2}\sigma_{p3} + \sigma_{p3}^2) \right.$$

$$+ \left. (\sigma_{p3}^2 - 2\sigma_{p3}\sigma_{p1} + \sigma_{p1}^2) \right]$$

$$= \frac{1 + \nu}{6E} \left[(\sigma_{p1} - \sigma_{p2})^2 + (\sigma_{p2} - \sigma_{p3})^2 + (\sigma_{p3} - \sigma_{p1})^2 \right] \qquad (a)$$

The maximum-distortion-energy theory of failure assumes that inelastic action will occur whenever the energy given by Eq. *a* exceeds the limiting value obtained from a tensile test. For this test, only one of the principal stresses will be nonzero. If this stress is called σ_f, the value of u_d becomes

$$(u_d)_f = \frac{1 + \nu}{3E} \sigma_f^2$$

and when this value is substituted in Eq. *a*, it becomes

$$2\sigma_f^2 = (\sigma_{p1} - \sigma_{p2})^2 + (\sigma_{p2} - \sigma_{p3})^2 + (\sigma_{p3} - \sigma_{p1})^2 \qquad (b)$$

or

$$\sigma_f^2 = \sigma_{p1}^2 + \sigma_{p2}^2 + \sigma_{p3}^2 - (\sigma_{p1}\sigma_{p2} + \sigma_{p2}\sigma_{p3} + \sigma_{p3}\sigma_{p1})$$

for failure by slip.

When a state of plane stress exists, assume σ_{p3} equals zero, Eq. *b* becomes

$$\sigma_{p1}^2 - \sigma_{p1}\sigma_{p2} + \sigma_{p2}^2 = \sigma_f^2$$

This last expression is the equation of an ellipse with its major axis along the line σ_{p1} equals σ_{p2} as shown in Fig. 8-11e.

The graphic representations of Figs. 8-11b, c, d, and e are superimposed in Fig. 8-11f for convenient comparison of the different theories. The maximum-distortion-energy theory has been found to agree best with available test data for ductile materials. The maximum-normal-strain theory is unsafe for ductile materials. The maximum-normal-stress theory is also unsafe for ductile materials with principal stresses of opposite sign. The maximum-shearing-stress theory is conservative for ductile materials and is used in some design codes. The maximum-normal-stress theory agrees with test data for brittle materials and is generally accepted in design practice for such materials.

The following example illustrates the application of the theories of failure in predicting the load-carrying capacity of a member.

EXAMPLE 8-7

The solid circular shaft of Fig. 8-14 has a proportional limit of 64 ksi and a Poisson's ratio of 0.30. Determine the value of the load R for failure by slip as predicted by each of the five given theories of failure. Assume that point A is the most severely stressed point.

Solution

When two equal opposite forces R are introduced along line EF of Fig. 8-14, it is evident that the bending moment at point A is $36R$ in.-kips, the torsional moment is $24R$ in.-kips, and the axial load is $12R$ kips. The flexural stress at A is

$$\sigma_1 = \frac{Mc}{I} = \frac{36R(2)}{\pi(2^4)/4} = \frac{18R}{\pi} \; T$$

The direct stress at A is

$$\sigma_2 = \frac{P}{A} = \frac{12R}{\pi(2^2)} = \frac{3R}{\pi} \; T$$

and the sum of the two tensile stresses at A becomes

$$\sigma = \sigma_1 + \sigma_2 = \frac{21R}{\pi}$$

The torsional shearing stress at A is

$$\tau = \frac{Tc}{J} = \frac{24R(2)}{\pi(2^4)/2} = \frac{6R}{\pi}$$

The maximum stresses at A are

$$\tau_{max} = \sqrt{\left(\frac{10.5R}{\pi}\right)^2 + \left(\frac{6R}{\pi}\right)^2} = \frac{12.1R}{\pi}$$

$$\sigma_{p1} = \frac{10.5R}{\pi} + \frac{12.1R}{\pi} = \frac{22.6R}{\pi} \; T$$

and

$$\sigma_{p2} = \frac{10.5R}{\pi} - \frac{12.1R}{\pi} = -\frac{1.6R}{\pi} = \frac{1.6R}{\pi} \; C$$

According to the maximum-normal-stress theory

$$64 = \sigma_f = \sigma_{p1} = 22.6R/\pi$$

from which

$$R = 64\pi/22.6 = \underline{8.90 \text{ kips}} \qquad \text{Ans.}$$

According to the maximum-shearing-stress theory

$$\frac{64}{2} = \tau_f = \tau_{\text{max}} = \frac{12.1R}{\pi}$$

from which

$$R = \frac{32\pi}{12.1} = \underline{8.31 \text{ kips}} \qquad\qquad \text{Ans.}$$

The maximum-normal-strain theory gives

$$\frac{\sigma_f}{E} = \frac{\sigma_{p1} - \nu\sigma_{p2}}{E}$$

or

$$64 = \frac{22.6R}{\pi} - 0.3\left(-\frac{1.6R}{\pi}\right)$$

which gives

$$R = \frac{64\pi}{23.1} = \underline{8.70 \text{ kips}} \qquad\qquad \text{Ans.}$$

According to the maximum-strain-energy theory

$$\sigma_f^2 = \sigma_{p1}^2 + \sigma_{p2}^2 - 2\nu\sigma_{p1}\sigma_{p2}$$

or

$$64^2 = \left(\frac{22.6R}{\pi}\right)^2 + \left(\frac{-1.6R}{\pi}\right)^2 - 2(0.30)\left(\frac{22.6R}{\pi}\right)\left(\frac{-1.6R}{\pi}\right)$$

which gives

$$R = \frac{64\pi}{23.1} = \underline{8.70 \text{ kips}} \qquad\qquad \text{Ans.}$$

According to the maximum-distortion-energy theory

$$\sigma_f^2 = \sigma_{p1}^2 + \sigma_{p2}^2 - \sigma_{p1}\sigma_{p2}$$

or

$$64^2 = \left(\frac{22.6R}{\pi}\right)^2 + \left(-\frac{1.6R}{\pi}\right)^2 - \frac{22.6R}{\pi}\left(-\frac{1.6R}{\pi}\right)$$

which gives

$$R = \frac{64\pi}{23.4} = \underline{8.59 \text{ kips}} \qquad\qquad \text{Ans.}$$

In this example, the maximum-shearing-stress theory is seen to be the most conservative, and the maximum-normal-stress theory gives the least conservative result.

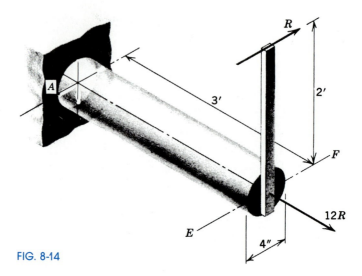

FIG. 8-14

PROBLEMS

> *Note:* In the following problems, all of the stresses are below the proportional limits of the materials.

8-85* A machine component fabricated from a material with a proportional limit in tension and compression of 60 ksi and a Poisson's ratio of 0.30 is subjected to a biaxial state of stress. The principal stresses are 30 ksi T and 50 ksi C. Determine which, if any, of the theories will predict failure by slip for this state of stress.

8-86* A machine component fabricated from a material with a proportional limit in tension and compression of 380 MPa and a Poisson's ratio of 0.33 is subjected to a biaxial state of stress. The principal stresses are 180 MPa T and 270 MPa C. Determine which, if any, of the theories will predict failure by slip for this state of stress.

8-87 At a point on the free surface of an alloy-steel ($E = 30,000$ ksi and $\nu = 0.30$) machine component the principal stresses are 45 ksi T and 25 ksi C. What minimum proportional limit is required according to each of the five theories if failure by slip is to be avoided?

8-88 At a point on the free surface of an aluminum alloy ($E = 73$ GPa and $\nu = 0.33$) machine component the principal stresses are 125 MPa T and 190 MPa C. What minimum proportional limit is required according to each of the five theories if failure by slip is to be avoided?

8-89* A material with a proportional limit of 36 ksi in tension and compression and a Poisson's ratio of 0.30 is subjected to a biaxial state of stress. The principal stresses are $\sigma_{p1} = 18$ ksi T, $\sigma_{p2} = 16$ ksi T, and $\sigma_{p3} = \sigma_z = 0$. Determine the factor of safety with respect to failure by slip according to each of the five theories of failure.

8-90* A material with a proportional limit of 36 ksi in tension and compression and a Poisson's ratio of 0.30 is subjected to a biaxial state of stress. The principal stresses are $\sigma_{p1} = 18$ ksi T, $\sigma_{p2} = 16$ ksi C, and $\sigma_{p3} = \sigma_z = 0$. Determine the factor of safety with respect to failure by slip according to each of the five theories of failure.

8-91 A material with a proportional limit of 430 MPa in tension and compression and a Poisson's ratio of 0.30 is subjected to a biaxial state of stress. The principal stresses are $\sigma_{p1} = 160$ MPa T, $\sigma_{p2} = 220$ MPa C, and $\sigma_{p3} = \sigma_z = 0$. Determine the factor of safety with respect to failure by slip according to each of the five theories of failure.

8-92* The shaft shown in Fig. P8-92 is made of steel having a proportional limit of 53 ksi in tension or compression and a Poisson's ratio of 0.30. If a factor of safety of 2.5 with respect to failure by slip is specified, determine the maximum permissible value for the axial load P according to

    (a) the maximum-normal-stress theory of failure.
    (b) the maximum-shearing-stress theory of failure.
    (c) the maximum-normal-strain theory of failure.

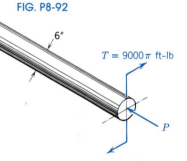

FIG. P8-92

$6''$

$T = 9000\pi$ ft-lb

P

8-93* The shaft shown in Fig. P8-92 is made of steel having a proportional limit of 53 ksi in tension or compression and a Poisson's ratio of 0.30. If a factor of safety of 2.0 with respect to failure by slip is specified, determine the maximum permissible value for the axial load P according to

    (a) the maximum-shearing-stress theory of failure.
    (b) the maximum-strain-energy theory of failure.
    (c) the maximum-distortion-energy theory of failure.

8-94 The shaft shown in Fig. P8-94 is made of steel having a proportional limit of 360 MPa in tension or compression and a Poisson's ratio of 0.30. If a factor of safety of 2.0 with respect to failure by slip is specified, determine the minimum permissible diameter D according to

    (a) the maximum-shearing-stress theory of failure.
    (b) the maximum-strain-energy theory of failure.
    (c) the maximum-distortion-energy theory of failure.

8-95* The shaft shown in Fig. 8-95 is made of steel that has a proportional limit of 76 ksi in tension or compression and a Poisson's ratio of 0.30. If a factor of safety of 2.0 with respect to failure by slip is specified, determine the maximum allowable value for the load R according to

    (a) the maximum-normal-stress theory of failure.

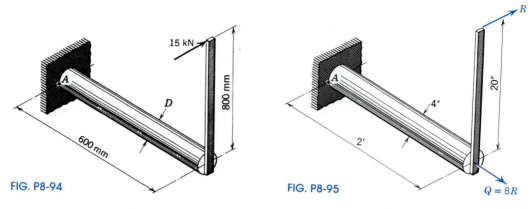

FIG. P8-94   FIG. P8-95

(b) the maximum-shearing-stress theory of failure.

(c) the maximum-normal-strain theory of failure.

8-96* The shaft shown in Fig. 8-95 is made of an aluminum alloy that has a proportional limit of 48 ksi in tension or compression and a Poisson's ratio of 0.33. If a factor of safety of 2.0 with respect to failure by slip is specified, determine the maximum allowable value for the load R according to

(a) the maximum-shearing-stress theory of failure.

(b) the maximum-strain-energy theory of failure.

(c) the maximum-distortion-energy theory of failure.

COMPUTER PROBLEMS

> *Note:* The following problems have been designed to be solved with a programmable calculator, microcomputer, or mainframe computer. Appendix F contains a description of a few numerical methods, together with a few simple programs in BASIC and FORTRAN, that can be modified for use in the solution of these problems.

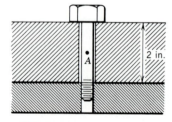

FIG. PC8-1

C8-1 A $\frac{3}{8}$-in.-diameter by 2.5-in.-long steel ($E = 29{,}000$ ksi) bolt is used to fasten two machine parts, as shown in Fig. PC8-1. Because of friction in the threads, the shank of the bolt is subjected to torsional stresses as well as axial tensile stresses. As the bolt is tightened, the 2-in. shank of the bolt is stretched 0.05-in. per turn of the head of the bolt. The torque (in.-lb) is about 6 percent of the tension (lb) in the bolt. For point A on the surface of the bolt, compute and plot:

(a) the axial stress σ_x as a function of the angle of twist θ ($0° \leqslant \theta \leqslant 30°$) of the head of the bolt.

(b) the shearing stress τ_{xy} as a function of the angle of twist θ ($0° \leqslant \theta \leqslant 30°$) of the head of the bolt.

(c) the principal stresses, σ_{p1} and σ_{p2}, as a function of the angle of twist θ ($0° \leqslant \theta \leqslant 30°$) of the head of the bolt.

(d) the maximum shearing stress τ_{max} as a function of the angle of twist θ ($0° \leqslant \theta \leqslant 30°$) of the head of the bolt.

C8-2 A hollow circular steel ($G = 80$ GPa) shaft 3 m long must be designed to withstand a torque of 300 kN-m and an axial load of 900 kN. If the angle of twist θ in the 3-m length must not exceed 0.22 deg and the maximum shearing stress τ_{max} in the shaft must not exceed 45 MPa, compute and plot the range of allowable outside diameters d_o as a function of the inside diameter d_i (0 mm $\leqslant d_i \leqslant$ 600 mm) of the shaft. Limit the wall thickness to values greater than 10 mm.

C8-3 As the C-clamp shown in Fig. PC8-3 is tightened, the web of the C-section is subjected to flexural stresses as well as axial tensile stresses. If the web of the section has a width h of 1 in. and a thickness b of 0.5 in., and the clamp has a capacity d of 3 in., compute and plot the maximum tensile and compressive stresses on section A–A as a function of the clamp force P (0 lb $\leqslant P \leqslant$ 500 lb).

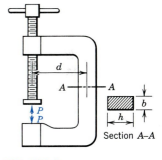

FIG. PC8-3

C8-4 For the clamp of Problem C8-3, compute and plot the maximum tensile and compressive stresses on section A–A as a function of the clamp capacity d (50 mm $\leqslant d \leqslant$ 150 mm). Use $P = 4000$ N, $h = 30$ mm, and $b = 15$ mm.

C8-5 For the clamp of Problem C8-3, the maximum allowable normal stress (tension or compression) is 19,000 psi. Compute and plot the required section width h as a function of the clamp force P (0 lb $\leqslant P \leqslant$ 500 lb). Use $d = 3$ in. and $b = 0.5$ in.

C8-6 A W356 × 179 wide-flange section (see Appendix C) is used for a simply supported beam that carries a concentrated load of 700 kN at the middle of a 3-m span. Compute and plot:

(a) the maximum normal stress σ_{max} at the bottom surface of the beam as a function of x (0 m $\leqslant x \leqslant$ 3 m).

(b) the maximum normal stress σ_{max} at the junction between the web and lower flange of the beam as a function of x (0 m $\leqslant x \leqslant$ 3 m).

(c) the maximum shearing stress τ_{max} at the neutral surface of the beam as a function of x (0 m $\leqslant x \leqslant$ 3 m).

C8-7 A WT4 × 29 T-section (see Appendix C) is used for a simply supported beam that carries a uniformly distributed load of 70 lb/ft for the full 25-ft span of the beam. The flange is placed at the bottom of the beam. Compute and plot the principal stresses σ_{p1} and σ_{p2} and the maximum shearing stress τ_{max} at the junction between the flange and stem of the section as functions of x (0 ft $\leqslant x \leqslant$ 25 ft).

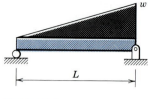

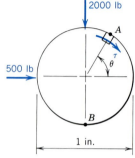

FIG. PC8-8

FIG. PC8-9

C8-8 A W102 × 19 wide-flange section (see Appendix C) is used for a beam that is loaded and supported as shown in Fig. PC8-8. The distributed load w increases linearly from 0 kN/m to 10 kN/m along the 2-m length of the beam. Plot the maximum shearing stress τ_{max} as a function of x (0 m $\leqslant x \leqslant$ 2 m) for a fiber

(a) on the outside surface of a flange.

(b) at the junction between a flange and the web.

(c) at the neutral surface of the beam.

C8-9 The circular shaft shown in Fig. PC8-9 has two cross shearing forces applied. Plot the shearing stress τ shown at point A as a function of angular position θ (0° $\leqslant \theta \leqslant$ 360°).

C8-10 The circular shaft shown in Fig. PC8-9 supports a bending moment of 500 in.-lb in addition to the two cross shearing forces shown in the figure. The bending moment produces a maximum tensile fiber stress at point B. Plot the principal stresses σ_{p1} and σ_{p2} at point A as a function of angular position θ (0° $\leqslant \theta \leqslant$ 360°).

C8-11 A 40-mm-diameter shaft supports two 100-mm-diameter spur gears as shown in Fig. PC8-11. The forces applied to the gears are in the yz plane. Plot the principal stresses σ_{p1} and σ_{p2} as a function of position x (0 mm $\leqslant x \leqslant$ 2800 mm) for points along the top surface of the shaft (points similar to point A).

C8-12 Compute and plot the principal stress σ at the boundary of a small hole as a function of angular position θ (0° $\leqslant \theta \leqslant$ 360°) measured counterclockwise from the x axis if the following stresses are applied at a distance far from the hole:

Case	σ_x	σ_y	τ_{xy}
a	0 ksi	0 ksi	10 ksi
b	10 ksi	10 ksi	0 ksi
c	0 psi	478 psi	509 psi
d	50 MPa	100 MPa	0 MPa
e	30 MPa	0 MPa	40 MPa
f	50 MPa	100 MPa	25 MPa

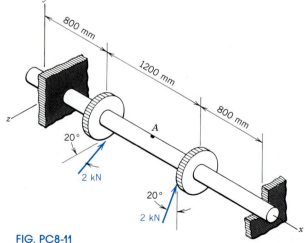

FIG. PC8-11

CHAPTER 9

COLUMNS

9-1

INTRODUCTION

In their simplest form, columns are long, straight, prismatic bars subjected to compressive, axial loads. As long as a column remains straight, it can be analyzed by the methods of Chapter 1; however, if a column begins to deform laterally, the deflection may become large and lead to catastrophic failure. This situation, called *buckling,* can be defined as the sudden large deformation of a structure due to a slight increase of an existing load under which the structure had exhibited little, if any, deformation before the load was increased. A yardstick illustrates this phenomenon. It will support a compressive load of several pounds without discernible lateral deformation, but once the load becomes large enough to cause the yardstick to ''bow out'' a slight amount, any further increase of load produces large lateral deflections.

Buckling of such a column is caused not by failure of the material of which the column is composed but by deterioration of what was a stable equilibrium to an unstable one. The three states of equilibrium can be illustrated with a ball at rest on a surface, as shown in Figure 9-1. The ball in Fig. 9-1a is in a *stable equilibrium* position at the bottom of the pit because gravity will cause it to return to its equilibrium position if perturbed. The ball in Fig. 9-1b is in a *neutral equilibrium*

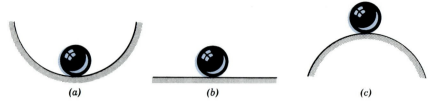

FIG. 9-1

position on the horizontal plane because it will remain at any new position to which it is displaced, tending neither to return to nor move farther from its original position. The ball in Fig. 9-1c, however, is in an *unstable equilibrium* position at the top of a hill because, if it is perturbed, gravity will cause it to move even farther from its original location until it eventually finds a stable equilibrium position at the bottom of another pit.

As the compressive load on a column is gradually increased from zero, the column is at first in a state of stable equilibrium. During this state, if the column is perturbed by inducing small lateral deflections, it will return to its straight configuration when the loads are removed. As the load is increased further, a critical value is reached at which the column is on the verge of experiencing a lateral deflection so, if it is perturbed, it will not return to its straight configuration. The load cannot be increased beyond this value unless the column is restrained laterally; should the lateral restraints be removed, the slightest perturbation will trigger large lateral deflections. For long, slender columns, the *critical buckling load* (the maximum load for which the column is in stable equilibrium) occurs at stress levels much less than the proportional limit for the material. This indicates that this type of buckling is an elastic phenomenon.

9-2

BUCKLING OF LONG STRAIGHT COLUMNS

The first solution for the buckling of long slender columns was published in 1757 by the Swiss mathematician Leonhard Euler (1707–1783). Although the results of this article can be used only for long slender columns, the analysis, similar to that used by Euler, is mathematically revealing and helps explain the behavior of columns.

The purpose of this analysis is to determine the minimum axial compressive load for which a column will experience lateral deflections. A straight, slender, pivot-ended column centrically loaded by axial compressive forces P at each end is shown in Fig. 9-2a. A pivot-ended column is supported such that the bending moment and lateral move-

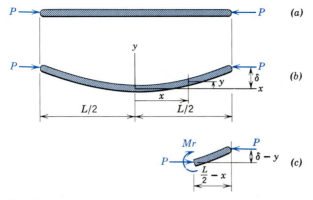

FIG. 9-2

ment are zero at the ends. In Fig. 9-2b, the load P has been increased sufficiently to cause a lateral deflection δ at the midpoint of the span. Axes are selected with the origin at the center of the span for convenience. If the column is sectioned at an arbitrary position x, a free-body diagram of the portion to the right of the section will appear as shown in Fig. 9-2c. The two forces constitute a couple of magnitude $P(\delta - y)$ that must equal the resisting moment M_r; thus, $M_r = P(\delta - y)$. The differential equation for the elastic curve, as given by Eq. 5-3, becomes

$$EI \frac{d^2y}{dx^2} = M_r = P(\delta - y)$$

or

$$\frac{d^2y}{dx^2} + \frac{P}{EI} y = \frac{P\delta}{EI} \qquad (a)$$

Equation a is an ordinary second-order linear differential equation with constant coefficients and a constant right-hand side. Established methods for the solution of such equations show it to be of the form

$$y = A \sin px + B \cos px + C \qquad (b)$$

where A, B, C, and p are constants. By differentiating Eq. b, substituting the results into Eq. a, and collecting like terms, the following expression is obtained for evaluating p and C:

$$(-p^2 + P/EI)(A \sin px + B \cos px) + (P/EI)C = (P/EI)\delta$$

from which it follows that

$$p^2 = P/EI \qquad \text{and} \qquad C = \delta$$

The constants A and B can be obtained from the two boundary conditions that the slope and deflection of the elastic curve are zero at the origin; that is, at $x = 0$, $y = 0$ and $dy/dx = 0$. Since the purpose of this analysis is to determine the minimum load P from which lateral deflections occur, it is necessary to have a third boundary condition; namely, at $x = L/2$, $y = \delta > 0$. The midspan deflection δ is required to be nonzero; otherwise; $\delta = 0$, $y \equiv 0$ could be a solution of Eq. a. When the boundary conditions are substituted in Eq. b and its first derivative, they give

$$0 = B + C$$

$$0 = Ap$$

from which

$$A = 0 \quad \text{and} \quad B = -C = -\delta$$

The solution to Eq. a thus becomes

$$y = \delta \left(1 - \cos \sqrt{\frac{P}{EI}} x \right) \tag{9-1}$$

In order to satisfy the condition that at $x = L/2$, $y = \delta$, the cosine term in Eq. 9-1 must vanish. Thus,

$$\cos \sqrt{\frac{P}{EI}} \frac{L}{2} = 0$$

This equation is satisfied when the argument of the cosine is an odd multiple of $\pi/2$, or

$$\sqrt{\frac{P}{EI}} \frac{L}{2} = \frac{\pi}{2}, \frac{3\pi}{2}, \frac{5\pi}{2}, \cdots$$

Only the first value has physical significance, since it determines the minimum value of P for a nontrivial solution. This value for P is called the critical buckling load, is designated by P_{cr}, and has the magnitude

$$P_{cr} = \frac{\pi^2 EI}{L^2} \tag{9-2}$$

The term P_{cr} is usually called the *Euler buckling load*[1] in honor of Leonhard Euler.

The moment of inertia I in Eq. 9-2 refers to the axis about which

---

[1] While the analysis predicts the buckling load, it does not determine the corresponding lateral deflection δ. This deflection can assume any nonzero value small enough that the nonlinear factor $[1 + (dy/dx)^2]^{3/2}$ in the curvature expression is approximately unity.

bending occurs. When I is replaced by Ar^2, where r is the radius of gyration about the axis of bending, Eq. 9-2 becomes

$$\frac{P_{cr}}{A} = \frac{\pi^2 E}{(L/r)^2} \tag{9-3}$$

The quantity L/r is the *slenderness ratio* and is determined for the axis about which bending tends to occur. For a pivot-ended, centrically loaded column with no intermediate bracing to restrain lateral motion, bending occurs about the axis of minimum moment of inertia (minimum radius of gyration).

The Euler buckling load as given by Eq. 9-2 or 9-3 agrees well with experiment only if the slenderness ratio is large, typically in excess of 140 for steel columns. Whereas short compression members can be treated as explained in Chapter 1, many columns lie between these extremes in which neither solution is applicable. These intermediate-length columns are analyzed by empirical formulas described in later sections.

EXAMPLE 9-1

A 2.5-m pivot-ended, air-dried Douglas fir column has a 50 × 100-mm rectangular cross section. Determine

(a) the slenderness ratio.
(b) the Euler buckling load.
(c) the ratio of the axial stress under the action of the buckling load to the elastic strength of the material.
(d) the maximum load the column can support with a factor of safety of 2.5.

$$r = \sqrt{\frac{I}{A}}$$

$$\frac{L}{r} = S.R.$$

Solution

(a) The length of the column is 2.5 m or 2500 mm. To determine the slenderness ratio, the minimum radius of gyration must be calculated. The moment of inertia of a rectangular cross section is $bh^3/12$ and the area is bh; therefore, the radius of gyration is $\sqrt{I/A}$ or $h/2\sqrt{3}$. The minimum radius of gyration is found by using the centroidal axis parallel to the longer side of the rectangle. Thus, $b = 100$ mm and $h = 50$ mm, so that

$$r = h/2\sqrt{3} = 50/2\sqrt{3} = 14.434 \text{ mm}$$

The slenderness ratio is then found to be

$$L/r = 2500/14.434 = \underline{173.2} \quad > 140 \qquad \text{Ans.}$$

(b) From Appendix A, the modulus of elasticity for air-dried Douglas fir is found to be 13 GPa. Thus, by using Eq. 9-3, the Euler buckling load is found to be

$$P_{cr} = \pi^2 EA(L/r)^2 = (\pi^2)(13)(10^9)(0.050)(0.100)/(173.2)^2$$
$$= 21.385(10^3) \text{ N} = \underline{21.4 \text{ kN}} \qquad \text{Ans.}$$

(c) The axial stress under the action of the buckling load is

$$\sigma_{cr} = P_{cr}/A = 21.385(10^3)/(0.050)(0.100)$$
$$= 4.277(10^6) \text{ N/m}^2 = 4.277 \text{ MPa}$$

From Appendix A, the elastic strength of air-dried Douglas fir is found to be 44 MPa. Thus, the stress ratio is

$$\sigma_{cr}/\sigma_y = 4.277/44 = 0.0972 = \underline{9.72 \text{ percent}} \qquad \text{Ans.}$$

(This demonstrates that buckling can occur at stresses well below the elastic limit of a material for sufficiently slender columns.)

(d) The maximum load that the column can support with a factor of safety of 2.5 must be based on the Euler buckling load. Therefore,

$$P_{max} = P_{cr}/2.5 = 21.385/2.5 = \underline{8.55 \text{ kN}} \qquad \text{Ans.}$$

EXAMPLE 9-2

Two $2 \times 2 \times \frac{1}{8}$-in. structural steel angles 9 ft long will be used as a pivot-ended column. Determine the slenderness ratio and the Euler buckling load if

(a) the two angles are not connected and each acts as an independent member.

(b) the two angles are fastened together as shown in Fig. 9-3 to act as a unit.

Solution

(a) If the angles are not connected and each acts independently, the slenderness ratio is determined by using the minimum radius of gyration of the individual cross sections. From Appendix C, the minimum radius of gyration r_{min} for this angle is 0.398 in. about the Z–Z axis. Thus the slenderness ratio is

$$\frac{L}{r} = \frac{L}{r_{min}} = \frac{9(12)}{0.398} = 271.4 \approx \underline{271} \qquad \text{Ans.}$$

The area for each angle is 0.484 in.^{2}; therefore, the cross-sectional area for the column is 0.968 in.^{2}. Since the modulus of elasticity for structural steel is $29(10^6)$ psi, the buckling load is

$$P_{cr} = \frac{\pi^2 EA}{(L/r)^2}$$

$$= \frac{\pi^2(29)(10^6)(0.968)}{(271.4)^2} = 3760 \text{ lb} \qquad \text{Ans.}$$

(b) With the two angles connected as shown in Fig. 9-3, both I_x and I_y or r_x and r_y must be known in order to determine the minimum radius of gyration. Cross-sectional properties for the angles are given in Appendix C. The value of I_y for the two angles is obtained by using the parallel axis theorem; thus,

$$I_y = 2I_c + 2A\ d^2 = 2Ar_c^2 + 2A\ d^2$$

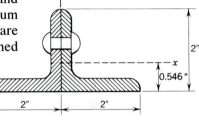

FIG. 9-3

and

$$r_y = \sqrt{I_y/(2A)} = \sqrt{\frac{2A(r_c^2 + d^2)}{2A}} = \sqrt{r_c^2 + d^2}$$

The above expression indicates that the radius of gyration for the two angles is the same as that for one angle, a fact obtained directly from the definition of radius of gyration. Then

$$r_y = \sqrt{(0.626)^2 + (0.546)^2} = 0.831 \text{ in.}$$

In a similar fashion, the radius of gyration about the x axis is the radius of gyration $r_x = 0.626$ in. for a single angle, since $d = 0$ for this axis. This means that the column tends to buckle about the x axis where the slenderness ratio is

$$\frac{L}{r} = \frac{L}{r_{min}} = \frac{9(12)}{0.626} = 172.52 \approx 172.5 \qquad \text{Ans.}$$

The corresponding Euler buckling load is

$$P_{cr} = \frac{\pi^2 EA}{(L/r)^2}$$

$$= \frac{\pi^2(29)(10^6)(0.968)}{(172.52)^2} = 9310 \text{ lb} \qquad \text{Ans.}$$

PROBLEMS

Note: In the following problems, assume that all columns are pivot-ended and that Euler's formula (Eq. 9-2) is applicable for steel with L/r greater than 140 and for aluminum or timber with L/r greater than 80.

9-1* A steel ($E = 30,000$ ksi) rod 1 in. in diameter and 50 in. long will be used to support an axial compressive load P. Determine

(a) the slenderness ratio.

(b) the Euler buckling load.

(c) the axial stress in the column when the Euler load is applied.

9-2* A hollow circular steel ($E = 200$ GPa) column 6 m long has an outside diameter of 125 mm and an inside diameter of 100 mm. Determine

(a) the slenderness ratio.

(b) the Euler buckling load.

(c) the axial stress in the column when the Euler load is applied.

9-3 A 15-ft column with the cross section shown in Fig. P9-3 is constructed from four pieces of timber. The timbers are nailed together so that they act as a unit. Determine

(a) the slenderness ratio.

(b) the Euler buckling load. Use $E = 1900$ ksi for the timber.

(c) the axial stress in the column when the Euler load is applied.

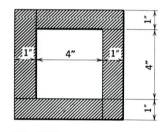

FIG. P9-3

9-4 A 3-m column with the cross section shown in Fig. P9-4 is constructed from three pieces of timber. The timbers are nailed together so that they act as a unit. Determine

(a) the slenderness ratio.

(b) the Euler buckling load. Use $E = 13$ GPa for the timber.

(c) the axial stress in the column when the Euler load is applied.

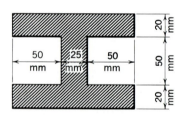

FIG. P9-4

9-5* A W24 × 62 structural steel section (see Appendix C for cross-sectional properties) is used for a 20-ft column. Determine

(a) the slenderness ratio.

(b) the Euler buckling load. Use $E = 29,000$ ksi for the steel.

(c) the axial stress in the column when the Euler load is applied.

9-6* A W356 × 64 structural steel section (see Appendix C for cross-sectional properties) is used for a 10-m column. Determine

(a) the slenderness ratio.

(b) the Euler buckling load. Use $E = 200$ GPa for the steel.

(c) the axial stress in the column when the Euler load is applied.

9-7 A WT6 × 36 structural steel section (see Appendix C for cross-sectional properties) is used for an 18-ft column. Determine

  (a) the slenderness ratio.

  (b) the Euler buckling load. Use $E = 29,000$ ksi for the steel.

  (c) the axial stress in the column when the Euler load is applied.

9-8 A WT152 × 89 structural steel section (see Appendix C for cross-sectional properties) is used for a 6-m column. Determine

  (a) the slenderness ratio.

  (b) the Euler buckling load. Use $E = 200$ GPa for the steel.

  (c) the axial stress in the column when the Euler load is applied.

9-9* Determine the maximum compressive load that a W36 × 160 structural steel column (see Appendix C for cross-sectional properties) can support if it is 36 ft long and a factor of safety of 2.24 is specified. Use $E = 29,000$ ksi.

9-10* Determine the maximum compressive load that a W610 × 155 structural steel column (see Appendix C for cross-sectional properties) can support if it is 12 m long and a factor of safety of 2.24 is specified. Use $E = 200$ GPa.

9-11 Determine the maximum compressive load that a WT9 × 38 structural steel column (see Appendix C for cross-sectional properties) can support if it is 32 ft long and a factor of safety of 1.92 is specified. Use $E = 29,000$ ksi.

9-12 Determine the maximum compressive load that a WT178 × 51 structural steel column (see Appendix C for cross-sectional properties) can support if it is 8 m long and a factor of safety of 1.92 is specified. Use $E = 200$ GPa.

9-13* Determine the maximum allowable compressive load for a 12-ft aluminum alloy ($E = 10,600$ ksi) column having the cross section shown in Fig. P9-13 if a factor of safety of 2.25 is specified.

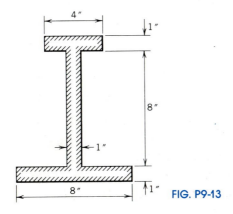

FIG. P9-13

9-14* Determine the maximum allowable compressive load for a 6.5-m steel (E = 200 GPa) column having the cross section shown in Fig. P9-14 if a factor of safety of 1.92 is specified.

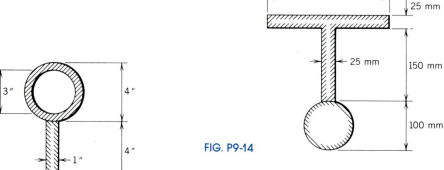

FIG. P9-14

9-15 Determine the maximum allowable compressive load for a 12-ft aluminum alloy (E = 10,600 ksi) column having the cross section shown in Fig. P9-15 if a factor of safety of 2.50 is specified.

9-16 Determine the maximum allowable compressive load for a 12-m steel (E = 200 GPa) column having the cross section shown in Fig. P9-16 if a factor of safety of 1.92 is specified.

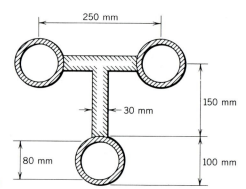

FIG. P9-15

FIG. P9-16

9-17* Determine the maximum total compressive load that three rectangular aluminum alloy (E = 10,600 ksi) bars 1 in. by 4 in. by 100 in. long will carry as columns if

    (a) each bar acts independently of the other two.

    (b) the three bars are welded together to form an H-column.

9-18* Two L127 × 76 × 12.7-mm structural steel angles (see Appendix C for cross-sectional properties) are used for a column that is 5 m long. Determine the total compressive load required to buckle the two members if

(a) they act independently of each other. Use $E = 200$ GPa.

(b) they are riveted together as shown in Fig. P9-18.

9-19 Two C9 × 20 structural steel channels (see Appendix C for cross-sectional properties) are used for a column that is 40 ft. long. Determine the total compressive load required to buckle the two members if

(a) they act independently of each other. Use $E = 29,000$ ksi.

(b) they are latticed 6 in. back-to-back as shown in Fig. P9-19.

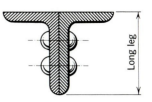

FIG. P9-18

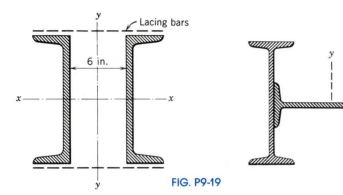

FIG. P9-19

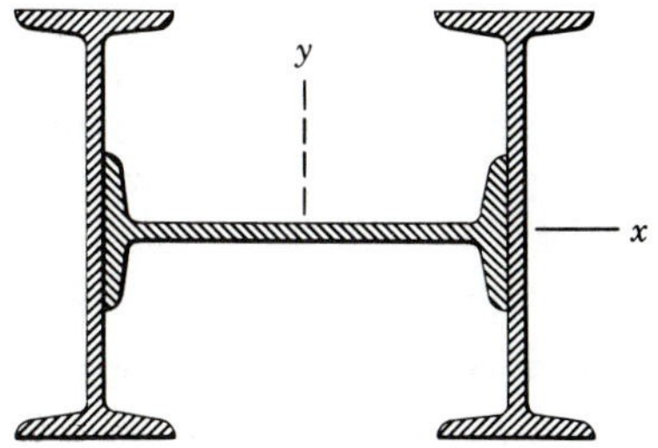

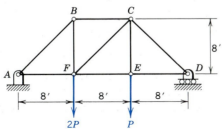

FIG. P9-20

9-20 A column 12 m long is made by riveting three S203 × 34 structural steel sections (see Appendix C for cross-sectional properties) together as shown in Fig. P9-20. Determine

(a) the maximum compressive load that this column can support. Use $E = 200$ GPa.

(b) the maximum compressive load that this column can support if the rivets are removed and the sections act as separate units.

9-21* A simple pin-connected truss is loaded and supported as shown in Fig. P9-21. All members of the truss have a hollow circular cross section with an inside diameter of 2 in. and a wall thickness of $\frac{1}{8}$ in. The members are made of structural steel with a modulus of elasticity of 29,000 ksi and a yield strength of 36 ksi. Determine the maximum load P that can be applied to the truss if a factor of safety of 1.75 with respect to failure by slip and a factor of safety of 4 with respect to failure by buckling is specified.

FIG. P9-21

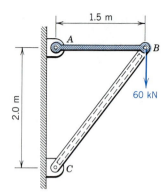

FIG. P9-22

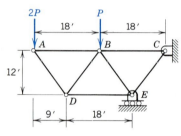

FIG. P9-23

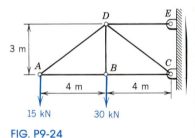

FIG. P9-24

9-22* A 60-kN load is supported by a tie rod *AB* and a pipe strut *BC* as shown in Fig. P9-22. The tie rod has a diameter of 30 mm and is made of steel with a modulus of elasticity of 210 GPa and a yield strength of 360 MPa. The pipe strut has an inside diameter of 50 mm and a wall thickness of 15 mm and is made of aluminum alloy with a modulus of elasticity of 73 GPa and a yield strength of 280 MPa. Determine the factor of safety with respect to failure by slip or buckling for the structure.

9-23 A simple pin-connected truss is loaded and supported as shown in Fig. P9-23. The members of the truss were fabricated by bolting two C10 × 30 channel sections (see Appendix C for cross-sectional properties) back-to-back to form an H-section. The channels are made of structural steel with a modulus of elasticity of 29,000 ksi and a yield strength of 36 ksi. Determine the maximum load *P* that can be applied to the truss if a factor of safety of 1.75 with respect to failure by slip and a factor of safety of 4 with respect to failure by buckling is specified.

9-24 A simple pin-connected truss is loaded and supported as shown in Fig. P9-24. All members of the truss are WT102 × 43 cut T-sections (see Appendix C for cross-sectional properties) made of structural steel with a modulus of elasticity of 200 GPa and a yield strength of 250 MPa. Determine

(a) the factor of safety with respect to failure by slip.

(b) the factor of safety with respect to failure by buckling.

9-25* A tie rod and a pipe strut are used to support a load *P* as shown in Fig. P9-25. Tie rod *AB* has a 1-in. diameter and is made of stainless steel with a modulus of elasticity of 28,000 ksi and a yield strength of 165 ksi. Pipe strut *BC* has an inside diameter of 2 in. and a wall thickness of $\frac{1}{8}$ in. and is made of an aluminum alloy with a modulus of elasticity of 10,600 ksi and a yield strength of 48 ksi. Determine the maximum load *P* that can be applied to the structure if a factor of safety of 1.50 with respect to failure by slip and a factor of safety of 3 with respect to failure by buckling is specified.

FIG. P9-25

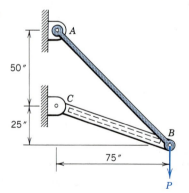

FIG. P9-26

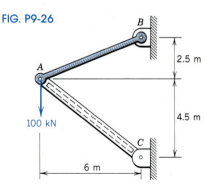

9-26 A 25-mm-diameter tie rod and a pipe strut with an inside diameter of 100 mm and a wall thickness of 25 mm are used to support a 100-kN load as shown in Fig. P9-26. Both the tie rod and the pipe strut are made of structural steel with a modulus of elasticity of 200 GPa and a yield strength of 250 MPa. Determine

(a) the factor of safety with respect to failure by slip.

(b) the factor of safety with respect to failure by buckling.

9-3

EFFECTS OF DIFFERENT IDEALIZED END CONDITIONS

The Euler buckling formula, as expressed by either Eq. 9-2 or Eq. 9-3, was derived for a column with pivoted ends. The Euler equation changes for columns with different end conditions such as the four common ones shown in Fig. 9-4.

While it is possible to set up the differential equation with the appropriate boundary conditions to determine the Euler equation for each new case, a more common approach makes use of the concept of an

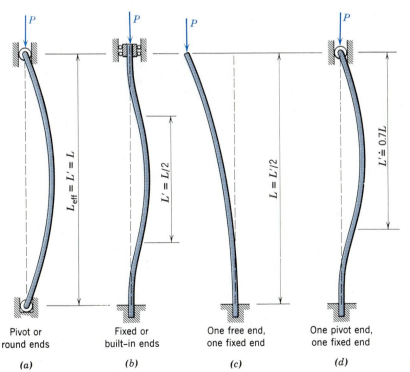

Pivot or round ends	Fixed or built-in ends	One free end, one fixed end	One pivot end, one fixed end
(a)	(b)	(c)	(d)

FIG. 9-4

effective length. The pivot-ended column, by definition, has zero bending moments at each end. The length L in the Euler equation, therefore, is the distance between successive points of zero bending moment. All that is needed to modify the Euler column formula for use with other end conditions is to replace L by L', where L' is defined as the *effective length* of the column (the distance between two successive inflection points or points of zero moment).

The ends of the column in Fig. 9-4b are built in or fixed; since the deflection curve is symmetrical, the distance between successive points of zero moment (inflection points) is half the length of the column. Thus, the effective length L' of a fixed-end column for use in the Euler column formula is half the true length ($L' = 0.5L$). The column in Fig. 9-4c, being fixed at one end and free at the other end, has zero moment only at the free end. If a mirror image of this column is visualized below the fixed end, however, the effective length between points of zero moment is seen to be twice the actual length of the column ($L' = 2L$). The column in Fig. 9-4d is fixed at one end and pinned at the other end. The effective length of this column cannot be determined by inspection, as could be done in the previous two cases; therefore, it is necessary to solve the differential equation to determine the effective length. This procedure yields $L' = 0.7L$.

A pin-ended column is usually loaded through a pin that, as a result of friction, is not completely free to rotate; therefore, there will always be an indeterminate moment at the ends of a pin-connected column that will reduce the distance between the inflection points to a value less than L. Also, it is impossible to support a column so that all rotation is eliminated, and so the effective length of the column in Fig. 9-4b will be somewhat greater than $L/2$. As a result, it is usually necessary to modify the effective column lengths indicated by the ideal end conditions. The amount of the corrections will depend on the individual application. In summary, *the term L/r in all column formulas in this book is interpreted to mean the effective slenderness ratio L'/r.* In the problems in this book, the length given for a member is assumed to be the effective length unless otherwise noted.

PROBLEMS

> *Note:* In the following problems, the Euler formula (Eq. 10-2) is applicable for steel columns with an effective slenderness ratio L'/r greater than 140 and for aluminum or timber columns with L'/r greater than 80.

9-27* A finished 2 × 4 (actual size $1\frac{5}{8}$ in. × $3\frac{1}{2}$ in. × 12 ft long) is used as a fixed-end, fixed-end column. If the modulus of elasticity for the timber is 1600 ksi and a factor of safety of 3 with respect to failure by buckling is specified, determine the maximum safe load for the column.

9-28* An L127 × 89 × 19.1-mm aluminum alloy (E = 73 GPa) angle is used for a fixed-end, pinned-end column having an actual length of 2.5 m. Determine the maximum safe load for the column if a factor of safety of 2.25 with respect to failure by buckling is specified. See Appendix C for cross-sectional properties; they are the same as those for a steel angle of the same size.

9-29 A W356 × 45 structural steel section (see Appendix C for cross-sectional properties) is used for a fixed-end, free-end column having an actual length of 3 m. Determine the maximum safe load for the column if a factor of safety of 1.95 with respect to failure by buckling is specified. Use E = 200 GPa.

9-30 Determine the maximum load that a 2-in. by 3-in. by 6-ft long aluminum alloy bar (E = 10,600 ksi) can support with a factor of safety of 3 with respect to failure by buckling if it is used as a fixed-end, pinned-end column.

9-31* A 6-in.-by-6-in.-by-20-ft-long timber (E = 1600 ksi) is used as a fixed-end, pinned-end column to support a 36,000-lb load. Determine the factor of safety based on the Euler buckling load.

9-32* A solid circular rod with diameter D, length L, and modulus of elasticity E will be used to support an axial compressive load P. The support system for the rod is shown in Fig. P9-32. Determine the critical buckling load in terms of D, L, and E.

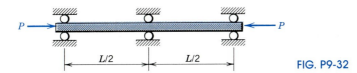

P ← → P

L/2   L/2

FIG. P9-32

9-33 A WT7 × 24 structural steel section (see Appendix C for cross-sectional properties) is used for a column with an actual length of 45 ft. If the modulus of elasticity for the steel is 29,000 ksi and a factor of safety of 2 with respect to failure by buckling is specified, determine the maximum safe load for the column under the following support conditions.

(a) Pivot-pivot.   (c) Pivot-fixed.
(b) Fixed-free.   (d) Fixed-fixed.

9-34 A WT229 × 57 structural steel section (see Appendix C for cross-sectional properties) is used for a column with an actual length of 20 m. If the modulus of elasticity for the steel is 200 GPa and a factor of safety of 2 with respect to failure by buckling is specified, determine the maximum safe load for the column under the following support conditions.

(a) Pivot-pivot.   (c) Pivot-fixed.
(b) Fixed-free.   (d) Fixed-fixed.

9-35* A structural steel ($E = 29,000$ ksi) column 20 ft long must support an axial compressive load of 200 kips. The column can be considered pinned at one end and fixed at the other end for bending about one axis and fixed at both ends for bending about the other axis. Select the lightest wide-flange or American standard section (see Appendix C) that can be used for the column.

9-36* A structural steel (E = 200 GPa) column 6 m long must support an axial compressive load of 500 kN. The column can be considered pinned at both ends for bending about one axis and fixed at both ends for bending about the other axis. Select the lightest wide-flange or American standard section (see Appendix C) that can be used for the column.

9-37 An S3 × 5.7 structural steel section (see Appendix C for cross-sectional properties) will be used for a 20-ft pivot-end, pivot-end column to support a 12.5-kip load. Equally spaced lateral braces, which offer no restraint to bending of the column, will be installed to prevent buckling about both the strong and the weak axes. Determine

(a) the spacing required for the lateral braces controlling buckling about the weak axis.
(b) the spacing required for the lateral braces controlling buckling about the strong axis.
(c) the maximum load the column can support once the lateral braces are installed. Use $E = 29,000$ ksi.

9-38 An S127 × 15 structural steel section (see Appendix C for cross-sectional properties) will be used for a 12-m pivot-end, pivot-end column to support a 60-kN load. Equally spaced lateral braces, which offer no restraint to bending of the column and no restraint to buckling about the strong axis, will be installed to prevent buckling about the weak axis. Determine

(a) the spacing required for the lateral braces.
(b) the maximum load the column can support once the lateral braces are installed. Use $E = 200$ GPa.

9-39 Derive the Euler formula, Eq. 9-2 or 9-3, for an axially loaded, pivot-ended column by solving the differential equation of the elastic curve written with the origin of coordinates at one end of the column.

9-40 Verify the effective length for the fixed-end, fixed-end column shown in Fig. 9-4*b* by solving the differential equation of the elastic curve and applying the appropriate boundary conditions.

9-41 Verify the effective length for the fixed-end, free-end column shown in Fig. 9-4*c* by solving the differential equation of the elastic curve and applying the appropriate boundary conditions.

9-42 A free-body diagram for a fixed-end, pivot-end column is shown in Fig. P9-42. Use the diagram to develop the differential equation of the elastic curve. Verify the effective length for the fixed-end, pivot-end column shown in Fig. 9-4*d* by solving the differential equation of the elastic curve and applying the appropriate boundary conditions.

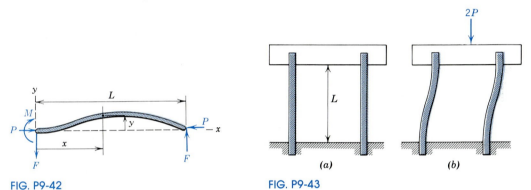

FIG. P9-42   FIG. P9-43

9-43 A rigid block is supported by two fixed-end, fixed-end columns as shown in Fig. P9-43*a*. Determine the effective lengths of the columns by solving the differential equation of the elastic curve and applying the appropriate boundary conditions. Assume that buckling occurs as shown in Fig. P9-43*b*.

9-4

THE SECANT FORMULA

Many real columns do not behave as predicted by the Euler formula because of imperfections in the alignment of the loading. In this section, the effect of imperfect alignment is examined by considering an eccentric loading. The pivot-ended column shown in Fig. 9-5*a* is subjected to compressive forces acting at a distance *e* from the centerline of the undeformed column. As the loading is increased, the column deflects laterally, as shown in Fig. 9-5*b*. If axes are chosen with the origin at the center of the span, as in Section 9-2, the bending moment at any section is

$$M = P(e + \delta - y)$$

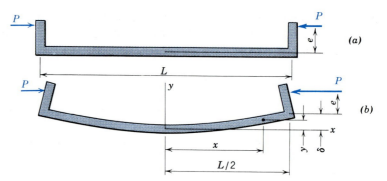

FIG. 9-5

and, if the stress does not exceed the proportional limit and deflections are small, the differential equation for the elastic curve becomes

$$EI \frac{d^2y}{dx^2} = P(e + \delta - y)$$

or

$$\frac{d^2y}{dx^2} + \frac{P}{EI} y = \frac{P}{EI} (e + \delta) \qquad (a)$$

This is the same differential equation as Eq. a of Section 9-2, except that the constant on the right-hand side is $P(e + \delta)/EI$ rather than $P\delta/EI$. The solution, therefore, is of the form

$$y = A \sin px + B \cos px + C \qquad (b)$$

where $p^2 = P/EI$, $C = (e + \delta)$, and A and B are constants to be determined from the boundary conditions, as was the case in Section 9-2. Since the slope and deflection of the elastic curve are zero at the origin (that is, when $x = 0$, $y = 0$ and $dy/dx = 0$), the constants A and B become

$$A = 0 \qquad \text{and} \qquad B = -(e + \delta)$$

The solution to Eq. a thus becomes

$$y = (e + \delta)[1 - \cos \sqrt{P/(EI)}x] \qquad (9\text{-}4)$$

In this case a relationship can be found between the maximum deflection δ, which is the value of y at $x = L/2$, and the load P. Thus,

$$\delta = e + \delta - e \cos \sqrt{P/(EI)}(L/2) - \delta \cos \sqrt{P/(EI)}(L/2)$$

or

$$\delta = e \left[\frac{1 - \cos \sqrt{P/(EI)}(L/2)}{\cos \sqrt{P/(EI)}(L/2)} \right] = e[\sec \sqrt{P/(EI)}(L/2) - 1] \qquad (c)$$

Equation c indicates that, for a given column in which E, I, and L are fixed and $e > 0$, the column exhibits lateral deflection for even small values of the load P. For any value of e, the quantity

$$\sec \sqrt{P/(EI)}(L/2)$$

approaches positive or negative infinity as the argument $\sqrt{P/(EI)}(L/2)$ approaches $\pi/2$, $3\pi/2$, $5\pi/2$, . . ., and the deflection δ increases without bound, indicating that the critical load corresponds to one of these angles. If $\pi/2$ is chosen, since this value yields the smallest load, then

$$\sqrt{P/(EI)}(L/2) = \pi/2$$

$$\sqrt{P/(EI)} = \pi/L$$

from which

$$P_{cr} = \frac{\pi^2 EI}{L^2} \tag{9-2}$$

which is the Euler formula discussed in Section 9-2.

Since it has been assumed that the stresses do not exceed the proportional limit (when the elastic curve equation was written), the maximum compressive stress can be obtained by superposition of the axial stress and the maximum bending stress. The maximum bending stress occurs on a section at the midspan of the column where the bending moment assumes its largest value, $M_{max} = P(e + \delta)$. Thus, the maximum stress is

$$\sigma_{max} = \frac{P}{A} + \frac{M_{max}c}{I} = \frac{P}{A} + \frac{P(e + \delta)c}{Ar^2} \tag{d}$$

in which r is the radius of gyration of the column cross section. From Eq. c,

$$e + \delta = e \sec \sqrt{\frac{P}{EI}} \frac{L}{2}$$

Therefore,

$$\sigma_{max} = \frac{P}{A} \left(1 + \frac{ec}{r^2} \sec \sqrt{\frac{P}{EI}} \frac{L}{2} \right)$$

or

$$\frac{P}{A} = \frac{\sigma_{max}}{1 + \dfrac{ec}{r^2} \sec \left(\dfrac{L}{2r} \sqrt{\dfrac{P}{EA}} \right)} \tag{9-5}$$

Equation 9-5 is known as the secant formula and relates the average stress P/A to the dimensions of the column, the properties of the column material, and the eccentricity e. The term L/r is the same slenderness ratio found in the Euler buckling formula, Eq. 9-3. For columns with different end conditions, the effective slenderness ratio L'/r should be used, as discussed in Section 9-3. The quantity ec/r^2 is called the *eccentricity ratio* and is seen to depend on the eccentricity of the load and the dimensions of the column. If the column was loaded axially, e would presumably be zero, and the maximum stress would be equal to P/A. It is virtually impossible, however, to eliminate all eccentricity that might result from various factors such as the initial crookedness of the column, minute flaws in the material, and a lack of uniformity of the cross section as well as accidental eccentricity of the load. A committee of the American Society of Civil Engineers[2] made an extensive study of the results of a great many column tests and came to the conclusion that a value of 0.25 for ec/r^2 would give results with the secant formula that would be in good agreement with experimental tests on axially loaded columns of structural steel in ordinary structural sizes. A more complete discussion of eccentrically loaded columns is included in Section 9-6. When Eq. 9-5 is used, σ_{max} is the proportional limit, although the yield point of structural steel may be used for convenience, and P is the load at which inelastic action would start to occur. This load should be divided by a factor of safety for design purposes. Note that P/A occurs on both sides of the equation, and thus it is not directly proportional to σ. Consequently, the load resulting from reducing σ to some proportion of the proportional limit will not be reduced by the same amount. In order to obtain a given factor of safety, the factor must be applied to the load each time it occurs in the formula and not to the value of σ; that is, in Eq. 9-5 kP_w should be substituted for P, where P_w is the working load (maximum safe load) and k is the factor of safety.

In order to make effective use of Eq. 9-5, curves showing P/A and L/r can be drawn for various values of ec/r^2 for any given material. Figure 9-6 is such a set of curves for a material with $\sigma = 40{,}000$ psi and $E = 30(10^6)$ psi (or $\sigma = 280$ MPa and $E = 210$ GPa). For these material properties, graphic solutions can be obtained from Fig. 9-6. Digital computers can also be programmed to directly solve the secant formula using iterative techniques.[3]

The outer envelope of Fig. 9-6, consisting of the horizontal line $\sigma = 40$ ksi and the Euler curve, corresponds to zero eccentricity. The

[2]Final Report of Special Committee on Steel Column Research, *Trans. Am. Soc. of Civil Engrs.*, Vol. 98, 1933.

[3]For a program for the secant formula, see *Numerical Methods in Fortran IV Case Studies*, W. S. Dorn and D. D. McCracken, Wiley, New York, 1972, pp. 43–49.

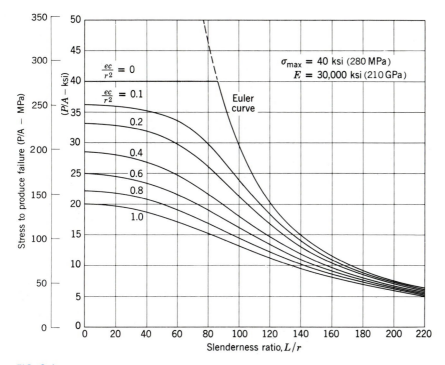

FIG. 9-6

Euler curve is truncated at 40 ksi since this is the maximum allowable stress for the material. With Young's modulus equal to 30,000 ksi, the truncation occurs when $L/r = 86$. From the data presented in Fig. 9-6, it is seen that eccentricity of loading plays a significant role in reducing the working load (maximum safe load) in the short and intermediate column ranges (slenderness ratios less than 150 for the steel in Fig. 9-6). For large slenderness ratios, the curves for the various eccentricity ratios tend to merge with the Euler curve. Consequently, the Euler formula can be used to analyze columns with large slenderness ratios. For a given problem, *the slenderness ratio must be computed* to determine whether or not the Euler equation is valid.

The quantities E, I, and L can be eliminated from Eq. *c* by using Eq. 9-2, in which case the maximum deflection becomes

$$\delta = e[\sec(\pi/2)\sqrt{P/P_{\mathrm{cr}}} - 1] \qquad (e)$$

From this equation it would appear that the deflection δ becomes infinite as P approaches P_{cr}; however, under these conditions, the slope of the deflected shape of the column is no longer sufficiently small to be neglected in the curvature expression. As a result, accurate deflections can only be obtained by using the nonlinear form of the differential equation of the elastic curve. Graphs of Eq. *e* for various values of *e*

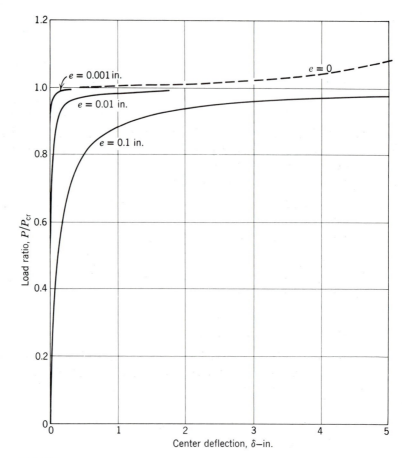

FIG. 9-7

are shown in Fig. 9-7. From these curves it can be seen that the deflection is extremely small as e approaches zero until the load is near P_{cr}, at which time the deflection increases very rapidly. In Fig. 9-7 the curve shown in dashed lines for $e = 0$ is a plot of the exact solution.[4] This limiting case represents an axially loaded column, and the remaining curves illustrate how the deflection under a supposedly axial load can be initiated by a small accidental eccentricity. The curve for $e = 0$ indicates that a load only slightly greater than P_{cr} would probably cause the column to collapse because the proportional limit of the material would have been exceeded.[5]

[4]See *Theory of Elastic Stability,* S. Timoshenko and J. Gere, 2nd ed. McGraw-Hill, New York, 1961, pp. 77–82.

[5]The curve is valid for any deflection only until the proportional limit of the material is exceeded, at which time the curve drops.

PROBLEMS

> *Note:* In the following problems, use the effective slenderness ratio for the specified end conditions when using either Fig. 9-6 or the secant formula.

9-44* The American Society of Civil Engineers recommends using a value of 0.25 for the eccentricity ratio for structural steel columns. What eccentricity does this imply for

(a) a solid circular bar of diameter d?

(b) a hollow circular bar with the inside diameter d_i equal to one-half the outside diameter d?

(c) a thin-walled tube of thickness t and diameter d?

9-45* The American Society of Civil Engineers recommends using a value of 0.25 for the eccentricity ratio for structural steel columns. What eccentricity does this imply for

(a) a W24 $\times$ 104 wide-flange section?

(c) an S18 $\times$ 70 American standard section?

9-46 A fixed-ended solid steel bar having a diameter of 50 mm and a length of 2 m will be used to support a compressive load. A factor of safety of 2.5 is specified and the eccentricity ratio $ec/r^2 = 0.50$. Determine

(a) the maximum safe compressive load (use Fig. 9-6).

(b) the implied value of the eccentricity.

9-47 A pivot-ended steel strut having a 2 $\times$ 4-in. cross section and a 50-in. length will be used to support a compressive load. A factor of safety of 2 is specified and the eccentricity ratio $ec/r^2 = 0.20$. Determine

(a) the maximum safe compressive load (use Fig. 9-6).

(b) the implied value of the eccentricity.

9-48* A fixed-ended steel strut having a 50 $\times$ 100-mm cross section and a length of 2.5 m will be used to support a compressive load. A factor of safety of 2.50 is specified. If the load may be applied with an eccentricity of 2 mm, determine

(a) the eccentricity ratio.

(b) the maximum safe compressive load (use Fig. 9-6).

9-49* A pivot-ended hollow circular steel tube having an outside diameter of 5 in., an inside diameter of 4 in., and a length of 8 ft will be used to support a compressive load. A factor of safety of 2.25 is specified. If the load may be applied with an eccentricity of 0.250 in., determine

(a) the eccentricity ratio.

(b) the maximum safe compressive load (use Fig. 9-6).

9-50* A W254 × 33 steel wide-flange section (see Appendix C for cross-sectional properties) is used for a free-end, fixed-end column that is 1.50 m long. Use Fig. 9-6 to estimate the maximum safe load this column could support if $ec/r^2 = 0.40$ and a factor of safety of 2 is specified.

9-51 A WT9 × 38 structural steel T-section (see Appendix C for cross-sectional properties) is used for a fixed-end, fixed-end column that is 20 ft long. Use Fig. 9-6 to estimate the maximum safe load this column could support if $ec/r^2 = 0.20$ and a factor of safety of 3 is specified.

9-52 A WT305 × 77 structural steel T-section (see Appendix C for cross-sectional properties) is used for a pivot-end, fixed-end column that is 6.50 m long. Use Fig. 9-6 to estimate the maximum safe load this column could support if $ec/r^2 = 0.40$ and a factor of safety of 2 is specified.

9-53* A hollow circular pivot-ended steel column 12 ft long has an inner diameter of 4 in. and an outer diameter of 5 in. When loaded in compression, the column buckles at a load of 150 kips. Use Fig. 9-6 to determine

(a) the value of the eccentricity ratio.

(b) the eccentricity of the applied load.

9-54* A fixed-ended steel strut has a 20-by-30-mm rectangular cross section and a 1.25-m length. The strut buckles about the long axis of the cross section at a load of 75 kN. Use Fig. 9-6 to determine

(a) the value of the eccentricity ratio.

(b) the eccentricity of the applied load.

9-55 A pivot-ended 4-in.-diameter 0.4%C hot-rolled steel ($E = 30,000$ ksi and $\sigma_{PL} = 53$ ksi) shaft is to support an axial compressive load of 80 kips with a factor of safety of 2.5. Use the secant formula to determine the maximum permissible length L if the eccentricity ratio $ec/r^2 = 0.20$.

9-56 A fixed-ended 50-mm-diameter by 2-m-long annealed stainless-steel bar ($E = 190$ GPa and $\sigma_{PL} = 250$ MPa) is to support an axial compressive load of 125 kN. Use the secant formula to determine the maximum eccentricity permitted if a factor of safety of 2.4 is specified.

9-57* A fixed-ended solid aluminum alloy ($E = 10,600$ ksi and $\sigma_{PL} = 48$ ksi) bar has a diameter of 3 in. and a length of 6 ft. Use the secant formula to determine the maximum safe compressive load if a factor of safety of 2.25 is specified and the eccentricity may be as much as 0.250 in.

9-58* A WT127 × 22 structural T-section (see Appendix C for cross-sectional properties) is used for a pivot-ended column. Use the secant formula to determine the allowable load P if the eccentricity ratio is 0.25, the modulus of elasticity is 200 GPa, the proportional limit is 250 MPa, the factor of safety is 2, and the length of the column is 2.75 m.

9-59 A W18 × 76 wide-flange section (see Appendix C for cross-sectional properties) is used for a fixed-end, pivot-end column. Use the secant formula to determine the allowable load P if the eccentricity ratio is 0.25, the modulus of elasticity is 29,000 ksi, the proportional limit is 36 ksi, the factor of safety is 2.25, and the actual length of the column is 25 ft.

9-60 A W610 × 125 wide-flange section (see Appendix C for cross-sectional properties) is used for a fixed-ended column. Use the secant formula to determine the allowable load P if the eccentricity ratio is 0.25, the modulus of elasticity is 200 GPa, the proportional limit is 250 MPa, the factor of safety is 1.95, and the actual length of the column is 7.50 m.

9-5

EMPIRICAL COLUMN FORMULAS: CENTRIC LOADING

Referring to Fig. 9-6, one observes that the envelope of the family of curves is the horizontal line $P/A = 40$ ksi up to L/r of 86 and the plot of the Euler formula (Eq. 9-3) for L/r greater than 86. This curve represents the failure stresses for *ideal* columns of any length under centric loading. However, for real columns (which will always have built-in eccentricities), the true curve will be an S-shaped curve similar to those shown in Fig. 9-6. Numerous column experiments indicate that the Euler formula (Eq. 9-2 or 9-3) is reliable for designing axially loaded columns, provided the slenderness ratio is within the range in which the eccentricity has relatively little effect. This range is called the *slender* range, which in Fig. 9-6 would begin in the region of L/r equal to 120–140 (the region of transition depends on the material). At the low end of the slenderness scale, the column would behave essentially as a compression block, and the failure stress would be the compressive strength of the material (for metals, usually the elastic strength σ_y). The extent of this range, the *compression block* range, is a matter of judgment, or is dictated by specifications. In Fig. 9-6, the small eccentricity curves seem to indicate that the range 0 to 40 could be considered the compression block range. The range between the compression block and the slender ranges is known as the *intermediate* range, and the only equation of a rational nature that applies to real columns in this range is the secant formula Eq. 9-5. Since the application of the secant formula to centric loading requires an estimate of the accidental eccentricity ratio, the equation acquires an empirical nature. Furthermore, the application is so involved that simpler empirical formulas have been developed that give results in reasonable agreement with experimental results within the intermediate range. Although a great many equations have been proposed, most of them come under one of three major classifications. These three general formulas are the *Gordon–Rankine,* the *straight-line,* and the *parabolic* formulas, described in the following paragraphs.

The *Gordon–Rankine* formula, one of the first empirical formulas, was developed by Gordon and Rankine, English engineers, during the middle of the nineteenth century. The formula is usually written as

$$\frac{P}{A} = \frac{\sigma_0}{1 + C_1(L/r)^2} \tag{9-6}$$

in which σ_0 and C_1 are experimentally determined constants. The equation plots as a curve similar to the secant curve.

The *straight-line* formula, proposed by T. H. Johnson about 1890, is written in general form as

$$\frac{P}{A} = \sigma_0 - C_2 \frac{L}{r} \tag{9-7}$$

where, again, σ_0 and C_2 are experimentally determined material-dependent constants.

At about the same time that T. H. Johnson proposed the straight-line formula, Professor J. B. Johnson proposed a *parabolic* formula of the general form

$$\frac{P}{A} = \sigma_0 - C_3 \left(\frac{L}{r}\right)^2 \tag{9-8}$$

where σ_0 and C_3 are the experimentally determined constants. This latter formula with the proper choice of constants can be made to agree quite closely in the intermediate range with the secant formula with certain prescribed numerical constants. See, for example, the close correlation between curves ② and ⑦ in Fig. 9-8.

For design purposes, the entire range of stresses for a given material is covered by an appropriate set of specifications known as a *column code.* The code will specify the empirical formula for the intermediate range and the limits of the intermediate range; it may also specify the Euler formula in the slender range. If the code is written for *safe* (or *allowable*) loads, the factor of safety k will be specified for the Euler formula and will either be specified or be included in the constants for the empirical formula. A few representative codes are listed in Table 9-1, in which the Euler formula is specified for the slender range, except

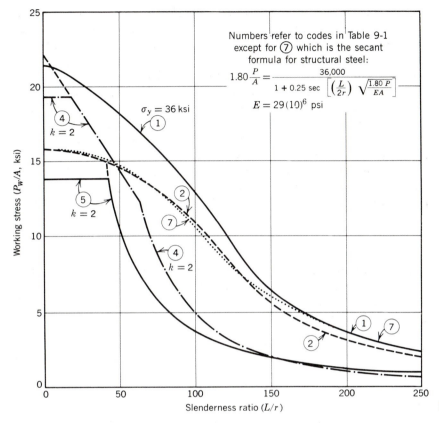

Numbers refer to codes in Table 9-1 except for $\textcircled{7}$ which is the secant formula for structural steel:

$$1.80\frac{P}{A} = \frac{36{,}000}{1 + 0.25\sec\left[\left(\frac{L}{2r}\right)\sqrt{\frac{1.80\,P}{EA}}\right]}$$

$$E = 29(10)^6 \text{ psi}$$

$\sigma_y = 36$ ksi

FIG. 9-8

for code 5. Some codes, such as 4 and 5, are for ultimate (buckling) loads. When these codes are used to determine safe loads, the values obtained from the formulas must be divided by an appropriate factor of safety. The data in Table 9-1 are taken or adapted from handbooks published by the American Institute of Steel Construction (A.I.S.C.), the American Association of State Highway Officials (A.A.S.H.O.), the Armed Forces Supply Support Center (Mil-Hdbk-5), the Aluminum Company of America (Alcoa Handbook), the Aluminum Association (Aluminum Construction Manual), and the Forest Products Laboratory (U.S.D.A. Handbook #72). A graphic comparison of selected codes is provided in Fig. 9-8.

In order to provide a smooth transition from the intermediate to the slender range, the constants may be adjusted so that the intermediate curve becomes tangent to the Euler curve, thus defining the upper limit of the intermediate range. For safe loads, the factor of safety to be included in the Euler formula can be adjusted to make this curve tangent to any of the formulas for safe loads in the intermediate range. This is done by equating the two expressions for P/A and equating the two

TABLE 9-1 Some Representative Column Codes

Code No.	Source	Material and Modulus of Elasticity	Intermediate Range Formulas and Limitations $\frac{L}{r}$ is the effective ratio $\frac{L'}{r}$	Slender Range Factor of Safety
1	A.I.S.C. Specification, 1970	Steel with a yield point σ_y	$\frac{P_w}{A} = \frac{1}{k}\left[\sigma_y - \frac{\sigma_y}{2C_c^2}\left(\frac{L}{r}\right)^2\right]$ $C_c^2 = \dfrac{2\pi^2 E}{\sigma_y}$ $k = \dfrac{5}{3} + \dfrac{3(L/r)}{8C_c} - \dfrac{(L/r)^3}{8C_c^3}$ $0 \leq \dfrac{L}{r} \leq C_c$	23/12
2	Adapted from A.A.S.H.O. Specification, 1973	A36 Structural steel $E = 29,000$ ksi or $E = 200$ GPa	$\dfrac{P_w}{A} = 16,000 - \dfrac{1}{2}\left(\dfrac{L}{r}\right)^2$, psi[a] $\dfrac{P_w}{A} = 110 - 0.00345\left(\dfrac{L}{r}\right)^2$, MPa[b] k included $0 \leq \dfrac{L}{r} \leq 130$	2.24[c]
3	Aluminum Construction Manual	Aluminum alloy 2014-T6 $E = 10,600$ ksi or $E = 73$ GPa	$\dfrac{P_w}{A} = 31,330 - 238\left(\dfrac{L}{r}\right)$, psi $\dfrac{P_w}{A} = 216 - 1.64\left(\dfrac{L}{r}\right)$, MPa[b] k included $11.3 \leq \dfrac{L}{r} \leq 55$	1.95

#	Material	Formula	Adapted from	Notes
4	Aluminum alloy 2024-T4 E = 10,600 ksi or E = 73 GPa	$\dfrac{P_u}{A} = 44{,}800 - 313\left(\dfrac{L}{r}\right)$, psi $\dfrac{P_u}{A} = 309 - 2.16\left(\dfrac{L}{r}\right)$, MPa[b] Ultimate load; use k as specified $18.5 \le \dfrac{L}{r} \le 64$	Adapted from Alcoa Structural Handbook, 1960	As specified
5	Magnesium alloy ZK60A-T5 E = 6500 ksi or E = 45 GPa	$\dfrac{P_u}{A} = \dfrac{7.6(10^6)}{(L/r)^{1.5}}$, psi $\dfrac{P_u}{A} = \dfrac{52.4(10^3)}{(L/r)^{1.5}}$, MPa[b] Ultimate load; use k as specified $43 < \dfrac{L}{r}$	Adapted from Mil-Hdbk-5, 1966	No Euler equation
6	Douglas fir (coast type) E = 1600 ksi or E = 11 GPa	$\dfrac{P_w}{A} = 1450 - 16.56(10^{-6})\left(\dfrac{L}{r}\right)^4$, psi $\dfrac{P_w}{A} = 10.0 - 114.2(10^{-9})\left(\dfrac{L}{r}\right)^4$, MPa[b] k included $38 \le \dfrac{L}{r} \le 73.5$ Use 1450 psi or 10.0 MPa for $\dfrac{L}{r} < 38$	Adapted from U.S.D.A. Wood Handbook #72, 1955 For square or round solid cross sections	3

[a]The formula in the A.A.S.H.O. specification is written in terms of the actual column length; changing the constant C_3 to 1/2 adjusts the formula for use with the effective length.

[b]Coverted to SI units.

[c]Not in the A.A.S.H.O. specification that specifies the secant formula for the slender range.

expressions for the first derivatives of P/A with respect to L/r. The resulting two equations can be solved for the factor of safety to be used in the Euler formula and the L/r at the transition from intermediate to slender range. The practice of making the intermediate and Euler curves tangent is not universal; many column codes plot as curves similar to that for curve ④ of Fig. 9-8.

The formulas in Table 9-1 are representative of a large number of column equations that have been incorporated in various design codes. The slenderness ratio in this table is always the effective slenderness ratio L'/r. If no end conditions are specified in the problems presented later, the stated length is the effective length. Note that the use of high-strength materials will increase the allowable (P/A) value for short columns but will have little, if any, effect on the strength of long columns, since the critical load (the Euler load) depends on Young's modulus, not on the elastic strength of the material. Note also that the use of fixed or restrained ends, which has the effect of reducing the length of the column, materially increases the strength of slender columns, but has much less influence on short compression members.

All of the discussion so far has been concerned with *primary* instability, in which the column deflects as a whole into a smooth curve. No discussion of compression loading is complete without reference to *local instability* in which the member fails locally by crippling of thin sections. Thin open sections such as angles, channels, and H-sections are particularly sensitive to crippling failure. The design of such members to avoid crippling failure is usually governed by specifications controlling the width-thickness ratios of outstanding flanges. Closed section members of thin material (thin-walled tubes, for example) must also be examined for crippling failure when the members are short (see Mil-Hdbk-5).

EXAMPLE 9-3

Two steel C10 × 25 channels are latticed 5 in. back-to-back, as shown in Figs. 9-9*a* and *b*, to form a column. Determine the maximum allowable axial load for effective lengths of 25 ft and 40 ft. Use code 1 for structural steel (see Appendix A for properties).

Solution

Both I_x and I_y (see Fig. 9-9*b*) or r_x and r_y must be known in order to determine the minimum radius of gyration. Properties of the channel section are given in Appendix C. The value of I_y for two channels is obtained by using the parallel axis theorem; thus,

$$I_y = 2I_c + 2A\,d^2 = 2Ar_c^2 + 2A\,d^2$$

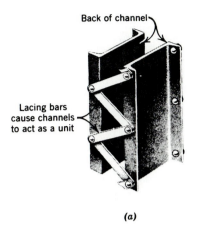

Back of channel

Lacing bars
cause channels
to act as a unit

(a)

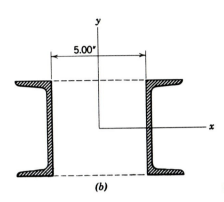

(b)

FIG. P9-9

and

$$r_y = \sqrt{I_y/(2A)} = \sqrt{\frac{2A(r_c^2 + d)^2}{2A}} = \sqrt{r_c^2 + d^2}$$
$$= \sqrt{(0.676)^2 + (2.50 + 0.617)^2} = 3.19 \text{ in.}$$

which is less than the tabular value of 3.52 in. for r_x. This means that the column tends to buckle with respect to the y axis, and the slenderness ratios are

$$\frac{L}{r} = \frac{12(25)}{3.19} = 94.0 \quad \text{and} \quad \frac{12(40)}{3.19} = 150.5$$

From code 1,

$$C_c^2 = 2\pi^2(29)(10^3)/36 = 15,900 \qquad C_c = 126$$

The slenderness ratios above indicate that the 25-ft column is in the intermediate range; hence,

$$k\frac{P}{A} = 36 - \frac{36(94^2)}{2(15,900)} = 26.0$$

The factor of safety is

$$k = \frac{5}{3} + \frac{3(94/126)}{8} - \frac{(94/126)^3}{8} = 1.89$$

Hence, the safe load for the 25-ft column is

$$P = \frac{26.0(2)(7.35)}{1.89} = \underline{202 \text{ kips}} \qquad \text{Ans.}$$

The 40-ft column having a slenderness ratio of 150.5 is in the slender range; hence,

$$P = \frac{\pi^2 EA}{k(L/r)^2} = \frac{\pi^2(29)(10^3)(2)(7.35)}{(23/12)(150.5^2)} = \underline{96.7 \text{ kips}} \qquad \text{Ans.}$$

EXAMPLE 9-4

Determine the dimensions necessary for a 500-mm rectangular strut to carry an axial load of 6.75 kN with a factor of safety of 2. The material is aluminum alloy 2024-T4, and the width of the strut is to be twice the thickness. Use code 4.

Solution

The code is represented by two different equations that depend on the value of L/r, which in turn will depend on the equation used. Thus, it will be necessary to assume that one of the equations applies and use it to obtain the dimensions of the column, after which the value of L/r must be calculated and used to check the validity of the equation used. Assume L/r is less than 64, in which case the straight-line equation is valid. The cross section is shown in Fig. 9-10, and the least moment of inertia I_x is equal to $bt^3/12$, the area A is equal to bt or $2t^2$, and the least radius of gyration is

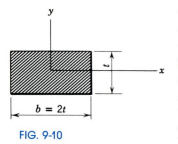

$b = 2t$

FIG. 9-10

$$r = \left(\frac{bt^3/12}{bt}\right)^{1/2} = \frac{t}{\sqrt{12}}$$

The slenderness ratio is

$$\frac{L}{r} = \frac{0.500 \sqrt{12}}{t}$$

and, when this value and the expression for the area are substituted in the straight-line formula of code 4, it becomes

$$k\left(\frac{P}{A}\right) = \frac{2(6.75)(10^3)}{2t^2} = 309(10^6) - 2.16(10^6)\left(\frac{0.500 \sqrt{12}}{t}\right)$$

from which

$$t = 13.70(10^{-3}) \text{ m} = 13.70 \text{ mm}$$

The value of L/r for this thickness is

$$\frac{L}{r} = \frac{0.500 \sqrt{12}}{0.0137} = 126.4$$

which is greater than 64 and indicates that the straight-line formula is not valid. The problem must be solved again using the Euler equation. Thus,

$$\frac{P}{A} = \frac{6.75(10^3)}{2t^2} = \frac{\pi^2 E}{2(L/r)^2} = \frac{(\pi^2)(73)(10^9)}{2[(0.500) \sqrt{12}/t]^2}$$

from which

$$t^4 = 28.1(10^{-9}) \text{ m}^4 \qquad \text{and} \qquad t = 12.95(10^{-3}) \text{ m} = 12.95 \text{ mm}$$

The value of L/r is 133.7 for this thickness which confirms the use of the Euler formula. The dimensions of the cross section are

$$t = 12.95 \text{ mm} \quad \text{and} \quad b = 25.9 \text{ mm} \quad \quad \text{Ans.}$$

EXAMPLE 9-5

Select the lightest wide-flange steel section listed in Appendix C to support a centric load of 150 kips as a 15-ft column. Use code 2.

Solution

If L/r was small, P/A would be 16,000 psi, and the area would be

$$A = 150,000/16,000 = 9.38 \text{ sq. in.}$$

A column should be selected from Appendix C with an area greater than 9.38 sq. in. for the first trial. In this case, try a W8 × 40 section for which A is 11.7 sq. in. and r_{min} is 2.04 in. The value of L/r for this column is

$$\frac{L}{r} = \frac{180}{2.04} = 88.2$$

and the load it can support is

$$P = A \left[16,000 - \frac{(L/r)^2}{2} \right] = 11.7 \left[16,000 - \frac{88.2^2}{2} \right] = 141,700 \text{ lb}$$

This load is less than the design load; therefore, a column with either a larger area, a larger radius of gyration, or both must be investigated. As a second trial value, use a W10 × 45 section for which A is 13.3 sq. in. and r is 2.01 in. For this section L/r is equal to 90, and the load is

$$P = 13.3(16,000 - 90^2/2) = 158,940 \text{ lb}$$

Since the 45-lb column is stronger than necessary and the 40-lb column is not strong enough, any other sections investigated should weigh between 40 and 45 lb/ft. The only wide flange section in Appendix C that might satisfy the requirements is a W14 × 43 section for which A is 12.6 sq. in. and r is 1.89 in. The slenderness ratio and allowable load for this column are

$$(L/r) = 95.2 \quad \text{and} \quad P = 144,530 \text{ lb}$$

which is less than the design load. Thus, the lightest column is a

$$\underline{\text{W10} \times \text{45 section}} \quad \quad \text{Ans.}$$

When a rolled section is to be selected to support a specified load, it is usually necessary to make several trial solutions, since there is no direct relationship between areas and radii of gyration for different structural shapes. The best section is usually the section with the least area (smallest mass) that will support the load. A minimum area can be obtained by assuming $L/r = 0$. The load-carrying capacity of various sections with areas larger than this minimum can then be calculated, using the proper column formula, to determine the lightest one that will carry the specified load.

The above is not necessarily the best procedure. Different designers have different approaches to the trial-and-error procedure, and for certain problems one approach may be better than another. The important point to make here is that the problem of design involving rolled shapes (other than simple geometric shapes) is, in general, solved by trial and error.

PROBLEMS

> *Note:* Refer to Appendix A for material properties and to Appendix C for cross-sectional properties of rolled structural shapes.

9-61* Three structural steel bars with a 1×4-in. rectangular cross section will be used for a 10-ft fixed-ended column. Determine the maximum compressive load permitted by code 1 if

  (a) the three bars act as independent axially loaded members.

  (b) the three bars are welded together to form an H-column.

9-62* A structural steel W356 $\times$ 122 wide-flange section will be used for a 9-m pivot-ended column. Determine the maximum compressive load permitted by code 1 if

  (a) the column is unbraced throughout its total length.

  (b) the column is braced such that the effective length for bending about the y axis is reduced to 6 m.

9-63 Two 4-in.-diameter solid circular cold-rolled 18-8 stainless-steel bars will be used for a 15-ft fixed-ended column. Determine the maximum compressive load permitted by code 1 if

  (a) the two bars act as independent axially loaded members.

  (b) the two bars are welded together as shown in Fig. P9-63.

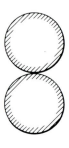

FIG. P9-63

9-64 Three hollow circular structural steel tubes with inside diameters of 50 mm and outside diameters of 80 mm will be used for a 3.5-m pivot-ended column. Determine the maximum compressive load permitted by code 2 if

(a) the three tubes act as independent axially loaded members.

(b) the three tubes are welded together as shown in Fig. P9-64.

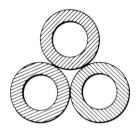

FIG. P9-64

9-65* An L6 × 6 × 1-in. structural steel angle will be used as a pivot-ended column to support an axial compressive load P. Determine the maximum length permitted by code 2

(a) if $P = 75$ kips.

(b) if $P = 120$ kips.

9-66* A WT178 × 36 structural steel T-section will be used as a pivot-ended column to support an axial compressive load P. Determine the maximum length permitted by code 2

(a) if $P = 200$ kN.

(b) if $P = 500$ kN.

9-67 Select the lightest structural steel wide-flange section listed in Appendix C that can be used as a 15-ft pivot-ended column to carry an axial compressive load of 250 kips according to code 1. Use 1.92 as the factor of safety for both ranges of column behavior.

9-68 Select the lightest structural steel angle listed in Appendix C that can be used as a 5-m fixed-ended column to carry an axial compressive load of 500 kN according to code 1. Use 1.92 as the factor of safety for both ranges of column behavior.

9-69 Select the lightest structural steel T-section listed in Appendix C that can be used as an 18-ft pivot-ended column to carry an axial compressive load of 175 kips according to code 2.

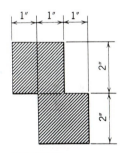

FIG. P9-71

9-70 Select the lightest pair of structural steel channels listed in Appendix C that can be used for a 16-m fixed-ended column to carry an axial compressive load of 750 kN. The channels are to be latticed back-to-back so that the resistance to bending is the same for both axes of symmetry. Specify the size and spacing to satisfy code 2.

9-71* A 5-ft strut having the cross section shown in Fig. P9-71 is made of aluminum alloy 2014-T6. If the strut is fixed at both ends, determine the maximum compressive load permitted by code 3.

9-72* A 3-m strut having the cross section shown in Fig. P9-72 is made of aluminum alloy 2024-T4. The strut is fixed at the bottom and pivoted at the top. Determine the maximum compressive load permitted by code 4 if a factor of safety of 2.5 is specified.

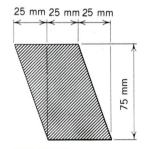

FIG. P9-72

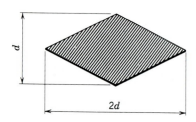

FIG. P9-74

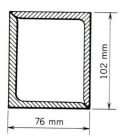

FIG. P9-77

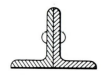

FIG. P9-78

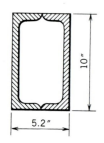

FIG. P9-79

9-73 A hollow circular aluminum alloy 2024-T4 fixed-ended column 100 in. long must support a load of 25 kips with a factor of safety of 2.50. If the ratio of inside to outside diameters is 0.75, determine the minimum outside diameter permitted by code 4.

9-74 A strut of aluminum alloy 2024-T4 having the cross section shown in Fig. P9-74 is to carry an axial compressive load of 600 kN with a factor of safety of 2. The strut is 1.25 m long and is fixed at the bottom and pivoted at the top. Determine the dimension d of the cross section by using code 4.

9-75* An L6 × 6 × ¾-in. aluminum alloy 2014-T6 angle will be used as a fixed-ended, pivot-ended column to support a load of 125 kips. The cross-sectional properties of aluminum and steel angles are the same; see Appendix C. Determine the maximum permissible length permitted by code 3.

9-76* A ZK60A-T5 magnesium alloy strut 1.75 m long must transmit an axial compressive load of 125 kN. Restraint at the ends of the strut makes the effective length one-third greater than that for completely fixed ends. The strut is to be rectangular in cross section with one dimension fixed at 75 mm. If a factor of safety of 2.50 is specified for the strut, determine the other dimension required to satisfy code 5.

9-77 Two L5 × 3½ × ½-in. structural steel angles 12 ft long are used as a pivot-ended column. Determine the maximum load permitted by code 2.

(a) if the angles are not connected and each acts as an independent axially loaded member.

(b) if the angles are fastened together as shown in Fig. P9-77.

9-78 Two L102 × 76 × 9.5-mm structural steel angles 7 m long are used as a pivot-ended column. Determine the maximum load permitted by code 2.

(a) if the angles are not connected and each acts as an independent axially loaded member.

(b) if the angles are welded together to form a 102 × 76-mm box section, as shown in Fig. P9-78.

9-79* Two C10 × 15.3 structural steel channels 12 ft long are used as a fixed-ended, pivot-ended column. Determine the maximum load permitted by code 2

(a) if the channels are not connected and each acts as an independent axially loaded member.

(b) if the channels are welded together to form a 10 × 5.2-in. box section, as shown in Fig. P9-79.

9-80* Two C305 × 45 structural steel channels 7 m long are used as a pivot-ended column. Determine the maximum load permitted by code 2

  (a) if the channels are not connected and each acts as an independent axially loaded member.

  (b) if the channels are latticed 150 mm back-to-back as shown in Fig. 9-9.

9-81 Four L3 × 3 × ½-in. structural steel angles 20 ft long are used as a fixed-ended column. Determine the maximum load permitted by-code 2

  (a) if the angles are not connected and each acts as an independent axially loaded member.

  (b) if the angles are fastened together as shown in Fig. P9-81.

FIG. P9-81

9-82 Four L102 × 76 × 9.5-mm structural steel angles 4 m long are used as a pivot-ended column. Determine the maximum load permitted by code 1

  (a) if the angles are not connected and each acts as an independent axially loaded member.

  (b) if the angles are welded together to form a 204 × 152-mm box section, as shown in Fig. P9-82.

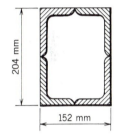

FIG. P9-82

9-83* A 25-ft plate and angle column consists of four L5 × 3½ × ½-in. structural steel angles riveted to a 10 × ½-in. structural steel plate as shown in Fig. P9-83. Determine the maximum safe load permitted by code 2

  (a) if the column is pinned at both ends.

  (b) if the column is fixed at the base and pinned at the top.

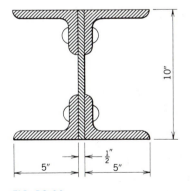

FIG. P9-83

9-84* Four C178 × 22 structural steel channels 12 m long are used to fabricate a column with the cross section shown in Fig. P9-84. The column is fixed at the base and pinned at the top. The pin at the top offers no restraint to bending about the y axis, but for bending about the x axis, the pin provides restraint sufficient to reduce the effective length to 7.5 m. Determine the maximum axial compressive load permitted by code 2.

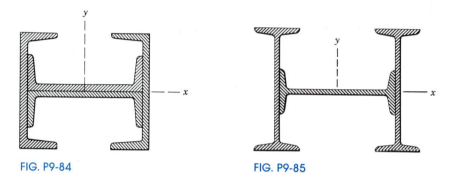

FIG. P9-84   FIG. P9-85

9-85 Three S10 × 25.4 structural steel sections 40 ft long are used to fabricate a column with the cross section shown in Fig. P9-85. The column is fixed at the base and pinned at the top. The pin at the top offers no restraint to bending about the y axis, but for bending about the x axis, the pin provides restraint sufficient to reduce the effective length to 24 ft. Determine the maximum axial compressive load permitted by code 2.

9-86 A connecting rod made of SAE 4340 heat-treated steel has the cross section shown in Fig. P9-86. The pins at the ends of the rod are parallel to the x axis and are 1250 mm apart. Assume that the pins offer no restraint to bending about the x axis, but provide essentially complete fixity for bending about the y axis. Determine the maximum axial compressive load permitted by code 1 if a factor of safety of 1.92 is specified for both ranges of column behavior.

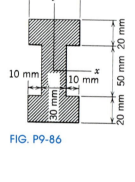

FIG. P9-86

9-87* The machine part shown in Fig. P9-87 is made of SAE 4340 heat-treated steel and carries an axial compressive load. The pins at the ends offer no restraint to bending about the axis of the pin, but restraint

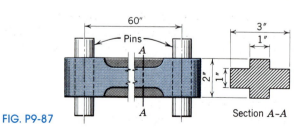

FIG. P9-87

about the perpendicular axis reduces the effective length to 36 in. Determine the maximum load permitted by code 1 if a factor of safety of 1.92 is specified for both ranges of column behavior.

9-88* The compression member AB of the truss shown in Fig. P9-88 is a structural steel W254 $\times$ 67 wide-flange section with the x–x axis lying in the plane of the truss. The member is continuous from A to B. Consider all connections to be the equivalent of pivot ends. If code 2 applies, determine

  (a) the maximum safe load for member AB.
  (b) the maximum safe load for member AB if the bracing member CD is removed.

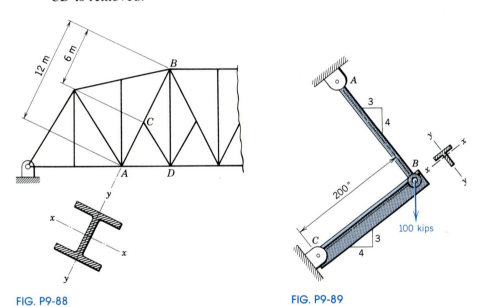

FIG. P9-88   FIG. P9-89

9-89 A tension bar AB and a compression strut BC are used to support a load $P = 100$ kips, as shown in Fig. P9-89. Assume that the pins at B and C offer no restraint to bending about the x axis, but provide end conditions that are essentially fixed at C and free at B for bending about the y axis. Select the lightest structural steel T-section listed in Appendix C that will satisfy code 2.

9-90 Douglas fir ($E = 11$ GPa) timber columns with square cross sections will be used to support axial compressive loads. If the effective lengths of the columns will be 3 m, determine the cross-sectional dimension b required to satisfy code 6 if

  (a) the axial compressive load is 10 kN.
  (b) the axial compressive load is 25 kN.
  (c) the axial compressive load is 100 kN.

9-6

ECCENTRICALLY LOADED COLUMNS

Although a given column will support the maximum load when the load is applied centrically, it is sometimes necessary to apply an eccentric load to a column. For example, a beam supporting a floor load in a building may in turn be supported by an angle riveted or welded to the side of a column, as shown in Fig. 9-11. Frequently, the floorbeam is framed into the column with a stiff connection, and the column is then subject to a bending moment due to the continuity of floorbeam and column. The eccentricity of the load, or the bending moment, will increase the stress in the column or reduce its load-carrying capacity. Three methods will be presented for computing the allowable load on a column subjected to an eccentric load (or an axial load combined with a bending moment).

1. The Secant Formula. The secant formula (Eq. 9-5) was derived on the assumption that the applied load had an initial eccentricity e; this value can be substituted in the equation to determine the failure load (the load to cause incipient inelastic action). As mentioned previously, there is usually a small amount of unavoidable eccentricity that must be approximated when using this formula for centric loads. This unavoidable eccentricity should be added to the intentional eccentricity, particularly if the intentional eccentricity is small. The use of the secant formula is quite time-consuming, unless a series of curves, such as those in Fig. 9-6, is available.[6] The secant formula is specified by the American Railway Engineering Association and the American Association of State Highway Officials.

2. Modified Column Formulas. The second solution is based on the arbitrary specification that the sum of the direct and bending stresses $(P/A + Mc/I)$ shall not exceed the average stress prescribed by the appropriate column formula. For example, when this approach is used with the parabolic formula, the equation becomes

$$\frac{P}{A} + \frac{Mc}{I} = \sigma_0 - C_3 \left(\frac{L}{r}\right)^2 \tag{9-9}$$

In effect, this merely replaces the low L/r end of the secant curve with a parabola (see Fig. 9-8), and if the constants in the parabola are properly chosen, the agreement between the two curves is quite good. Since the parabola will deviate very rapidly from the second curve beyond the point of reverse curvature, Eq. (9-9) is limited to the intermediate

[6]See, for example, Appendix C of the A.A.S.H.O. Standard Specifications for Highway Bridges.

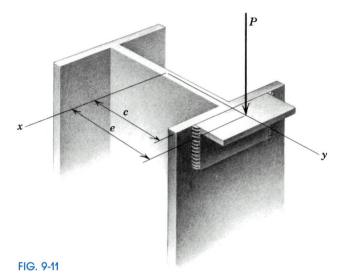

FIG. 9-11

range. This leaves only the secant formula valid for columns in the slender range with eccentric loading. Although it is possible to use other types of intermediate formulas for the right side of Eq. 9-9, the use of the parabola only is recommended here, since it is an attempt to approximate a formula known to be rational.

The application of the secant formula or Eq. 9-9 to an eccentrically loaded column such as that of Fig. 9-11 must be accompanied by an investigation of the stability of the column as an axially loaded member. When the load is applied in the plane containing the axis about which the moment of inertia is least (the *y* axis of Fig. 9-11), the column might fail by buckling about either axis, and both possibilities should be checked.

1. It can fail by buckling about the axis of least radius of gyration (the *y* axis in Fig. 9-11) because L/r is largest about this axis. In this case, all formulas developed for centric loading apply[7] and the radius of gyration is r_y.

2. The column can fail by buckling about the axis of maximum radius of gyration (the *x* axis) due to the combined effect of the bending moment and the direct load. In this case, the values required are r_x, I_x, and the distance *c*, as shown in Fig. 9-11. Note that *c* is the maximum distance to the compression side from the centroidal axis. The smaller of the two loads is, of course, the limiting load.

[7]The load is assumed to be axial; that is, the eccentricity in the *y* direction does not decrease resistance to buckling about the *y* axis of Fig. 9-11. For columns with I-sections in the slender range, it can be shown that such an assumption may be unwarranted. See "Lateral Buckling of I-Section Columns with Eccentric End Loads in Plane of the Web," B. Johnston, *J. of Appl. Mech.*, December 1941, p. A-176.

3. The Interaction Formula. One of the modern expressions for treating combined loads is known as the interaction formula, of which several types are in use. The analysis for compression members subjected to bending and direct stress may be derived as follows. It is assumed that the stress in the column can be written as

$$\frac{P}{A} + \frac{Mc}{I} = \sigma_w$$

and, when this expression is divided by σ_w, it becomes

$$\frac{P/A}{\sigma_w} + \frac{Mc/I}{\sigma_w} = 1 \tag{9-10}$$

When considering eccentrically loaded columns, the value of σ_w will, in general, be different for the two terms. In the first term P/A represents an axial stress on a column; therefore, the value of σ_w should be the average working stress on an axially loaded column as obtained by an empirical formula such as those presented in Section 9-5 or the Euler formula. In the second term, Mc/I represents the bending stress induced in the member as a result of the eccentricity of the load or an applied bending moment; therefore, the corresponding value of σ_w should be the allowable flexural stress. Since the two values of σ_w are different, one recommended form of Eq. 9-10 is the following inter-action formula[8]

$$\frac{P/A}{\sigma_a} + \frac{Mc/I}{\sigma_b} = 1 \tag{9-11}$$

in which P/A = the average stress on the eccentrically loaded column, σ_a = the *allowable* average stress for an axially loaded column (note that the greatest value of L/r should be used to calculate σ_a), Mc/I = the bending stress in the column, and σ_b = the *allowable* bending stress.[9]

When Eq. 9-11 is used to select the most economic section, it will usually not be possible to obtain a section that will exactly satisfy the equation. Any section that makes the sum of the terms on the left side of the equation less than unity is considered safe, and the *safe* section giving the largest sum (less than unity) is the most efficient section.

---

[8] The 1970 A.I.S.C. specifications permit this formula if $(P/A)/\sigma_a \le 0.15$; otherwise, the second term of the equation is expanded to provide for secondary moments due to lateral displacement of any cross section.

[9] The 1970 A.I.S.C. specifications gives $\sigma_b = 0.66\sigma_y$ with a reduced value for compression flanges in some cases. See *The Steel Construction Manual*.

EXAMPLE 9-6

A 6-m-wide-flange steel column with the following properties $A = 18\,210$ mm^2, $r_{min} = 68.8$ mm, $r_{max} = 196$ mm, and $c = 230.7$ mm supports an eccentric load applied 125 mm from the center of the section measured along the web. Use code 1 with $\sigma_y = 290$ MPa, $C_c^2 = 13\,610$, and $k = 1.90$ and 1.76 for maximum and miminum L/r, respectively. Determine the maximum safe load

(a) according to the modified column formula.

(b) according to the interaction formula with σ_b equal to 190 MPa.

Solution

(a) Assume first that the column tends to buckle about the axis parallel to the web. Note that there is no eccentricity about this axis. The value of L/r is

$$\frac{L}{r} = \frac{6(10^3)}{68.8} = 87.2$$

which is less than C_c of 116.7. The average stress P/A is

$$\frac{P}{A} = \frac{290}{1.90}\left[1 - \frac{(87.2)^2}{2(13610)}\right] = 110.0 \text{ MPa}$$

For bending about the axis perpendicular to the web, the radius of gyration is 196 mm (L/r will be much less than 87.2), and the modified column formula becomes

$$\frac{P}{A} + \frac{Mc}{I} = \frac{\sigma_y}{k}\left[1 - \frac{(L/r)^2}{2C_c^2}\right]$$

The bending moment is Pe, and when numerical values are substituted in the equation, it becomes

$$\frac{P}{A} + \frac{P(0.125)(0.2307)}{A(0.196)^2} = \frac{290}{1.76}\left[1 - \frac{(6/0.196)^2}{2(13610)}\right]$$

from which

$$\frac{P}{A} = 90.9 \text{ MPa}$$

Since this last stress is less than that for buckling about the web, the maximum allowable load is

$$P = 90.9(10^6)(18210)(10^{-6}) = 1655(10^3) \text{ N} = \underline{1655 \text{ kN}} \quad \text{Ans.}$$

(b) The interaction formula for eccentrically loaded columns is

$$\frac{P/A}{\sigma_a} + \frac{Mc/I}{\sigma_b} = \frac{P/A}{\sigma_a} + \frac{Pec/(Ar^2)}{\sigma_b} = 1$$

The value of σ_a was found to be 110.0 MPa in part (a), and when numerical data are substituted in the expression, it becomes

$$\frac{P}{A}\left[\frac{1}{110.0} + \frac{(0.125)(0.2307)/(0.196)^2}{190}\right] = 1$$

which gives

$$\frac{P}{A} = 76.7 \text{ MPa}$$

The allowable load using this specification is

$$P = 76.7(10^6)(18210)(10^{-6}) = 1397(10^3) \text{ N} = \underline{1397 \text{ kN}} \quad \text{Ans.}$$

The two results are about 18 percent apart in this case.

PROBLEMS

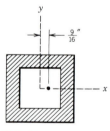

FIG. P9-91

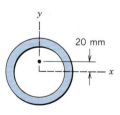

FIG. P9-92

> *Note:* Use the table of codes (Table 9-1) of Section 9-5 modified for eccentric loads unless otherwise specified. Refer to Appendix A for material properties and to Appendix C for cross-sectional properties of rolled structural shapes.

9-91* A hollow square steel member with outside and inside dimensions of 5 in. and 4 in. (walls are $\frac{1}{2}$ in. thick) functions as a pivot-ended column 12 ft long. Determine the maximum load that the column can carry if the load is applied with a known eccentricity of $\frac{9}{16}$ in., as shown in Fig. P9-91. Assume an accidental eccentricity ratio of 0.20. Use Fig. 9-6 and a factor of safety of 2.

9-92* A hollow circular steel member with outside and inside diameters of 150 mm and 120 mm (walls are 15 mm thick) functions as a pivot-ended column 4 m long. Determine the maximum load that the column can carry if the load is applied 20 mm from the axis of the member as shown in Fig. P9-92. Assume an accidental eccentricity ratio of 0.15. Use Fig. 9-6 and a factor of safety of 2.50.

9-93 A 2-in.-diameter steel strut is subjected to an eccentric compressive load as shown in Fig. P9-93. The effective length for bending about the x axis is 75 in., but for bending about the y axis, the end conditions reduce the effective length to 50 in. Determine the maximum load the strut can carry with a factor of safety of 1.5. Use Fig. 9-6 and assume an accidental eccentricity ratio of 0.10.

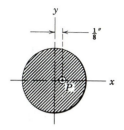

FIG. P9-93

9-94 A hollow square steel member with outside and inside dimensions of 100 mm and 70 mm (walls are 15 mm thick) functions as a pivot-ended column 4 m long. Determine the maximum load that the column can carry if the load is applied with a known eccentricity of 15 mm along a diagonal of the square as shown in Fig. P9-94. Assume an accidental eccentricity ratio of 0.15. Use Fig. 9-6 and a factor of safety of 2.

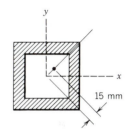

FIG. P9-94

9-95* A W14 $\times$ 82 structural steel section is used for a 24-ft pivot-ended column. Determine the maximum load permitted by code 2 if the load is applied at a point on the centerline of the web 5 in. from the axis of the column.

9-96* A W356 $\times$ 64 structural steel section is used for a 12-m fixed-ended column. Determine the maximum load permitted by code 1 (use $k = 1.92$ for both ranges) if the load is applied at a point on the centerline of the web 150 mm from the axis of the column.

9-97 A WT8 $\times$ 25 structural steel T-section is to be used as a compression member to transmit an eccentric load of 100 kips. The effective length of the member is 12 ft. If code 1 (use $k = 1.92$ for both ranges) is specified, determine the maximum eccentricity e permitted if the point of load application is on the centerline of the stem between the outer edge of the flange and the centroidal axis of the column.

9-98 A WT203 $\times$ 37 structural steel T-section is to be used as a compression member to transmit an eccentric load of 450 kN. The effective length of the member is 3.5 m. If code 1 (use $k = 1.92$ for both ranges) is specified, determine the maximum eccentricity e permitted if the point of load application is on the centerline of the stem between the end of the stem and the centroidal axis of the column.

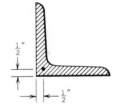

FIG. P9-99

9-99* An L8 $\times$ 8 $\times$ 1 structural steel angle will be used for a 12-ft pivot-ended column. Determine the maximum load permitted by code 2 if the load is applied at the point on the cross section indicated in Fig. P9-99.

9-100* Two 50 $\times$ 150-mm structural steel bars 5 m long will be welded together as shown in Fig. P9-100 and used for a pivot-ended column. Determine the maximum load permitted by code 2 if the load is applied at the point on the cross section indicated in Fig. P9-100.

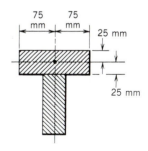

FIG. P9-100

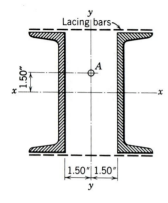

FIG. P9-101

9-101 Two C8 × 18.75 structural steel sections 25 ft long are latticed 3 in. back-to-back as shown in Fig. P9-101 to form a pivot-ended column. Determine the maximum load permitted by code 2 if the load is applied at point A of the cross section.

9-102 A WT203 × 30 structural steel T-section with an effective length of 7 m is used to support an axial compressive load of 100 kN. If code 1 (use $k = 1.92$ for both ranges) is specified, determine the maximum bending moment, applied to the ends of the member in the yz plane (the z axis is the axis of the member), that can be superimposed on the 100 kN load if the moment induces compression in the flange.

9-103* Two L6×3½ × ½-in. structural steel angles are welded together, as shown in Fig. P9-103, to form a column with an effective length of 15 ft. Determine

 (a) the maximum axial compressive load P permitted by code 2.

 (b) the maximum and minimum values for the distance d when the load of part (a) is applied.

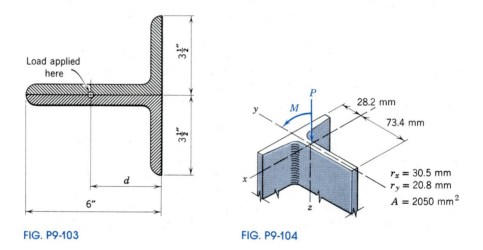

FIG. P9-103   FIG. P9-104

9-104* A 2014-T6 aluminum alloy compression member with an effective length of 1.25 m has the T-cross section shown in Fig. P9-104. Determine

 (a) the maximum axial compressive load P permitted by code 3.

 (b) the maximum bending moment M in the yz plane that can be superimposed as shown on the axial load of part (a).

9-105 Determine the maximum load permitted by code 6 for a 12-ft Douglas fir pivot-ended column with the cross section shown in Fig. P9-105 if the load is applied at the point indicated on the cross section.

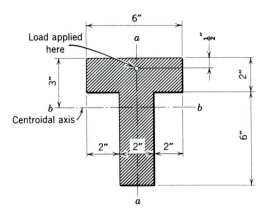

Load applied here

6"

a

$\frac{1}{2}$"

3"

2"

b

b

Centroidal axis

6"

2" 2" 2"

a

FIG. P9-105

9-106 A W533 × 92 structural steel section is used as a 7.5-m column. The column is fixed at the base, but can be considered free at the top for bending with respect to the x axis (see Appendix C) and pivot-ended at the top for bending with respect to the y axis. The load is applied at a point on the centerline of the web 80 mm from the axis of the column. Determine the maximum load permitted by code 2.

9-107* A W8 × 40 structural steel compression member has an effective length of 25 ft. The member is subjected to an axial load P and a bending moment of 10 ft-kips with respect to the x axis of the cross section (see Appendix C). Determine the maximum safe load P. Use code 1 (with $k = 1.92$ for both ranges) and the interaction formula with an allowable flexural stress of $0.66\sigma_y$.

9-108* Two C229 × 30 structural steel channels are latticed 120 mm back-to-back with the flanges turned out to form a column with an effective length of 10 m. The load is applied at a point on the axis of symmetry parallel to the backs of the channels 50 mm from the axis of the column. Determine the maximum allowable load. Use code 2 and the interaction formula with an allowable flexural stress of 160 MPa.

COMPUTER PROBLEMS

Note: The following problems have been designed to be solved with a programmable calculator, microcomputer, or mainframe computer. Appendix F contains a description of a few numerical methods, together with a few simple programs in BASIC and FORTRAN, that can be modified for use in the solution of these problems

C9-1 Plot the average stress P/A as a function of the slenderness ratio L/r ($0 \leqslant L/r \leqslant 250$) permitted by the secant formula for eccentricity ratios ec/r^2 of 0, 0.1, 0.2, 0.4, 0.6, 0.8, and 1.0. Use $\sigma_{max} = 4$ ksi and $E = 1800$ ksi.

C9-2 Plot the average stress P/A as a function of the slenderness ratio L/r ($0 \leqslant L/r \leqslant 250$) permitted by the secant formula for eccentricity ratios ec/r^2 of 0, 0.1, 0.2, 0.4, 0.6, 0.8, and 1.0. Use $\sigma_{max} = 280$ MPa and $E = 73$ GPa.

C9-3 Repeat Problem C9-1 using $\sigma_{max} = 8$ ksi and a factor of safety k of 2.

C9-4 Repeat Problem C9-2 using $\sigma_{max} = 330$ MPa and a factor of safety k of 2.25.

C9-5 A W4 $\times$ 13 structural steel ($E = 29,000$ ksi) wide-flange section (see Appendix C) will be used for a column. Assume an eccentricity e of $L'/1000$, where L' is the effective length of the column. Use iterative techniques and the secant formula (Eq. 9-5) to determine the critical load for

(a) a column with an effective length L' of 6 ft.

(b) a column with an effective length L' of 12 ft.

(c) a column with an effective length L' of 24 ft.

C9-6 A W102 $\times$ 19 structural steel ($E = 200$ GPa) wide-flange section (see Appendix C) is being used for a pivot-ended column. Assume that the eccentricity ratio ec/r^2 resulting from manufacturing and loading imperfections equals 0.25. Prepare a plot (similar to Fig. 9-7) showing the load ratio P/P_{cr} versus center deflection δ of the column.

C9-7 Two C18 $\times$ 58 structural steel ($E = 29,000$ ksi) channels (see Appendix C) will be laced back-to-back and used for a 30-ft column to support an axial compressive load. Prepare a plot showing the working load P_w as a function of the lacing distance d. Use code 1.

C9-8 A W102 $\times$ 19 structural steel ($E = 200$ GPa) wide-flange section (see Appendix C) is being used for a column with an effective length L' of 4 m. Assume that the eccentricity ratio ec/r^2 resulting from manufacturing imperfections equals 0.25. Use iterative techniques and the secant formula (Eq. 9-5) to determine the critical load if the load is applied at a point on the centerline of the web at

(a) a distance $e = 0$ mm from the axis of the column.

(b) a distance $e = 10$ mm from the axis of the column.

(c) a distance $e = 100$ mm from the axis of the column.

(d) a distance $e = 200$ mm from the axis of the column.

CHAPTER 10
INELASTIC BEHAVIOR

10-1
INTRODUCTION

The preceding chapters in this book covered elastic analyses of axial, torsional, and flexural members. Stress was proportional to strain (Hooke's law applied) and the maximum normal and shearing stresses in the members were not permitted to exceed the corresponding proportional limits of the materials. In some design situations, the restriction on elastic behavior is not required and a limited amount of inelastic action can be permitted. The theory of inelastic behavior in axial, torsional, and flexural members is introduced in this chapter.

10-2
INELASTIC BEHAVIOR OF AXIALLY LOADED MEMBERS

The stress-strain diagram used to represent the behavior of a specific material was introduced in Section 1-13. Typical diagrams for structural steel and a magnesium alloy were presented in Figs. 1-31a and 1-31b. These diagrams were used to define a number of material properties.

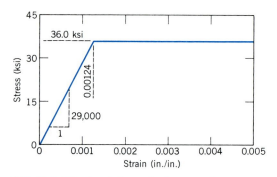

FIG. 10-1 Idealized stress-strain curve for structural steel.

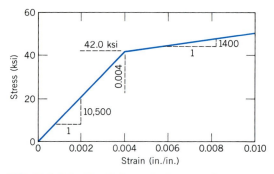

FIG. 10-2 Idealized stress-strain curve for an aluminum alloy.

The useful small-deformation region of the stress-strain diagram for structural steel can be approximated by the idealized diagram shown in Fig. 10-1. The elastic portion of the diagram is a straight line whose slope is the modulus of elasticity E of the material. The inelastic portion of the diagram is a straight line of zero slope beginning at the yield stress. A material for which the slope is zero in the initial portion of the inelastic region is called an *elastoplastic material*. The stress-strain diagram for structural steel, shown in Fig. 1-31*a*, indicates that the yield stress is approximately 36 ksi, and the strain at yield is approximately 0.00124, which gives a modulus of elasticity of approximately 29,000 ksi. Figure 1-31*a* indicates that plastic flow continues with no increase in stress beyond the yield stress until the strain reaches a value of approximately 0.015. Beyond this point (which is beyond the useful small-deformation region of the material) strain hardening begins to occur and an increase in the stress level is required to produce additional strain.

The useful small-deformation region of the stress-strain diagram for an aluminum alloy can be approximated by the idealized diagram shown in Fig. 10-2. The elastic portion of the diagram is a straight line. The inelastic portion of the diagram is a straight line of different slope beginning at the yield stress. A material for which the slope in the initial portion of the inelastic region is not zero is called a *strain-hardening material*. The data presented in Fig. 10-2 indicates that the yield stress for the aluminum alloy is approximately 42 ksi and the strain at yield is approximately 0.0040. Thus, the modulus of elasticity for the aluminum alloy is approximately 10,500 ksi. The slope of the diagram in the initial portion of the inelastic region is 1400 ksi; therefore, the increment of stress required to produce a specified increment of strain in the inelastic region is less than it is in the elastic region. It is obvious from the diagram that a strain-hardening material does not permit an increase in strain without an increase in stress.

The useful small-deformation region of the stress-strain diagram for

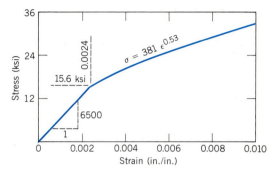

FIG. 10-3 Idealized stress-strain curve for a magnesium alloy.

a magnesium alloy can be approximated by the idealized diagram shown in Fig. 10-3. The elastic portion of the diagram is a straight line. The inelastic portion of the diagram cannot be represented adequately by a straight line of different slope beginning at the yield stress; therefore, this nonlinear region is defined by an appropriate mathematical function. The data presented in Fig. 10-3 indicates that the yield stress for the magnesium alloy is approximately 15.6 ksi and the strain at yield is approximately 0.0024. Thus, the modulus of elasticity for the magnesium alloy is approximately 6500 ksi. It is obvious from the diagram that this magnesium alloy is a strain-hardening material that does not permit an increase in strain without an increase in stress.

A statically determinate, axially loaded member will deform elastically until stresses in the member reach the yield point of the material. Once the yield stress for the member is exceeded, the strain in the member must be obtained by referring to the stress-strain diagram for the material. Failure of such members is assumed when a prescribed level of strain is reached.

The elastic analysis of statically indeterminate, axially loaded members was discussed in Section 2-4. The procedure outlined in that section for solving statically indeterminate problems consisted of the following steps:

1. Draw a free-body diagram.
2. Identify the unknown forces.
3. List the independent equations of equilibrium.
4. Write the required number of deformation equations.
5. Solve the equilibrium and deformation equations for the unknown forces.

When the stresses in some members extend into the inelastic range, stress-strain diagrams such as those shown in Figs. 10-1, 10-2, and 10-3 must be used to relate the loads and the deflections before the equations can be solved for the unknown forces. For the case of an elastoplastic material (Fig. 10-1), the force in the member will have a constant value after the yield stress is reached. Once the forces in these members are known, the remaining forces can be determined by using statical methods. For the case of strain-hardening materials (Figs. 10-2 and 10-3), trial-and-error solutions are required, since the stress in each of the members depends on the deflection imposed on the member.

The procedure for analyzing axially loaded members with stresses extending into the inelastic range is illustrated in the following example.

EXAMPLE 10-1

The rigid plate C in Fig. 10-4a is fastened to the 0.50-in.-diameter steel rod A and to the aluminum alloy pipe B. The other ends of A and B are fastened to rigid supports. When the force P is zero, there are no stresses in A and B. Rod A is made of low-carbon steel (assumed to be elastoplastic) with a proportional limit and yield point of 40 ksi and a modulus of elasticity of 30,000 ksi. Pipe B has a cross-sectional area of 2.0 sq. in. and is made of an aluminum alloy with the stress-strain diagram shown in Fig. 10-4b. A load of 30 kips is applied to plate C as shown. Determine the axial stresses in rod A and pipe B and the displacement of plate C.

Solution

Figure 10-4c is a free-body diagram of plate C and portions of members A and B. The free-body diagram contains two unknown forces. Only one equation of equilibrium is available; therefore, the problem is statically indeterminate. As the load is applied to the plate C, it moves downward an amount δ, which represents the total deformation in members A and B (see Fig. 10-4d). This observation provides a displacement equation that can be used with the equilibrium equation to solve the problem. The two equations are

$$\overset{+}{\circlearrowleft} \qquad \Sigma F_y = 0 = P_A + P_B - 30$$

or

$$(\pi/4)(0.50)^2 \sigma_A + 2\sigma_B = 30 \qquad (a)$$

and

$$\delta_A = \delta_B$$

or

$$10\epsilon_A = 20\epsilon_B$$

In order to solve these equations, the stresses and strains must be related. If the stresses are less than the corresponding proportional

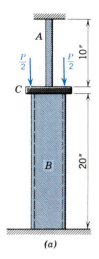

(a)

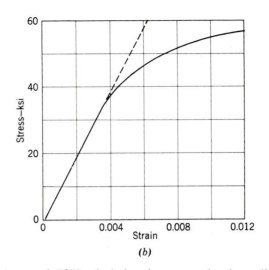

(b)

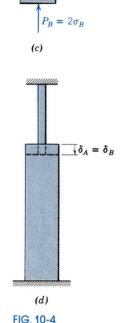

$P_A = (\pi/4)(0.5)^2 \, \sigma_A$

15 k   15 k

$P_B = 2\sigma_B$

(c)

$\delta_A = \delta_B$

(d)

FIG. 10-4

limits, Hooke's law can be used. If Hooke's law is assumed to be valid. the strain equation becomes

$$\frac{10\sigma_A}{30(10^3)} = \frac{20\sigma_B}{10(10^3)}$$

from which

$$\sigma_A = 6\sigma_B \qquad\qquad (b)$$

The value of $E_B = 10(10^3)$ ksi is obtained as the slope of the stress-strain diagram in Fig. 10-4b. The simultaneous solution of Eqs. a and b gives $\sigma_A = 56.6$ ksi T and $\sigma_B = 9.45$ ksi C.

These results indicate that bar A is stressed beyond the porportional limit of the steel and that Hooke's law does not apply. Since the material is assumed to be perfectly plastic beyond the proportional limit, the stress in bar A must be 40 ksi. When this value is substituted in equilibrium Eq. a, the equation becomes

$$(\pi/16)(40) + 2\sigma_B = 30$$

from which

$$\sigma_B = \underline{11.07 \text{ ksi } C} \qquad \text{and} \qquad \sigma_A = \underline{40 \text{ ksi } T} \qquad \text{Ans.}$$

The stress in B is less than the proportional limit for aluminum; therefore, Hooke's law can be used to determine the movement of the plate, which is

$$\delta = \delta_B = 20(11.07)/(10^4) = \underline{0.0221 \text{ in. downward}} \qquad \text{Ans.}$$

If the stress in B had been more than the proportional limit of the aluminum, it would have been necessary to determine the strain from the curve in Fig. 10-4b. If the material in bar A were not perfectly plastic, a trial-and-error solution would be required, since the stresses in both A and B depend on the movement of C.

PROBLEMS

10-1* The rigid bar *CD* of Fig. P10-1 is horizontal and bars *A* and *B* are unstressed before the load *P* is applied. Bar *A* has a cross-sectional area of 2 in.² and is made of a low-carbon steel (elastoplastic) that has a proportional limit and yield point of 40 ksi and a modulus of elasticity of 30,000 ksi. Bar *B* has a cross-sectional area of 1.5 in.² and is made of an aluminum alloy that has a proportional limit of 48 ksi and a modulus of elasticity of 10,600 ksi. Determine

(a) the axial stress in each of the bars after the load *P* is applied.

(b) the vertical displacement (deflection) of pin *C*.

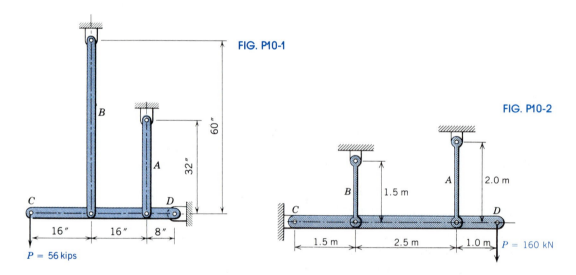

FIG. P10-1

FIG. P10-2

10-2* The rigid bar *CD* of Fig. P10-2 is horizontal and bars *A* and *B* are unstressed before the load *P* is applied. Bar *A* has a cross-sectional area of 500 mm² and is made of an aluminum alloy that has a proportional limit of 330 MPa and a modulus of elasticity of 73 GPa. Bar *B* has a cross-sectional area of 750 mm² and is made of a low-carbon steel (elastoplastic) that has a proportional limit and yield point of 275 MPa and a modulus of elasticity of 210 GPa. Determine

(a) the axial stress in each of the bars after the load *P* is applied.

(b) the vertical displacement (deflection) of pin *D*.

10-3 Bar *A* of Fig. P10-3 is made of an aluminum alloy that has a proportional limit of 55 ksi and a modulus of elasticity of 10,500 ksi. Bars *B* are made of structural steel (elastoplastic) that has a proportional limit and yield point of 36 ksi and a modulus of elasticity of 29,000 ksi. All bars have a 2.5-by-1.0-in. rectangular cross section. Determine

(a) the axial stress in each of the bars after a 250-kip load *P* is applied.

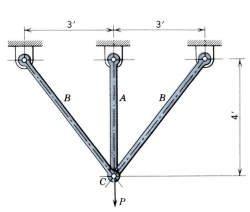

FIG. P10-3

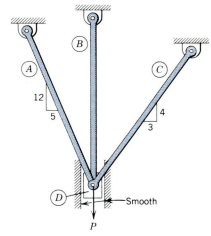

FIG. P10-5

    (b) the vertical displacement (deflection) of pin C produced by the 250-kip load.

10-4 The three bars shown in Fig. P10-3 are made of structural steel that has a proportional limit and yield point of 36 ksi and a modulus of elasticity of 29,000 ksi. All bars have a 2.5-by-1.0-in. rectangular cross section. Determine

    (a) the maximum load P that the structure will support if elastoplastic action is assumed (sometimes called the "limit load").

    (b) the displacement of pin C when the yield point is reached in the last bar to change from elastic to plastic action.

    (c) the maximum deflection of pin C if strain hardening begins when the strain level in a bar reaches 0.018 in./in.

10-5* Three bars, each 50 mm wide by 25 mm thick by 4 m long, are connected and loaded as shown in Fig. P10-5. Bar A is made of Monel that has a proportional limit of 400 MPa and a modulus of elasticity of 180 GPa. Bar B is made of a magnesium alloy that has a proportional limit of 100 MPa and a modulus of elasticity of 40 GPa. Bar C is made of structural steel (elastoplastic) that has a proportional limit and yield point of 240 MPa and a modulus of elasticity of 200 GPa. Determine

    (a) the axial stress in each of the bars after a 650-kN load P is applied.

    (b) the vertical displacement (deflection) of pin D produced by the 650-kN load.

10-6* The three bars shown in Fig. P10-5 are made of structural steel that has a proportional limit and yield point of 240 MPa and a modulus of elasticity of 200 GPa. All bars have a 75-by-25-mm rectangular cross section and are 4 m long. Determine

(a) the maximum load P that the structure will support if elastoplastic action is assumed (sometimes called the "limit load").

(b) the displacement of pin D when the yield point is reached in the last bar to change from elastic to plastic action.

(c) the maximum deflection of pin D if strain hardening begins when the strain level in a bar reaches 0.015 m/m.

10-7 Bar A of Fig. P10-7 is made of an aluminum alloy with a proportional limit of 55 ksi and a modulus of elasticity of 10,500 ksi. Bar B is made of structural steel (elastoplastic) with a proportional limit and yield point of 36 ksi and a modulus of elasticity of 29,000 ksi. Both bars have a 1.5-by-1.0-in. rectangular cross section. Determine

(a) the axial stress in each of the bars after a 50-kip load P is applied.

(b) the vertical displacement (deflection) of pin D produced by the 50-kip load.

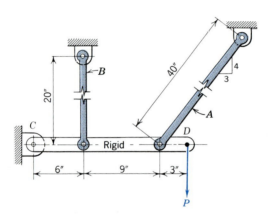

FIG. P10-7

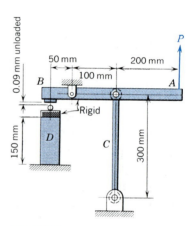

FIG. P10-8

10-8 The rigid bar AB of Fig. P10-8 is horizontal and bar C and post D are unstressed before the load P is applied. Bar C has a cross-sectional area of 600 mm² and is made of a low-carbon steel (elastoplastic) that has a proportional limit and yield point of 240 MPa and a modulus of elasticity of 200 GPa. Post D has a cross-sectional area of 2000 mm² and is made of cold-rolled brass that has a proportional limit of 410 MPa and a modulus of elasticity of 100 GPa. Determine

(a) the axial stresses in bar C and post D after a 100-kN load P is applied.

(b) the vertical displacement (deflection) of pin A produced by the 100-kN load.

10-9* The rigid bar CD of Fig. P10-9 is horizontal and bars A and B are unstressed before the load P is applied. Bar A has a cross-sectional area of 2 in.² and is made of an aluminum alloy that has the stress-strain diagram shown in Fig. 10-4b. Bar B has a cross-sectional area of 2.5 in.² and is made of a low-carbon steel (elastoplastic) that has a proportional limit and yield point of 36 ksi and a modulus of elasticity of 30,000 ksi. Determine

(a) the axial stress in each of the bars after a 40-kip load P is applied.

(b) the axial stress in each of the bars after a 60-kip load P is applied.

(c) the vertical displacement (deflection) of pin C after a 65-kip load P is applied.

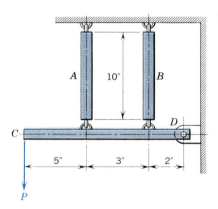

FIG. P10-9

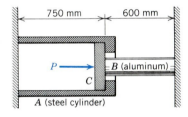

FIG. P10-10

10-10* The assembly shown in Fig. P10-10 consists of a steel cylinder A that has an inside diameter of 160 mm and a wall thickness of 15 mm, a rigid bearing plate C, and an aluminum alloy bar B that has an outside diameter of 50 mm. The aluminum alloy has a proportional limit of 430 MPa and a modulus of elasticity of 73 GPa. The steel (elastoplastic) has a proportional limit and yield point of 275 MPa and a modulus of elasticity of 210 GPa. Determine

(a) the axial stresses in cylinder A and bar B after a 2000-kN load P is applied.

(b) the axial stresses in cylinder A and bar B after a 3000-kN load P is applied.

(c) the horizontal displacement of the bearing plate produced by the 3000-kN load P.

10-3

INELASTIC BEHAVIOR OF TORSIONAL MEMBERS

One of the limitations of the torsion formula (Eq. 3-5) is that stresses must be less than the proportional limit of the material. In some design situations, a limited amount of inelastic action can be permitted; there-

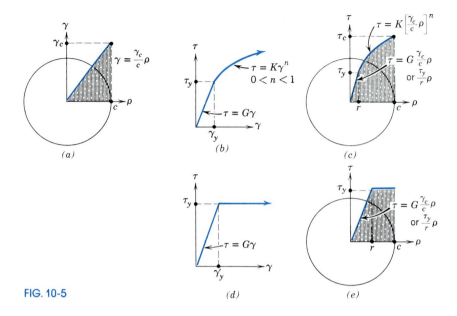

FIG. 10-5

fore, the analysis presented in Section 3-3 will be extended in this section to include some inelastic action in the material. In Section 3-2, the assumption was made that a plane section remains plane and a diameter remains straight for circular sections under pure torsion. This assumption led to the development of Eq. 3-3.

$$\gamma_\rho = \frac{\gamma_c}{c}\rho \qquad (3\text{-}3)$$

which states that the shearing strain at any point on a transverse cross section of a circular shaft is proportional to the distance of the point from the axis of the shaft. Equation 3-3 is valid for elastic or inelastic action provided the strains are not too large ($\tan \gamma \approx \gamma$). Once the shearing strain distribution is known, the law of variation of the shearing stress τ (meaning $\tau_{z\theta}$) with radius ρ on the transverse plane can be determined from a shear stress-strain diagram for the material. As an aid to the thought sequence, it is suggested that sketches be drawn of γ vs. ρ, then τ vs. γ, and finally, τ vs. ρ, as illustrated for two typical examples in Fig. 10-5. The resisting torque (or other unknown if the torque is known) can then be evaluated by substituting the expression for τ in terms of ρ into Eq. 3-1.

$$T = T_r = \int_{\text{area}} \rho\tau_\rho\, dA \qquad (3\text{-}1)$$

The procedure is illustrated in the following examples.

EXAMPLE 10-2

A solid circular steel shaft 4 in. in diameter is subjected to a pure torque of 7π ft-kips. Assume the steel is elastoplastic, having a yield point in shear of 18 ksi and a modulus of rigidity of 12,000 ksi. Determine the maximum shearing stress in the shaft and the magnitude of the angle of twist in a 10-ft length.

Solution

Assume first that the maximum stress is less than the proportional limit. In this case, if the material is homogeneous and isotropic, the expression $\tau = G\gamma$ applies over the entire cross section (see Fig. 10-5d), and the τ vs. ρ diagram would appear like the γ vs. ρ diagram (Fig. 10-5a), with γ_c replaced by τ_c, making the equation of the curve $\tau = (\tau_c/c)\rho$. Substituting this expression for τ into Eq. 3-1 yields

$$7\pi(12) = \int_0^2 \int_0^{2\pi} \rho\left(\frac{\tau_c}{c}\rho\right)(\rho\,d\rho\,d\theta) = \int_0^2 \rho\left(\frac{\tau_c}{2}\rho\right)(2\pi\rho\,d\rho) = \pi\tau_c\left(\frac{2^4}{4}\right)$$

from which τ_c is 21 ksi, which is greater than 18 ksi. Therefore, the assumption is not valid, and the stress distribution curve should be as in Fig. 10-5e with the maximum shearing stress

$$\tau_{max} = \underline{18 \text{ ksi}} \qquad\qquad \text{Ans.}$$

Observe that the first integration above replaces the original dA with an annular area of width $d\rho$. In the future dA will be written in this form.

To obtain the angle of twist, Eq. 3-2 will be rewritten thus

$$\theta = \frac{\gamma L}{\rho} = \frac{\gamma_y L}{r} \qquad\qquad (a)$$

where the shearing strain at the yield point γ_y equals $\tau_y/G = 18/12,000 = 1.5(10)^{-3}$ rad and r is the radius of the part of the cross section that is deforming elastically (Fig. 10-5e). The value of r will be found by means of Eq. 3-1. Thus,

$$T_r = \int_0^r \rho\left(\frac{18}{r}\rho\right)(2\pi\rho\,d\rho) + \int_r^2 \rho(18)(2\pi\rho\,d\rho)$$

$$7\pi(12) = 2\pi\left(\frac{18}{r}\right)\left(\frac{r^4}{4}\right) + 2\pi(18)(1/3)(2^3 - r^3)$$

from which $r^3 = 4$ and $r = 1.587$ in. Now, from Eq. a, the magnitude of the angle of twist is

$$\theta = \frac{1.5(10)^{-3}(10)(12)}{1.587} = \underline{0.1134 \text{ rad}} \qquad\qquad \text{Ans.}$$

EXAMPLE 10-3

A straight shaft with a hollow circular cross section with outside and inside diameters of 3 and 2 in., respectively, is made of a magnesium alloy having the shear stress-strain diagram of Fig. 10-6. Determine the torque required to twist a 5-ft length of the shaft through 0.240 rad.

Solution

From Eq. 3-2 the shear strains at the outside and inside surfaces are obtained. Thus,

$$\gamma_{1.5} = \frac{\rho\theta}{L} = \frac{(3/2)(0.240)}{5(12)} = 6(10)^{-3} \text{ rad}$$

$$\gamma_{1.0} = (2/3)\gamma_{1.5} = 4(10)^{-3} \text{ rad}$$

Since both the minimum and maximum shear strains lie within the right-hand portion of the curve of Fig. 10-6, the shear stress-strain function given for this curve applies for the entire cross section. Thus, the γ vs. ρ function is

$$\gamma = \frac{\gamma_c}{c} \rho = \frac{0.006}{1.5} \rho = 0.004\rho$$

The τ vs. ρ function then becomes

$$\tau = 77.4 \, \gamma^{0.44} = 77.4(0.004\rho)^{0.44} = 6.818\rho^{0.44}$$

Substituting this expression into Eq. 3-1 yields

$$T_r = \int_{1.0}^{1.5} \rho \, (6.818 \, \rho^{0.44})(2\pi\rho \, d\rho) = 12.453\rho^{3.44} \Big]_{1.0}^{1.5}$$

from which

$$T = T_r = \underline{37.8 \text{ in.-kips}} \qquad \text{Ans.}$$

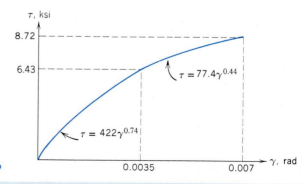

FIG. 10-6

PROBLEMS

10-11* A solid circular 4-in.-diameter shaft is made of an elastoplastic steel (G = 12,000 ksi) that has a shearing yield point of 24 ksi. Determine

    (a) the applied torque when the plastic zone is $\frac{1}{2}$-in. deep.

    (b) the applied torque when the plastic zone is 1-in. deep.

10-12* A solid circular 100-mm-diameter shaft is made of an elastoplastic steel (G = 80 GPa) that has a shearing yield point of 120 MPa. Determine

    (a) the torque required to initiate plastic action in the shaft.

    (b) the percent increase in torque required to produce plastic action over the entire cross section of the shaft.[1]

10-13 A solid circular 3-in.-diameter shaft is made of an elastoplastic steel (G = 12,000 ksi) that has a shearing yield point of 24 ksi. When the plastic zone in the shaft is $\frac{3}{4}$ in. deep, determine

    (a) the shearing strain at the surface of the shaft.

    (b) the torque applied to the shaft.

10-14 A hollow shaft has an inside diameter of 50 mm and an outside diameter of 100 mm. The shaft is made of an elastoplastic steel (G = 80 GPa) that has a shearing yield point of 140 MPa. Determine

    (a) the applied torque when the plastic zone is 10-mm deep.

    (b) the applied torque when the shearing stress at a point on the inside surface of the shaft reaches the yield point.

10-15* A solid circular 4-in.-diameter shaft is twisted through an angle of 5° in a 4-ft length. The shaft is made of an elastoplastic steel (G = 12,000 ksi) that has a yield point in shear of 18 ksi. Determine

    (a) the maximum shearing strain in the shaft.

    (b) the magnitude of the applied torque.

    (c) the shearing stress at a point on a cross section $\frac{1}{2}$ in. from the axis of the shaft.

10-16* A hollow shaft with an outside diameter of 100 mm and an inside diameter of 50 mm is twisted through an angle of 8° in a 3-m length. The shaft is made of an elastoplastic steel (G = 80 GPa) that has a yield point in shear of 140 MPa. Determine

    (a) the maximum shearing strain in the shaft.

    (b) the magnitude of the applied torque.

    (c) the shearing stress at a point on a cross section at the inside surface of the shaft.

[1]*Note:* There will always be some elastic action near the axis of the shaft; however, this effect is so small that it can be neglected without producing significant error.

10-17 A hollow shaft has an inside diameter of 2.5 in., an outside diameter of 4 in., and is 3 ft long. The shaft is made of an aluminum alloy that has a shearing stress-strain diagram that can be approximated by the two straight lines shown in Fig. P10-17. Determine

   (a) the torque required to develop a shearing stress of 30 ksi at a point on a cross section at the inner surface of the shaft.

   (b) the angle of twist when the torque of part (a) is applied.

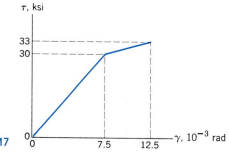

FIG. P10-17

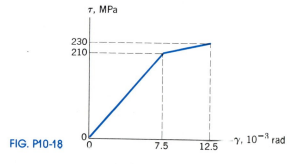

FIG. P10-18

10-18 A solid circular 100-mm-diameter shaft is 2 m long. The shaft is made of an aluminum alloy that has a shearing stress-strain diagram that can be approximated by the two straight lines shown in Fig. P10-18. Determine

   (a) the torque required to develop a maximum shearing stress of 230 MPa in the shaft.

   (b) the angle of twist when the torque of part (a) is applied.

10-19* The composite shaft shown in Fig. P10-19a consists of a hollow steel cylinder with a 6-in. outside diameter and a 4-in. inside diameter over a solid 2-in.-diameter cold-rolled brass core. The two members are rigidly connected to bar AB at the right end and to the wall at the left end. The shearing stress-strain diagram for the steel is shown in Fig. P10-19b. The brass ($G = 5000$ ksi) has a proportional limit in shear of 30 ksi. Determine the torque required to rotate bar AB through an angle of 0.30 rad.

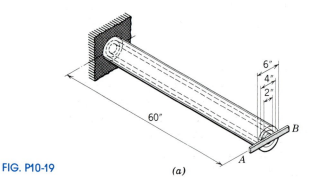

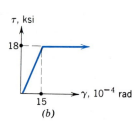

FIG. P10-19

(a)

(b)

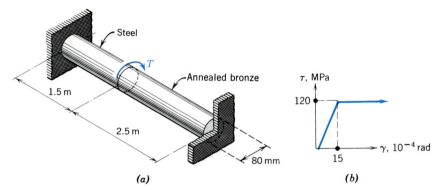

FIG. P10-20

10-20* Two 80-mm-diameter steel ($G = 80$ GPa) and bronze ($G = 45$ GPa) shafts are rigidly connected and supported as shown in Fig. P10-20a. The shearing stress-strain diagram for the steel is shown in Fig. P10-20b. The bronze has a proportional limit in shear of 84 MPa. Determine the torque required to produce a maximum shearing stress of 60 MPa in the bronze.

10-21 The steel shaft of Fig. P10-21 is rigidly secured at the ends. The left 2 ft of the shaft is solid and the right 3 ft is hollow with an inside diameter of 2 in. The steel ($G = 12,000$ ksi) has a yield point in shear of 18 ksi. Determine the minimum torque, applied as shown, that will produce completely plastic action in the hollow segment.

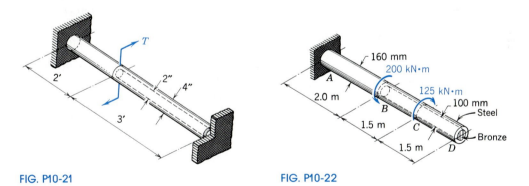

FIG. P10-21   FIG. P10-22

10-22 The 160-mm-diameter steel shaft of Fig. P10-22 has a 100-mm bronze core inserted and securely bonded to the steel ($G = 80$ GPa) in 3 m of the right end. The steel is elastoplastic and has a yield point in shear of 120 MPa. The bronze has a modulus of rigidity G of 40 GPa and a proportional limit in shear of 240 MPa. Determine

  (a) the maximum shearing stress in each of the materials.

  (b) the rotation of the free end of the shaft.

10-23* The circular shaft of Fig. P10-23 consists of a steel (G = 12,000 ksi) segment ABC securely connected to a bronze (G = 6000 ksi) segment CD. Ends A and D of the shaft are securely fastened to rigid supports. The steel is elastoplastic and has a yield point in shear of 18 ksi. The bronze has a proportional limit in shear of 35 ksi. When the two equal torques T are applied, the section at B rotates 0.072 rad in the direction of T. Determine

 (a) the magnitude of the applied torque T.

 (b) the maximum shearing stress in each of the materials.

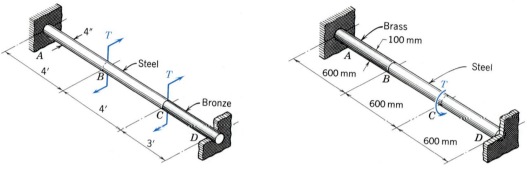

FIG. P10-23   FIG. P10-24

10-24* The 100-mm-diameter shaft shown in Fig. P10-24 is composed of a brass (G = 40 GPa) segment and a steel (G = 80 GPa) segment. The steel is elastoplastic and has a yield point in shear of 120 MPa. The brass has a proportional limit in shear of 60 MPa. Determine the magnitude of the torque, applied as shown, that is required to produce a maximum shearing stress in the brass of 50 MPa.

10-25 The stress-strain curve for a polymeric material used for small gears and shafts is shown in Fig. P10-25. If the material is used for a circular shaft of radius R that is to transmit a torque T, develop expressions for

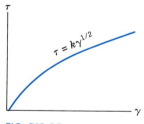

FIG. P10-25

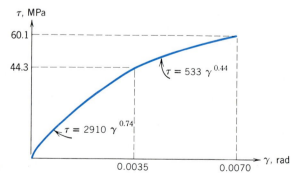

FIG. P10-26

(a) the angle of twist in terms of T, R, k, and length L.

(b) the maximum shearing stress in the shaft in terms of T and R.

10-26 A solid circular 80-mm-diameter shaft is made of a magnesium alloy that has the shearing stress-strain diagram shown in Fig. P10-26. Determine the torque required to twist a 2-m length of the shaft through 0.300 rad.

10-4

INELASTIC BEHAVIOR OF FLEXURAL MEMBERS

A large proportion of structural and machine designs are based on elastic analysis, for which the flexure formula is applicable. However, for some designs, particularly when the weight of the structure is important (aircraft design, for example), the limitation requiring stresses to remain below the proportional limit of the material results in uneconomical or inefficient designs. Therefore, this limitation is sometimes discarded and higher stress levels are tolerated in the design. This section is concerned with the analysis of stresses in instances in which the proportional limit of the material is exceeded or the material does not exhibit a linear stress-strain relationship. In both of these cases, Hooke's law does not apply.

The basic approach to problems involving inelastic action is the same as that outlined in Sections 4-2 and 4-3 for linearly elastic action. A plane section is assumed to remain plane; therefore, a linear distribution of strain exists. If a stress-strain diagram is available for the material, it can be used with the strain distribution to obtain a stress distribution for the beam. The stress distribution shown in Fig. 10-7a is for a beam of symmetrical cross section made of a material having the stress-strain

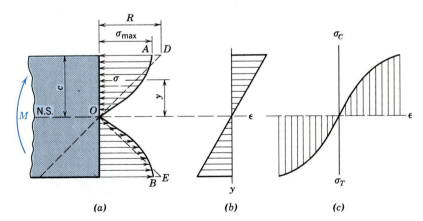

(a)   (b)   (c)   FIG. 10-7

diagram shown in Fig. 10-7c. If an equation can be written for the stress distribution, the resisting moment can be obtained by using Eq. 4-1. If it is not practical to obtain an equation for the stress distribution, the beam depth can be divided into layers of finite thickness and the stress associated with each layer determined. The method of Section 4-3 can then be used to evaluate the resisting moment. As a simplifying technique, the stress-strain diagram can be approximated by a series of straight lines, resulting in a simplified stress distribution diagram and a reduction in the amount of work involved.

An alternate approach to the problem of inelastic action involves replacing the stress distribution *AOB* of Fig. 10-7a by a linear distribution *DOE* (shown dotted) with a maximum stress *R*. The stress *R* (known as the modulus of rupture in bending) is selected to give the same resisting moment as that obtained from the actual stress distribution *AOB*. For the assumed linear distribution *DOE*, the flexure formula is valid; therefore, $R = Mc/I$. The value of *R* is obtained by loading a beam to rupture and substituting the magnitude of the bending moment at rupture into the flexure formula. The modulus of rupture is useful only for predicting the rupture load. It is also valid only for a particular material and cross section and, in general, may be dependent on some shape parameter (as, for example, the diameter to thickness ratio for tubing).

If the material is elastoplastic (see Section 10-2), the large strains associated with yielding will permit plastic action to take place as indicated in Fig. 10-8. The stress distribution for fully elastic action is shown in Fig. 10-8a, for partially plastic action in Fig. 10-8b, and for essentially complete plastic action in Fig. 10-8c. There must always be a slight amount of elastic action if the strain at the neutral surface is zero. The plastic action shown in Fig. 10-8c is idealized in Fig. 10-8d; this is the assumed distribution used for plastic analysis of structural steel beams.

In members subjected to inelastic action, the neutral axis does not necessarily coincide with a centroidal axis as it does when the action is elastic. If the section is symmetrical and the stress-strain diagrams

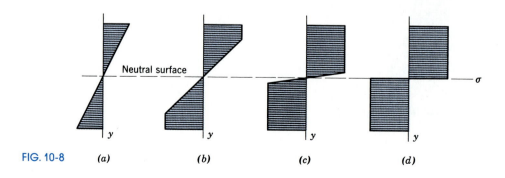

FIG. 10-8   *(a)*   *(b)*   *(c)*   *(d)*

for tension and compression are identical, the neutral axis for plastic action conicides with the centroidal axis. However, if the section is unsymmetrical (for example, a T-section) or if the stress-strain relations for tension and compression differ appreciably (for example, cast iron), the neutral axis shifts away from the fibers that first experience inelastic action, and it is necessary to locate the axis before the resisting moment can be evaluated. The neutral axis is located by using the equation

$$\Sigma F_x = \int_{\text{Area}} \sigma \, dA = 0 \qquad\qquad (a)$$

and solving for the location of the axis where σ is zero.

The following examples illustrate the concepts of inelastic analysis as applied to the pure bending of beams that are loaded in a plane of symmetry.

EXAMPLE 10-4

A beam having the T cross section shown in Fig. 10-9 is made of elastoplastic steel ($E = 200$ GPa) with a proportional limit (equal to the yield point) of 240 MPa. Determine

(a) the bending moment (applied in the vertical plane of symmetry) that will produce a longitudinal strain of -0.0012 at point B on the lower face of the flange.

(b) the bending moment required to produce completely plastic action in the beam.

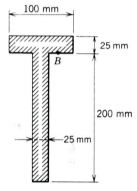

FIG. 10-9

Solution

(a) The strain distribution, stress-strain, and stress distribution diagrams are shown in Figs. 10-10a, b, and c. Because of the inelastic action and unsymmetrical section, the location of the neutral axis must first be determined by using Eq. a. Thus,

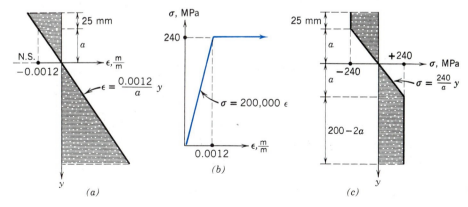

FIG. 10-10

$$\int_A \sigma \, dA = -240(10^6)(0.025)(0.100) - \frac{240(10^6)}{2}(a)(0.025)$$

$$+ \frac{240(10^6)}{2}(a)(0.025)$$

$$+ 240(10^6)(0.200 - 2a)(0.025) = 0$$

from which

$$a = 0.050 \text{ m} = 50 \text{ mm}$$

Once the neutral axis is located, the easiest way to find the moment is by using the free-body diagram method of Example 4-1. Thus, from Fig. 10-11,

$$M_r = 240(10^6)(0.100)(0.025)(0.0625)$$

$$+ \tfrac{1}{2}(240)(10^6)(0.050)(0.025)(2/3)(0.050)(2)$$

$$+ 240(10^6)(0.100)(0.025)(0.100) = 107.5(10^3) \text{ N·m}$$

from which

$$M = M_r = \underline{107.5 \text{ kN·m as shown}} \qquad \text{Ans.}$$

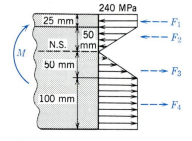

FIG. 10-11

(b) In this case, the stress distribution diagram would be similar to the diagram shown in Fig. 10-8d. If the distance from the neutral surface to the bottom of the stem is designated as distance a, Eq. a yields

$$\int_A \sigma \, dA = \sigma(0.025)(a) - \sigma(0.025)(0.200 - a)$$

$$- \sigma(0.100)(0.025) = 0$$

from which $a = 150$ mm; therefore, the neutral surface is 75 mm below the top of the flange. The plastic moment obtained by using the free-body diagram method is

$$M_r = 240(10^6)(0.100)(0.025)(0.0625)$$

$$+ 240(10^6)(0.050)(0.025)(0.025)$$

$$+ 240(10^6)(0.150)(0.025)(0.075) = 112.5(10^3)\text{N·m}$$

$$M_p = M_r = \underline{112.5 \text{ kN·m}} \qquad \text{Ans.}$$

EXAMPLE 10-5

A beam having the cross section of Fig. 10-12a is made of magnesium alloy having the approximate stress-strain diagram of Fig. 10-12b. Determine the magnitude of the bending moment applied in the vertical plane of symmetry necessary to produce a flexural stress of magnitude of 14.4 ksi at point A, which is at the lower surface of the top flange.

Solution

Because of symmetry, the neutral axis coincides with the centroidal z axis; therefore, the strain and stress distribution diagrams are as shown in Figs. 10-12c and d. Since the specified stress is 14.4 ksi at the lower surface of the top flange ($y = 2$ in.), the strains (from Fig. 10-12b), at $y = 2$ in. and $y = 3$ in. are 0.0035 in./in. and 0.00525 in./in., respectively. Thus, the first stress-strain function is valid for the entire web and the second function is valid for the entire flange. Once the stress distributions are known, Eq. 4-1 can be used to calculate the bending moment. Thus,

$$M_r = \int_A \sigma y \, dA = 2 \left[\int_0^2 405 \left(\frac{0.0035}{2} y \right)^{0.59} (y)(1 \, dy) + \int_2^3 93.1 \left(\frac{0.0035}{2} y \right)^{0.33} (y)(4 \, dy) \right]$$

$$= 2 \left[\int_0^2 9.569 y^{1.59} \, dy + \int_2^3 45.837 y^{1.33} \, dy \right] = 2 \left[\frac{9.569 \, y^{2.59}}{2.59} \right]_0^2 + 2 \left[\frac{45.836 \, y^{2.33}}{2.33} \right]_2^3$$

$$= 44.49 + 311.00 = \underline{355 \text{ in.-kips}} \qquad \text{Ans.}$$

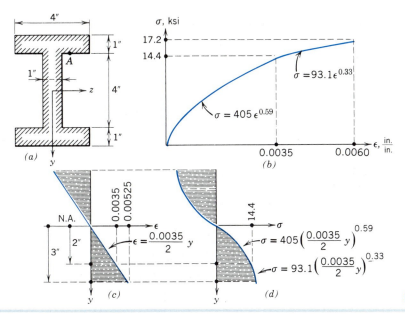

(a)

(b)

(c)

(d)

FIG. 10-12

EXAMPLE 10-6

A beam is to be fabricated from three 6-by-1-in. plates that will be welded together to form a symmetrical I-section. Determine the maximum elastic and plastic bending moments that the beam will sustain if the material is

(a) elastoplastic steel with a proportional limit (equal to the yield point) of 40 ksi.

(b) an aluminum alloy with the stress-strain diagram shown in Fig. 10-13a. Assume the diagram is the same in tension and compression and limit the strain to 0.010.

Solution

(a) For the elastic moment, the flexure formula $\sigma = Mc/I$ applies. The moment of inertia I about the neutral axis is 166 in.^{4}. Therefore,

$$M_e = \frac{\sigma I}{c} = \frac{40(166)}{4} = \underline{1660 \text{ in.-kips}} \qquad \text{Ans.}$$

For this material, the ideal stress distribution of Fig. 10-8d may be assumed when computing the plastic moment, which is

$$M_p = \left[40(6)(1)\,\frac{7}{2} + 40(3)(1)\,\frac{3}{2} \right] [2] = \underline{2040 \text{ in.-kips}} \qquad \text{Ans.}$$

To check the degree of approximation involved in assuming the ideal stress distribution for this section, the maximum strain before strain hardening takes place will be assumed to be 16 times the elastic strain (a reasonable value for structural steel). When this strain is reached in the outer fibers, the maximum elastic strain occurs at $(\frac{1}{16})$ (4) or $\frac{1}{4}$ in. above and below the neutral surface; this means that the actual stress distribution diagram is like that of Fig. 10-8c, the distance from the neutral surface to the point where the stress is constant being $\frac{1}{4}$ in. This distribution makes the moment of 2040 in.-kips computed above too large by the amount

$$\Delta M_r = \frac{40}{2}\,\frac{1}{4}\,(1)\,\frac{1}{3}\,\frac{1}{4}\,(2) = 0.833 \text{ in.-kip}$$

or 0.04 percent.

(b) The elastic moment is given by the flexure formula, where σ is the proportional limit that, from Fig. 10-13a, is 30 ksi. Then,

$$M_e = 30(166/4) = \underline{1245 \text{ in.-kips}} \qquad \text{Ans.}$$

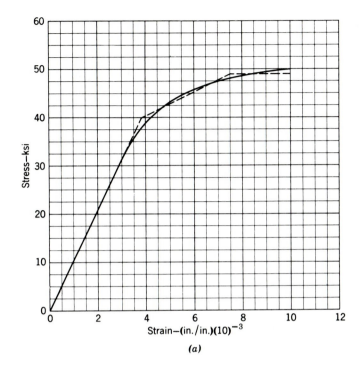

(a)

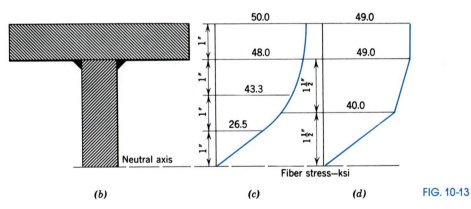

(b)   *(c)*   *(d)*   FIG. 10-13

The plastic moment will first be determined with the aid of the stress distribution diagram of Fig. 10-13c that was obtained from Fig. 10-13a by assuming the maximum strain in the beam to be 0.010, and since the strains are proportional to the distance from the neutral surface of the beam, the stress for any point in the cross section may be read from the diagram of Fig. 10-13a. With the assumption that the curve in Fig. 10-13c is a series of chords connecting the points at 1-in. vertical intervals, the following solution is obtained:

$$M_p = 2 \left[6(1)(48)\frac{7}{2} + (50 - 48)\frac{1}{2}(6)(1)\left(3 + \frac{2}{3}\right) \right.$$

$$+ 43.3(1)(1)\frac{5}{2} + (48 - 43.3)\frac{1}{2}\left(2 + \frac{2}{3}\right)$$

$$+ 26.5(1)(1)\frac{3}{2} + (43.3 - 26.5)\frac{1}{2}\left(1 + \frac{2}{3}\right)$$

$$\left. + \frac{26.5}{2}\frac{2}{3}\right]$$

$$= 2414 \text{ in.-kips} \hspace{3cm} \text{Ans.}$$

This result will be checked by assuming the curve below the flange to be a parabola. The flange moment will be the same as above and equals 2060 in.-kips; then,

$$M_p = 2060 + \frac{2}{3}(48)(3)(1)\frac{5}{8}(3)(2) = 2420 \text{ in.-kips} \hspace{0.5cm} \text{Ans.}$$

One more technique will be applied to this problem. The stress-strain diagram of Fig. 10-13a will be approximated by the dotted lines shown, and the resulting stress distribution diagram for the beam will be as in Fig. 10-13d. With the use of this diagram, the plastic moment becomes

$$M_p = 2 \left[49(6)(1)\left(\frac{7}{2}\right) + 40\left(\frac{3}{2}\right)(1)\left(\frac{3}{2} + \frac{3}{4}\right) \right.$$

$$+ (49 - 40)\left(\frac{1}{2}\right)\left(\frac{3}{2}\right)(1)\left(\frac{3}{2} + 1\right)$$

$$\left. + 40\left(\frac{1}{2}\right)\left(\frac{3}{2}\right)(1)\left(\frac{2}{3}\right)\left(\frac{3}{2}\right) \right]$$

$$= 2422 \text{ in.-kips} \hspace{3cm} \text{Ans.}$$

The last three results indicate that any of the three techniques is adequate for this problem.

From the preceding discussion and examples, one may observe that, if the load capacity of a beam is based on the plastic moment rather than the elastic moment, a considerable saving in material may be realized (with the same factor of safety). Furthermore, in the design of statically indeterminate beams and frames, the plastic method of analysis is considerably less time-consuming than the elastic analysis.

This is not to say that the elastic method of analysis should be discarded as outmoded, for there are many designs that must be based on elastic action, and for some situations, particularly when repeated loading is involved, the application of plastic analysis may be dangerous.

PROBLEMS

Note: In the following problems, all beams are loaded in a plane of symmetry parallel to the largest cross-sectional dimension, except as noted otherwise.

10-27* Determine the maximum elastic and plastic bending moments for a W33 × 201 steel beam having a proportional limit (equal to the yield point) of 36 ksi.

10-28* Determine the maximum elastic and plastic bending moments for a W762 × 196 steel beam having a proportional limit (equal to the yield point) of 250 MPa.

10-29 Determine the maximum elastic and plastic bending moments for a W14 × 120 steel beam having a proportional limit (equal to the yield point) of 36 ksi.

10-30 Determine the maximum elastic and plastic bending moments for a W203 × 60 steel beam having a proportional limit (equal to the yield point) of 250 MPa.

10-31* Determine the ratio of the plastic moment to the maximum elastic moment for a beam of elastoplastic material with a rectangular cross section.

10-32* Determine the ratio of the plastic moment to the maximum elastic moment for a beam of elastoplastic material with a circular cross section.

10-33 Determine the ratio of the plastic moment to the maximum elastic moment for a beam of elastoplastic material with the cross section shown in Fig. P10-33.

10-34 Determine the ratio of the plastic moment to the maximum elastic moment for a beam of elastoplastic material with a square cross section if

    (a) the neutral axis is located as shown in Fig. P10-34*a*.

    (b) the neutral axis is located as shown in Fig. P10-34*b*.

10-35* A beam of elastoplastic material (yield point of 36 ksi) has the cross section shown in Fig. P10-35. Determine

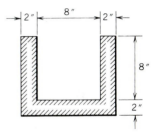

FIG. P10-33

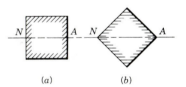

(a)   (b)

FIG. P10-34

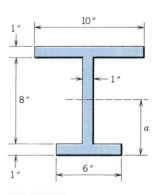

FIG. P10-35

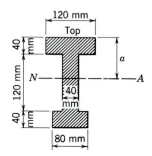

FIG. P10-36

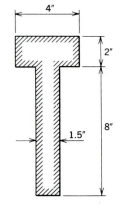

FIG. P10-37

(a) the location of the neutral axis when the stress in the outer fibers of the top flange reaches the yield point.

(b) the moment required to produce the condition of part (a).

(c) the ratio of the plastic moment to the maximum elastic moment for this cross section.

10-36* A beam of elastoplastic material (yield point of 240 MPa) has the cross section shown in Fig. P10-36. Determine

(a) the location of the neutral axis when the stress in the outer fibers of the top flange reaches the yield point.

(b) the moment required to produce the condition of part (a).

(c) the ratio of the plastic moment to the maximum elastic moment for this cross section.

10-37 A beam of elastoplastic material (yield point of 36 ksi) has the cross section shown in Fig. P10-37. Determine

(a) the location of the neutral axis when the stress in the outer fibers of the top flange reaches the yield point.

(b) the moment required to produce the condition of part (a).

(c) the ratio of the plastic moment to the maximum elastic moment for this cross section.

10-38 A WT305 × 70 steel beam of elastoplastic material has a proportional limit (equal to the yield point) of 250 MPa. Determine

(a) the location of the neutral axis when the stress in the outer fibers of the flange reaches the yield point.

(b) the moment required to produce the condition of part (a).

(c) the ratio of the plastic moment to the maximum elastic moment for this cross section.

10-39* A beam of rectangular cross section is made of a material for which the stress-strain diagram in tension can be represented by the expression $\sigma = K\epsilon^{1/2}$. The shape of the diagram is the same in tension and compression. Develop an expression similar to the flexure formula for relating fiber stress and applied moment.

10-40 A beam having the T cross section shown in Fig. P10-40a is made of a magnesium alloy that has the stress-strain diagram shown in Fig. P10-40b. The beam is subjected to a bending moment that produces a maximum flexural stress of 99.3 MPa T. When this moment is applied, the neutral surface is located at the junction of the flange and stem. Determine

(a) the dimension c.

(b) the bending moment applied to the beam.

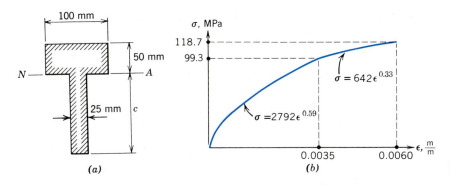

FIG. P10-40

10-5

PLASTIC ANALYSIS OF STATICALLY INDETERMINATE BEAMS

Because of increasing emphasis on structural analysis based on plastic action, a brief introduction to plastic analyses of statically indeterminate beams is presented in this section. The analysis will be limited to ideal elastoplastic materials as defined in Section 10-2. For practical purposes, ordinary structural steel will be satisfactory, and for this material the following assumptions are reasonable and may be verified (within practical limits) by experiment:

1. As the load on the beam increases, the fiber stress will increase proportionately until the yield point for the material is attained at the section of maximum moment, after which yielding at this section will progress over the entire section (see Section 10-4 and Fig. 10-8*d*). The resisting moment will increase with the load until the stress in each fiber has reached the yield point, after which the moment will remain constant. In other words, a hinge exerting a constant moment forms at the section, producing a discontinuity in the slope of the elastic curve. The moment at the hinge is termed the *plastic moment* M_p.
2. As the loading is increased, a second hinge will form and additional hinges will form until the beam is reduced to what has been termed a *mechanism;* that is, a system of members, connected at their ends by pins (each pin exerting a constant moment), which will continue to deflect with no additional load.

The load producing a mechanism is considered the *ultimate load,* although strain hardening of the material might necessitate the application of a larger load to produce complete collapse. The hinge, which actually extends over a finite length of beam, is assumed to form at a discrete point in the beam.

One procedure for analyzing statically indeterminate beams consists of arbitrarily assigning the plastic moment value to the redundant moments; this procedure makes the beam statically determinate. Additional hinges are introduced at points of high moment until a mechanism develops. Finally, a check of the moments should show that nowhere does the moment exceed the magnitude of the plastic moment M_p. This is not the only method used in plastic analysis,[2] but for the simpler cases it is probably the easiest to apply. The method is illustrated by the following example.

EXAMPLE 10-7

The continuous beam of Fig. 10-14a is a steel W36 $\times$ 160 section. Consider the steel to be elastoplastic with a yield point of 36 ksi and determine the ultimate load w (which for span BC includes the weight of the beam) for the beam.

Solution

The free-body diagram of Fig. 10-14b indicates three unknown reactions, and for this force system there are only two independent equations of equilibrium; hence, there is one redundant quantity. If one moment, shear, or reaction is determined, the beam becomes statically determinate. However, the ultimate load is not attained until a mechanism forms; therefore, it becomes necessary to assume more than one possible hinge location. To assist in estimating the location of possible hinges, the shape of the moment diagram by parts and the composite moment diagram are shown in Fig. 10-14c, where the solid line indicates the composite diagram and the dashed lines represent the diagrams for two simple beams with a moment at B. This diagram indicates that hinges may possibly form at B, D, and E. Possible deformations are indicated (greatly exaggerated) in Fig. 10-14d, the top curve indicating the shape of the elastic curve when the action is entirely elastic. As the magnitude of the loading is increased beyond that necessary to produce the yield point stress in the outer fibers, a hinge will form at the section where the flexural stress is the highest. The first hinge is assumed to form at B, and under this condition the beam would assume the shape indicated by the middle curve of Fig. 10-14d. Since a mechanism has not yet formed, the loading is increased until a second hinge

[2]For additional material, see *Plastic Design of Steel Frames*, L. S. Beedle, Wiley, New York, 1958.

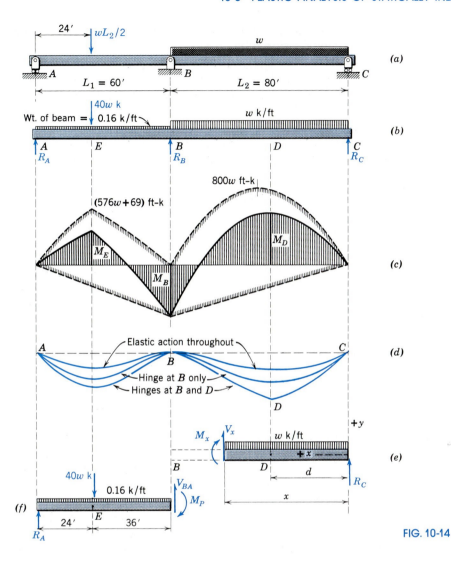

FIG. 10-14

forms. This hinge is assumed to form at D, and from the lower curve of Fig. 10-14d it is apparent that a mechanism has developed since, with hinges at B and D and with end C supported on rollers, deflection can continue in the span BC without an increase in the load. The second hinge may have formed at E instead of at D; hence, the moment at E must be evaluated as a check on the hinge assumptions.

The magnitude of the load necessary to produce hinges at B and D will first be determined, after which the moment at E will be checked. It should be noted that when the plastic moment M_p is used in either an equilibrium or a moment equation, it must have the correct sense or sign. With reference to the free-body diagram of Fig. 10-14e, which represents the right portion of span BC, the equation for the moment in span BC is

$$M_x = R_C x - wx^2/2 \qquad (a)$$

When x is 80 ft,

$$M_B = -M_p = R_C(80) - w(80)^2/2$$

from which

$$R_C = 40w - M_p/80 \qquad (b)$$

The negative sign on M_B is obtained from Fig. 10-14c. The section of maximum positive moment in span BC is at D, where the shear is zero and is located from the shear equation

$$V_x = -R_C + wx$$

When x is equal to d,

$$V_x = 0 = -R_C + wd$$

from which

$$d = R_C/w \qquad (c)$$

From Eq. a the moment at D (the plastic moment) is

$$M_D = +M_p = R_C d - wd^2/2$$

Since R_C is equal to wd from Eq. c, the plastic moment becomes

$$M_p = wd^2 - \frac{wd^2}{2} = \frac{w}{2}\left(40 - \frac{M_p}{80w}\right)^2 \qquad (d)$$

where the plus sign on M_D is obtained from Fig. 10-14c and R_C is obtained from Eq. b. Solving Eq. d for w gives

$$w = \frac{5.8284 M_p}{2300} \quad \text{or} \quad \frac{0.1716 M_p}{3200}$$

When these expressions are substituted into Eq. b, only the larger value yields a positive result for d; therefore, this expression will give the correct magnitude for w unless the bending moment at E exceeds M_p. The plastic moment for the W36 $\times$ 160 section is

$$M_p = [\overbrace{\sigma_{yp}(12)(1.02)(18 - 0.51)}^{\text{one flange}} + \overbrace{\sigma_{yp}(0.650)(18 - 1.02)^2(1/2)]}^{\text{one-half web}}(2)$$

$$= 616\sigma_{yp}$$

$$= 616(36)/12 = 1848 \text{ ft-kips}$$

Substituting this value for M_p in the expression for w yields

$$w = 5.828(616)(3)/3200 = 3.366 \text{ kips/ft}$$

To check the bending moment at E, the reaction at A will first be obtained by summing moments with respect to B on the free-body diagram of Fig. 10-14f. The equation

$$60R_A - 40w(36) - 0.16(60)(30) + M_p = 0$$

gives

$$R_A = 24(3.366) + 4.80 - 616(3)/60 = 54.8 \text{ kips}$$

The bending moment at E is

$$M_E = 54.8(24) - 0.16(24)(12) = 1269 \text{ ft-kips}$$

which is less than 1848; therefore, a hinge does not form at E, the moment in the beam is nowhere greater than M_p, and the ultimate load is

$$w = \underline{3.37 \text{ kips/ft}} \qquad \text{Ans.}$$

Some economy in the design could be effected by reducing the size of the beam in the left span to such a section that a hinge might form at E; that is, a beam section could be selected for which the plastic moment is at least 1269 ft-kips. If a table of plastic section moduli was available,[3] the section could be determined from the expression $M_p = Z \sigma_{yp}$, where Z is the plastic section modulus.[4] If such a table is not available, the following method may be used. It has been shown[5] that the ratio Z/S (where S is the elastic section modulus) for wide flange and American standard sections ranges from 1.10 to 1.23 with 1.14 as the average for all wide flange sections and 1.18 as the average for

[3] See AISC *Manual of Steel Construction* and BCSA-CONSTRADO *Structural Steel-work Handbook* for standard metric sections.

[4] The plastic modulus is the arithmetic sum of the absolute values of the first moments, with respect to the neutral axis, of the cross-sectional areas on each side of the neutral axis.

[5] See Fig. 7.1 of *Plastic Design of Steel Frames*, L. S. Beedle, Wiley, New York, 1958.

American standard sections. Using the average value of 1.14 gives

$$S = \frac{Z}{1.14} = \frac{M_E}{1.14\sigma_{yp}} = \frac{1269(12)}{1.14(36)} = 371 \text{ in.}^3$$

From Appendix C, a W33 × 130 section with a section modulus of 406 appears to be a likely choice, and it will be checked as follows. When the cross-sectional dimensions given by Appendix C are used, the plastic moment is

$$\overbrace{[11.51(0.855)(16.12)}^{\text{one flange}} + \overbrace{15.69(0.580)(15.69/2)]}^{\text{one-half web}}(2)(36/12) = 1379 \text{ ft-kips}$$

which is more than adequate. Therefore, one could splice the W33 × 130 beam to the W36 × 160 beam at some section between B and E. Theoretically, the splice could be made where the bending moment in span AB has a value of -1379 ft-kips (which is approximately 5.3 ft left of B). Practically, to avoid splicing at a section of large moment, the joint might be located some distance to the left, the distance depending on the judgment of the designer or a pertinent specification. The resulting beam will develop hinges at B and D and will just begin to yield at E under a 3.37 kips/ft.

The maximum value for w, based on elastic analysis and with the assumption that the moment of inertia I_{AB} for the full length of 60 ft is $(7/10)I_{BC}$, is found to be 2.76 kips/ft; thus, the plastic analysis for this problem gives a loading approximately 22 percent higher than that obtained from elastic analysis. Part of this increase is due to the plastic moment being greater (about 14 percent) than the elastic moment, and part is due to the increase in load necessary to cause formation of the second plastic hinge. These loads will be considered failure loads that, when divided by an appropriate factor of safety,[6] will yield working loads.

The discussion presented in this section is very limited in scope and hence does no more than present some of the fundamental concepts of plastic design involving an ideal elastoplastic material. A complete analysis would involve the rotation capacity of the beam sections used, to ascertain if the first hinge to form will sustain sufficient plastic deformation to permit the formation of successive hinges. Furthermore, shearing stresses or local buckling may be quite significant for some designs. These and other design problems are beyond the scope of this book.

[6]Here, the factor of safety is defined as the ratio of the ultimate (failure) load to the computed (or specified) applied load. This definition, which differs from that of Section 1-16, will be used in all plastic analysis problems in this book.

PROBLEMS

Note: The following problems are to be solved using plastic analysis, and for simplification the weight of the beam is to be neglected. All beams are made of structural steel having a yield point and proportional limit of 36 ksi or 250 MPa. For wide-flange beam design, use $Z = 1.14 \, S$.

10-41* A beam is loaded and supported as shown in Fig. P10-41. Determine the ultimate load P in terms of M_p and L

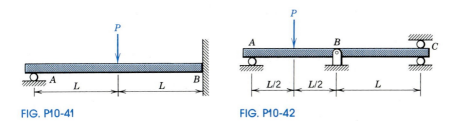

FIG. P10-41   FIG. P10-42

10-42* A beam is loaded and supported as shown in Fig. P10-42. Determine the ultimate load P in terms of M_p and L.

10-43 A beam is loaded and supported as shown in Fig. P10-43. Determine the ultimate load M in terms of M_p and L.

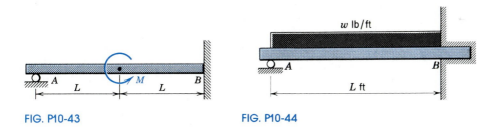

FIG. P10-43   FIG. P10-44

10-44 A beam is loaded and supported as shown in Fig. P10-44. Determine the ultimate load w in terms of M_p and L.

10-45* The beam shown in Fig. P10-45 is a structural steel wide-flange section that has a plastic section modulus $Z = 57.6 \, \text{in.}^3$ and a length $L = 15$ ft. Determine the magnitude of the ultimate load w.

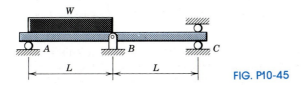

FIG. P10-45

10-46* The beam shown in Fig. P10-46 is a structural steel wide-flange section that has a plastic section modulus $Z = 900(10^3)$ mm³ and a length $L = 6$ m. Determine the magnitude of the ultimate load P.

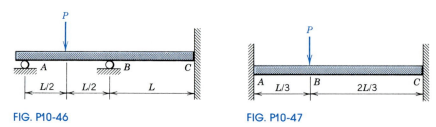

FIG. P10-46   FIG. P10-47

10-47 The beam shown in Fig. P10-47 is a structural steel wide-flange section that has a plastic section modulus $Z = 123$ in.³ and a length $L = 18$ ft. Determine the magnitude of the ultimate load P.

10-48 The beam shown in Fig. P10-48 is a structural steel wide-flange section that has a plastic section modulus $Z - 2018(10^3)$ mm³ and a length $L = 8$ m. Determine the magnitude of the ultimate load w.

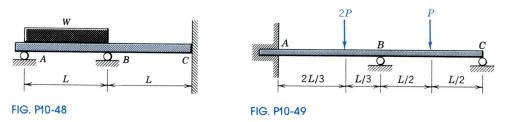

FIG. P10-48   FIG. P10-49

10-49* The beam shown in Fig. P10-49 has a length $L = 10$ ft and is carrying a load $P = 65$ kips. Determine the lightest structural steel wide-flange section required if a factor of safety of 2 is specified.

10-50* The beam shown in Fig. P10-50 has a length $L = 6$ m and is carrying a load $w = 40$ kN/m. Determine the lightest structural steel wide-flange section required if a factor of safety of 1.75 is specified.

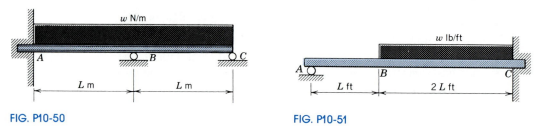

FIG. P10-50   FIG. P10-51

10-51 The beam shown in Fig. P10-51 has a length $L = 8$ ft and is carrying a load $w = 1500$ lb/ft. Determine the lightest structural steel wide-flange section required if a factor of safety of 2.25 is specified.

10-52 The beam shown in Fig. P10-52 has a length $L = 8$ m and is carrying a load $w = 100$ kN/m. Determine the lightest structural steel wide-flange section required if a factor of safety of 2.5 is specified.

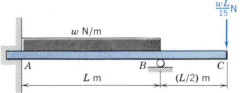

FIG. P10-52

10-53 A two-span continuous beam with simple supports is subjected to a uniformly distributed load of 1000 lb/ft on both spans. The span lengths are 32 ft and 40 ft, respectively. Determine the lightest structural steel wide-flange section required if a factor of safety of 2 is specified. The beam is to have a constant section throughout its length.

10-54* A structural steel wide-flange section having a plastic section modulus $Z = 3667(10^3)$ mm^3 is used as a two-span, simply supported, continuous beam with spans of 8 m and 14 m, respectively. The beam carries uniformly distributed loads of $2w$ kN/m on the 8-m span and w kN/m on the 14-m span. Determine

(a) the ultimate load w for the beam.

(b) the lightest wide-flange beam required for the 8-m span.

(c) the theoretical splice point.

10-6

INELASTIC COLUMN THEORY

The column formulas discussed in Sections 9-2, 9-4, and 9-5 were based on the assumption that the action was elastic; that is, the stresses were assumed to be below the proportional limit. The Euler formula can thus be used for an *ideal, axially* loaded column for any value of P/A less than the proportional limit of the material.

The first theory for predicting the buckling loads of columns loaded in the inelastic range was proposed by F. R. Engesser about 1890 and is known as the tangent-modulus theory. It is based on the assumption that the Euler formula can be used in the inelastic range, provided that Young's modulus is replaced by the tangent modulus of the material. The tangent modulus is defined as the slope of the stress-strain diagram for a material at *a particular* stress, and it is thus a function of the stress (and strain) for stresses greater than the proportional limit. For stresses less than the proportional limit, the tangent modulus, of course, is the same as Young's modulus. For an ideal elastoplastic material the tangent modulus becomes zero at the proportional limit since the stress-strain diagram becomes horizontal. If E_T represents the tangent modulus, the buckling load, according to the tangent-modulus theory, is

$$\frac{P}{A} = \frac{\pi^2 E_T}{(L/r)^2} \tag{10-1}$$

In order to use Eq. 10-1, values of E_T corresponding to various values of P/A must be obtained from a stress-strain diagram. It is difficult to determine precise values of the slope at specific points on the curve; therefore, it is usually desirable to draw a supplementary curve showing E_T plotted against P/A and use this curve to obtain values for E_T. When corresponding values of E_T and P/A are available, values of L/r can be computed from Eq. 10-1 and used to plot a graph indicating the load-carrying capacity of columns when loaded in the plastic region of the stress-strain diagram. The Euler formula was derived on the assumption that the modulus of elasticity was constant over the cross section of the column, whereas the tangent-modulus formula assumes a variable modulus; therefore, the latter formula cannot be considered to provide a rigorous solution. If the assumption is made that for any given value of the load the *variation* of strain across any section of the column is small, the slope of the stress-strain diagram (the tangent modulus) may be considered a constant for the material at all points in any given section. Since the strains will vary linearly (plane section remains plane), the result will be a linear stress distribution, thus justifying the use of the differential equation for bending. On the basis of this reasoning, it appears that the modified Euler (tangent-modulus) formula should give reasonable results, particularly for materials for which the tangent modulus does not vary too abruptly.

The shape of the tangent-modulus curve is influenced to a considerable degree by the shape of the stress-strain diagram near the proportional limit. Three idealized stress-strain diagrams are shown in Fig. 10-15 for three hypothetical materials with identical diagrams to the proportional limit. Material A is elastoplastic, B has a diagram represented by two straight lines, and C has a gradual curve at the knee. The variation of the tangent modulus with the stress is shown on the same figure, with the modulus represented by horizontal distances. The buckling loads, as predicted by the tangent modulus, are shown in Fig. 10-16. Note that the buckling load predicted by the tangent-modulus formula (Eq. 10-1) is the same as that predicted by the Euler formula (Eq. 9-3) in the elastic range, and that this load is usually greater than loads obtained by actual experiment. However, when sufficient care is exercised to assure a centric load, the buckling loads indicated by

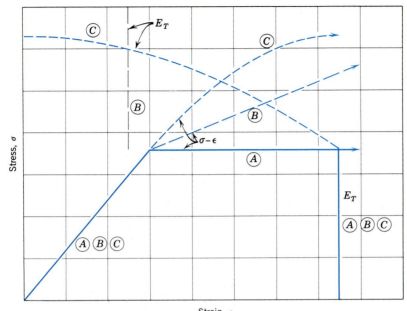

Strain, ϵ
Tangent modulus, E_T

FIG. 10-15

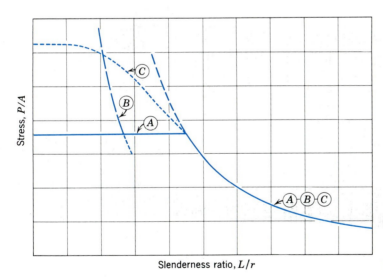

Slenderness ratio, L/r

FIG. 10-16

the tangent-modulus formula are fairly close to experimental results[7]
for columns loaded in the elastic and plastic range.

[7]"Evaluation of Aeroplane Metals," J. A. Van den Broek, *J. R. Aeronaut. Soc.*,
1946.

The application of the Euler-Engesser equation (Eq. 10-1) in the inelastic range may require an iterative solution. Such a solution can be obtained with the aid of the digital computer if the time and expense are justified. However, many problems can be solved using successive approximations with a precision consistent with ordinary engineering applications. The following paragraphs outline three of the possible approaches to such solutions.

1. The actual compression stress-strain diagram can be approximated by a series of straight lines. One then solves a given problem by assuming the stress range is entirely within a particular segment of the diagram, using the value of E_T for this segment in Eq. 10-1, solving for σ_{cr} (or P_{cr}/A), and checking this value against the values of σ for which the curve segment is valid. If the computed value of σ_{cr} does not lie in the range selected, the process is repeated using a different segment of the diagram. Since Eq. 10-1 is identical with Eq. 9-3 in the elastic range, the curve segments to be considered include the elastic range. Note that at the junction of two segments, there are two different values of E_T, leading to an abrupt discontinuity in the P/A vs. L/r curve, as indicated for material B in Fig. 10-16. The Euler-Engesser equation is not valid at such points of discontinuity, since the derivation is based on the assumption that the variation of the tangent modulus over the cross section of the member is small.

2. The nonlinear part of the actual stress-strain diagram can be approximated by one or more continuous curves, each of which is represented by a differentiable function of stress and strain. Within certain limits, the tangent modulus ($E_T = d\sigma/d\epsilon$) is a continuous function of σ, which can be substituted in Eq. 10-1, resulting in an expression for σ_{cr} as a function of L/r (where σ_{cr} is P/A). Here again, one should assume a certain segment of the curve is applicable, then check the value of the computed stress against the range of stresses for which the curve segment is applicable. Note that although the stress-strain functions may provide a reasonably continuous curve, the derivatives of these functions may not be continuous through the change from one function to another. In other words, at the junction of two curves, there can be two different values for E_T, thus invalidating Eq. 10-1, as discussed in the paragraph above.

3. A solution can be obtained by making use of a stress vs. tangent modulus curve, such as the one shown in Fig. 10-17. The solution can be an iterative process, by assuming values for E_T, solving for σ_{cr}, and checking with the diagram. Some time might be saved by solving Eq. 10-1 for σ/E_T for the given problem and reading the corresponding value of σ_{cr} from a σ vs. σ/E_T curve (such as the one shown in Fig. 10-17) if such a curve is available. This latter curve

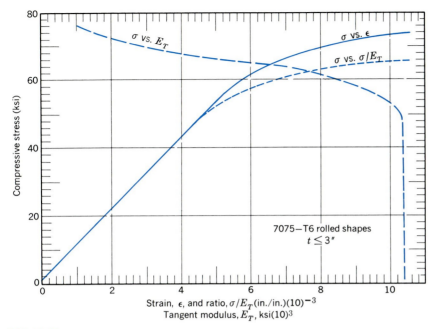

FIG. 10-17

is usually not provided; hence, it would be necessary to construct the σ vs. σ/E_T curve from the given curves or data. However, in actual design practice, it is more practical to construct a P/A vs. L/r curve by solving Eq. 10-1 for values of L/r corresponding to values of σ and E_T selected from given curves or data. This curve can then be used as follows: if for a given problem, L/r is known and the critical buckling load is wanted, the value of P/A corresponding to the L/r can be read directly from the curve. If the problem is one of selecting a section to carry a specified load, the trial-solution approach is necessary. In other words, the techniques of Section 9-5 are followed, with the formulas replaced by graphs. Since there are usually no abrupt discontinuities in the σ vs. E_T curves, the ambiguity referred to in the above paragraphs does not, in general, exist.

EXAMPLE 10-8

A compression strut is fabricated from an aluminum alloy 2024-T4 tube having an outside diameter of 30 mm, a wall thickness of 2.00 mm, and a length between fittings of 250 mm. End conditions at the fittings reduce the effective length to 170 mm. The material has the stress-strain diagram shown in Fig. 10-18. Determine the axial buckling load.

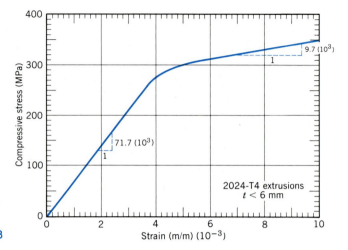

FIG. 10-18

Solution

The cross-sectional area A and moment of inertia I for the tube are

$$A = \left(\frac{\pi}{4}\right)(30^2 - 26^2) = 175.93 \text{ mm}^2$$

$$I = \left(\frac{\pi}{64}\right)(30^4 - 26^4) = 17{,}329 \text{ mm}^4$$

Thus, the radius of gyration for the cross section is

$$r = \sqrt{\frac{I}{A}} = \sqrt{\frac{17{,}329}{175.93}} = 9.925 \text{ mm}$$

The slenderness ratio for the strut is

$$\frac{L'}{r} = \frac{170}{9.925} = 17.128$$

The tangent modulus E_T for the stress range 300 MPa $\leq \sigma \leq$ 350 MPa is 9.7 GPa (see Fig. 10-18). Thus, from Eq. 10-1, the buckling load P is

$$P = \frac{\pi^2 E_T A}{(L'/r)^2}$$

$$= \frac{\pi^2 (9.7)(10^9)(175.93)(10^{-6})}{(17.128)^2}$$

$$= 54.7(10^3) \text{ N} = \underline{54.7 \text{ kN}} \qquad \text{Ans.}$$

PROBLEMS

10-55* A tube with an outside diameter of 2.50 in. and an inside diameter of 2.00 in. is to be used for a compression strut. End conditions at the fittings reduce the effective length to one-half of the true length. The stress-strain diagram for the material can be approximated by the three straight lines shown in Fig. P10-55. Determine the axial load at which buckling would probably occur if

    (a) the length between fittings is 100 in.

    (b) the length between fittings is 46 in.

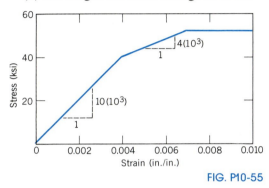

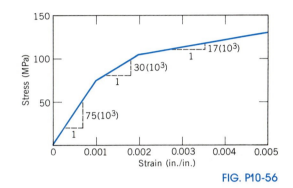

FIG. P10-55   FIG. P10-56

10-56* A tube with an outside diameter of 75 mm and a wall thickness of 5 mm is to be used for a compression strut. End conditions at the fittings reduce the effective length to three-quarters of the true length. The stress-strain diagram for the material can be approximated by the three straight lines shown in Fig. P10-56. Determine the maximum allowable length for the strut if

    (a) the axial load the strut must support is 55 kN.

    (b) the axial load the strut must support is 100 kN.

    (c) the axial load the strut must support is 140 kN.

10-57 A compressive strut with a length of 30 in. between end fittings is composed of two L2 × 2 × ¼-in. angles (see Appendix C for cross-sectional properties) fastened together to form a T-section. End conditions at the fittings reduce the effective length to 21 in. Determine the maximum permissible axial load if the strut is made from the aluminum alloy of Fig. 10-17 and a factor of safety of 2 is specified.

10-58 A compressive strut is fabricated from an aluminum alloy 2024-T4 tube having an outside diameter of 30 mm and a wall thickness of 2.00 mm. End conditions at the fittings reduce the effective length to three-quarters of the actual length between the fittings. The material has the stress-strain diagram shown in Fig. 10-18. Determine the length between fittings at which buckling would probably occur under an axial compressive load of 57 kN.

10-59* A single L2 × 2 × ¼-in. angle (see Appendix C for cross-sectional properties) is to support an axial compressive load of 30 kips with a factor of safety of 2. The angle is made from the aluminum alloy of Fig. 10-17. Determine the maximum allowable total length if the end fittings reduce the effective length to three-quarters of the total length.

10-60 A 3-in.-diameter tube with a wall thickness of 0.25 in. is to be used as a compression strut with a length of 50 in. between fittings. End conditions at the fittings reduce the effective length to 34 in. Determine the axial buckling load if the strut is made from an aluminum alloy with the stress-strain diagram shown in Fig. P10-60.

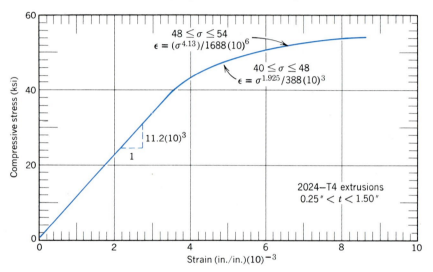

FIG. P10-60

10-61 A compression strut with a length of 16 in. between end fittings is to have a solid rectangular cross section 2.00 × b in. The strut must carry an axial working load of 160 kips with a factor of safety of 1.5. End conditions at the fittings reduce the effective length to seven-tenths of the true length. Determine the dimension b if the strut is made from an aluminum alloy with the stress-strain diagram shown in Fig. P10-60.

COMPUTER PROBLEMS

Note: The following problems have been designed to be solved with a programmable calculator, microcomputer, or mainframe computer. Appendix F contains a description of a few numerical methods, together with a few simple programs in BASIC and FORTRAN, that can be modified for use in the solution of these problems.

C10-1 A rigid platform suspended by two rods supports a load P as shown in Fig. PC10-1. The $\frac{3}{8}$-in.-diameter aluminum alloy rods are 24-in. long and have a modulus of elasticity E, which can be approximated by

$$E = 10,500 \text{ ksi}, \qquad 0 \text{ ksi} \leqslant |\sigma| \leqslant 42.0 \text{ ksi}$$
$$E = 1400 \text{ ksi}, \qquad 42.0 \text{ ksi} \leqslant |\sigma|$$

(a) Compute and plot the deflection δ at point E of the platform as a function of the load P ($0 \text{ kip} \leqslant P \leqslant 30 \text{ kips}$).

(b) Compute and plot the rotation θ (the angle between the platform and the horizontal) as a function of the load P ($0 \text{ kip} \leqslant P \leqslant 30 \text{ kips}$).

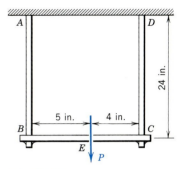

FIG. PC10-1

C10-2 A magnesium alloy post supports a load P, as shown in Fig. PC10-2. The magnesium alloy has a stress-strain relationship that can be approximated by the expressions

$$\sigma = 45(10^9)\epsilon, \qquad 0 \text{ MPa} \leqslant |\sigma| \leqslant 108 \text{ MPa}$$
$$\sigma = 2.642(10^9)\epsilon^{0.53}, \qquad 108 \text{ MPa} \leqslant |\sigma|$$

Compute and plot the deflection δ_A of the top end of the post as a function of the load P ($0 \text{ kN} \leqslant P \leqslant 28 \text{ kN}$).

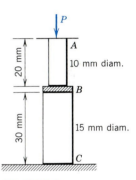

FIG. PC10-2

C10-3 A weight W is being raised with a cable and pulley, as shown in Fig. PC10-3. The aluminum alloy support bars AB and BC are 12-in. long and have a cross-sectional area of 1 in.^{2}. The modulus of elasticity of the aluminum alloy can be approximated as

$$E = 10,500 \text{ ksi}, \qquad 0 \text{ ksi} \leqslant |\sigma| \leqslant 42.0 \text{ ksi}$$
$$E = 1400 \text{ ksi}, \qquad 42.0 \text{ ksi} \leqslant |\sigma|$$

Compute and plot the horizontal and vertical components δ_x and δ_y of the displacement of pin B as a function of the load W ($0 \text{ kip} \leqslant W \leqslant 30 \text{ kips}$).

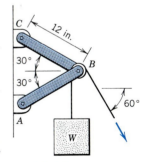

FIG. PC10-3

C10-4 A 20-mm-diameter aluminum alloy shaft and a 25-mm-diameter steel shaft are rigidly connected and supported, as shown in Fig. PC10-4a. The shearing stress-strain diagrams for the aluminum alloy and the steel are shown in Figs. PC10-4b and c, respectively. Compute and plot the rotation θ for a section at C as a function of the torque T ($0 \text{ N·m} \leqslant T \leqslant 475 \text{ N·m}$) applied at C.

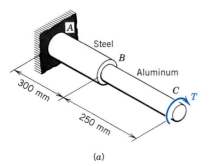

(a)

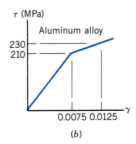

(b)

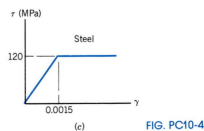

(c)   FIG. PC10-4

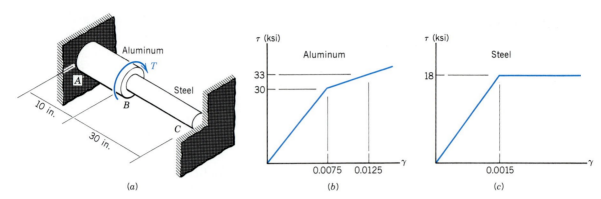

FIG. PC10-5

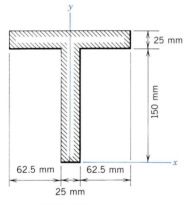

FIG. PC10-6

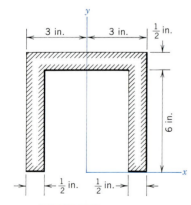

FIG. PC10-7

C10-5 A 1-in.-diameter aluminum alloy shaft and a $\frac{1}{2}$-in.-diameter steel shaft are rigidly connected and supported as shown in Fig. PC10-5a. The shearing stress-strain diagrams for the aluminum alloy and the steel are shown in Figs. PC10-5b, and c, respectively. Compute and plot the rotation θ for a section at B as a function of the torque T (0 ft-lb $\leqslant T \leqslant$ 700 ft-lb) applied at B.

C10-6 The cross section for a T-beam is shown in Fig. PC10-6. The beam is made of an elastoplastic low-carbon steel that has a modulus of elasticity of 200 GPa and a yield strength of 250 MPa. Assume that the properties are the same in tension and compression. Compute and plot,

(a) the location of the neutral axis y_{NA} as a function of the bending moment M (0 kN·m $\leqslant M \leqslant$ 80 kN·m) supported by the beam.

(b) the location of the elastic–plastic boundary as a function of the bending moment M (0 kN·m $\leqslant M \leqslant$ 80 kN·m) supported by the beam.

C10-7 The cross section for a U-beam is shown in Fig. PC10-7. The beam is made of an elastoplastic low-carbon steel that has a modulus of elasticity of 28,800 ksi and a yield strength of 36 ksi. Assume that the properties are the same in tension and compression. Compute and plot,

(a) the location of the neutral axis y_{NA} as a function of the bending moment M (0 ft-kip $\leqslant M \leqslant$ 48 ft-kips) supported by the beam.

(b) the location of the elastic–plastic boundary as a function of the bending moment M (0 ft-kip $\leqslant M \leqslant$ 48 ft-kips) supported by the beam.

CHAPTER 11

ENERGY METHODS

11-1

INTRODUCTION

The concept of strain energy was introduced in Section 8-8 by considering the work done by a slowly applied axial load P in elongating a bar of uniform cross section by an amount δ. From the load deformation diagram of Fig. 8-12, it was observed that the work done in elongating the bar is

$$W_k = \int_0^\delta P \, d\delta \qquad (a)$$

Since the work done on the bar must equal the strain energy stored in the bar, the expression for strain energy in terms of axial stress and axial strain became

$$W_k = U = \int_0^\epsilon (\sigma)(A)(L) \, d\epsilon = AL \int_0^\epsilon \sigma \, d\epsilon \qquad (b)$$

When the stress remains below the elastic limit of the material, Hooke's law applies and $d\epsilon$ may be expressed as $d\sigma/E$. Equation b then becomes

$$U = \left(\frac{AL}{E}\right) \int_0^\sigma \sigma \, d\sigma$$

or

$$U = AL \left(\frac{\sigma^2}{2E} \right) \tag{11-1}$$

Equation 11-1 gives the *elastic strain energy* (which is, in general, recoverable) for axial loading of a material obeying Hooke's law. The quantity in parentheses, $\sigma^2/(2E)$, is the elastic strain energy u in tension or compression per unit volume, or strain energy intensity, for a particular value of σ. For shear loading the expression would be identical except that σ would be replaced by τ and E by G.

The integral $\int \sigma d\epsilon$ of Eq. *b* represents the area under the stress-strain curve and, if evaluated from zero to the elastic limit (for practical purposes, the proportional limit), yields a property known as the *modulus of resilience*. The modulus is defined as the maximum strain energy per unit volume that a material will absorb without inelastic deformation. Customary units are inch-pounds per cubic inch or newton-meters per cubic meter. The area under the entire stress-strain curve from zero to rupture gives the property known as the *modulus of toughness* and denotes the energy per unit volume necessary to rupture the material. The values of these moduli for the steel of Fig. 1-31*a* are

$$u_R = (35,000)(0.001,17)/2 = 22 \text{ in.-lb/in.}^3$$

and

$$u_T = 57,000(0.3) = 17,000 \text{ in.-lb/in.}^3,$$

where the 57,000 is an estimated average value of stress for a rectangle having approximately the same area as that under the upper curve of Fig. 1-31*a*.

When the state of stress in a body can be represented by a nonzero normal stress, Eq. 11-1 can be written as

$$U = \int_{\text{volume}} \frac{\sigma^2}{2E} \, dx \, dy \, dz \tag{c}$$

Equation *c* is valid for a beam subjected to pure bending, and if the energy (usually small) resulting from the shearing stresses is neglected, the expression applies to any beam with transverse loading. The stress in Eq. *c* must be expressed as a function of one or more of the other variables before integration.

For the case of pure bending in a beam of constant cross section A and length L, the value of σ_x is

$$\sigma_x = \frac{My}{I}$$

and the strain energy (when x is measured along the axis of the beam) is

$$U = \frac{1}{2E} \int \int \left(\frac{My}{I}\right)^2 dA \, dx = \frac{1}{2E} \int_0^L \frac{M^2}{I^2} dx \int_{\text{area}} y^2 \, dA$$

which reduces to

$$U = \int_0^L \frac{M^2}{2EI} dx \qquad (11\text{-}2)$$

The procedure leading to Eq. c for a normal stress can be applied in exactly the same manner to a shearing stress to give

$$U = \int_{\text{volume}} \frac{\tau^2}{2G} dx \, dy \, dz \qquad (d)$$

where τ must be expressed as a function of x, y, and z before integration. For the case of a circular shaft of length L and cross section A subjected to a torque T, the expression for τ is

$$\tau = \frac{T\rho}{J}$$

and the energy becomes

$$U = \frac{1}{2G} \int \int \left(\frac{T\rho}{J}\right)^2 dA \, dx = \frac{1}{2G} \int_0^L \frac{T^2}{J^2} dx \int_{\text{area}} \rho^2 \, dA$$

which gives

$$U = \int_0^L \frac{T^2 \, dx}{2GJ} \qquad (11\text{-}3)$$

One of the important concepts disclosed by Eqs. c and d is that the energy-absorbing capacity of a member (that is, the resistance to failure) is a function of the *volume* of material available, in contrast to the resistance to failure under static loading, which is a function of cross-sectional area or section modulus.

The concept of elastic strain energy can be extended to include biaxial and triaxial loadings by writing the expression for strain energy intensity u as $\sigma\epsilon/2$ and adding the energies due to each of the stresses. Since energy is a positive scalar quantity, the addition is the arithmetic sum of the energies. For a system of triaxial principal stresses σ_{p1}, σ_{p2}, σ_{p3}, the total elastic strain energy intensity is

$$u = \left(\frac{1}{2}\right)(\sigma_{p1}\epsilon_{p1} + \sigma_{p2}\epsilon_{p2} + \sigma_{p3}\epsilon_{p3}) \qquad (e)$$

When the strains are expresed in terms of the stresses, this equation becomes

$$u = \frac{1}{2E} [\sigma_{p1}^2 + \sigma_{p2}^2 + \sigma_{p3}^2 - 2\nu(\sigma_{p1}\sigma_{p2} + \sigma_{p2}\sigma_{p3} + \sigma_{p3}\sigma_{p1})] \quad (f)$$

If σ_x, σ_y, σ_z are not principal stresses, the strain energy due to the shearing stresses on the x, y, z planes must be added to the energy contributed by the normal stresses to obtain the total elastic strain energy intensity; thus,

$$u = \frac{1}{2} (\sigma_x \epsilon_x + \sigma_y \epsilon_y + \sigma_z \epsilon_z + \tau_{xy}\gamma_{xy} + \tau_{yz}\gamma_{yz} + \tau_{zx}\gamma_{zx}) \quad (11\text{-}4)$$

$$u = \frac{1}{2E} [\sigma_x^2 + \sigma_y^2 + \sigma_z^2 - 2\nu(\sigma_x\sigma_y + \sigma_y\sigma_z + \sigma_z\sigma_x)]$$

$$+ \frac{1}{2G} (\tau_{xy}^2 + \tau_{yz}^2 + \tau_{zx}^2) \quad (11\text{-}5)$$

PROBLEMS

Note: In the following problems, the mass of the loaded members is to be neglected, and unless stated otherwise, all of the applied energy is to be assumed to be absorbed elastically by the loaded members. Neglect stress concentrations.

11-1* A square bar with cross-sectional area A and length L is made from a homogeneous material which has a specific weight γ and a modulus of elasticity E. Determine the elastic strain energy stored in the bar (as a result of its own weight) if it hangs vertically while suspended from one end. Express the result in terms of γ, L, A, and E.

11-2* A cantilever beam with a rectangular cross section has a width b, a depth $2c$, and a span L. The beam has a couple M applied at the free end. Compare the elastic strain energy stored in the beam with the elastic strain energy stored in a bar of the same size that is axially loaded to the same maximum tensile stress level.

11-3 A simply supported beam with a rectangular cross section has a width b, a depth $2c$, and a span L. The beam is subjected to end couples that produce a constant bending moment for the full length of the beam. Compare the elastic strain energy stored in the beam with the elastic strain energy stored in a bar of the same size that is axially loaded to the same maximum tensile stress level.

11-4 A cantilever beam with a rectangular cross section has a width b, a depth $2c$, and a span L. The beam carries a concentrated load P at the free end. Compare the elastic strain energy stored in the beam with the elastic strain energy stored in a bar of the same size that is axially loaded to the same maximum tensile stress level. Neglect the energy resulting from the shearing stresses.

11-5* A simply supported beam with a rectangular cross section has a width b, a depth $2c$, and a span L. The beam carries a concentrated load P at midspan. Compare the elastic strain energy stored in the beam with the elastic strain energy stored in a bar of the same size that is axially loaded to the same maximum tensile stress level. Neglect the energy resulting from the shearing stresses.

11-6* A cantilever beam with a rectangular cross section has a width b, a depth $2c$, and a span L. The beam carries a uniformly distributed load w over its entire length. Compare the elastic strain energy stored in the beam with the elastic strain energy stored in a bar of the same size that is axially loaded to the same maximum tensile stress level. Neglect the energy resulting from the shearing stresses.

11-7 A simply supported beam with a rectangular cross section has a width b, a depth $2c$, and a span L. The beam carries a uniformly distributed load w over its entire length. Compare the elastic strain energy stored in the beam with the elastic strain energy stored in a bar of the same size that is axially loaded to the same maximum tensile stress level. Neglect the energy resulting from the shearing stresses.

11-8 A solid circular shaft with diameter d and length L is subjected to a constant torque T. Compare the elastic strain energy stored in this shaft with the elastic strain energy stored in a bar of the same size that is axially loaded to the same maximum tensile stress level.

11-9* A hollow circular shaft with outside diameter d, inside diameter $d/2$, and length L is subjected to a constant torque T. Compare the elastic strain energy stored in this shaft with the elastic strain energy stored in a bar of the same size that is axially loaded to the same maximum tensile stress level.

11-10 A cantilever beam with a rectangular cross section has a width b, a depth $2c$, and a span L. The beam carries a concentrated load P at the free end. Compare the elastic strain energy produced in the beam by the transverse shearing stresses with the elastic strain energy produced in the beam by the fiber stresses.

11-2

DYNAMIC LOADING—IMPACT

When the motion of a body is changed (accelerated), the force necessary to produce this acceleration is called a *dynamic force* or *load*. For example, the force an elbow in a pipeline exerts on the fluid in the pipe to change its direction of flow, the pressure on the wings of an airplane pulling out of a dive, the collision of two automobiles, and a man jumping on a diving board are all examples of dynamic loading. A suddenly applied load is called an *impact load*. The last two of the preceding examples are considered impact loads. Under impact loading, if there is elastic action, the loaded system will vibrate until equilibrium is established. A dynamic load may be expressed in terms of mass times the acceleration of the mass center, in terms of the rate of change of the momentum, or in terms of the change of the kinetic energy of the body.

Energy methods can be used to obtain solutions for many problems in mechanics of materials. When an energy approach is used, the magnitude of the load is expressed in terms of the kinetic energy delivered to the loaded system; hence, it is often referred to as an *energy load*. For example, a particle of mass m moving with a speed v possesses a kinetic energy $mv^2/2$; if this particle is stopped by a body, the energy absorbed by the body (the loaded system) is some fractional part of $mv^2/2$, the balance of the energy being converted into sound, heat, and permanent deformation of the striking particle.

In the loaded system, dynamic loading produces stresses and strains, the magnitude and distribution of which will depend not only on the usual parameters encountered previously but also on the velocity of propagation of the strain waves through the material of which the system is composed. This latter consideration, although very important when loads are applied with high velocities, may often be neglected when the velocity of application of the load is low. The velocities are considered to be low when the loading time permits the material to act in the same manner as it does under static loading; that is, the relations between stress and strain and between load and deflection are essentially the same as those already developed for static loading. The use of energy methods is illustrated in the following example.

EXAMPLE 11-1

A weight of 30 lb is dropped from a height of 4 ft onto the center of a small rigid platform, as shown in Fig. 11-1. The two steel ($E = 30,000$ ksi) rods supporting the platform have 1-by-2-in. rectangular cross sections and are 8 ft long. Determine

(a) the maximum tensile stress developed in the rods.

(b) the maximum deflection of the platform.

Solution

The energy applied to the platform by the falling weight W is

$$U = W(h + \delta)$$

The elastic strain energy stored in the rods is given by Eq. 11-1 as

$$U = \frac{\sigma^2 AL}{2E}$$

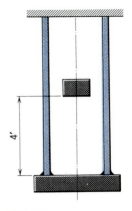

FIG. 11-1

(a) If energy losses in the system are negligible,

$$\frac{\sigma^2 AL}{2E} = W(h + \delta) = W\left(h + \frac{\sigma L}{E}\right)$$

or

$$AL\sigma^2 - 2WL\sigma - 2WhE = 0$$

which has the solution

$$\sigma = \frac{W}{A} + \sqrt{\left(\frac{W}{A}\right)^2 + \frac{2WhE}{AL}}$$

$$= \frac{30}{4} + \sqrt{\left(\frac{30}{4}\right)^2 + \frac{2(30)(4)(12)(30{,}000{,}000)}{4(8)(12)}}$$

$$= 15{,}008 \text{ psi} = \underline{15.01 \text{ ksi}} \qquad \text{Ans.}$$

Note that the static stress produced by the 30-lb weight would be

$$\sigma = \frac{W}{A} = \frac{30}{4} = 7.50 \text{ psi}$$

(b) The maximum deflection experienced by the platform would be

$$\delta = \frac{\sigma L}{E} = \frac{15{,}008(8)(12)}{30{,}000{,}000} = \underline{0.048 \text{ in.}} \qquad \text{Ans.}$$

Note that the static deflection produced by the 30-lb weight would be

$$\delta = \frac{\sigma L}{E} = \frac{7.50(8)(12)}{30{,}000{,}000} = 0.000024 \text{ in.}$$

PROBLEMS

> *Note:* In the following problems, the mass of the loaded members is to be neglected, and unless stated otherwise, all of the applied energy is to be assumed to be absorbed elastically by the loaded members. Neglect stress concentrations.

11-11* A solid circular steel (E = 30,000 ksi) bar 50 in. long is to be used as a tension member subject to axial energy loads. The allowable axial tensile stress is 25 ksi. Determine the minimum diameter required for an axial energy load of 100 ft-lb.

11-12* A rolled bronze (E = 100 GPa) bar with a cross-sectional area of 2500 mm² is to be used as a tension member subject to axial energy loads. The allowable axial tensile stress is 70 MPa. Determine the minimum length of bar required for an axial energy load of 200 N·m.

11-13 A hot-rolled Monel (E = 26,000 ksi) bar with a cross-sectional area of 3.5 in.² is to be used as a tension member subject to axial energy loads. The allowable axial tensile stress is 10 ksi. Determine the minimum length of bar required for an axial energy load of 50 ft-lb.

11-14 A 25-mm-diameter steel (E = 200 GPa) rod 900 mm long is supported at the top end and fitted with a loading flange (stop) at the bottom end, as shown in Fig. P11-14. A collar W slides freely on the rod. Determine the maximum mass that the collar may have without exceeding the proportional limit (250 MPa) of the rod if

(a) the collar is dropped from a height of 600 mm.

(b) the collar is suddenly applied to the stop (dropped from a very small height).

(c) the collar is slowly applied to the stop (static load).

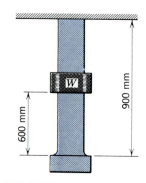

FIG. 11-14

11-15* A 10-ft-steel (E = 30,000 ksi and G = 12,000 ksi) shaft 3 in. in diameter is subjected to an energy load of 100 ft-lb. Determine

(a) the maximum normal stress in the shaft when the load is applied as an axial energy load.

(b) the maximum shearing stress in the shaft when the load is applied as a torsional energy load.

11-16* A timber (E = 8.2 GPa) beam 150 mm wide, 100 mm deep, and 3 m long is simply supported at the ends. From what maximum height can a 14-kg mass be dropped onto the center of the beam without causing the maximum fiber stress to exceed 10 MPa?

11-17 A weight of 10 lb is dropped from a height of 8 in. onto the free end of a steel (E = 30,000 ksi) cantilever beam. The beam is 2 in. wide, 3 in. deep, and 30 in. long. Determine

(a) the maximum deflection produced by the falling weight.

(b) the static load needed to produce the same deflection.

(c) the maximum fiber stress caused by the falling weight.

11-18 An elevator cage with a mass of 1400 kg is being lowered at a rate of 1.5 m/s. The elevator cable has an effective cross-sectional area of 650 mm² and an effective modulus of elasticity of 170 GPa. If the cage is suddenly stopped by the cable drum (not the brakes on the side of the cage) after descending 60 m, determine the maximum tensile stress developed in the cable. Neglect the mass of the cable.

11-19* A torsional energy load of 20 ft-lb is applied to the steel ($G = 12,000$ ksi) shaft shown in Fig. P11-19. Determine

(a) the maximum shearing stress in the shaft.

(b) the maximum angle of rotation at the right end of the shaft.

(c) the maximum shearing stress in the shaft if its diameter is reduced to a uniform 2 in. over the complete length of the shaft.

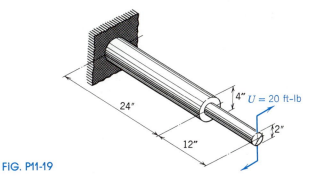

FIG. P11-19

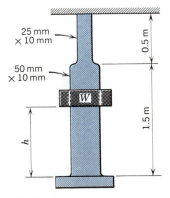

FIG. P11-20

11-20 A flat piece of steel ($E = 200$ GPa) 10 mm thick and 2 m long has a width of 25 mm for 0.5 m of its length and a width of 50 mm for the remaining 1.5 m. A 2-kg collar W is dropped onto a loading flange at the end of the bar, as shown in Fig. P11-20. Determine the maximum height h from which the collar can be dropped if the axial tensile stress in the rod is not to exceed 200 MPa.

11-3

THE EQUIVALENT STATIC LOAD

The stresses and strains produced by dynamic loading, as noted in Section 11-2, depend on the velocity of propagation of the strain waves through the material. However, for many engineering applications (where the velocities of the applied load are small with respect to the velocities of propagation of the strain waves), such complications can be neglected and it can be assumed that the material behaves as it does under slowly applied loads. This will be the approach taken in this book, and with such a simplifying assumption there is no need to evaluate the integrals of Eqs. *c* and *d* of Section 11.1. Instead, an *equivalent static load* (a slowly applied load that will produce the same maximum deflection as the dynamic load) is applied to the machine component or structure of interest. If the assumption of identical material behavior under the two loadings is valid (most mechanical-type loadings are applied slowly), the stress-strain diagram for any point in the loaded system does not change and the stress and strain distributions produced by the equivalent static load will be the same as those produced by the dynamic load.

As an illustration, consider the steel cantilever beam loaded with a falling weight W as shown in Fig. 11-2. The variation of the end deflection of the beam with respect to time is shown in Fig. 11-2*b*, where Δ, equal to δ_{max}, is the maximum deflection resulting from the falling weight transferring part of its kinetic energy ($mv^2/2$) to the beam and where δ_{stat} is the static deflection resulting when the beam and weight finally reach equilibrium. The static load P, applied as in Fig. 11-2*c*, produces the same maximum deflection Δ; the deflection curve and stress-strain diagrams are assumed to be the same in both cases. Hence, the strain energy in the beam is the same in both cases, and since the strain energy is equal to the effective applied energy (work done on the beam), the following relation may be stated:

$$\begin{pmatrix} \text{effective} \\ \text{energy} \\ \text{applied} \end{pmatrix} = \begin{pmatrix} \text{work done by} \\ \text{equivalent} \\ \text{static load} \end{pmatrix} \qquad (a)$$

The effective energy is that portion of the applied energy that produces the maximum deflection. In addition to energy lost in the form of sound, heat, and local distortion, some is absorbed by the supports.

From Section 11-1, the strain energy in terms of the static load equals $\int P \, d\delta$ or $P_{avg}(\delta_{max})$. Equation *a* may then be written as

$$U = P_{avg}(\Delta)$$

and if stress and strain are proportional (load-deflection diagram is a straight line), P_{avg} is equal to one-half P_{max}. Hence, for a material possessing a linear stress-strain relationship,

$$U = P\Delta/2 \qquad (11\text{-}6)$$

The right-hand side of Eq. 11-6 involves the relations between load, stress, and strain previously encountered under static loading and applies to any type of loading. If the effective energy applied, U, is due

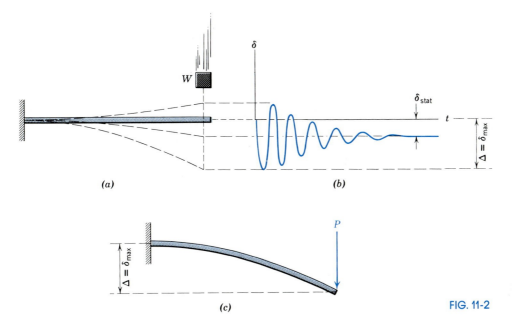

FIG. 11-2

to a change of velocity of a translating body, the left side of Eq. 11-6 becomes

$$\eta m v^2 / 2$$

where η is the effective portion of the applied energy, m is the mass of the body, and v is the velocity of the moving body just before impact. If the effective energy applied comes from a body falling from rest and being stopped by the resisting structure, the left-hand side of Eq. 11-6 comes from the change in potential energy of the body and is

$$\eta m g h$$

where g is the local acceleration due to gravity and h is the total vertical distance through which the body falls.

For torsional energy loading, Eq. a can be written as

$$U = T\theta/2 \tag{11-7}$$

where T is the slowly applied torque that will produce the same maximum angle of twist θ, as the suddenly applied torque if the maximum shearing stress is less than the proportional limit.

A term sometimes encountered in working with dynamic loading is load factor (or impact factor), defined as the *ratio of the equivalent static load to the magnitude of the force the loading mass would exert on the loaded system if slowly applied.* For example, the equivalent static loading on the wings of an airplane pulling out of a dive divided by the gravitational force mg on the plane (loading on horizontal tail surfaces neglected) is the load factor.

EXAMPLE 11-2

The upper 900 mm of the steel ($E = 200$ GPa) bar AB in Fig. 11-3a is 50 mm wide by 12 mm thick, and the lower 600 mm is 25 mm wide by 12 mm thick. When the collar W is released from rest from a height of 500 mm onto the end A of the bar, the axial stress in the bar (neglect stress concentration) is not to exceed 100 MPa. Assume that 80 percent of the energy of the falling collar is absorbed elastically by member AB. Determine

(a) the maximum permissible mass for collar W.

(b) the maximum permissible mass for W if the bar AB is 25 mm wide through its entire length.

Solution

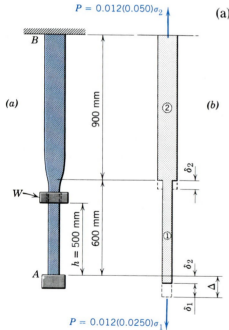

(a)

(b)

FIG. 11-3

(a) Figure 11-3b indicates the static load applied to the bar and the resulting deflections. When Eq. 11-6 is used, the following expression is obtained:

$$\eta mgh = (0.8)(mg)(0.500 + \Delta) = P\Delta/2 \qquad (b)$$

The quantities P and Δ can be calculated directly from the given data. From Fig. 11-3b,

$$0.012(0.050)(\sigma_2) = 0.012(0.025)(\sigma_1) \quad \text{or} \quad \sigma_1 = 2\sigma_2$$

This expression indicates that σ_1 (equal to 100 MPa) is the limiting stress and that σ_2 must be 50 MPa. The load P is

$$P = \sigma_1 A_1 = 100(10^6)(0.012)(0.025) = 30(10^3)\ N = 30\ \text{kN}$$

and the deflection Δ is

$$\Delta = \delta_2 + \delta_1 = 0.900\epsilon_2 + 0.600\epsilon_1 + (0.900\sigma_2 + 0.600\sigma_1)/200(10^9)$$
$$= \frac{0.900(50)(10^6) + 0.600(100)(10^6)}{200(10^9)}$$
$$= 0.525(10^{-3})\text{m} = 0.525\ \text{mm}$$

Substituting these values of P and Δ into Eq. b gives the expression

$$0.8(m)(9.81)(0.500 + 0.000525) = 30(10^3)(0.525)(10^{-3})/2$$

from which

$$m = 2.00 \text{ kg} \qquad \text{Ans.}$$

(b) If the bar is 25 mm wide for its entire length,

$$\Delta = \frac{(0.900 + 0.600)(100)(10^6)}{200(10^9)} = 0.750(10^{-3})\text{m} = 0.750 \text{ mm}$$

and

$$0.8(m)(9.81)(0.500 + 0.000750) = 30(10^3)(0.750)(10^{-3})/2$$

from which

$$m = 2.86 \text{ kg} \qquad \text{Ans.}$$

In view of the development of Section 11-1, which indicates that the energy-absorbing capacity of a member is a function of the amount of material present, the results in parts (a) and (b) seem paradoxical, as the strength of the member is increased by removing some material. The apparent discrepancy is probably best explained by referring to the stress-strain diagram of Fig. 11-4. For part (a) of the problem, the energy absorbed per unit volume by the lower 600 mm of the bar is represented by the area *ocd*, which gives

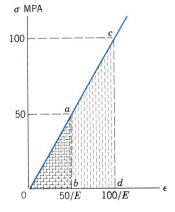

$$u = \frac{1}{2}\left[\frac{100(10^6)}{200(10^9)}\right](100)(10^6) = 25(10^3)\text{N} \cdot \text{m/m}^3 = 25 \text{ kJ/m}^3$$

FIG. 11-4

The energy absorbed per unit volume by the upper 900 mm of the bar is represented by area *oab* of Fig. 11-4 and is seen to be only one-fourth as much because both the stress and the strain are one-half as much as in the lower 600 mm. When one-half of the volume of the upper 900 mm is removed, the energy absorbed by each unit volume of the remaining material is quadrupled (area *ocd* as compared to area *oab*), resulting in a net gain in energy-absorbing capacity. For this reason, bolts designed to carry dynamic loads are sometimes turned down so that the diameter of the shank of the bolt is the same as the diameter at the root of the threads. In this way the stress distribution in the bolt is approximately uniform, and the capacity of the bolt to absorb energy is a maximum.

EXAMPLE 11-3

The simply supported 6061-T6 aluminum alloy beam A of Fig. 11-5a is 3 in. wide by 1 in. deep. The center support is a helical spring with a modulus of 100 lb/in. The spring is initially unstressed and in contact with the beam. When the 50-lb block drops 2 in. onto the top of the beam, 90 percent of the energy is absorbed elastically by the system consisting of beam A and spring B. Determine

(a) the maximum flexural stress in the beam for this loading.

(b) the load factor.

(c) the proportion of the effective energy absorbed by each of the components A and B.

Solution

A deflection diagram for the system and free-body diagrams for the component parts are shown in Figs. 11-5b, c, and d. Since the action is assumed to be elastic and the materials obey Hooke's law, the load-deflection relation will be linear, and Eq. 11-6 is valid; hence,

$$50(2 + \Delta)(0.9) = P\Delta/2 \qquad (c)$$

The fiber stress is wanted; therefore, the forces P and Q will be evaluated, and from these forces and Fig. 11-5c the fiber stress can be determined. From Figs. 11-5c and d and Appendix D,

$$\delta_A = \frac{(P - Q)L^3}{48EI} = \frac{(P - Q)(10^3)(12^3)}{48(10^7)(3)(1^3)/12} = \frac{(P - Q)(144)}{(10^4)}$$

and

$$\delta_B = \frac{Q}{100}$$

From Fig. 11-5b,

$$\delta_A = \delta_B = \Delta$$

or

$$\frac{(P - Q)(144)}{10,000} = \frac{Q}{100}$$

from which

$$Q = \frac{36P}{61}$$

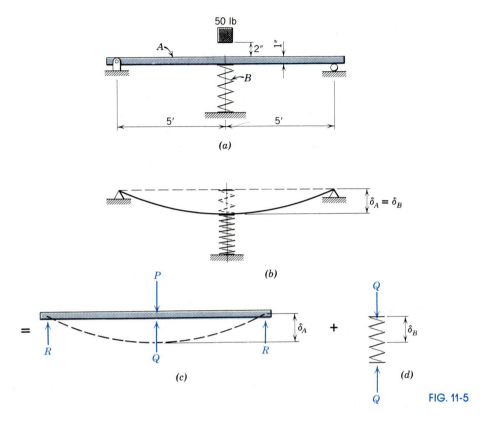

FIG. 11-5

Then,

$$\Delta = \frac{Q}{100} = \frac{36P}{6100}$$

and Eq. *c* becomes

$$50\left(2 + \frac{36P}{6100}\right)0.9 = \left(\frac{1}{2}\right)\frac{36P^2}{6100}$$

or

$$P^2 - 90P = 30,500$$

from which

$$P = 225.3 \quad \text{and} \quad -135.3$$

The positive root is the magnitude of P necessary to produce the maximum downward deflection of the beam, whereas the negative root is the force necessary to produce the maximum upward deflection (see Fig. 11-2*b*) of the vibrating system.

When the value of 225.3 lb for P is used,

$$Q = \frac{36(225.3)}{61} = 133.0 \text{ lb}$$

and

$$P - Q = 92.3 \text{ lb}$$

(a) From Fig. 11-5*c*, R is 46.15 lb, and the maximum bending moment occurs at the center of the beam and is 46.15(5)(12) in.-lb; therefore, the maximum flexural stress is

$$\sigma = \frac{Mc}{I} = \frac{46.15(60)(0.50)}{3(1^3)/12}$$

$$= \underline{5540} \text{ psi } T \text{ and } C \qquad \qquad \text{Ans.}$$

which is well below the proportional limit, and therefore the action is elastic.

(b) The load factor is

$$\frac{225.3}{50} = \underline{4.51} \qquad \qquad \text{Ans.}$$

(c) The resultant equivalent static load at the center of A is $(P - Q)$ and is 92.3 lb, and the energy absorbed elastically by the beam is

$$U_A = \frac{(92.3)(\Delta)}{2}$$

The elastic energy absorbed by the helical spring is

$$U_B = \frac{(133.3)(\Delta)}{2}$$

The total energy absorbed elastically by the system is

$$U = \frac{(225.3)(\Delta)}{2}$$

Therefore, the proportions absorbed by each component are

$$\frac{92.3}{225.3}(100) = \underline{41 \text{ percent for } A} \qquad \qquad \text{Ans.}$$

and

$$\frac{133.0}{225.3}(100) = \underline{59 \text{ percent for } B} \qquad \text{Ans.}$$

To recapitulate, Eq. 11-6 is applicable only to systems having a linear relationship between load and deformation, and only when the velocity of application of the dynamic load is not too high.

PROBLEMS

Note: In the following problems, the mass of the loaded members is to be neglected, and unless stated otherwise, all of the applied energy is to be assumed to be absorbed elastically by the loaded members. Neglect stress concentrations.

11-21* A load of 1000 lb compresses a given railroad car spring 0.10 in. What static load would be required to produce the same deflection as would be produced by a load of 350 lb dropped from a height of 1 ft?

11-22* A mass of 24 kg is dropped from a height of 100 mm onto a helical spring that has a modulus of 18 kN/m. Determine

(a) the maximum deflection of the spring.

(b) the static load that would produce the same deflection.

11-23 A collar W is dropped from a height of 10 in. onto the end of the steel ($E = 30,000$ ksi) bar shown in Fig. P11-23. If 80 percent of the energy is effective in producing stress in the bar and if the maximum stress is not to exceed 18 ksi, determine the maximum allowable weight of the collar if

(a) the bar has the cross section shown in Fig. P11-23.

(b) the bar has a uniform cross section of 2 in.² throughout its entire length.

11-24 A 10-kg collar W is dropped from a height h onto the end of the steel ($E = 200$ GPa) bar shown in Fig. P11-24. If 75 percent of the energy is effective in producing stress in the bar and if the maximum stress is not to exceed 200 MPa, determine the maximum height h from which the collar can be dropped if

(a) the bar has the cross section shown in Fig. P11-24.

(b) the bar has a 25-by-10-mm uniform cross section throughout its entire length.

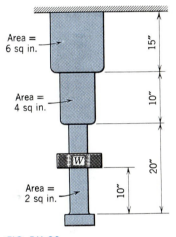

FIG. P11-23

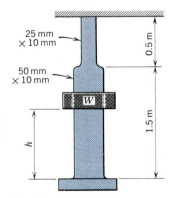

FIG. P11-24

11-25* A weight of 80 lb is dropped from a height of 2 in. onto the free end of a steel ($E = 30{,}000$ ksi) cantilever beam. The beam is 2 in. wide, 3 in. deep, and 30 in. long. If 50 percent of the energy of the falling weight is effective in producing stress in the beam, determine

(a) the maximum fiber stress produced by the falling weight.

(b) the maximum deflection due to the falling weight.

(c) the static load needed to produce the same deflection.

11-26 A timber ($E = 8.2$ GPa) beam 150 mm wide, 100 mm deep, and 3 m long is simply supported at the ends. A 14-kg mass is dropped onto the center of the beam from a height of 250 mm. If 80 percent of the energy of the falling body is effective in producing stress in the beam, determine

(a) the maximum fiber stress produced by the falling weight.

(b) the maximum deflection due to the falling weight.

(c) the static load needed to produce the same deflection.

11-27 A steel ($G = 12{,}000$ ksi) shaft 15 ft long consists of a 5-ft solid section with a diameter of 10 in. and a 10-ft hollow section with an outside diameter of 10 in. and an inside diameter of 8 in. Determine

(a) the torsional energy load required to produce a maximum shearing stress of 10 ksi in the shaft.

(b) the angle of twist produced by the energy load of part (a).

11-28* A torsional energy load of 35 J is applied to the steel ($G = 80$ GPa) shaft shown in Fig. P11-28. Determine

(a) the maximum shearing stress produced by the energy load.

(b) the torsional energy load required to produce the same maximum shearing stress if the shaft is turned to a uniform 50-mm diameter except for a 50-mm length at the support.

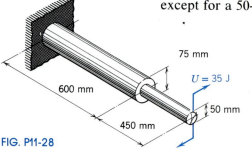

FIG. P11-28

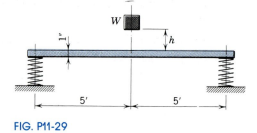

FIG. P11-29

11-29* The beam of Fig. P11-29 is 3 in. wide and 1 in. deep. Each of the supporting coil springs has a modulus of 50 lb/in. When the block W is slowly applied to the center of the beam, the ends of the beam move downward 0.10 in. and the center of the beam moves downward an additional 0.12 in. Determine

(a) the weight of block W.

(b) the height h from which the block W must be dropped to produce a fiber stress of 2400 psi in the beam.

11-30 The beam of Fig. P11-30 is made of an aluminum alloy ($E = 70$ GPa). The 25-kg block B is dropped from a height of 12 mm onto the top of the coil spring C, which has a modulus of 35 kN/m. The resulting load factor is 2. Determine the percentage of the energy of block B that is absorbed by

(a) the coil spring C.

(b) the beam A.

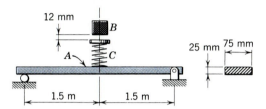

FIG. P11-30

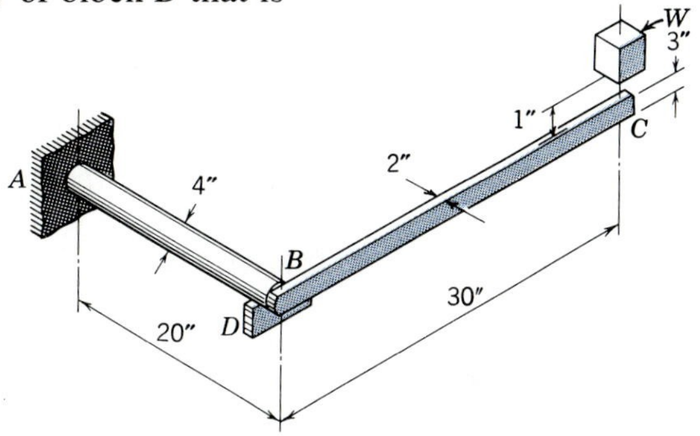

FIG. P11-31

11-31* The structure shown in Fig. P11-31 consists of a solid steel ($G = 12{,}000$ ksi) shaft AB rigidly connected to the steel ($E = 30{,}000$ ksi) beam BC. The shaft is fixed at A and supported by a smooth bearing D at B. The 100-lb weight W is dropped from a height of 1 in. onto the beam. If 80 percent of the energy of the falling weight is absorbed elastically by the shaft and the beam, determine the maximum fiber stress in the beam and the maximum shearing stress in the shaft.

11-32* Beams A and B of Fig. P11-32 are made of wood ($E = 8$ GPa) and have cross sections that are 50 mm deep by 150 mm wide. When the 12-kg block W is dropped on beam A the maximum deflection is 38 mm. If 90 percent of the energy is absorbed elastically by the beams, determine

(a) the height h from which W is dropped.

(b) the load factor.

(c) the amount of energy absorbed by each beam.

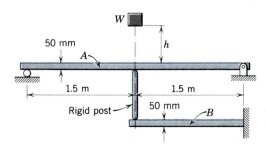

FIG. P11-32

11-33 Beams A and B of Fig. P11-33 are made of wood (E = 1200 ksi) and have cross sections that are 2 in. deep by 6 in. wide. When the 20-lb block W is dropped onto beam A, 90 percent of the energy of the falling block is absorbed elastically by the beams. If the maximum fiber stress developed in beam B is 1200 psi, determine

(a) the height h from which W is dropped.

(b) the load factor.

(c) the amount of energy absorbed by each beam.

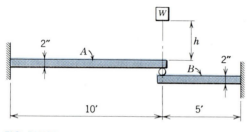

FIG. P11-33   FIG. P11-34

11-34 The aluminum alloy (E = 70 GPa) beams shown in Fig. P11-34 have cross sections that are 25 mm deep by 100 mm wide. The helical spring C between beams A and B has a modulus of 20 kN/m. The 5-kg block is dropped onto beam A with a load factor of 4. If 80 percent of the energy of the falling block is absorbed elastically by the beams and spring, determine

(a) the height h from which the block is dropped.

(b) the amount of energy absorbed by each component.

11-4

SOME DESIGN CONSIDERATIONS FOR BEAMS SUBJECTED TO ENERGY LOADS

Some design considerations for beams subjected to energy loads are readily obtained from the equivalent static load procedure of Section 11-3. For example, consider the energy absorbed by a cantilever beam of length L with a concentrated load P at the free end. The strain energy (see Appendix D) is

$$U = P\Delta/2 = \frac{P(PL^3)}{2(3EI)}$$

With the load P replaced by its value in terms of the maximum fiber stress in the beam, σ equals PLc/I, the expression for the energy becomes

$$U = \frac{\sigma^2 IL}{6c^2 E} = \frac{\sigma^2 L}{6E}\left(\frac{I}{c^2}\right)$$

For a rectangular beam I/c^2 is equal to one-third the cross-sectional area, regardless of whether the long or short dimension of the rectangle is parallel to the load. When the parenthetical quantity is replaced by its value in terms of the area of the rectangle, the expression for the energy becomes

$$U = (AL)\frac{\sigma^2}{18E} \qquad\qquad (a)$$

which shows that for a rectangular cross section the capacity to absorb energy is independent of the moment of inertia of the cross section. Comparison of Eq. a with Eq. 11-1 of Section 11-1 reveals that the rectangular cantilever beam considered has only one-ninth as much capacity to absorb or store elastic energy as an axially loaded member of the same volume. Too much of the volume of the cantilever beam has low stresses for efficient energy load capacity. The situation can be improved by using a tapered beam whose width varies directly as the distance from the free end.

For this purpose consider the cantilever beam of constant depth h of Fig. 11-6. The maximum fiber stress at any section is

$$\sigma = \frac{Mc}{I} = \frac{Pxh/2}{2zh^3/12} = \frac{6PL}{bh^2}$$

which shows that the maximum fiber stress is independent of the distance x along the beam. The differential equation of the elastic curve for the beam is

$$E\frac{d^2y}{dx^2} = -\frac{Px}{I} = -\frac{12PL}{bh^3}$$

After integration twice and evaluation of the constants of integration, the magnitude of the maximum deflection at the free end of the beam is found to be

$$\Delta = \frac{6PL^3}{bh^3E}$$

The energy U then becomes

$$U = \frac{P\Delta}{2} = \frac{P}{2}\frac{6PL^3}{bh^3E}$$

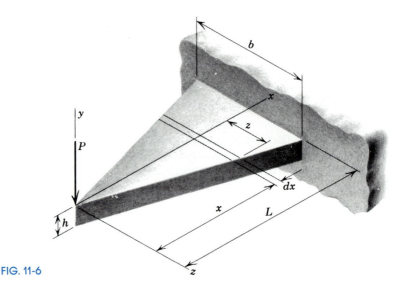

FIG. 11-6

or in terms of the maximum fiber stress the energy is

$$U = (\text{volume})[\sigma^2/(6E)] \qquad (b)$$

Comparison of Eqs. *a* and *b* shows that the tapered cantilever beam of Fig. 11-6 can absorb three times as much elastic energy as a cantilever beam of the same volume with a constant rectangular cross section. However, comparison with Eq. 11-1 of Section 11-1 shows that the beam of Fig. 11-6 is still only one-third as effective as an axially loaded member of the same volume.

For torsion members subjected to energy loads, a hollow shaft will provide higher energy-absorbing capacity per unit volume than a solid circular shaft.

11-5

DEFLECTIONS BY ENERGY METHODS— CASTIGLIANO'S THEOREM

Strain energy techniques are frequently used to analyze the deflections of beams and structures. Of the many available methods, the application of Castigliano's theorem, to be developed here, is one of the most widely used. It was presented in 1873 by the Italian engineer Alberto Castigliano (1847–1884). Although the theorem will be derived by considering the strain energy stored in beams, it is applicable to any structure for which the force-deformation relations are linear.

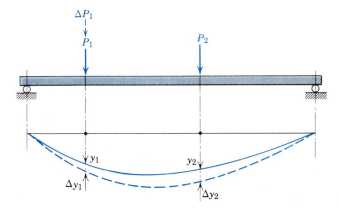

FIG. 11-7

If the beam shown in Fig. 11-7 is slowly and simultaneously loaded by the two forces P_1 and P_2 with resulting deflections y_1 and y_2, the strain energy U of the beam is, by Clapeyron's theorem (see Section 8-8), equal to the work done by the forces. Therefore,

$$U = \frac{1}{2} P_1 y_1 + \frac{1}{2} P_2 y_2 \qquad (a)$$

Let the force P_1 be increased by a small amount ΔP_1, and Δy_1 and Δy_2 be the changes in deflection due to this incremental load. Since the forces P_1 and P_2 are already present, the increase in the strain energy is

$$\Delta U = \frac{1}{2}\Delta P_1 \Delta y_1 + P_1 \Delta y_1 + P_2 \Delta y_2 \qquad (b)$$

If the order of loading was reversed so that the incremental force, ΔP_1, was applied first, followed by P_1 and P_2, the resulting strain energy would be

$$U + \Delta U = \frac{1}{2}\Delta P_1 \Delta y_1 + \Delta P_1 y_1 + \frac{1}{2} P_1 y_1 + \frac{1}{2} P_2 y_2 \qquad (c)$$

The resulting strain energy must be independent of the order of loading; hence, by combining Eqs. *a, b,* and *c,* one obtains

$$\Delta P_1 y_1 = P_1 \Delta y_1 + P_2 \Delta y_2 \qquad (d)$$

Equations *b* and *d* can be combined to give

$$\frac{\Delta U}{\Delta P_1} = y_1 + \frac{1}{2} \Delta y_1$$

or upon taking the limit as ΔP_1 approaches zero[1]

$$\frac{\partial U}{\partial P_1} = y_1 \qquad (e)$$

For the general case in which there are many loads involved, Eq. *e* is written as

$$\frac{\partial U}{\partial P_i} = y_i \qquad (11\text{-}8)$$

The following is a statement of Castigliano's theorem. *If the strain energy of a linearly elastic structure is expressed in terms of the system of external loads, the partial derivative of the strain energy with respect to a concentrated external load is the deflection of the structure at the point of application and in the direction of that load.* By a similar development, Castigliano's theorem can also be shown to be valid for applied moments and the resulting rotations (or changes in slope) of the structure, Thus,

$$\frac{\partial U}{\partial M_i} = \theta_i \qquad (11\text{-}9)$$

If the deflection is required either at a point where there is no unique point load or in a direction not aligned with the applied load, a dummy load is introduced at the desired point acting in the proper direction. The deflection is obtained by first differentiating the strain energy with respect to the dummy load and then taking the limit as the dummy load approaches zero. Also, for the application of Eq. 11-9, either a unique point moment or a dummy moment must be applied at point *i*. The moment will be in the direction of rotation at the point. Note that if the loading consists of a number of point loads, all expressed in terms of a single parameter (e.g., P, $2P$, $3P$, wL, $2wL$) and if the deflection is wanted at one of the applied loads, one must either write the moment equation with this load as a separate identifiable term or add a dummy load at the point, so that the partial derivative can be taken with respect to this load only.

It was shown in Section 8-8 that the strain energy per unit volume for uniaxial stress is $\sigma^2/(2E)$; hence, the total strain energy under uniaxial stress is

$$U = \int_{\text{vol.}} \frac{\sigma^2}{2E}\, dV$$

For a beam of constant (or slowly varying) cross section subjected to pure bending, the principal stresses are parallel to the axis of the beam; therefore, using Eq. 4-7, the strain energy becomes

---

[1] The partial derivative is used because the strain energy is a function of both P_1 and P_2.

$$U = \frac{1}{2E} \int_{\text{vol.}} \left(\frac{My}{I}\right)^2 dV \qquad (f)$$

By writing dV as $dA\ dx$ (where x is measured along the axis of the beam), Eq. f becomes

$$U = \frac{1}{2E} \int_0^L \left(\frac{M^2}{I^2} \int_{\text{area}} y^2\ dA\right) dx = \frac{1}{2E} \int_0^L \frac{M^2}{I}\ dx \qquad (11\text{-}10)$$

Equation 11-10 was developed for a beam loaded in pure bending. However, most real beams will be subjected to transverse loads that induce shearing stresses and, in the case of distributed loading, transverse normal stresses.

In Section 5-12, it was shown that deflections due to transverse shearing stresses in beams are small enough to neglect except for short deep beams. It can also be shown that transverse normal stresses are small in long beams. Therefore, Eq. 11-10, which neglects strain energy due to these stresses, is applicable for most real beams encountered in engineering practice. When Eqs. 11-8, 11-9, or 11-10 are used to solve a particular problem, it is usually much simpler to apply Leibnitz's rule to differentiate under the integral sign, so that

$$y_i = \frac{\partial U}{\partial P_i} = \frac{1}{E} \int_0^L \frac{M}{I} \frac{\partial M}{\partial P_i}\ dx \qquad (11\text{-}11)$$

The use of Castigliano's theorem to solve deflection problems is illustrated in the following examples.

EXAMPLE 11-4

Determine the deflection at the free end of a cantilever beam of constant cross section and length L, loaded with a force P at the free end and a distributed load that varies linearly from zero at the free end to w at the support.

Solution

From the free-body diagram of Fig. 11-8, the moment equation is

$$M = -Px - \frac{wx^3}{6L}$$

The partial derivative of M with respect to P is $-x$, and Eq. 11-11 becomes

$$EIy = \int_0^L \left(Px^2 + \frac{wx^4}{6L}\right) dx = +\frac{PL^3}{3} + \frac{wL^3}{30}$$

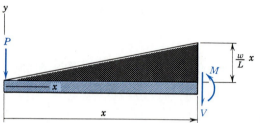

FIG. 11-8

The positive signs indicate deflection in the direction of the force P; hence,

$$y = \frac{PL^3}{3EI} + \frac{wL^4}{30EI} \quad \text{downward} \qquad \text{Ans.}$$

The result may be verified by reference to Appendix D.

EXAMPLE 11-5

Determine the deflection at the center of a simply supported beam of constant cross section and span L carrying a uniformly distributed load w lb/ft over its entire length.

Solution

In the free-body diagram of Fig. 11-9, the dashed force P represents a fictitious load applied at the location and in the direction of the desired deflection. The moment equation is

$$M = M_w + M_P = \frac{wLx}{2} - \frac{wx^2}{2} + \frac{Px}{2} - P\left\langle x - \frac{L}{2} \right\rangle^1$$

where the quantity $\left\langle x - (L/2) \right\rangle^1$ is zero for all $x \leqslant L/2$ (see Section 5-6). The partial derivative of M with respect to P is

$$\frac{\partial M}{\partial P} = \frac{x}{2} - \left\langle x - \frac{L}{2} \right\rangle^1$$

The dummy force P is equated to zero and the deflection is given by

$$EIy = \int_0^L \left(\frac{wLx}{2} - \frac{wx^2}{2} \right)\left(\frac{x}{2} - \left\langle x - \frac{L}{2} \right\rangle^1 \right) dx$$

which, for ease of integration, can be written as

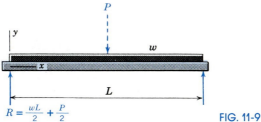

$$R = \frac{wL}{2} + \frac{P}{2}$$

FIG. 11-9

$$Ely = \frac{w}{4} \int_0^L (Lx^2 - x^3)dx + \frac{w}{4} \int_{L/2}^L (L^2x - 3Lx^2 + 2x^3)dx = \frac{5wL^4}{384EI}$$

Since the deflection is positive in the direction of the force, the deflection is downward and

$$y = \frac{5wL^4}{384EI} \quad \text{downward} \qquad \text{Ans.}$$

PROBLEMS

> *Note:* Use Castigliano's theorem to solve the following problems.

11-35* A beam is loaded and supported as shown in Fig. P11-35. Determine the deflection at the concentrated load P in terms of P, L, E, and I.

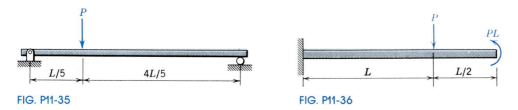

FIG. P11-35 FIG. P11-36

11-36* A beam is loaded and supported as shown in Fig. P11-36. Determine the deflection at the concentrated load P in terms of P, L, E, and I.

11-37 For the beam shown in Fig. P11-37, determine the slope at the section where the couple M is applied in terms of M, L, E, and I.

11-38 For the beam shown in Fig. P11-37, determine the deflection at the section where the couple M is applied in terms of M, L, E, and I.

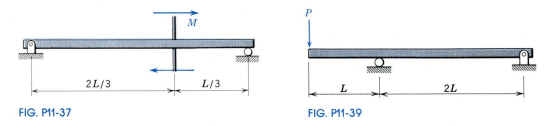

FIG. P11-37

FIG. P11-39

11-39* For the beam shown in Fig. P11-39, determine the deflection at the concentrated load P in terms of P, L, E, and I.

11-40* For the beam shown in Fig. P11-39, determine the slope at the left end of the beam in terms of P, L, E, and I.

11-41 For the beam shown in Fig. P11-39, determine the deflection at the section midway between the supports in terms of P, L, E, and I.

11-42 A beam is loaded and supported as shown in Fig. P11-42. Determine the deflection at the left end when $E = 10,000$ ksi and $I = 100$ in.^{4}.

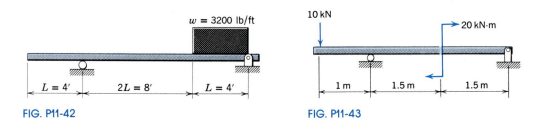

FIG. P11-42

FIG. P11-43

11-43* A beam is loaded and supported as shown in Fig. P11-43. Determine the deflection at the left end when $E = 70$ GPa and $I = 20(10^6)$ mm^4.

11-44* A beam is loaded and supported as shown in Fig. P11-43. Determine the deflection at a section midway between the supports when $E = 200$ GPa and $I = 10(10^6)$ mm^4.

11-45 Determine the deflection at a section midway between the supports in terms of w, L, E, and I if the beam is loaded and supported as shown in Fig. P11-45.

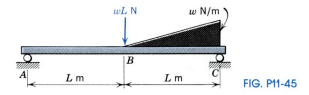

FIG. P11-45

11-46* Determine the vertical component of the deflection due to flexure in terms of P, R, E, and I at point B of the curved bar shown in Fig. P11-46. Assume that the radius of curvature of the bar is large relative to the depth of the cross section so that the elastic flexure formula applies.

11-47 Determine the horizontal component of the deflection due to flexure in terms of P, R, E, and I at point B of the curved bar shown in Fig. P11-46. Assume that the radius of curvature of the bar is large relative to the depth of the cross section so that the elastic flexure formula applies.

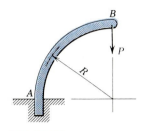

FIG. P11-46

11-48* Determine the vertical component of the deflection due to flexure in terms of P, R, E, and I at point B of the curved bar shown in Fig. P11-48. Assume that the radius of curvature of the bar is large relative to the depth of the cross section so that the elastic flexure formula applies.

11-49 Determine the horizontal component of the deflection due to flexure in terms of P, R, E, and I at point B of the curved bar shown in Fig. P11-48. Assume that the radius of curvature of the bar is large relative to the depth of the cross section, so that the elastic flexure formula applies.

11-50 Determine the slope due to flexure at point B of the curved bar shown in Fig. P11-48 in terms of P, R, E, and I. Assume that the radius of curvature of the bar is large relative to the depth of the cross section, so that the elastic flexure formula applies.

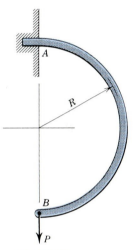

FIG. P11-48

11-51* Use Castigliano's theorem to solve Problem 5-160.

11-52* Use Castigliano's theorem to solve Problem 5-161.

11-53 Use Castigliano's theorem to solve Problem 5-162.

11-54 Use Castigliano's theorem to solve Problem 5-163.

11-55 Use Castigliano's theorem to solve Problem 5-164.

11-6

STATICALLY INDETERMINATE BEAMS— CASTIGLIANO'S THEOREM

Castigliano's theorem (see Section 11-5) is an effective supplement to the equations of equilibrium in the solution of statically indeterminate structures. If the deflection at some point in a structure is known (a boundary condition), Eq. 11-8 may be applied at this point and set equal to the known deflection. The resulting equation will contain one or more unknown forces. If the equation contains more than one unknown, additional equations must be written, such as equilibrium equations or the application of Eq. 11-9. However, there are certain restrictions necessary to ensure that the equations are independent; also, some remarks regarding procedure may be in order. The following outline may be found helpful:

1. For any system of loads and reactions, as many independent energy equations (Eq. 11-8 or 11-9) can be written as there are redundant unknowns—defined as unknowns not necessary to maintain equilibrium of the structure. This restriction is equivalent to the restriction of Section 6-2 requiring one boundary condition for each extra unknown.

2. If only one independent energy equation can be written and the moment equation contains two unknowns (forces or couples, including dummy loads), one of the unknowns must be expressed in terms of the other by means of an equilibrium equation. This case is illustrated in the alternate solution of Example 11-6 which follows. If two independent energy equations can be written, these can contain two unknowns that can be obtained from a solution of the two energy equations, as illustrated by Example 11-8.

3. If the loading changes along the beam, the moment equation can be written for the entire span using the singularity notation of Section 5-6 and, after multiplication by $\partial M/\partial P_i$ or $\partial M/\partial M_i$ (also using singularity notation), integration can be performed over the entire span. However, this procedure requires integration by parts of such expressions as $x\langle x - L\rangle^1\, dx$. Sometimes it may be more convenient to write an ordinary algebraic moment equation for each interval, multiply by $\partial M/\partial P_i$ or $\partial M/\partial M_i$ (different for each interval), and integrate over each interval; then add the results of all the integrations.

EXAMPLE 11-6

A beam is loaded and supported as shown in Fig. 11-10. Use Castigliano's theorem to determine the reactions in terms of w and L.

Solution

The free-body diagram of Fig. 11-10b indicates that the problem is statically indeterminate, since there are three unknown reaction components and only two independent equations of equilibrium available.

If the portion of the beam to the right of the reaction R was removed and replaced by an equivalent shearing force and bending moment at

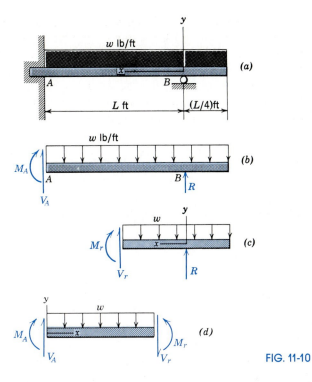

FIG. 11-10

the transverse section above R, neither R nor the elastic curve in the interval $0 \leqslant x \leqslant L$ would be changed. Hence, it is necessary to deal with only the strain energy of the beam in the interval $0 \leqslant x \leqslant L$. With the coordinate system placed as shown, the resulting moment equation obtained from the free-body diagram in Fig. 11-10c is

$$M = Rx - \frac{w}{2}\left(x + \frac{L}{4}\right)^2$$

$$= Rx - \frac{wx^2}{2} - \frac{wLx}{4} - \frac{wL^2}{32}$$

From Eq. 11-8, the deflection at the right support is given by

$$y_O = \frac{\partial U}{\partial R} = \frac{1}{EI}\int_0^L M\frac{\partial M}{\partial R}\,dx$$

Since the partial derivative of M with respect to R is x, the expression for the deflection is

$$y_O = \frac{1}{EI}\int_0^L \left(Rx^2 - \frac{wx^3}{2} - \frac{wLx^2}{4} - \frac{wL^2x}{32}\right)dx$$

which upon integration and substitution of limits becomes

$$y_O = \frac{1}{EI} \left(\frac{RL^3}{3} - \frac{wL^4}{8} - \frac{wL^4}{12} - \frac{wL^4}{64} \right)$$

Since the support is unyielding ($y = 0$ when $x = 0$), the above expression can be equated to zero and solved for R, yielding

$$R = \frac{43wL}{64} \text{ lb upward} \qquad\qquad \text{Ans.}$$

The other reaction components (M_A and V_A) are obtained from the equilibrium equations $\Sigma M_A = 0$ and $\Sigma F_y = 0$ and are

$$M_A = \frac{7wL^2}{64} \text{ ft-lb counterclockwise} \qquad \text{Ans.}$$

and

$$V_A = \frac{37wL}{64} \text{ lb upward} \qquad\qquad \text{Ans.}$$

Alternate Solution

This solution will make use of the free-body diagram of Fig. 11-10d, from which the moment equation is

$$M = M_A + V_A x - \frac{wx^2}{2}$$

Since there are two unknowns in the moment equation and only one independent energy equation can be written, the equilibrium equation $\Sigma M_B = 0$ will be used to obtain a relation between the two unknowns. Thus,

$$M_A + V_A L - \frac{wL^2}{2} + \frac{wL^2}{32} = 0$$

from which

$$M_A = \frac{15wL^2}{32} - V_A L \qquad \text{or} \qquad V_A = \frac{15wL}{32} - \frac{M_A}{L}$$

Eliminating M_A from the moment equation gives

$$M = V_A(x - L) + \frac{15wL^2}{32} - \frac{wx^2}{2}$$

and

$$\frac{\partial M}{\partial V_A} = x - L$$

From Eq. 11-8, the deflection at the left support is given by

$$y_A = \frac{\partial U}{\partial V_A} = \frac{1}{EI} \int_0^L M \frac{\partial M}{\partial V_A} dx$$

$$= \int_0^L \left[V_A(x - L)^2 - \frac{wx^2}{2}(x - L) + \frac{15wL^2}{32}(x - L) \right] dx = 0$$

from which

$$V_A = \frac{37wL}{64} \text{ lb upward} \qquad\qquad \text{Ans.}$$

The other reaction components (M_A and R) can be obtained from equilibrium equations.

EXAMPLE 11-7

Beam BC of Fig. 11-11a is a rolled steel W203 × 22 beam having a cross-sectional moment of inertia of $20(10^6)$ mm^4. The beam is fixed at B and supported at C by an aluminum alloy tie rod having a cross-sectional area of 100 mm^2. Determine the force in the tie rod due to the distributed load of 12 kN/m. Use 200 GPa and 70 GPa for the moduli of elasticity of the steel and aluminum, respectively.

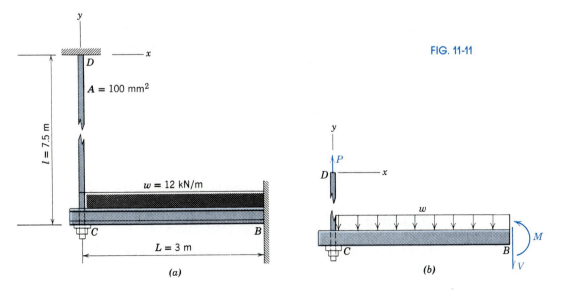

FIG. 11-11

(a)

(b)

Solution

The free-body diagram of the beam and tie rod is shown in Fig. 11-11b with the origin of coordinates at the top of the tie rod where the boundary condition is $x = 0$, $y = 0$. The total strain energy in the system is the sum of the energies in the tie rod and the beam; that is,

$$U = \frac{\sigma^2}{2E_a} Al + \int_0^L \frac{M^2}{2E_s I}\, dx = \frac{P^2 l}{2AE_a} + \int_0^L \frac{M^2}{2E_s I}\, dx$$

The deflection at D is

$$y_O = \frac{\partial U}{\partial P} = \frac{Pl}{AE_a} + \frac{1}{E_s I}\int_0^L M \frac{\partial M}{\partial P}\, dx = 0$$

The moment in the beam is

$$M = Px - \frac{wx^2}{2}$$

and

$$\frac{\partial M}{\partial P} = x$$

Therefore,

$$y_O = \frac{Pl}{AE_a} + \frac{1}{E_s I}\int_0^L \left(Px^2 - \frac{wx^3}{2}\right) dx = \frac{Pl}{AE_a} + \frac{PL^3}{3E_s I} - \frac{wL^4}{8E_s I}$$

Since the structure is fixed at D, y_O is zero and substitution of the given data yields

$$\frac{P(7.5)}{100(10^{-6})(70)(10^9)} + \frac{P(3)^3}{3(200)(10^9)(20)(10^{-6})} = \frac{12(10^3)(3)^4}{8(200)(10^9)(20)(10^{-6})}$$

from which

$$P = 9145 \text{ N} = \underline{9.15 \text{ kN } T} \qquad \text{Ans.}$$

EXAMPLE 11-8

Determine the reactions at the left end of the beam of Fig. 11-12a in terms of P, L, and a.

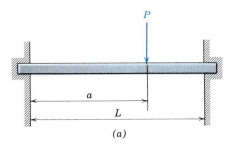

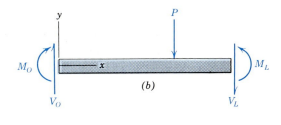

FIG. 11-12

Solution

The free-body diagram of Fig. 11-12b indicates four unknown reaction components, and since there are only two independent equations of equilbrium, two supplementary equations are necessary. These are obtained by applying Eq. 11-8 and 11-9 and solving the resulting simultaneous equations for M_O and V_O. With the origin at the left support, the boundary conditions are $y = 0$ and $\theta = 0$ at $x = 0$. Since the moment equation is

$$M = M_O + V_O x - P\langle x - a \rangle^1$$

the deflection at the left end is given by Eq. 11-8 as

$$y_O = \frac{\partial U}{\partial V_O} = \frac{1}{EI} \int_0^L M \frac{\partial M}{\partial V_O} dx$$

and

$$EIy_O = \left[\frac{M_O x^2}{2} + \frac{V_O x^3}{3} - P \left(\frac{x}{2} \langle x - a \rangle^2 - \frac{1}{6} \langle x - a \rangle^3 \right) \right]_0^L$$

where the last term was integrated by parts. Substitution of the limits gives

$$EIy_O = \frac{M_O L^2}{2} + \frac{V_O L^3}{3} - \frac{PL}{2} (L - a)^2 + \frac{P}{6} (L - a)^3$$

The application of the boundary condition $y = 0$ at $x = 0$ gives

$$\frac{M_O L^2}{2} + \frac{V_O L^3}{3} - \frac{PL^3}{3} + \frac{PL^2 a}{2} - \frac{Pa^3}{6} = 0 \qquad (a)$$

When Eq. 11-9 is applied, the rotation at the left end is given by the expression

$$\theta_O = \frac{\partial U}{\partial M_O} = \frac{1}{EI} \int_0^L M \frac{\partial M}{\partial M_O} \, dx$$

and

$$EI\theta_O = \int_0^L (M_O + V_O x - P\langle x - a\rangle^1) \, dx$$

$$= \left[M_O x + \frac{V_O x^2}{2} - \frac{P}{2} \langle x - a \rangle^2 \right]_0^L$$

which reduces to

$$M_O L + \frac{V_O L^2}{2} - \frac{PL^2}{2} + PaL - \frac{Pa^2}{2} = 0 \qquad (b)$$

Equations a and b are solved to obtain the following results:

$$V_O = P - \frac{3Pa^2}{L^2} + \frac{2Pa^3}{L^3} \qquad \text{Ans.}$$

and

$$M_O = -Pa + \frac{2Pa^2}{L} - \frac{Pa^3}{L^2} \qquad \text{Ans.}$$

PROBLEMS

Note: Use Castigliano's theorem to solve the following problems.

11-56* A beam is loaded and supported as shown in Fig. P11-56. Determine

(a) the reactions at supports A and B in terms of w and L.
(b) the deflection at midspan in terms of w, L, E, and I.

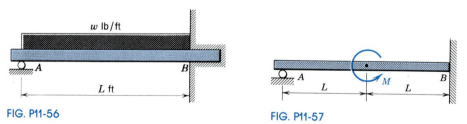

FIG. P11-56

FIG. P11-57

11-57* A beam is loaded and supported as shown in Fig. P11-57. Determine

(a) the reactions at supports A and B in terms of M and L.
(b) the deflection at the middle of span AB in terms of M, L, E, and I.

11-58 A beam is loaded and supported as shown in Fig. P11-58. Determine

(a) the reactions at supports A and B in terms of P.

(b) the deflection at the middle of the span in terms of P, L, E, and I.

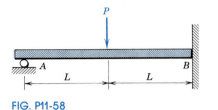

FIG. P11-58

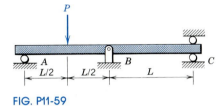

FIG. P11-59

11-59 A beam is loaded and supported as shown in Fig. P11-59. Determine

(a) the reactions at supports A, B, and C in terms of P.

(b) the deflection at the middle of span AB in terms of P, L, E, and I.

11-60* A beam is loaded and supported as shown in Fig. P11-60. Determine

(a) the reactions at supports A, B, and C in terms of w and L.

(b) the deflection at the middle of span BC in terms of w, L, E, and I.

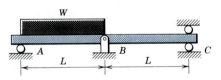

FIG. P11-60

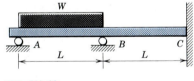

FIG. P11-61

11-61* A beam is loaded and supported as shown in Fig. P11-61. Determine

(a) the reactions at supports A, B, and C in terms of w and L.

(b) the deflection at the middle of span AB in terms of w, L, E, and I.

11-62 A beam is loaded and supported as shown in Fig. P11-62. Determine

(a) the reactions at supports A and B in terms of w and L.

(b) the deflection at C in terms of w, L, E, and I.

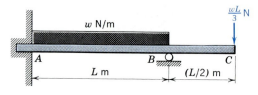

FIG. P11-62

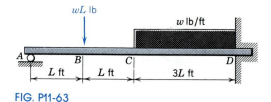

FIG. P11-63

11-63 A beam is loaded and supported as shown in Fig. P11-63. Determine

(a) the reactions at supports A and D in terms of w and L.
(b) the deflection at B in terms of w, L, E, and I.

11-64* A beam is loaded and supported as shown in Fig. P11-64. Determine

(a) the reactions at supports A, B, and C in terms of w and L.
(b) the deflection at the middle of span AB in terms of w, L, E, and I.

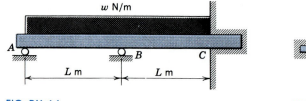

FIG. P11-64

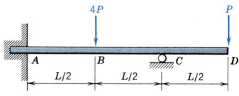

FIG. P11-65

11-65* A beam is loaded and supported as shown in Fig. P11-65. Determine

(a) the reactions at supports A and C in terms of P.
(b) the deflection at B in terms of P, L, E, and I.

11-66 The curved beam shown in Fig. P11-66 has a constant radius and constant E and I. Neglect the energy due to the axial stress and assume that the ratio R/c is large enough for the elastic flexure formula to apply. Determine the reaction at B in terms of P.

11-67* The frame ABC shown in Fig. P11-67 is a continuous member of constant E and I. Assume the corner at B remains a right angle and neglect the energy due to the axial stress in leg BC. Determine

(a) the reaction at support A in terms of w and L.
(b) the horizontal deflection of end A in terms of w, L, E, and I.

11-68 A frame ABC, similar to the one shown in Fig. P11-67, is pinned at end A as shown in Fig. P11-68. The frame supports a uniformly

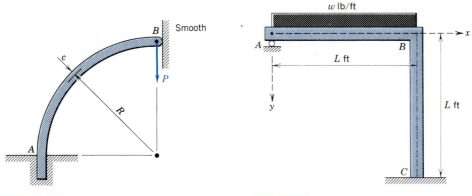

FIG. P11-66

FIG. P11-67

distributed load w as shown in Fig. P11-67. Neglect the energy due to axial stresses in AB and BC. Determine the horizontal and vertical components of the pin reaction at A on member AB.

11-69* Use Castigliano's theorem to solve Problem 6-92.

11-70 Use Castigliano's theorem to solve Problem 6-93.

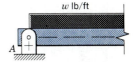

FIG. P11-68

11-7

UNIT LOAD METHOD

A method for determining deflections or rotations at a point on a beam or structure, which is simple to use in practice, is the Maxwell–Mohr method. This method is also commonly known as the unit-load method or the dummy-load method. In Section 11-5 it was shown that the deflection (or rotation) at a point on a beam or structure can be determined by using Castigliano's theorem. Thus, from Equation 11-11 for the general case where a number of loads are applied to a beam or structure, the deflection y_i in the direction of force P_i at point i is

$$y_i = \frac{\partial U}{\partial P_i} = \frac{1}{E} \int_0^L \frac{M}{I} \frac{\partial M}{\partial P_i} \, dx \qquad (11\text{-}11)$$

or for rotations

$$\theta_i = \frac{\partial U}{\partial M_i} = \frac{1}{E} \int_0^L \frac{M}{I} \frac{\partial M}{\partial M_i} \, dx \qquad (11\text{-}12)$$

Since the moment equation consists of the forces P_i multiplied by appropriate lengths, the differentiation process leaves just the length

TABLE 11-1 Values of Integrals $\int_0^L Mm\,dx$

	a (rectangle)	a (triangle ↗)	a (triangle ↘)	a_1, a_2 (trapezoid)
b (rectangle)	abL	$\tfrac{1}{2}abL$	$\tfrac{1}{2}abL$	$\tfrac{1}{2}bL(a_1+a_2)$
b (triangle ↗)	$\tfrac{1}{2}abL$	$\tfrac{1}{3}abL$	$\tfrac{1}{6}abL$	$\tfrac{1}{6}bL(a_1+2a_2)$
b (triangle ↘)	$\tfrac{1}{2}abL$	$\tfrac{1}{6}abL$	$\tfrac{1}{3}abL$	$\tfrac{1}{6}bL(2a_1+a_2)$
b_1, b_2 (trapezoid)	$\tfrac{1}{2}aL(b_1+b_2)$	$\tfrac{1}{6}aL(b_1+2b_2)$	$\tfrac{1}{6}aL(2b_1+b_2)$	$\tfrac{1}{6}b_1L(2a_1+a_2)+\tfrac{1}{6}b_2L(a_1+2a_2)$
* b	$\tfrac{1}{3}abL$	$\tfrac{1}{4}abL$	$\tfrac{1}{12}abL$	$\tfrac{1}{12}bL(a_1+3a_2)$
** b	$\tfrac{2}{3}abL$	$\tfrac{5}{12}abL$	$\tfrac{1}{4}abL$	$\tfrac{1}{12}bL(3a_1+5a_2)$
*** b	$\tfrac{1}{4}abL$	$\tfrac{1}{5}abL$	$\tfrac{1}{20}abL$	$\tfrac{1}{20}bL(a_1+4a_2)$

*Spandrel of second-degree parabola.   **Segment of second-degree parabola.   ***Spandrel of third-degree parabola.

terms when the derivative $\partial M/\partial P_i$ is evaluated. Thus, the term $\partial M/\partial P_i$ is equivalent to a moment m_y produced by a unit load P_i and Equation 11-11 can be written as

$$y_i = \frac{\partial U}{\partial P_i} = \frac{1}{E}\int_0^L \frac{M}{I} m_y\,dx \tag{11-13}$$

or for rotations

$$\theta_i = \frac{\partial U}{\partial M_i} = \frac{1}{E}\int_0^L \frac{M}{I} m_\theta\,dx \tag{11-14}$$

A positive result means that the deflection (or rotation) is in the direction assumed for the unit load. Similarly, a negative result indicates that the deflection (or rotation) is in a direction opposite to that assumed for the unit load. The integral $\int Mm\,dx$ often involves moment expressions that can be represented by simple moment diagrams. Values for a number of the frequently encountered integrals are given in Table 11-1.

The following examples illustrate the use of the unit load method and the integrals listed in Table 11-1.

EXAMPLE 11-9

A beam with constant EI is loaded and supported as shown in Fig. 11-13a. Use the unit-load method together with the values for the moment products listed in Table 11-1 to determine the rotation at the left end of the beam.

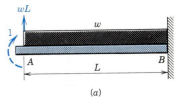

(a)

Solution

The moment diagram (by parts) for the beam shown in Fig. 11-13a together with the moment m_θ produced by a unit moment applied to the left end of the beam are shown in Fig. 11-13b. From the values for the integrals $\int Mm\,dx$ listed in Table 11-1, the angle θ is

$$\theta = \frac{1}{EI}\left(\frac{1}{2}abL - \frac{1}{3}abL\right) = \frac{1}{EI}\left(\frac{1}{2}wL^3 - \frac{1}{6}wL^3\right) = \frac{wL^3}{3EI} \qquad \text{Ans.}$$

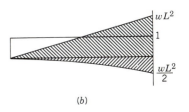

(b)

FIG. 11-13

EXAMPLE 11-10

A simple frame with constant EI is loaded and supported as shown in Fig. 11-14a. Use the unit-load method together with the values for the moment products listed in Table 11-1 to determine

(a) the horizontal component of the deflection at point A.
(b) the rotation at point A.

Solution

(a) The moment diagram for the frame is shown in Fig. 11-14b. The moment m_y produced in the frame by a unit load applied in a horizontal direction at point A is shown in Fig. 11-14c. From the values for the integrals $\int Mm\,dx$ listed in Table 11-1, the horizontal component of the deflection at point A is

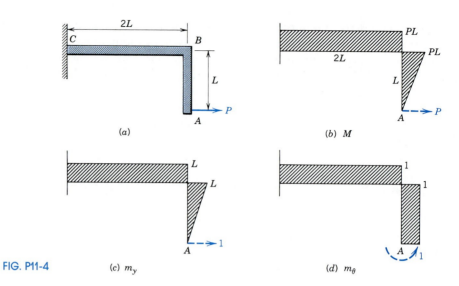

FIG. P11-4

(a)

(b) M

(c) m_y

(d) m_θ

$$\delta_H = \frac{1}{EI}\left[\frac{1}{3}abL + ab(2L)\right]$$

$$= \frac{1}{EI}\left[\frac{1}{3}PL^3 + PL^2(2L)\right] = \frac{7PL^3}{3EI} \qquad \text{Ans.}$$

(b) The moment m_θ produced in the frame by a unit moment applied in a counterclockwise direction at point A is shown in Fig. 11-14d. From the values for the integrals $\int Mm\ dx$ listed in Table 11-1, the rotation at point A is

$$\theta = \frac{1}{EI}\left[\frac{1}{2}abL + ab(2L)\right]$$

$$= \frac{1}{EI}\left[\frac{1}{2}PL^2 + PL(2L)\right] = \frac{5PL^2}{2EI} \qquad \text{Ans.}$$

PROBLEMS

Note: Use the unit-load method and values from Table 11-1 to solve the following problems.

1-71* A beam with constant EI is loaded and supported as shown in Fig. P11-71. Determine the deflection at the middle of the span.

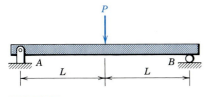

FIG. P11-71

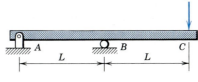

FIG. P11-73

11-72* A beam with constant EI is loaded and supported as shown in Fig. P11-71. Determine the rotation at the left end of the beam.

11-73 A beam with constant EI is loaded and supported as shown in Fig. P11-73. Determine

(a) the deflection at the right end of the beam.

(b) the rotation at the left end of the beam.

11-74 A beam with constant EI is loaded and supported as shown in Fig. P11-74. Determine

(a) the deflection at the middle of span AC.

(b) the rotation at the left end of the beam.

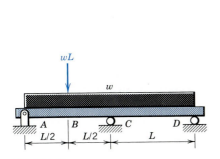

FIG. P11-74

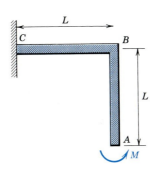

FIG. P11-75

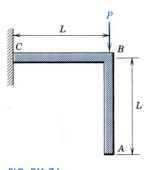

FIG. P11-76

11-75* A simple frame with constant EI is loaded and supported as shown in Fig. P11-75. Determine

(a) the horizontal component of the deflection at point A.

(b) the vertical component of the deflection at point A.

11-76* A simple frame with constant EI is loaded and supported as shown in Fig. P11-76. Determine

(a) the horizontal component of the deflection at point A.

(b) the vertical component of the deflection at point A.

(c) the rotation at point A.

11-77 A simple frame with constant EI is loaded and supported as shown in Fig. P11-77. Determine

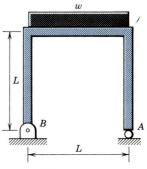

FIG. P11-77

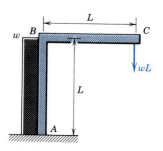

(a) the deflection at point A.

(b) the rotation at point B.

11-78 A simple frame with constant EI is loaded and supported as shown in Fig. P11-78. Determine

(a) the horizontal component of the deflection at point C.

(b) the vertical component of the deflection at point C.

(c) the rotation at point B.

11-79* Use the unit-load method to solve Problem 5-161.

11-80* Use the unit-load method to solve Problem 5-162.

11-81 Use the unit-load method to solve Problem 5-163.

11-82 Use the unit-load method to solve Problem 5-164.

COMPUTER PROBLEMS

Note: The following problems have been designed to be solved with a programmable calculator, microcomputer, or mainframe computer. Appendix F contains a description of a few numerical methods, together with a few simple programs in BASIC and FORTRAN, that can be modified for use in the solution of these problems.

C11-1 A hollow circular shaft and a solid circular shaft have the same length L and the same cross-sectional area A. If the shafts are made of the same material and are stressed to the same elastic, maximum shearing stress level, compute and plot the ratio U_H/U_S versus r_o/r_s where

U_H is the elastic strain energy stored in the hollow shaft.

U_S is the elastic strain energy stored in the solid shaft.

r_o is the outside radius of the hollow shaft.

r_s is the radius of the solid shaft.

C11-2 A 2-kg mass is dropped from a height h onto the free end of a cantilever beam. The beam has a 10-by-60-mm rectangular cross section and is 500 mm long. The beam is made of cold-rolled bronze ($E = 100$ GPa), and bending occurs about the weak axis of the cross section. Compute and plot

(a) the deflection δ of the free end of the beam as a function of the drop height h (0 mm $< h <$ 200 mm).

(b) the maximum fiber stress σ in the beam as a function of the drop height h (0 mm $< h <$ 200 mm).

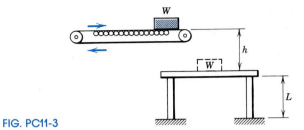

FIG. PC11-3

C11-3 A conveyer belt drops a 40-lb package onto the center of a table, as shown in Fig. PC11-3. Each of the four structural steel ($E = 30{,}000$ ksi) legs of the table has a 1-in.-square cross section and is 48 in. long. Assume that 25 percent of the energy of the falling weight is absorbed by the legs and that the effective length for buckling of the legs is 48 in. Compute and plot the axial stress σ in the legs of the table as a function of the drop height h for stresses up to one-half of the critical buckling stress.

APPENDIX A

AVERAGE PROPERTIES OF SELECTED MATERIALS

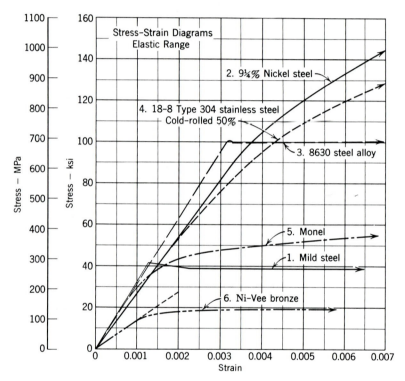

FIG. A-1 Stress-strain diagrams in the elastic range for a selection of engineering materials.

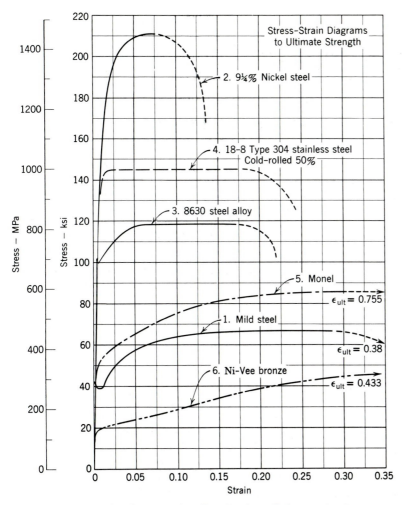

FIG. A-2 Stress-strain diagrams to ultimate strength for a selection of engineering materials.

TABLE A-1a Average Properties of Selected Engineering Materials (U.S. Customary System of Units)

Exact values may vary widely with changes in composition, heat treatment, and mechanical working. More precise information can be obtained from manufacturers.

Materials	Specific Weight (lb/in.³)	Elastic Strength[a] Tension (ksi)	Comp. (ksi)	Shear (ksi)	Ultimate Strength Tension (ksi)	Comp. (ksi)	Shear (ksi)	Endurance Limit[c] (ksi)	Modulus of Elasticity (1000 ksi)	Modulus of Rigidity (1000 ksi)	Percent Elongation in 2 in.	Coefficient of Thermal Expansion (10⁻⁶/°F)
Ferrous metals												
Wrought iron	0.278	30	b		48	b	25	23	28		30[d]	6.7
Structural steel	0.284	36	b		66	b		28	29	11.0	28[d]	6.6
Steel, 0.2% C hardened	0.284	62	b		90	b			30	11.6	22	6.6
Steel, 0.4% C hot-rolled	0.284	53	b		84	b		38	30	11.6	29	
Steel, 0.8% C hot-rolled	0.284	76	b		122	b			30	11.6	8	
Cast iron—gray	0.260				25	100		12	15		0.5	6.7
Cast iron—malleable	0.266	32	b		50	b			25		20	6.6
Cast iron—nodular	0.266	70			100				25		4	6.6
Stainless steel (18-8) annealed	0.286	36	b		85	b		40	28	12.5	55	9.6
Stainless steel (18-8) cold-rolled	0.286	165	b		190	b		90	28	12.5	8	9.6
Steel, SAE 4340, heat-treated	0.283	132	145		150	b	95	76	29	11.0	19	
Nonferrous metal alloys												
Aluminum, cast, 195-T6	0.100	24	25		36	b	30	7	10.3	3.8	5	
Aluminum, wrought, 2014-T4	0.101	41	41	24	62	b	38	18	10.6	4.0	20	12.5
Aluminum, wrought, 2024-T4	0.100	48	48	28	68	b	41	18	10.6	4.0	19	12.5
Aluminum, wrought, 6061-T6	0.098	40	40	26	45	b	30	13.5	10.0	3.8	17	12.5
Magnesium, extrusion, AZ80X	0.066	35	26		49	b	21	19	6.5	2.4	12	14.4
Magnesium, sand cast, AZ63-HT	0.066	14	14		40	b	19	14	6.5	2.4	12	14.4
Monel, wrought, hot-rolled	0.319	50	b		90	b		40	26	9.5	35	7.8
Red brass, cold-rolled	0.316	60			75				15	5.6	4	9.8
Red brass, annealed	0.316	15	b		40	b			15	5.6	50	9.8
Bronze, cold-rolled	0.320	75			100				15	6.5	3	9.4
Bronze, annealed	0.320	20	b		50	b			15	6.5	50	9.4
Titanium alloy, annealed	0.167	135	b		155	b			14	5.3	13	9.4
Invar, annealed	0.292	42	b		70	b			21	8.1	41	0.6
Nonmetallic materials												
Douglas fir, green[e]	0.022	4.8	3.4			3.9	0.9		1.6			
Douglas fir, air dry[e]	0.020	8.1	6.4			7.4	1.1		1.9			
Red oak, green[e]	0.037	4.4	2.6			3.5	1.2		1.4			1.9
Red oak, air dry[e]	0.025	8.4	4.6			6.9	1.8		1.8			
Concrete, medium strength	0.087		1.2			3.0			3.0			6.0
Concrete, fairly high strength	0.087		2.0			5.0			4.5			6.0

[a] Elastic strength may be represented by proportional limit, yield point, or yield strength at a specified offset (usually 0.2 percent for ductile metals).
[b] For ductile metals (those with an appreciable ultimate elongation), it is customary to assume the properties in compression have the same values as those in tension.   [c] Rotating beam.   [d] Elongation in 8 in.   [e] All timber properties are parallel to the grain.

TABLE A-1b Average Properties of Selected Engineering Materials (International System of Units)

Exact values may vary widely with changes in composition, heat treatment, and mechanical working. More precise information can be obtained from manufacturers.

Materials	Density (Mg/m³)	Elastic Strength[a] Tension (MPa)	Elastic Strength[a] Comp. (MPa)	Elastic Strength[a] Shear (MPa)	Ultimate Strength Tension (MPa)	Ultimate Strength Comp. (MPa)	Ultimate Strength Shear (MPa)	Endurance Limit[c] (MPa)	Modulus of Elasticity (GPa)	Modulus of Rigidity (GPa)	Percent Elongation in 50 mm	Coefficient of Thermal Expansion (10^{-6}/°C)
Ferrous metals												
Wrought iron	7.70	210	b		330	b	170	160	190		30[d]	12.1
Structural steel	7.87	250	b		450	b		190	200	76	28[d]	11.9
Steel, 0.2% C hardened	7.87	430	b		620	b			210	80	22	11.9
Steel, 0.4% C hot-rolled	7.87	360	b	160	580	b		260	210	80	29	
Steel, 0.8% C hot-rolled	7.87	520	b	190	840	b			210	80	8	
Cast iron—gray	7.20			180	170	690		80	100		0.5	12.1
Cast iron—malleable	7.37	220	b		340	b			170		20	11.9
Cast iron—nodular	7.37	480			690				170		4	11.9
Stainless steel (18-8) annealed	7.92	250	b		590	b		270	190	86	55	17.3
Stainless steel (18-8) cold-rolled	7.92	1140	b		1310	b		620	190	86	8	17.3
Steel, SAE 4340, heat-treated	7.84	910	1000		1030	b	650	520	200	76	19	
Nonferrous metal alloys												
Aluminum, cast 195-T6	2.77	160	170		250	b	210	50	71	26	5	22.5
Aluminum, wrought, 2014-T4	2.80	280	280		430	b	260	120	73	28	20	22.5
Aluminum, wrought, 2024-T4	2.77	330	330		470	b	280	120	73	28	19	22.5
Aluminum, wrought, 6061-T6	2.71	270	270		310	b	210	93	70	26	17	22.5
Magnesium, extrusion, AZ80X	1.83	240	180		340	b	140	130	45	16	12	25.9
Magnesium, sand cast, AZ63-HT	1.83	100	96		270	b	130	100	45	16	12	25.9
Monel, wrought, hot-rolled	8.84	340	b		620	b		270	180	65	35	14.0
Red brass, cold-rolled	8.75	410			520				100	39	4	17.6
Red brass, annealed	8.75	100	b		270	b			100	39	50	17.6
Bronze, cold-rolled	8.86	520			690				100	45	3	16.9
Bronze, annealed	8.86	140	b		340	b			100	45	50	16.9
Titanium alloy, annealed	4.63	930	b		1070	b			96	36	13	
Invar, annealed	8.09	290	b		480	b			140	56	41	1.1
Nonmetallic materials												
Douglas fir, green[e]	0.61	33	23			27	6.2		11			
Douglas fir, air dry[e]	0.55	56	44			51	7.6		13			
Red oak, green[e]	1.02	30	18			24	8.3		10			3.4
Red oak, air dry[e]	0.69	58	32			48	12.4		12			
Concrete, medium strength	2.41		8			21			21			10.8
Concrete, fairly high strength	2.41		14			34			31			10.8

[a] Elastic strength may be represented by proportional limit, yield point, or yield strength at a specified offset (usually 0.2 percent for ductile metals). [b] For ductile metals (those with an appreciable ultimate elongation), it is customary to assume the properties in compression have the same values as those in tension. [c] All timber properties are parallel to the grain.

[c] Rotating beam.   [d] Elongation in 200 mm.   [e] All timber properties are parallel to the grain.

PROPERTIES OF SELECTED AREAS

Many engineering formulas used to determine stresses or deflections in Mechanics of Materials problems require knowledge of the cross-sectional area of a section, the location of the centroid of an area, or the second moment of an area (moment of inertia) about a centroidal axis. Such properties for rectangles, circles, semicircles, triangles, and parabolas are listed in Table B-1.

TABLE B-1 Properties of Selected Areas

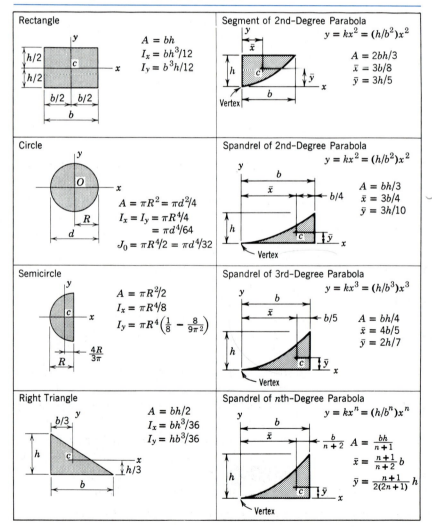

Rectangle

$$A = bh$$
$$I_x = bh^3/12$$
$$I_y = b^3h/12$$

Segment of 2nd-Degree Parabola

$$y = kx^2 = (h/b^2)x^2$$
$$A = 2bh/3$$
$$\bar{x} = 3b/8$$
$$\bar{y} = 3h/5$$

Circle

$$A = \pi R^2 = \pi d^2/4$$
$$I_x = I_y = \pi R^4/4$$
$$= \pi d^4/64$$
$$J_0 = \pi R^4/2 = \pi d^4/32$$

Spandrel of 2nd-Degree Parabola

$$y = kx^2 = (h/b^2)x^2$$
$$A = bh/3$$
$$\bar{x} = 3b/4$$
$$\bar{y} = 3h/10$$

Semicircle

$$A = \pi R^2/2$$
$$I_x = \pi R^4/8$$
$$I_y = \pi R^4\left(\frac{1}{8} - \frac{8}{9\pi^2}\right)$$

Spandrel of 3rd-Degree Parabola

$$y = kx^3 = (h/b^3)x^3$$
$$A = bh/4$$
$$\bar{x} = 4b/5$$
$$\bar{y} = 2h/7$$

Right Triangle

$$A = bh/2$$
$$I_x = bh^3/36$$
$$I_y = hb^3/36$$

Spandrel of nth-Degree Parabola

$$y = kx^n = (h/b^n)x^n$$
$$A = \frac{bh}{n+1}$$
$$\bar{x} = \frac{n+1}{n+2}b$$
$$\bar{y} = \frac{n+1}{2(2n+1)}h$$

PROPERTIES OF ROLLED-STEEL SHAPES

In the following tables,[1] the properties of a few structural-steel rolled shapes are listed for use in solving some of the problems presented in the text. The data listed in these tables were obtained from publications of the American Institute of Steel Construction.

The notation used in the tables is

I = moment of inertia

S = section modulus

r = radius of gyration

x = distance to the centroid

y = distance to the centroid

The designations used to describe the various structural shapes have the following meanings.

C12 × 30	a 12-in. rolled channel section with a weight of 30 lb/ft
C305 × 45	a 305-mm rolled channel section with a mass of 45 kg/m
L6 × 4 × 1/2	a rolled angle section with 6-in. and 4-in. legs $\frac{1}{2}$ in. thick
L102 × 89 × 13	a rolled angle section with 102-mm and 89-mm legs 13 mm thick
S10 × 35	a 10-in. American standard rolled section (I-beam) with a weight of 35 lb/ft
S127 × 22	a 127-mm American standard rolled section (I-beam) with a mass of 22 kg/m
W16 × 100	a 16-in. wide-flange rolled section with a weight of 100 lb/ft
W914 × 342	a 914-mm wide-flange rolled section with a mass of 342 kg/m
WT8 × 50	an 8-in. structural T-section (cut from a W-section) with a weight of 50 lb/ft
WT457 × 171	a 457-mm structural T-section (cut from a W-section) with a mass of 171 kg/m

[1]These tables are courtesy of the American Institute of Steel Construction.

Wide-Flange Shapes: Properties for Designing (*W*-Shapes) (U.S. Customary Units)

Desig-nation	Area (in.²)	Depth (in.)	Flange Width (in.)	Flange Thick-ness (in.)	Web Thick-ness (in.)	Axis X–X I (in.⁴)	Axis X–X S (in.³)	Axis X–X r (in.)	Axis Y–Y I (in.⁴)	Axis Y–Y S (in.³)	Axis Y–Y r (in.)
W36 × 230	67.6	35.90	16.470	1.260	0.760	15000	837	14.9	940	114	3.73
× 160	47.0	36.01	12.000	1.020	0.650	9750	542	14.4	295	49.1	2.50
W33 × 201	59.1	33.68	15.745	1.150	0.715	11500	684	14.0	749	95.2	3.56
× 152	44.7	33.49	11.565	1.055	0.635	8160	487	13.5	273	47.2	2.47
× 130	38.3	33.09	11.510	0.855	0.580	6710	406	13.2	218	37.9	2.39
W30 × 132	38.9	30.31	10.545	1.000	0.615	5770	380	12.2	196	37.2	2.25
× 108	31.7	29.83	10.475	0.760	0.545	4470	299	11.9	146	27.9	2.15
W27 × 146	42.9	27.38	13.965	0.975	0.605	5630	411	11.4	443	63.5	3.21
× 94	27.7	26.92	9.990	0.745	0.490	3270	243	10.9	124	24.8	2.12
W24 × 104	30.6	24.06	12.750	0.750	0.500	3100	258	10.1	259	40.7	2.91
× 84	24.7	24.10	9.020	0.770	0.470	2370	196	9.79	94.4	20.9	1.95
× 62	18.2	23.74	7.040	0.590	0.430	1550	131	9.23	34.5	9.80	1.38
W21 × 101	29.8	21.36	12.290	0.800	0.500	2420	227	9.02	248	40.3	2.89
× 83	24.3	21.43	8.355	0.835	0.515	1830	171	8.67	81.4	19.5	1.83
× 62	18.3	20.99	8.240	0.615	0.400	1330	127	8.54	57.5	13.9	1.77
W18 × 97	28.5	18.59	11.145	0.870	0.535	1750	188	7.82	201	36.1	2.65
× 76	22.3	18.21	11.035	0.680	0.425	1330	146	7.73	152	27.6	2.61
× 60	17.6	18.24	7.555	0.695	0.415	984	108	7.47	50.1	13.3	1.69
W16 × 100	29.4	16.97	10.425	0.985	0.585	1490	175	7.10	186	35.7	2.52
× 67	19.7	16.33	10.235	0.665	0.395	954	117	6.96	119	23.2	2.46
× 40	11.8	16.01	6.995	0.505	0.305	518	64.7	6.63	28.9	8.25	1.57
× 26	7.68	15.69	5.500	0.345	0.250	301	38.4	6.26	9.59	3.49	1.12
W14 × 120	35.3	14.48	14.670	0.940	0.590	1380	190	6.24	495	67.5	3.74
× 82	24.1	14.31	10.130	0.855	0.510	882	123	6.05	148	29.3	2.48
× 43	12.6	13.66	7.995	0.530	0.305	428	62.7	5.82	45.2	11.3	1.89
× 30	8.85	13.84	6.730	0.385	0.270	291	42.0	5.73	19.6	5.82	1.49
W12 × 96	28.2	12.71	12.160	0.900	0.550	833	131	5.44	270	44.4	3.09
× 65	19.1	12.12	12.000	0.605	0.390	533	87.9	5.28	174	29.1	3.02
× 50	14.7	12.19	8.080	0.640	0.370	394	64.7	5.18	56.3	13.9	1.96
× 30	8.79	12.34	6.520	0.440	0.260	238	38.6	5.21	20.3	6.24	1.52
W10 × 60	17.6	10.22	10.080	0.680	0.420	341	66.7	4.39	116	23.0	2.57
× 45	13.3	10.10	8.020	0.620	0.350	248	49.1	4.33	53.4	13.3	2.01
× 30	8.84	10.47	5.810	0.510	0.300	170	32.4	4.38	16.7	5.75	1.37
× 22	6.49	10.17	5.750	0.360	0.240	118	23.2	4.27	11.4	3.97	1.33
W8 × 40	11.7	8.25	8.070	0.560	0.360	146	35.5	3.53	49.1	12.2	2.04
× 31	9.13	8.00	7.995	0.435	0.285	110	27.5	3.47	37.1	9.27	2.02
× 24	7.08	7.93	6.495	0.400	0.245	82.8	20.9	3.42	18.3	5.63	1.61
× 15	4.44	8.11	4.015	0.315	0.245	48.0	11.8	3.29	3.41	1.70	0.876
W6 × 25	7.34	6.38	6.080	0.455	0.320	53.4	16.7	2.70	17.1	5.61	1.52
× 16	4.74	6.28	4.030	0.405	0.260	32.1	10.2	2.60	4.43	2.20	0.967
W5 × 16	4.68	5.01	5.000	0.360	0.240	21.3	8.51	2.13	7.51	3.00	1.27
W4 × 13	3.83	4.16	4.060	0.345	0.280	11.3	5.46	1.72	3.86	1.90	1.00

Courtesy of The American Institute of Steel Construction.

Wide-Flange Shapes: Properties for Designing (*W*-Shapes) (Converted to SI Units)

Designation	Area (mm²)	Depth (mm)	Flange Width (mm)	Flange Thickness (mm)	Web Thickness (mm)	Axis X–X I (10⁶ mm⁴)	Axis X–X S (10³ mm³)	Axis X–X r (mm)	Axis Y–Y I (10⁶ mm⁴)	Axis Y–Y S (10³ mm³)	Axis Y–Y r (mm)
W914 × 342	43610	912	418	32.0	19.3	6245	13715	378	391	1870	94.7
× 238	30325	915	305	25.9	16.5	4060	8880	366	123	805	63.5
W838 × 299	38130	855	400	29.2	18.2	4785	11210	356	312	1560	90.4
× 226	28850	851	294	26.8	16.1	3395	7980	343	114	775	62.7
× 193	24710	840	292	21.7	14.7	2795	6655	335	90.7	620	60.7
W762 × 196	25100	770	268	25.4	15.6	2400	6225	310	81.6	610	57.2
× 161	20450	758	266	19.3	13.8	1860	4900	302	60.8	457	54.6
W686 × 217	27675	695	355	24.8	15.4	2345	6735	290	184	1040	81.5
× 140	17870	684	254	18.9	12.4	1360	3980	277	51.6	406	53.8
W610 × 155	19740	611	324	19.1	12.7	1290	4230	257	108	667	73.9
× 125	15935	612	229	19.6	11.9	985	3210	249	39.3	342	49.5
× 92	11750	603	179	15.0	10.9	645	2145	234	14.4	161	35.1
W533 × 150	19225	543	312	20.3	12.7	1005	3720	229	103	660	73.4
× 124	15675	544	212	21.2	13.1	762	2800	220	33.9	320	46.5
× 92	11805	533	209	15.6	10.2	554	2080	217	23.9	228	45.0
W457 × 144	18365	472	283	22.1	13.6	728	3080	199	83.7	592	67.3
× 113	14385	463	280	17.3	10.8	554	2395	196	63.3	452	66.3
× 89	11355	463	192	17.7	10.5	410	1770	190	20.9	218	42.9
W406 × 149	18970	431	265	25.0	14.9	620	2870	180	77.4	585	64.0
× 100	12710	415	260	16.9	10.0	397	1915	177	49.5	380	62.5
× 60	7615	407	178	12.8	7.7	216	1060	168	12.0	135	39.9
× 39	4950	399	140	8.8	6.4	125	629	159	3.99	57.2	28.4
W356 × 179	22775	368	373	23.9	15.0	574	3115	158	206	1105	95.0
× 122	15550	363	257	21.7	13.0	367	2015	154	61.6	480	63.0
× 64	8130	347	203	13.5	7.7	178	1025	148	18.8	185	48.0
× 45	5710	352	171	9.8	6.9	121	688	146	8.16	95.4	37.8
W305 × 143	18195	323	309	22.9	14.0	347	2145	138	112	728	78.5
× 97	12325	308	305	15.4	9.9	222	1440	134	72.4	477	76.7
× 74	9485	310	205	16.3	9.4	164	1060	132	23.4	228	49.8
× 45	5670	313	166	11.2	6.6	99.1	633	132	8.45	102	38.6
W254 × 89	11355	260	256	17.3	10.7	142	1095	112	48.3	377	65.3
× 67	8580	257	204	15.7	8.9	103	805	110	22.2	218	51.1
× 45	5705	266	148	13.0	7.6	70.8	531	111	6.95	94.2	34.8
× 33	4185	258	146	9.1	6.1	49.1	380	108	4.75	65.1	33.8
W203 × 60	7550	210	205	14.2	9.1	60.8	582	89.7	20.4	200	51.8
× 46	5890	203	203	11.0	7.2	45.8	451	88.1	15.4	152	51.3
× 36	4570	201	165	10.2	6.2	34.5	342	86.7	7.61	92.3	40.9
× 22	2865	206	102	8.0	6.2	20.0	193	83.6	1.42	27.9	22.3
W152 × 37	4735	162	154	11.6	8.1	22.2	274	68.6	7.12	91.9	38.6
× 24	3060	160	102	10.3	6.6	13.4	167	66.0	1.84	36.1	24.6
W127 × 24	3020	127	127	9.1	6.1	8.87	139	54.1	3.13	49.2	32.3
W102 × 19	2470	106	103	8.8	7.1	4.70	89.5	43.7	1.61	31.1	25.4

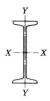

American Standard Beams: Properties for Designing (S-Shapes) (U.S. Customary Units)

Desig-nation	Area (in.²)	Depth (in.)	Flange Width (in.)	Flange Thick-ness (in.)	Web Thick-ness (in.)	Axis X–X I (in.⁴)	Axis X–X S (in.³)	Axis X–X r (in.)	Axis Y–Y I (in.⁴)	Axis Y–Y S (in.³)	Axis Y–Y r (in.)
S24 × 121	35.6	24.50	8.050	1.090	0.800	3160	258	9.43	83.3	20.7	1.53
× 106	31.2	24.50	7.870	1.090	0.620	2940	240	9.71	77.1	19.6	1.57
× 100	29.3	24.00	7.245	0.870	0.745	2390	199	9.02	47.7	13.2	1.27
× 90	26.5	24.00	7.125	0.870	0.625	2250	187	9.21	44.9	12.6	1.30
× 80	23.5	24.00	7.000	0.870	0.500	2100	175	9.47	42.2	12.1	1.34
S20 × 96	28.2	20.30	7.200	0.920	0.800	1670	165	7.71	50.2	13.9	1.33
× 86	25.3	20.30	7.060	0.920	0.660	1580	155	7.89	46.8	13.3	1.36
× 75	22.0	20.00	6.385	0.795	0.635	1280	128	7.62	29.8	9.32	1.16
× 66	19.4	20.00	6.255	0.795	0.505	1190	119	7.83	27.7	8.85	1.19
S18 × 70	20.6	18.00	6.251	0.691	0.711	926	103	6.71	24.1	7.72	1.08
× 54.7	16.1	18.00	6.001	0.691	0.461	804	89.4	7.07	20.8	6.94	1.14
S15 × 50	14.7	15.00	5.640	0.622	0.550	486	64.8	5.75	15.7	5.57	1.03
× 42.9	12.6	15.00	5.501	0.622	0.411	447	59.6	5.95	14.4	5.23	1.07
S12 × 50	14.7	12.00	5.477	0.659	0.687	305	50.8	4.55	15.7	5.74	1.03
× 40.8	12.0	12.00	5.252	0.659	0.462	272	45.4	4.77	13.6	5.16	1.06
× 35	10.3	12.00	5.078	0.544	0.428	229	38.2	4.72	9.87	3.89	0.980
× 31.8	9.35	12.00	5.000	0.544	0.350	218	36.4	4.83	9.36	3.74	1.00
S10 × 35	10.3	10.00	4.944	0.491	0.594	147	29.4	3.78	8.36	3.38	0.901
× 25.4	7.46	10.00	4.661	0.491	0.311	124	24.7	4.07	6.79	2.91	0.954
S8 × 23	6.77	8.00	4.171	0.426	0.441	64.9	16.2	3.10	4.31	2.07	0.798
× 18.4	5.41	8.00	4.001	0.426	0.271	57.6	14.4	3.26	3.73	1.86	0.831
S7 × 20	5.88	7.00	3.860	0.392	0.450	42.4	12.1	2.69	3.17	1.64	0.734
× 15.3	4.50	7.00	3.662	0.392	0.252	36.7	10.5	2.86	2.64	1.44	0.766
S6 × 17.25	5.07	6.00	3.565	0.359	0.465	26.3	8.77	2.28	2.31	1.30	0.675
× 12.5	3.67	6.00	3.332	0.359	0.232	22.1	7.37	2.45	1.82	1.09	0.705
S5 × 14.75	4.34	5.00	3.284	0.326	0.494	15.2	6.09	1.87	1.67	1.01	0.620
× 10	2.94	5.00	3.004	0.326	0.214	12.3	4.92	2.05	1.22	0.809	0.643
S4 × 9.5	2.79	4.00	2.796	0.293	0.326	6.79	3.39	1.56	0.903	0.646	0.569
× 7.7	2.26	4.00	2.663	0.293	0.193	6.08	3.04	1.64	0.764	0.574	0.581
S3 × 7.5	2.21	3.00	2.509	0.260	0.349	2.93	1.95	1.15	0.586	0.468	0.516
× 5.7	1.67	3.00	2.330	0.260	0.170	2.52	1.68	1.23	0.455	0.390	0.522

Courtesy of The American Institute of Steel Construction.

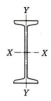

American Standard Beams: Properties for Designing (*S*-Shapes) (Converted to SI Units)

Desig-nation	Area (mm²)	Depth (mm)	Flange Width (mm)	Flange Thick-ness (mm)	Web Thick-ness (mm)	Axis X–X I (10^6 mm⁴)	Axis X–X S (10^3 mm³)	Axis X–X r (mm)	Axis Y–Y I (10^6 mm⁴)	Axis Y–Y S (10^3 mm³)	Axis Y–Y r (mm)
S610 × 180	22970	622.3	204.5	27.7	20.3	1315	4225	240	34.7	339	38.9
× 158	20130	622.3	199.9	27.7	15.7	1225	3935	247	32.1	321	39.9
× 149	18900	609.6	184.0	22.1	18.9	995	3260	229	19.9	216	32.3
× 134	17100	609.6	181.0	22.1	15.9	937	3065	234	18.7	206	33.0
× 119	15160	609.6	177.8	22.1	12.7	874	2870	241	17.6	198	34.0
S508 × 143	18190	515.6	182.9	23.4	20.3	695	2705	196	20.9	228	33.8
× 128	16320	515.6	179.3	23.4	16.8	658	2540	200	19.5	218	34.5
× 112	14190	508.0	162.2	20.2	16.1	533	2100	194	12.4	153	29.5
× 98	12520	508.0	158.9	20.2	12.8	495	1950	199	11.5	145	30.2
S457 × 104	13290	457.2	158.8	17.6	18.1	358	1690	170	10.0	127	27.4
× 81	10390	457.2	152.4	17.6	11.7	335	1465	180	8.66	114	29.0
S381 × 74	9485	381.0	143.3	15.8	14.0	202	1060	146	6.53	91.3	26.2
× 64	8130	381.0	139.7	15.8	10.4	186	977	151	5.99	85.7	27.2
S305 × 74	9485	304.8	139.1	16.7	17.4	127	832	116	6.53	94.1	26.2
× 61	7740	304.8	133.4	16.7	11.7	113	744	121	5.66	84.6	26.9
× 52	6645	304.8	129.0	13.8	10.9	95.3	626	120	4.11	63.7	24.1
× 47	6030	304.8	127.0	13.8	8.9	90.7	596	123	3.90	61.3	25.4
S254 × 52	6645	254.0	125.6	12.5	15.1	61.2	482	96.0	3.48	55.4	22.9
× 38	4815	254.0	118.4	12.5	7.9	51.6	408	103	2.83	47.7	24.2
S203 × 34	4370	203.2	105.9	10.8	11.2	27.0	265	78.7	1.79	33.9	20.3
× 27	3490	203.2	101.6	10.8	6.9	24.0	236	82.8	1.55	30.5	21.1
S178 × 30	3795	177.8	98.0	10.0	11.4	17.6	198	68.3	1.32	26.9	18.6
× 23	2905	177.8	93.0	10.0	6.4	15.3	172	72.6	1.10	23.6	19.5
S152 × 26	3270	152.4	90.6	9.1	11.8	10.9	144	57.9	0.961	21.3	17.1
× 19	2370	152.4	84.6	9.1	5.9	9.20	121	62.2	0.758	17.9	17.9
S127 × 22	2800	127.0	83.4	8.3	12.5	6.33	99.8	47.5	0.695	16.6	15.7
× 15	1895	127.0	76.3	8.3	5.4	5.12	80.6	52.1	0.508	13.3	16.3
S102 × 14	1800	101.6	71.0	7.4	8.3	2.83	55.6	39.6	0.376	10.6	14.5
× 11	1460	101.6	67.6	7.4	4.9	2.53	49.8	41.7	0.318	9.41	14.8
S76 × 11	1425	76.2	63.7	6.6	8.9	1.22	32.0	29.2	0.244	7.67	13.1
× 8.5	1075	76.2	59.2	6.6	4.3	1.05	27.5	31.2	0.189	6.39	13.3

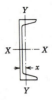

Standard Channels: Properties for Designing (*C*-Shapes) (U.S. Customary Units)

Designation	Area (in.²)	Depth (in.)	Flange Width (in.)	Flange Average Thickness (in.)	Web Thickness (in.)	Axis X–X I (in.⁴)	Axis X–X S (in.³)	Axis X–X r (in.)	Axis Y–Y I (in.⁴)	Axis Y–Y S (in.³)	Axis Y–Y r (in.)	x (in.)
*C18 × 58	17.1	18.00	4.200	0.625	0.700	676	75.1	6.29	17.8	5.32	1.02	0.862
× 51.9	15.3	18.00	4.100	0.625	0.600	627	69.7	6.41	16.4	5.07	1.04	0.858
× 45.8	13.5	18.00	4.000	0.625	0.500	578	64.3	6.56	15.1	4.82	1.06	0.866
× 42.7	12.6	18.00	3.950	0.625	0.450	554	61.6	6.64	14.4	4.69	1.07	0.877
C15 × 50	14.7	15.00	3.716	0.650	0.716	404	53.8	5.24	11.0	3.78	0.867	0.798
× 40	11.8	15.00	3.520	0.650	0.520	349	46.5	5.44	9.23	3.37	0.886	0.777
× 33.9	9.96	15.00	3.400	0.650	0.400	315	42.0	5.62	8.13	3.11	0.904	0.787
C12 × 30	8.82	12.00	3.170	0.501	0.510	162	27.0	4.29	5.14	2.06	0.763	0.674
× 25	7.35	12.00	3.047	0.501	0.387	144	24.1	4.43	4.47	1.88	0.780	0.674
× 20.7	6.09	12.00	2.942	0.501	0.282	129	21.5	4.61	3.88	1.73	0.799	0.698
C10 × 30	8.82	10.00	3.033	0.436	0.673	103	20.7	3.42	3.94	1.65	0.669	0.649
× 25	7.35	10.00	2.886	0.436	0.526	91.2	18.2	3.52	3.36	1.48	0.676	0.617
× 20	5.88	10.00	2.739	0.436	0.379	78.9	15.8	3.66	2.81	1.32	0.692	0.606
× 15.3	4.49	10.00	2.600	0.436	0.240	67.4	13.5	3.87	2.28	1.16	0.713	0.634
C9 × 20	5.88	9.00	2.648	0.413	0.448	60.9	13.5	3.22	2.42	1.17	0.642	0.583
× 15	4.41	9.00	2.485	0.413	0.285	51.0	11.3	3.40	1.93	1.01	0.661	0.586
× 13.4	3.94	9.00	2.433	0.413	0.233	47.9	10.6	3.48	1.76	0.962	0.669	0.601
C8 × 18.75	5.51	8.00	2.527	0.390	0.487	44.0	11.0	2.82	1.98	1.01	0.599	0.565
× 13.75	4.04	8.00	2.343	0.390	0.303	36.1	9.03	2.99	1.53	0.854	0.615	0.553
× 11.5	3.38	8.00	2.260	0.390	0.220	32.6	8.14	3.11	1.32	0.781	0.625	0.571
C7 × 14.75	4.33	7.00	2.299	0.366	0.419	27.2	7.78	2.51	1.38	0.779	0.564	0.532
× 12.25	3.60	7.00	2.194	0.366	0.314	24.2	6.93	2.60	1.17	0.703	0.571	0.525
× 9,8	2.87	7.00	2.090	0.366	0.210	21.3	6.08	2.72	0.968	0.625	0.581	0.540
C6 × 13	3.83	6.00	2.157	0.343	0.437	17.4	5.80	2.13	1.05	0.642	0.525	0.514
× 10.5	3.09	6.00	2.034	0.343	0.314	15.2	5.06	2.22	0.866	0.564	0.529	0.499
× 8.2	2.40	6.00	1.920	0.343	0.200	13.1	4.38	2.34	0.693	0.492	0.537	0.511
C5 × 9	2.64	5.00	1.885	0.320	0.325	8.90	3.56	1.83	0.632	0.450	0.489	0.478
× 6.7	1.97	5.00	1.750	0.320	0.190	7.49	3.00	1.95	0.479	0.378	0.493	0.484
C4 × 7.25	2.13	4.00	1.721	0.296	0.321	4.59	2.29	1.47	0.433	0.343	0.450	0.459
× 5.4	1.59	4.00	1.584	0.296	0.184	3.85	1.93	1.56	0.319	0.283	0.449	0.457
C3 × 6	1.76	3.00	1.596	0.273	0.356	2.07	1.38	1.08	0.305	0.268	0.416	0.455
× 5	1.47	3.00	1.498	0.273	0.258	1.85	1.24	1.12	0.247	0.233	0.410	0.438
× 4.1	1.21	3.00	1.410	0.273	0.170	1.66	1.10	1.17	0.197	0.202	0.404	0.436

Courtesy of The American Institute of Steel Construction.

*Not part of the American Standard Series.

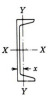

Standard Channels: Properties for Designing (*C*-Shapes) (Converted to SI Units)

Desig-nation	Area (mm²)	Depth (mm)	Flange Width (mm)	Flange Thick-ness (mm)	Thick-ness (mm)	Axis X–X I (10^6 mm⁴)	Axis X–X S (10^3 mm³)	Axis X–X r (mm)	Axis Y–Y I (10^6 mm⁴)	Axis Y–Y S (10^3 mm³)	Axis Y–Y r (mm)	x (mm)
C457 × 86	11030	457.2	106.7	15.9	17.8	281	1230	160	7.41	87.2	25.9	21.9
× 77	9870	457.2	104.1	15.9	15.2	261	1140	163	6.83	83.1	26.4	21.8
× 68	8710	457.2	101.6	15.9	12.7	241	1055	167	6.29	79.0	26.9	22.0
× 64	8130	457.2	100.3	15.9	11.4	231	1010	169	5.99	76.9	27.2	22.3
C381 × 74	9485	381.0	94.4	16.5	18.2	168	882	133	4.58	61.9	22.0	20.3
× 60	7615	381.0	89.4	16.5	13.2	145	762	138	3.84	55.2	22.5	19.7
× 50	6425	381.0	86.4	16.5	10.2	131	688	143	3.38	51.0	23.0	20.0
C305 × 45	5690	304.8	80.5	12.7	13.0	67.4	442	109	2.14	33.8	19.4	17.1
× 37	4740	304.8	77.4	12.7	9.8	59.9	395	113	1.86	30.8	19.8	17.1
× 31	3930	304.8	74.7	12.7	7.2	53.7	352	117	1.61	28.3	20.3	17.7
C254 × 45	5690	254.0	77.0	11.1	17.1	42.9	339	86.9	1.64	27.0	17.0	16.5
× 37	4740	254.0	73.3	11.1	13.4	38.0	298	89.4	1.40	24.3	17.2	15.7
× 30	3795	254.0	69.6	11.1	9.6	32.8	259	93.0	1.17	21.6	17.6	15.4
× 23	2895	254.0	66.0	11.1	6.1	28.1	221	98.3	0.949	19.0	18.1	16.1
C229 × 30	3795	228.6	67.3	10.5	11.4	25.3	221	81.8	1.01	19.2	16.3	14.8
× 22	2845	228.6	63.1	10.5	7.2	21.2	185	86.4	0.803	16.6	16.8	14.9
× 20	2540	228.6	61.8	10.5	5.9	19.9	174	88.4	0.733	15.7	17.0	15.3
C203 × 28	3555	203.2	64.2	9.9	12.4	18.3	180	71.6	0.824	16.6	15.2	14.4
× 20	2605	203.2	59.5	9.9	7.7	15.0	148	75.9	0.637	14.0	15.6	14.0
× 17	2180	203.2	57.4	9.9	5.6	13.6	133	79.0	0.549	12.8	15.9	14.5
C178 × 22	2795	177.8	58.4	9.3	10.6	11.3	127	63.8	0.574	12.8	14.3	13.5
× 18	2320	177.8	55.7	9.3	8.0	10.1	114	66.0	0.487	11.5	14.5	13.3
× 15	1850	177.8	53.1	9.3	5.3	8.87	99.6	69.1	0.403	10.2	14.8	13.7
C152 × 19	2470	152.4	54.8	8.7	11.1	7.24	95.0	54.1	0.437	10.5	13.3	13.1
× 16	1995	152.4	51.7	8.7	8.0	6.33	82.9	56.4	0.360	9.24	13.4	12.7
× 12	1550	152.4	48.8	8.7	5.1	5.45	71.8	59.4	0.288	8.06	13.6	13.0
C127 × 13	1705	127.0	47.9	8.1	8.3	3.70	58.3	46.5	0.263	7.37	12.4	12.1
× 10	1270	127.0	44.5	8.1	4.8	3.12	49.2	49.5	0.199	6.19	12.5	12.3
C102 × 11	1375	101.6	43.7	7.5	8.2	1.91	37.5	37.3	0.180	5.62	11.4	11.7
× 8	1025	101.6	40.2	7.5	4.7	1.60	31.6	39.6	0.133	4.64	11.4	11.6
C76 × 9	1135	76.2	40.5	6.9	9.0	0.862	22.6	27.4	0.127	4.39	10.6	11.6
× 7	948	76.2	38.0	6.9	6.6	0.770	20.3	28.4	0.103	3.82	10.4	11.1
× 6	781	76.2	35.8	6.9	4.6	0.691	18.0	29.7	0.082	3.31	10.3	11.1

Angles Equal Legs: Properties for Designing (*L*-Shapes) (U.S. Customary Units)

Size and Thickness (in.)	Weight (lb/ft)	Area (in.²)	Axis X–X or Y–Y				Axis Z–Z
			I (in.⁴)	S (in.³)	r (in.)	x or y (in.)	r (in.)
L8 × 8 × 1	51.0	15.0	89.0	15.8	2.44	2.37	1.56
× 7/8	45.0	13.2	79.6	14.0	2.45	2.32	1.57
× 3/4	38.9	11.4	69.7	12.2	2.47	2.28	1.58
× 5/8	32.7	9.61	59.4	10.3	2.49	2.23	1.58
× 1/2	26.4	7.75	48.6	8.36	2.50	2.19	1.59
L6 × 6 × 1	37.4	11.0	35.5	8.57	1.80	1.86	1.17
× 7/8	33.1	9.73	31.9	7.63	1.81	1.82	1.17
× 3/4	28.7	8.44	28.2	6.66	1.83	1.78	1.17
× 5/8	24.2	7.11	24.2	5.66	1.84	1.73	1.18
× 1/2	19.6	5.75	19.9	4.61	1.86	1.68	1.18
× 3/8	14.9	4.36	15.4	3.53	1.88	1.64	1.19
L5 × 5 × 7/8	27.2	7.98	17.8	5.17	1.49	1.57	0.973
× 3/4	23.6	6.94	15.7	4.53	1.51	1.52	0.975
× 5/8	20.0	5.86	13.6	3.86	1.52	1.48	0.978
× 1/2	16.2	4.75	11.3	3.16	1.54	1.43	0.983
× 3/8	12.3	3.61	8.74	2.42	1.56	1.39	0.990
L4 × 4 × 3/4	18.5	5.44	7.67	2.81	1.19	1.27	0.778
× 5/8	15.7	4.61	6.66	2.40	1.20	1.23	0.779
× 1/2	12.8	3.75	5.56	1.97	1.22	1.18	0.782
× 3/8	9.8	2.86	4.36	1.52	1.23	1.14	0.788
× 1/4	6.6	1.94	3.04	1.05	1.25	1.09	0.795
L3½ × 3½ × 1/2	11.1	3.25	3.64	1.49	1.06	1.06	0.683
× 3/8	8.5	2.48	2.87	1.15	1.07	1.01	0.687
× 1/4	5.8	1.69	2.01	0.794	1.09	0.968	0.694
L3 × 3 × 1/2	9.4	2.75	2.22	1.07	0.898	0.932	0.584
× 3/8	7.2	2.11	1.76	0.833	0.913	0.888	0.587
× 1/4	4.9	1.44	1.24	0.577	0.930	0.842	0.592
L2½ × 2½ × 1/2	7.7	2.25	1.23	0.724	0.739	0.806	0.487
× 3/8	5.9	1.73	0.984	0.566	0.753	0.762	0.487
× 1/4	4.1	1.19	0.703	0.394	0.769	0.717	0.491
L2 × 2 × 3/8	4.7	1.36	0.479	0.351	0.594	0.636	0.389
× 1/4	3.19	0.938	0.348	0.247	0.609	0.592	0.391
× 1/8	1.65	0.484	0.190	0.131	0.626	0.546	0.398

Courtesy of The American Institute of Steel Construction.

Angles Equal Legs: Properties for Designing (*L*-Shapes) (Converted to SI Units)

Size and Thickness (mm)	Mass (kg/m)	Area (mm²)	Axis X–X or Y–Y				Axis Z–Z
			I (10^6 mm⁴)	S (10^3 mm³)	r (mm)	x or y (mm)	r (mm)
L203 × 203 × 25.4	75.9	9675	37.0	259	62.0	60.2	39.6
× 22.2	67.0	8515	33.1	229	62.2	58.9	39.9
× 19.1	57.9	7355	29.0	200	62.7	57.9	40.1
× 15.9	48.7	6200	24.7	169	63.2	56.6	40.1
× 12.7	39.3	5000	20.2	137	63.5	55.6	40.4
L152 × 152 × 25.4	55.7	7095	14.8	140	45.7	47.2	29.7
× 22.2	49.3	6275	13.3	125	46.0	46.2	29.7
× 19.1	42.7	5445	11.7	109	46.5	45.2	29.7
× 15.9	36.0	4585	10.1	92.8	46.7	43.9	30.0
× 12.7	29.2	3710	8.28	75.5	47.2	42.7	30.0
× 9.5	22.2	2815	6.61	57.8	47.8	41.7	30.2
L127 × 127 × 22.2	40.5	5150	7.41	84.7	37.8	39.9	24.7
× 19.1	35.1	4475	6.53	74.2	38.4	38.6	24.8
× 15.9	29.8	3780	5.66	63.3	38.6	37.6	24.8
× 12.7	24.1	3065	4.70	51.8	39.1	36.3	25.0
× 9.5	18.3	2330	3.64	39.7	39.6	35.3	25.1
L102 × 102 × 19.1	27.5	3510	3.19	46.0	30.2	32.3	19.8
× 15.9	23.4	2975	2.77	39.3	30.5	31.2	19.8
× 12.7	19.0	2420	2.31	32.3	31.0	30.0	19.9
× 9.5	14.6	1845	1.81	24.9	31.2	29.0	20.0
× 6.4	9.8	1250	1.27	17.2	31.8	27.7	20.2
L89 × 89 × 12.7	16.5	2095	1.52	24.4	26.9	26.9	17.3
× 9.5	12.6	1600	1.19	18.8	27.2	25.7	17.4
× 6.4	8.6	1090	0.837	13.0	27.7	24.6	17.6
L76 × 76 × 12.7	14.0	1775	0.924	17.5	22.8	23.7	14.8
× 9.5	10.7	1360	0.732	13.7	23.2	22.6	14.9
× 6.4	7.3	929	0.516	9.46	23.6	21.4	15.0
L64 × 64 × 12.7	11.5	1450	0.512	11.9	18.8	20.5	12.4
× 9.5	8.8	1115	0.410	9.28	19.1	19.4	12.4
× 6.4	6.1	768	0.293	6.46	19.5	18.2	12.5
L51 × 51 × 9.5	7.0	877	0.199	5.75	15.1	16.2	9.88
× 6.4	4.75	605	0.145	4.05	15.5	15.0	9.93
× 3.2	2.46	312	0.079	2.15	15.9	13.9	10.1

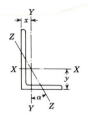

L

Angles Unequal Legs: Properties for Designing (L-Shapes) (U.S. Customary Units)

Size and Thickness	Weight (lb/ft)	Area (in.²)	Axis X–X				Axis Y–Y				Axis Z–Z	
			I (in.⁴)	S (in.³)	r (in.)	y (in.)	I (in.⁴)	S (in.³)	r (in.)	x (in.)	r (in.)	Tan α
L9 × 4 × 5/8	26.3	7.73	64.9	11.5	2.90	3.36	8.32	2.65	1.04	0.858	0.847	0.216
1/2	21.3	6.25	53.2	9.34	2.92	3.31	6.92	2.17	1.05	0.810	0.854	0.220
L8 × 6 × 1	44.2	13.0	80.8	15.1	2.49	2.65	38.8	8.92	1.73	1.65	1.28	0.543
× 3/4	33.8	9.94	63.4	11.7	2.53	2.56	30.7	6.92	1.76	1.56	1.29	0.551
× 1/2	23.0	6.75	44.3	8.02	2.56	2.47	21.7	4.79	1.79	1.47	1.30	0.558
L8 × 4 × 1	37.4	11.0	69.6	14.1	2.52	3.05	11.6	3.94	1.03	1.05	0.846	0.247
× 3/4	28.7	8.44	54.9	10.9	2.55	2.95	9.36	3.07	1.05	0.953	0.852	0.258
× 1/2	19.6	5.75	38.5	7.49	2.59	2.86	6.74	2.15	1.08	0.859	0.865	0.267
L7 × 4 × 3/4	26.2	7.69	37.8	8.42	2.22	2.51	9.05	3.03	1.09	1.01	0.860	0.324
× 1/2	17.9	5.25	26.7	5.81	2.25	2.42	6.53	2.12	1.11	0.917	0.872	0.335
× 3/8	13.6	3.98	20.6	4.44	2.27	2.37	5.10	1.63	1.13	0.870	0.880	0.340
L6 × 4 × 3/4	23.6	6.94	24.5	6.25	1.88	2.08	8.68	2.97	1.12	1.08	0.860	0.428
× 1/2	16.2	4.75	17.4	4.33	1.91	1.99	6.27	2.08	1.15	0.987	0.870	0.440
× 3/8	12.3	3.61	13.5	3.32	1.93	1.94	4.90	1.60	1.17	0.941	0.877	0.446
L6 × 3½ × 1/2	15.3	4.50	16.6	4.24	1.92	2.08	4.25	1.59	0.972	0.833	0.759	0.344
× 3/8	11.7	3.42	12.9	3.24	1.94	2.04	3.34	1.23	0.988	0.787	0.767	0.350
L5 × 3½ × 3/4	19.8	5.81	13.9	4.28	1.55	1.75	5.55	2.22	0.977	0.996	0.748	0.464
× 1/2	13.6	4.00	9.99	2.99	1.58	1.66	4.05	1.56	1.01	0.906	0.755	0.479
× 3/8	10.4	3.05	7.78	2.29	1.60	1.61	3.18	1.21	1.02	0.861	0.762	0.486
× 1/4	7.0	2.06	5.39	1.57	1.62	1.56	2.23	0.830	1.04	0.814	0.770	0.492
L5 × 3 × 1/2	12.8	3.75	9.45	2.91	1.59	1.75	2.58	1.15	0.829	0.750	0.648	0.357
× 3/8	9.8	2.86	7.37	2.24	1.61	1.70	2.04	0.888	0.845	0.704	0.654	0.364
× 1/4	6.6	1.94	5.11	1.53	1.62	1.66	1.44	0.614	0.861	0.657	0.663	0.371
L4 × 3½ × 1/2	11.9	3.50	5.32	1.94	1.23	1.25	3.79	1.52	1.04	1.00	0.722	0.750
× 3/8	9.1	2.67	4.18	1.49	1.25	1.21	2.95	1.17	1.06	0.955	0.727	0.755
× 1/4	6.2	1.81	2.91	1.03	1.27	1.16	2.09	0.808	1.07	0.909	0.734	0.759
L4 × 3 × 1/2	11.1	3.25	5.05	1.89	1.25	1.33	2.42	1.12	0.864	0.827	0.639	0.543
× 3/8	8.5	2.48	3.96	1.46	1.26	1.28	1.92	0.866	0.879	0.782	0.644	0.551
× 1/4	5.8	1.69	2.77	1.00	1.28	1.24	1.36	0.599	0.896	0.736	0.651	0.558
L3½ × 3 × 1/2	10.2	3.00	3.45	1.45	1.07	1.13	2.33	1.10	0.881	0.875	0.621	0.714
× 3/8	7.9	2.30	2.72	1.13	1.09	1.08	1.85	0.851	0.897	0.830	0.625	0.721
× 1/4	5.4	1.56	1.91	0.776	1.11	1.04	1.30	0.589	0.914	0.785	0.631	0.727
L3½ × 2½ × 1/2	9.4	2.75	3.24	1.41	1.09	1.20	1.36	0.760	0.704	0.705	0.534	0.486
× 3/8	7.2	2.11	2.56	1.09	1.10	1.16	1.09	0.592	0.719	0.660	0.537	0.496
× 1/4	4.9	1.44	1.80	0.755	1.12	1.11	0.777	0.412	0.735	0.614	0.544	0.506
L3 × 2½ × 1/2	8.5	2.50	2.08	1.04	0.913	1.00	1.30	0.744	0.722	0.750	0.520	0.667
× 3/8	6.6	1.92	1.66	0.810	0.928	0.956	1.04	0.581	0.736	0.706	0.522	0.676
× 1/4	4.5	1.31	1.17	0.561	0.945	0.911	0.743	0.404	0.753	0.661	0.528	0.684
L3 × 2 × 1/2	7.7	2.25	1.92	1.00	0.924	1.08	0.672	0.474	0.546	0.583	0.428	0.414
× 3/8	5.9	1.73	1.53	0.781	0.940	1.04	0.543	0.371	0.559	0.539	0.430	0.428
× 1/4	4.1	1.19	1.09	0.542	0.957	0.993	0.392	0.260	0.574	0.493	0.435	0.440
L2½ × 2 × 3/8	5.3	1.55	0.912	0.547	0.768	0.813	0.514	0.363	0.577	0.581	0.420	0.614
× 1/4	3.62	1.06	0.654	0.381	0.784	0.787	0.372	0.254	0.592	0.537	0.424	0.626

L

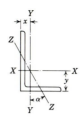

Angles Unequal Legs: Properties for Designing (L-Shapes) (Converted to SI Units)

Size and Thickness	Mass (kg/m)	Area (mm²)	Axis X–X				Axis Y–Y				Axis Z–Z	
			I (10⁶ mm⁴)	S (10³ mm³)	r (mm)	y (mm)	I (10⁶ mm⁴)	S (10³ mm³)	r (mm)	x (mm)	r (mm)	Tan α
L229 × 102 × 15.9	39.1	4985	27.0	188	73.7	85.3	3.46	43.4	26.4	21.8	21.5	0.216
× 12.7	31.7	4030	22.1	153	74.2	84.1	2.88	35.6	26.7	20.6	21.7	0.220
L203 × 152 × 25.4	65.8	8385	33.6	247	63.2	67.3	16.1	146	43.9	41.9	32.5	0.543
× 19.1	50.3	6415	26.4	192	64.3	65.0	12.8	113	44.7	39.6	32.8	0.551
× 12.7	34.2	4355	18.4	131	65.0	62.7	9.03	78.5	45.5	37.3	33.0	0.558
L203 × 102 × 25.4	55.7	7095	29.0	231	64.0	77.5	4.83	64.6	26.2	26.7	21.5	0.247
× 19.1	42.7	5445	22.9	179	64.8	74.9	3.90	50.3	26.7	24.2	21.6	0.258
× 12.7	29.2	3710	16.0	123	65.8	72.6	2.81	35.2	27.4	21.8	22.0	0.267
L178 × 102 × 19.1	39.0	4960	15.7	138	56.4	63.8	3.77	49.7	27.7	25.7	21.8	0.324
× 12.7	26.6	3385	11.1	95.2	57.2	61.5	2.72	34.7	28.2	23.3	22.1	0.335
× 9.5	20.2	2570	8.57	72.8	57.7	60.2	2.12	26.7	28.7	22.1	22.4	0.340
L152 × 102 × 19.1	35.1	4475	10.2	102	47.8	52.8	3.61	48.7	28.4	27.4	21.8	0.428
× 12.7	24.1	3065	7.24	71.0	48.5	50.5	2.61	34.1	29.2	25.1	22.1	0.440
× 9.5	18.3	3230	5.62	54.4	49.0	49.3	2.04	26.2	29.7	23.9	22.3	0.446
L152 × 89 × 12.7	22.8	2905	6.91	69.5	48.8	52.8	1.77	26.1	24.7	21.2	19.3	0.344
× 9.5	17.4	2205	5.37	53.1	49.3	51.8	1.39	20.2	25.1	20.0	19.5	0.350
L127 × 89 × 19.1	29.5	3750	5.79	70.1	39.4	44.5	2.31	36.4	24.8	25.3	19.0	0.464
× 12.7	20.2	2580	4.16	49.0	40.1	42.2	1.69	25.6	25.7	23.0	19.2	0.479
× 9.5	15.5	1970	3.24	37.5	40.6	40.9	1.32	19.8	25.9	21.9	19.4	0.486
× 6.4	10.4	1330	2.24	25.7	41.1	39.6	0.928	13.6	26.4	20.7	19.6	0.492
L127 × 76 × 12.7	19.0	2420	3.93	47.7	40.4	44.5	1.07	18.8	21.1	19.1	16.5	0.357
× 9.5	14.6	1845	3.07	36.7	40.9	43.2	0.849	14.6	21.5	17.9	16.6	0.364
× 6.4	9.82	1250	2.13	25.1	41.1	42.2	0.599	10.1	21.9	16.7	16.8	0.371
L102 × 89 × 12.7	17.7	2260	2.21	31.8	31.2	31.8	1.58	24.9	26.4	25.4	18.3	0.750
× 9.5	13.5	1725	1.74	24.4	31.8	30.7	1.23	19.2	26.9	24.3	18.5	0.755
× 6.4	9.22	1170	1.21	16.9	32.3	29.5	0.870	13.2	27.2	23.1	18.6	0.759
L102 × 76 × 12.7	16.5	2095	2.10	31.0	31.8	33.8	1.01	18.4	21.9	21.0	16.2	0.543
× 9.5	12.6	1600	1.65	23.9	32.0	32.5	0.799	14.2	22.3	19.9	16.4	0.551
× 6.4	8.63	1090	1.15	16.4	32.5	31.5	0.566	9.82	22.8	18.7	16.5	0.558
L89 × 76 × 12.7	15.2	1935	1.44	23.8	27.2	28.7	0.970	18.0	22.4	22.2	15.8	0.714
× 9.5	11.8	1485	1.13	18.5	27.7	27.4	0.770	13.9	22.8	21.1	15.9	0.721
× 6.4	8.04	1005	0.795	12.7	28.2	26.4	0.541	9.65	23.2	19.9	16.0	0.727
L89 × 64 × 12.7	14.0	1775	1.35	23.1	27.7	30.5	0.566	12.5	17.9	17.9	13.6	0.486
× 9.5	10.7	1360	1.07	17.9	27.9	29.5	0.454	8.70	18.3	16.8	13.6	0.496
× 6.4	7.29	929	0.749	12.4	28.4	28.2	0.323	6.75	18.7	15.6	13.8	0.506
L76 × 64 × 12.7	12.6	1615	0.866	17.0	23.2	25.4	0.541	12.2	18.3	19.1	13.2	0.667
× 9.5	9.82	1240	0.691	13.3	23.6	24.3	0.433	9.52	18.7	17.9	13.3	0.676
× 6.4	6.70	845	0.487	9.19	24.0	23.1	0.309	6.62	19.1	16.8	13.4	0.684
L76 × 51 × 12.7	11.5	1450	0.799	16.4	23.5	27.4	0.280	7.77	13.9	14.8	10.9	0.414
× 9.5	8.78	1115	0.637	12.8	23.9	26.4	0.226	6.08	14.2	13.7	10.9	0.428
× 6.4	6.10	768	0.454	8.88	24.3	25.2	0.163	4.26	14.6	12.5	11.0	0.440
L64 × 51 × 9.5	7.89	1000	0.380	8.96	19.5	20.7	0.214	5.95	14.7	14.8	10.7	0.614
× 6.4	5.39	684	0.272	6.24	19.9	20.0	0.155	4.16	15.0	13.6	10.8	0.626

Structural Tees Cut from W Shapes:
Properties for Designing
(*WT*-Shapes)
(U.S. Customary Units)

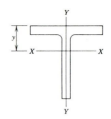

Structural Tees Cut from W Shapes: Properties for Designing (*WT*-Shapes) (U.S. Customary Units)

Desig-nation	Area (in.²)	Depth of Tee (in.)	Flange Width (in.)	Flange Thick-ness (in.)	Stem Thick-ness (in.)	Axis X–X I (in.⁴)	S (in.³)	r (in.)	y (in.)	Axis Y–Y I (in.⁴)	S (in.³)	r (in.)
WT18 × 115	33.8	17.950	16.470	1.260	0.760	934	67.0	5.25	4.01	470	57.1	3.73
× 80	23.5	18.005	12.000	1.020	0.650	740	55.8	5.61	4.74	147	24.6	2.50
WT15 × 66	19.4	15.155	10.545	1.000	0.615	421	37.4	4.66	3.90	98.0	18.6	2.25
× 54	15.9	14.915	10.475	0.760	0.545	349	32.0	4.69	4.01	73.0	13.9	2.15
WT12 × 52	15.3	12.030	12.750	0.750	0.500	189	20.0	3.51	2.59	130	20.3	2.91
× 47	13.8	12.155	9.065	0.875	0.515	186	20.3	3.67	2.99	54.5	12.0	1.98
× 42	12.4	12.050	9.020	0.770	0.470	166	18.3	3.67	2.97	47.2	10.5	1.95
× 31	9.11	11.870	7.040	0.590	0.430	131	15.6	3.79	3.46	17.2	4.90	1.38
WT9 × 38	11.2	9.105	11.035	0.680	0.425	71.8	9.83	2.54	1.80	76.2	13.8	2.61
× 30	8.82	9.120	7.555	0.695	0.415	64.7	9.29	2.71	2.16	25.0	6.63	1.69
× 25	7.33	8.995	7.495	0.570	0.355	53.5	7.79	2.70	2.12	20.0	5.35	1.65
× 20	5.88	8.950	6.015	0.525	0.315	44.8	6.73	2.76	2.29	9.55	3.17	1.27
WT8 × 50	14.7	8.485	10.425	0.985	0.585	76.8	11.4	2.28	1.76	93.1	17.9	2.51
× 25	7.37	8.130	7.070	0.630	0.380	42.3	6.78	2.40	1.89	18.6	5.26	1.59
× 20	5.89	8.005	6.995	0.505	0.305	33.1	5.35	2.37	1.81	14.4	4.12	1.57
× 13	3.84	7.845	5.500	0.345	0.250	23.5	4.09	2.47	2.09	4.80	1.74	1.12
WT7 × 60	17.7	7.240	14.670	0.940	0.590	51.7	8.61	1.71	1.24	247	33.7	3.74
× 41	12.0	7.155	10.130	0.855	0.510	41.2	7.14	1.85	1.39	74.2	14.6	2.48
× 34	9.99	7.020	10.035	0.720	0.415	32.6	5.69	1.81	1.29	60.7	12.1	2.46
× 24	7.07	6.985	8.030	0.595	0.340	24.9	4.48	1.87	1.35	25.7	6.40	1.91
× 15	4.42	6.920	6.730	0.385	0.270	19.0	3.55	2.07	1.58	9.79	2.91	1.49
× 11	3.25	6.870	5.000	0.335	0.230	14.8	2.91	2.14	1.76	3.50	1.40	1.04
WT6 × 60	17.6	6.560	12.320	1.105	0.710	43.4	8.22	1.57	1.28	172	28.0	3.13
× 48	14.1	6.355	12.160	0.900	0.550	32.0	6.12	1.51	1.13	135	22.2	3.09
× 36	10.6	6.125	12.040	0.670	0.430	23.2	4.54	1.48	1.02	97.5	16.2	3.04
× 25	7.34	6.095	8.080	0.640	0.370	18.7	3.79	1.60	1.17	28.2	6.97	1.96
× 15	4.40	6.170	6.520	0.440	0.260	13.5	2.75	1.75	1.27	10.2	3.12	1.52
× 8	2.36	5.995	3.990	0.265	0.220	8.70	2.04	1.92	1.74	1.41	0.706	0.773
WT5 × 56	16.5	5.680	10.415	1.250	0.755	28.6	6.40	1.32	1.21	118	22.6	2.68
× 44	12.9	5.420	10.265	0.990	0.605	20.8	4.77	1.27	1.06	89.3	17.4	2.63
× 30	8.82	5.110	10.080	0.680	0.420	12.9	3.04	1.21	0.884	58.1	11.5	2.57
× 15	4.42	5.235	5.810	0.510	0.300	9.28	2.24	1.45	1.10	8.35	2.87	1.37
× 6	1.77	4.935	3.960	0.210	0.190	4.35	1.22	1.57	1.36	1.09	0.551	0.785
WT4 × 29	8.55	4.375	8.220	0.810	0.510	9.12	2.61	1.03	0.874	37.5	9.13	2.10
× 20	5.87	4.125	8.070	0.560	0.360	5.73	1.69	0.988	0.735	24.5	6.08	2.04
× 12	3.54	3.965	6.495	0.400	0.245	3.53	1.08	0.999	0.695	9.14	2.81	1.61
× 9	2.63	4.070	5.250	0.330	0.230	3.41	1.05	1.14	0.834	3.98	1.52	1.23
× 5	1.48	3.945	3.940	0.205	0.170	2.15	0.717	1.20	0.953	1.05	0.532	0.841
WT3 × 10	2.94	3.100	6.020	0.365	0.260	1.76	0.693	0.774	0.560	6.64	2.21	1.50
× 6	1.78	3.015	4.000	0.280	0.230	1.32	0.564	0.861	0.677	1.50	0.748	0.918
WT2 × 6.5	1.91	2.080	4.060	0.345	0.280	0.526	0.321	0.524	0.440	1.93	0.950	1.00

Courtesy of The American Institute of Steel Construction.

Structural Tees Cut from W Shapes:
Properties for Designing
(*WT*-Shapes)
(Converted to SI Units)

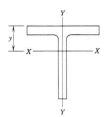

Structural Tees Cut from W Shapes: Properties for Designing (*WT*-Shapes) (Converted to SI Units)

Desig- nation	Area (mm²)	Depth of Tee (mm)	Flange Width (mm)	Flange Thick- ness (mm)	Stem Thick- ness (mm)	Axis X–X I (10^6 mm⁴)	Axis X–X S (10^3 mm³)	Axis X–X r (mm)	Axis X–X y (mm)	Axis Y–Y I (10^6 mm⁴)	Axis Y–Y S (10^3 mm³)	Axis Y–Y r (mm)
WT457 × 171	21805	455.9	418.3	32.0	19.3	389	1098	133	102	196	936	94.7
× 119	15160	457.3	304.8	25.9	16.5	308	914	142	120	61.2	403	63.5
WT381 × 98	12515	384.9	267.8	25.4	15.6	175	613	118	99.1	40.8	305	57.2
× 80	10260	378.8	266.1	19.3	13.8	145	524	119	102	30.4	228	54.6
WT305 × 77	9870	305.6	323.9	19.1	12.7	78.7	328	89.2	65.8	54.1	333	73.9
× 70	8905	308.7	230.3	22.2	13.1	77.4	333	93.2	75.9	22.7	197	50.3
× 63	8000	306.1	229.1	19.6	11.9	69.1	300	93.2	75.4	19.6	172	49.5
× 46	5875	301.5	178.8	15.0	10.9	54.5	256	96.3	87.9	7.16	80.3	35.1
WT229 × 57	7225	231.3	280.3	17.3	10.8	29.9	161	64.5	45.7	31.7	226	66.3
× 45	5690	231.6	191.9	17.7	10.5	26.9	152	68.8	54.9	10.4	109	42.9
× 37	4730	228.5	190.4	14.5	9.0	22.3	128	68.6	53.8	8.32	87.7	41.9
× 30	3795	227.3	152.8	13.3	8.0	18.6	110	70.1	58.2	3.98	51.9	32.3
WT203 × 74	9485	215.5	264.8	25.0	14.9	32.0	187	57.9	44.7	38.8	293	63.8
× 37	4755	206.5	179.6	16.0	9.7	17.6	111	61.0	48.0	7.74	86.2	40.4
× 30	3800	203.3	177.7	12.8	7.7	13.8	87.7	60.2	46.0	5.99	67.5	39.9
× 19	2475	199.3	139.7	8.8	6.4	9.78	67.0	62.7	53.1	2.00	28.5	28.4
WT178 × 89	11420	183.9	372.6	23.9	15.0	21.5	141	43.4	31.5	103	552	95.0
× 61	7740	181.7	257.3	21.7	13.0	17.1	117	47.0	35.3	30.9	239	63.0
× 51	6445	178.3	254.9	18.3	10.5	13.6	93.2	46.0	32.8	25.3	198	62.5
× 36	4560	177.4	204.0	15.1	8.6	10.4	73.4	47.5	34.3	10.7	105	48.5
× 22	2850	175.8	170.9	9.8	6.9	7.91	58.2	52.6	40.1	4.07	47.7	37.8
× 16	2095	174.5	127.0	8.5	5.8	6.16	47.7	54.4	44.7	1.46	22.9	26.4
WT152 × 89	11355	166.6	312.9	28.1	18.0	18.1	135	39.9	32.5	71.6	459	79.5
× 71	9095	161.4	308.9	22.9	14.0	13.3	100	38.4	28.7	56.2	364	78.5
× 54	6840	155.6	305.8	17.0	10.9	9.66	74.4	37.6	25.9	40.6	265	77.2
× 37	4735	154.8	205.2	16.2	9.4	7.78	62.1	40.6	29.7	11.7	114	49.8
× 22	2840	156.7	165.6	11.2	6.6	5.62	45.1	44.5	32.3	4.25	51.1	38.6
× 12	1525	152.3	101.3	6.7	5.6	3.62	33.4	48.8	44.2	0.587	11.6	19.6
WT127 × 83	10645	144.3	264.5	31.8	19.2	11.9	105	33.5	30.7	49.1	370	68.1
× 65	8325	137.7	260.7	25.1	15.4	8.66	78.2	32.3	26.9	37.2	285	66.8
× 45	5690	129.8	256.0	17.3	10.7	5.37	49.8	30.7	22.5	24.2	188	65.3
× 22	2850	133.0	147.6	13.0	7.6	3.86	36.7	36.8	27.9	3.48	47.0	34.8
× 9	1140	125.3	100.6	5.3	4.8	1.81	20.0	39.9	34.5	0.454	9.03	19.9
WT102 × 43	5515	111.1	208.8	20.6	13.0	3.80	42.8	26.2	22.2	15.6	150	53.3
× 30	3785	104.8	205.0	14.2	9.1	2.39	27.7	25.1	18.7	10.2	99.6	51.8
× 18	2285	100.7	165.0	10.2	6.2	1.47	17.7	25.4	17.7	3.80	46.0	40.9
× 13	1695	103.4	133.4	8.4	5.8	1.42	17.2	29.0	21.2	1.66	24.9	31.2
× 7	955	100.2	100.1	5.2	4.3	0.895	11.7	30.5	24.2	0.437	8.72	21.4
WT76 × 15	1895	78.7	152.9	9.3	6.6	0.733	11.4	19.7	14.2	2.76	3.62	38.1
× 9	1150	76.6	101.6	7.1	5.8	0.549	9.24	21.9	17.2	0.624	12.3	23.3
WT51 × 10	1230	52.8	103.1	8.8	7.1	0.219	5.26	13.3	11.2	0.803	15.6	25.4

APPENDIX D

TABLE OF BEAM DEFLECTIONS AND SLOPES

The differential equation of the deflection curve for a beam is a linear differential equation; therefore, solutions of the equation for different loading conditions can be superimposed. The method of superposition for beam deflections consists of finding the resultant effect of several loads acting simultaneously on a member as the sum of the contributions from each of the loads applied individually. The deflections for the separate loads can be determined by using any of the methods described in Chapter 5 of the text or by using the table of beam deflections published in any of the structural engineering handbooks. The equations of the elastic curves presented in Table D-1 and the deflections and slopes presented in Table D-2 are representative of the results found in any of the handbooks. The particular cases presented are provided solely for the purpose of providing a means for solving a selection of beam-deflection problems by using the superposition method.

TABLE D-1 Equations of the Elastic Curve for Selected Beams

Case	Load and Support (Length L)	Equation of Elastic Curve
1		$y = \dfrac{P}{6EI}(x^3 - 3Lx^2)$
2		$y = \dfrac{w}{24EI}(-x^4 + '4Lx^3 - 6L^2x^2)$
3		$y = \dfrac{w}{120EIL}(x^5 - 5Lx^4 + 10L^2x^3 - 10L^3x^2)$
4		$y = -\dfrac{M}{2EI}x^2$
5		$y = \dfrac{Pb}{6EIL}(x^3 + b^2x - L^2x)$
6		$y = \dfrac{P}{48EI}(4x^3 - 3L^2x)$
7		$y = \dfrac{w}{24EI}(-x^4 + 2Lx^3 - L^3x)$
8		$y = \dfrac{M}{6EIL}(x^3 - L^2x)$

TABLE D-2 Beam Deflections and Slopes

Case	Load and Support (Length L)	Slope at End ($+ \measuredangle \theta$)	Maximum Deflection (+ upward)
1		$\theta = -\dfrac{PL^2}{2EI}$ at $x = L$	$y_{max} = -\dfrac{PL^3}{3EI}$ at $x = L$
2		$\theta = -\dfrac{wL^3}{6EI}$ at $x = L$	$y_{max} = -\dfrac{wL^4}{8EI}$ at $x = L$
3		$\theta = -\dfrac{wL^3}{24EI}$ at $x = L$	$y_{max} = \dfrac{wL^4}{30EI}$ at $x = L$
4		$\theta = +\dfrac{ML}{EI}$ at $x = L$	$y_{max} = +\dfrac{ML^2}{2EI}$ at $x = L$
5		$\theta_1 = -\dfrac{Pb(L^2 - b^2)}{6LEI}$ at $x = 0$ $\theta_2 = +\dfrac{Pa(L^2 - a^2)}{6LEI}$ at $x = L$	$y_{max} = -\dfrac{Pb(L^2 - b^2)^{3/2}}{9\sqrt{3}LEI}$ at $x = \sqrt{(L^2 - b^2)/3}$ $y_{\substack{center \\ not\ max}} = -\dfrac{Pb(3L^2 - 4b^2)}{48EI}$
6		$\theta_1 = -\dfrac{PL^2}{16EI}$ at $x = 0$ $\theta_2 = +\dfrac{PL^2}{16EI}$ at $x = L$	$y_{max} = -\dfrac{PL^3}{48EI}$ at $x = L/2$
7		$\theta_1 = -\dfrac{wL^3}{24EI}$ at $x = 0$ $\theta_2 = +\dfrac{wL^3}{24EI}$ at $x = L$	$y_{max} = -\dfrac{5wL^4}{384EI}$ at $x = L/2$
8		$\theta_1 = -\dfrac{ML}{6EI}$ at $x = 0$ $\theta_2 = +\dfrac{ML}{3EI}$ at $x = L$	$y_{max} = -\dfrac{ML^2}{9\sqrt{3}\,EI}$ at $x = L/\sqrt{3}$ $y_{\substack{center \\ not\ max}} = -\dfrac{ML^2}{16EI}$

THE INTERNATIONAL SYSTEM OF UNITS (SI)

E-1

INTRODUCTION

The importance of the regulation of weights and measures was recognized as early as 1787 when Article 1, Section 8, of the United States Constitution was written. The metric system was legalized in the United States in 1866; in 1893, the international meter and kilogram became the fundamental standards of length and mass both for metric and customary weights and measures. International standardization began with an International Metric Convention in 1875, which established a permanent International Bureau of Weights and Measures. The National Bureau of Standards represents the United States in this international body.

The original metric system provided a set of units for the measurement of length, area, volume, capacity, and mass based on two fundamental units: the meter and the kilogram. With the addition of a unit of time, practical measurements began to be based on the meter-kilogram-second (MKS) system. In 1960, The Eleventh General Conference on Weights and Measures formally adopted the *International System of Units,* for which the abbreviation is *SI* in all languages, as the international standard. Thirty-six countries, including the United States, participated in this conference.

E-2

SI UNITS

The Internationl System of Units adopted by the conference includes three classes of units: (1) base units, (2) supplementary units, and (3) derived units. The system is founded on the seven base units listed in Table E-1.

TABLE E-1

Quantity	Name of Base SI Unit	Symbol
Length	meter	m
Mass	kilogram	kg
Time	second	s
Electric current	ampere	A
Thermodynamic temperature	kelvin	K
Amount of substance	mole	mol
Luminous intensity	candela	cd

TABLE E-2

Quantity	Name of Supplementary SI Unit	Symbol
Plane angle	radian	rad
Solid angle	steradian	sr

TABLE E-3

Quantity	Derived SI Unit	Special Name	Symbol
Area	square meter	—	m^2
Volume	cubic meter	—	m^3
Linear velocity	meter per second	—	m/s
Angular velocity	radian per second	—	rad/s
Linear acceleration	meter per second squared	—	m/s^2
Frequency	(cycle) per second	hertz	Hz
Density	kilogram per cubic meter	—	kg/m^3
Force	kilogram·meter per second squared	newton	N
Moment of force	newton·meter	—	N·m
Pressure	newton per meter squared	pascal	Pa
Stress	newton per meter squared	pascal	Pa or N/m^2
Work, energy	newton·meter	joule	J
Power	joule per second	watt	W

Certain units of the International System have not been classified under either base units or derived units. These units, listed in Table E-2, are called supplementary units and may be regarded either as base units or as derived units.

Derived units are expressed algebraically in terms of base units and/or supplementary units. Their symbols are obtained by means of the mathematical signs of multiplication and division. For example, the SI unit for velocity is meter per second (m/s) and the SI unit for angular velocity is radian per second (rad/s). For some of the derived units, special names and symbols exist; those of interest in mechanics are listed in Table E-3.

E-3

MULTIPLES OF SI UNITS

Prefixes are used to form names and symbols of multiples (decimal multiples and submultiples) of SI units. The choice of the appropriate multiple is governed by convenience and should usually be chosen so that the numerical values will be between 0.1 and 1000. Only one prefix should be used in forming a multiple of a compound SI unit, and prefixes in the denominator should be avoided. Approved prefixes with their names and symbols are listed in Table E-4.

TABLE E-4

Factor by Which Unit is Multiplied	Prefix	
	Name	Symbol
10^{12}	tera	T
10^{9}	giga	G
10^{6}	mega	M
10^{3}	kilo	k
10^{2}	hecto[a]	h
10	deca[a]	da
10^{-1}	deci[a]	d
10^{-2}	centi[a]	c
10^{-3}	milli	m
10^{-6}	micro	μ
10^{-9}	nano	n
10^{-12}	pico	p
10^{-15}	femto	f
10^{-18}	atto	a

[a]To be avoided when possible.

E-4

CONVERSION BETWEEN THE SI AND U.S. CUSTOMARY SYSTEMS

As the use of SI becomes more commonplace in the United States, engineers will be required to be familiar with both SI and the U.S. Customary system in common use today. As an aid to interpreting the physical significance of answers in SI units for those more accustomed to the U.S. Customary system, the following conversion factors are provided:

TABLE E-5 Conversion Factors

Quantity	U.S. Customary to SI	SI to U.S. Customary
Length	1 in. = 25.40 mm	1 m = 39.37 in.
	1 ft = 0.3048 m	1 m = 3.281 ft
Area	1 in.^{2} = 645.2 mm^2	1 m^2 = 1550 in.^{2}
	1 ft = 0.09290 m^2	1 m^2 = 10.76 ft^2
Volume	1 in.^{3} = 16.39(10^3) mm^3	1 mm^3 = 61.02(10^{-6}) in.^{3}
	1 ft^3 = 0.02832 m^3	1 m^3 = 35.31 ft^3
Moment of inertia	1 in.^{4} = 0.4162 (10^6) mm^4	1 mm^4 = 2.402(10^{-6}) in.^{4}
Force	1 lb = 4.448 N	1 N = 0.2248 lb
Distributed load	1 lb/ft = 14.59 N/m	1 kN/m = 68.53 lb/ft
Pressure or stress	1 psi = 6.895 kPa	1 kPa = 0.145 psi
	1 ksi = 6.895 MPa	1 MPa = 145.0 psi
Bending moment or torque	1 ft·lb = 1.356 N·m	1 N·m = 0.7376 ft·lb
Work or energy	1 ft·lb = 1.356 J	1 J = 0.7376 ft·lb
Power	1 hp = 745.7 W	1 kW = 1.341 hp

APPENDIX F

COMPUTATIONAL METHODS

The material in this appendix was prepared by
Leroy D. Sturges of Iowa State University.

F-1

INTRODUCTION

The purpose of this appendix is to provide a few simple numerical methods that can be used to help solve (to take some of the drudgery out of the solution of) mechanics problems. No attempt has been made to provide a collection of computer programs to be used for the solution of the various types of problems encountered in a course in mechanics of materials. The problems designated as computer problems are basically simple applications of the elementary principles. They simply require solving the same problem over and over again as some parameter in the problem varies. Although they could be solved by hand or with a programmable calculator, these problems are most conveniently solved using a computer—either a microcomputer programmed in BASIC or a mainframe computer programmed in FORTRAN. In any event, the parametric study is intended to display characteristics of the problems that cannot be seen by solving for just one particular value. For example, Problem C1-1 investigates the effectiveness of different sizes of washers in distributing the stress under a bolt head or a nut.

This appendix is not designed to teach the student all there is to know about numerical methods. The numerical methods presented here are purposely kept simple—simple for understanding and simple for use. Much more sophisticated methods exist. Students interested in these more sophisticated methods should take a course in numerical methods and/or see some of the references listed at the end of this appendix.

This appendix addresses four types of problems encountered in various mechanics problems:

1. Nonlinear equations (root solving)
2. Systems of linear equations
3. Numerical integration
4. Ordinary differential equations

One or two simple methods are presented for the solution of each of these types of problems. The methods presented are purposely kept simple so that they can be used in hand calculation with a calculator or with a programmable calculator as well as with a computer.

In each case, a simple program is included to demonstrate the use of the numerical method. Versions are supplied in both BASIC and FORTRAN. A companion disk is not included, as the programs are short enough to type in without too much effort. The programs are not elegant; like the numerical methods they illustrate, they have been kept as simple as possible so that they will be easy to understand, easy to modify, and easy to customize, and also so that they will run on the widest possible selection of computers. Students will probably want to modify and enhance these programs to improve the transfer of data into the programs and to improve the format of the results output by the programs. Students may also want to modify the programs to take advantage of special features of their individual computers, such as graphical output.

F-2

NONLINEAR EQUATIONS

Problems in mechanics often require the solution of nonlinear equations, such as

$$x^3 - 40x^2 - 2000x + 48,000 = 0 \qquad \text{(F-1)}$$

These problems are sometimes stated in the form, find the zeros or roots of the function

$$f(x) = x^3 - 40x^2 - 2000x + 48{,}000$$

(that is, find the values of x that make $f(x) = 0$). Therefore, they are sometimes called root solving problems. Equation F-1 is a typical equation encountered in the problem of finding the maximum stress at a point (see Chapter 7).

While such equations can be solved by trial and error (simply guessing values until the left-hand side of the equation is nearly zero), there exist simple, systematic ways to solve such problems. Two such methods—the Method of False Position and the Newton–Raphson method—will be discussed here.

F-2-1 METHOD OF FALSE POSITION

The Method of False Position is a systematic method of narrowing down the region in which a root exists. For example, Fig. F-1 shows a graph of $f(x)$ as a function of x. Point L lies on the left side of the point where $f(x) = 0$ and point R lies on the right side. Note that for two such points, L and R, which bracket a simple root, $f(x_L)f(x_R) < 0$, always. A rough graph of the function $f(x)$ may be needed to find the initial points L and R.

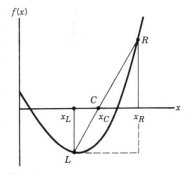

FIG. F-1

Construct point C—the point where the straight line joining point L and point R goes through zero—and use it as an estimate of the root. Point C can be found by similar triangles

$$\frac{f(x_R)}{x_R - x_C} = \frac{f(x_R) - f(x_L)}{x_R - x_L}$$

Rearranging and solving for x_C gives

$$x_C = x_R - \frac{f(x_R)\,(x_R - x_L)}{f(x_R) - f(x_L)} \qquad \text{(F-2)}$$

However, since $f(x)$ is not a straight line, $f(x_C)$ will not be zero. If $f(x_C)f(x_R) < 0$, then x_C is on the left side of the root. In this case, point L is moved to point C. If $f(x_C)f(x_L) < 0$, then x_C is on the right side of the root. In this case, point R is moved to point C. The process is then repeated until point C becomes the same as one of the end points (within the limits of numerical roundoff error). Then point C is the desired root.

EXAMPLE PROBLEM F-1

Solve Eq. F-1 using the Method of False Position to a relative accuracy of 0.001 percent.

Solution

Based on a rough sketch of the function

$$f(x) = x^3 - 40x^2 - 2000x + 48,000$$

the initial points $x_L = 40$ and $x_R = 70$ are chosen. Point C is constructed using Eq. F-2, giving

$$x_C = 70 - \frac{55,000\,(70 - 40)}{55,000 - (-32,000)} = 51.0344$$

Since $f(51.0345) = -25,329 < 0$, point L is moved to point C and Eq. F-2 is used to compute a new point C

$$x_C = 70 - \frac{55,000\,(70 - 51.0344)}{55,000 - (-25,329)} = 57.0146$$

Enter XL = 40
Enter XR = 70
Enter Error = .00001

XL	XC	XR	
40.00000	51.03440	70.00000	Move XL to XC
51.03440	57.01460	70.00000	Move XL to XC
57.01460	59.13280	70.00000	Move XL to XC
59.13280	59.75920	70.00000	Move XL to XC
59.75920	59.93400	70.00000	Move XL to XC
59.93400	59.98190	70.00000	Move XL to XC
59.98190	59.99500	70.00000	Move XL to XC
59.99500	59.99860	70.00000	Move XL to XC
59.99860	59.99960	70.00000	Move XL to XC
59.99960	59.99990	70.00000	Move XL to XC

The root is 59.9999

FIG. F-2 Typical output listing for Eq. F-1 showing the procedure of the Method of False Position and the rate of convergence to the correct answer.

The process is repeated until the relative error, either

$$\frac{x_R - x_C}{x_C} \quad \text{or} \quad \frac{x_C - x_L}{x_C}$$

is less than or equal to 0.00001. The result of this process is shown in Fig. F-2. After ten iterations, the root is located as $x = 59.9999$.

Simple programs in BASIC and FORTRAN to solve nonlinear equations are given in Programs F-1a and F-1b, respectively. The function statement in line 100 must be changed for the particular problem being solved.

```
100 DEF FNY(X)=X*X*X-40*X*X-2000*X+48000
110 PRINT "Enter XL = ";
120 INPUT XL
130 PRINT "Enter XR = ";
140 INPUT XR
150 PRINT "Enter Error = ";
160 INPUT ER
170 XC=XR-FNY(XR)*(XR-XL)/(FNY(XR)-FNY(XL))
180    IF FNY(XC)*FNY(XL)<=0 THEN 250
190    IF FNY(XC)*FNY(XR)<=0 THEN 220
200 PRINT "**** ERROR ****"
210 END
220 IF ABS((XC-XL)/XC)<ER THEN 280
230    XL=XC
240    GOTO 170
250 IF ABS((XC-XR)/XC)<ER THEN 280
260    XR=XC
270    GOTO 170
280 PRINT
290 PRINT "The root is ";XC
```

PROGRAM F-1a. BASIC program listing for performing the method of false position.

```
100 Y(X) = X*X*X-40*X*X-2000*X+48000
    PRINT *, 'Enter XL = '
    ACCEPT *, XL
    PRINT *, 'Enter XR = '
    ACCEPT *, XR
    PRINT *, 'Enter Error = '
    ACCEPT *, ER
170 XC = XR-Y(XR)*(XR-XL)/(Y(XR)-Y(XL))
    IF (Y(XC)*Y(XL).LE.0) GOTO 250
    IF (Y(XC)*Y(XR).LE.0) GOTO 220
    PRINT *, '**** ERROR ****'
    CALL EXIT
220 IF (ABS((XC-XL)/XC).LT.ER) GOTO 280
    XL = XC
    GOTO 170
250 IF (ABS((XC-XR)/XC).LT.ER) GOTO 280
    XR = XC
    GOTO 170
280 PRINT *, 'The root is  ',XC
    CALL EXIT
    END
```

PROGRAM F-1b. FORTRAN program listing for performing the method of false position.

F-2-2 NEWTON–RAPHSON METHOD

The Newton–Raphson method of root solving is more sophisticated than the Method of False Position. It uses the slope of the function to estimate the location of the root. For most functions this method converges much more quickly to the solution than does the Method of False Position, and the Newton–Raphson method is the method of choice in most cases. However, there are certain types of functions that are not solved very well by the Newton–Raphson method, so the Method of False Position is a good method to use on these functions.

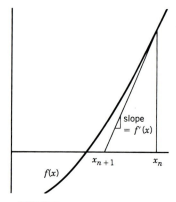

FIG. F-3

The Newton–Raphson method uses the tangent line to the curve at the point x_n (Fig. F-3) to estimate the location of the root. The slope of the tangent line at x_n is just the derivative of the function evaluated at x_n

$$\text{slope} = f'(x_n)$$

But from the geometry of Fig. F-3, the slope is also given by

$$\text{slope} = \frac{f(x_n) - 0}{x_n - x_{n+1}}$$

Setting these two expressions equal and solving for x_{n+1} gives

$$x_{n+1} = x_n - \frac{f(x_n)}{f'(x_n)} \tag{F-3}$$

This formula is used iteratively to get improved estimates of the location of the root. A rough graph of the function $f(x)$ should be used to get a first estimate of the root location, x_0.

EXAMPLE PROBLEM F-2

Solve Eq. F-1 using the Newton–Raphson method to a relative accuracy of 0.001 percent.

Solution

Based on a rough sketch of the function,

$$f(x) = x^3 - 40x^2 - 2000x + 48{,}000$$

the initial point $x_0 = 70$ is chosen. Equation F-3 is then used to generate the next several points

$$x_1 = x_0 - \frac{x_0^3 - 40x_0^2 - 2000x_0 + 48{,}000}{3x_0^2 - 80x_0 - 2000}$$

$$= 70 - \frac{55{,}000}{7100} = 62.2535$$

$$x_2 = 62.2535 - \frac{9736.4987}{4646.2210} = 60.1579$$

```
Enter Xo = 70
Enter Error = .00001
```

Xo	DL	ABS(DL/X1)
70.00000	− 7.74648	0.12443
62.25350	− 2.09558	0.03483
60.15790	− 0.15708	0.00262
60.00080	− 0.00087	0.00001
60.00000	1.00000	0.01639
61.00000	− 0.96685	0.01611
60.03310	− 0.03311	0.00055
60.00000	− 0.00004	0.00000

The root is 60

FIG. F-4 Typical output listing for Eq. F-1 showing the procedure of the Newton–Raphson method and the rate of convergence to the correct answer.

etc. The process is repeated until the relative error

$$\left| \frac{x_{n+1} - x_n}{x_{n+1}} \right|$$

(the ratio of the difference between successive approximations and the current approximation) is less than or equal to 0.00001. The result of this process is shown in Fig. F-4. The column labeled DL is just the second term in the iteration formula

$$DL = -\frac{f(x_n)}{f'(x_n)} = x_{n+1} - x_n$$

After seven iterations, the root is located as $x = 60.0000$.

Simple programs in BASIC and FORTRAN to solve nonlinear equations using the Newton–Raphson method are given in Programs F-2a and F-2b, respectively. The function statements defining the function and its derivative in lines 100 and 110 must be changed for the particular problem being solved.

```
100 DEF FNY(X)=X*X*X-40*X*X-2000*X+48000
110 DEF FNYP(X)=3*X*X-80*X-2000
120 PRINT "Enter Xo = ";
130 INPUT XO
140 PRINT "Enter Error = ";
150 INPUT ER
160     DL=-FNY(XO)/FNYP(XO)
170     X1=XO+DL
180     IF ABS(DL/X1)<ER THEN 210
190         XO=X1
200         GOTO 160
210 PRINT "The root is ";X1
```

PROGRAM F-2*a*. BASIC program listing for performing the Newton–Raphson method.

```
100 Y(X) = X*X*X-40*X*X-2000*X+48000
110 YP(X) = 3*X*X-80*X-2000
    PRINT *, 'Enter Xo = '
    ACCEPT *, XO
    PRINT *, 'Enter Error = '
    ACCEPT *, ER
160     DL = -Y(XO)/YP(XO)
        X1 = XO+DL
        IF (ABS(DL/X1).LT.ER) THEN
            GOTO 210
        ELSE
            XO = X1
            GOTO 160
        END IF
210 PRINT *, 'The root is ',X1
    CALL EXIT
    END
```

PROGRAM F-2*b*. FORTRAN program listing for performing the Newton–Raphson method.

F-3

SYSTEMS OF LINEAR EQUATIONS

Many problems in mechanics require the solution of a system of linear equations, such as

$$5x + 3y + 4z = 23 \qquad \text{(F-4}a)$$
$$2x + 1y + 1z = 7 \qquad \text{(F-4}b)$$
$$1x + 3y + 5z = 22 \qquad \text{(F-4}c)$$

One possible scheme for solving such equations would be to:

1. Use Eq. F-4*a* to eliminate *x* from Eqs. F-4*b* and F-4*c*. For example, subtract (2/5) times Eq. F-4*a* from eq. F-4*b* and subtract (1/5) times Eq. F-4*a* from Eq. F-4*c*. (If the coefficient of *x* in Eq. F-4*a* were zero, the equations must first be reordered so that the coefficient is not zero.)
2. Similarly, use the resulting Eq. F-4*b* to eliminate *y* from Eqs. F-4*a* and F-4*c*.
3. Finally, use Eq. F-4*c* to eliminate *z* from Eqs. F-4*a* and F-4*b*.
4. At this point, Eq. F-4*a* will give the value of *x*; Eq. F-4*b*, the value of *y*; and Eq. F-4*c*, the value of *z*

The procedure described above is called the Gauss–Jordan method for the solution of systems of linear equations. Since the letters representing the unknowns (x, y, and z) serve no function in the solution scheme except to hold the coefficients in their proper places, the procedure is conveniently carried out on a matrix of the coefficients

$$\begin{bmatrix} 5 & 3 & 4 & 23 \\ 2 & 1 & 1 & 7 \\ 1 & 3 & 5 & 22 \end{bmatrix} \qquad \text{(F-5)}$$

where the first column contains the coefficients of x, the second column contains the coefficients of y, the third column contains the coefficients of z, and the last column contains the right-hand sides of the equations. The rows of the matrix are to be multiplied by constants and added to other rows in the same manner that the equations were in the procedure described above.

EXAMPLE PROBLEM F-3

Solve the system of linear equations (Eqs. F-4a, b, and c) using the Gauss–Jordan method.

Solution

The equations are first written in matrix form as in Eq. F-5. Then (2/5) times the first row is subtracted from the second row, and (1/5) times the first row is subtracted from the third row to give

$$\begin{bmatrix} 5 & 3 & 4 & 23 \\ 0 & -0.2 & -0.6 & -2.2 \\ 0 & 2.4 & 4.2 & 17.4 \end{bmatrix}$$

Then, 15 times the second row is added to the first row and 12 times the second row is added to the third row to give

$$\begin{bmatrix} 5 & 0 & -5.0 & -10.0 \\ 0 & -0.2 & -0.6 & -2.2 \\ 0 & 0 & -3.0 & -9.0 \end{bmatrix}$$

Finally, (5/3) times the third row is subtracted from the first row and (0.2) times the third row is subtracted from the second row to give

$$\begin{bmatrix} 5 & 0 & 0 & 5 \\ 0 & -0.2 & 0 & -0.4 \\ 0 & 0 & -3.0 & -9.0 \end{bmatrix}$$

The answer is more easily interpreted if the first row is divided by 5; the second row by (-0.2), and the third row by (-3), which gives the matrix

$$\begin{bmatrix} 1 & 0 & 0 & 1 \\ 0 & 1 & 0 & 2 \\ 0 & 0 & 1 & 3 \end{bmatrix}$$

The rows of this matrix represent the three equations

$$x = 1$$
$$y = 2$$
$$z = 3$$

which is the solution to the original system of equations.

Simple programs in BASIC and FORTRAN to solve systems of linear equations using the Gauss–Jordan elimination procedure are given in Programs F-3*a* and F-3*b*, respectively. The number of equations being solved and the coefficients of the matrix are included in data statements (lines 100–130) and must be changed for the specific problem being solved. A desirable modification of these programs would be to have these values entered directly from the keyboard for small numbers of equations or read from a file for large numbers of equations.

PROGRAM F-3*a*. BASIC program listing for solving a system of linear equations using the Gauss–Jordan elimination method.

```
100 DATA 3
110 DATA 5,3,4,23
120 DATA 2,1,1,7
130 DATA 1,3,5,22
140 DIM A(20,21)
150 GOSUB 310
160 FOR I=1 TO N
170    GOSUB 500
180    GOSUB 720
190    NEXT I
200 PRINT
210 PRINT
220 PRINT "The solution is:"
230 PRINT
240 FOR I=1 TO N
250    PRINT "x(";I;") = ";A(I,N+1)
260    NEXT I
270 END
280 REM
290 REM    Read the input matrix from DATA statements
300 REM
310 READ N
320 FOR I=1 TO N
```

```
330    FOR J=1 TO N+1
340        READ A(I,J)
350        NEXT J
360    NEXT I
370 CLEAR
380 PRINT "The input matrix is:"
390 PRINT
400 FOR I=1 TO N
410    FOR J=1 TO N+1
420        PRINT A(I,J),
430        NEXT J
440    PRINT
450    NEXT I
460 RETURN
470 REM
480 REM     Search for row with largest element in column I
490 REM
500 TEMP=ABS(A(I,I))
510 KT=I
520 FOR K=I TO N
530    TT=ABS(A(K,I))
540    IF TT<=TEMP THEN 570
550        KT=K
560        TEMP=TT
570    NEXT K
580 IF (KT=I) THEN 680
590 REM
600 REM     Interchange rows if necessary to make 'pivot' element
610 REM     as large as possible
620 REM
630 FOR K=1 TO N+1
640    TEMP=A(I,K)
650    A(I,K)=A(KT,K)
660    A(KT,K)=TEMP
670    NEXT K
680 RETURN
690 REM
700 REM     'Normalize' the pivot row
710 REM
720 PV=A(I,I)
730 FOR K=I TO N+1
740    A(I,K)=A(I,K)/PV
750    NEXT K
760 REM
770 REM     Eliminate all entries from the Ith column
780 REM     except the pivot element which has been
790 REM     normalized to 1
800 REM
810 FOR K=1 TO N
820    IF K=I THEN 870
830    PV=A(K,I)
840    FOR KK=I TO N+1
850        A(K,KK)=A(K,KK)-PV*A(I,KK)
860        NEXT KK
870    NEXT K
880 RETURN
```

PROGRAM F-3*b*. FORTRAN
program listing for solving
a system of linear
equations using the
Gauss–Jordan elimination
method.

```
      REAL  A(20,21)
100 DATA  N/3/
110 DATA  A(1,1),A(1,2),A(1,3),A(1,4)/5,3,4,23/
120 DATA  A(2,1),A(2,2),A(2,3),A(2,4)/2,1,1, 7/
130 DATA  A(3,1),A(3,2),A(3,3),A(3,4)/1,3,5,22/
      PRINT *, 'The input matrix is:'
      PRINT *, '   '
      DO 360  I = 1, N
         PRINT *, (A(I,J), J = 1, N+1)
360   CONTINUE
      DO 190  I = 1, N
         CALL PIVOT(N,A,I)
         CALL ELIM(N,A,I)
190   CONTINUE
      PRINT *, '   '
      PRINT *, '   '
      PRINT *, 'The solution is:'
      PRINT *, '   '
      DO 260  I = 1, N
         PRINT *, 'X(',I,') = ',A(I,N+1)
260   CONTINUE
      CALL EXIT
      END
C
      SUBROUTINE PIVOT(N,A,I)
      REAL A(20,21)
C
C         Search for row with largest element in column I
C
      TEMP = ABS(A(I,I))
      KT = I
      DO 570  K = I, N
         TT = ABS(A(K,I))
         IF (TT.GT.TEMP)  THEN
            KT = K
            TEMP = TT
            END IF
570   CONTINUE
      IF (KT.EQ.I) GOTO 680
C
C         Interchange rows if necessary to make 'pivot' element
C         as large as possible
C
      DO 670  K = 1, N+1
         TEMP = A(I,K)
         A(I,K) = A(KT,K)
         A(KT,K) = TEMP
670   CONTINUE
680 RETURN
      END
C
      SUBROUTINE ELIM(N,A,I)
      REAL  A(20,21)
C
C         'Normalize' the pivot row
```

```
C
      PV = A(I,I)
      DO 750  K = I, N+1
         A(I,K) = A(I,K)/PV
  750    CONTINUE
C
C         Eliminate all entries from the Ith column
C         except the pivot element which has been
C         normalized to 1
C
      DO 870  K = 1, N
         IF (K.NE.I)  THEN
            PV = A(K,I)
            DO 860  KK = I, N+1
               A(K,KK) = A(K,KK)-PV*A(I,KK)
  860          CONTINUE
            END IF
  870    CONTINUE
      RETURN
      END
```

F-4

NUMERICAL INTEGRATION

Most of the functions to be integrated in mechanics of materials problems are polynomials or other simple functions and are easily evaluated analytically. Sometimes, however, the function to be integrated will be complex enough to require advanced integration techniques. Other times, the function to be integrated will not be given explicitly. Instead, the function may be given by experimentally determined values at a few points. In these last two cases, numerical methods may be useful in evaluating the integrals.

In the simplest form, numerical integration dervies from the physical interpretation of an integral as the area under a curve. The area may be approximated using several rectangles, trapezoids, or other simple shapes whose area is easily determined. The approximate value of the integral is obtained by adding together the areas of the several pieces. The method to be described here uses trapezoids to approximate the area under the curve; hence the name—Trapezoidal Rule.

For example, the value of the integral

$$I = \int_a^b f(x)\, dx$$

is represented by the shaded area under the curve in Fig. F-5a. Approximating this area by one large trapezoid of width $h = b - a$, as in Fig. F-5b, gives

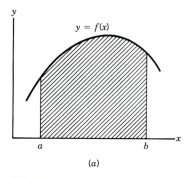
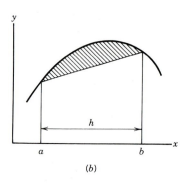
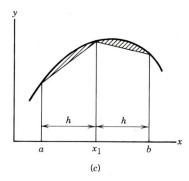

FIG. F-5

$$I \approx T_1 = \frac{h}{2}[f_a + f_b]$$

where $f_a = f(a)$ and $f_b = f(b)$. This approximation is obviously in error by the amount of the shaded area in Fig. F-5b between the top of the trapezoid and the curve.

The amount of error can be reduced, however, by using two trapezoids of width $h = (b-a)/2$, as in Fig. F-5c, and adding their areas together

$$I \approx T_2 = \frac{h}{2}[f_a + f_1] + \frac{h}{2}[f_1 + f_b]$$

$$= \frac{h}{2}[f_a + f_b + 2f_1]$$

where, $f_1 = f(x_1)$ and $x_1 = a + h$. The tops of the two trapezoids follow the curve more closely than did the single trapezoid. The error in the approximation T_2 (represented by the shaded area in Fig. F-5c) is less than the error in the approximation T_1 (represented by the shaded area in Fig. F-5b).

Continuing with this logic and dividing the interval into N panels (trapezoids) of equal width, h, gives the Trapezoidal Rule approximation

$$I \approx T_N = \frac{h}{2}[f_0 + f_n + 2\Sigma f_i] \qquad \text{(F-6)}$$

where

$$f_0 = f(a)$$
$$f_N = f(b)$$
$$f_i = f(x_i) = f(a + ih); \qquad i = 1, 2, \ldots, N-1$$
$$h = \frac{b-a}{N}$$

As the number of panels used is increased, the width of the panels will decrease, the tops of the trapezoids will more closely fit the function being integrated, and the total error of the approximation will be reduced. It can be shown that the error in using the Trapezoidal Rule is approximately proportional to h^2. Therefore, reducing h by a factor of 2 should reduce the error by a factor of 4. Typically, the integral is evaluated several times using smaller and smaller panel widths until the value for two different panel widths is nearly the same. This value then is taken to be the value of the integral.

For experimental data given only at discrete points (possibly not equispaced), the Trapezoidal Rule is applied to each pair of points and the sum is then added together. An alternate approach would be to use an interpolation routine to generate additional points at values required for the Trapezoidal Rule.

EXAMPLE PROBLEM F-4

Evaluate the integral

$$\int_0^2 xe^{-x^2} \, dx$$

using the Trapezoidal Rule to an estimated relative error of 0.1 percent.

Solution

Evaluating the integral using Eq. F-6 and a single panel gives

$$T_1 = \frac{2}{2}[0 + 2e^{-4}] = 0.0366$$

while using two panels gives

$$T_2 = \frac{1}{2}[0 + 2e^{-4} + 2(e^{-1})] = 0.3862$$

The absolute error (just the difference between the current estimate and the correct value) is estimated to be

$$E_{abs} = |T_2 - T_1| = 0.3496$$

and the relative error (the ratio of the absolute error to the correct value) is estimated to be

$$E_{rel} = \left| \frac{T_2 - T_1}{T_2} \right| \times 100 = 90.5\%$$

TABLE F-1 Evaluation of the integral $\int_0^2 xe^{-x^2}\, dx$, using the Trapezoidal Rule (Eq. F-6) and Varying Numbers of Panels

Number of Panels	Value of Integral	% Relative Error
1	0.036631	
2	0.386195	90.514800
4	0.466847	17.275900
8	0.484937	3.730330
16	0.489371	0.906144
32	0.490475	0.224984
64	0.490750	0.056155
128	0.490819	0.014032

Note: The result is 0.4908 with an estimated accuracy of 0.1 percent.

Using four panels gives

$$T_4 = \frac{0.5}{2}\,[0 + 2e^{-4}$$
$$+\ 2(0.5e^{-0.25} + e^{-1}) + 1.5e^{-2.25}] = 0.4668$$

and the relative error is estimated to be

$$E_{\text{rel}} = \left| \frac{0.4668 - 0.3862}{0.4668} \right| \times 100 = 17.28\%$$

Continuing using more and more panels until the desired accuracy has been reached gives the results in Table F-1. Using 64 panels, the integral is evaluated as 0.4908 with an estimated error of 0.06 percent. (Notice in Table F-1 that reducing the width of the panels by a factor of 2 reduced the error by approximately a factor of 4 in accordance with the error estimate mentioned earlier).

A listing of the BASIC and FORTRAN programs used to generate the output of Table F-1 are given in Programs F-4*a* and F-4*b*, respectively. The function statement and integration limits in lines 100–120 need to be changed for the function being integrated.

```
100 DEF FNY(X)=X*EXP(-X*X)
110 A=0
120 B=2
130 CLEAR
140 N=1
150 H=B-A
160 T=H*(FNY(A)+FNY(B))/2
170 PRINT "no  of     value of      % rel"
180 PRINT "panels     integral      error"
190 PRINT
200 PRINT USING "  ####    ###.######",N,T
210 FOR K=1 TO 7
220    TOLD=T
230    H=H/2
240    T=FNY(A)+FNY(B)
250    N=(B-A)/H
260    FOR I=1 TO N-1
270       X1=A+I*H
280       T=T+2*FNY(X1)
290       NEXT I
300    T=T*H/2
310    ER=100*ABS((T-TOLD)/T)
320    PRINT USING "  ####    ###.######   ###.######",N,T,ER
330    NEXT K
```

PROGRAM F-4*a*. BASIC program listing for evaluating integrals using the trapezoidal rule.

```
100 Y(X) = X*EXP(-X*X)
110 A = 0
120 B = 2
    N = 1
    H = B-A
    T = H*(Y(A)+Y(B))/2
    PRINT *, 'no  of     value of      % rel'
    PRINT *, 'panels     integral      error'
    PRINT *, '   '
    PRINT 200, N, T
200 FORMAT (2X,I4,2(3X,F10.6))
    DO 330  K = 1, 7
       TOLD = T
       H = H/2
       T = Y(A)+Y(B)
       N = (B-A)/H
       DO 290  I = 1, N-1
          X1 = A+I*H
          T = T+2*Y(X1)
290       CONTINUE
       T = T*H/2
       ER = 100*ABS((T-TOLD)/T)
       PRINT 200, N,T,ER
330    CONTINUE
    CALL EXIT
    END
```

PROGRAM F-4*b*. FORTRAN program listing for evaluating integrals using the trapezoidal rule.

EXAMPLE PROBLEM F-5

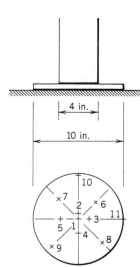

FIG. F-6

A circular plate is used to distribute the weight carried by a column (Fig. F-6). Because the plate is not rigid, the pressure between the plate and the floor is not constant. If pressure sensors on the bottom of the plate indicate the following radial pressure distribution,

Radial Position (in.)	Pressure (psi)	Radial Position (in.)	Pressure (psi)
0.0	21.6	3.0	2.4
0.5	21.8	3.5	1.8
1.0	21.7	4.0	1.4
1.5	21.8	4.5	1.0
2.0	13.6	5.0	0.9
2.5	3.5		

determine the total weight carried by the plate.

Solution

The total weight W carried by the plate is

$$W = \int_0^5 p(2\pi r \, dr) = 2\pi \int_0^5 (pr) \, dr$$

Since the data are given at equispaced values of r, the Trapezoidal Rule (Eq. F-6) will be used directly with $N = 10$ and $h = (5-0)/10 = 0.5$ to get

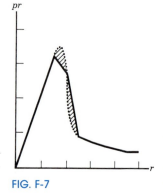

FIG. F-7

$$\begin{aligned}
W = \ & 2\pi(0.52) \, \{(0)(21.6) + (5.0)(0.9) \\
& + 2 \, [(0.5)(21.8) + (1.0)(21.7) + (1.5)(21.8) \\
& + (2.0)(13.6) + (2.5)(3.5) + (3.0)(2.4) \\
& \quad (3.5)(1.8) + (4.0)(1.4) + (4.5)(1.0)]\} \\
= \ & 399.3 \text{ lb}
\end{aligned}$$

The function being integrated, pr, is plotted in Fig. F-7 as a function of r. The solid line indicates the area used by the Trapezoidal Rule, while the dotted line indicates the expected true function shape. The expected error is the shaded region between the two curves. Because of cancellation between positive and negative error areas, the resulting answer is expected to be quite accurate.

F-5

ORDINARY DIFFERENTIAL EQUATIONS

Many phenomena in mechanics require the solution of differential equations. Many of these problems are simple, standard equations that have well-known solutions. Some differential equations, however, are difficult to solve analytically but are easily solved numerically using a simple procedure called Euler's method. The procedure will be described in three parts

1. First-order ordinary differential equations
2. General initial-value problems in which the order of the ordinary differential equation is greater than one, but for which all data are given at a single point
3. Boundary-value problems in which the order of the ordinary differential equation is greater than one and the data are given at two or more points

F-5-1 FIRST-ORDER ORDINARY DIFFERENTIAL EQUATIONS

A problem governed by a first-order, ordinary differential equation is of the form

$$\frac{dy}{dt} = f(t, y) \qquad a \leqslant t \leqslant b \qquad \text{(F-7a)}$$

$$y(a) = y_0 \qquad\qquad \text{(F-7b)}$$

the solution of the differential equation is the function $y(t)$, which satisfies Eq. F-7a. But since $f(t, y) = dy/dt$ is the slope of the function $y(t)$, it seems reasonable to expect that $f(a, y_0)$ can be used to predict the value of $y_1 = y(t_1)$, $t_1 = a + \Delta t$, (see Fig. F-8). Then knowing the value y_1, $f(t_1, y_1)$ can be used to predict $y_2 = y(t_2)$, $t_2 = a + 2\Delta t$, etc. The solution, $y(t)$, is generated in a marching fashion

$$y_{n+1} = y_n + f(t_n, y_n) \Delta t \qquad \text{(F-8)}$$

where $t_n = a + n\Delta t$ and $y_n = y(t_n)$. A more theoretical justification of Eq. F-8 could be obtained from the Taylor's series for $y(t)$. In fact, rewriting Eq. F-8 in the form

$$y(t_n + \Delta t) = y(t_n) + y'(t_n) \Delta t$$

shows that it is just the first two terms in the Taylor series expansion of $y(t)$ about the point (t_n).

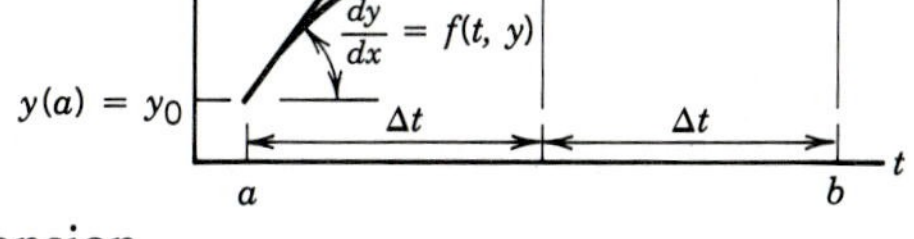

FIG. F-8

The Euler method for solving differential equation F-7 consists of using the recursion relation, Eq. F-8, to generate the points (t_n, y_n) starting from (a, y_o) and marching along the curve to $t = b$. The points generated are connected by straight-line segments to define the function $y(t)$. Enough points should be generated so that the resulting curve looks smooth.

Besides generating enough points so that the curve looks smooth, points must be generated close enough together so that the method does not stray too far from the solution curve. If the step size, Δt, is too large, Euler's method will not follow the solution curve very closely (see Example Problem F-6). The smaller the step size, the better Euler's method works, but more steps and more computation time are required to generate the solution.

It can be shown that the error in using Euler's method is proportional to the step size. Therefore, reducing the step size by a factor of 2 should reduce the error by a factor of 2 also. The normal procedure is to solve the problem with successively smaller step sizes until the solution no longer changes.

EXAMPLE PROBLEM F-6

Solve the differential equation

$$\frac{dy}{dt} = y^2 e^{-t}$$

for $y(t)$, $1 < t < 5$ and $y(1) = 0.5$.

Solution

Starting from the initial condition, $t_0 = 1$ and $y_0 = 0.5$, use Eq. F-8 and $\Delta t = 1$ to generate the points

$$y_1 = y_0 + (y_0^2 e^{-t_0})\Delta t$$
$$= 0.5 + (0.5)^2 \, e^{-1} \, (1) = 0.5920$$
$$t_1 = t_0 + \Delta t = 1 + 1 = 2$$
$$y_2 = (y_1 + y_1^2 e^{-t_1})\Delta t$$
$$= 0.5920 + (0.5920)^2 \, e^{-2} \, (1) = 0.6394$$
$$t_2 = t_1 + \Delta t = 2 + 1 = 3$$
$$y_3 = 0.6394 + (0.6394)^2 \, e^{-3} \, (1) = 0.6597$$
$$t_3 = t_2 + \Delta t = 3 + 1 = 4$$
$$y_4 = 0.6597 + (0.6597)^2 \, e^{-4} \, (1) = 0.6677$$
$$t_4 = t_3 + \Delta t = 4 + 1 = 5$$

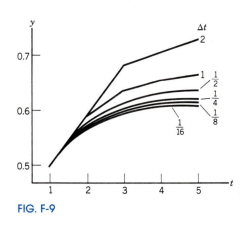

FIG. F-9

These points are plotted in Fig. F-9 and connected by straight-line segments. The resulting curve is very coarse and probably does not follow the true solution very closely. The calculations were repeated using smaller step sizes ($\frac{1}{2}$, $\frac{1}{4}$, $\frac{1}{8}$, and $\frac{1}{16}$) until the curve became smooth and no longer changed. These results are also shown in Fig. F-9.

The programs in BASIC and FORTRAN used to generate the output of Fig. F-9 are given as Programs F-5a and F-5b, respectively. The function $f(t,y)$ that defines the problem and the initial data, t_0 and y_0, in statements 110–130 must be changed for the particular problem being solved.

```
100 DIM X(100),Y(100)
110 DEF FNF(X,Y)=Y*Y*EXP(-X)
120 XI=1
130 YI=.5
140 H=1
150 FOR M=1 TO 5
160    NS=4/H
170    X=XI
180    Y=YI
190    X(0)=X
200    Y(0)=Y
210    FOR N=1 TO NS
220       Y=Y+H*FNF(X,Y)
230       X=X+H
240       X(N)=X
250       Y(N)=Y
260    NEXT N
270    PRINT
280    PRINT
290    PRINT
300    A$="  ###.####   ##.#####"
310    FOR N=0 TO NS
320       PRINT USING A$,X(N),Y(N)
330    NEXT N
340    H=H/2
350 NEXT M
360 END
```

PROGRAM F-5a. BASIC program listing for solving ordinary differential equations using Euler's method (see Example Problem F-6).

```
100 REAL  FX(100),FY(100)
110 F(X,Y) = Y*Y*EXP(-X)
120 XI = 1
130 YI = .5
140 H = 1
    DO 350  M = 1, 5
       NS = 4/H
       X = XI
       Y = YI
       FX(1) = X
       FY(1) = Y
       DO 260  N = 1, NS
          Y = Y+H*F(X,Y)
          X = X+H
          FX(N+1) = X
          FY(N+1) = Y
260    CONTINUE
       PRINT *, '      '
       PRINT *, '      '
       PRINT *, '      '
300 FORMAT(F12.4,F12.5)
       DO 330  N = 1, NS+1
          PRINT 300, FX(N), FY(N)
330    CONTINUE
       H = H/2
350    CONTINUE
    CALL EXIT
    END
```

PROGRAM F-5b. FORTRAN program listing for solving ordinary differential equations using Euler's method (see Example Problem F-6).

F-5-2 GENERAL INITIAL-VALUE PROBLEMS

Euler's method can also be applied to higher order differential equations such as

$$\frac{d^4y}{dt^4} + t\left(\frac{d^2y}{dt^2}\right)^2 + y^2\frac{dy}{dt} = t\sin y \tag{F-9}$$

$$y(1) = 1.0$$
$$y'(1) = 0.0$$
$$y''(1) = -1.0$$
$$y'''(1) = 0.5$$

In order to apply Euler's method to this type of problem, the differential equation (Eq. F-9) must first be reduced to a system of first-order differential equations. One way of doing this is to define $n-1$ new variables (the dependent variable and its first $n-1$ derivatives where n is the order of the differential equation)

$$x_1 = y$$

$$x_2 = \frac{dy}{dt} = \frac{dx_1}{dt}$$

$$x_3 = \frac{d^2y}{dt^2} = \frac{dx_2}{dt}$$

$$x_4 = \frac{d^3y}{dt^3} = \frac{dx_3}{dt}$$

The last three equations are three first-order differential equations relating the variables t, x_1, x_2, x_3, and x_4. A fourth equation comes from the original differential equation (Eq. F-9). Thus,

$$x_1' = x_2 = f_1(t, x_1, x_2, x_3, x_4) \tag{F-10a}$$
$$x_2' = x_3 = f_2(t, x_1, x_2, x_3, x_4) \tag{F-10b}$$
$$x_3' = x_4 = f_3(t, x_1, x_2, x_3, x_4) \tag{F-10c}$$
$$x_4' = t\sin x_1 - tx_3^2 + x_1^2 x_2$$
$$= f_4(t, x_1, x_2, x_3, x_4) \tag{F-10d}$$

In terms of these four new variables, the boundary conditions are

$$x_1(1) = 1.0$$
$$x_2(1) = 0.0$$
$$x_3(1) = -1.0$$
$$x_4(1) = 0.5$$

After the nth order differential equation (Eq. F-9) is reduced to a system of n first-order differential equations (Eqs. F-10), Euler's method (Eq. F-8) is applied to each of the first-order differential equations, in turn generating the sequence of points

$$x_{11} = x_{10} + f_1(t_0, x_{10}, x_{20}, x_{30}, x_{40}) \, \Delta t$$
$$x_{21} = x_{20} + f_2(t_0, x_{10}, x_{20}, x_{30}, x_{40}) \, \Delta t$$
$$x_{31} = x_{30} + f_3(t_0, x_{10}, x_{20}, x_{30}, x_{40}) \, \Delta t$$
$$x_{41} = x_{40} + f_4(t_0, x_{10}, x_{20}, x_{30}, x_{40}) \, \Delta t$$
$$x_{12} = x_{11} + f_1(t_1, x_{11}, x_{21}, x_{31}, x_{41}) \, \Delta t$$
$$x_{22} = x_{21} + f_2(t_1, x_{11}, x_{21}, x_{31}, x_{41}) \, \Delta t$$

etc., where $x_{mn} = x_m(t_n)$. Once the solution to the system of first-order equations (Eqs. F-10) is complete, the solution to the original differential equation (Eq. F-9) is obtained simply as

$$y(t) = x_1(t)$$

EXAMPLE PROBLEM F-7

The equation used in simple beam theory to generate the elastic curve

$$EI \frac{d^2y}{dx^2} = M(x) \qquad (a)$$

is an approximation that is valid for small deflections (small slope dy/dx). For large deflections (slopes greater than about 0.1 radians), the exact differential equation

$$EI \frac{d^2y/dx^2}{(1 + (dy/dx)^2)^{3/2}} = M(x) \qquad (b)$$

should be solved. For the cantilever beam of Fig. F-10 with $L = 10$ ft, $P = 2500$ lb, $E = 10.6 \times 10^6$ psi, and $I = 5.0$ in.⁴:

(a) Compute (using Eq. a) and plot the deflection of the beam (plot y as a function of x; $0 \le x \le L$).

(b) Compute (using Eq. b and Euler's method for solving differential equations) and plot the deflection of the beam (plot y as a function of x; $0 \le x \le L$).

(c) Determine the minimum moment of inertia, I, such that the error resulting from using Eq. a is less than 5 percent.

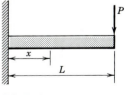

FIG. F-10

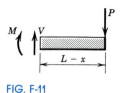

FIG. F-11

Solution

(a) The moment equation for the cantilever beam is determined by considering a free-body diagram of a segment of the end of the beam (Fig. F-11). Summing moments about the cut (left) end of the segment gives

$$M(x) = -P(L-x)$$

Substituting the moment relationship into Eq. *a* and integrating twice gives

$$EI\theta(x) = \frac{P}{2}[(L-x)^2 - L^2]$$

$$EIy(x) = \frac{P}{6}[-(L-x)^3 - 3L^2x + L^3]$$

where the constants of integration were determined using the initial conditions $\theta(0) = y(0) = 0$. Then the deflection of the beam as given by simple beam theory is

$$y(x) = \frac{2500[-(120-x)^3 - 3(120)^2x + (120)^3]}{6(10.6)(10^6)(5.0)}$$

where *x* and *y* are both in inches. The results are tabulated in Table F-2 and plotted in Fig. F-12.

(b) Equation *b* is first transformed into a pair of first-order differential equations through the change of variables

$$y_1 = y$$

$$y_2 = \frac{dy}{dx} = y_1'$$

Then, the pair of differential equations to be solved is

$$y_1' = y_2$$

$$y_2' = \frac{M(x)}{EI}[1 + (y_2)^2]^{3/2}$$

$$= \frac{-P(L-x)}{EI}[1 + (y_2)^2]^{3/2}$$

$$= \frac{-2500(120-x)}{(10.6)(10^6)(5)}[1 + (y_2)^2]^{3/2}$$

subject to the initial conditions

$$y_1(0) = 0$$

$$y_2(0) = 0$$

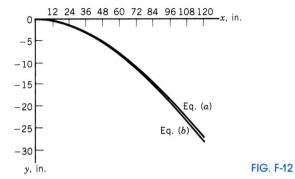

FIG. F-12

TABLE F-2 Comparison of the Deflection of a Cantilever Beam Using Simple Beam Theory Y(SBT) and Integrating the Complete Differential Equation Using Euler's Method Y(Euler).

Example Problem F-7

$L = 120$	$X_i = 0.00$
$EI = 53,000,000$	$Y_i = 0.00$
$P = 2500.00$	$X_f = 120$
	$TH_i = 0.00000$

X	Y(Euler)	Theta	Y(SBT)	Error
0.00	0.0000	0.00000	0.0000	0.000
6.00	−0.0983	−0.03315	−0.1002	1.947
12.00	−0.3907	−0.06470	−0.3940	0.837
18.00	−0.8679	−0.09471	−0.8711	0.376
24.00	−1.5208	−0.12325	−1.5215	0.048
30.00	−2.3406	−0.15033	−2.3349	0.244
36.00	−3.3187	−0.17596	−3.3011	0.529
42.00	−4.4462	−0.20013	−4.4100	0.815
48.00	−5.7144	−0.22281	−5.6513	1.105
54.00	−7.1143	−0.24397	−7.0149	1.397
60.00	−8.6365	−0.26357	−8.4905	1.690
66.00	−10.2716	−0.28155	−10.0681	1.981
72.00	−12.0097	−0.29785	−11.7374	2.267
78.00	−13.8405	−0.31243	−13.4882	2.546
84.00	−15.7535	−0.32521	−15.3103	2.814
90.00	−17.7379	−0.33614	−17.1935	3.069
96.00	−19.7822	−0.34517	−19.1277	3.308
102.00	−21.8750	−0.35225	−21.1027	3.531
108.00	−24.0044	−0.35734	−23.1082	3.734
114.00	−26.1585	−0.36041	−25.1341	3.916
120.00	−28.3251	−0.36145	−27.1702	4.077

Note: The maximum error in using simple beam theory is about 4 percent (for the given properties and loading) and occurs at the free end of the beam.

PROGRAM F-6*a*. BASIC program listing for solving ordinary differential equations using Euler's method (see Example Problem F-7).

```
100 L=120
110 EI=.53E+08
120 P=2500
130 DEF FNM(X)=-P*(L-X)
140 DEF FNY(X)=(-P/(6*EI))*((L-X)*(L-X)*(L-X)+3*L*L*X-L*L*L)
150 XI=0
160 XF=L
170 YI=0
180 TH=0
190 DIM X(25),Y1(25),Y2(25),FY(25),ER(25)
200 NS=1000
210 H=(XF-XI)/NS
220 X=XI
230 Y1=YI
240 Y2=TH
250 GOSUB 480
260 FOR N=0 TO 20
270     FY(N)=FNY(X)
280     X(N)=X
290     Y1(N)=Y1
300     Y2(N)=Y2
310     IF ABS(Y1)<.0001 THEN 330
320     ER(N)=100*ABS((Y1(N)-FY(N))/Y1(N))
330     GOSUB 390
340     NEXT N
350 GOSUB 600
360 END
370 REM
380 REM
390 FOR K=1 TO NS/20
400     Y1=Y1+H*Y2
410     TEMP=SQR(1+Y2*Y2)
420     Y2=Y2+H*(FNM(X)*TEMP*TEMP*TEMP/EI)
430     X=X+H
440     NEXT K
450 RETURN
460 REM
470 REM
480 PRINT
490 PRINT
500 PRINT ,"   Example Problem  F-7"
510 PRINT
520 PRINT USING "    L  =           ####   ",L
530 PRINT USING "    EI = ##,##########   ",EI
540 PRINT USING "    P  =          ##,###.##",P
550 PRINT USING "    Xi = ####.##               Xf = ####",XI,XF
560 PRINT USING "    Yi = ####.##               THi = ####.#####",YI,
570 PRINT
580 RETURN
590 REM
600 PRINT "     X        Y(Euler)    Theta       Y(SBT)        Erro
610 A$="  ####.##    ####.####   ###.#####   ####.####   ####.###"
620 FOR N=0 TO 20
630     PRINT USING A$,X(N),Y1(N),Y2(N),FY(N),ER(N)
640     NEXT N
650 RETURN
```

```
190 REAL  FX(25), FY1(25), FY2(25), FY(25), ER(25), L
    COMMON  L, P, EI
100 L = 120
110 EI = .53E+08
120 P = 2500
150 XI = 0
160 XF = L
170 YI = 0
180 TH = 0
    NS = 1000
    H = (XF-XI)/NS
    X = XI
    Y1 = YI
    Y2 = TH
    PRINT *, '     '
    PRINT *, '     '
    PRINT *, '    Example Problem  F-7'
    PRINT *, '    '
    PRINT *, '   L  =           ', L
    PRINT *, '   EI = ', EI
    PRINT *, '   P  =         ', P
    PRINT *, '   Xi = ', XI, '              Xf = ', XF
    PRINT *, '   Yi = ', YI, '              THi = ', TH
    PRINT *, '    '
    DO 340  N = 1, 21
       FY(N) = Y(X)
       FX(N) = X
       FY1(N) = Y1
       FY2(N) = Y2
       IF (ABS(Y1).GT.0.0001)  ER(N)=100*ABS((FY1(N)-FY(N))/FY1(N))
       CALL EULER(NS,Y1,Y2,X,H)
340    CONTINUE
    PRINT *, '    X        Y(Euler)     Theta',
   +        '        Y(SBT)        Error'
610 FORMAT(F10.2,2(F12.4,F12.5))
    DO 640  N = 1, 21
       PRINT 610, FX(N), FY1(N), FY2(N), FY(N), ER(N)
640    CONTINUE
    CALL EXIT
    END
C
    FUNCTION Y(X)
    REAL  L
    COMMON  L, P, EI
140 Y = (-P/(6.*EI))*((L-X)*(L-X)*(L-X)+3.*L*L*X-L*L*L)
    RETURN
    END
C
    FUNCTION FM(X)
    REAL  L
    COMMON  L, P, EI
130 FM = -P*(L-X)
    RETURN
    END
```

PROGRAM F-6*b*. FORTRAN program listing for solving ordinary differential equations using Euler's method (see Example Problem F-7).

```
C
        SUBROUTINE EULER(NS,Y1,Y2,X,H)
        REAL   L
        COMMON   L, P, EI
        DO 440   K = 1, NS/20
            Y1 = Y1+H*Y2
            TEMP = SQRT(1+Y2*Y2)
            Y2 = Y2+H*(FM(X)*TEMP*TEMP*TEMP/EI)
            X = X+H
    440     CONTINUE
        RETURN
        END
```

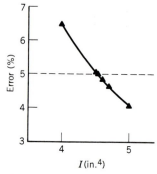

FIG. F-13

These equations were solved using the programs in BASIC and FOR-TRAN listed as Programs F-6a and F-6b, respectively, and a step size $\Delta x = L/1000$ in. These results are also tabulated in Table F-2 and plotted in Fig. F-12. The maximum difference between the solutions using Eqs. a and b is about 4 percent.

(c) Equation b was solved for several different values of the moment of inertia from $I = 4$ in.^{4} to $I = 5$ in.^{4} The resulting error at $x = L$ is plotted as a function of I in Fig. F-13 and indicates a minimum moment of inertia of about $I = 4.53$ in.^{4} is required to keep the maximum error less than 5 percent.

F-5-3   BOUNDARY-VALUE PROBLEMS

A boundary-value problem consists of a differential equation of order 2 or higher (such as Eq. F-9) that has the data given at two or more points. Euler's method, however, requires all of the data be given at the initial point, so that it can march along the solution. Therefore, Euler's method cannot be used directly.

One approach in solving boundary-value problems is to guess values for the derivatives needed by the Euler method, solve the problem using the Euler method, and see how close the solution comes to the second boundary value. If necessary, the guesses for the initial derivatives are adjusted and the problem solved again until the solution meets the second boundary value. This is called a "shooting method" by analogy with the method of aiming an artillery gun—guess a barrel angle, fire the gun, see if the shell goes beyond the target or falls short, adjust the angle, fire again, etc.

EXAMPLE PROBLEM F-8

Repeat Example Problem F-7 for the simply supported beam shown in Fig. F-14, if $L = 5$ m, $w = 6000$ N/m, $E = 100$ GPa, and $I = 10^6$ mm⁴.

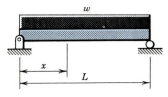

FIG. F-14

Solution

(a) The moment equation for the beam is determined by considering a free-body diagram of a segment of the end of the beam (Fig. F-15). Summing moments about the cut (right) end of the segment gives

$$M(x) = \frac{w}{2} [Lx - x^2]$$

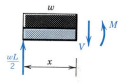

FIG. F-15

Substituting the moment relationship into Eq. *a* and integrating twice gives

$$EI\theta(x) = \frac{w}{24} [6Lx^2 - 4x^3 - L^3]$$

$$EIy(x) = \frac{w}{24} [2Lx^3 - x^4 - L^3x]$$

where the constants of integration were determined using the boundary conditions $y(0) = y(L) = 0$. Then the deflection of the beam as given by simple beam theory is

$$y(x) = \frac{6000[(2)(5)x^3 - x^4 - (5)^3x]}{24(100)(10^9)(10^{-6})}$$

where x and y are both given in meters. The results are tabulated in Table F-3 and plotted in Fig. F-16.

(b) Equation *b* is first transformed into a pair of first-order differential equations through the change of variables

$$y_1 = y$$

$$y_2 = \frac{dy}{dx} = y_1'$$

Then the pair of differential equations to be solved is

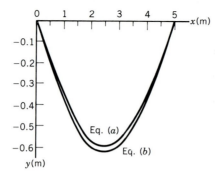

FIG. F-16

TABLE F-3 Comparison of the Deflection of a Simply Supported Beam Using Simple Beam Theory Y(SBT) and Integrating the Complete Differential Equation Using Euler's Method Y(Euler).

Example Problem F-8

$L = 5$	$X_i = 0.00$
$EI = 100{,}000$	$Y_i = 0.00$
$W = 6000.00$	$X_f = 5$
	$TH_i = -0.32829$

X	Y(Euler)	Theta	Y(SBT)	Error
0.00	0.0000	−0.32829	0.0000	0.000
0.25	−0.0817	−0.32312	−0.0777	4.790
0.50	−0.1608	−0.30826	−0.1533	4.677
0.75	−0.2352	−0.28510	−0.2246	4.494
1.00	−0.3029	−0.25512	−0.2900	4.264
1.25	−0.3625	−0.21971	−0.3479	4.016
1.50	−0.4126	−0.18013	−0.3970	3.776
1.75	−0.4525	−0.13749	−0.4363	3.566
2.00	−0.4814	−0.09272	−0.4650	3.406
2.25	−0.4989	−0.04660	−0.4824	3.310
2.50	−0.5049	0.00017	−0.4883	3.286
2.75	−0.4991	0.04694	−0.4824	3.339
3.00	−0.4817	0.09308	−0.4650	3.465
3.25	−0.4529	0.13790	−0.4363	3.656
3.50	−0.4131	0.18060	−0.3970	3.899
3.75	−0.3631	0.22025	−0.3479	4.176
4.00	−0.3035	0.25577	−0.2900	4.464
4.25	−0.2358	0.28590	−0.2246	4.739
4.50	−0.1613	0.30922	−0.1533	4.977
4.75	−0.0820	0.32427	−0.0777	5.163
5.00	−0.0000	0.32965	0.0000	0.000

Note: The error in using simple beam theory is about 3.3 percent (for the given properties and loading) at the middle of the beam (where the deflection is the greatest).

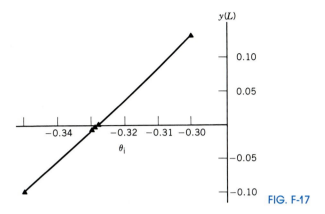

FIG. F-17

$$y_1' = y_2$$

$$y_2' = \frac{M(x)}{EI} [1 + (y_2)^2]^{3/2}$$

$$= \frac{w(Lx - x^2)}{2EI} [1 + (y_2)^2]^{3/2}$$

$$= \frac{6000(5x - x^2)}{2(100)(10^9)(10^{-6})} [1 + (y_2)^2]^{3/2}$$

subject to the boundary conditions

$$y_1(0) = 0$$

$$y_1(L) = 0$$

Guessing that $y_2(0) = -0.3$ and solving the above equations using Euler's method gives $y_1(L) = 0.132$. Trying again and guessing $y_2(0) = -0.35$ gives $y_1(L) = -0.0994$. The second guess came closer to the desired solution $y_1(L) = 0$, but overcorrected, so $y_2(0) = -0.33$ is used for the third guess and gives $y_1(L) = -0.00791$ (see Fig. F-17). Continuing in this fashion, a few more guesses led to the final choice $y_2(0) = -0.32829$. The solution for this value of $y_2(0)$ is also tabulated in Table F-3 and plotted in Fig. F-16.

The difference between the solutions using Eqs. *a* and *b* is about 3.4 percent over the middle segment of the beam where the deflection is the greatest.

(c) Equation *b* was solved for different values of the moment of inertia from $I = 0.7(10^6)$ mm⁴ to $I = 1.0(10^6)$ mm⁴. The error at the center of the beam ($x = L/2$) is plotted as a function of I in Fig. F-18 and indicates a minimum moment of inertia of about $0.825(10^6)$ mm⁴ is required to keep the error less than 5 percent in the middle of the beam where the deflection is the greatest.

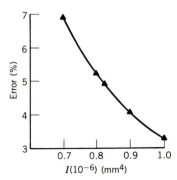

FIG. F-18

F-6

FURTHER READING

1. Chapra, S. C., and Canale, R. P. (1985). *Numerical Methods for Engineers with Personal Computer Applications*, McGraw-Hill, New York.

2. Cheney, W., and Kincaid, D. (1980). *Numerical Mathematics and Computing,* Brooks/Cole, Monterey, Calif.

3. James, M. L., Smith, G. M. and Wolford, J. C. (1985). *Applied Numerical Methods for Digital Computation,* 3rd ed., Harper & Row, New York.

4. Johnston, R. L. (1982). *Numerical Methods: A Software Approach,* Wiley, New York.

5. Mathews, J. H. (1987). *Numerical Methods for Computer Science, Engineering and Mathematics*. Prentice-Hall, Englewood Cliffs, N.J.

APPENDIX G

SECOND MOMENT OF PLANE AREAS

G-1

DEFINITIONS

The second moment of the area in Fig. G-1 with respect to the x axis is defined as the integral $\int_A y^2 \, dA$ where the integration is carried out over the area A. The quantity is called the second moment of the area to distinguish it from the first moment about the axis, which is defined as $\int_A y \, dA$. The second moment is frequently referred to as *moment of inertia* of the area because of the similarity of the defining equation to that used to express the moment of inertia of the mass of a body and the terms *second moment* and *moment of inertia* are used interchangeably.

From the definition in the preceding paragraph, it may be noted that a similar equation can be written for the second moment of the area with respect to any axis in the plane of the area. The symbol I is used for the second moment with respect to any axis in the plane of the area and is sometimes called the rectangular moment of inertia; thus,

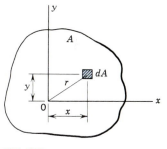

FIG. G-1

$$I_x = \int_A y^2 \, dA \qquad I_y = \int_A x^2 \, dA \qquad \text{(G-1)}$$

When the reference axis is normal to the plane of the area, for example through 0 in Fig. G-1, the integral is called the polar second moment or polar moment of inertia J_0 and can be written as

$$J_0 = \int_A r^2 \, dA = \int_A (x^2 + y^2) \, dA = \int_A x^2 \, dA + \int_A y^2 \, dA$$

Thus,

$$J_0 = I_y + I_x \qquad \text{(G-2)}$$

Note that the x and y axes can be any two mutually perpendicular axes intersecting at 0.

From the definition, the second moment of an area is the product of a distance square multiplied by an area and thus is always positive and has the dimension of a length raised to the fourth power (L^4). Common units are mm^4 and in.^{4}.

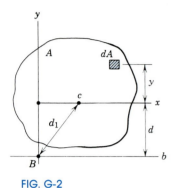

FIG. G-2

G-2

PARALLEL AXIS THEOREM

When the second moment of an area has been determined with respect to a given axis, the second moment with respect to a parallel axis can be obtained by means of the *parallel axis theorem* (sometimes called the transfer formula), provided one of the axes passes through the centroid of the area.

The second moment of the area in Fig. G-2 about the b axis is

$$I_b = \int_A (y + d)^2 \, dA = \int_A y^2 \, dA + 2d \int_A y \, dA + d^2 \int_A dA$$

$$= I_x + 2d \int_A y \, dA + Ad^2 \tag{a}$$

The integral $\int_A y \, dA$ is the first moment of the area with respect to the x axis. If the x axis passes through the centroid of the area, the first moment is zero and Eq. a becomes

$$I_b = I_c + Ad^2 \tag{G-3}$$

where I_c is the second moment of the area with respect to an axis parallel to b and passing through the centroid C, and d is the distance between the two axes. In a similar manner it can be shown that

$$J_B = J_c + Ad_1^2 \tag{G-4}$$

The parallel-axis theorem states that the *second moment of an area with respect to any axis is equal to the second moment of the area with respect to a parallel axis through the centroid of the area added to the product of the area and the square of the distance between the two axes*. The theorem demonstrates that the second moment of an area with respect to an axis through the centroid of the area is less than that for any parallel axis. Thus, solving Eq. G-3 for I_c gives

$$I_c = I_b - Ad^2$$

G-3

SECOND MOMENTS OF AREAS BY INTEGRATION

When determining the second moment of an area by integration, it is first necessary to select an element of area. If all parts of the element are the same distance from the axis, the second moment can be determined directly from the definition (see Section G-1). When any other element is selected, the second moment of the *element* with respect to the axis must either be known or be obtainable from a known result by the parallel-axis theorem.

Either single or double integration may be involved, depending on the element selected. When double integration is used, all parts of the element will be the same distance from the moment axis, and the second moment of the element can be written directly. Special care must be taken in establishing the limits for the two integrations to see that the correct area is included. If a strip element is selected, the second moment can usually be obtained by a single integration, but the element must be properly selected in order for its second moment about the reference axis to be either known or readily calculated. Either Cartesian or polar coordinates can be used for some problems. The choice of a coordinate system and the selection of an element is a matter of either personal preference or previous experience. The following examples illustrate the procedure for determining the second moments of areas by integration.

EXAMPLE G-1

A rectangle has a base b and a height h. Determine the second moment of the area with respect to

(a) the base of the rectangle.

(b) an axis through the centroid parallel to the base.

(c) an axis through the centroid normal to the area (polar second moment).

Solution

(a) The rectangular area is shown in Fig. G-3, together with an element of area parallel to the reference axis. The second moment of the area of the element about the x axis is

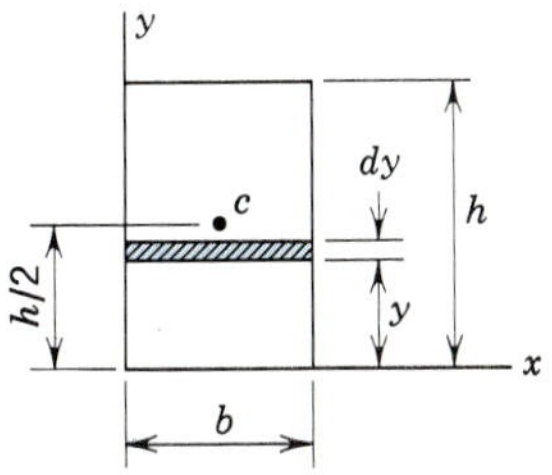

$$dI_x = y^2 \, dA = y^2 b \, dy$$

and the second moment of the entire area is

$$I_x = b \int_0^h y^2 \, dy = \frac{bh^3}{3} \qquad \text{Ans.}$$

(b) The parallel-axis theorem can be used to determine the second moment about an axis through c parallel to the x axis. Thus,

FIG. G-3

$$I_{xc} = I_x - Ad^2 = \frac{bh^3}{3} - (bh)\left(\frac{h}{2}\right)^2 = \frac{bh^3}{12} \qquad \text{Ans.}$$

(c) The second moment with respect to the vertical axis through c can be obtained from the preceding solution by interchanging b and h; that is,

$$I_{yc} = \frac{hb^3}{12}$$

and the polar second moment of the area about an axis at c is

$$J_c = I_{xc} + I_{yc} = \frac{bh^3}{12} + \frac{hb^3}{12} = \frac{bh}{12}(h^2 + b^2) \qquad \text{Ans.}$$

EXAMPLE G-2

Determine the second moment of a circular area with a radius R with respect to a diameter of the circle.

Solution

Figure G-4 shows a circular area with a radius R. The x axis is the moment axis. Polar coordinates are convenient for this problem. The second moment of the element about the x axis is

$$dI_x = (r \sin \theta)^2 \, dr(r \, d\theta) = r^3 \sin^2\theta \, dr \, d\theta$$

The second moment of the entire area is

$$I_x = \int_0^{2\pi} \int_0^R r^3 \sin^2\theta \, dr \, d\theta = \int_0^{2\pi} \frac{R^4}{4} \sin^2\theta \, d\theta$$

$$= \frac{R^4}{4} \int_0^{2\pi} \frac{1 - \cos 2\theta}{2} \, d\theta = \frac{R^4}{4}\left[\frac{\theta}{2} - \frac{\sin 2\theta}{4}\right]_0^{2\pi} = \frac{\pi R^4}{4} \qquad \text{Ans.}$$

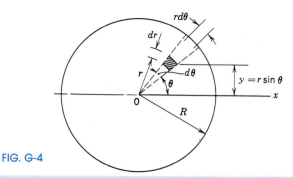

FIG. G-4

EXAMPLE G-3

The equation of the parabolic curve in Fig. G-5 is $cy = x^2$ where c is a constant. Determine the second moment of the shaded area with respect to the x and y axes.

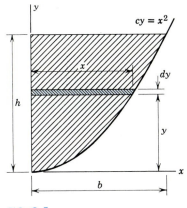

FIG. G-5

Solution

The constant in the equation can be obtained by substituting $x = b$ and $y = h$ in the equation, which gives $c = b^2/h$ and the equation becomes

$$b^2y = hx^2$$

A horizontal element (Fig. G-5) is selected for this example. The area of the element is

$$dA = x\, dy = bh^{-1/2}y^{1/2}dy$$

and the second moment of the element about the x axis is

$$dI_x = y^2\, dA = bh^{-1/2}y^{5/2}dy$$

The second moment of the area is

$$I_x = bh^{-1/2}\int_0^h y^{5/2}dy = bh^{-1/2}\left[\frac{2}{7}y^{7/2}\right]_0^h = \frac{2}{7}bh^3 \qquad \text{Ans.}$$

Since all parts of the element are *not* the same distance from the y axis, the horizontal element can be used if its second moment with respect to the y axis is shown. The element is a rectangle of height x and width dy and the reference axis is the base of the rectangle. From Example G-1, the second moment of the element about the y axis is

$$dI_y = \frac{1}{3}(dy)x^3 = \frac{1}{3}\left(\sqrt{\frac{b^2y}{h}}\right)^3 dy = \frac{b^3y^{3/2}}{3h^{3/2}}dy$$

and the second moment of the area is

$$I_y = \frac{b^3}{3h^{3/2}}\int_0^h y^{3/2}\, dy = \frac{2}{15}b^3h \qquad \text{Ans.}$$

EXAMPLE G-4

Determine the second moment of the area in Fig. G-6 with respect to the x axis.

Solution

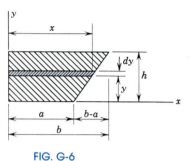

FIG. G-6

An element of area parallel to the x axis is selected as shown in Fig. G-6. The area of the element is $dA = x \, dy$ where

$$x = a + \frac{b-a}{h} y$$

and the second moment of the element with respect to the x axis is

$$dI_x = y^2 \, dA = \left(ay^2 + \frac{b-a}{h} y^3 \right) dy$$

The second moment of the area about the x axis is

$$I_x = \int_0^h \left(ay^2 + \frac{b-a}{h} y^3 \right) dy = \frac{(a+3b)h^3}{12} \qquad \text{Ans.}$$

G-4

RADIUS OF GYRATION OF AREAS

It is sometimes convenient to express the second moment of area (moment of inertia) as a function of the area and a length. Since the second moment of an area has the dimensions of length to the fourth power, it can be expressed as the product of the area multiplied by a length squared. *The radius of gyration of an area with respect to a specified axis is defined as the length that when squared and multiplied by the area, will give the second moment of the area with respect to the specified axis.* The definition can be expressed by the equations

$$Ak_b^2 = I_b \qquad \text{and} \qquad Ak_0^2 = J_0 \qquad (G-5)$$

where k_b and k_0 are the radii of gyration with respect to the b axis and the polar axis through 0, respectively. The radius of gyration is *not* the distance from the reference axis to a fixed point in the area (such as the centroid), but it is a useful property of the area and the specified axis.

The radius of gyration of an area with respect to any axis is always greater than the distance from the axis to the centroid of the area. From Fig. G-7, the moment of inertia of the area with respect to the b axis is

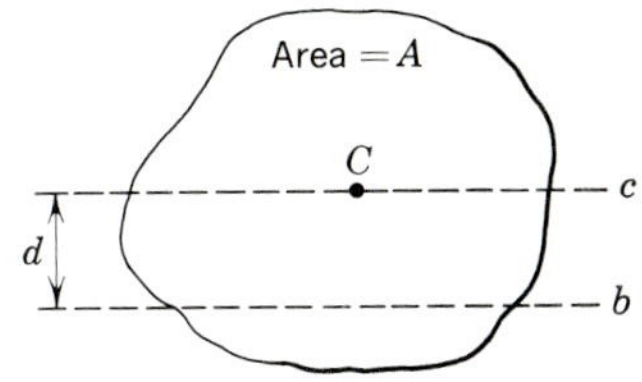

$$I_b = I_c + Ad^2 = Ak_b^2 = Ak_c^2 + Ad^2$$

or

$$k_b^2 = k_c^2 + d^2$$

from which

$$k_b = (k_c^2 + d^2)^{1/2}$$

This equation indicates that k_b will always be greater than the distance d from the b axis to the centroid. If the distance d is increased, the value of k_b will also be increased, but the increase in k_b will be less than the increase in d, indicating that k_b is *not* the distance from the b axis to a fixed point in the area.

The radius of gyration of an area is a convenient means of expressing a property of an area and is often useful in formulas developed in mechanics—for example, formulas giving the strength of columns.

FIG. G-7

G-5

SECOND MOMENT OF COMPOSITE AREAS

It is often necessary to calculate the second moment of an irregularly shaped area. When such an area is made up of a series of simple shapes such as rectangles, triangles, and circles, the second moment can be determined by use of the parallel-axis theorem rather than by integration. The second moment of the irregular area, the *composite area,* with respect to any axis is equal to the sum of the second moments of the separate parts of the area with respect to the specified axis. When an area such as a hole is removed from a larger area, its second moment must be subtracted from the second moment of the larger area to obtain the resulting second moment.

It is useful to have formulas available giving second moments of areas frequently encountered in routine work. Table B-1 lists properties of several geometric shapes frequently encountered. Tables listing second moments of the cross-sectional areas of various structural shapes are found in engineering handbooks and in data books prepared by industrial organizations such as the American Institute of Steel Construction and the Aluminum Company of America. Properties of a few selected shapes are included in Appendix C for use in solving problems.

EXAMPLE G-5

Determine the second moment of the shaded area in Fig. G-8 with respect to the x axis.

Solution

The shaded area can be divided into a 10×6-in. rectangle A and a 10×6-in. triangle B with a 4-in. radius semicircle C removed. The location of the centroids of the three areas, as determined from Table B-1, is shown to the right of the figure. The second moments of the three areas with respect to the x axis can be obtained, by using information from Table B-1, as follows:

For Area A,

$$I_x = \frac{bh^3}{3} = \frac{10(6)^3}{3} = 720 \text{ in.}^4$$

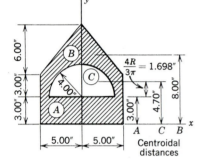

FIG. G-8

For area B,

$$I_x = I_c + Ad^2 = \frac{bh^3}{36} + Ad^2$$

$$= \frac{10(6)^3}{36} + \frac{10(6)}{2}(8)^2 = 1980 \text{ in.}^4$$

For area C, although I_c for the semicircle is given in Table B-1, the development of the expression here will expand the coverage of the subject matter. Since the area is half of a circle, its second moment with respect to a diameter is half that of a full circle, and from the parallel axis theorem

$$I_c = I_{\text{dia}} - Ad^2 = \frac{1}{2}\frac{\pi R^4}{2} - \frac{\pi R^2}{2}\left(\frac{4R}{3\pi}\right)^2$$

$$= \frac{\pi(4)^4}{8} - \frac{8(4)^4}{9\pi} = 28.10 \text{ in.}^4$$

and

$$I_x = I_c + Ad^2 = 28.10 + \frac{\pi(4)^2}{2}(4.698)^2 = 582.8 \text{ in.}^4$$

The second moment of the composite area is

$$I_x = 720 + 1980 - 583 = \underline{2117 \text{ in.}^4} \qquad \text{Ans.}$$

G-6

PRODUCTS OF INERTIA OF AREAS

The product of inertia dI_{xy} of the element of area dA in Fig. G-9 with respect to the x and y axes is defined as the product of the two coordinates of the element multiplied by the area of the element; the product of inertia of the total area A about the x and y axes is the sum of the products of inertia of the elements of the area; thus,

$$I_{xy} = \int_A xy \, dA \qquad \text{(G-6)}$$

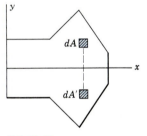

FIG. G-9

The dimensions of the product of inertia are the same as for the moment of inertia (second moment), but since the product xy can be either positive or negative, the product of inertia can be positive, negative, or zero as contrasted with the second moment, which is always positive.

The product of inertia of an area with respect to any two orthogonal axes is zero when either of the axes is an axis of symmetry. This statement can be demonstrated by means of Fig. G-10, which is symmetrical with respect to the x axis. The products of inertia of the elements dA and dA' on opposite sides of the axis of symmetry will be equal in magnitude and opposite in sign and will thus cancel each other in the summation. Therefore, the resulting product of inertia will be zero.

The parallel-axis theorem for products of inertia can be derived from Fig. G-11 in which the x' and y' axes pass through the centroid C and are parallel to the x and y axes. The product of inertia with respect to the x and y axes is

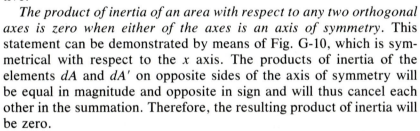

FIG. G-10

$$I_{xy} = \int_A xy \, dA = \int_A (x_c + x')(y_c + y') \, dA$$

$$= x_c y_c \int_A dA + x_c \int_A y' \, dA + y_c \int_A x' \, dA + \int_A x'y' \, dA$$

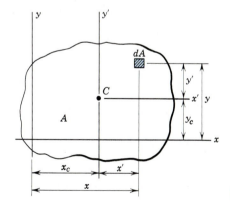

FIG. G-11

The second and third integrals in the preceding equation are zero since x' and y' are centroidal axes and the last integral is the product of inertia with respect to the centroidal axes. Consequently, the product of inertia is

$$I_{xy} = I_{x'y'} + x_c y_c A \qquad \text{(G-7)}$$

The parallel axis theorem for products of inertia can be stated as follows: *The product of inertia of an area with respect to any two perpendicular axes x and y is equal to the product of inertia of the area with respect to a pair of centroidal axes parallel to the x and y axes added to the product of the area and the two centroidal distances from the x and y axes.*

The product of inertia is useful in determining principal axes of inertia, as discussed in the following section. The determination of the product of inertia is illustrated in the next two examples.

EXAMPLE G-6

Determine the product of inertia of the triangular area shown in Fig. G-12 with respect to

(a) the x and y axes.

(b) a pair of centroidal axes parallel to the x and y axes.

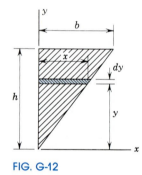

FIG. G-12

Solution

(a) The horizontal element in Fig. G-12 has a length x and a height dy. From symmetry, the product of inertia of the element with respect to axes through the centroid of the element parallel to the x and y axes is zero. Thus, the product of inertia of the element with respect to the x and y axes is

$$dI_{xy} = \frac{x}{2} y \, dA = \frac{x}{2} y (x \, dy)$$

From similar triangles, x is equal to $(b/h)y$ and the product of inertia of the element becomes

$$dI_{xy} = \frac{b^2}{2h^2} y^3 \, dy$$

The product of inertia of the triangular area is

$$I_{xy} = \frac{b^2}{2h^2} \int_0^h y^3 \, dy = \frac{b^2 h^2}{8} \qquad \text{Ans.}$$

(b) The product of inertia with respect to the centroidal axes can be obtained from the parallel axis theorem; thus,

$$I_{x'y'} = I_{xy} - Ax_c y_c = \frac{b^2 h^2}{8} - \frac{bh}{2} \left(\frac{b}{3}\right)\left(\frac{2h}{3}\right) = \underline{\frac{b^2 h^2}{72}} \quad \text{Ans.}$$

EXAMPLE G-7

Determine the product of inertia of the quadrant of the circle in Fig. G-13 with respect to the x and y axes.

Solution

The horizontal element selected is shown in Fig. G-13. As in Example G-6, the product of inertia of the element with respect to its centroidal axes is zero and its product of inertia with respect to the x and y axes is

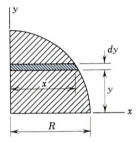

FIG. G-13

$$dI_{xy} = \frac{x}{2}(y)(x \; dy) = \frac{y}{2} x^2 \; dy$$

Since $R^2 = x^2 + y^2$, the preceding expression becomes

$$dI_{xy} = \frac{y}{2}(R^2 - y^2) \; dy$$

and the product of inertia of the area is

$$I_{xy} = \frac{1}{2}\int_0^R (R^2 y - y^3) \; dy = \underline{\frac{R^4}{8}} \quad \text{Ans.}$$

G-7

PRINCIPAL MOMENTS OF INERTIA

The moment of inertia of the area A in Fig. G-14 with respect to the x' axis through 0 will, in general, vary with the angle θ. The x and y axes used to obtain Eq. G-2 were any pair of orthogonal axes in the plane of the area passing through 0; therefore,

$$J_0 = I_x + I_y = I_{x'} + I_{y'}$$

where x' and y' are any pair of orthogonal axes through 0. Since the

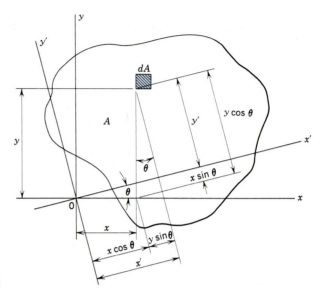

FIG. G-14

sum of $I_{x'}$ and $I_{y'}$ is a constant, $I_{x'}$ will be maximum and the corresponding $I_{y'}$ will be minimum for one particular value of θ.

The set of axes for which the second moments are maximum and minimum are called the *principal axes* of the area through point 0 and are designated as the u and v axes. The second moments of the area with respect to these axes are called the *principal moments of inertia* of the area and are designated I_u and I_v. There is only one set of principal axes for any point in an area unless all axes have the same second moment, such as the diameters of a circle.

A convenient way to determine the principal moments of inertia of an area is to express $I_{x'}$ as a function of I_x, I_y, I_{xy}, and θ and then set the derivative of $I_{x'}$ with respect to θ equal to zero to obtain the value of θ giving the maximum and minimum second moments. From Fig. G-14,

$$dI_{x'} = y'^2 \, dA = (y \cos \theta - x \sin \theta)^2 \, dA$$

and

$$I_{x'} = \cos^2 \theta \int_A y^2 \, dA - 2 \sin \theta \cos \theta \int_A xy \, dA + \sin^2 \theta \int_A x^2 \, dA$$

$$= I_x \cos^2 \theta - 2I_{xy} \sin \theta \cos \theta + I_y \sin^2 \theta$$

$$= \frac{1}{2}(I_x + I_y) + \frac{1}{2}(I_x - I_y)\cos 2\theta - I_{xy} \sin 2\theta \qquad \text{(G-8)}$$

The angle 2θ for which $I_{x'}$ is maximum can be obtained by setting the derivative of $I_{x'}$ with respect to θ equal to zero; thus,

$$\frac{dI_{x'}}{d\theta} = -2(1/2)(I_x - I_y)\sin 2\theta - 2I_{xy}\cos 2\theta = 0$$

from which

$$\tan 2\beta = -\frac{I_{xy}}{(I_x - I_y)/2} \tag{G-9}$$

where β represents the two values of θ that locate the principal axes u and v. Equation G-9 gives two values of 2β that are 180° apart, and thus two values of β that are 90° apart. The principal moments of inertia can be obtained by substituting these values of β in Eq. G-8. From Eq. G-9,

$$\cos 2\beta = \mp \frac{(I_x - I_y)/2}{\{[(I_x - I_y)/2]^2 + I_{xy}^2\}^{1/2}}$$

and

$$\sin 2\beta = \pm \frac{I_{xy}}{\{[(I_x - I_y)/2]^2 + I_{xy}^2\}^{1/2}}$$

When these expressions are substituted in Eq. G-8, the principal moments of inertia reduce to

$$I_{u,v} = \frac{I_x + I_y}{2} \pm \sqrt{\left(\frac{I_x - I_y}{2}\right)^2 + I_{xy}^2} \tag{G-10}$$

The product of inertia of the element of area in Fig. G-14 with respect to the x' and y' axes is

$$dI_{x'y'} = x'y'\, dA = (x\cos\theta + y\sin\theta)(y\cos\theta - x\sin\theta)\, dA$$

and the product of inertia of the area is

$$I_{x'y'} = (\cos^2\theta - \sin^2\theta)\int_A xy\, dA + \sin\theta\cos\theta\int_A (y^2 - x^2)\, dA$$

$$= I_{xy}\cos 2\theta + \frac{1}{2}(I_x - I_y)\sin 2\theta \tag{G-11}$$

The product of inertia $I_{x'y'}$ will be zero when

$$\tan 2\theta = -\frac{I_{xy}}{(I_x - I_y)/2}$$

which is the same as Eq. G-9, which locates the principal axes. Consequently, *the product of inertia is zero with respect to the principal axes*. Since the product of inertia is zero with respect to any axis of symmetry, it follows that any axis of symmetry must be a principal axis for any point on the axis of symmetry.

The following example illustrates the procedure for determining second moments of areas with respect to the principal axes.

EXAMPLE G-8

Determine the maximum and minimum second moments of the area of the unequal-leg angle in Fig. G-15 with respect to axes through the centroid of the area.

Solution

The area is divided into two rectangles A and B, and the location of the centroid of each area is indicated by the dimensions. The centroid of the composite area is at C. The values of I_x and I_y for the two areas are obtained by applying the parallel-axis theorem to the expression for I_c from Table B-1; thus,

$$I_x = \frac{bh^3}{12} + Ay_c^2$$

$$I_x = I_{Ax} + I_{Bx} = \frac{1(8)^3}{12} + 1(8)(1.346)^2 + \frac{5(1)^3}{12} + 5(1)(2.154)^2$$

$$= 80.8 \text{ in.}^4$$

$$I_y = I_{Ay} + I_{By} = \frac{8(1)^3}{12} + 8(1)(1.154)^2 + \frac{1(5)^3}{12} + 1(5)(1.846)^2$$

$$= 38.8 \text{ in.}^4$$

The products of inertia are determined as indicated in Section G-6; thus,

$$I_{xy} = I_{x'y'} + Ax_cy_c = 0 + Ax_cy_c$$

$$I_{xy} = I_{Axy} + I_{Bxy} = 1(8)(-1.154)(1.346) + 5(1)(1.846)(-2.154)$$

$$= -32.3 \text{ in.}^4$$

From Eq. G-9,

$$\tan 2\beta = -\frac{-32.3}{(80.8 - 38.8)/2} = 1.538$$

$$2\beta = 57.0° \quad \text{or} \quad 237°$$

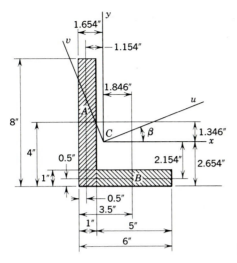

FIG. G-15

Therefore,

$$\beta = 28.5° \quad \text{or} \quad 118.5°$$

From Eq. G-8 with $\beta = 28.5°$, the maximum second moment is

$$I_u = 80.8(0.879)^2 - 2(-32.3)(0.477)(0.879) + 38.8(0.477)^2$$
$$= \underline{98.3 \text{ in.}^4} \qquad\qquad\qquad \text{Ans.}$$

With $\beta = 118.5°$, the minimum second moment is

$$I_v = 80.8(-0.477)^2 - 2(-32.3)(0.879)(-0.477) + 38.8(0.879)^2$$
$$= \underline{21.3 \text{ in.}^4} \qquad\qquad\qquad \text{Ans.}$$

The principal moments of inertia can also be determined by means of Eq. G-10.

$$I_{u,v} = \frac{80.8 + 38.8}{2} \pm \sqrt{\left(\frac{80.8 - 38.8}{2}\right)^2 + (-32.3)^2}$$
$$= \underline{98.3 \text{ in.}^4} \quad \text{and} \quad \underline{21.3 \text{ in.}^4} \qquad \text{Ans.}$$

G-8

MOHR'S CIRCLE FOR SECOND MOMENTS OF AREAS

The use of Mohr's circle for determining principal stresses was discussed in Section 7-4. A comparison of Eqs. 7-1a and 7-1b with Eq.

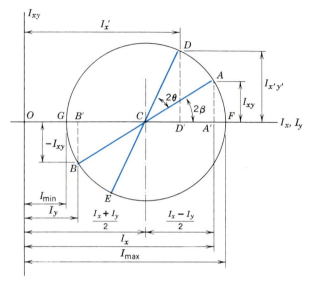

G-8 indicates that a similar procedure can be used to obtain the principal moments of inertia of an area.

Figure G-16 illustrates the use of Mohr's circle for second moments. Assume that I_x is greater than I_y and that I_{xy} is positive. Moments of inertia are plotted along the horizontal axis, and products of inertia are plotted along the vertical axis. Second moments are always positive and are plotted to the right of the origin. Products of inertia can be either positive or negative; positive values are plotted above the horizontal axis. The distance OA' along the I axis is equal to I_x and $A'A$, parallel to the I_{xy} axis, is equal to I_{xy}. Similarly, OB' is laid off equal to I_y and $B'B$ is equal to $-I_{xy}$. The line AB intersects the I axis at C and AB is the diameter of Mohr's circle for second moments. Each point on the circle represents $I_{x'}$ and $I_{x'y'}$ for one particular orientation of the x' and y' axes. To demonstrate this statement, the diameter DE is drawn at an angle 2θ counterclockwise from line AB. From the figure

$$OD' = OC + CD\cos(2\beta + 2\theta)$$

which, since CD is equal to CA, reduces to

$$OD' = OC + CA\cos 2\beta \cos 2\theta - CA \sin 2\beta \sin 2\theta$$

Since $CA \cos 2\beta$ and $CA \sin 2\beta$ are equal to CA' and $A'A$, respectively, the distance OD' becomes

$$OD' = \frac{1}{2}(I_x + I_y) + \frac{1}{2}(I_x - I_y)\cos 2\theta - I_{xy}\sin 2\theta = I_{x'}$$

from Eq. G-8. In a similar manner,

$$D'D = CD \sin(2\beta + 2\theta)$$

$$= \frac{1}{2}(I_x - I_y)\sin 2\theta + I_{xy} \cos 2\theta = I_{x'y'}$$

from Eq. G-11.

Since the horizontal coordinate of each point on the circle represents a particular value of $I_{x'}$, the maximum second moment is represented by OF and its value is

$$I_{max} = OF = OC + CF = OC + CA$$

$$= \frac{I_x + I_y}{2} + \sqrt{\left(\frac{I_x - I_y}{2}\right)^2 + I_{xy}^2}$$

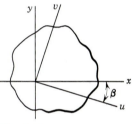

which agrees with Eq. G-10. Figure G-17 represents an area and a set of axes for which the data used to construct Fig. G-16 are valid. In deriving Eq. G-8, the angle between the x and x' axes was θ. In obtaining the same equation from Mohr's circle, the angle between the radii to points (I_x, I_{xy}), line CA, and $(I_x, I_{x'y'})$, line CD, is 2θ. Thus, all angles on Mohr's circle are twice the corresponding angles for the given area.

FIG. G-17

EXAMPLE G-9

Solve Example G-8 by means of Mohr's circle.

Solution

Values of I_x, I_y, and I_{xy} must be computed in the same manner as in Example G-8. These values are

$$I_x = 80.8 \text{ in.}^4 \qquad I_y = 38.8 \text{ in.}^4 \qquad I_{xy} = -32.3 \text{ in.}^4$$

The circle in Fig. G-18 is constructed as indicated from these values. Point A is located at I_x and I_{xy} (which is negative) and B is at I_y and $-I_{xy}$. Point C is the center of the circle and is 59.8 units from O. The radius of the circle is

$$CA = \sqrt{(21.0)^2 + (32.3)^2} = 38.5 \text{ in.}^4$$

and the principal moments of inertia are

$$I_{max} = OC + CF = 59.8 + 38.5 = \underline{98.3 \text{ in.}^4} \qquad \text{Ans.}$$

$$I_{min} = OC - CG = 59.8 - 38.5 = \underline{21.3 \text{ in.}^4} \qquad \text{Ans.}$$

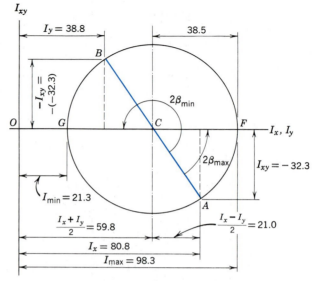

FIG. G-18

Twice the angle from the x axis to the principal axis with the maximum moment of inertia is shown on the sketch as $2\beta_{max}$ and

$$\beta_{max} = \frac{1}{2} \tan^{-1} \frac{32.3}{21.0} = 28.5° \text{ counterclockwise from the } x \text{ axis.}$$

The angle from the x axis to the principal axis with the minimum moment of inertia is β_{min} and is

$$\beta_{min} = \beta_{max} + 90° = 118.5° \text{ counterclockwise from the } x \text{ axis.}$$

These values were obtained analytically from the geometry of Mohr's circle. They could also have been measured directly from the figure; however, the accuracy of the results obtained by scaling the distances from the figure would depend on the scale used and the care employed in constructing the figure.

ANSWERS

CHAPTER 1

1-1 d_s = 1.382 in.
      d_b = 0.651 in.
      d_a = 1.349 in.

1-2   (a) σ = 175.0 MPa T
      (b) σ = 50.0 MPa T
      (c) σ = 90.0 MPa T

1-3   (a) σ = 23.1 ksi C
      (b) σ = 30.0 ksi C
      (c) σ = 20.0 ksi C

1-4 A_{AB} = 4770 mm²
      A_{BC} = 2600 mm²
      A_{CD} = 1167 mm²

1-7   (a) τ = 11.32 ksi
      (b) L = 6.37 in.

1-8 L = 175.4 mm

1-11 L_1 = 2.55 in.
      L_2 = 3.40 in.

1-12   (a) σ_b = 23.6 MPa C
      (b) d = 242 mm

1-15 d = 0.1270 in.

1-16 P = 589 kN

1-19   (a) d = 1.160 in.
      (b) d = 1.641 in.
      (c) d = 1.690 in.

1-20   (a) P = 67.5 kN
      (b) d = 25.2 mm
      (c) d = 21.2 mm

1-23   (a) σ = 11.88 ksi C
      (b) d = 1.630 in.
      (c) d = 1.289 in.

1-24   (a) σ = 117.7 MPa T

(b) A = 589 mm²
(c) A = 117.7 mm²

1-27 P = 4.88 kips

1-29 d = 1.596 in.

1-30 σ = 5.37 MPa C
      τ = 8.27 MPa

1-33 P = 9.38 kips

1-34   (a) P = 695 kN
      (b) τ = 16.52 MPa
      (c) σ_{max} = 34.7 MPa C
          τ_{max} = 17.37 MPa

1-37   7.17 × 7.17 in.

1-38 t = 40.6 mm

1-41 σ = 12.31 ksi T
      τ = 1.075 ksi

1-42 σ = 121.5 MPa T
      τ = 9.64 MPa

1-45 σ = 4.37 ksi T
      τ = 14.24 ksi

1-46 σ = 163.8 MPa T
      τ = 9.85 MPa

1-49   (a) τ_h = 3.25 ksi
          τ_v = 3.25 ksi
      (b) τ = 4.75 ksi

1-50   (a) σ_x = 98.4 MPa T
      (b) τ = 76.3 MPa

1-53 σ = 5.47 ksi C

1-54   (a) σ = 99.0 MPa C
          τ = 168.0 MPa
      (b) σ = 1.000 MPa C
          τ = 168.0 MPa

1-59 ϵ = 2080 μin./in.

1-60 ΔL = 0.480 mm

1-63 $\gamma = 1396 \ \mu\text{rad}$

1-64 $\delta = 0.01000 \ \text{mm}$

1-69 (a) $\epsilon = 1543 \ \mu\text{in./in.}$
(b) $\epsilon = 1400 \ \mu\text{in./in.}$
(c) $\epsilon = 1638 \ \mu\text{in./in.}$

1-70 (a) $\epsilon = 2000 \ \mu\text{m/m}$
(b) $\epsilon = 1583 \ \mu\text{m/m}$
(c) $\epsilon = 2750 \ \mu\text{m/m}$

1-73 (a) $\Delta L = \gamma L^2/6E$
(b) $\epsilon_{\text{avg}} = \gamma L/6E$
(c) $\epsilon_{\text{max}} = \gamma L/3E$

1-74 (a) $\Delta L = 5.00 \ \text{mm}$
(b) $\epsilon_{\text{avg}} = 1667 \ \mu\text{m/m}$
(c) $\epsilon_{\text{max}} = 5000 \ \mu\text{m/m}$

1-77 $\gamma = 2820 \ \mu\text{rad}$

1-78 $\gamma_{xy} = 1560 \ \mu\text{rad}$

1-81 $\epsilon_n = 1536 \ \mu\text{in./in.}$

1-82 $\epsilon_n = -2420 \ \mu\text{m/m}$

1-85 (a) $E = 15.79(10^6) \ \text{psi}$
(b) $\nu = \frac{1}{3}$
(c) $\sigma = 23.7 \ \text{ksi}$

1-86 (a) $E = 78.7 \ \text{GPa}$
(b) $\nu = 0.326$
(c) $\sigma = 354 \ \text{MPa}$

1-91 $d = 0.921 \ \text{in.}$

1-92 (a) $\epsilon_A = -58.2 \ \mu\text{m/m}$
(b) $\epsilon_B = -131.0 \ \mu\text{m/m}$
(c) $\Delta d = 3.93 \ \mu\text{m}$

1-95 (a) $V = 2.00 \ \text{kips}$
(b) $\delta = 0.000347 \ \text{in.}$

1-96 (a) $h = 104.2 \ \text{mm}$
(b) $P = 360 \ \text{kN}$

1-97 $\sigma_x = 8.95 \ \text{ksi} \ T$
$\sigma_y = 0.1753 \ \text{ksi} \ C$
$\tau_{xy} = -1.520 \ \text{ksi}$

1-98 $\sigma_x = 191.2 \ \text{MPa} \ T$
$\sigma_y = 130.4 \ \text{MPa} \ T$
$\tau_{xy} = -42.6 \ \text{MPa}$

1-101 $\epsilon_x = 375 \ \mu\text{in./in.}$
$\epsilon_y = -58.3 \ \mu\text{in./in.}$
$\epsilon_z = 50.0 \ \mu\text{in./in.}$
$\gamma_{xy} = 477 \ \mu\text{rad}$
$\gamma_{yz} = 412 \ \mu\text{rad}$
$\gamma_{zx} = 277 \ \mu\text{rad}$

1-102 $\epsilon_x = 1825 \ \mu\text{m/m}$

$\epsilon_y = -1910 \ \mu\text{m/m}$
$\epsilon_z = 458 \ \mu\text{m/m}$
$\gamma_{xy} = 1276 \ \mu\text{rad}$
$\gamma_{yz} = 1749 \ \mu\text{rad}$
$\gamma_{zx} = 2770 \ \mu\text{rad}$

1-107 $\Delta L = 48.7 \ \text{in.}$

1-108 $\Delta L = -72.0 \ \text{mm}$

1-109 $\Delta L = 4.39 \ \text{in.}$
$\Delta d = 0.234 \ \text{in.}$

1-110 $\Delta L = 0.0238\%$

1-113 $\delta = 0.0472 \ \text{in.} \ \uparrow$

1-114 $\delta = 2.12 \ \text{mm} \ \leftarrow$

1-117 (a) $d_i = 5.994 \ \text{in.}$
(b) $\sigma = 14.10 \ \text{ksi} \ T$

1-118 (a) $T = 81.9°\text{C}$
(b) $\sigma = 83.3 \ \text{MPa} \ T$

1-121 $\sigma_x = 9.34 \ \text{ksi} \ C$
$\sigma_y = 5.58 \ \text{ksi} \ C$

1-122 $\sigma_x = 81.3 \ \text{MPa} \ T$
$\sigma_y = 304 \ \text{MPa} \ T$

1-123 (a) $k = 3.53$
(b) $k = 6.48$

1-124 (a) $k = 2.50$
(b) $k = 4.50$

1-127 (a) $k_{AC} = 1.932$
$k_{BC} = 2.39$
(b) $k_{AC} = 7.73$
$k_{BC} = 8.54$

1-128 (a) $k_{AC} = 1.913$
$k_{BC} = 1.566$
(b) $k_{AC} = 3.95$
$k_{BC} = 3.13$

1-131 (a) $D = 0.623 \ \text{in.}$
(b) $D = 0.600 \ \text{in.}$
(c) $D = 0.747 \ \text{in.}$

1-132 (a) $D = 48.0 \ \text{mm}$
(b) $D = 60.5 \ \text{mm}$
(c) $D = 54.7 \ \text{mm}$

1-135 (a) $t = 6.67 \ \text{hr}$
(b) $t = 16.67 \ \text{hr}$
(c) $t = 23.8 \ \text{hr}$
(d) $t = \text{unlimited}$

1-137 (a) $\sigma_w = 66 \ \text{ksi} \ (455 \ \text{MPa})$
(b) $\sigma_w = 59 \ \text{ksi} \ (407 \ \text{MPa})$
(c) $\sigma_w = 54 \ \text{ksi} \ (372 \ \text{MPa})$

1-139 $k = 2.06$

1-140 $\sigma_{max} = 60.1$ ksi T
$\sigma_{min} = 36.1$ ksi C

CHAPTER 2

2-1 (a) $\delta = 0.01405$ in.
(b) $k = 3.18$

2-2 (a) $\delta = 3.89$ mm
(b) $k = 4.61$

2-3 $d = 1.596$ in.

2-4 $d = 12.06$ mm

2-7 (a) $\Delta L = 0.00905$ in.
(b) $\delta = 0.0727$ in.

2-8 (a) $\delta = 1.804$ mm
(b) $\delta = 1.404$ mm

2-11 $t = 0.428$ in.

2-12 $d = 6.21$ mm
$L = 2.21$ m

2-15 $\delta = 90.6$ μin.

2-16 $\delta = 0.0545$ mm

2-17 (a) $\sigma_A = 28.1$ ksi T
$\sigma_B = 22.5$ ksi T
(b) $\delta_C = 0.0375$ in.
(c) $P = 31.5$ kips

2-18 (a) $\sigma_A = 228$ MPa T
$\sigma_B = 87.6$ MPa T
(b) $\delta_D = 1.800$ mm
(c) $P = 150.5$ kN

2-21 $d = 1.932$ in.

2-22 $d = 42.1$ mm

2-25 (a) $A_A = 0.316$ in.^{2}
$A_B = 0.508$ in.^{2}
(b) $\delta_A = 0.1857$ in.
$\delta_B = 0.0894$ in.

(c) $\delta_v = 0.1615$ in. $\downarrow$
$\delta_h = 0.1010$ in. $\rightarrow$

2-26 (a) $A_{AB} = 348$ mm^2
$A_{BC} = 779$ mm^2
(b) $\delta_{AB} = 3.41$ mm
$\delta_{BC} = 2.22$ mm
(c) $\delta_v = 4.52$ mm $\downarrow$
$\delta_h = 3.40$ mm $\rightarrow$

2-29 (a) $\sigma_T = 8.42$ ksi T
$\sigma_B = 14.73$ ksi C
(b) $\delta_C = 0.0236$ in. $\downarrow$

2-30 (a) $\sigma_A = 56.6$ MPa C
$\sigma_B = 84.9$ MPa T
(b) $\delta_C = 0.849$ mm

2-32 (a) $\sigma_A = 0$
$\sigma_{BL} = 12.73$ ksi T
$\sigma_{BR} = 12.73$ ksi C
(b) $\delta = 0.0306$ in.

2-34 $P = 90.3$ N

2-35 (a) $\sigma_s = 8.16$ ksi C
$\sigma_c = 1.224$ ksi C
(b) $\delta = 0.00653$ in.

2-36 (a) $P = 3.36$ MN
(b) $\delta = 0.375$ mm

2-39 (a) $\sigma_s = 9.57$ ksi C
$\sigma_c = 1.166$ ksi C
(b) $\Delta L = +0.00818$ in.

2-40 (a) $\sigma_p = 1.074$ MPa T
$\sigma_b = 11.93$ MPa T
(b) $a = 200.1023$ mm

2-43 (a) $\sigma_A = 1.113$ ksi C
$\sigma_B = 25.7$ ksi C
(b) $k_A = 55.7$
$k_B = 2.92$
(c) $\delta_C = 0.0448$ in.

2-44 (a) $A_A = 1884$ mm^2
(b) $\delta_C = 0.321$ mm

2-45 (a) $\sigma_A = 12.03$ ksi T
$\sigma_B = 15.89$ ksi T
(b) $\delta_D = 0.0681$ in.

2-46 (a) $\sigma_A = 110.0$ MPa T
$\sigma_B = 78.5$ MPa T
(b) $\delta_C = 0.524$ mm

2-49 (a) $\sigma_A = 14.93$ ksi T
$\sigma_B = 5.96$ ksi T
(b) $\delta_D = 0.1408$ in.

2-50 (a) $\sigma_A = 143.5$ MPa T
$\sigma_B = 68.6$ MPa T
(b) $\delta_C = 1.279$ mm

2-53 (a) $\sigma_A = 24.6$ ksi T
$\sigma_B = 31.5$ ksi T
(b) $\delta_C = 0.0788$ in.

2-54 (a) $\sigma = 59.0$ MPa T
(b) $\delta_D = 2.53$ mm

2-57 (a) $P = 64.3$ kips
(b) $\delta = 0.0408$ in.

2-58 (a) $\sigma_A = 157.3$ MPa T
(b) $\delta_A = 1.180$ mm

2-61 (a) $P = 62.8$ kN
(b) $\delta = 0.968$ mm

2-62 (a) $P = 71.8$ kN
(b) $\delta = 0.939$ mm

2-65 $L = 6.94$ in.

2-66 $L = 130.0$ mm

2-67 $L_1' = 8.22$ in.
$L_2' = 2.08$ in.

2-68 $L_1' = 248$ mm
$L_2' = 72.5$ mm

2-71 (a) $L_1' = 11.20$ in.
$L_2' = 4.57$ in.
(b) $L_1' = 12.82$ in.
$L_2' = 3.95$ in.

2-72 (a) $L_1' = 327$ mm
$L_2' = 131.0$ mm
(b) $L_1' = 390$ mm
$L_2' = 106.8$ mm

2-75 $P = 16.63$ kips

2-76 $P = 332$ kN

2-79 $P = 2.05$ kips

2-80 $r = 12.00$ mm

2-83 $\sigma = 720$ psi T

2-84 $t = 2.50$ in.

2-85 $t = 0.600$ in.

2-86 $P = 1.235$ MPa

2-89 $t = 0.474$ in.

2-90 (a) $\sigma = 58.0$ MPa T
(b) $p = 3.13$ MPa

2-93 $t = 0.312$ in.

2-94 (a) $t = 16.68$ mm
(b) $t = 8.34$ mm

2-98 $N = 62.3$ kN
$V = 15.43$ kN

2-100 (a) $p = 3.03$ MPa
(b) $\sigma_a = 150$ MPa T
$\sigma_h = 300$ MPa T
(c) $\epsilon_h = 1275\ \mu$m/m

2-101 $p = 3.75$ MPa

2-105 $\sigma_m = 7\gamma r^2/18t\ T$
$\sigma_t = \gamma r^2/9t\ T$

2-106 $\sigma_m = \gamma r^2/6t\ T$
$\sigma_t = 5\gamma r^2/6t\ T$

2-109 $\sigma_m = 75.0$ psi T
$\sigma_t = 158.3$ psi T

CHAPTER 3

3-1 (a) $\tau = 11.46$ ksi
(b) $\theta = 0.0688$ rad

3-2 (a) $\tau_o = 65.8$ MPa
(b) $\tau_i = 43.9$ MPa
(c) $\theta = 0.0274$ rad

3-5 (a) $d_{AB} = 3.37$ in.
$d_{BC} = 2.48$ in.
(b) $\theta_{B/A} = 0.0204$ rad
(c) $\theta_{C/A} = 0.0490$ rad

3-6 (a) $d_{AB} = 76.1$ mm
$d_{AC} = 61.9$ mm
(b) $\theta_{B/A} = 0.0458$ rad
(c) $\theta_{C/B} = 0.0218$ rad

3-9 (a) $\tau = 4.75$ ksi
(b) $\theta = 0.00335$ rad

3-11 (a) $\tau = 81.5$ MPa
(b) $\tau = 80.7$ MPa
(c) $\theta = 0.644$ rad

3-13 (a) $d_{AB} = 0.546$ in.
(b) $d_{CD} = 0.402$ in.
(c) $L = 2.69$ ft

3-14 (a) $d = 65.4$ mm
(b) $\theta_{D/A} = 0.0724$ rad

3-17 (a) $\tau_{AB} = 7.64$ ksi
(b) $\tau_{BC} = 15.09$ ksi
(c) $\theta_{B/C} = 0.0252$ rad
(d) $\theta_{A/C} = 0.0557$ rad

3-18 (a) $d_{AB} = 68.8$ mm
$d_{BC} = 93.4$ mm
(b) $d = 74.9$ mm

3-21 (a) $T = 149.6$ in.-kips
(b) $k = 3.60$

3-22 (a) $d = 99.3$ mm
(b) $\theta_{D/A} = 0.0288$ rad

3-25 (a) $T_3 = 983$ in.-kips
(b) $\theta_B = 0.0540$ rad
(c) $\theta_D = 0.0670$ rad

3-26 (a) $\tau = 86.6$ MPa

(b) $\theta_3 = 0.0729$ rad
(c) $\theta_R = 0.0748$ rad

3-29 (a) $d_{AB} = 1.503$ in.
  $d_{CD} = 1.894$ in.
  (b) $\theta_A = 0.1749$ rad

3-30 (a) $\tau_{AB} = 67.9$ MPa
  (b) $\tau_{CE} = 79.6$ MPa
  (c) $\theta_{E/D} = 0.0295$ rad
  (d) $\theta_{D/A} = 0.1006$ rad

3-33 $\theta = 3TL/16\pi Gtr^3$

3-35 $\theta = qL^2/\pi Gc^4$

3-37 $P_w = 9345$ hp

3-38 $d = 85.9$ mm

3-39 $P_w = 323$ hp

3-40 $d = 106.8$ mm

3-43 (a) $\tau = 3.96$ ksi
  (b) $\tau = 4.95$ ksi
  (c) $\theta = 0.00660$ rad

3-44 (a) $d = 79.8$ mm
  (b) $N = 576$ rpm

3-47 (a) $\tau_{AB} = 19.26$ ksi
  (b) $\tau_{BC} = 3.80$ ksi
  (c) $\theta_{C/A} = 0.0355$ rad

3-48 (a) $d_1 = 80.4$ mm
  (b) $d_2 = 57.9$ mm
  (c) $\theta_{C/A} = 0.0538$ rad

3-51 (a) $d = 1.608$ in.
  (b) $d = 3.35$ in.
  (c) $\theta = 0.1104$ rad

3-52 (a) $d = 95.4$ mm
  (b) $d = 55.8$ mm
  (c) $\theta = 0.246$ rad

3-55 $\tau = 8.91$ ksi

3-56 (a) $\tau_s = 13.49$ MPa
  $\tau_m = 15.35$ MPa
  (b) $\theta = 0.01349$ rad

3-59 (a) $\tau = 16.04$ ksi
  (b) $\theta = 0.0321$ rad

3-60 (a) $T = 15.94$ kN·m
  (b) $\theta = 0.0556$ rad

3-63 (a) $\theta = 0.01801$ rad
  (b) $\theta = 0.01698$ rad
  (c) $\Delta\theta = 5.75\%$

3-64 (a) $T = 28.1$ kN·m
  (b) $\theta = 0.0438$ rad

3-67 $\tau_b = 5.26$ ksi
  $\tau_s = 12.72$ ksi

3-68 $d = 77.5$ mm

3-71 (a) $T = 38.2$ in.-kips
  (b) $\tau = 600$ psi
  (c) $T = 674$ in.-kips

3-72 (a) $T = 15.81$ kN·m
  (b) $\tau = 158.3$ MPa
  (c) $\tau = 32.6$ MPa

3-75 $T = 201$ in.-kips

3-76 (a) $\tau = 87.3$ MPa
  (b) $\tau = 116.4$ MPa
  (c) $\theta = 0.0582$ rad

3-79 (a) $T = 3.41$ in.-kips
  (b) $\tau = 462$ psi
  (c) $\theta = 0.00418$ rad

3-80 $\tau = 18.11$ ksi

3-81 $\tau = 81.5$ MPa

3-84 $d = 60.0$ mm

3-85 $\tau = 13.47$ ksi

3-88 (a) $T = 110.0$ in.-kips
  (b) $T = 66.5$ in.-kips
  (c) $T = 86.4$ in.-kips
  (d) $T = 76.8$ in.-kips

3-89 (a) $T = 8.95$ kN·m
  (b) $T = 5.41$ kN·m
  (c) $T = 7.03$ kN·m
  (d) $T = 6.75$ kN·m

3-92 $\tau = 12.73$ ksi

3-93 $\tau = 40.5$ MPa

3-95 $t = 0.0638$ in.

3-96 $t = 1.750$ mm

CHAPTER 4

4-1 $h = 0.0249$ in.

4-2 $R = 4.55$ m

4-5 $M_r = 64.0$ in.-kips

4-6 $M_r = 33.8$ kN·m

4-9 (a) $M_r = 53.3$ in.-kips
  (b) $M_r = 26.7$ in.-kips

4-10 (a) $\sigma = 252$ MPa

(b) $M_r = 5.25$ kN·m

4-13 $M_F = 90.4\%$

4-14 (a) $M_r = 10.00$ kN·m
(b) $\Delta M_r = 21.1\%$

4-17 $F = 40.8$ kips T

4-18 $M = 121.5$ kN·m

4-23 $\sigma = 2.21$ ksi C

4-24 $\sigma = 17.65$ MPa T

4-27 (a) $M_r = 26.7$ in.-kips
(b) $M_r = 53.3$ in.-kips

4-28 (a) $M_r = 10.00$ kN·m
(b) $\Delta M_r = 21.1\%$

4-31 $\sigma = 7.85$ ksi T

4-32 $\sigma = 87.5$ MPa T

4-35 (a) $\sigma = 8.51$ ksi T
(b) $\sigma = 5.41$ ksi C

4-36 (a) $\sigma = 116.7$ MPa T
(b) $\sigma = 85.3$ MPa C

4-39 (a) $\sigma_A = 10.48$ ksi C
(b) $\sigma_B = 7.86$ ksi T

4-40 (a) $M = 21.0$ kN·m
(b) $P = 21.0$ kN
(c) $w = 10.48$ kN/m

4-43 (a) $\sigma = 10.33$ ksi C
(b) $\sigma = 10.92$ ksi T or C

4-44 (a) $\sigma = 50.7$ MPa T
(b) $\sigma = 53.2$ MPa T or C

4-47 $L = 43.2$ ft

4-48 $h = 212$ mm

4-53 $V_x = 1000$ lb
$M_x = 1000x + 8000$ ft-lb

4-54 $V_x = -15x + 50$ kN
$M_x = -7.5x^2 + 50x$ kN·m

4-57 $V_x = -200x + 1200$ lb
$M_x = -100x^2 + 1200x -$
3000 ft-lb

4-58 $V_x = -8x + 36$ kN
$M_x = -4x^2 + 36x -$
56 kN·m

4-61 (a) $V_x = 1400$ lb
$M_x = 1400x + 8000$ ft-lb
(b) $V_x = -500x + 5400$ lb
$M_x = -250x^2 +$
$5400x - 8000$ ft-lb

4-62 (a) $V_x = 32$ kN
$M_x = 32x - 6$ kN·m

(b) $V_x = -30x + 92$ kN
$M_x = -15x^2 + 92x -$
66 kN·m

4-65 (a) $V_x = -62.5x^2 + 3000$ lb
$M_x = -20.83x^3 +$
$3000x$ ft-lb
(b) $V_{max} = -6000$ lb at
$x = 12$ ft
$M_{max} = 13.86$ ft-kips at
$x = 6.93$ ft

4-66 (a) $V_x = 2.5x^2 - 25x +$
41.67 kN
$M_x = 0.833x^3 -$
$12.5x^2 + 41.7x$ kN·m
(b) $V_{max} = 41.7$ at $x = 0$
$M_{max} = 40.1$ kN·m at
$x = 2.113$ m

4-69 (a) $V_x = (10,000/\pi)$ cos
$(\pi x/10)$ lb
$M_x = (100,000/\pi^2)$ sin
$(\pi x/10)$ ft-lb
(b) $V_{max} = 3.18$ kips
$M_{max} = 10.13$ ft-kips

4-70 (a) $V_x = (400/\pi^2) - (200/\pi)$
sin $(\pi x/8)$ kN
$M_x = (400/\pi^2)(x - 4) +$
$(1600/\pi^2)$ cos $(\pi x/8)$
kN·m
(b) $V_{max} = 40.5$ kN
$M_{max} = 34.1$ kN·m

4-73 Partial Results
$R_A = 5750$ lb
$R_B = 7250$ lb
$V_{max} = -7250$ lb
$M_{max} = 36,250$ ft-lb

4-74 Partial Results
$R_A = 4.5$ kN
$R_C = 30.5$ kN
$V_{max} = 15.5$ kN
$M_{max} = -37.5$ kN·m

4-77 (a) Partial Results:
$R_A = 11,000$ lb
$R_C = 21,000$ lb
$V_{max} = -13,000$ lb
$M_{max} = -64,000$ ft-lb
(b) $V_x = -2000x +$
11,000 lb
$M_x = -1000x^2 +$
$11,000x$ ft-lb

4-78 (a) Partial Results:
$R_A = 45.0$ kN
$R_D = 95.0$ kN
$V_{max} = -95.0$ kN
$M_{max} = 150.4$ kN·m
(b) $V_x = -30x + 145$ kN
$M_x = -15x^2 + 145x - 200$ kN·m

4-81 (a) Partial Results:
$R_A = 3083$ lb
$R_D = 3417$ lb
$V_{max} = -3417$ lb
$M_{max} = 13,753$ ft-lb
(b) $V_x = -600x + 4583$ lb
$M_x = -300x^2 + 4583x - 3750$ ft-lb

4-82 (a) Partial Results:
$R_A = 16.00$ kN
$R_D = 20.00$ kN
$V_{max} = -20.0$ kN
$M_{max} = 40.6$ kN·m
(b) $V_x = -5x + 25$ kN
$M_x = -2.5x^2 + 25x - 22.5$ kN·m

4-85 Partial Results:
$R_L = 3600$ lb
$R_R = 4400$ lb
$V_{max} = -4400$ lb
$M_{max} = 19,360$ ft-lb

4-86 Partial Results:
$R_L = 75.0$ kN
$R_R = 21.0$ kN
$V_{max} = 51.0$ kN
$M_{max} = 84.0$ kN·m

4-89 Use a W8 × 15 section

4-90 Use a W254 × 33 section

4-95 (a) $\sigma = 5.44$ ksi T
(b) $\sigma = 8.16$ ksi C

4-96 (a) $\sigma = 64.6$ MPa T
(b) $\sigma = 93.8$ MPa C

4-99 $w = 10.43$ kips/ft

4-100 $w = 11.13$ kN/m

4-103 (a) Partial Results:
$R_B = -2000$ lb
$V_{max} = -2480$ lb
$M_{max} = 9840$ ft-lb
(b) $M_x = -60x^2 - 1280x + 9840$ ft-lb

4-104 (a) Partial Results:
$R_B = 215$ kN
$R_C = -35$ kN
$R_E = 50$ kN
$V_{max} = -120$ kN
$M_{max} = -90$ kN·m
(b) $M_x = -20x^2 + 95x - 90$ kN·m
(c) $M_x = (-\frac{1}{9})(40x^3 - 300x^2 + 480x + 320)$ kN·m

4-109 (a) $M_2 = 1620$ ft-lb
$F_1 = 2880$ lb
$F_2 = 4320$ lb
(b) $V = 1440$ lb
(c) $\tau = 40.0$ psi

4-110 (a) $\tau = 112.5$ kPa
(b) $\tau = 150.0$ kPa

4-113 (a) $V = 9.00$ kips
(b) $\tau = 397$ psi
(c) $\tau = 265$ psi

4-114 (a) $F_1 = 8.41$ kN
$F_2 = 13.21$ kN
(b) $V = 4.80$ kN
(c) $\tau = 353$ kPa

4-117 (a) $V = 519$ lb
(b) $s = 2.31$ in.

4-119 Use 5 welds on each side

4-120 $s = 808$ mm

4-123 (a) $\tau = 152.9$ psi
(b) $\tau = 179.6$ psi
(c) $\tau = 214$ psi at N.A.

4-124 (a) $\tau = 815$ kPa
(b) $\tau = 982$ kPa
(c) $\tau = 1141$ kPa at N.A.

4-125 (a) $\tau = 140.6$ psi
(b) $\tau = 187.5$ psi
(c) $\sigma = 4.50$ psi T

4-126 (a) $\tau = 1800$ kPa
(b) $\tau = 645$ kPa
(c) $\sigma = 21.6$ Mpa C

4-129 (a) $\tau = 110.6$ psi
(b) $\tau = 179.3$ psi at N.A. near the supports
(c) $\sigma = 3.36$ ksi T

4-130 (a) $\tau = 2.31$ MPa
(b) $\tau = 3.25$ MPa at N.A. near the supports

(c) $\tau = 2.88$ MPa near the supports

4-133 (a) $w = 768$ lb/ft
(b) $\tau = 48.9$ psi
(c) $\sigma = 1800$ psi

4-134 (a) $\tau = 1.957$ MPa
(b) $\tau = 1.671$ MPa
(c) $\tau = 3.17$ MPa
(d) $\sigma = 30.8$ MPa

4-137 (a) $\sigma = 1600$ psi
(b) $\tau = 83.3$ psi

4-138 $P = 2.67$ kN

4-141 (a) $\sigma = 6.73$ ksi T
(b) $\sigma = 5.05$ ksi C
(c) $\tau = 580$ psi
(d) $\tau = 545$ psi

4-142 (a) $\sigma = 71.6$ MPa T
(b) $\tau = 19.00$ MPa
(c) $\tau = 12.13$ MPa

4-145 $r = 0.0625$ in.

4-146 $M = 882$ N·m

4-149 $P = 1.961$ kips

4-150 $b = 300$ mm

4-154 $\sigma_w = 9.19$ MPa C

4-155 $\sigma_s = 10.35$ ksi T

4-158 $\sigma_w = 8.21$ MPa
$\sigma_s = 138.9$ MPa

4-159 $\sigma_w = 0.669$ ksi
$\sigma_s = 14.42$ ksi

4-162 b = 81.1 mm

4-163 b = 1.110 in.

4-166 $\sigma_c = 3.84$ MPa C
$\sigma_s = 85.3$ MPa T

4-168 $w = 12.62$ kN/m

4-169 (a) $A_s = 2.23$ in.^{2}
(b) $d = 23.2$ in.

4-170 (a) $A_s = 1689$ mm^2
(b) $M = 91.4$ kN·m

4-171 $M = 46.8$ ft-kips

CHAPTER 5

5-1 $\sigma = 6.60$ ksi

5-2 $\sigma = 333$ MPa

5-5 $d = 11.72$ in.

5-6 $d = 600$ mm

5-9   (a) $\rho = 17,800$ in.
(b) $\delta = 0.1456$ in.
(c) $\sigma = 5.38$ ksi

5-10 (a) $\rho = 598$ m
(b) $\delta = 0.836$ mm
(c) $\sigma = 62.1$ MPa

5-13 $\sigma = 1667$ psi

5-14 (a) $I = 82.2(10^3)$ mm^4
(b) $b = 17.48$ mm
$h = 38.4$ mm

5-27 (a) $y = (P/6EI)(x^3 - 3Lx^2)$
(b) $\theta = -PL^2/2EI$
(c) $\delta = PL^3/3EI \downarrow$

5-28 (a) $y = (P/6EI)(-x^3 + 3Lx^2 - 2L^3)$
(b) $\theta = +PL^2/2EI$
(c) $\delta = PL^3/3EI \downarrow$

5-31 (a) $y = (w/120EIL)(-x^5 + 5L^4x - 4L^5)$
(b) $\theta = +wL^3/24EI$
(c) $\delta = wL^4/30EI \downarrow$

5-32 (a) $y = (w/120EIL)(-x^5 + 10L^2x^3 - 20L^3x^2)$
(b) $\theta = -wL^3/8EI$
(c) $\delta = 11wL^4/120EI \downarrow$

5-35 (a) $y = (w/24EI)(-x^4 + 2Lx^3 - L^3x)$
(b) $\theta = -wL^3/24EI$
(c) $\delta = 5wL^4/384EI \downarrow$

5-36 (a) $y = (w/360EIL)(-3x^5 + 10L^2x^3 - 7L^4x)$
(b) $\theta = +wL^3/45EI$
(c) $\delta = 5wL^4/768EI \downarrow$

5-39 (a) $y = (w/24EI)(-x^4 + 2Lx^3 - 2L^3x + L^4)$
(b) $\delta = 0.292$ in. $\uparrow$

5-40 (a) $y = (w/16EI)(Lx^3 - 4L^2x^2 + 3L^3x)$
(b) $\delta = 7.03$ mm $\uparrow$

5-43 (a) $\delta = 0.270$ in. $\uparrow$
(b) $\delta = 0.324$ in. $\downarrow$

5-44 (a) $y = (w/120EIL)(-x^5 + 10L^2x^3 - 25L^4x)$
(b) $\delta = 54.0$ mm $\downarrow$

5-47 (a) $\theta = -PL^2/16EI$
(b) $\delta = PL^3/48EI$ ↓

5-48 (a) $\theta = -4PL^2/9EI$
(b) $\delta = 23PL^3/48EI$ ↓
(c) $\delta = 0.484PL^3/EI$ ↓

5-55 (a) $\delta = 7PL^3/12EI$ ↓
(b) $\delta = 23PL^3/12EI$ ↓

5-56 (a) $\delta = 31PL^3/96EI$ ↓
(b) $\delta = 35PL^3/96EI$ ↓

5-59 (a) $\delta = 5wL^4/24EI$ ↓
(b) $\delta = 2wL^4/3EI$ ↓

5-60 (a) $\delta = 59wL^4/72EI$ ↓
(b) $\delta = 57wL^4/64EI$ ↓

5-61 (a) $y = (w/840EIL^3)(-x^7 + 7L^6x - 6L^7)$
(b) $\delta = wL^4/140EI$ ↓
(c) $V_B = wL/4$ ↑
$M_B = -wL^2/20$

5-62 (a) $y = (w/840EIL^3)(-x^7 + 7L^4x^3 - 6L^6x)$
(b) $\delta = 13wL^4/5120EI$ ↓
(c) $\delta = 0.00256wL^4/EI$ ↓
(d) $R_A = wL/20$ ↑
$R_B = wL/5$ ↑

5-67 $\delta = 11PL^3/6EI$ ↓

5-68 $\delta = 7PL^3/2EI$ ↓

5-71 (a) $\delta = 5PL^3/6EI$ ↓
(b) $\delta = 23PL^3/24EI$ ↓

5-72 (a) $\delta = PL^3/2EI$ ↓
(b) $\delta = PL^3/64EI$ ↑

5-75 (a) $\delta = 2wL^4/3EI$ ↓
(b) $\delta = 23wL^4/12EI$ ↓

5-76 (a) $\delta = 7wL^4/24EI$ ↓
(b) $\delta = 7wL^4/8EI$ ↓

5-79 (a) $\delta = 11wa^4/24EI$ ↓
(b) $\delta = 205wa^4/384EI$ ↓

5-80 (a) $\delta = 123wL^4/720EI$ ↓
(b) $\delta = 0.1709wL^4/EI$ ↓

5-101 (a) $\delta = PL^3/3EI$ ↓
(b) $\delta = 7PL^3/12EI$ ↓

5-102 (a) $\delta = wL^4/8EI$ ↓
(b) $\delta = 5wL^4/24EI$ ↓

5-103 (a) $\delta = wL^4/30EI$ ↓
(b) $\delta = 13wL^4/240EI$ ↓

5-104 (a) $\delta = 3wL^4/8EI$ ↑
(b) $\delta = 11wL^4/12EI$ ↑

5-107 (a) $\delta = 2wL^4/3EI$ ↓
(b) $\delta = 23wL^4/12EI$ ↓

5-108 (a) $\delta = 41wL^4/24EI$ ↓
(b) $\delta = 23wL^4/8EI$ ↓

5-111 (a) $\delta = 3PL^3/256EI$ ↓
(b) $\delta = 11PL^3/768EI$ ↓

5-112 (a) $\delta = 2PL^3/3EI$ ↓
(b) $\delta = PL^3/16EI$ ↑

5-115 $\delta = 5wL^4/48EI$ ↓

5-116 $\delta = 147wL^4/720EI$ ↓

5-119 $\delta = 0.1376$ in. ↓

5-120 $\delta = 52.0$ mm ↓

5-123 $\delta = 25PL^3/18EI$ ↓

5-124 $\delta = 43PL^3/144EI$ ↓

5-127 $\delta = 61wL^4/72EI$ ↓

5-129 $\delta = 0.484\, PL^3/EI$ ↓

5-130 $\delta = 5.82PL^3/EI$ ↓

5-133 $\delta = 0.1679$ in. ↓

5-134 $\delta = 85.8$ mm ↓

5-137 Use an S24 × 80 section

5-138 Use a WT229 × 30 section

5-141 (a) $\delta = 0.1116$ in. ↓
(b) $\Delta\delta = 0.538\%$

5-143 $\delta = 0.1356$ in. ↓

5-144 $\delta = 12.58$ mm ↓

5-145 $\delta = PL^3/2EI$ ↑

5-146 $\delta = 71wL^4/24EI$ ↓

5-149 $\delta = 5wL^4/128EI$ ↑

5-150 $\delta = 5wL^4/128EI$ ↑

5-155 $\delta = 0.243$ in. ↓

5-156 $\delta = 14.77$ mm ↓

5-160 (a) $\delta_v = ML^2/2EI$ ↑
(b) $\delta_h = 5ML^2/8EI$ →

5-161 (a) $\delta_v = 7wL^4/EI$ ↓
(b) $\delta_h = 5wL^4/2EI$ ←

5-165 $\Delta\delta = 0.458\%$

5-166 $\Delta\delta = 0.580\%$

CHAPTER 6

6-1 (a) $M = PL/2$
(b) $\delta = PL^3/12EI$ ↑

6-2 (a) $M = wL^2/6$
   (b) $M = wL^2/4$

6-5 (a) $R_A = 3M/2L \uparrow$
   $V_B = 3M/2L \downarrow$
   $M_B = +M/2$
   (b) $\theta = +ML^2/32EI$

6-6 (a) $R_A = 3wL/8 \uparrow$
   $V_B = 5wL/8 \uparrow$
   $M_B = -wL^2/8$
   (b) $\delta = wL^4/192EI \downarrow$

6-9 $R_A = wL/24 \uparrow$
   $V_B = 7wL/24 \uparrow$
   $M_B = -wL^2/24$
   (b) $\delta = 0.001267wL^4/EI \downarrow$

6-11 (a) $R_A = 13wL/32 \uparrow$
   $V_B = 19wL/32 \uparrow$
   $M_B = -3wL^2/32$
   (b) $\delta = 0.00726wL^4/EI \downarrow$
   (c) $M = -3wL^2/32$

6-12 (a) $P = wL/6$
   (b) $R_A = 5wL/12 \uparrow$
   $R_B = 3wL/4 \uparrow$
   (c) $\delta = 0.00788wL^4/EI \downarrow$

6-15 (a) $R_A = 9M/16L \uparrow$
   $V_B = 9M/16L \downarrow$
   $M_B = +M/8$
   (b) $\delta = ML^2/32EI \downarrow$

6-16 (a) $R_A = 5P/16 \uparrow$
   $V_B = 11P/16 \uparrow$
   $M_B = -3PL/8$
   (b) $\delta = 7PL^3/96EI \downarrow$

6-19 (a) $R_B = 0.791$ kips $\uparrow$
   $R_C = 3.87$ kips $\uparrow$
   $R_D = 3.34$ kips $\uparrow$
   (b) $M = -3.26$ ft-kips
   (c) $M = +10.00$ ft-kips

6-20 (a) $R_A = 26.7$ kN $\uparrow$
   $V_D = 45.3$ kN $\uparrow$
   $M_D = -111.4$ kN·m
   (b) $\delta = 9.98$ mm $\downarrow$

6-23 (a) $V_A = wL/2 \uparrow$
   $R_B = wL \uparrow$
   $V_C = wL/2 \uparrow$
   $M_A = -wL^2/12$
   $M_C = -wL^2/12$
   (b) $M_B = -wL^2/12$

6-25 $R_B = 30.1$ kips

6-26 $V_A = 7.83$ kN $\uparrow$
   $V_B = 7.53$ kN $\downarrow$
   $M_A = -9.66$ kN·m
   $M_B = +9.54$ kN·m

6-31 $M = wL^2/4$

6-32 $P = wL/3$

6-35 (a) $V_A = P \uparrow$
   $V_B = P \uparrow$
   $M_A = -2PL/9$
   $M_B = -2PL/9$
   (b) $\delta = 5PL^3/648EI \downarrow$

6-36 (a) $V_A = wL/2 \uparrow$
   $V_B = wL/2 \uparrow$
   $M_A = -wL^2/12$
   $M_B = -wL^2/12$
   (b) $\delta = wL^4/384EI \downarrow$

6-39 (a) $V_A = 2P \uparrow$
   $M_A = -PL/2$
   $R_C = 3P \uparrow$
   (b) $\delta = PL^3/48EI \downarrow$

6-41 (a) $V_A = 13wL/16 \uparrow$
   $V_C = 3wL/16 \uparrow$
   $M_A = -11wL^2/48$
   $M_C = -5wL^2/48$
   (b) $\delta = wL^4/48EI \downarrow$

6-42 (a) $R_A = 10wL/27 \uparrow$
   $V_C = 44wL/27 \uparrow$
   $M_C = -8wL^2/9$
   (b) $\delta = 22wL^4/81EI \downarrow$

6-45 (a) $R_A = 13P/32 \uparrow$
   $R_B = 11P/16 \uparrow$
   $R_C = 3P/32 \downarrow$
   (b) $\theta = +PL^2/32EI$
   (c) $\delta = 23PL^3/1536EI \downarrow$

6-46 (a) $R_A = 11wL/32 \uparrow$
   $R_B = 21wL/16 \uparrow$
   $R_C = 11wL/32 \uparrow$
   (b) $\theta = -wL^3/96EI$
   (c) $\delta = 17wL^4/1536EI \downarrow$

6-49 (a) $R_A = 11P/28 \uparrow$
   $R_B = 43P/56 \uparrow$
   $V_C = 9P/56 \downarrow$
   $M_C = 3PL/56$
   (b) $\theta = +3PL^2/112EI$
   (c) $\delta = 19PL^3/1344EI \downarrow$

6-51 (a) $M = 11PL/12$
    (b) $R_B = 9P/4$ ↑
        $R_C = 5P/4$ ↓
    (c) $\theta = +PL^2/4EI$
    (d) $\delta = PL^3/648EI$ ↑

6-53 (a) $R_B = 3.42$ kips ↑
        $R_C = 5.15$ kips ↑
        $R_E = 1.425$ kips ↑
    (b) $\theta = -0.000472$ rad
    (c) $\delta = 0.0372$ in. ↓

6-55 $R_A = 29wL/48$ ↑
     $R_B = 19wL/24$ ↑
     $R_C = 29wL/48$ ↑

6-56 $R_A = 11wL/48$ ↑
     $R_B = 11wL/8$ ↑
     $R_C = 35wL/48$ ↑

6-57 $R_A = 9wL/16$ ↑
     $V_B = 7wL/16$ ↑
     $M_B = +wL^2/16$

6-59 (a) $V_A = 65wL/72$ ↑
        $M_A = -11wL^2/36$
        $R_C = 7wL/72$
    (b) $\delta = 19wL^4/864EI$ ↓

6-60 (a) $V_A = 3P/10$ ↑
        $M_A = -3PL/5$
        $R_C = 7P/10$ ↑
    (b) $\delta = PL^3/12EI$ ↓

6-63 (a) $M = wL^2/6$
    (b) $M = wL^2/4$

6-64 (a) $P = wL/3$
    (b) $P = 3wL/8$

6-65 (a) $R_A = 5wL/12$ ↑
        $R_B = 7wL/6$ ↑
        $R_C = 5wL/12$ ↑
    (b) $\delta = 5wL^4/1152EI$ ↓

6-66 (a) $R_A = 10wL/27$ ↑
        $V_C = 44wL/27$ ↑
        $M_C = -8wL^2/9$
    (b) $\delta = 22wL^4/81EI$ ↓

6-69 (a) $M = PL/12$
    (b) $M = PL/9$

6-71 (a) $P = 2wL/9$
    (b) $V_A = 117wL/216$ ↑
        $M_A = -7wL^2/72$
        $R_B = 147wL/216$ ↑

6-73 (a) $V_A = 13wL/16$ ↑
        $M_A = -11wL^2/48$
        $V_C = 3wL/16$ ↑
        $M_C = -5wL^2/48$
    (b) $\delta = wL^4/48EI$ ↓

6-74 (a) $V_A = 13wL/28$ ↑
        $M_A = -wL^2/14$
        $R_B = 8wL/7$ ↑
        $R_C = 11wL/28$ ↑
    (b) $\theta = +wL^3/42EI$

6-77 (a) $R_A = 3.32$ kips ↑
        $R_B = 5.36$ kips ↑
        $R_C = 3.32$ kips ↑
    (b) $\sigma = 6.48$ ksi

6-79 (a) $R_A = P - Pa(5L^2 - 4a^2)/2L^3$
        $R_B = Pa(3L^2 - 4a^2)/L^3$
        $R_C = Pa(4a^2 - L^2)/2L^3$
    (b) $M = 0.1037PL$

6-81 $R_A = 2.66$ kips ↑
     $R_B = 4.69$ kips ↑
     $R_C = 2.66$ kips ↑

6-82 $W = 13.89$ kN

6-85 $R_B = 28.3$ kN

6-86 $R_B = 17.83$ kN

6-88 $P = 5.51$ kN

6-89 $R = 750$ lb

6-90 (a) $R_A = 3P/20$ ↑
    (b) $\delta = 17PL^3/60EI$ →

6-91 (a) $A_x = wL/10$ ←
    (b) $A_y = 9wL/20$ ↑

CHAPTER 7

7-1 $\sigma_n = 11.01$ ksi T
    $\tau_{nt} = 9.59$ ksi

7-2 $\sigma_n = 60.8$ MPa T
    $\tau_{nt} = 14.40$ MPa

7-5 $\sigma_n = 30.9$ ksi T
    $\tau_{nt} = -1.794$ ksi

7-6 $\sigma_n = 46.6$ MPa C
    $\tau_{nt} = -47.4$ MPa

7-9 $\sigma_n = 10.37$ ksi T
$\tau_{nt} = -8.73$ ksi

7-10 $\sigma_n = 155.2$ MPa C
$\tau_{nt} = 53.8$ MPa

7-13 $\sigma_n = 21.8$ ksi T
$\sigma_t = 9.23$ ksi T
$\tau_{nt} = 1.724$ ksi

7-14 $\sigma_n = 5.98$ MPa T
$\sigma_t = 86.0$ MPa C
$\tau_{nt} = 19.64$ MPa

7-17 $\sigma_n = 11.89$ ksi T
$\sigma_t = 1.110$ ksi T
$\epsilon_n = 385$ μin./in.

7-19 (a) $\sigma_{p1} = 18.77$ ksi T
$\sigma_{p2} = 2.77$ ksi C
$\sigma_{p3} = 0$
$\tau_p = 10.77$ ksi
$\tau_{max} = 10.77$ ksi
(b) $\theta_p = 34.1°$

7-20 (a) $\sigma_{p1} = 77.7$ MPa T
$\sigma_{p2} = 7.72$ MPa C
$\sigma_{p3} = 0$
$\tau_p = 42.7$ MPa
$\tau_{max} = 42.7$ MPa
(b) $\theta_p = 34.7°$

7-21 $\sigma_{p1} = 14.00$ ksi T
$\sigma_{p2} = 6.00$ ksi C
$\sigma_{p3} = 0$
$\tau_p = 10.00$ ksi
$\tau_{max} = 10.00$ ksi
$\theta_p = -18.43°$

7-22 $\sigma_{p1} = 86.0$ MPa T
$\sigma_{p2} = 36.0$ MPa C
$\sigma_{p3} = 0$
$\tau_p = 61.0$ MPa
$\tau_{max} = 61.0$ MPa
$\theta_p = -17.50°$

7-25 $\sigma_{p1} = 31.0$ ksi T
$\sigma_{p2} = 1.000$ ksi T
$\sigma_{p3} = 0$
$\tau_p = 15.00$ ksi
$\tau_{max} = 15.50$ ksi
$\theta_p = 26.6°$

7-26 $\sigma_{p1} = 44.0$ MPa T
$\sigma_{p2} = 18.00$ MPa T
$\sigma_{p3} = 0$
$\tau_p = 13.00$ MPa

$\tau_{max} = 22.0$ MPa
$\theta_p = 33.7°$

7-31 $\sigma_x = 9.86$ ksi T
$\sigma_y = 17.86$ ksi C
$\theta_p = 15.00°$

7-32 $\sigma_{p1} = 105.0$ MPa T
$\sigma_{p2} = 145.0$ MPa C
$\sigma_{p3} = 0$
$\tau_{xy} = -75.0$ MPa
$\theta_p = -18.43°$

7-35 $\sigma_{p1} = 27.0$ ksi T
$\sigma_{p2} = 37.0$ ksi C
$\sigma_{p3} = 0$
$\tau_{xy} = -19.97$ ksi

7-36 $\sigma_{p1} = 200$ MPa T
$\sigma_{p2} = 30.0$ MPa T
$\sigma_{p3} = 0$
$\tau_{max} = 100.0$ MPa
$\sigma_x = 155.0$ MPa
$\tau_{xy} = -75.0$ MPa

7-37 $\sigma_n = 6.62$ ksi T
$\tau_{nt} = 11.03$ ksi

7-38 $\sigma_n = 23.6$ ksi T
$\tau_{nt} = -32.9$ ksi

7-41 (a) $\sigma_{p1} = 16.45$ ksi T
$\sigma_{p2} = 10.45$ ksi C
$\sigma_{p3} = 0$
$\tau_p = 13.45$ ksi
$\tau_{max} = 13.45$ ksi
(b) $\sigma_n = 1.160$ ksi C
$\tau_{nt} = 12.79$ ksi

7-42 (a) $\sigma_{p1} = 98.2$ MPa T
$\sigma_{p2} = 118.2$ MPa C
$\sigma_{p3} = 0$
$\tau_p = 108.2$ MPa
$\tau_{max} = 108.2$ MPa
(b) $\sigma_n = 60.3$ MPa C
$\tau_{nt} = -95.8$ MPa

7-45 (a) $\sigma_{p1} = 30.4$ ksi T
$\sigma_{p2} = 6.57$ ksi T
$\sigma_{p3} = 0$
$\tau_p = 11.93$ ksi
$\tau_{max} = 15.22$ ksi
(b) $\sigma_n = 10.37$ ksi T
$\tau_{nt} = -8.73$ ksi

7-46 (a) $\sigma_{p1} = 14.22$ MPa C
$\sigma_{p2} = 175.8$ MPa C

$\sigma_{p3} = 0$

$\tau_p = 80.8$ MPa

$\tau_{max} = 87.9$ MPa

(b) $\sigma_n = 155.2$ MPa

$\tau_{nt} = 53.8$ MPa

7-49 (a) $\sigma_{p1} = 21.0$ ksi T

$\sigma_{p2} = 6.00$ ksi T

$\sigma_{p3} = 0$

$\tau_p = 7.50$ ksi

$\tau_{max} = 10.50$ ksi

(b) $\sigma_n = 20.9$ ksi T

$\tau_{nt} = 0.897$ ksi

7-50 (a) $\sigma_{p1} = 63.0$ MPa T

$\sigma_{p2} = 18.00$ MPa T

$\sigma_{p3} = 0$

$\tau_p = 22.5$ MPa

$\tau_{max} = 31.5$ MPa

(b) $\sigma_n = 60.6$ MPa T

$\tau_{nt} = -10.18$ MPa

7-53 $\sigma_n = 21.8$ ksi T

$\sigma_t = 9.23$ ksi T

$\tau_{nt} = 1.724$ ksi

7-54 $\sigma_n = 5.98$ MPa T

$\sigma_t = 86.0$ MPa C

$\tau_{nt} = 19.64$ MPa

7-57 $\epsilon_n = 610\ \mu$

$\epsilon_t = -810\ \mu$

$\gamma_{nt} = 1259\ \mu$rad

7-58 $\epsilon_n = 937\ \mu$

$\epsilon_t = 613\ \mu$

$\gamma_{nt} = -217\ \mu$rad

7-61 $\epsilon_n = -398\ \mu$

$\epsilon_t = -402\ \mu$

$\gamma_{nt} = 721\ \mu$rad

7-62 $\epsilon_n = 695\ \mu$

$\epsilon_t = 855\ \mu$

$\gamma_{nt} = -356\ \mu$rad

7-65 $\epsilon_{p1} = 672\ \mu$

$\epsilon_{p2} = -272\ \mu$

$\epsilon_{p3} = 0$

$\gamma_p = 943\ \mu$rad

$\gamma_{max} = 943\ \mu$rad

$\theta_p = 16.00°$

7-66 $\epsilon_{p1} = 1004\ \mu$

$\epsilon_{p2} = -364\ \mu$

$\epsilon_{p3} = 0$

$\gamma_p = 1367\ \mu$rad

$\gamma_{max} = 1367\ \mu$rad

$\theta_p = -10.28°$

7-69 $\epsilon_{p1} = 1009\ \mu$

$\epsilon_{p2} = -759\ \mu$

$\epsilon_{p3} = 0$

$\gamma_p = 1768\ \mu$rad

$\gamma_{max} = 1768\ \mu$rad

$\theta_p = 4.07°$

7-70 $\epsilon_{p1} = 777\ \mu$

$\epsilon_{p2} = -437\ \mu$

$\epsilon_{p3} = 0$

$\gamma_p = 1215\ \mu$rad

$\gamma_{max} = 1215\ \mu$rad

$\theta_p = 8.62°$

7-73 $\epsilon_{p1} = 908\ \mu$

$\epsilon_{p2} = 388\ \mu$

$\epsilon_{p3} = 0$

$\gamma_p = 519\ \mu$rad

$\gamma_{max} = 908\ \mu$rad

$\theta_p = -16.85°$

7-74 $\epsilon_{p1} = 970\ \mu$

$\epsilon_{p2} = 580\ \mu$

$\epsilon_{p3} = 0$

$\gamma_p = 391\ \mu$rad

$\gamma_{max} = 970\ \mu$

$\theta_p = -25.1°$

7-77 $\epsilon_x = 500\ \mu$

$\epsilon_y = -300\ \mu$

$\gamma_{xy} = 600\ \mu$rad

$\gamma_p = 1000\ \mu$rad

$\gamma_{max} = 1000\ \mu$rad

7-78 $\epsilon_x = 640\ \mu$

$\epsilon_y = -800\ \mu$

$\gamma_{xy} = 960\ \mu$rad

$\gamma_p = 1730\ \mu$rad

$\gamma_{max} = 1730\ \mu$rad

7-81 $\epsilon_{p1} = 966\ \mu$

$\epsilon_{p2} = -241\ \mu$

$\epsilon_{p3} = 0$

$\gamma_p = 1207\ \mu$rad

$\gamma_{max} = 1207\ \mu$rad

$\theta_p = 6.59°$

7-82 $\epsilon_{p1} = 1072\ \mu$

$\epsilon_{p2} = -505\ \mu$

$\epsilon_{p3} = 0$

$\gamma_p = 1576\ \mu$rad

$\gamma_{max} = 1576\ \mu$rad

$\theta_p = 19.27°$

7-85 $\gamma_{xy} = \pm 641 \ \mu\text{rad}$
$\epsilon_{p2} = -774 \ \mu$
$\gamma_p = 1188 \ \mu\text{rad}$
$\gamma_{\max} = 1188 \ \mu\text{rad}$
$\theta_p = \pm 16.34°$

7-86 $\gamma_{xy} = \pm 912 \ \mu\text{rad}$
$\epsilon_{p2} = -180 \ \mu$
$\gamma_p = 960 \ \mu\text{rad}$
$\gamma_{\max} = 960 \ \mu\text{rad}$
$\theta_p = \pm 35.9°$

7-89 (a) $\epsilon_x = 750 \ \mu$
$\epsilon_y = -250 \ \mu$
$\gamma_{xy} = -750 \ \mu\text{rad}$
(b) $\epsilon_{p1} = 875 \ \mu$
$\epsilon_{p2} = -375 \ \mu$
$\epsilon_{p3} = -214 \ \mu$
$\gamma_p = 1250 \ \mu\text{rad}$
$\gamma_{\max} = 1250 \ \mu\text{rad}$

7-90 (a) $\epsilon_x = 780 \ \mu$
$\epsilon_y = -251 \ \mu$
$\gamma_{xy} = -782 \ \mu\text{rad}$
(b) $\epsilon_{p1} = 911 \ \mu$
$\epsilon_{p2} = -383 \ \mu$
$\epsilon_{p3} = -260 \ \mu$
$\gamma_p = 1294 \ \mu\text{rad}$
$\gamma_{\max} \ 1294 \ \mu\text{rad}$

7-93 $\epsilon_a = 748 \ \mu$
$\epsilon_b = 958 \ \mu$
$\epsilon_c = 172.5 \ \mu$

7-94 $\epsilon_a = 840 \ \mu$
$\epsilon_b = 840 \ \mu$
$\epsilon_c = 120.0 \ \mu$

7-97 (a) $\epsilon_x = -350 \ \mu$
$\epsilon_y = 1150 \ \mu$
$\gamma_{xy} = 520 \ \mu\text{rad}$
(b) $\sigma_x = 0.514 \ \text{ksi} \ T$
$\sigma_y = 33.5 \ \text{ksi} \ T$
$\tau_{xy} = 5.72 \ \text{ksi}$
(c) $\sigma_{p1} = 34.5 \ \text{ksi} \ T$
$\sigma_{p2} = 0.449 \ \text{ksi} \ C$
$\sigma_{p3} = 0$
$\tau_{\max} = 17.46 \ \text{ksi}$

7-98 (a) $\epsilon_x = 525 \ \mu$
$\epsilon_y = 1350 \ \mu$
$\gamma_{xy} = -975 \ \mu\text{rad}$
(b) $\sigma_x = 75.2 \ \text{MPa} \ T$
$\sigma_y = 121.4 \ \text{MPa} \ T$
$\tau_{xy} = -27.3 \ \text{MPa}$

(c) $\sigma_{p1} = 134.0 \ \text{MPa} \ T$
$\sigma_{p2} = 62.5 \ \text{MPa} \ T$
$\sigma_{p3} = 0$
$\tau_{\max} = 67.0 \ \text{MPa}$

7-101 (a) $\sigma_x = 18.96 \ \text{ksi} \ T$
$\sigma_y = 1.813 \ \text{ksi} \ C$
$\tau_{xy} = 6.35 \ \text{ksi}$
(b) $\epsilon_{p1} = 727 \ \mu$
$\epsilon_{p2} = -327 \ \mu$
$\epsilon_{p3} = -171.4 \ \mu$
$\gamma_{\max} = 1055 \ \mu\text{rad}$
(c) $\sigma_{p1} = 20.7 \ \text{ksi} \ T$
$\sigma_{p2} = 3.60 \ \text{ksi} \ C$
$\sigma_{p3} = 0$
$\tau_{\max} = 12.17 \ \text{ksi}$

7-102 (a) $\sigma_x = 64.7 \ \text{MPa} \ T$
$\sigma_y = 2.49 \ \text{MPa} \ T$
$\tau_{xy} = -42.8 \ \text{MPa}$
(b) $\epsilon_{p1} = 1272 \ \mu$
$\epsilon_{p2} = -655 \ \mu$
$\epsilon_{p3} = -304 \ \mu$
$\gamma_{\max} = 1927 \ \mu\text{rad}$
(c) $\sigma_{p1} = 86.5 \ \text{MPa} \ T$
$\sigma_{p2} = 19.27 \ \text{MPa} \ C$
$\sigma_{p3} = 0$
$\tau_{\max} = 52.9 \ \text{MPa}$

7-106 (a) $\sigma_n = 71.9 \ \text{MPa} \ T$
$\tau_{nt} = 10.98 \ \text{MPa}$
(b) $\sigma_{p1} = 73.8 \ \text{MPa} \ T$
$\sigma_{p2} = 9.41 \ \text{MPa} \ T$
$\sigma_{p3} = 43.2 \ \text{MPa} \ C$
$\tau_{\max} = 58.5 \ \text{MPa}$

7-107 (a) $\sigma_n = 14.29 \ \text{ksi} \ T$
$\tau_{nt} = 4.17 \ \text{ksi}$
(b) $\sigma_{p1} = 18.00 \ \text{ksi} \ T$
$\sigma_{p2} = 12.00 \ \text{ksi} \ T$
$\sigma_{p3} = 6.00 \ \text{ksi} \ T$
$\tau_{\max} = 6.00 \ \text{ksi}$

7-110 (a) $\sigma_n = 41.3 \ \text{MPa} \ T$
$\tau_{nt} = 53.0 \ \text{MPa}$
(b) $\sigma_{p1} = 81.3 \ \text{MPa} \ T$
$\sigma_{p2} = 4.75 \ \text{MPa} \ C$
$\sigma_{p3} = 36.6 \ \text{MPa} \ C$
$\tau_{\max} = 58.9 \ \text{MPa}$

7-111 (a) $\sigma_n = 105.0 \ \text{MPa} \ T$
$\tau_{nt} = 20.5 \ \text{MPa}$
(b) $\sigma_{p1} = 110.9 \ \text{MPa} \ T$
$\sigma_{p2} = 36.5 \ \text{MPa} \ T$

$\sigma_{p3} = 97.4$ MPa C

$\tau_{max} = 104.1$ MPa

7-113 (a) $\sigma_{p1} = 106.2$ MPa T

$\sigma_{p2} = 10.00$ MPa T

$\sigma_{p3} = 3.77$ MPa T

$\tau_{max} = 51.2$ MPa

(b) $\theta_{1x} = 42.8°$

$\theta_{1y} = 56.0°$

$\theta_{1z} = 67.4°$

7-114 (a) $\sigma_{p1} = 28.0$ ksi T

$\sigma_{p2} = 14.22$ ksi T

$\sigma_{p3} = 6.24$ ksi C

$\tau_{max} = 17.13$ ksi

(b) $\theta_{1x} = 35.1°$

$\theta_{1y} = 57.1°$

$\theta_{1z} = 79.2°$

CHAPTER 8

8-1 $\sigma_{p1} = 5.10$ ksi T

$\sigma_{p2} = 1.118$ ksi C

$\sigma_{p3} = 0$

$\tau_{max} = 3.11$ ksi

$\theta_p = 25.1°$

8-2 $\sigma_{p1} = 56.6$ MPa T

$\sigma_{p2} = 29.3$ MPa C

$\sigma_{p3} = 0$

$\tau_{max} = 43.0$ MPa

$\theta_p = 35.7°$

8-5 $P = 182.0$ kips

8-6 $T = 4.02$ kN·m

8-9 $T = 62.4$ kips

8-10 $P = 954$ kN

8-13 $P = 11.11$ kips

$T = 4.16$ in.-kips

8-15 (a) $\tau = 19.60$ ksi

(b) $\Delta\tau = 23.0\%$

8-16 (a) $\tau = 255$ MPa

(b) $\Delta\tau = 19.83\%$

8-19 $R = 2.33$ in.

$n = 6$ turns

8-20 $R = 116.5$ mm

$n = 5$ turns

8-23 $d = 0.292$ in.

8-25 $\sigma = 3.58$ ksi T

$\sigma = 1.989$ ksi C

8-26 $\sigma = 15.25$ MPa T

$\sigma = 44.3$ MPa C

8-29 $\sigma_A = 1.250$ ksi T

$\sigma_B = 1.750$ ksi C

8-30 $\sigma_{CD} = 89.8$ MPa T

$\sigma_{EF} = 126.0$ MPa C

8-33 $\sigma_A = 1000$ psi T

$\sigma_B = 200$ psi C

$\sigma_C = 1800$ psi C

$\sigma_D = 600$ ksi C

8-34 $\sigma_A = 8.75$ MPa T

$\sigma_B = 1.250$ MPa T

$\sigma_C = 13.75$ MPa C

$\sigma_D = 6.25$ MPa C

8-37 $h = 1.035$ in.

8-38 (a) $\sigma = 125.0$ MPa C

(b) $\sigma = 125.3$ MPa C

8-41 $\sigma = 52.6$ ksi

8-42 $W = 90.5$ kips

8-45 $P = 49.1$ lb

$Q = 393$ lb

$x = 7.00$ in.

8-47 $\sigma_{p1} = 14.96$ ksi T

$\sigma_{p2} = 2.85$ ksi C

$\sigma_{p3} = 0$

$\tau_p = 8.90$ ksi

$\tau_{max} = 8.90$ ksi

$\theta_p = -23.6°$

8-48 $\sigma_{p1} = 120.7$ MPa T

$\sigma_{p2} = 17.86$ MPa C

$\sigma_{p3} = 0$

$\tau_p = 69.3$ MPa

$\tau_{max} = 69.3$ MPa

$\theta_p = -21.0°$

8-51 $\sigma_{max} = 20.0$ ksi T or C

$\tau_{max} = 10.00$ ksi

8-52 $\sigma_{max} = 208$ MPa T or C

$\tau_{max} = 110.1$ MPa

8-55 $\sigma_{p1} = 32.3$ ksi T

$\sigma_{p2} = 0.279$ ksi C

$\sigma_{p3} = 0$

$\tau_p = 16.28$ ksi

$\tau_{max} = 16.28$ ksi

$\theta_p = 5.31°$

8-57 $\sigma_{p1} = 13.83$ ksi T
$\sigma_{p2} = 2.37$ ksi C
$\sigma_{p3} = 0$
$\tau_p = 8.10$ ksi
$\tau_{max} = 8.10$ ksi
$\theta_p = 22.5°$

8-58 $\sigma_{p1} = 101.9$ MPa T
$\sigma_{p2} = 36.7$ MPa C
$\sigma_{p3} = 0$
$\tau_p = 69.3$ MPa
$\tau_{max} = 69.3$ MPa
$\theta_p = 31.0°$

8-61 $\sigma_{p1} = 32.3$ ksi T
$\sigma_{p2} = 8.15$ ksi C
$\sigma_{p3} = 0$
$\tau_p = 20.2$ ksi
$\tau_{max} = 20.2$ ksi

8-62 $\sigma_{p1} = 144.4$ MPa T
$\sigma_{p2} = 14.54$ MPa C
$\sigma_{p3} = 0$
$\tau_p = 79.5$ MPa
$\tau_{max} = 79.5$ MPa

8-65 $d = 4.12$ in.

8-68 $\sigma_{p1} = 5.07$ MPa T
$\sigma_{p2} = 75.8$ MPa C
$\sigma_{p3} = 0$
$\tau_p = 40.4$ MPa
$\tau_{max} = 40.4$ MPa

8-69 $\sigma_{p1} = 8.14$ ksi T
$\sigma_{p2} = 1.089$ ksi C
$\sigma_{p3} = 0$
$\tau_p = 4.61$ ksi
$\tau_{max} = 4.61$ ksi

8-71 $V = 189.4$ lb

8-72 $V = 21.8$ kN

8-76 $\sigma_{max} = 166.4$ MPa T

8-77 $\sigma_{max} = 2.73$ ksi T

8-80 $\sigma_{max} = 99.3$ ksi T
$\tau_{max} = 49.7$ ksi

8-82 $\sigma_{max} = 20.2$ MPa T
$\tau_{max} = 10.08$ MPa

8-83 $\sigma_{max} = 19.10$ ksi T

8-85 Failure predicted by:
Maximum-shearing-stress
theory
Maximum-strain-energy
theory

Maximum-distortion-energy
theory

8-86 Failure predicted by:
Maximum-shearing-stress
theory
Maximum-distortion-energy
theory

8-89 Maximum-normal-stress
$k = 2.00$
Maximum-shearing-stress
$k = 2.00$
Maximum-normal-strain
$k = 2.73$
Maximum-strain-energy
$k = 1.78$
Maximum-distortion-energy
$k = 2.11$

8-90 Maximum-normal-stress
$k = 2.00$
Maximum-shearing-stress
$k = 1.059$
Maximum-normal-strain
$k = 1.579$
Maximum-strain-energy
$k = 1.312$
Maximum-distortion-energy
$k = 1.222$

8-92 (a) $P = 514$ kips
(b) $P = 393$ kips
(c) $P = 483$ kips

8-93 (a) $P = 597$ kips
(b) $P = 655$ kips
(c) $P = 639$ kips

8-95 (a) $R = 7.65$ kips
(b) $R = 6.94$ kips
(c) $R = 7.42$ kips

8-96 (a) $R = 4.38$ kips
(b) $R = 4.65$ kips
(c) $R = 4.58$ kips

CHAPTER 9

9-1 (a) $L/r = 200$
(b) $P_{cr} = 5.81$ kips
(c) $\sigma = 7.40$ ksi

9-2 (a) $L/r = 149.9$

(b) $P_{cr} = 388$ kN
(c) $\sigma = 87.8$ MPa

9-5   (a) $L/r = 173.9$
(b) $P_{cr} = 172.2$ kips
(c) $\sigma = 9.46$ ksi

9-6   (a) $L/r = 208$
(b) $P_{cr} = 370$ kN
(c) $\sigma = 45.5$ MPa

9-9 $P_w = 201$ kips

9-10 $P_w = 660$ kN

9-13 $P_w = 109.1$ kips

9-14 $P_w = 916$ kN

9-17   (a) $P_{cr} = 10.46$ kips
(b) $P_{cr} = 115.1$ kips

9-18   (a) $P_{cr} = 104.0$ kN
(b) $P_{cr} = 310$ kN

9-21 $P = 0.779$ kips

9-22 $k = 2.62$

9-25 $P = 1.668$ kips

9-27 $P_w = 1.271$ kips

9-28 $P_w = 141.5$ kN

9-31 $k = 1.678$

9-32 $P_{cr} = \pi^3 ED^4/16L^2$

9-35   Use a W10 × 22 section

9-36   Use a W254 × 33 section

9-44   (a) $e = d/32$
(b) $e = 5d/128$
(c) $e = d/16$

9-45   (a) $e = 0.332$ in.
(b) $e = 0.0933$ in.

9-48   (a) $ec/r^2 = 0.240$
(b) $P_w = 336$ kN

9-49   (a) $ec/r^2 = 0.244$
(b) $P_w = 91.1$ kips

9-50 $P_w = 293$ kN

9-53   (a) $ec/r^2 = 0.35$
(b) $e = 0.359$ in.

9-54   (a) $ec/r^2 = 0.30$
(b) $e = 1.000$ mm

9-57 $P_w = 64.2$ kips

9-58 $P_w = 227$ kN

9-61   (a) $P_w = 41.5$ kips
(b) $P_w = 206$ kips

9-62   (a) $P_w = 784$ kN
(b) $P_w = 1461$ kN

9-65   (a) $L = 13.35$ ft
(b) $L = 9.84$ ft

9-66   (a) $L = 6.73$ m
(b) $L = 0.479$ m

9-71 $P_w = 151.7$ kips

9-72 $P_w = 43.3$ kN

9-75 $L = 100.6$ in.

9-76 $h = 53.0$ mm

9-79   (a) $P_w = 57.4$ kips
(b) $P_w = 133.2$ kips

9-80   (a) $P_w = 77.0$ kN
(b) $P_w = 1035$ kN

9-83   (a) $P_w = 139.7$ kips
(b) $P_w = 237$ kips

9-84 $P_w = 390$ kN

9-87 $P_w = 95.8$ kips

9-88   (a) $P_w = 536$ kN
(b) $P_w = 137.1$ kN

9-91 $P_w = 87.8$ kips

9-92 $P_w = 293$ kN

9-95 $P_{max} = 181.2$ kips

9-96 $P_{max} = 459$ kN

9-99 $P_{max} = 37.9$ kips

9-100 $P_{max} = 551$ kN

9-103   (a) $P_w = 58.2$ kips
(b) $0.669 < d < 2.83$ in.

9-104   (a) $P_w = 210$ kN
(b) $M = 3.14$ kN·m

9-107 $P_{max} = 69.1$ kips

9-108 $P_{max} = 307$ kN

CHAPTER 10

10-1   (a) $\sigma_A = 40.0$ ksi
$\sigma_B = 44.4$ ksi
(b) $\delta_C = 0.419$ in.

10-2   (a) $\sigma_A = 245$ MPa
$\sigma_B = 275$ MPa
(b) $\delta_D = 8.40$ mm

10-5   (a) $\sigma_A = 282$ MPa
$\sigma_B = 67.9$ MPa
$\sigma_C = 240$ MPa
(b) $\delta_C = 6.79$ mm

10-6   (a) $P = 1225$ kN

(b) $\delta_D = 6.00$ mm
(c) $\delta_D = 60.0$ mm

10-9 (a) $\sigma_A = 25.0$ ksi
$\sigma_B = 30.0$ ksi
(b) $\sigma_A = 42.0$ ksi
$\sigma_B = 36.0$ ksi
(c) $\delta_C = 0.1300$ in.

10-10 (a) $\sigma_A = 220$ MPa
$\sigma_B = 95.5$ MPa
(b) $\sigma_A = 275$ MPa
$\sigma_B = 373$ MPa
(c) $\delta_C = 3.07$ mm

10-11 (a) $T = 360$ in.-kips
(b) $T = 390$ in.-kips

10-12 (a) $T = 23.6$ kN·m
(b) $\Delta T = 33.4\%$

10-15 (a) $\gamma = 0.00364$ rad
(b) $T = 296$ in.-kips
(c) $\tau = 10.91$ ksi

10-16 (a) $\gamma = 0.00233$ rad
(b) $T = 30.5$ kN·m
(c) $\tau = 93.1$ MPa

10-19 $T = 756$ in.-kips

10-20 $T = 20.9$ kN·m

10-23 (a) $T = 296$ in.-kips
(b) $\tau_b = 23.8$ ksi
$\tau_s = 18.00$ ksi

10-24 $T = 37.2$ kN·m

10-27 $M_e = 24{,}600$ in.-kips
$M_p = 27{,}500$ in.-kips

10-28 $M_e = 1556$ kN·m
$M_p = 1771$ kN·m

10-31 $M_p/M_e = 1.500$

10-32 $M_p/M_e = 1.698$

10-35 (a) $a = 6.38$ in.
(b) $M = 2750$ in.-kips
(c) $M_p/M_e = 1.363$

10-36 (a) $a = 88.3$ mm
(b) $M = 148.2$ kN·m
(c) $M_p/M_e = 1.489$

10-39 $\sigma = 5\sqrt{2}M(y/h)^{0.5}/bh^2$

10-41 $P = 3M_p/L$

10-42 $P = 6M_p/L$

10-45 $w = 8.95$ kips/ft

10-46 $P = 225$ kN

10-49 Use a W18 $\times$ 60 section

10-50 Use a W406 $\times$ 60 section

10-54 (a) $w = 54.5$ kN/m
(b) Use a W457 $\times$ 89 section
(c) $x = 7.19$ m

10-55 (a) $P = 44.7$ kips
(b) $P = 84.5$ kips

10-56 (a) $L = 4.02$ m
(b) $L = 1.888$ m
(c) $L = 1.201$ m

10-59 $L = 17.13$ in.

CHAPTER 11

11-1 $U = \gamma^2 A L^3/6E$

11-2 $U_{bar} = 3U_{beam}$

11-5 $U_{bar} = 9U_{beam}$

11-6 $U_{bar} = 15U_{beam}$

11-9 $U_{shaft} = 5(1 + \nu)/4\ U_{bar}$

11-11 $d = 1.713$ in.

11-12 $L = 3.27$ m

11-15 (a) $\sigma = 9.21$ ksi
(b) $\tau = 8.24$ ksi

11-16 $h = 204$ mm

11-19 (a) $\tau = 16.48$ ksi
(b) $\theta = 0.01854$ rad
(c) $\tau = 10.09$ ksi

11-21 $P_s = 9.52$ kips

11-22 (a) $\Delta = 65.9$ mm
(b) $P_s = 1186$ N

11-25 (a) $\sigma = 15.90$ ksi
(b) $\Delta = 0.1060$ in.
(c) $P_s = 1590$ lb

11-28 (a) $\tau = 100.2$ MPa
(b) $U = 62.1$ J

11-29 (a) $W = 10.00$ lb
(b) $h = 0.880$ in.

11-31 $\sigma = 12.08$ ksi
$\tau = 2.88$ ksi

11-32 (a) $h = 189.2$ mm
(b) $LF = 10.76$
(c) $U_A = 16.05$ J
$U_B = 8.02$ J

11-35 $\delta = 16PL^3/1875EI \downarrow$

11-36 $\delta = PL^3/6EI \uparrow$

11-39 $\delta = PL^3/EI \downarrow$

11-40 $\theta = +7PL^2/6EI$

11-43 $\delta = 11.31$ mm $\downarrow$

11-44 $\delta = 2.81$ mm $\uparrow$

11-46 $\delta_v = \pi PR^3/4EI \downarrow$

11-48 $\delta_v = \pi PR^3/2EI \downarrow$

11-51 (a) $\delta_v = ML^2/2EI \uparrow$
       (b) $\delta_h = 5ML^2/8EI \rightarrow$

11-52 (a) $\delta_v = 7wL^4/EI \downarrow$
       (b) $\delta_h = 5wL^4/2EI \leftarrow$

11-56 (a) $R_A = 3wL/8 \uparrow$
           $V_B = 5wL/8 \uparrow$
           $M_B = -wL^2/8$
       (b) $\delta = wL^4/192EI \downarrow$

11-57 (a) $R_A = 9M/16L \uparrow$
           $V_B = 9M/16L \downarrow$
           $M_B = +M/8$
       (b) $\delta = ML^2/32EI \downarrow$

11-60 (a) $R_A = 7wL/16 \uparrow$
           $R_B = 5wL/8 \uparrow$
           $R_C = wL/16 \downarrow$
       (b) $\delta = wL^4/256EI \uparrow$

11-61 (a) $R_A = 3wL/7 \uparrow$
           $R_B = 19wL/28 \uparrow$

$V_C = 3wL/28 \downarrow$
$M_C = +wL^2/28$
       (b) $\delta = 23wL^4/2688EI \downarrow$

11-64 (a) $R_A = 11wL/28 \uparrow$
           $R_B = 8wL/7 \uparrow$
           $V_C = 13wL/28 \uparrow$
           $M_C = -wL^2/14$
       (b) $\delta = 17wL^4/2688EI \downarrow$

11-65 (a) $V_A = 2P \uparrow$
           $M_A = -PL/2$
           $R_C = 3P \uparrow$
       (b) $\delta = PL^3/48EI \downarrow$

11-67 (a) $R_A = 15wL/32 \uparrow$
       (b) $\delta_h = wL^4/64EI \leftarrow$

11-69 (a) $R_A = 3P/20 \uparrow$
       (b) $\delta_h = 17PL^3/60EI \rightarrow$

11-71 $\delta = PL^3/6EI \downarrow$

11-72 $\theta = +PL^2/4EI$

11-75 (a) $\delta_h = 3ML^2/2EI \rightarrow$
       (b) $\delta_v = ML^2/2EI \uparrow$

11-76 (a) $\delta_h = PL^3/2EI \leftarrow$
       (b) $\delta_v = PL^3/3EI \downarrow$
       (c) $\theta = -PL^2/2EI$

11-79 (a) $\delta_v = 7wL^4/EI \downarrow$
       (b) $\delta_h = 5wL^4/2EI \leftarrow$

11-80 (a) $\delta_v = 3wL^4/EI \downarrow$
       (b) $\delta_h = wL^4/3EI \rightarrow$

NAME INDEX

Beedle, L.S., 582, 585
Boresi, A.P., 124

Castigliano, A., 620
Clapeyron, B.P.E., 496
Clebsch, A., 315
Coulomb, C.A., 145, 208, 494
Culmann, K., 412

Dally, J.W., 47
Dorn, W.S., 526
Duleau, A., 145

Engesser, F.R., 589
Euler, L., 508

Galilei, Galileo, 208
Gere, J.M., 528
Goodier, J.N., 126
Guest, J.J., 494

Hayden, H.W., 52
Heaviside, O., 315
Hencky, H., 497
Hetenyi, M., 47
Hooke, Robert, 47
Huber, M.T., 497

Johnson, J.B., 532
Johnson, T.H., 532
Johnston, B., 547
Jourawski, D.J., 252

Kirsch, G., 124

Lynn, P.P., 124

McCracken, D.D., 526
Mariotte, E., 208
Moffatt, W.G., 52
Mohr, O.C., 412

Osgood, W.R., 50

Parent, A., 208
Pilkey, W.D., 315
Poisson, S.D., 51

Ramberg, W., 50
Rankine, W.J.M., 492
Riley, W.F., 47

Saint-Venant, Barre de, 126, 494
Soderberg, C.R., 81
Sokolnikoff, I.S., 196

Timoshenko, S.P., 126, 145, 208, 528
Troelsch, H.W., 117

Van den Broek, J.A., 591
Van Vlack, L.H., 52
von Mises, R., 497

Wahl, A.M., 460
Wulff, J., 52

Young, Thomas, 47

SUBJECT INDEX

AVERAGE PROPERTIES OF SELECTED ENGINEERING MATERIALS (INTERNATIONAL SYSTEM OF UNITS)

Exact values may vary widely with changes in composition, heat treatment, and mechanical working. More precise information can be obtained from manufacturers.

Materials	Density (Mg/m³)	Elastic Strength[a]			Ultimate Strength			Endurance Limit[c] (MPa)	Modulus of Elasticity (GPa)	Modulus of Rigidity (GPa)	Percent Elongation in 50 mm	Coefficient of Thermal Expansion (10⁻⁶/°C)
		Tension (MPa)	Comp. (MPa)	Shear (MPa)	Tension (MPa)	Comp. (MPa)	Shear (MPa)					
Ferrous metals												
Wrought iron	7.70	210	b		330	b	170	160	190		30[d]	12.1
Structural steel	7.87	250	b		450	b		190	200	76	28[d]	11.9
Steel, 0.2% C hardened	7.87	430	b		620	b			210	80	22	11.9
Steel, 0.4% C hot-rolled	7.87	360	b		580	b		260	210	80	29	
Steel, 0.8% C hot-rolled	7.87	520	b		840	b			210	80	8	
Cast iron—gray	7.20				170	690		80	100		0.5	12.1
Cast iron—malleable	7.37	220	b		340	b			170		20	11.9
Cast iron—nodular	7.37	480			690				170		4	11.9
Stainless steel (18-8) annealed	7.92	250	b		590	b		270	190	86	55	17.3
Stainless steel (18-8) cold-rolled	7.92	1140	b		1310	b		620	190	86	8	17.3
Steel, SAE 4340, heat-treated	7.84	910	1000		1030	b	650	520	200	76	19	
Nonferrous metal alloys												
Aluminum, cast 195-T6	2.77	160	170		250		210	50	71	26	5	
Aluminum, wrought, 2014-T4	2.80	280	280	160	430	b	260	120	73	28	20	22.5
Aluminum, wrought, 2024-T4	2.77	330	330	190	470	b	280	120	73	28	19	22.5
Aluminum, wrought, 6061-T6	2.71	270	270	180	310	b	210	93	70	26	17	22.5